Graduate Texts in Mathematics

37

J. Donald Monk

Mathematical Logic

Springer-Verlag
New York Heidelberg Berlin
1976

J. Donald Monk
Department of Mathematics
University of Colorado
Boulder, Colorado 80302

AMS Subject Classifications
Primary: 02-xx
Secondary: 10N-xx, 06-XX, 08-XX, 26A98

Library of Congress Cataloging in Publication Data

Monk, James Donald, 1930–
Mathematical logic.

(Graduate texts in mathematics ; 37)
Bibliography
Includes indexes.
1. Logic, Symbolic and mathematical. I. Title. II. Series.
QA9.M68 511'.3 75–42416

Printed in the United States of America

ISBN 0–387–90170–1 Springer-Verlag New York

ISBN 3–540–90170–1 Springer-Verlag Berlin Heidelberg

to Dorothy

Preface

This book is a development of lectures given by the author numerous times at the University of Colorado, and once at the University of California, Berkeley. A large portion was written while the author worked at the Forschungsinstitut für Mathematik, Eidgennössische Technische Hochschule, Zürich.

A detailed description of the contents of the book, notational conventions, etc., is found at the end of the introduction.

The author's main professional debt is to Alfred Tarski, from whom he learned logic. Several former students have urged the author to publish such a book as this; for such encouragement I am especially indebted to Ralph McKenzie.

I wish to thank James Fickett and Stephen Comer for invaluable help in finding (some of the) errors in the manuscript. Comer also suggested several of the exercises.

J. Donald Monk
October, 1975

Contents

Part III

Decidable and Undecidable Theories 231

Part IV

Model Theory 309

Part V

Unusual Logics 471

Introduction

Leafing through almost any exposition of modern mathematical logic, including this book, one will note the highly technical and purely mathematical nature of most of the material. Generally speaking this may seem strange to the novice, who pictures logic as forming the foundation of mathematics and expects to find many difficult discussions concerning the philosophy of mathematics. Even more puzzling to such a person is the fact that most works on logic presuppose a substantial amount of mathematical background, in fact, usually more set theory than is required for other mathematical subjects at a comparable level. To the novice it would seem more appropriate to begin by assuming nothing more than a general cultural background. In this introduction we want to try to justify the approach used in this book and similar ones. Inevitably this will require a discussion of the philosophy of mathematics. We cannot do full justice to this topic here, and the interested reader will have to study further, for example in the references given at the end of this introduction. We should emphasize at the outset that the various possible philosophical viewpoints concerning the nature or purpose of mathematics do not effect one way or the other the correctness of mathematical reasoning (including the technical results of this book). They do effect how mathematical results are to be intuitively interpreted, and which mathematical problems are considered as more significant.

We shall discuss first a possible definition of mathematics, and then turn to a deeper discussion of the meaning of mathematics. After this we can in part justify the methods of modern logic described in this book. The introduction closes with an outline of the contents of the book and some comments on notation.

As a tentative definition of mathematics, we may say it is an *a priori*, *exact*, *abstract*, *absolute*, *applicable*, and *symbolic* scientific discipline. We now consider these defining characteristics one by one. To say that mathematics is *a priori* is to say that it is independent of experience. Unlike physics or chemistry, the laws of mathematics are not laws of nature or dependent upon laws of nature. Theorems would remain valid in other possible worlds, where the laws of physics might be entirely different. If we take mathematical knowledge to mean a body of theorems and their formal proofs, then we can say that such knowledge is independent of all experience except the very rudimentary process of mechanically checking that the proofs are really proofs in the logical sense—lists of formulas subject to rules of inference. Of course this is a very limited conception of mathematical knowledge, but there can be little doubt that, so conceived, it is *a priori* knowledge. Depending on one's attitude towards mathematical truth, one might wish to broaden this view of mathematical knowledge; we shall discuss this later. Under broadened views, it is certainly possible to challenge the *a priori* nature of mathematics; see, e.g., Kalmar [6] (bibliography at the end of this introduction).

Mathematics is *exact* in the sense that all its terms, definitions, rules of proof, etc. have a precise meaning. This is especially true when mathematics is based upon logic and set theory, as it is customary to do these days. This aspect of mathematics is perhaps the main thing that distinguishes it from other scientific disciplines. The possibility of being exact stems partially from its *a priori* nature. It is of course difficult to be very precise in discussing empirical evidence, because nature is so complex, difficult to classify, observations are subject to experimental error, etc. But in the realm of ideas divorced from experience it is possible to be precise, and in mathematics one is precise. Of course some parts of philosophical speculation are concerned with *a priori* matters also, but such speculation differs from mathematics in not being exact.

Another distinguishing feature of mathematical discourse is that it is generally much more *abstract* than ordinary language. One of the hallmarks of modern mathematics is its abstractness, but even classical mathematics is very abstract compared to other disciplines. Number, line, plane, etc. are not concrete concepts compared to chairs, cars, or planets. There are different levels of abstractness in mathematics, too; one may contemplate a progression like numbers, groups, universal algebras, categories. This characteristic of mathematics is shared by many other disciplines. In physics, for example, discussion may range from very concrete engineering problems to possible models for atomic nuclei. But in mathematics the concepts are *a priori*, already implying some degree of abstractness, and the tendency toward abstractness is very rampant.

Next, mathematical results are *absolute*, not revisable on the basis of experience. Again, viewing mathematics just as a collection of theorems and formal proofs, there is little to quarrel with in this statement. Thus we see once more a difference between mathematics and experimental evidence; the latter is certainly subject to revision as measurements become more exact.

Of course the appropriateness of a mathematical discipline for a given empirical study is highly subject to revision. Experimental evidence and *a posteriori* reasoning hence play a role in motivation for studying parts of mathematics and in the directions for mathematical research. One's attitude toward the absoluteness of mathematics is also colored by differing commitments to the nature of mathematical truth (see below).

A feature of mathematics which is probably not inherent in its nature is its *applicability*. A very great portion of mathematics arises by trying to give a precise mathematical theory for some concrete, perhaps even nonmathematical, situation. Of course geometry and much of classical mathematics arose in this way from special intuition derived from actual sense evidence. Also, logic owes much to this means of development; formal languages arose from less formal mathematical discourse, the notion of Turing machine from the intuitive notion of computability, etc. Many very abstract mathematical disciplines arose from an analysis of less abstract parts of mathematics, and may hence be subsumed under this facet of the discipline; group theory and algebraic topology may be mentioned as examples. This aspect of mathematics is emphasized in Rogers [12], for example.

Finally, the use of *symbolic* notation is a main characteristic of mathematics. This is connected with its exact nature, but even more connected with the development of mathematics as a kind of language. In fact, mathematics is often said just to be a language of a special kind. Most linguists would reject this claim, for mathematics fails to satisfy many of their criteria for a language, e.g., that of universality (capability of expressing usual events, emotions, ideas, etc. which occur in ordinary life). But mathematics does have many features in common with ordinary languages. It has proper names, such as π and e, and many mathematical statements have a subject–predicate form. In fact, almost all mathematical statements can be given an entirely nonsymbolic rendering, although this may be awkward in many cases. Thus mathematics can be considered as embedded in the particular natural language—English, Russian, etc.—in which it is partially expressed. But also mathematics can, in principle, be expressed purely symbolically; indeed, a large portion of mathematics was so expressed in Russell and Whitehead's *Principia Mathematica*.

Now we turn to a discussion of the nature of mathematical truth. We shall briefly mention three opposed views here: platonism, formalism, and intuitionism. The views of most mathematicians as to what their subject is all about are combinations of these three. On a subjective evaluation, we would estimate the mathematical world as populated with 65% platonists, 30% formalists, and 5% intuitionists. We describe here the three extremes. There are (perhaps) more palatable versions of all three.

According to extreme *platonism*, mathematical objects are real, as real as any things in the world we live in. For example, infinite sets exist, not just as a mental construct, but in a real sense, perhaps in a "hyperworld." Similarly, nondenumerable sets, real numbers, choice functions, Lebesgue measure, and the category of all categories have a real existence. Since all of the mathematical objects are real, the job of a mathematician is as empirical as that of a geologist or physicist; the mathematician looks at a special aspect of nature and tries to discover some of the facts. The various mathematical statements, like the Riemann hypothesis or the continuum hypothesis, are either true or false in the real world. The axioms of set theory are axioms in the Greek sense—self-evident statements which form a partial basis to deductively arrive at other truths. Hence such results as the independence of the continuum hypothesis relative to the usual set-theoretical axioms force the platonist into a search for new insights and intuitions into the nature of sets so as to decide the truth or falsity of those statements which cannot be decided upon the basis of already accepted facts. Thus for him the independence results are not results about mathematics, but just about the formalization of mathematics. This view of mathematics leads to some revisions of the "definition" of mathematics we gave earlier. Thus it no longer is independent of empirical facts, but is as empirical as physics or chemistry. But since a platonist will still insist upon the absolute, immutable nature of mathematics, it still has an *a priori* aspect. For more detailed accounts of platonism see Mostowski [10] or Gödel [3].

In giving the definition of mathematics we have implicitly followed the view of formalists. A *formalist* does not believe that any mathematical objects have a real existence. For him, mathematics is just a collection of axioms, theorems, and formal proofs. Of course, the activity of mathematics is not just randomly writing down formal proofs for random theorems. The choices of axioms, of problems, of research directions, are influenced by a variety of considerations—practical, artistic, mystical—but all really nonmathematical. A revised version of platonism is to think of mathematical concepts not as actually existing but as mental constructs. A very extensive understructure for much of formalism is very close to this version of platonism—the formal development of a mathematical theory to correspond to certain mental constructions. Good examples are geometry and set theory, both of which have developed in this way. And all concept analysis (e.g., analyzing the intuitive notion of computability) can be viewed as philosophical bases for much formal mathematics. Another motivating principle behind much formalism is the desire to inter-relate different parts of mathematics; for example, one may cite the ties among sentential logic, Boolean algebra, and topology. Thus while mathematics itself is precise and formal, a mathematician is more of an artist than an experimental scientist. For more on formalism, see Hilbert [5], A. Robinson [11], and P. Cohen [2]. For another discussion of platonism and formalism see Monk [9].

Intuitionism is connected with the constructivist trend in mathematics: a mathematical object exists only if there is a (mental) construction for it. This philosophy implies that much ordinary mathematics must be thrown out, while platonism and formalism can both be used to justify present day mathematics. Even logical principles themselves must be modified on the basis of intuitionism. Thus the law of excluded middle—for any statement A, either A holds or (*not* A) holds—is rejected. The reasoning here goes as follows. Let A, for example, be the statement that there are infinitely many primes p such that $p + 2$ is also a prime. Then A does not presently hold, for we do not possess a construction which can go from any integer m given to us and produce primes p and $p + 2$ with $m < p$. But (*not* A) also does not hold, since we do not possess a construction which can go from any hypothetical construction proving A and produce a contradiction. One may say that intuitionism is the only branch of mathematics dealing directly with real, constructible objects. Other parts of mathematics introduce idealized concepts which have no constructive counterpart. For most mathematicians this idealism is fully justified, since one can make contact with verifiable, applicable mathematics as an offshoot of idealistic mathematics. See Heyting [4] and Bishop [1].

Now from the point of view of these brief comments on the nature of mathematics let us return to the problem of justifying our purely technical approach to logic. First of all, we do want to consider logic as a branch of mathematics, and subject this branch to as severe and searching an analysis as other branches. It is natural, from this point of view, to take a no-holds-barred attitude. For this reason, we shall base our discussion on a set-theoretical foundation like that used in developing analysis, or algebra, or topology. We may consider our task as that of giving a mathematical analysis of the basic concepts of logic and mathematics themselves. Thus we treat mathematical and logical practice as given empirical data and attempt to develop a purely mathematical theory of logic abstracted from these data. Our degree of success, that is, the extent to which this abstraction corresponds to the reality of mathematical practice, is a matter for philosophers to discuss. It will be evident also that many of our technical results have important implications in the philosophy of mathematics, but we shall not discuss these. We shall make some comments concerning an application of technical logic within mathematics, namely to the precise development of mathematics. Indeed, mathematics, formally developed, starts with logic, proceeds to set theory, and then branches into its several disciplines. We are not in the main concerned with this development, but a proper procedure for such a development will be easy to infer from the easier portions of our discussion in this book.

Inherent in our treatment of logic, then, is the fact that our whole discussion takes place within ordinary intuitive mathematics. Naturally, we do not develop this intuitive mathematics formally here. Essentially all that we presuppose is elementary set theory, such as it is developed in Monk [8] for example. (See the end of this introduction for a description of set-theoretic notation we use that is not standard.) Since our main concern in the book is certain formal languages, we thus are confronted with two levels of language in the book: the informal metalanguage, in which the whole discussion takes place, and the object languages which we discuss. The latter will be defined, in due course, as certain sets (!), in keeping with the foundation of all mathematics upon set theory. It is important to keep sharply in mind this distinction between language and metalanguage. But it should also be emphasized that many times we take ordinary metalanguage arguments and "translate" them into a given formal language; see Chapter 17, for example.

Briefly speaking, the book is divided up as follows. Part I is devoted to the elements of recursive function theory—the mathematical theory of effective, machine-like processes. The most important things in Part I are the various equivalent definitions of recursive functions. In Part II we give a short course in elementary logic, covering topics frequently found in undergraduate courses in mathematical logic. The main results are the completeness and compactness theorems. The heart of the book is in the remaining three parts. Part III treats one of the two basic questions of mathematical logic: given a theory T, is there an automatic method for determining the validity of sentences in T? Aside from general results, the chapter treats this question for many ordinary theories, with both positive and negative results. For example, there is no such method for set theory, but there is for ordinary addition of integers. As corollaries we present celebrated results of Gödel concerning the incompleteness of strong theories and the virtual impossibility of giving convincing consistency proofs for strong theories. The second basic question of logic is treated in Part IV: what is the relationship between semantic properties of languages (truth of sentences, denotations of words, etc.) and formal characteristics of them (form of sentences, etc.)? Some important results of this chapter are Beth's completeness theorem for definitions, Lindstrom's abstract characterization of languages, and the Keisler–Shelah mathematical characterization of the formal definability of classes of structures. In both of these chapters the languages studied are of a comprehensive type known as first-order languages. Other popular languages are studied in Part V, e.g., the type theory first extensively developed by Russell and Whitehead and the languages with infinitely long expressions.

Optional chapters in the book are marked with an asterisk *. For the interdependence of the chapters, see the graph following this introduction. The book is provided with approximately 320 exercises. Difficult or lengthy ones are marked with an asterisk *. Most of the exercises are not necessary for further work in the book; those that are are marked with a prime ′. The end of a proof is signaled by the symbol □.

As already mentioned, we will be following the set-theoretical notation found in [8]. For the convenience of the reader we set out here the notation from [8] that is not in general use. For informal logic we use "$\Rightarrow$" for "implies," "$\Leftrightarrow$" or "iff" for "if and only if," "$\neg$" for "not," "$\forall$" for "for all," and "$\exists$" for "there exists." We distinguish between classes and sets in the usual fashion. The notation $\{x : \varphi(x)\}$ denotes the class of all sets x such that $\varphi(x)$. Inclusion and proper inclusion are denoted by $\subseteq$ and $\subset$ respectively. The empty set is denoted by 0, and is the same as the ordinal number 0. We let $A \sim B = \{x : x \in A, x \notin B\}$. The ordered pair (a, b) is defined by $(a, b) = \{\{a\}, \{a, b\}\}$; and $(a, b, c) = ((a, b), c)$, $(a, b, c, d) = ((a, b, c), d)$, etc. A binary relation is a set of ordered pairs; ternary, quaternary relations are defined similarly. The domain and range of a binary relation R are denoted by Dmn R and Rng R respectively. *The value of a function f at an argument a is denoted variously by* ${}^af$, ${}_af$, f^a, f_a, fa, $f(a)$; *and we may change notation frequently, especially for typographical reasons.* The symbol $\langle \tau(i) : i \in I\rangle$ denotes a function f with domain I such that $fi = \tau(i)$ for all $i \in I$. The sequence $\langle x_0, \ldots, x_{m-1}\rangle$ is the function with domain m and value x_i for each $i \in m$. The set ${}^AB$ is the set of all functions mapping A into B. An m-ary relation is a subset of ${}^mA$, for some A. Thus a 2-ary relation is a set of ordered pairs, $\langle x, y\rangle$. By abuse of notation we shall sometimes identify the two kinds of ordered pairs, of binary relations, ternary relations, etc. We write f^*A for $\{fa : a \in A\}$. The notations $f\colon A \to B$, $f\colon A \twoheadrightarrow B$, $f\colon A \rightarrowtail B$, and $f\colon A \rightarrowtail\!\!\!\twoheadrightarrow B$ mean that f is a function mapping A into (onto, one-one into, one-one onto respectively) B. The identity function (on the class of all sets) is denoted by $\boldsymbol{I}$. The restriction of a function F to a set A is denoted by $F \restriction A$. The class of all subsets of A is denoted by $\boldsymbol{S}A$. Given an equivalence relation R on a set A, the equivalence class of $a \in A$ is denoted by $[a]_R$ or $[a]$, while the set of all equivalence classes is denoted by A/R. Ordinals are denoted by small Greek letters $\alpha, \beta, \gamma, \ldots$, while cardinals are denoted by small German letters $\mathfrak{m}, \mathfrak{n}, \ldots$. The cardinality of a set A is denoted by $|A|$. The least cardinal greater than a cardinal $\mathfrak{m}$ is denoted by $\mathfrak{m}^+$. *For typographical reasons we sometimes write* (exp $(\mathfrak{m}, \mathfrak{n})$ *for* $\mathfrak{m}^{\mathfrak{n}}$ *and* exp $\mathfrak{m}$ *for* $2^{\mathfrak{m}}$.

One final remark on our notation throughout the book: *in various symbolisms introduced with superscripts or subscripts, we will omit the latter when no confusion is likely* (e.g., $[a]_R$ and $[a]$ above).

BIBLIOGRAPHY

1. Bishop, E. *Foundations of Constructive Analysis.* New York: McGraw-Hill (1967).
2. Cohen, P. Comments on the foundations of set theory. In: *Axiomatic Set Theory*. Providence: Amer. Math. Soc. (1971), 9–16.
3. Gödel, K. What is Cantor's continuum problem? *Amer. Math. Monthly*, **54** (1947), 515–525.
4. Heyting, A. *Intuitionism.* Amsterdam: North-Holland (1966).
5. Hilbert, D. Die logischen Grundlagen der mathematik. *Math. Ann.*, *88* (1923), 151–165.
6. Kalmar, L. Foundations of mathematics—whither now. In: *Problems in the Philosophy of Mathematics.* Amsterdam: North-Holland (1967), 187–194.
7. Kreisel, G. Observations on popular discussions of foundations. In: *Axiomatic Set Theory.* Providence: Amer. Math. Soc. (1971), 189–198.
8. Monk, J. D. *Introduction to Set Theory.* New York: McGraw-Hill (1969).

9. Monk, J. D. On the foundations of set theory. *Amer. Math. Monthly*, *77* (1970), 703–711.

10. Mostowski, A. Recent results in set theory. In: *Problems in the Philosophy of Mathematics*. Amsterdam: North-Holland (1967), 82–96.

11. Robinson, A. Formalism 64. In: *Logic, Methodology, and the Philosophy of Science*. Amsterdam: North-Holland (1964), 228–246.

12. Rogers, R. Mathematical and philosophical analyses. *Philos. Sci.*, *31* (1964), 255–264.

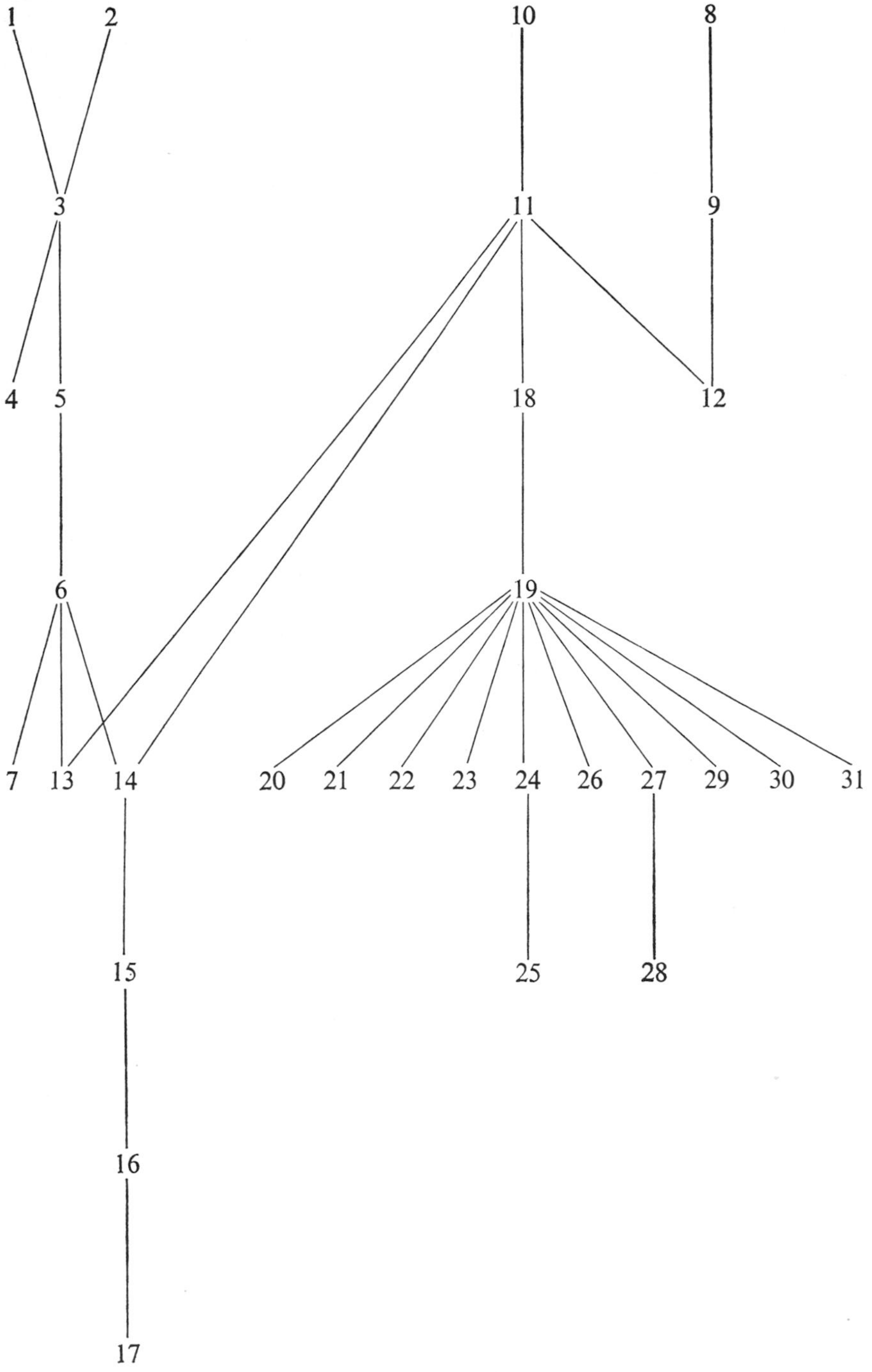

Interdependence of Chapters

PART I

Recursive Function Theory

This first part is of a purely mathematical nature, not involving notions of formal logic. We wish to give precise versions of such intuitive notions as *effective procedure*, *calculable function*, *algorithm*, *effective list*, etc. Thus our topic is in the field of computer science; but it has many applications in logic, and much of its deeper theory uses concepts from logic. We find it convenient to have these notions available when we begin the study of logic itself. Much of logic, particularly the more classical portions, is concerned with problems of effectiveness, e.g., to try to have an effective method for recognizing theorems in a given mathematical theory. This justifies treating effectiveness before logic.

We shall briefly talk about these intuitive notions of effectiveness now, and in the course of this part we shall try to give intuitive versions for some of our main theorems and proofs. In fact, such intuitive versions are frequently enlightening and can be translated into rigorous proofs so easily that, with practice, they can take the place of some of the usual rigorous proofs. Underlying all of the notions we shall discuss is the main notion of an *effective procedure*. By an effective procedure we understand a purely mechanical, step by step process which begins with a finite amount of data and proceeds to an end result, or perhaps proceeds forever, producing output as it goes. An effective procedure must be specified in advance, in all details, in a cookbook fashion: first do this, then do that, etc. The specifications must be given in a finite amount of space and must be completely unambiguous. All the data are given in a discrete, finite form, and the operating procedure is also to occupy just a finite amount of physical space. To avoid complications, however, we put no limit on the (finite) sizes of the data, operating specifications, or the procedure itself. Many examples of effective procedures are familiar to us all. There are the usual algorithms for adding and multiplying decimal numbers; for finding the roots of quadratic equations; for approximating π to any desired degree of accuracy, etc. It is quite clear that each of these algorithms can be specified in enough detail to qualify as an effective procedure in our sense. For most purposes it is not necessary to be any more explicit or precise than this about what an effective procedure really is. Mathematics abounds in algorithms, and usually they are easily recognizable as such. With the advent of large scale computers, a great number of mathematical algorithms have been worked out explicitly as effective procedures in the hardware or software of computers. However, once we start considering the possibility of negative results—showing that some procedure is *not* effective—it becomes essential to have a precise mathematical notion of effective procedure.

Using the notion of effective procedure, one can subject common mathematical notions to a process of effectivization. Thus a function $f: A \to B$ is *effective* if there is an effective procedure which, given any element $a \in A$, mechanically produces fa in a finite amount of time. A set A is *effective* if there is an effective procedure which, given any object a, will decide in a finite amount of time whether $a \in A$ or $a \notin A$. A set A is *effectively enumerable* if there is an effective function f mapping ω onto A, or, equivalently, if there is an effective procedure which generates all of the members of A, one after another. A real number is *effective* if the function giving its decimal expansion is effective. We shall see several concrete and detailed examples of the effectivization of mathematical notions in this part.

To give a rigorous definition of effective procedure requires considerable abstraction from the intuitive notion. We can, however, simplify the problem of giving such a definition in several ways. First, because all such procedures act upon finite and discrete (not continuous) data, one can restrict attention to the natural numbers as input and output. Indeed, the natural numbers can be used to code any finite and discrete data, simply by effectively assigning numbers to the data; this idea is called *Gödel numbering*, or *arithmetization*, and we shall see many applications of it. Second, among the various mathematical notions applicable to natural numbers, it is reasonable to concentrate our efforts on effectivizing just one of them, that of a function. Most of the other effectiveness notions can be easily expressed using the notion of an effective function, as we shall see.

These two restrictions then enable us to pose a well-defined problem of applied mathematics: single out rigorously a class of number-theoretic functions which coincides with the intuitive class of effective number-theoretic functions. This part is devoted to this problem and to the exposition of some deeper aspects of the theory of effective procedures. There are many rigorous versions of the notion of effective function. For each of these versions one can give persuasive arguments that the intuitive notion has been faithfully captured. All of the versions turn out to be equivalent, which again makes it plausible that they form adequate rigorous versions of the intuitive notion. In this part we base our exposition on the version involving Turing machines.

Roughly speaking, with them effectivity is identified with calculability by a machine. We also discuss definitions of effectivity involving closure conditions—explicitly defining the class of effective functions. Another verson is that of Markov algorithms, corresponding closely to the intuitive notion of (written) algorithm. In the exercises we also discuss a version involving generalized digital computers, and the Herbrand–Gödel–Kleene calculus of functions. In Part III another important version—syntactic definability in number theory—is discussed. Thus there will be available to the reader a variety of equivalent rigorous versions of effectiveness. Different ones of these versions appeal to different mathematicians. Each one seems a little arbitrary, and additions to each are natural to suggest. It turns out, though, that almost all of these additions are superfluous in the sense that they can be effected in some way by the original notion. We shall give a few remarks on this topic as we proceed.

1 Turing Machines

In this chapter we shall present a popular mathematical version of effectiveness, Turing computability, which will form our main rigorous basis for the mathematical discussion of effectivity. Actually in this section we present only some of the basic definitions concerning Turing machines and some elementary results which both illuminate these definitions and form a basis for later work. The definition of Turing computability itself is found in Chapter 3. After giving the formal definition of a Turing machine we discuss briefly the motivation behind the definition.

In our exposition of Turing machines we follow Hermes [2] rather closely. A Turing machine (intuitively) consists of a mass of machinery, a reading head, and a tape infinite in both directions. The machine may be in any of finitely many *internal states*. The tape is divided up into squares called *fields* of the tape (see figure).

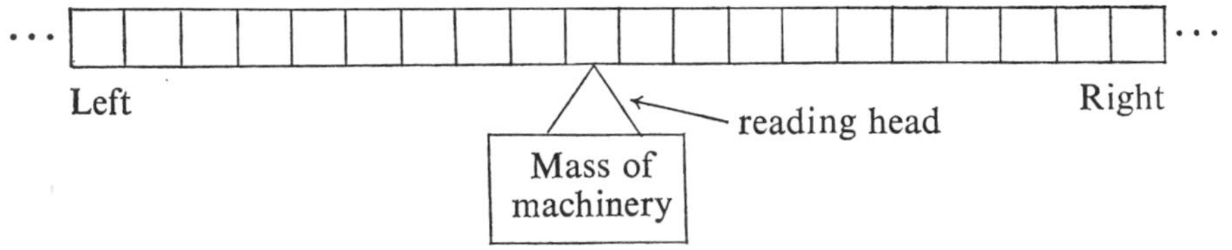

The machine proceeds step by step. At a given step it takes an action depending on what state it is in and upon what it finds on the field that the reading head is on. We allow only two symbols, 0 and 1, to be on a given field, and all but finitely many of the fields have 0 on them. These are the actions the machine can take:

(1) Write 0 on the given field (first erasing what is there).
(2) Write 1 on the given field (first erasing what is there).

(3) Move tape one square to the right.
(4) Move tape one square to the left.
(5) Stop.

We now want to make this rigorous.

Definition 1.1. A *Turing machine* is a matrix of the form

$$\begin{matrix} c_1 & 0 & v_1 & d_1 \\ c_1 & 1 & v_2 & d_2 \\ c_2 & 0 & v_3 & d_3 \\ c_2 & 1 & v_4 & d_4 \\ \vdots & \vdots & \vdots & \vdots \\ c_n & 0 & v_{2n-1} & d_{2n-1} \\ c_n & 1 & v_{2n} & d_{2n} \\ \vdots & \vdots & \vdots & \vdots \\ c_m & 0 & v_{2m-1} & d_{2m-1} \\ c_m & 1 & v_{2m} & d_{2m} \end{matrix}$$

where: $c_1, \ldots, c_m$ are distinct members of ω, $v_1, \ldots, v_{2m} \in \{0, 1, 2, 3, 4\}$ and $d_1, \ldots, d_{2m} \in \{c_1, \ldots, c_m\}$. $c_1, \ldots, c_m$ are called *states*. c_1 is called the *initial state* of the machine.

We think of a row $c_i\ \varepsilon\ v_j\ d_j$ of this matrix as giving the following information: when the machine is in state c_i and scans the symbol ε on the tape, it takes action v_j and then moves into state d_j. Here the action given by v_j is as follows:

$v_j = 0$: write 0 on scanned square;
$v_j = 1$: write 1 on scanned square;
$v_j = 2$: move tape one square to the right;
$v_j = 3$: move tape one square to the left;
$v_j = 4$: stop.

To make *this* precise, we proceed as follows:

Definition 1.2. Let $\mathbb{Z}$ be the set of all (negative and nonnegative) integers. A *tape description* is a function F mapping $\mathbb{Z}$ into $\{0, 1\}$ which is 0 except for finitely many values. A *configuration* of a given Turing machine T is a triple (F, d, e) such that F is a tape description, d is a state, and e is an integer (which tells us, intuitively, where the reading head is). A *computation step* of T is a pair $((F, d, e), (F', d', e'))$ of configurations such that: if the line of the Turing machine beginning with (d, Fe) is (d, Fe, w, f), then:

$$\begin{array}{lllll} \text{if } w = 0 & \text{then} & F' = F_0^e, & d' = f, & e' = e; \\ \text{if } w = 1 & \text{then} & F' = F_1^e, & d' = f, & e' = e; \\ \text{if } w = 2 & \text{then} & F' = F, & d' = f, & e' = e - 1; \\ \text{if } w = 3 & \text{then} & F' = F, & d' = f, & e' = e + 1. \end{array}$$

Here F_ε^e is the function $(F \sim \{(e, Fe)\}) \cup \{(e, \varepsilon)\}$. Thus F_ε^e is the tape description acting like F except possibly at e, and $F_\varepsilon^e e = \varepsilon$. A *computation* of T is a finite sequence $\langle (F_0, d_0, e_0), \ldots, (F_m, d_m, e_m) \rangle$ of configurations such that $d_0 = c_1$, $((F_i, d_i, e_i), (F_{i+1}, d_{i+1}, e_{i+1}))$ is a computation step for each $i < m$, and the row of the Turing machine beginning (d_m, Fe_m) has 4 as its third entry.

The way a Turing machine runs has now been described. To compute a function f, roughly speaking we hand the machine a number x and it produces fx as an output. Since only zeros and ones appear on a tape, we cannot literally hand x to the machine; it must be coded by zeros and ones. The mathematically most obvious way of coding x is to use its binary representation as a "decimal" with base 2. However, this is inconvenient, in view of the very primitive operations which a Turing machine can perform. We elect instead to represent x by a sequence of $x + 1$ one's. (This is sometimes called the *tally* notation.) The extra "one" is added in order to be able to recognize the code of the number zero as different from a zero entry on the tape whose purpose is just as a blank. The precise way in which functions are computed by a Turing machine will be defined in Chapter 3. In this chapter we want to see how these rather primitive looking machines can nevertheless perform some intricate operations on strings of zeros and ones. These results will be useful in Chapter 3 and later work.

Using the intuitive notion of coding we can argue as follows that Turing machines are really quite powerful: We have seen informally how to represent any number on a tape. A sequence of numbers can be represented by putting blanks (zeros) between the strings of ones representing the numbers. By using two blanks one can code several blocks of numbers, or one can use the two blanks to recognize a portion of the tape set aside for a special purpose. By repeated adjoining of a one, it is possible to add with a Turing machine; and by repeated addition, one can multiply. Since a new state depends on the currently scanned symbol, it is possible to set up different actions depending upon what is on the tape. And we are not really restricted to just one square in this decision making, since by using several states we can examine any restricted portion of the tape.

In the general theory of Turing machines, one allows several symbols instead of just 0 and 1 (see, e.g., [2]). However, it is clearly possible to code these different symbols by different strings of 1's. Several tapes may also be allowed. Again such a modification can be coded within our machines; in the case of two tapes, for example, one may instead use odd and even numbered squares on a single tape.

These intuitive comments on the strength of Turing machines of course would require proof. Some of them will be proved later, and we hope that they will all seem plausible after we have worked with Turing machines a while. For a more detailed argument on the strength of Turing machines see the introduction to [2].

Definition 1.3. T_{right} is the following machine:

0	0	2	1
0	1	2	1
1	0	4	1
1	1	4	1

Proposition 1.4. *For any tape description F and any $e \in \mathbb{Z}$, $\langle(F, 0, e), (F, 1, e - 1)\rangle$ is a computation of T_{right}.*

Thus T_{right} merely moves the tape one square to the right, and then stops.

Definition 1.5. T_{left} is the following machine:

0	0	3	1
0	1	3	1
1	0	4	1
1	1	4	1

Proposition 1.6. *For any tape description F and any $e \in \mathbb{Z}$, $\langle(F, 0, e), (F, 1, e + 1)\rangle$ is a computation of T_{left}.*

Thus T_{left} moves the tape one square to the left and then stops.

Definition 1.7. T_0 is the following machine:

0	0	4	0
0	1	0	0

Proposition 1.8. *For any tape description F and any $e \in \mathbb{Z}$, (i) if $Fe = 0$, then $\langle(F, 0, e)\rangle$ is a computation of T_0; (ii) if $Fe = 1$, then $\langle(F, 0, e), (F^e_0, 0, e)\rangle$ is a computation of T_0. Thus T_0 writes a 0 if a zero is not here, but does not move the tape.*

Definition 1.9. T_1 is the following machine:

$$\begin{array}{cccc} 0 & 0 & 1 & 0 \\ 0 & 1 & 4 & 0 \end{array}$$

Proposition 1.10. *For any tape description F and any $e \in \mathbb{Z}$, (i) if $Fe = 0$, then $\langle (F, 0, e), (F^e_1, 0, e)\rangle$ is a computation of T_1; (ii) if $Fe = 1$, then $\langle (F, 0, e)\rangle$ is a computation of T_1. T_1 writes a 1 if a 1 is not there, but does not move the tape.*

Definition 1.11. If a is any set and $m \in \omega$, let $a^{(m)}$ be the unique element of ${}^m\{a\}$. Thus $a^{(m)}$ is an m-termed sequence of a's, $a^{(m)} = \langle a, a, \ldots, a\rangle$ (m times). If x and y are finite sequences, say $x = \langle x_0, \ldots, x_{m-1}\rangle$ and $y = \langle y_0, \ldots, y_{n-1}\rangle$, we let $xy = \langle x_0, \ldots, x_{m-1}, y_0, \ldots, y_{n-1}\rangle$. Frequently we write a for $\langle a\rangle$.

Definition 1.12. $T_{l\,\mathrm{seek}\,0}$ is the following machine:

$$\begin{array}{cccc} 0 & 0 & 2 & 1 \\ 0 & 1 & 2 & 1 \\ 1 & 0 & 4 & 1 \\ 1 & 1 & 1 & 0 \end{array}$$

A computation with $T_{l\,\mathrm{seek}\,0}$ can be indicated as follows, where we use an obvious notation:

```
—— 0 1^(m)       a ————————
                 ∧
—— 0 1^(m-1)     1 a ——————
                 ∧
—— 0 1^(m-2)     1 1 a ————
                 ∧
                 ⋮
———————— 0 1 1^(m-1) a ————
               ∧
———————————— 0 1^(m) a ————
             ∧
             ↑
        Reading head
```

Thus $T_{l\,\mathrm{seek}\,0}$ finds the first 0 to the left of the square it first looks at and stops at that 0. In this and future cases we shall not formulate an exact theorem describing such a fact; we now feel the reader can in principle translate such informal statements as the above into a rigorous form.

Definition 1.13. $T_{r\,\text{seek}\,0}$ is the following machine:

0	0	3	1
0	1	3	1
1	0	4	1
1	1	1	0

$T_{r\,\text{seek}\,0}$ finds the first 0 to the right of the square it first looks at and stops at that 0.

Definition 1.14. $T_{l\,\text{seek}\,0}$ is the following machine:

0	0	2	1
0	1	2	1
1	0	0	0
1	1	4	1

$T_{l\,\text{seek}\,1}$ finds the first 1 to the left of the square it first looks at and stops at that 1. It may be that no such 1 exists; then the machine continues forever, and no computation exists.

Definition 1.15. $T_{r\,\text{seek}\,1}$ is the following machine:

0	0	3	1
0	1	3	1
1	0	0	0
1	1	4	1

$T_{r\,\text{seek}\,1}$ finds the first 1 to the right of the square it first looks at and stops at that 1. But again, it may be that no such 1 exists.

Definition 1.16. Suppose M, N, and P are Turing machines with pairwise disjoint sets of states. By $M \to N$ we mean the machine obtained by writing down N after M, after *first* replacing all rows of M of the forms $(c\ \ 0\ \ 4\ \ d)$ or $(c'\ \ 1\ \ 4\ \ d')$ by the rows $(c,\ \ 0\ \ 0\ \ e)$ or $(c'\ \ 1\ \ 1\ \ e)$ respectively, where e is the initial state of N. By

$$\begin{array}{ccc} M & \xrightarrow{\text{if } 0} & N \\ {\scriptstyle \text{if } 1}\downarrow & & \\ P & & \end{array}$$

we mean the machine obtained by writing down M, then N, then P, after first replacing all rows of M of the forms $(c\ \ 0\ \ 4\ \ d)$ or $(c'\ \ 1\ \ 4\ \ d')$ by the rows (c 0 0 e) or (c' 1 1 e') respectively, where e is the initial state of N and e' is the initial state of P.

Obviously we can change the states of a Turing machine by a one-one mapping without effecting what it does to a tape description. Hence we can apply the notation just introduced to machines even if they do not have pairwise disjoint sets of states. Furthermore, the above notation can be combined into large "flow charts" in an obvious way.

Definition 1.17. $T_{\text{seek 1}}$ is the following machine:

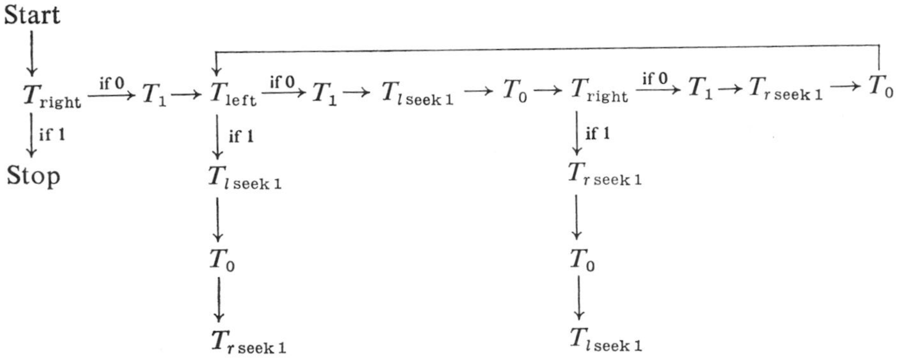

(Here by $T_{\text{right}} \overset{\text{if 1}}{\rightarrow}$ Stop we mean that the row (1 1 4 1) of T_{right} is not to be changed.)

This machine just finds a 1 and stops there. It must look both left and right to find such a 1; 1's are written (but later erased) to keep track of how far the search has gone, so that the final tape description is the same as the initial one. If the tape is blank initially the computation continues forever.

Since this is a rather complicated procedure we again indicate in detail a computation using $T_{\text{seek 1}}$. First we have two trivial cases:

```
Starting with  1 a        Starting with  0 1
                 ∧                         ∧

    1 a                        0 1
      ∧                          ∧
      1 a                        0 1
      ∧                          ∧
                                 1 1
                                 ∧
                               1 1
                                 ∧
                                 1 1
                                 ∧
                                 0 1
                                 ∧
                               0 1
                                 ∧
```

In the nontrivial case we start with $— 0^{(m)}\, \underset{\wedge}{0}\, 0^{(n)} —$; $m > 0$:

$$
\begin{array}{llllllll}
 & & & \textit{Start} & & & \\
 & & 0^{(m)} & \underset{\wedge}{0} & 0^{(n)} & & \\
 & & 0^{(m-1)} & \underset{\wedge}{0} & 0^{(n+1)} & & \\
 & & 0^{(m-1)} & \underset{\wedge}{1} & 0^{(n+1)} & & \\
 & 0^{(m-1)} & 1 & \underset{\wedge}{0} & 0^{(n)} & & \\
 & & & \vdots & & & \\
0^{(m-i)} & 1 & 0^{(2i-2)} & \underset{\wedge}{0} & 0^{(n-i+1)} & & \\
0^{(m-i)} & 1 & 0^{(2i-2)} & \underset{\wedge}{1} & 0^{(n-i+1)} & & \\
 & & 0^{(m-i)} & \underset{\wedge}{1} & 0^{(2i-2)} & 1 & 0^{(n-i+1)} \\
 & & 0^{(m-i)} & \underset{\wedge}{0} & 0^{(2i-2)} & 1 & 0^{(n-i+1)} \\
 & & 0^{(m-i-1)} & \underset{\wedge}{0} & 0^{(2i-1)} & 1 & 0^{(n-i+1)} \\
 & & 0^{(m-i-1)} & \underset{\wedge}{1} & 0^{(2i-1)} & 1 & 0^{(n-i+1)} \\
0^{(m-i-1)} & 1 & 0^{(2i-1)} & \underset{\wedge}{1} & 0^{(n-i+1)} & & \\
0^{(m-i-1)} & 1 & 0^{(2i-1)} & \underset{\wedge}{0} & 0^{(n-i+1)} & & \\
0^{(m-i-1)} & 1 & 0^{(2i)} & \underset{\wedge}{0} & 0^{(n-i)} & &
\end{array}
$$

Here $i = 1$ initially, and the portion beyond $0^{(m-1)}\, 1\, 0^{(2i-2)}\, \underset{\wedge}{0}\, 0^{(n-i+1)}$ takes place only if $i < m$ and $i \leq n$. Thus, if we start with $— 1\, 0^{(m)}\, \underset{\wedge}{0}\, 0^{(n)} —$, and $n + 1 \geq m$, we end as follows (setting $i = m$):

$$
\begin{array}{rllll}
1\ 1\ 0^{(2m-2)} & \underset{\wedge}{0} & 0^{(n-m+1)} & & \\
1\ 1\ 0^{(2m-2)} & \underset{\wedge}{1} & 0^{(n-m+1)} & & \\
1 & \underset{\wedge}{1} & 0^{(2m-2)} & 1 & 0^{(n-m+1)} \\
1 & \underset{\wedge}{0} & 0^{(2m-2)} & 1 & 0^{(n-m+1)} \\
 & \underset{\wedge}{1} & 0^{(2m-1)} & 1 & 0^{(n-m+1)} \\
1\ 0^{(2m-1)} & \underset{\wedge}{1} & 0^{(n-m+1)} & & \\
1\ 0^{(2m-1)} & \underset{\wedge}{0} & 0^{(n-m+1)} & & \\
 & \underset{\wedge}{1} & 0^{(n+m+1)} & &
\end{array}
$$

On the other hand, if we start with — $0^{(m)}\,\underset{\wedge}{0}\,0^{(n)}\,1$ —, and $n + 1 < m$ we end as follows (setting $i = n + 1$):

$$0^{(m-n-1)}\ 1\ 0^{(2n)}\quad \underset{\wedge}{0}\ 1$$

$$0^{(m-n-1)}\ 1\ 0^{(2n)}\quad \underset{\wedge}{1}\ 1$$

$$0^{(m-n-1)}\ \underset{\wedge}{1}\ 0^{(2n)}\quad 1\ 1$$

$$0^{(m-n-1)}\ \underset{\wedge}{0}\ 0^{(2n)}\quad 1\ 1$$

$$0^{(m-n-2)}\ \underset{\wedge}{0}\ 0^{(2n+1)}\ 1\ 1$$

$$0^{(m-n-2)}\ \underset{\wedge}{1}\ 0^{(2n+1)}\ 1\ 1$$

$$0^{(m-n-2)}\ 1\ 0^{(2n+1)}\quad \underset{\wedge}{1}\ 1$$

$$0^{(m-n-2)}\ 1\ 0^{(2n+1)}\quad \underset{\wedge}{0}\ 1$$

$$0^{(m-n-2)}\ 1\ 0^{(2n+2)}\quad \underset{\wedge}{1}$$

$$0^{(m-n-2)}\ \underset{\wedge}{1}\ 0^{(2n+2)}\ 1$$

$$0^{(m-n-2)}\ \underset{\wedge}{0}\ 0^{(2n+2)}\ 1$$

$$0^{(m+n+1)}\ \underset{\wedge}{1}$$

Definition 1.18. $T_{l\,\text{end}}$ is the following machine:

$$\text{Start} \longrightarrow T_{l\,\text{seek}\,0} \longrightarrow T_{\text{right}} \xrightarrow{\text{if } 0} T_{\text{left}} \qquad (T_{\text{right}} \xrightarrow{\text{if } 1} T_{l\,\text{seek}\,0})$$

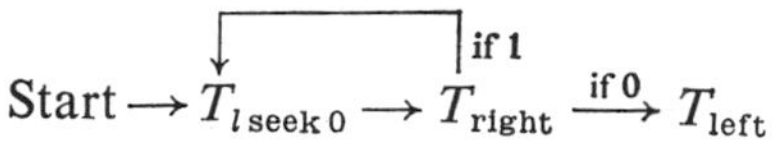

$T_{l\,\text{end}}$ moves the tape to the right until finding 00, and stops on the right-most of these two zeros. $T_{l\,\text{end}}$ does not start counting zeros until moving the tape.

Definition 1.19. $T_{r\,\text{end}}$ is the following machine:

$$\text{Start} \longrightarrow T_{r\,\text{seek}\,0} \longrightarrow T_{\text{left}} \xrightarrow{\text{if } 0} T_{\text{right}} \qquad (T_{\text{left}} \xrightarrow{\text{if } 1} T_{r\,\text{seek}\,0})$$

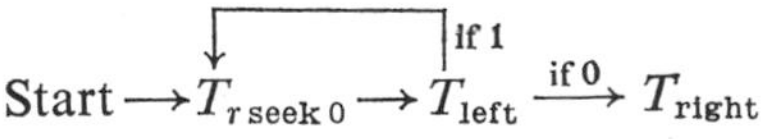

$T_{r\,\text{end}}$ moves the tape to the left until finding 00, and stops on the left-most of these two zeros. $T_{r\,\text{end}}$ does not start counting zeros until moving the tape.

Definition 1.20. $T_{l\,\text{trans}}$ is the following machine:

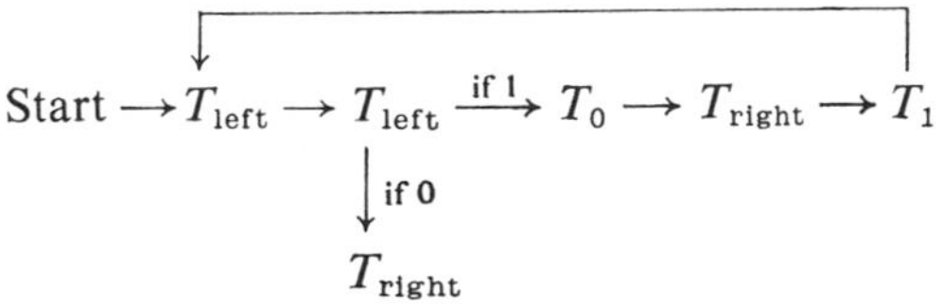
Start
T_{left}
T_{left}
if 1
T_0
T_{right}
T_1
if 0
T_{right}

The action of $T_{l\,\text{trans}}$ is indicated thus, in the case of interest to us:

$$\text{———————}\ \underset{\wedge}{a}\ 0\ 1^{(b+1)}\ 0\ \text{—}$$

$$\vdots$$

$$\text{——}\ a\ 1^{(b+1)}\ \underset{\wedge}{0}\ 0\ \text{————}$$

The tape is otherwise unchanged.

Definition 1.21. $T_{l\,\text{shift}}$ is the following machine:

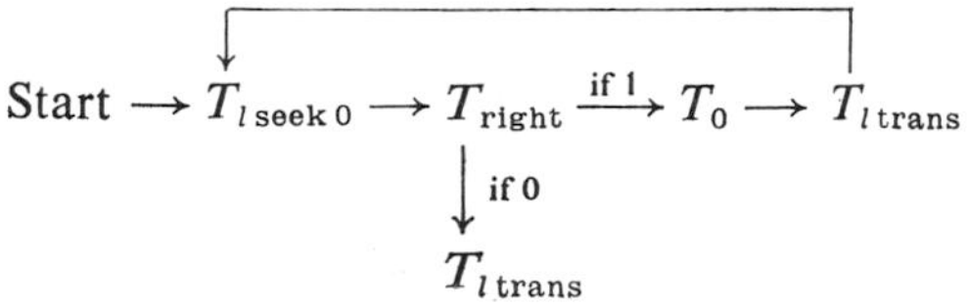
Start
$T_{l\,\text{seek}\,0}$
T_{right}
if 1
T_0
$T_{l\,\text{trans}}$
if 0
$T_{l\,\text{trans}}$

$T_{l\,\text{shift}}$ acts as follows in the case of interest to us:

$$\text{——}\ 0\ 1^{(x+1)}\ 0\ 1^{(y+1)}\ \underset{\wedge}{0}\ \text{————}$$

$$\vdots$$

$$\text{———————}\ 0\ 1^{(y+1)}\ \underset{\wedge}{0}\ 0^{(x+2)}\ \text{———}$$

The tape to the left and right of this portion of $x + y + 5$ symbols is unchanged.

Definition 1.22. T_{fin} is the following machine:

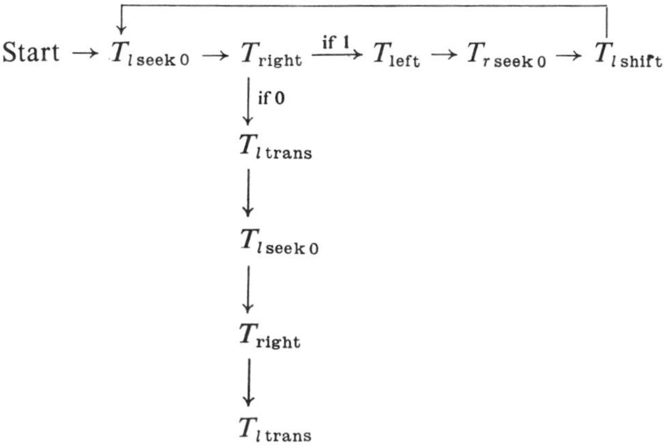
Start
$T_{l\,\text{seek}\,0}$
T_{right}
if 1
T_{left}
$T_{r\,\text{seek}\,0}$
$T_{l\,\text{shift}}$
if 0
$T_{l\,\text{trans}}$
$T_{l\,\text{seek}\,0}$
T_{right}
$T_{l\,\text{trans}}$

T_{fin} acts as follows in the case of interest to us:

$$\begin{array}{l} \text{———} \; 0^{(2)} \; 1^{(x0+1)} \; 0 \; 1^{(x1+1)} \; 0 \; \cdots \; 0 \; 1^{(x(m-1)+1)} \; 0 \; 1^{(y+1)} \; \underset{\wedge}{0} \; \text{———} \\ \qquad \vdots \\ \text{—————————————} \; 1^{(y+1)} \; \underset{\wedge}{0} \; 0^{(p)} \; \text{———} \end{array}$$

where $p = x_0 + x_1 + \cdots + x_{m-1} + 2m + 2$. In case $m = 0$, it works like this:

$$\begin{array}{l} \text{———} \; 0^{(2)} \; 1^{(y+1)} \; \underset{\wedge}{0} \; \text{———} \\ \qquad \vdots \\ \text{———} \; 1^{(y+1)} \; \underset{\wedge}{0} \; 0^{(2)} \; \text{———} \end{array}$$

In each case the tape is otherwise unchanged. Here "fin" abbreviates "finish."

This machine will be used at the end of computations to erase scratchwork.

Definition 1.23. T_{copy} is the following machine:

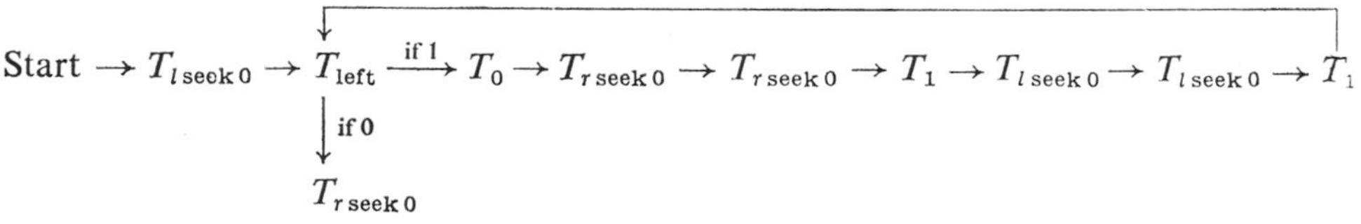

T_{copy} acts as follows:

$$\begin{array}{l} \text{———} \; 0 \; 1^{(x+1)} \; \underset{\wedge}{0} \; 0^{(x+2)} \; \text{———} \\ \qquad \vdots \\ \text{———} \; 0 \; 1^{(x+1)} \; 0 \; 1^{(x+1)} \; \underset{\wedge}{0} \; \text{———} \end{array}$$

The tape is otherwise unchanged.

A machine M repeated $m > 0$ times will be indicated by M^m in our diagrams.

Definition 1.24. For $n > 0$, $T_{n\,\text{copy}}$ is the following machine:

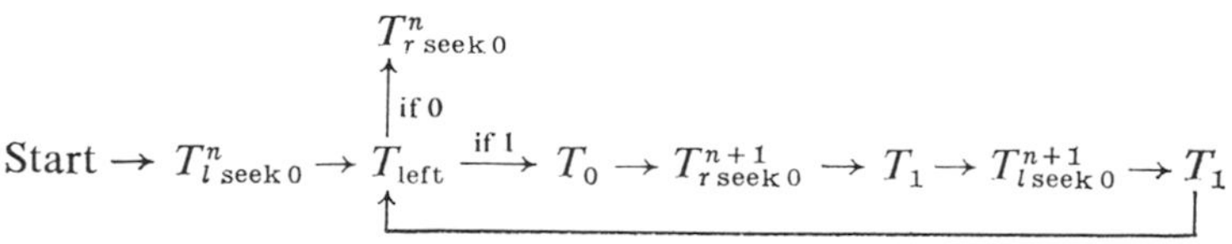

$T_{n\,\mathrm{copy}}$ acts as follows:

$$0\ 1^{(x0+1)}\ 0\ 1^{(x1+1)}\ 0\ \cdots\ 1^{(x(n-1)+1)}\ \underset{\wedge}{0}\ 0^{(x0+2)}$$

$$0\ 1^{(x0+1)}\ 0\ 1^{(x1+1)}\ 0\ \cdots\ 0\ 1^{(x(n-1)+1)}\ 0\ \cdots\ 1^{(x0+1)}\ \underset{\wedge}{0}$$

The tape is otherwise unchanged. This machine copies the nth block to the left.

These are all the basic machines needed to compute functions. We shall return to Turing machines after discussing some classes of number-theoretic functions.

BIBLIOGRAPHY

1. Davis, M. *Computability and Unsolvability.* New York: McGraw-Hill (1958).
2. Hermes, H. *Enumerability, Decidability, Computability*, 2nd ed. New York: Springer (1969).
3. Minsky, M. *Computation.* Englewood Cliffs: Prentice-Hall (1967).

EXERCISES

1.25. Give an example of a Turing machine which gets in a loop—repeats some configurations over and over.

1.26. Give an example of a Turing machine which never stops, but doesn't get in a loop.

1.27′. Prove rigorously that $T_{l\,\mathrm{trans}}$ does what is said in the text.

1.28′. Prove rigorously that $T_{l\,\mathrm{shift}}$ does what is said in the text.

1.29′. Prove rigorously that T_{fin} does what is said in the text.

1.30′. Prove rigorously that T_{copy} does what is said in the text.

1.31′. Prove rigorously that $T_{n\,\mathrm{copy}}$ does what is said in the text.

1.32. Show that there is no Turing machine which, started at an arbitrary position, will find the left-most 1 on the tape.

1.33. Construct a Turing machine which will print the sequence 11001100. . . .

1.34. Construct a Turing machine that stops iff there are at least two one's on the tape.

2 Elementary recursive and primitive recursive functions

To show that many number-theoretic functions are Turing computable, it is convenient to distinguish some functions by closure conditions.

The class of elementary recursive functions which we shall now define in this way is a class of intuitively effective functions which contains most of the effective functions actually encountered in practice. However, not every effective function is elementary recursive. Toward the end of the chapter we introduce the wider class of primitive recursive functions, which still does not cover all kinds of intuitively effective functions. In the next chapter we go from primitive recursive functions to a class of functions, the recursive functions, intuitively corresponding to the entire class of effective functions. An elementary recursive function is just a function obtainable from the usual arithmetic operations of addition, subtraction, multiplication, and division by composition, summation, and multiplication. Most of this chapter is concerned with listing out some elementary functions and with giving operations which lead from elementary functions to elementary functions. This is necessary in order to be able to easily recognize that some of the rather complicated intuitively effective functions are, in fact, elementary recursive. A more detailed treatment of the topics of this section can be found in Péter [2].

Definition 2.1. A *number-theoretic function* is a function which is, for some positive integer m, an m-ary operation on ω. The class of *elementary recursive*, or for brevity *elementary functions*, is the intersection of all classes A of number-theoretic functions such that, first of all, the following specific functions are in A:

(1) $+$, the usual 2-ary operation of addition;
(2) $\cdot$, the usual 2-ary operation of multiplication;

(3) the 2-ary operation f such that $f(m, n) = |m - n|$ for all $m, n \in \omega$;
(4) the 2-ary operation f such that $f(m, n)$ is the greatest nonnegative integer $\leq m/n$ (if $n \neq 0$), 0 if $n = 0$; we denote $f(m, n)$ by $[m/n]$;
(5) for each positive integer n and each $i < n$, the n-ary operation f on ω such that for all $x_0, \ldots, x_{n-1} \in \omega, f(x_0, \ldots, x_{n-1}) = x_i$; f is denoted by U_i^n; it is called an *identity* or *projection* function.

Second, and last, A is required to be closed under the following operations upon number-theoretic functions:

(a) The operation of composition. If f is an m-ary function, and $g_0, \ldots, g_{m-1}$ are n-ary functions, then the *composition* of f with $g_0, \ldots, g_{m-1}$ is denoted by $\mathrm{K}_n^m(f; g_0, \ldots, g_{m-1})$; it is defined to be the n-ary function h such that for all $x_0, \ldots, x_{n-1} \in \omega$,

$$h(x_0, \ldots, x_{n-1}) = f(g_0(x_0, \ldots, x_{n-1}), \ldots, g_{m-1}(x_0, \ldots, x_{n-1})).$$

(b) The operation of summation. If f is an m-ary function, then g (m-ary) is obtained from f by *summation*, in symbols $g = \sum f$, if for all $x_0, \ldots, x_{m-1} \in \omega$,

$$g(x_0, \ldots, x_{m-1}) = \sum \{f(x_0, \ldots, x_{m-2}, y) : y < x_{m-1}\}$$

[note that if $m = 1$ the definition reads

$$gx = \sum_{y<x} fy;$$

for any m, we have $g(x_0, \ldots, x_{m-2}, 0) = 0$ by convention].
(c) The operation of multiplication. If f is an m-ary function, then g (m-ary) is obtained from f by *multiplication*, in symbols $g = \prod f$, if for all $x_0, \ldots, x_{m-1} \in \omega$,

$$g(x_0, \ldots, x_{m-1}) = \prod \{f(x_0, \ldots, x_{m-2}, y) : y < x_{m-1}\}$$

[if $x_{m-1} = 0$, the right hand side is 1 by convention].

It should be evident that each elementary function is effectively calculable in the intuitive sense. To convince oneself of this, it is enough to argue that each of the functions (1)–(5) above is effectively calculable, and that the class of effectively calculable functions is closed under the operations (a)–(c). For (1)–(5), the ordinary school algorithms suffice for this argument. As to (a)–(c), suppose, for example, that f, an m-ary function, is effectively calculable, and we wish to show that $\sum f$ also is. Given $x_0, \ldots, x_{m-1} \in \omega$, we merely calculate $f(x_0, \ldots, x_{m-2}, 0), f(x_0, \ldots, x_{m-2}, 1), \ldots, f(x_0, \ldots, x_{m-2}, x_{m-1} - 1)$, which we can do since f is effectively calculable, and then we add them all up by the school process, giving us $(\sum f)(x_0, \ldots, x_{m-1})$.

Proposition 2.2. *Suppose f is m-ary, $g_0, \ldots, g_{m-1}$ are n-ary, and $h_0, \ldots, h_{n-1}$ are p-ary. Then*

$$\mathrm{K}_p^n(\mathrm{K}_n^m(f; g_0, \ldots, g_{m-1}); h_0, \ldots, h_{n-1}) = \mathrm{K}_p^m(f; \mathrm{K}_p^n(g_0; h_0, \ldots, h_{n-1}), \ldots, \mathrm{K}_p^n(g_{m-1}; h_0, \ldots, h_{n-1})).$$

PROOF. If $x_0, \ldots, x_{p-1} \in \omega$, then, with l = left hand side and r = right hand side,

$$\begin{aligned} l(x_0, \ldots, x_{p-1}) &= (\mathrm{K}_n^m(f; g_0, \ldots, g_{m-1}))(h_0(x_0, \ldots, x_{p-1}), \ldots, h_{n-1}(x_0, \ldots, x_{p-1})) \\ &= f(g_0(h_0(x_0, \ldots, x_{p-1}), \ldots, h_{n-1}(x_0, \ldots, x_{p-1})), \ldots, \\ &\quad g_{m-1}(h_0(x_0, \ldots, x_{p-1}), \ldots, h_{n-1}(x_0, \ldots, x_{p-1}))) \\ &= f((\mathrm{K}_p^n(g_0; h_0, \ldots, h_{n-1}))(x_0, \ldots, x_{p-1}), \ldots, \\ &\quad (\mathrm{K}_p^n(g_{m-1}; h_0, \ldots, h_{n-1}))(x_0, \ldots, x_{p-1})) \\ &= r(x_0, \ldots, x_{p-1}). \end{aligned}$$

□

The following theorem is the usual set-theoretical consequence of a definition like 2.1.

Proposition 2.3. *A number-theoretic function f is elementary iff there is a finite sequence $\langle g_0, \ldots, g_{k-1}\rangle$ of number-theoretic functions such that $g_{k-1} = f$, and for each $i < k$ one of the following conditions holds:*

(*i*) $g_i = +$,
(*ii*) $g_i = \cdot$,
(*iii*) g_i = *subtraction* (*in the sense of* 2.1(3)),
(*iv*) g_i = *division* (*in the sense of* 2.1(4)),
(*v*) $g_i = \mathrm{U}_j^n$ *for some* $n > 0$, *some* $j < n$,
(*vi*) g_i *is* n-*ary, and for some* $m > 0$ *there exist* $j < i$ *and* $k_0, \ldots, k_{m-1} < i$ *such that* g_j *is* m-*ary*, $g_{k0}, \ldots, g_{k(m-1)}$ *are* n-*ary, and* $g_i = \mathrm{K}_n^m(g_j; g_{k0}, \ldots, g_{k(m-1)})$ (g_i *is obtained from earlier functions by composition*),
(*vii*) *there is a* $j < i$ *such that* $g_i = \sum(g_j)$,
(*viii*) *there is a* $j < i$ *such that* $g_i = \prod(g_j)$.

PROOF. Let A be the set of all f such that there is a finite sequence of the kind described in the theorem. By considering 1-termed sequences it is easy to see that $+$, $\cdot$, subtraction, division, and U_j^n are all in A (for any $n > 0$ and $j < n$). Suppose $f \in A$, f is m-ary, $h_0, \ldots, h_{m-1} \in A$, all of $h_0, \ldots, h_{m-1}$ are n-ary. Choose a finite sequence $\langle g_0, \ldots, g_{k-1}\rangle$ such that $g_{k-1} = f$ and or each $i < k$ one of the conditions (*i*)–(*viii*) holds for g_i. For each $j < m$ choose a finite sequence $\langle l_{j,0}, \ldots, l_{j,aj-1}\rangle$ such that $l_{j,aj-1} = h_j$ and for each $i < a_j$ one of the conditions (*i*)–(*viii*) holds for l_{ji}. Then consideration of the sequence

$$\langle g_0, \ldots, g_{k-1}, l_{0,0}, \ldots, l_{0,a0-1}, \ldots, l_{m-1,0}, \ldots, \\ l_{m-1,a(m-1)}, \mathrm{K}_n^m(f; h_0, \ldots, h_{m-1})\rangle$$

shows that $\mathrm{K}_n^m(f; h_0, \ldots, h_{m-1}) \in A$. Thus A is closed under composition. If $f \in A$, so that a sequence $\langle g_0, \ldots, g_{k-1}\rangle$ exists as in the theorem, then consideration of $\langle g_0, \ldots, g_{k-1}, \sum f\rangle$ and $\langle g_0, \ldots, g_{k-1}, \prod f\rangle$ show that $\sum f$, $\prod f \in A$. Hence every elementary function appears in A. This proves $\Rightarrow$. If $f \in A$, with $\langle g_0, \ldots, g_{k-1}\rangle$ given as in the theorem, then it is easily shown by induction on i that g_i is elementary for each $i < k$. In particular, $f = g_{k-1}$ is elementary; this proves $\Leftarrow$. □

We now proceed to show that many garden-variety number-theoretic functions are elementary and that simple operations on elementary functions again give elementary functions.

For later purposes it is convenient to formulate results of the second kind in a more general way. A class A of number-theoretic functions is said to be *closed under elementary recursive operations* provided A contains all the elementary functions 2.1(1)–(5) and is closed under composition, summation, and multiplication. Obviously the class of all elementary functions is closed under elementary recursive operations. So will be all of the wider classes of effective functions which we discuss later.

Proposition 2.4. *Let A be closed under elementary recursive operations. If f is m-ary and $f \in A$, and π is a permutation of $\{0, \ldots, m-1\}$, then the m-ary function g such that $g(x_0, \ldots, x_{m-1}) = f(x_{\pi 0}, \ldots, x_{\pi(m-1)})$ for all $x_0, \ldots, x_{m-1} \in \omega$ is also in A.*

PROOF. $g = \mathrm{K}^m_m(f; \mathrm{U}^m_{\pi 0}, \ldots, \mathrm{U}^m_{\pi m-1)})$. □

Proposition 2.5 (Identification of variables). *Let A be closed under elementary recursive operations. If f is m-ary, $m > 1$, and $f \in A$, then the $(m-1)$-ary function g such that $g(x_0, \ldots, x_{m-2}) = f(x_0, \ldots, x_{m-2}, x_0)$ for all $x_0, \ldots, x_{m-2} \in \omega$ is in A.*

PROOF. $g = \mathrm{K}^m_{m-1}(f; \mathrm{U}^{m-1}_0, \ldots, \mathrm{U}^{m-1}_{m-2}, \mathrm{U}^{m-1}_0)$. □

By means of 2.4 and 2.5 variables can be identified in an arbitrary number of places. Thus if f is 3-ary elementary, so is the function g with $g(x, y) = f(x, y, y)$, for if $h(x, y, z) = f(y, x, z)$, h is elementary by 2.4; letting $k(x, y) = h(x, y, x)$ for all $x, y \in \omega$, k is elementary by 2.5, and $g(x, y) = k(y, x)$ for all $x, y \in \omega$, so g is elementary by 2.4. Usually it is just as easy in cases like this to use the method of proof of 2.4 and 2.5.

Proposition 2.6 (Adjoining apparent variables). *Let A be closed under elementary recursive operations. If f is m-ary and $f \in A$, then the $(m+1)$-ary function g such that $g(x_0, \ldots, x_m) = f(x_0, \ldots, x_{m-1})$ for all $x_0, \ldots, x_m \in \omega$, is in A.*

PROOF. $g = \mathrm{K}^m_{m+1}(f; \mathrm{U}^{m+1}_0, \ldots, \mathrm{U}^{m+1}_{m-1})$. □

Definition 2.7

(*i*) For $n > 0$, $m \in \omega$, C^n_m is the n-ary function such that $\mathrm{C}^n_m(x_0, \ldots, x_{n-1}) = m$ for all $x_0, \ldots, x_{n-1} \in \omega$.

(*ii*) sg and $\overline{\mathrm{sg}}$ are unary functions; for $x \in \omega$,

$$\mathrm{sg}\, x = \begin{cases} 0 & \text{if } x = 0, \\ 1 & \text{if } x \neq 0, \end{cases}$$

$$\overline{\mathrm{sg}}\, x = \begin{cases} 1 & \text{if } x = 0, \\ 0 & \text{if } x \neq 0. \end{cases}$$

(*iii*) $\not{p}$ is a unary function:

$$\not{p}x = \begin{cases} 0 & \text{if } x = 0 \\ x - 1 & \text{if } x \neq 0 \end{cases} \text{ for all } x \in \omega.$$

(*iv*) By convention, $0^0 = 1$, $0^x = 0$ for $x \neq 0$; $0! = 1$.

(*v*) σ is a unary function: $\sigma x = x + 1$ for all $x \in \omega$.

Thus C^n_m is the n-ary constant function with value m. The functions sg and $\overline{\text{sg}}$ are of a technical usefulness. $\not{p}$ is the predecessor function and σ the successor function.

Proposition 2.8. *The following functions are elementary:*

(*i*) C^n_m (*for* $n \neq 0$)
(*ii*) σ
(*iii*) sg
(*iv*) $\overline{\text{sg}}$
(*v*) *exponentiation*
(*vi*) *factorial*
(*vii*) $\not{p}$

PROOF

(1) C^1_0 is elementary: $C^1_0\, x = |x - x|$ for all $x \in \omega$.
(2) $\overline{\text{sg}}$ is elementary: $\overline{\text{sg}}\, x = \prod_{y<x} C^1_0\, y$, for all $x \in \omega$.
(3) sg is elementary: $\text{sg}\, x = \overline{\text{sg}}\, \overline{\text{sg}}\, x$ for all $x \in \omega$.
(4) C^1_1 is elementary: $C^1_1\, x = \overline{\text{sg}}\, C^1_0\, x$ for all $x \in \omega$.
(5) C^1_m is elementary: (by induction on m) $C^1_{m+1}\, x = C^1_m\, x + C^1_1\, x$ for all $x \in \omega$.
(6) C^n_m is elementary: $C^n_m\,(x_0, \ldots, x_{n-1}) = C^1_m\, U^n_0\,(x_0, \ldots, x_{n-1})$ for all $x_0, \ldots, x_{n-1} \in \omega$.
(7) $\sigma : \sigma x = x + C^1_1\, x$.
(8) exponentiation: $x^y = \prod_{z<y} U^2_0\,(x, z)$.
(9) factorial: $x! = \prod_{z<x} \sigma z$.
(10) $\not{p} : \not{p}x = |x - C^1_1 x| \cdot \text{sg}\, x$. □

Definition 2.9

(*i*) By an m-ary *number-theoretic relation* ($m > 0$) we mean a set of m-tuples of natural numbers. ${}^m\omega$ is the set of *all* m-tuples of natural numbers. As usual, we identify ${}^1\omega$ and ω, in an informal way.

(*ii*) If R is an m-ary number-theoretic relation, its *characteristic function* χ_R, is the m-ary number-theoretic function such that for all $x_0, \ldots, x_{m-1} \in \omega$,

$$\chi_R(x_0, \ldots, x_{m-1}) = \begin{cases} 0 & \text{if } \langle x_0, \ldots, x_{m-1}\rangle \notin R, \\ 1 & \text{if } \langle x_0, \ldots, x_{m-1}\rangle \in R. \end{cases}$$

(*iii*) An m-ary number-theoretic relation R is *elementary* if χ_R is elementary.

The definition 2.9(*iii*) is motivated by our intuitive feeling that a relation R is effective iff χ_R is an effective function. In fact, if we have an effective procedure for determining membership in R, then we can effectively calculate χ_R as follows. Given any object a, determine whether $a \in R$ or $a \notin R$. In the first case, $\chi_R a = 1$, and in the second case, $\chi_R a = 0$. Conversely, suppose we have an effective procedure for calculating values of χ_R. Given any object a, calculate $\chi_R a$. If $\chi_R a = 1$, then $a \in R$. If $\chi_R a = 0$, then $a \notin R$.

Given any class A of number-theoretic functions, an m-ary number-theoretic relation R is said to be an *A-relation* if $\chi_R \in A$.

Proposition 2.10. *0 and ω are elementary; if $x \in \omega$ then $\{x\}$ is elementary.*

PROOF. $\chi_0 = \mathrm{C}_0^1$ and $\chi_\omega = \mathrm{C}_1^1$. If $x \in \omega$, then for any $y \in \omega$, $\chi_{\{x\}} y = \overline{\mathrm{sg}}\,(|x - y|)$; hence $\chi_{\{x\}} y = \overline{\mathrm{sg}}\,(|\mathrm{C}_x^1\, y - \mathrm{U}_0^1\, y|)$. □

By 2.10, $\{x\}$ is always on effectively decidable set. Intuitively speaking, to check whether $y \in \{x\}$ we simply check if $y = x$ (surely an effective matter). As an example, let $B = \{0\}$ if Fermat's last theorem is true, while $B = 0$ if it is false. B is an effectively decidable set, although we do not know now whether $0 \in B$ or not. Thus there is a decision procedure for membership in B, but we don't know what it is (it is either the obvious one for $\{0\}$ or the obvious one for 0).

Proposition 2.11. *Let A be closed under elementary recursive operations. If R and S are A-relations, then so are $R \cap S$, $R \cup S$, and ${}^m\omega \sim R$.*

PROOF. For all $x_0, \ldots, x_{m-1}$, $\chi_{R\cap S}(x_0, \ldots, x_{m-1}) = \chi_R(x_0, \ldots, x_{m-1}) \cdot \chi_S(x_0, \ldots, x_{m-1})$, $\chi_T(x_0, \ldots, x_{m-1}) = \overline{\mathrm{sg}}\, \chi_R(x_0, \ldots, x_{m-1})$, with $T = {}^m\omega \sim R$, $R \cup S = {}^m\omega \sim [({}^m\omega \sim R) \cap ({}^m\omega \sim S)]$. □

Corollary 2.12. *Every finite subset of ω is elementary, and so is every cofinite set.*

Proposition 2.13. *The binary relations $\leq$, $<$, $\geq$, $=$, $\neq$ are elementary.*

PROOF. For any $x, y \in \omega$,

$$\chi_<(x, y) = \overline{\mathrm{sg}}\, [\sigma x / \sigma y] = \overline{\mathrm{sg}}\, [\sigma\, \mathrm{U}_0^2\,(x, y) / \sigma\, \mathrm{U}_1^2\,(x, y)].$$

Thus $<$ is elementary. Further

$$\chi_{\neq}(x, y) = \mathrm{sg}\,(|x - y|),$$

so $\neq$ is elementary. Finally, $\leq\ = (<\ \cup =)$, $\geq\ = ({}^2\omega \sim <)$, $>\ = ({}^2\omega \sim \leq)$, $=\ = ({}^2\omega \sim \neq)$. □

Proposition 2.14 (Bounded existential quantifier). *Let A be closed under elementary recursive operations. Suppose R is an m-ary A-relation. Let*

$S = \{\langle x_0, \ldots, x_{m-1}\rangle$: there is a $y < x_{m-1}$ such that $\langle x_0, \ldots, x_{m-2}, y\rangle \in R\}$. Then S is an A-relation.

PROOF. $\chi_S(x_0, \ldots, x_{m-1}) = \mathrm{sg} \sum \{\chi_R(x_0, \ldots, x_{m-2}, y) : y < x_{m-1}\}$. □

Proposition 2.15 (Bounded universal quantifier). *Let A be closed under elementary recursive operations. Suppose R is an m-ary A-relation. Let $T = \{\langle x_0, \ldots, x_{m-1}\rangle$: for every $y < x_{m-1}$ we have $\langle x_0, \ldots, x_{m-2}, y\rangle \in R\}$. Then T is an A-relation.*

PROOF. Let S be as in 2.14, with R replaced by ${}^m\omega \sim R$. Then $T = {}^m\omega \sim S$. □

Definition 2.16 (Bounded minimum). Let R be an m-ary relation. For all $x_0, \ldots, x_{m-1} \in \omega$, let

$$f(x_0, \ldots, x_{m-1}) = \begin{cases} \text{the least } y < x_{m-1} \text{ such that } \langle x_0, \ldots, x_{m-2}, y\rangle \in R, \\ \quad \text{if there is such a } y, \\ 0 \quad \text{otherwise.} \end{cases}$$

$f(x_0, \ldots, x_{m-1})$ is denoted by $\mu y < x_{m-1} R(x_0, \ldots, x_{m-2}, y)$.

Proposition 2.17. *Let A be closed under elementary recursive operations. If R is an m-ary A-relation, then the function f of 2.16 is a member of A.*

PROOF. Note that

$$\text{(1)} \quad \overline{\mathrm{sg}} \sum_{y<i} \chi_R(x_0, \ldots, x_{m-2}, y) = \begin{cases} 1 & \text{if } \langle x_0, \ldots, x_{m-1}, y\rangle \notin R \text{ for all } y < i, \\ 0 & \text{otherwise.} \end{cases}$$

Let $g(x_0, \ldots, x_{m-2}, i) = \overline{\mathrm{sg}} \sum_{y<i} \chi_R(x_0, \ldots, x_{m-2}, y)$ for all $x_0, \ldots, x_{m-2}, i \in \omega$. Thus $g \in A$. From (1) we see that

$$\sum \{g(x_0, \ldots, x_{m-2}, \mathfrak{s}i) ; i < x_{m-1}\} = \begin{cases} f(x_0, \ldots, x_{m-1}) & \text{if there is a } y < \\ \quad x_{m-1} \text{ such that } \langle x_0, \ldots, x_{m-2}, y\rangle \in R, \\ x_{m-1} & \text{otherwise.} \end{cases}$$

Hence

$$f(x_0, \ldots, x_{m-1}) = \overline{\mathrm{sg}}\, g(x_0, \ldots, x_{m-1}) \cdot \sum \{g(x_0, \ldots, x_{m-2}, \mathfrak{s}i) : i < x_{m-1}\},$$

so $f \in A$. □

The rather technical proof of 2.17 may be compared with a proof of the intuitive version of the proposition, which goes: if R is an m-ary effective relation, then the function f of 2.16 is effective. In fact, to calculate $f(x_0, \ldots, x_{m-1})$, we test successively whether $\langle x_0, \ldots, x_{m-2}, 0\rangle \in R$, $\langle x_0, \ldots, x_{m-2}, 1\rangle \in R, \ldots, \langle x_0, \ldots, x_{m-2}, x_{m-1}\rangle \in R$. If at some point we reach an i such that $\langle x_0, \ldots, x_{m-2}, i\rangle \in R$, we set $f(x_0, \ldots, x_{m-1}) = i$ and stop testing. If we complete our testing without finding such an i we set $f(x_0, \ldots, x_{m-1}) = 0$.

Proposition 2.18 (Definition by cases). *Let A be closed under elementary recursive operations. Suppose $g_0, \ldots, g_{m-1}$ are n-ary members of A,*

$R_0, \ldots, R_{m-1}$ are pairwise disjoint n-ary A-relations with $\bigcup_{i<m} R_i = {}^n\omega$, and f is the n-ary function such that, for all $x_0, \ldots, x_{n-1} \in \omega$,

$$f(x_0, \ldots, x_{n-1}) = \begin{cases} g_0(x_0, \ldots, x_{n-1}) & \textit{if } \langle x_0, \ldots, x_{n-1}\rangle \in R_0, \\ g_1(x_0, \ldots, x_{n-1}) & \textit{if } \langle x_0, \ldots, x_{n-1}\rangle \in R_1, \\ \cdots\cdots & \vdots \\ g_{m-1}(x_0, \ldots, x_{n-1}) & \textit{if } \langle x_0, \ldots, x_{n-1}\rangle \in R_{m-1}. \end{cases}$$

Then $f \in A$.

Proof. For any $x_0, \ldots, x_{n-1} \in \omega$,

$$f(x_0, \ldots, x_{n-1}) = \chi_{R0}(x_0, \ldots, x_{n-1}) \cdot g_0(x_0, \ldots, x_{n-1}) + \cdots + \chi_{R(m-1)}(x_0, \ldots, x_{n-1}) \cdot g_{m-1}(x_0, \ldots, x_{n-1}). \quad \square$$

Definition 2.19

(*i*) for $x, y \in \omega$, let

$$x \dot{-} y = \begin{cases} x - y & \text{if } x \geq y, \\ 0 & \text{if } x < y. \end{cases}$$

(*ii*)

$$\min(x, y) = \begin{cases} x & \text{if } x \leq y, \\ y & \text{if } x > y. \end{cases}$$

(*iii*) (by induction). For $m > 2$, $\min_m(x_0, \ldots, x_{m-1}) = \min(\min_{m-1}(x_0, \ldots, x_{m-2}), x_{m-1})$, with $\min_2(x, y) = \min(x, y)$.

(*iv*) $\max(x, y)$, $\max_m(x_0, \ldots, x_{m-1})$ similarly.

(*v*) $\mathrm{rm}(x, y) =$ remainder upon dividing x by y, if $y \neq 0$; $\mathrm{rm}(x, 0) = 0$.

(*vi*) $| = \{(x, y) : x \text{ divides } y\} = \{(x, y) : \text{there is a } z \text{ such that } y = x \cdot z\}$.

(*vii*) $\mathrm{PM} = \{x : x \text{ is a positive prime}\}$.

Proposition 2.20. *All of the functions and relations of* 2.19 *are elementary.*

Proof. Obvious, as concerns (*i*)–(*iv*). For (*v*),

$$\mathrm{rm}(x, y) = \begin{cases} x - (y \cdot [x/y]) & \text{if } y \neq 0, \\ 0 & \text{if } y = 0. \end{cases}$$

For (*vi*), note that $x | y$ iff there is a z such that $y = x \cdot z$ iff there is a $z < y$ such that $y = x \cdot z$; now see 2.14. Finally, $p \in \mathrm{PM}$ iff for every $x < p$, either not $x | p$ or $x = 1$, and $p \neq 0$, $p \neq 1$; cf. 2.15. $\square$

Definition 2.21. For every k let p_k be the $(k + 1)$st prime; thus $\mathrm{p}_0 = 2$, $\mathrm{p}_1 = 3$, $\mathrm{p}_2 = 5, \ldots$.

Proposition 2.22 (Number-theoretic). *For every k, $\mathrm{p}_k \leq \exp(2, 2^k)$.*

Proof. By induction on k. Trivial for $k = 0, 1$. Induction step, $k > 0$:

$$\begin{aligned} \mathrm{p}_{k+1} &\leq \mathrm{p}_0 \cdot \ldots \cdot \mathrm{p}_k - 1 && \text{(Euclid)} \\ &\leq \exp(2, 2^0) \cdot \ldots \cdot \exp(2, 2^k) - 1 && \text{(induction hypothesis)} \\ &= 2^{\Sigma\{\exp(2,i) : i \leq k\}} - 1 \\ &= \exp(2^{k+1} - 1) - 1 \leq \exp(2^{k+1}). \end{aligned} \quad \square$$

Proposition 2.23. p *is elementary.*

PROOF. Let $N = \{(x, y) : x, y \in \mathrm{PM}, x < y$, and y is the next prime after $x\}$. Thus $N = \{(x, y) : x, y \in \mathrm{PM}$ and $x < y$ and for all $z < y$, either $z \leq x$ or not $z \notin \mathrm{PM}\}$, so N is elementary. Let $Pr = \{(x, k) : x$ is the $(k + 1)$st prime$\}$. Thus $(x, k) \in Pr$ if $x \in \mathrm{PM}$ and $\sum_{y<x} \chi_{\mathrm{PM}} y = k$, so Pr is elementary. Finally, $\mathrm{p}_k = \mu x < \exp(2, 2^k) + 1((x, k) \in Pr)$, so p is elementary. □

Definition 2.24. If $a = 0$ or $a = 1$, let $(a)_i = 0$. If $a > 1$ let $(a)_i$ be the exponent of p_i in the prime decomposition of a. Sometimes we write $(a)i$ instead of $(a)_i$.

Proposition 2.25. ( ) *is elementary.*

PROOF. $(a)_i = \mu x < a(\mathrm{p}_i^x | a$ and not $\mathrm{p}_i^{x+1} | a)$. □

Definition 2.26. $\mathrm{l}a$ = greatest i such that $\mathrm{p}_i | x$ ($=0$ if $x = 0$ or 1).

Proposition 2.27. l *is elementary.*

PROOF. $\mathrm{l}a = \mu i < a[\mathrm{p}_i | x$ and $\forall j \leq a(i < j \Rightarrow \mathrm{p}_j \nmid a)]$. □

We now proceed to study a larger class of functions, the class of primitive recursive functions. Most of the effective functions encountered in the literature were actually shown to be primitive recursive. Actually most of them are even elementary, and usually this can easily be shown. We feel that it is only an historical accident that elementary functions are not more widely discussed than primitive recursive functions.

Definition 2.28. The class of *primitive recursive functions* is the intersection of all classes A of functions such that $\mathfrak{s}$, $\mathrm{U}_i^n \in A$ for all $n > 0$ and $i < n$, and such that A is closed under composition and under the following two operations:

(*i*) *The parameterized operation of primitive recursion:* if f is m-ary and h $(m + 2)$-ary, $m > 0$, then define g recursively as follows:

$$g(x_0, \ldots, x_{m-1}, 0) = f(x_0, \ldots, x_{m-1}),$$
$$g(x_0, \ldots, x_{m-1}, \mathfrak{s}y) = h(x_0, \ldots, x_{m-1}, y, g(x_0, \ldots, x_{m-1}, y)),$$

for all $x_0, \ldots, x_{m-1}, y \in \omega$. Then g *is obtained from* f *and* h *by primitive recursion*, in symbols $g = \mathrm{R}^m(f, h)$.

(*ii*) *The no-parameter operation of primitive recursion:* if $a \in \omega$ and h is 2-ary, define g:

$$g0 = a,$$
$$g\mathfrak{s}y = h(y, gy),$$

for all $y \in \omega$. In symbols $g = \mathrm{R}^0(a, h)$.

A relation R is *primitive recursive* iff χ_R is a primitive recursion function.

Note that the operations of primitive recursion are rather special kinds of recursive or inductive definitions. Many recursive definitions can be reduced to primitive recursive ones; see, e.g., the important course-of-values recursion, 2.33. But there are recursive definitions which cannot be reduced to primitive recursion. See, e.g., Theorem 3.6. The class of general recursive functions introduced in the next section encompasses all of the natural notions of recursive definitions.

Clearly the primitive recursive schema affords an effective procedure for calculating values of $\mathrm{R}^m(f, h)$, if f and h are effectively calculable. Similarly for $\mathrm{R}^0(a, h)$. Thus every primitive recursive function is effectively calculable in the intuitive sense.

Analogously to 2.3 we have:

Proposition 2.29. *A number-theoretic function f is primitive recursive iff there is a finite sequence $\langle g_0, \ldots, g_{k-1}\rangle$ of functions such that $g_{k-1} = f$, and for each $i < k$ one of the following conditions holds:*

(*i*) $g_i = \mathfrak{s}$,
(*ii*) $g_i = \mathrm{U}^n_j$ *for some* $n > 0, j < n$,
(*iii*) *as in* 2.3 (*vi*) (*composition*),
(*iv*) *there exist* j, $h < i$ *and* $m \in \omega$, $m \neq 0$, *such that* g_j *is* m-*ary*, g_h *is* $(m + 2)$-*ary, and* $g_i = \mathrm{R}^m(g_j, g_h)$,
(*v*) *there exist* $j < i$ *and* $a \in \omega$ *such that* g_j *is* 2-*ary and* $g_i = \mathrm{R}^0(a, g_j)$.

A class A of number-theoretic functions is said to be *closed under primitive recursive operations* provided A contains all the primitive recursive functions $\mathfrak{s}$, U^n_i and is closed under the primitive recursive operations given in 2.28, including composition.

Theorem 2.30. *If A is closed under primitive recursive operations, then A is closed under elementary recursive operations. In particular, every elementary function is primitive recursive.*

PROOF

(1) $\mathfrak{p}$ is primitive recursive. For,

$$\mathfrak{p}0 = 0,$$
$$\mathfrak{p}\mathfrak{s}y = \mathrm{U}^2_0(y, \mathfrak{p}y).$$

(2) $\dot{-}$ is primitive recursive. For,

$$x \dot{-} 0 = \mathrm{U}^1_0\, x,$$
$$x \dot{-} \mathfrak{s}y = \mathfrak{p}\, \mathrm{U}^3_2\, (x, y, x \dot{-} y).$$

(3) C^1_0 is primitive recursive:

$$\mathrm{C}^1_0\, x = \mathrm{U}^1_0\, x \dot{-} \mathrm{U}^1_0\, x.$$

(4) $+$ is primitive recursive:

$$x + 0 = U_0^1\, x,$$
$$x + \sigma y = \sigma\, U_2^3\,(x, y, x + y).$$

(5) $\cdot$ is primitive recursive:

$$x \cdot 0 = C_0^1\, x,$$
$$x \cdot \sigma y = x \cdot y + x = +(U_2^3\,(x, y, x \cdot y), U_0^3\,(x, y, x \cdot y)).$$

(6) $|x - y| = (x \dot{-} y) + (y \dot{-} x)$.
(7) $\overline{\mathrm{sg}}$ is primitive recursive: $\overline{\mathrm{sg}}\, x = 1 \dot{-} x$.
(8) $\mathrm{sg}\,(x) = \overline{\mathrm{sg}}\, \overline{\mathrm{sg}}\, x$.
(9) rm is primitive recursive. Define

$$f(x, 0) = 0,$$
$$f(x, \sigma y) = \sigma f(x, y) \cdot \mathrm{sg}\, |x - \sigma f(x, y)|.$$

Then $\mathrm{rm}\,(x, y) = f(y, x)$.

(10) Division is primitive recursive. Define

$$f(x, 0) = 0,$$
$$f(x, \sigma y) = f(x, y) + \overline{\mathrm{sg}}\, |x - \sigma\, \mathrm{rm}\,(x, y)|.$$

Then $[x/y] = f(y, x)$.

Now assume that A is closed under primitive recursive operations. In particular, A is closed under composition.

(11) A is closed under summation. For, suppose $f \in A$, f m-ary, and $g = \sum f$. Then

$$g(x_0, \ldots x_{m-2}, 0) = 0,$$
$$g(x_0, \ldots, x_{m-2}, \sigma z) = \sum_{y < \sigma z} f(x_0, \ldots, x_{m-2}, y)$$
$$= \sum_{y < z} f(x_0, \ldots, x_{m-2}, y) + f(x_0, \ldots, x_{m-2}, z)$$
$$= g(x_0, \ldots, x_{m-2}, z) + f(x_0, \ldots, x_{m-2}, z).$$

Hence $g \in A$.
(12) A is closed under multiplication.
This is proved similarly. The proof of 2.30 is complete. □

The converse of 2.30 fails; see 2.45.

To express another important property of primitive recursion, we need a new coding device. Given a finite sequence $\langle x_0, \ldots, x_{m-1} \rangle$ of natural numbers, it is natural to code it by the single integer $\prod_{i<m} \mathrm{p}_i^{xi+1}$. The added one in the exponents is essential for uncoding, to distinguish between the codes for $\langle 2, 3, 0 \rangle$ and $\langle 2, 3 \rangle$, for example. The mapping that assigns to each finite sequence of natural numbers its code as above is a one-one function into ω. From the code y the original sequence x is easily extracted:

$$x = \langle (y)_0 \dot{-} 1, \ldots, (y)_{\mathrm{l}y} \dot{-} 1 \rangle.$$

The following definition gives a special instance of this coding device:

Definition 2.31. If f is an m-ary number-theoretic function, we define $\tilde{f}$, the *course-of-values function* for f, as follows: $\tilde{f}$ is again m-ary, and for any $x_0, \ldots, x_{m-1} \in \omega$,

$$\tilde{f}(x_0, \ldots, x_{m-1}) = \prod \{\mathrm{p}_i^{f(x0,\ldots,x(m-2),i)+1} : i < x_{m-1}\}.$$

Thus $\tilde{f}(x_0, \ldots, x_{m-1})$ codes the whole sequence $\langle f(x_0, \ldots, x_{m-2}, 0), \ldots, f(x_0, \ldots, x_{m-2}, x_{m-1} - 1)\rangle$. Note that $\tilde{f}(x_0, \ldots, x_{m-2}, 0) = 1$.

Proposition 2.32. *Let A be closed under primitive recursive operations. Then $f \in A$ iff $\tilde{f} \in A$.*

PROOF. Assume first that $f \in A$. Then

$$\begin{aligned}
\tilde{f}(x_0, \ldots, x_{m-2}, 0) &= 1,\\
\tilde{f}(x_0, \ldots, x_{m-2}, \sigma y) &= \prod_{i<\sigma y} \mathrm{p}_i^{f(x0,\ldots,x(m-2),i)+1}\\
&= \tilde{f}(x_0, \ldots, x_{m-2}, y)\cdot \mathrm{p}_y^{f(x0,\ldots,x(m-2),y)+1}\\
&= h(x_0, \ldots, x_{m-2}, y, \tilde{f}(x_0, \ldots, x_{m-2}, y)),
\end{aligned}$$

where $h(z_0, \ldots, z_m) = z_m \cdot \mathrm{p}_{z(m-1)}^{f(z0,\ldots,z(m-1))+1}$ for all $z_0, \ldots, z_m \in \omega$. Conversely, if $\tilde{f} \in A$, then

$$f(x_0, \ldots, x_{m-1}) = (\tilde{f}(x_0, \ldots, x_{m-2}, \sigma x_{m-1}))_{x(m-1)},$$

so $f \in A$. □

The next proposition shows that recursion in which the successor step depends on several preceding values can still be reduced to primitive recursion.

Proposition 2.33 (Course-of-values recursion). *Let A be closed under primitive recursive operations. Suppose f is an m-ary function and h is an $(m + 1)$-ary member of A such that, for all $x_0, \ldots, x_{m-1} \in \omega$,*

$$f(x_0, \ldots, x_{m-1}) = h(x_0, \ldots, x_{m-1}, \tilde{f}(x_0, \ldots, x_{m-1})).$$

Then $f \in A$.

PROOF

$$\begin{aligned}
\tilde{f}(x_0, \ldots, x_{m-2}, 0) &= 1,\\
\tilde{f}(x_0, \ldots, x_{m-2}, \sigma y) &= \prod_{i<\sigma y} \mathrm{p}_i^{f(x0,\ldots,x(m-2),i)+1}\\
&= \tilde{f}(x_0, \ldots, x_{m-2}, y)\cdot \mathrm{p}_y^{f(x0,\ldots,x(m-2),y)+1}\\
&= \tilde{f}(x_0, \ldots, x_{m-2}, y)\cdot \mathrm{p}_y^{h(x0,\ldots,x(m-2),y,\tilde{f}(x0,\ldots,x(m-2),y))+1}
\end{aligned}$$

Thus $\tilde{f} \in A$. By 2.32, $f \in A$. □

Next we show how close elementary functions are to primitive recursive functions—the class of elementary functions is closed under a restricted kind of primitive recursion.

Proposition 2.34 (Bounded primitive recursion)

(*i*) *Suppose* $m > 0$, f *and* h *are elementary, m-ary and* $(m + 2)$*-ary respectively,* $g = \mathrm{R}^m(f, h)$, k *is elementary, and* $g(x_0, \ldots, x_m) \leq k(x_0, \ldots, x_m)$ *for all* $x_0, \ldots, x_m \in \omega$. *Then* g *is elementary.*

(*ii*) *Suppose* h *is a binary elementary function,* $g = \mathrm{R}^0(a, h)$ (*with* $a \in \omega$), *and* $gx \leq kx$ *for all* $x \in \omega$, *with* k *elementary. Then* g *is elementary.*

PROOF. (*i*) For any $x_0, \ldots, x_m, z \in \omega$ let

$$s(x_0, \ldots, x_m) = (x_m + 1) \cdot \sum_{z < xm} k(x_0, \ldots, x_{m-1}, z).$$

Let R consist of all $(m + 2)$-tuples $\langle x_0, \ldots, x_m, y\rangle$ such that there is a $q \leq \mathrm{p}_{xm}^{s(x0,\ldots,xm)}$ so that

(1) $$(q)_0 = f(x_0, \ldots, x_{m-1})$$

and, for all $z < x_m$,

(2) $$(q)_{z+1} = h(x_0, \ldots, x_{m-1}, z, (q)_z)$$

and, finally, $y = (q)_{xm}$. Obviously R is elementary. Now (*i*) follows from

(3) $$g(x_0, \ldots, x_m) = \mu y \leq k(x_0, \ldots, x_m)[\langle x_0, \ldots, x_m, y\rangle \in R].$$

To prove (3), assume that $x_0, \ldots, x_m \in \omega$, let t be the sequence $\langle g(x_0, \ldots, x_{m-1}, 0), \ldots, g(x_0, \ldots, x_{m-1}, x_m)\rangle$, and let

$$q = \prod_{i<xm} \mathrm{p}_i^{ti}.$$

Then for each $i \leq x_m$ we have

$$t_i \leq k(x_0, \ldots, x_{m-1}, i) \leq \sum_{z \leq xm} k(x_0, \ldots, x_{m-1}, z)$$

and so

$$q \leq \mathrm{p}_{xm}^{s(x0,\ldots,xm)}.$$

Furthermore, q satisfies the conditions (1), (2). Thus $\langle x_0, \ldots, x_m, g(x_0, \ldots, x_m)\rangle \in R$. It is also clear that $\langle x_0, \ldots, x_m, y\rangle \in R$ implies that $y = g(x_0, \ldots, x_m)$, so (3) holds.

Condition (*ii*) is proved similarly. □

As our final result of this chapter we shall give an example of a primitive recursive function which is not elementary.

Definition 2.35. $\mathscr{a}$ is the binary operation on ω given by the following conditions: for any $m, n \in \omega$,

$$\mathscr{a}(m, 0) = m,$$
$$\mathscr{a}(m, n + 1) = m^{\mathscr{a}(m,n)}.$$

Thus $a(m, n)$ is the iterated exponential, m raised to the m power n times. Although exponentiation is elementary by 2.8(v), we shall see that iterated exponentiation is not. The reason is that it grows faster than any elementary function; see 2.44. Obviously, we have:

Lemma 2.36. *a is primitive recursive.*

Lemma 2.37. *$m \leq a(m, n)$ for all m, n.*

PROOF. We may assume that $m \neq 0$. Now we prove 2.37 by induction on n: $a(m, 0) = m$. Assuming $m \leq a(m, n)$,

$$a(m, n + 1) = m^{a(m,n)} \geq m^m \geq m. \qquad \square$$

Lemma 2.38. *$a(m, n) < a(m, n + 1)$ for all $m > 1$ and all $n \in \omega$.*

PROOF. $a(m, n + 1) = m^{a(m,n)} > a(m, n)$. $\square$

Lemma 2.39. *$a(m, n) < a(m + 1, n)$ for all $m \neq 0$ and all $n \in \omega$.*

PROOF. We proceed by induction on n: $a(m, 0) = m < m + 1 = a(m + 1, 0)$. Assuming our result for n,

$$\begin{aligned} a(m, n + 1) &= m^{a(m,n)} \leq (m + 1)^{a(m,n)} \\ &< (m + 1)^{a(m+1,n)} = a(m + 1, n + 1). \end{aligned} \qquad \square$$

Lemma 2.40. *$a(m, n) + a(m, p) \leq a(m, \max(n, p) + 1)$ for all $m > 1$ and all $n, p \in \omega$.*

PROOF.
$$\begin{aligned} a(m, n) + a(m, p) &\leq 2a(m, \max(n, p)) \text{ by 2.38} \\ &\leq 2^{a(m,\max(n,p))} \leq m^{a(m,\max(n,p))} \\ &= a(m, \max(n, p) + 1). \end{aligned} \qquad \square$$

Lemma 2.41. *$a(m, n) \cdot a(m, p) \leq a(m, \max(n, p) + 1)$ for all $m > 1$ and all $n, p \in \omega$.*

PROOF. If $n = p = 0$ then the inequality is obvious. Hence assume that $n \neq 0$ or $p \neq 0$. Then

$$\begin{aligned} a(m, n) \cdot a(m, p) &\leq a(m, \max(n, p))^2 \quad \text{by 2.38} \\ &= (m^{a(m, \max(n,p)-1)})^2 = m^{2a(m, \max(n,p)-1)} \\ &\leq m^{\exp(2, a(m,\max(n,p)-1))} \leq m^{a(m,\max(n,p))} \\ &= a(m, \max(n, p) + 1). \end{aligned} \qquad \square$$

Lemma 2.42. *$a(m, n)^{a(m,p)} \leq a(m, \max(p + 2, n + 1))$ for all $m > 1$ and all $n, p \in \omega$.*

PROOF. For $n = 0$ we have

$$a(m, n)^{a(m,p)} = m^{a(m,p)} = a(m, p + 1) \leq a(m, \max(p + 2, n + 1))$$

(using 2.38). If $n \neq 0$ we have

$$\begin{aligned} a(m, n)^{a(m,p)} &= m^{a(m,n-1)\cdot a(m,p)} \leq m^{a(m,\max(n-1,p)+1)} \quad \text{by 2.41} \\ &= a(m, \max(p + 2, n + 1)). \end{aligned} \qquad \square$$

Lemma 2.43. *$\mathcal{a}(\mathcal{a}(m, n), p) \leq \mathcal{a}(m, n + 2p)$ for all $m > 1$ and all $n, p \in \omega$.*

PROOF. We proceed by induction on p:

$$\mathcal{a}(\mathcal{a}(m, n), 0) = \mathcal{a}(m, n) = \mathcal{a}(m, n + 2\cdot 0).$$

Assuming our result for p, we then have

$$\begin{aligned} \mathcal{a}(\mathcal{a}(m, n), p + 1) &= \mathcal{a}(m, n)^{\mathcal{a}(\mathcal{a}(m,n),p)} \leq \mathcal{a}(m, n)^{\mathcal{a}(m,n+2p)} \\ &\leq \mathcal{a}(m, \max(n + 2p + 2, n + 1)) \qquad \text{by 2.42} \\ &= \mathcal{a}(m, n + 2(p + 1)). \end{aligned}$$ □

Lemma 2.44. *If g is a k-ary elementary function then there is an $m \in \omega$ such that for all $x_0, \ldots, x_{k-1} \in \omega$, if $\max(x_0, \ldots, x_{k-1}) > 1$ then $g(x_0, \ldots, x_{k-1}) < \mathcal{a}(\max(x_0, \ldots, x_{k-1}), m)$.*

PROOF. Let A be the set of all functions g (of any rank) for which there is such an m. To prove the lemma it suffices to show that A is closed under elementary recursive operations.

(1) $$+ \in A.$$

In fact, let $m = 2$: for any $x_0, x_1 \in \omega$ with $\max(x_0, x_1) > 1$,

$$\begin{aligned} x_0 + x_1 &\leq \max(x_0, x_1) + \max(x_0, x_1) \\ &= \mathcal{a}(\max(x_0, x_1), 0) + \mathcal{a}(\max(x_0, x_1), 0) \\ &< \mathcal{a}(\max(x_0, x_1), 1) + \mathcal{a}(\max(x_0, x_1), 1) \qquad \text{by 2.38} \\ &\leq \mathcal{a}(\max(x_0, x_1), 2) \qquad \text{by 2.40} \end{aligned}$$

Thus (1) holds, Analogously,

(2) $$\cdot \in A.$$

(3) $f \in A$, where $f(m, n) = |m - n|$ for all $m, n \in \omega$.

For if $\max(x_0, x_1) > 1$, then $|x_0 - x_1| \leq \max(x_0, x_1) = \mathcal{a}(\max(x_0, x_1), 0) < \mathcal{a}(\max(x_0, x_1), 1)$. Similarly, the next two statements hold:

(4) $f \in A$, where $f(m, n) = [m/n]$ for all $m, n \in \omega$.
(5) $U_i^n \in A$, for any positive $n \in \omega$ and any $i < n$.
(6) A is closed under composition.

For, suppose f is m-ary, $g_0, \ldots, g_{m-1}$ are n-ary, and $f, g_0, \ldots, g_{m-1} \in A$. Choose $p, q_0, \ldots, q_{m-1} \in \omega$ such that $\max(x_0, \ldots, x_{m-1}) > 1$ implies that $f(x_0, \ldots, x_{m-1}) < \mathcal{a}(\max(x_0, \ldots, x_{m-1}), p)$, and such that for each $i < m$, $\max(x_0, \ldots, x_{n-1}) > 1$ implies that $g_i(x_0, \ldots, x_{n-1}) < \mathcal{a}(\max(x_0, \ldots, x_{n-1}), q_i)$. Let $h = K_n^m(f; g_0, \ldots, g_{m-1})$. Let

$$s = \max\{q_i : i < m\} + 2p + \max\{f(x_0, \ldots, x_{m-1}) : x_0, \ldots, x_{m-1} \leq 1\} + 1.$$

Now suppose that $\max(x_0, \ldots, x_{n-1}) > 1$. Then if $g_0(x_0, \ldots, x_{n-1}), \ldots, g_{n-1}(x_0, \ldots, x_{n-1}) \leq 1$, we obviously have

$$\begin{aligned} h(x_0, \ldots, x_{n-1}) &= f(g_0(x_0, \ldots, x_{n-1}), \ldots, g_{n-1}(x_0, \ldots, x_{n-1})) \\ &< s \leq \mathcal{a}(\max(x_0, \ldots, x_{n-1}), s) \qquad \text{by 2.38} \end{aligned}$$

Assume now that $\max\{g_i(x_0, \ldots, x_{n-1}) : i < m\} > 1$. Then

$$\begin{aligned} h(x_0, \ldots, x_{n-1}) &= f(g_0(x_0, \ldots, x_{n-1}), \ldots, g_{n-1}(x_0, \ldots, x_{n-1})) \\ &< \alpha(\max\{g_i(x_0, \ldots, x_{n-1}) : i < m\}, p) \\ &< \alpha(\max\{\alpha(\max\{x_0, \ldots, x_{n-1}\}, q_i) : i < m\}, p) && \text{by 2.39} \\ &= \alpha(\alpha(\max\{x_0, \ldots, x_{n-1}\}, \max\{q_0, \ldots, q_{m-1}\}), p) && \text{by 2.38} \\ &\leq \alpha(\max\{x_0, \ldots, x_{n-1}\}, \max\{q_0, \ldots, q_{m-1}\} + 2p) && \text{by 2.43} \\ &< \alpha(\max\{x_0, \ldots, x_{n-1}\}, s) && \text{by 2.38} \end{aligned}$$

(7) A is closed under $\sum$.

In fact, suppose $f \in A$, say f is m-ary, and let $g = \sum f$. Since $f \in A$, choose $p \in \omega$ such that $\max(x_0, \ldots, x_{m-1}) > 1$ implies that $f(x_0, \ldots, x_{m-1}) < \alpha(\max(x_0, \ldots, x_{m-1}), p)$. Let

$$q = p + 1 + \max\{f(x_0, \ldots, x_{m-1}) : x_0, \ldots, x_{m-1} \leq 1\}.$$

Then for any $x_0, \ldots, x_{m-1} \in \omega, f(x_0, \ldots, x_{m-1}) < \alpha(\max(x_0, \ldots, x_{m-1}, 2), q)$, using 2.38. Thus if $\max(x_0, \ldots, x_{m-1}) > 1$ we have

$$\begin{aligned} g(x_0, \ldots, x_{m-1}) &= \sum_{y < x(m-1)} f(x_0, \ldots, x_{m-2}, y) \\ &< \sum_{y < x(m-1)} \alpha(\max(x_0, \ldots, x_{m-2}, y, 2), q) \\ &\leq \sum_{y < x(m-1)} \alpha(\max(x_0, \ldots, x_{m-1}), q) && \text{by 2.39} \\ &= \alpha(\max(x_0, \ldots, x_{m-1}), q) \cdot x_{m-1} \\ &\leq \alpha(\max(x_0, \ldots, x_{m-1}), q) \cdot \alpha(\max(x_0, \ldots, x_{m-1}), q) && \text{by 2.37} \\ &\leq \alpha(\max(x_0, \ldots, x_{m-1}), q + 1) && \text{by 2.41} \end{aligned}$$

Similarly, using 2.42,

(8) A is closed under $\prod$.

This completes the proof of 2.44. □

Theorem 2.45. *There are primitive recursive functions which are not elementary in fact, α is such a function.*

PROOF. By 2.36, α is primitive recursive. Suppose α is elementary. Let $fm = \alpha(m, m)$ for all $m \in \omega$. Thus f is elementary. By 2.44 choose $m \in \omega$ such that $x > 1$ implies that $fx < \alpha(x, m)$. Then

$$\begin{aligned} \alpha(m + 2, m + 2) &= f(m + 2) < \alpha(m + 2, m) \\ &< \alpha(m + 2, m + 2) && \text{by 2.38} \end{aligned}$$

contradiction. □

BIBLIOGRAPHY

1. Grzegorczyk, A. Some classes of recursive functions. *Rozprawy Matematyczne*, *4* (1953).
2. Péter, R. *Recursive Funktionen*. Berlin: Akademie-Verlag (1957).

EXERCISES

2.46. Show that the following functions are elementary:

$$(1)\ f(x_0, \ldots, x_{m-2}, z) = \begin{cases} \max y \leqslant z((x_0, \ldots, x_{m-2}, y) \in R), \\ =0 \quad \text{if there is no such } y, \end{cases}$$

where R is elementary.

(2) $g(x_0, \ldots, x_{m-2}, y) = \max \{f(x_0, \ldots, x_{m-2}, z) : z \leq y\}$, with f elementary.

(3) $g(x_0, \ldots, x_{m-2}, y) = \min \{f(x_0, \ldots, x_{m-2}, z) : z \leq y\}$, with f elementary.

2.47. Show that the following functions and relations are elementary:

(1) $(a, b) =$ gcd (greatest common divisor) of a and b, $=0$ if $a = 0$ or $b = 0$.
(2) $sa =$ sum of positive divisors of a.
(3) the set of perfect numbers, i.e., numbers a with $sa = 2a$.
(4) the Euler φ function: $\varphi a =$ the number of elements of $\{x : 1 \leqslant x \leqslant a\}$ with $(x, a) = 1$.

2.48. Let $fn = [e \cdot n] =$ greatest integer $\leqslant e \cdot n$, for every $n \in \omega$, where e is the base of the natural system of logarithms. Show that f is elementary. *Hint:* write

$$e = 1 + \frac{1}{1!} + \frac{1}{2!} + \cdots + \frac{1}{n!} + \frac{1}{(n+1)!} + \cdots,$$

$$= \frac{1}{n!}\left(n! + \frac{n!}{1!} + \cdots + \frac{n!}{n!}\right) + \frac{1}{(n+1)!} + \cdots.$$

Let $Sn = n! + n!/1! + \cdots + n!/n!$. Define S primitive recursively, but show that is bounded by an elementary function. Let $Rn = 1/(n+1)! + \cdots$ (Note: R is *not* a number-theoretic function, since its values are actually transcendental.) Show that for $n > 1$, $Rn < 1/n!$. Hence conclude that $[e \cdot n] = [Sn/(n-1)!]$ for $n > 1$, as desired.

2.49. Show that $\binom{n}{m}$ (combinatorial symbol) is elementary.

The purpose of the following two exercises is to show how one can be rigorous in applying the results of this section in showing that functions or relations are elementary. However, later we shall not use these exercises, since the application of results of this section are obvious anyway. Both exercises have to do with certain formal languages which are special cases of languages which will be discussed in detail later.

2.50 (Explicit definition). Let A be a class of number-theoretic functions closed under composition, and such that $U_i^n \in A$ whenever $n > 0$ and $i < n$. For each $f \in A$ introduce a symbol R_f. Allow, in addition, variables v_0, v_1, $v_2, \ldots$. We define *term*: any variable standing alone is a term. If $f \in A$, f m-ary ($m > 0$), and $\sigma_0, \ldots, \sigma_{m-1}$ are terms, then so is $R_f\sigma_0, \ldots, \sigma_{m-1}$. These are all the terms.

Let i be such that all the variables appearing in a certain term τ are in the list $v_0, \ldots, v_i$. Define g_τ^i:

$$g_\tau^i(v_0, \ldots, v_i) = \tau$$

for all $v_0, \ldots, v_i \in \omega$, where each R_f occurring in τ is interpreted as f. Show that $g_\tau \in A$. [Try induction on how τ is built up.]

2.51 (COMPLEX EXPLICIT DEFINITION). For each elementary function f introduce a symbol F_f, and for each elementary relation R a symbol $\mathscr{R}_R$. Also let N_0, $N_1, \ldots$ be some more symbols, and $v_0, v_1, \ldots$ variables. For logical symbols we take $\exists, \forall, \mu, \neg, \vee, \wedge, \rightarrow, \leftrightarrow, =$. Special symbols: (,), $<$. We define terms and formulas simultaneously and recursively:

(1) v_i is a term;
(2) if f is an m-ary elementary function and $\sigma_0, \ldots, \sigma_{m-1}$ are terms, then $F_f(\sigma_0, \ldots, \sigma_{m-1})$ is a term;
(3) N_i is a term;
(4) if R is an m-ary elementary relation and $\sigma_0, \ldots, \sigma_{m-1}$ are terms, then $\mathscr{R}_R(\sigma_0, \ldots, \sigma_{m-1})$ is a formula;
(5) if σ, τ are terms then $\sigma = \tau$ is a formula;
(6) if φ and ψ are formulas then so are $\neg\varphi$, $\varphi \vee \psi$, $\varphi \wedge \psi$, $\varphi \rightarrow \psi$, $\varphi \leftrightarrow \psi$;
(7) if v_i does not occur in a term σ, and if φ is a formula, then $\exists v_i < \sigma, \varphi$ and $\forall v_i < \sigma, \varphi$ are formulas;
(8) under the assumptions of (7), $\mu v_i < \sigma, \varphi$ is a term.

These are all the terms and formulas. Now show:

(9) *if σ is a term whose variables are in the list $v_0, \ldots, v_i$, and if $f_\sigma^i(v_0, \ldots, v_i) = \sigma$ for all $v_0, \ldots, v_i \in \omega$, then f_σ^i is elementary;*
(10) *if φ is a formula whose variables are in the list $v_0, \ldots, v_i$ and if $R_\varphi^i = \{\langle v_0, \ldots, v_i\rangle : v_0, \ldots, v_i \in \omega$ and $\varphi\}$, then R_φ^i is elementary.*

In (9) and (10), the symbol F_f is to be interpreted as f; $\mathscr{R}_R$ as R; N_i as i, and the other symbols are to have their natural meanings.
Suggestion: prove (9), (10) simultaneously by induction on how σ and φ are built up.

2.52. Suppose g and g' are 1-ary primitive recursive and h and h' are 3-ary primitive recursive. Define f and f' simultaneously:

$$f(x, 0) = g(x), \qquad f(x, y + 1) = h(f(x, y), f'(x, y), x)$$
$$f'(x, 0) = g'(x), \qquad f'(x, y + 1) = h'(f(x, y), f'(x, y), x).$$

Show that f and f' are primitive recursive. Hint: define $f''(x, y) = 2^{f(x,y)} \cdot 3^{f'(x,y)}$.

2.53. Suppose that g is 1-ary primitive recursive, h is 4-ary primitive recursive and f is defined as follows:

$$f(0, n) = f(1, n) = gn$$
$$f(m + 1, n) = h(f(m - 1, n), f(m, n), m, n) \qquad \text{for } m > 0.$$

Show that f is primitive recursive.

2.54. Show that there are exactly $\aleph_0$ primitive recursive functions. Show that there is a number-theoretic function which is not primitive recursive.

3 Recursive Functions; Turing Computability

In this chapter we shall give three versions of the notion of effectively calculable function: recursive functions (defined explicitly by means of closure conditions), an analogous but less redundant version due to Julia Robinson, and the notion of Turing computable function, based upon Turing machines. These three notions will be shown to be equivalent; here the results of Chapters 1 and 2 serve as essential lemmas. In the exercises, three further equivalent notions are outlined: a variant of our official definition of recursiveness, the Gödel–Herbrand–Kleene calculus, and a generalized computer version which is even closer to actual computers than Turing machines. As stated in the introduction to this part, none of these different versions stands out as overwhelmingly superior to the others in any reasonable way. The versions involving closure conditions are mathematically the simplest. The ones using generalized machines seem the most intuitively appealing. The Kleene calculus and the Markov algorithms of the next section are closest to the kinds of symbol manipulations and algorithmic procedures that one works out on paper or within natural languages. Take your pick.

Definition 3.1. Let $m > 1$. An m-ary number-theoretic function f is called *special* if for all $x_0, \ldots, x_{m-2} \in \omega$ there is a y such that $f(x_0, \ldots, x_{m-2}, y) = 0$. If f is a special function, we let

$$k(x_0, \ldots, x_{m-2}) = \text{the least } y \text{ such that } f(x_0, \ldots, x_{m-2}, y) = 0.$$

We write "$\mu y(f(x_0, \ldots, x_{m-2}, y) = 0)$" for "$k(x_0, \ldots, x_{m-2})$". The operation of passing from f to k is called the operation of (unbounded) *minimalization.*

The class of *general recursive functions* is the intersection of all classes A of functions such that $\mathscr{s}$, $U_i^n \in A$ for all $n > 0$ and $i < n$, and such that A is closed under composition, primitive recursion, and minimalization (applied to special functions). A relation R is *general recursive* iff χ_R is general recursive. Frequently, both for functions and relations, we shall say merely *recursive* instead of *general recursive*. A class A of number-theoretic functions is said to be *closed under general recursive operations* provided that A contains all the functions $\mathscr{s}$, U_i^n and is closed under composition, primitive recursion, and minimalization (applied to special functions).

Several comments on Definition 3.1 should be made before we proceed. First, the minimalization operator used in 3.1 is somewhat different from the one in 2.16, and the difference in their notations reflects this. We shall see later that this difference is essential (see, e.g., 3.6). To see that all general recursive functions are effectively calculable it suffices to assume that f is an m-ary special effectively calculable function with $m > 1$ and that k is obtained from f by minimalization and argue that k is effectively calculable. In fact, given $x_0, \ldots, x_{m-2} \in \omega$, start computing $f(x_0, \ldots, x_{m-2}, 0)$, $f(x_0, \ldots, x_{m-2}, 1), \ldots$. Since f is special, 0 eventually appears in this sequence. The first y for which $f(x_0, \ldots, x_{m-2}, y) = 0$ is the desired value of k at $\langle x_0, \ldots, x_{m-2}\rangle$, and the calculation can then terminate. Thus the assumption that f is special is very crucial. Otherwise, for some arguments this procedure would continue forever without yielding an output.

We can argue as follows, intuitively, that every effectively calculable function is general recursive. Let f, m-ary, be effectively calculable. We then have a finitary procedure P to calculate it. Given an argument $\langle x_0, \ldots, x_{m-1}\rangle$, from P we make a calculation c; the last step of the calculation has the value $f(x_0, \ldots, x_{m-1}\rangle$ coded in it. Let T consist of all sequences $\langle P, x_0, \ldots, x_{m-1}, c\rangle$ of this sort. Presumably T itself is effectively calculable and probably more easily calculable than f. By a coding device we may assume that $P \in \omega$ and $c \in \omega$. Let V be the function that finds the output $f(x_0, \ldots, x_{m-1})$ within c. Now it is reasonable to suppose that both T and V are simple enough that they are recursive, for no matter how complicated f is, T and V must be very routinely calculable. Also, it is reasonable to assume that c is uniquely determined by P and $x_0, \ldots, x_{m-1}$. Hence

$$f(x_0, \ldots, x_{m-1}) = V\mu c(\overline{\text{sg}}\,\chi_T(P, x_0, \ldots, x_{m-1}, c) = 0),$$

so f is recursive. We shall see that this intuitive argument is very close to the rigorous argument that every Turing computable function is recursive.

Church's thesis is the philosophical principle that every effectively calculable function is recursive. This principle is important in supplying motivation for our notion of recursiveness. We shall not use it, however, in our formal development. Later, especially in Part III, we shall use what we will call the *weak Church's thesis*, which is just that certain definite arguments and constructions which we shall make are to be seen to be recursive (or even elementary) without a detailed proof. The weak Church's thesis rests on the same foot as the common feeling that most mathematics can be formalized within set theory. Of course we can take extensive practice with checking the weak Church's thesis as strong evidence for Church's thesis itself.

Theorem 3.2. *If A is closed under recursive operations, then A is closed under primitive recursive operations. In particular, every primitive recursive function is recursive.*

Now we want to see that there is a recursive function which is not primitive recursive. The argument which we shall use for this purpose is of some independent interest, so we shall first formulate it somewhat abstractly.

Definition 3.3. Let A be a collection of number-theoretic functions. A binary number-theoretic function f is said to be *universal for unary members of A* provided that for every unary $g \in A$ there is an $m \in \omega$ such that for every $n \in \omega$, $f(m, n) = gn$.

Theorem 3.4. *Let A be a set of number-theoretic functions closed under elementary recursive operations. If f is universal for unary members of A, then $f \notin A$.*

PROOF. Assume that $f \in A$. Let $gm = f(m, m) + 1$ for all $m \in \omega$. Thus $g \in A$. Since f is universal for unary members of A, choose $m \in \omega$ such that $f(m, n) = gn$ for all $n \in \omega$. Then $gm = f(m, m) = f(m, m) + 1$, contradiction. □

The proof just given is an instance of the Cantor diagonal argument. Other instances will play an important role in this part as well as in Part III; see, e.g., 15.18 and 15.20.

Lemma 3.5. *There is a general recursive function which is universal for unary primitive recursive functions.*

PROOF. We first define an auxiliary binary function h by a kind of recursion which is not primitive recursion, and afterwards we will show that h is actually general recursive. We accompany the recursive definition with informal comments. We think of a number x as coding information about an associated primitive recursive function f: $(x)_0$ is the number of arguments of f, and the next prime factor of x indicates in which case of the construction of 2.29 we are in. The definition of $h(x, y)$ for arbitrary $x, y \in \omega$ breaks into the following cases depending upon x:

Case 1 (Successor). $x = 2$. Let $h(x, y) = (y)_0 + 1$ for all y.

Case 2 (Identity functions). $x = 2^n \cdot 3^{i+1}$, where $i < n$. Let $h(x, y) = (y)_i$ for all y.

Case 3 (Composition). $x = 2^n \cdot 5^m \cdot \mathrm{p}_3^q \cdot \mathrm{p}_4^{r0} \cdot \ldots \cdot \mathrm{p}_{m+3}^{r(m-1)}$, with $n, m > 0$. For any y, let

$$h(x, y) = h(q, \mathrm{p}_0^{h(r0,y)} \cdot \ldots \cdot \mathrm{p}_{m-1}^{h(r(m-1),y)}).$$

Note here that $q < x$ and $r0, \ldots, r(m-1) < x$, so the recursion is legal.

Case 4 (Primitive recursion without parameters). $x = 2 \cdot 7^q \cdot 11^a$ with $q > 0$. We define $h(x, y)$ by recursion on y:

$$\begin{aligned} h(x, 1) &= a, \\ h(x, 2^{y+1}) &= h(q, 2^y \cdot 3^{h(x, \exp(2,y))}), \\ h(x, z) &= 0 \qquad \text{for } z \text{ not of the form } 2^u. \end{aligned}$$

Case 5 (Primitive recursion with parameters). $x = 2^{m+1} \cdot 11^q \cdot 13^r$ with $m > 0$ and $q > 0$. We define $h(x, y)$ by recursion on y. First let y be given with $(y)_m = 0$. We set

$$\begin{aligned} h(x, y) &= h(q, y) \\ h(x, y \cdot \mathrm{p}_m^{z+1}) &= h(r, y \cdot \mathrm{p}_m^z \cdot \mathrm{p}_{m+1}^{h(x, y \cdot \exp(pm, z))}). \end{aligned}$$

Case 6. For x not of one of the above forms, let $h(x, y) = 0$ for all y.

This completes the recursive definition of h. We first claim:

(1) for every $m \in \omega \sim 1$ and for every m-ary primitive recursive function f there is an $x \in \omega \sim 1$ such that, for all $y_0, \ldots, y_{m-1} \in \omega$,

$$f(y_0, \ldots, y_{m-1}) = h(x, \mathrm{p}_0^{y0} \cdot \ldots \cdot \mathrm{p}_{m-1}^{y(m-1)}).$$

Indeed, let Γ be the set of all f such that an x exists. Then, for all y,

$$h(2, 2^y) = y + 1,$$

so $\mathfrak{s} \in \Gamma$. Next, suppose $i < n$. Then for any $y_0, \ldots, y_{n-1} \in \omega$,

$$h(2^n \cdot 3^{i+1}, \mathrm{p}_0^{y0} \cdot \ldots \cdot \mathrm{p}_{n-1}^{y(n-1)}) = y_i,$$

so $\mathrm{U}_i^n \in \Gamma$. To show that Γ is closed under composition, suppose that $f \in \Gamma$, $g_0, \ldots, g_{m-1} \in \Gamma$, f m-ary, and $g_0, \ldots, g_{m-1}$ each n-ary. Choose $u \in \omega$ for f and $v_0, \ldots, v_{m-1} \in \omega$ for $g_0, \ldots, g_{m-1}$ respectively so that (1) holds for f, u; g_0, v_0; $\ldots$; g_{m-1}, v_{m-1}. Let $x = 2^n \cdot 5^m \cdot \mathrm{p}_3^u \cdot \mathrm{p}_4^{v0} \cdot \ldots \cdot \mathrm{p}_{m+3}^{v(m-1)}$. Then for any $y_0, \ldots, y_{n-1} \in \omega$ we have, with $z = \mathrm{p}_0^{y0} \cdot \ldots \cdot \mathrm{p}_{n-1}^{y(n-1)}$, $g_i(y_0, \ldots, y_{n-1}) = t_i$ for each $i < m$,

$$\begin{aligned} h(x, z) &= h(u, \mathrm{p}_0^{h(v0,z)} \cdot \ldots \cdot \mathrm{p}_{m-1}^{h(v(m-1),z)}) \\ &= h(u, \mathrm{p}_0^{t0} \cdot \ldots \cdot \mathrm{p}_{m-1}^{t(m-1)}) \\ &= f(g_0(y_0, \ldots, y_{n-1}), \ldots, g_{m-1}(y_0, \ldots, y_{n-1})). \end{aligned}$$

Thus Γ is closed under composition. To show that Γ is closed under primitive recursion without parameters, suppose $f \in \Gamma$, f binary, with associated number q so that (1) works, and suppose that $x = 2^1 \cdot 7^q \cdot 11^a$. Let $k0 = a$, $k(n+1) = f(n, kn)$ for all $n \in \omega$. Then we show that $ky = h(x, 2^y)$ for all $y \in \omega$ by induction on y:

$$\begin{aligned} h(x, 2^0) &= a = k0, \\ h(x, 2^{y+1}) &= h(q, 2^y \cdot 3^{h(x, \exp(2,y))}) \\ &= h(q, 2^y \cdot 3^{ky}) \\ &= f(y, ky) = k(y+1). \end{aligned}$$

It is similarly show that Γ is closed under primitive recursion with parameters. Thus (1) holds.

Now let $f(x, y) = h(x, 2^y)$ for all $x, y \in \omega$. Then by (1), f is universal for unary primitive recursive functions. Hence it only remains to show that h (and hence f) is general recursive. This proof can easily be modified to show that almost any legal kind of recursion leads to a general recursive function. This kind of proof is, however, very laborious. There is a much easier way of proving this kind of thing; see the comments following the recursion theorem in Chapter 5.

The computation of $h(x, y)$ can be done in finitely many steps, in which we compute successively certain other values of h: $h(a_0, b_0), \ldots, h(a_{m-1}, b_{m-1})$. We identify this sequence of computations with the number $\mathrm{p}_0^{c0} \cdot \ldots \cdot \mathrm{p}_{m-1}^{c(m-1)}$, where, for each $i < m$, $ci = 2^{ai} \cdot 3^{bi} \cdot 5^{h(ai,bi)}$. This intuitive idea should be kept in mind in checking the following statement, which clearly shows that h is general recursive. For brevity, we write $(a)_{ij}$ (or $(a)_{i,j}$ or $(a)(i, j)$) in place of $((a)_i)_j$; similar abbreviations hold for $(((a)_i)_j)_k$, etc.

Statement. For any $x, y \in \omega$, $h(x, y) = (z)_{\mathrm{l}z,2}$, where z is the least u such that $u \geq 2$, $(u)_{\mathrm{l}u,0} = x$, $(u)_{\mathrm{l}u,1} = y$, and for each $i \leq \mathrm{l}u$ one of the following holds:

(2) $(u)_{i0} = 2$ and $(u)_{i2} = (u)_{i10} + 1$;
(3) $\mathrm{l}(u)_{i0} = 1$ and $(u)_{i01} - 1 < (u)_{i00}$ and $(u)_{i2} = (u)(i, 1, (u)_{i01} - 1)$;
(4) $(u)_{i00} \neq 0$, $(u)_{i01} = 0$, $(u)_{i02} \neq 0$, $\mathrm{l}(u)_{i0} \leq (u)_{i02} + 3$, and there is a $j < i$ such that $(u)_{j0} = (u)_{i03}$, $\mathrm{l}(u)_{j1} \leq (u)_{i02} - 1$, for all $k < (u)_{i02}$ there is a $q < i$ such that $(u)_{q0} = (u)_{i,0,k+4}$, $(u)_{q1} = (u)_{i1}$, and $(u)_{q2} = (u)_{j1k}$, and, finally, $(u)_{j2} = (u)_{i2}$;
(5) $(u)_{i00} = 1$, $(u)_{i01} = (u)_{i02} = 0$, $(u)_{i03} \neq 0$, $\mathrm{l}(u)_{i0} \leq 4$, and one of the following three cases holds:
   (5′) $(u)_{i1} = 1$ and $(u)_{i2} = (u)_{i04}$;
   (5″) there is a $w < (u)_i$ such that $(u)_{i1} = 2^{w+1}$, and there is a $j < i$ such that $(u)_{j0} = (u)_{i03}$ and for some $k < i$, $(u)_{k0} = (u)_{i0}$, $(u)_{k1} = 2^w$, $(u)_{j1} = 2^w \cdot 3^{(u)k2}$, and $(u)_{j2} = (u)_{i2}$;
   (5‴) there is no $w < (u)_i$ such that $(u)_{i1} = 2^w$, and $(u)_{i2} = 0$;
(6) $(u)_{i00} > 1$, $(u)_{i01} = (u)_{i02} = (u)_{i03} = 0$, $(u)_{i04} \neq 0$, $\mathrm{l}(u)_{i0} \leq 5$, and one of the following conditions holds (with $(u)_{i00} - 1 = m$ for brevity):
   (6′) $(u)_{i1m} = 0$ and there is a $j < i$ such that $(u)_{j0} = (u)_{i04}$, $(u)_{j1} = (u)_{i1}$, and $(u)_{j2} = (u)_{i2}$;
   (6″) $(u)_{i1m} \neq 0$, say $(u)_{i1} = t \cdot \mathrm{p}_m$, and there exist $j, k < i$ such that $(u)_{j0} = (u)_{i05}$, $(u)_{j1} = t \cdot \exp(\mathrm{p}_{m+1}, (u)_{k2})$, $(u)_{k0} = (u)_{i0}$, $(u)_{k1} = t$, and $(u)_{j2} = (u)_{i2}$;
(7) none of the above, and $(u)_{i2} = 0$.

To check this statement carefully, let A be the set of all $u \geq 2$ satisfying the condition above beginning "for each $i \leq \mathrm{l}u$". Then the following condition is clear:

(8) if $u, v \in A$, then $u \cdot \prod_{i \leqslant \mathrm{l}v} \mathrm{p}_{\mathrm{l}u+i+1}^{(v)_i} \in A$.
(9) for all $x, y \in \omega$, there is a $u \in A$ with $(u)_{\mathrm{l}u,0} = x$ and $(u)_{\mathrm{l}u,1} = y$; for any such u, $(u)_{\mathrm{l}u,2} = h(x, y)$.

Condition (9) is established by induction on x.

This completes the proof of 3.5. □

Theorem 3.6. *There is a recursive function which is not primitive recursive.*

Theorem 3.7. *There are exactly $\aleph_0$ recursive functions.*

PROOF. Let A_0 consist of all of the functions $\mathfrak{s}$, U_i^n with $i < n$. Thus $|A_0| = \aleph_0$. Having defined A_n, let A_{n+1} consist of all members of A_n together with all functions obtainable from members of A_n by one application of composition, primitive recursion, or minimalization (applied to special functions). Thus if $|A_n| = \aleph_0$, then $|A_{n+1}| = \aleph_0$. Clearly, then, $|\bigcup_{n\in\omega} A_n| = \aleph_0$. Obviously $\bigcup_{n\in\omega} A_n$ is exactly the set of all recursive functions. □

Theorem 3.8. *There is a number-theoretic function which is not recursive.*

Although Theorem 3.8 follows from 3.7 purely on grounds of cardinality, we can also explicitly exhibit a nonrecursive function. Let $f_0, f_1, \ldots$ be an enumeration of all unary recursive functions (by 3.7). Define $gm = f_m m + 1$ for all $m \in \omega$. Then g is obviously not in our enumeration, so g is not recursive. We are really just repeating the proof of Theorem 3.4 here in a special case.

We now turn to the notion of a Turing computable function.

Definition 3.9

(*i*) If $g = \langle g_0, \ldots, g_{m-1}\rangle$ is a finite sequence of 0's and 1's and F is a tape description (recall Definition 1.2), then we say that g *lies on F beginning at q and ending at n* (where $q, n \in \mathbb{Z}$), provided that $Fq = g0$, $F(q + 1) = g_1, \ldots, F_n = g_{m-1}$ (thus $n = q + m - 1$).

(*ii*) An m-ary number-theoretic function f is *Turing computable* iff there is a Turing machine M, with notation as in Definition 1.1, such that for every tape description F, all $q, n \in \mathbb{Z}$, and all $x_0, \ldots, x_{m-1} \in \omega$, if $01^{(x0+1)} 0\cdots01^{(x(m-1)+1)}$ lies on F beginning at q and ending at n, and if $Fi = 0$ for all $i > n$, then there is a computation $\langle(F, c_1, n + 1), (G_1, a_1, b_1), \ldots, (G_{p-1}, a_{p-1}, b_{p-1})\rangle$ of M having the following properties:

(1) $G_{p-1}i = Fi$ for all $i \leq n + 1$;
(2) $1^{(f(x0,\ldots,x(m-1))+1)}$ lies on G_{p-1} beginning at $n + 2$ and ending at $b_{p-1} - 1$;
(3) $G_{p-1}i = 0$ for all $i \geq b_{p-1}$.

We then say that f is *computed by M*.

There are, of course, several arbitrary aspects in this definition of computable function. Many details could be changed without modifying in an essential way the power of the notion. We have simply specified in a detailed way how an input for the machine is to be presented and how the output is to be located. The condition (1) is particularly useful in combining several computations. Now we show that *every recursive function is Turing computable.*

Lemma 3.10. *$\mathscr{s}$ is Turing computable.*

PROOF. A machine for $\mathscr{s}$ is:

$$T_{\text{copy}} \to T_1 \to T_{\text{left}}. \qquad \square$$

Lemma 3.11. *U_i^n is Turing computable.*

PROOF. The machine is $T_{(n-i)\text{copy}}$. $\square$

Lemma 3.12. *The class of Turing computable functions is closed under composition.*

PROOF. Suppose f m-ary, $g_0, \ldots, g_{m-1}$ n-ary. Suppose $f, g_0, \ldots, g_{m-1}$ are computed by $M, N_0, \ldots, N_{m-1}$ respectively. Then the following machine computes $\mathrm{K}_n^m(f; g_0, \ldots, g_{m-1})$:

$$\begin{array}{l} T_{\text{left}} \to T_1 \to T_{\text{left}} \to T^n_{(n+1)\text{copy}} \to T^n_{l\,\text{seek}\,0} \to T_{\text{right}} \to T_0 \to T_{\text{rend}} \to \\ N_0 \to T^n_{(n+1)\text{copy}} \to N_1 \to \cdots \to T^n_{(n+1)\text{copy}} \to N_{m-1} \to T_{(m+(m-1)n)\text{copy}} \to \\ T_{(m+(m-2)n)\text{copy}} \to \cdots \to T_{m\,\text{copy}} \to M \to T_{\text{fin}}. \end{array} \qquad \square$$

Lemma 3.13. *The class of Turing computable functions is closed under primitive recursion without a parameter.*

PROOF. Suppose that f is a binary operation on ω, computed by a machine M, and $a \in \omega$. Let $g0 = a$, $g(n+1) = f(n, gn)$ for all $n \in \omega$. Then the following machine computes g:

$$T_{\text{left}} \to T_1 \to T_{\text{left}} \to T_{2\,\text{copy}} \to T_{l\,\text{seek}\,0} \to T_{\text{right}} \to T_0 \to T_{\text{rend}} \to$$

$$\begin{array}{l} T_{\text{left}} \to (T_1 \to T_{\text{left}})^{a+1} \to T_{2\,\text{copy}} \to T_{\text{right}} \to T_0 \to T_{\text{right}} \xrightarrow{\text{if } 1} T_{\text{left}} \to \\ \qquad\qquad\qquad\qquad\qquad\qquad\qquad\qquad\quad \downarrow \text{if } 0 \\ \qquad\qquad\qquad\qquad\qquad\qquad\qquad\qquad\quad T_{\text{fin}} \end{array}$$

$$\begin{array}{l} T_{\text{left}} \to \overleftarrow{T_1 \to T_{\text{left}} \to T_{3\,\text{copy}} \to M \to T_{4\text{copy}} \to T_{\text{right}} \to T_0 \to T_{\text{right}} \xrightarrow{\text{if } 1} T_{\text{left}} \to T_{4\text{copy}}} \\ \qquad\qquad\qquad\qquad\qquad\qquad\qquad\qquad\qquad\qquad\qquad\qquad \downarrow \text{if } 0 \\ \qquad\qquad\qquad\qquad\qquad\qquad\qquad\qquad\qquad\qquad\qquad\qquad T_{\text{fin}} \end{array} \qquad \square$$

Lemma 3.14. *The class of Turing computable functions is closed under primitive recursion with parameters.*

PROOF. Suppose that f is m-ary, $m > 0$, g is $(m+2)$-ary and that they are computed by M and N respectively. Let $h(x_0, \ldots, x_{m-1}, 0) = f(x_0, \ldots, x_{m-1})$,

$h(x_0, \ldots, x_{m-1}, y + 1) = g(x_0, \ldots, x_{m-1}, y, g(x_0, \ldots, x_{m-1}, y))$. Then the following machine computes h:

$$T_{\text{left}} \to T_1 \to T_{\text{left}} \to T_{2\,\text{copy}} \to T^m_{(m+3)\text{copy}} \to T^{m+1}_{l\,\text{seek}\,0} \to T_{\text{right}} \to$$

$$T_0 \to T_{\text{rend}} \to M \to T_{(m+2)\,\text{copy}} \to T_{\text{right}} \to T_0 \to T_{\text{right}} \xrightarrow{\text{if } 1} T_{\text{left}} \to$$

$$\left\downarrow \text{if } 0 \right.$$

$$T_{\text{fin}}$$

$$T^m_{(m+2)\,\text{copy}} \to T_{\text{left}} \to T_1 \to T_{\text{left}} \to T_{(m+3)\,\text{copy}} \to N \to T_{(m+4)\,\text{copy}} \to T_{\text{right}} \xrightarrow{\text{if } 1}$$

$$\left\downarrow \text{if } 0 \right.$$

$$T_{\text{fin}}$$

$$T_{\text{left}} \to T^{m+1}_{(m+4)\,\text{copy}}$$

$\square$

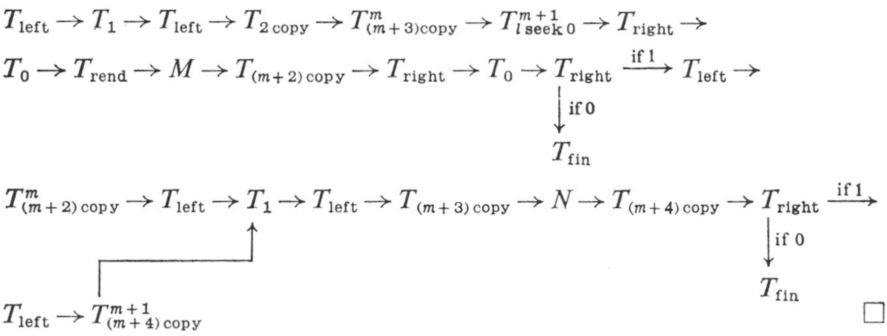

Lemma 3.15. *The class of Turing computable functions is closed under minimalization (applied to special functions).*

PROOF. Let f be an m-ary special function, $m > 1$, and suppose that f is computed by a machine M. Let $g(x_0, \ldots, x_{m-2}) = \mu y[f(x_0, \ldots, x_{m-2}, y) = 0]$ for all $x_0, \ldots, x_{m-2} \in \omega$. Then the following machine computes g:

$$T_{\text{left}} \to T_1 \to T_{\text{left}} \to M \to T_{\text{right}} \to T_0 \to T_{\text{right}} \xrightarrow{\text{if } 1} T_0 \to T_{\text{right}}$$

$$\left\downarrow \text{if } 0 \right. \qquad\qquad \text{if } 1 \left\uparrow\right.$$

$$\text{Stop} \qquad\qquad \square$$

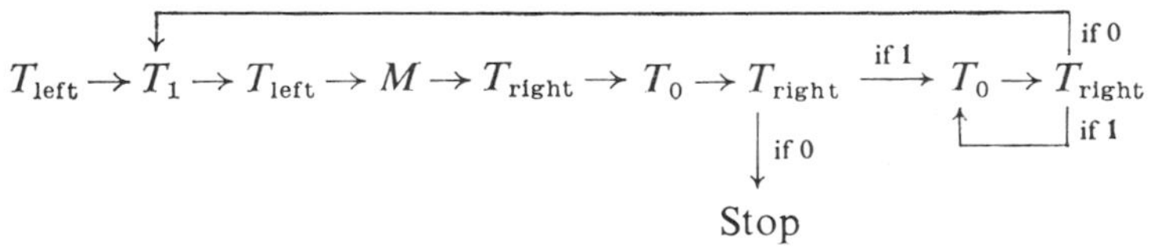

Summarizing Lemmas 3.10–3.15, we have:

Lemma 3.16. *Every general recursive function is Turing computable.*

We now want to get the converse of 3.16. This requires Gödel numbering. This process, whose name is just a catch-word for the process of number-theoretically effectivizing nonnumber-theoretic concepts (already hinted at in the introduction to this part), has already been used twice in less crucial contexts. In discussing course-of-values recursion, we numbered finite sequences of numbers; see 2.31. And in constructing a function universal for unary primitive recursive functions essentially we numbered construction sequences for primitive recursive functions; see 3.5. Now we want to effectively number various of the concepts surrounding the notion of Turing machine. Besides our immediate purpose of proving the equivalence of Turing computability and recursiveness, this effectivization will be important for our later discussion of general recursion theory.

The script letter $\mathscr{g}$ will be used for Gödel numbering functions throughout this book; we will usually just depend on the context to distinguish the various particular uses of "$\mathscr{g}$."

Definition 3.17. Let $\mathbb{E}$ be the set of even numbers. Let T be the class of all Turing machines. If M is a Turing machine, with notation as in 1.1, we let the *Gödel number* of M, $\mathcal{g}M$, be the number

$$\prod_{i<2m} \mathrm{p}_i^{ti},$$

where, for each $i < 2m$, $ti = 2^{c[i+2/2]} \cdot 3^{(\chi\mathbb{E})(i+1)} \cdot 5^{v(i+1)} \cdot 7^{d(i+1)}$.

Lemma 3.18. *$\mathcal{g}^*T$ is elementary.*

PROOF. For any $x \in \omega$, $x \in \mathcal{g}^*T$ if $\mathrm{l}x$ is odd, $x > 1$, for every $i \leq \mathrm{l}x$ we have $((x)_i)_2 < 5$, for every $i \leq \mathrm{l}x$ there is a $j \leq \mathrm{l}x$ such that $((x)_i)_3 = ((x)_j)_0$, for every $i \leq \mathrm{l}x$, if i is even then $((x)_i)_0 = ((x)_{i+1})_0$, and for all $i, j \leq \mathrm{l}x$, if $i + 2 \leq j$, then $((x)_i)_0 \neq ((x)_j)_0$, and if i is even then $((x)_i)_1 = 0$, while if i is odd, $((x)_i)_1 = 1$. □

Definition 3.19. If F is a tape description, then the *Gödel number* of F, $\mathcal{g}F$, is the number

$$\prod_{i=0}^{\infty} \mathrm{p}_i^{ki},$$

where

$$k_i = \begin{cases} F(i/2) & \text{if } i \text{ is even,} \\ F(-(i+1)/2) & \text{if } i \text{ is odd.} \end{cases}$$

Note that a natural number m is the Gödel number of some tape description iff $\forall x < \mathrm{l}m((m)_x < 2)$ and $m \neq 0$.

Definition 3.20. A *complete configuration* is a quadruple (M, F, d, e) such that (F, d, e) is a configuration in the Turing machine M. $\mathbb{C}$ is the set of all complete configurations. The *Gödel number* $\mathcal{g}(M, F, d, e)$ of such a complete configuration is the number

$$2^{\mathcal{g}M} \cdot 3^{\mathcal{g}F} \cdot 5^d \cdot 7^n,$$

where

$$n = \begin{cases} 2e & \text{if } e \geq 0, \\ -2e - 1 & \text{if } e < 0. \end{cases}$$

Lemma 3.21. *$\mathcal{g}^*\mathbb{C}$ is elementary.*

PROOF. For any $x \in \omega$, $x \in \mathcal{g}^*\mathbb{C}$ iff $\forall i \leq \mathrm{l}(x)_1(((x)_1)_i < 2)$, $(x)_1 \neq 0$, $(x)_0 \in \mathcal{g}^*T$, and there is an $i \leq \mathrm{l}(x)_0$ such that $(x)_2 = (((x)_0)_i)_0$, and $\mathrm{l}x \leq 3$. □

Definition 3.22. (*i*) For any $e \in \mathbb{Z}$, let

$$\mathcal{g}e = \begin{cases} 2e & \text{if } e \geq 0, \\ -2e - 1 & \text{if } e < 0. \end{cases}$$

For any $x \in \omega$, let

$$f_0 x = \begin{cases} x + 2 & \text{if } x \text{ is even,} \\ 0 & \text{if } x = 1, \\ x - 2 & \text{if } x \text{ is odd and } x > 1, \end{cases}$$

$$f_1 x = \begin{cases} x - 2 & \text{if } x \text{ is even and } x > 0, \\ 1 & \text{if } x = 0, \\ x + 2 & \text{if } x \text{ is odd.} \end{cases}$$

Lemma 3.23. *f_0 and f_1 are elementary. For any $e \in \mathbb{Z}$ we have $f_0 \mathscr{g} e = \mathscr{g}(e + 1)$ and $f_1 \mathscr{g} e = \mathscr{g}(e - 1)$.*

Proof

$$f_0 \mathscr{g} e = \begin{Bmatrix} f_0 2e & e \geq 0 \\ f_0(-2e - 1) & e < 0 \end{Bmatrix} = \begin{Bmatrix} 2(e + 1) & e \geq 0 \\ 0 & e = -1 \\ -2e - 3 & e < -1 \end{Bmatrix} = \mathscr{g}(e + 1);$$

$$f_1 \mathscr{g} e = \begin{Bmatrix} f_1 2e & e \geq 0 \\ f_1(-2e - 1) & e < 0 \end{Bmatrix} = \begin{Bmatrix} 2(e - 1) & e > 0 \\ 1 & e = 0 \\ -2e + 1 & e < 0 \end{Bmatrix} = \mathscr{g}(e - 1). \qquad \square$$

Lemma 3.24. *Let $R_0 = \{(x, n, \varepsilon, y) : x = \mathscr{g}F$ for some tape description F, $n = \mathscr{g}e$ for some $e \in \mathbb{Z}$, $\varepsilon = 0$ or $\varepsilon = 1$, and $y = \mathscr{g}(F^e_\varepsilon)\}$. Then R_0 is elementary.*

Proof. $(x, n, \varepsilon, y) \in R_0$ iff $\forall i \leq \text{l}x((x)_i < 2)$, $x \neq 0$, $\varepsilon < 2$, and $y = [x/\text{p}_n^{(x)n}] \cdot \text{p}_n^\varepsilon$. $\square$

Lemma 3.25. *Let $R_1 = \{(x, y) : x$ is the Gödel number of a complete configuration (M, F, d, e), y is the Gödel number of a complete configuration (M, F', d', e') (same M), and $((F, d, e), (F', d', e'))$ is a computation step$\}$. Then R_1 is elementary.*

Proof. For any x, y, $(x, y) \in R_1$ iff $x \in \mathscr{g}^*\mathbb{C}$, $y \in \mathscr{g}^*\mathbb{C}$, $(x)_0 = (y)_0$, and there is an $i \leq \text{l}((x)_0)$ such that $(x)_2 = (((x)_0)_i)_0$, $(((x)_0)_i)_1 = ((x)_1)_{(x)3}$, and one of the following conditions holds:

(a) $(((x)_0)_i)_2 = 0$, $((x)_1, (x)_3, 0, (y)_1) \in R_0$, $(y)_2 = (((x)_0)_i)_3$, and $(y)_3 = (x)_3$;
(b) $(((x)_0)_i)_2 = 1$, $((x)_1, (x)_3, 1, (y)_1) \in R_0$, $(y)_2 = (((x)_0)_i)_3$, and $(y)_3 = (x)_3$;
(c) $(((x)_0)_i)_2 = 2$, $(y)_1 = (x)_1$, $(y)_2 = (((x)_0)_i)_3$, and $(y)_3 = f_1((x)_3)$;
(d) $(((x)_0)_i)_2 = 3$, $(y)_1 = (x)_1$, $(y)_2 = (((x)_0)_i)_3$, and $(y)_3 = f_0((x)_3)$. $\square$

Definition 3.26. A *complete computation* is a sequence $\mathfrak{M} = \langle (M, F_0, d_0, e_0), \ldots, (M, F_m, d_m, e_m) \rangle$ such that $\langle (F_0, d_0, e_0), \ldots, (F_m, d_m, e_m) \rangle$ is a computation in M. The *Gödel number* of such a complete computation is the number

$$\prod_{i < m} \text{p}_i^{\mathscr{g}(M, F_i, d_i e_i)}.$$

Let R_2 be the set of all Gödel numbers of complete computations.

Lemma 3.27. *R_2 is elementary.*

PROOF. For any x, $x \in R_2$ iff for every $i \leq \mathrm{l}x$, $(x)_i \in \mathcal{g}^*\mathbb{C}$, and $((((x)_0)_0)_0)_0 = ((x)_0)_2$, and for every $i < \mathrm{l}x$, $((x)_i, (x)_{i+1}) \in R_1$, and there is an $i \leq \mathrm{l}((x)_0)_0$ such that $((((x)_0)_0)_i)_0 = ((x)_{\mathrm{l}x})_2$, $((((x)_0)_0)_i)_1 = (((x)_{\mathrm{l}x})_1)_{((x)\mathrm{l}x)3}$, and

$$((((x)_0)_0)_i)_2 = 4. \qquad \square$$

Definition 3.28. If h is a finite sequence of 0's and 1's, we let

$$\mathcal{g}h = \prod_{i < \mathrm{Dmn}h} \mathrm{p}_i^{hi+1}.$$

For any $x \in \omega$, let $f_2x = \prod_{i \leq x} \mathrm{p}_i^2$.

Lemma 3.29. *f_2 is elementary, and $f_2x = \mathcal{g}1^{(x+1)}$ for any x.*

Definition 3.30. For any $x, y \in \omega$, $\mathrm{Cat}\,(x, y) = x \cdot \prod_{i \leq \mathrm{l}y} \mathrm{p}_{\mathrm{l}x+i+1}^{(y)i}$.

Lemma 3.31. *If h and k are finite sequences of 0's and 1's, then $\mathcal{g}(hk) = \mathrm{Cat}\,(\mathcal{g}h, \mathcal{g}k)$. (Recall the definition of hk from 1.11.)*

Definition 3.32. $f_3^1x = \mathrm{Cat}\,(2, f_2x)$. For $m > 1$,

$$f_3^m(x_0, \ldots, x_{m-1}) = \mathrm{Cat}\,(f_3^{m-1}(x_0, \ldots, x_{m-2}), \mathrm{Cat}\,(2, f_2x_{m-1})).$$

Lemma 3.33. *f_3^m is elementary for each m, and $f_3^m(x_0, \ldots, x_{m-1}) = \mathcal{g}(0\ 1^{(x0+1)}\ 0 \cdots 0\ 1^{(x(m-1)+1)})$.*

Lemma 3.34. *Let $R_3 = \{(x, y, m, n)$: x is the Gödel number of a tape description F, y is the Gödel number of a finite sequence h of 0's and 1's, $m = \mathcal{g}e$ and $n = \mathcal{g}e'$ for certain $e, e' \in \mathbb{Z}$, and h lies on F beginning at e and ending at $e'\}$. Then R_3 is elementary.*

PROOF. For any x, y, m, n $(x, y, m, n) \in R_3$ iff $y \neq 0$, $\forall i \leq \mathrm{l}x((x)_i < 2)$, $x \neq 0$, either $y = 1$ and $m = n$, or else $y > 1$, for every $i \leq \mathrm{l}y[(y)_i = 1$ or $(y)_i = 2]$, and there is a $z \leq (m + 2y)^y$ such that $(z)_0 = m$, $f_0((z)_i) = (z)_{i+1}$ for each $i < \mathrm{l}z$, $\mathrm{l}z = \mathrm{l}y$, $(z)_{\mathrm{l}z} = n$, and for each $i \leq \mathrm{l}z$, $(x)_{(z)i} = (y)_i \dot{-} 1$. $\square$

The notations $f_0, f_1, f_2, f_3^m, R_0, R_1, R_2, R_3$ will not be used beyond the present section. The relations T_m introduced next, however, are fundamental for the aspects of recursion theory dealt with in Chapters 5 and 6.

Definition 3.35. For $m > 0$ let $\mathrm{T}_m = \{(e, x_0, \ldots, x_{m-1}, u)$: e is the Gödel number of a Turing machine M, and u is the Gödel number of a complete computation $\langle(M, F_0, d_0, v_0), \ldots, (M, F_n, d_n, v_n)\rangle$ such that $01^{(x0+1)}\ 0 \cdots 0\ 1^{(x(m-1)+1)}$ lies on F_0 ending at -1, F_0 is zero otherwise, $v_0 = 0$, and $F_n1 = 1.\}$

Note that for any $e, x_0, \ldots, x_{m-1} \in \omega$ *there is at most one* u *such that* $(e, x_0, \ldots, x_{m-1}, u) \in \mathrm{T}_m$.

Lemma 3.36. T_m *is elementary.*

PROOF. For any $e, x_0, \ldots, x_{m-1}, u$, $(e, x_0, \ldots, x_{m-1}, u) \in \mathrm{T}_m$ iff $e \in \mathcal{g}^*\mathbb{T}$, $u \in R_2$, $((u)_0)_0 = e$, $(((u)_0)_1, f_3^m(x_0, \ldots, x_{m-1}), s, 1) \in R_3$ and $\forall t \leq ((u)_0)_1$ (t odd and $t > s \Rightarrow (((u)_0)_1)_t = 0$) and $\forall t \leq ((u)_0)_1$ (t even $\Rightarrow (((u)_0)_1)_t = 0$) for some $s \leq ((u)_0)_1$, $((u)_0)_3 = 0$, and $(((u)_{1u})_1)_2 = 1$. □

Definition 3.37. For any $x \in \omega$ let

$$\mathrm{V}x = \mu y \leq x[(((x)_{1x})_1, \mathrm{Cat}\,(f_2 y, 2), 2, 2y + 4) \in R_3]$$

Obviously V is elementary.

Lemma 3.38. *Every Turing computable function is recursive.*

PROOF. Let M be a Turing machine which computes f as described in Definition 3.9(*ii*), and let $e = \mathcal{g}M$. Then for any $x_0, \ldots, x_{m-1} \in \omega$,

$$f(x_0, \ldots, x_{m-1}) = \mathrm{V}\mu u[(e, x_0, \ldots, x_{m-1}, u) \in \mathrm{T}_m],$$

as desired. □

Theorem 3.39. *A function is Turing computable iff it is recursive.*

We close this chapter with a variant of the notion of recursiveness due to Julia Robinson [3]. It will be useful to us later on. The idea is to simplify the definition of recursive function by using rather complicated initial functions but very simple recursive operations.

Definition 3.40. $[\surd]$ is the function such that $[\sqrt{x}]$ = greatest integer $\leq \sqrt{x}$ for each $x \in \omega$. Also, for any $x \in \omega$ we let $\mathrm{Exc}\, x = x - [\sqrt{x}]^2$; this is the *excess* of x over a square.

Lemma 3.41. $[\surd]$ *and* Exc *are elementary.*

PROOF. $[\sqrt{x}] \leq x$ for all $x \in \omega$. Further,

$$[\sqrt{0}] = 0$$

$$[\sqrt{(n + 1)}] = \begin{cases} [\sqrt{n}] & \text{if } n + 1 \neq ([\sqrt{n}] + 1)^2, \\ [\sqrt{n}] + 1 & \text{otherwise} \end{cases}$$

$$= [\sqrt{n}] + \overline{\mathrm{sg}}\, |n + 1 - ([\sqrt{n}] + 1)^2|$$

Thus we may use 2.34. Finally, $\mathrm{Exc}\, n = n \dot{-} [\sqrt{n}]^2$. □

The next definition and theorem introduce special cases of the important device of *pairing functions*, extensively used in recursive function theory.

Definition 3.42. (*i*) $J(a, b) = ((a + b)^2 + b)^2 + a$ for all $a, b \in \omega$. (*ii*) $Lx = \text{Exc}\,[\sqrt{x}]$ for all $x \in \omega$.

Theorem 3.43

(*i*) J *and* L *are elementary;*
(*ii*) $\text{Exc}\,0 = 0$ *and* $L0 = 0$;
(*iii*) *if* $\text{Exc}\,(a + 1) \neq 0$, *then* $\text{Exc}\,(a + 1) = \text{Exc}\,a + 1$ *and* $L(a + 1) = La$;
(*iv*) $\text{Exc}\,J(a, b) = a$;
(*v*) $LJ(a, b) = b$;
(*vi*) J *is* $1 - 1$.

Proof. (*i*) and (*ii*) are obvious. As to (*iii*), choose x such that $a = x^2 + \text{Exc}\,a < (x + 1)^2$. Since $\text{Exc}\,(a + 1) \neq 0$, it is then clear that $\text{Exc}\,(a + 1) = \text{Exc}\,a + 1$. Furthermore, clearly $x^2 \leq a < (x + 1)^2$ and $x^2 \leq a + 1 < (x + 1)^2$, so $x = [\sqrt{a}] = [\sqrt{(a + 1)}]$ and hence $La = L(a + 1)$.

To prove (*iv*), note that

$$\begin{aligned} ((a + b)^2 + b)^2 &\leq J(a, b) \\ &< ((a + b)^2 + b)^2 + 2(a + b)^2 + 2b + 1 \\ &= ((a + b)^2 + b + 1)^2. \end{aligned}$$

Hence $\text{Exc}\,J(a, b) = a$, and (*iv*) holds. Furthermore, clearly from the above $[\sqrt{J(a, b)}] = (a + b)^2 + b$; since

$$\begin{aligned} (a + b)^2 &\leq [\sqrt{J(a, b)}] \\ &< (a + b)^2 + 2a + 2b + 1 \\ &= (a + b + 1)^2, \end{aligned}$$

we infer that $LJ(a, b) = \text{Exc}\,[\sqrt{J(a, b)}] = b$, as desired in (*v*). Finally, (*vi*) is a purely set-theoretical consequence of (*iv*) and (*v*). □

For the next results we assume a very modest acquaintance with number theory; see any number theory textbook.

Theorem 3.44 (Number-theoretic: The Chinese remainder theorem). *Let* $m_0, \ldots, m_{r-1}$ *be natural numbers* > 1, *with* $r > 1$, *the* m_i*'s pairwise relatively prime. Let* $a_0, \ldots, a_{r-1}$ *be any r natural numbers. Then there is an* $x \in \omega$ *such that* $x \equiv a_i (\text{mod } m_i)$ *for all* $i < r$.

Proof. By induction on r; we first take the case $r = 2$. Since m_0 and m_1 are relatively prime, there exist integers (positive, negative, or zero) s and t such that $1 = m_0 s + m_1 t$. Then $a_0 - a_1 = m_0 s(a_0 - a_1) + m_1 t(a_0 - a_1)$. Choose $u \in \omega$ such that $a_0 - m_0 s(a_0 - a_1) + u m_0 m_1 > 0$, and let $x = a_0 - m_0 s(a_0 - a_1) + u m_0 m_1$. Then $x \equiv a_0 (\text{mod } m_0)$, and $x = a_1 + m_1 t(a_0 - a_1) + u m_0 m_1 \equiv a_1 (\text{mod } m_1)$, as desired.

Now we assume the theorem true for r and prove it with "r" replaced by

"$r + 1$". With s, t, u as above, choose $x \in \omega$ such that $x \equiv a_0 - m_0 s(a_0 - a_1) + u m_0 m_1 (\text{mod } m_0 m_1)$, $x \equiv a_2 (\text{mod } m_2), \ldots, x \equiv a_r (\text{mod } m_r)$. Then $x \equiv a_0$ (mod m_0) and, since $a_0 - m_0 s(a_0 - a_1) = a_1 + m_1 t(a_0 - a_1)$, $x \equiv a_1 (\text{mod } m_1)$ as desired. □

Definition 3.45. For all $x, i \in \omega$ let $\beta(x, i) = \text{rm}(\text{Exc } x, 1 + (i + 1)\text{L}x)$.

Theorem 3.46 (Number-theoretic: Gödel's β-function lemma). *For any finite sequence $y_0, \ldots, y_{n-1}$ of natural numbers there is an $x \in \omega$ such that $\beta(x, i) = y_i$ for each $i < n$.*

Proof. Let s be the maximum of $y_0, \ldots, y_{n-1}, n$. For each $i < n$ let $m_i = 1 + (i + 1) \cdot s!$ Then for $i < j < n$ the integers m_i and m_j are relatively prime. For, if a prime p divides both m_i and m_j, it also divides $m_j - m_i = (j + 1) \cdot s! - (i + 1) \cdot s! = (j - i) \cdot s!$ Now $p \nmid s!$, since $p | 1 + (i + 1)s!$. Hence $p | j - i$. But $j - i < n \leq s$, and hence this would imply that $p | s!$, which we know is impossible. Thus indeed m_i and m_j are relatively prime.

Hence by the Chinese remainder theorem choose v such that

$$v \equiv y_i (\text{mod } m_i) \qquad \text{for each } i < n.$$

Let $x = \text{J}(v, s!)$. Then $\text{Exc } x = v$ by 3.43(*iv*), and $\text{L}x = s!$ by 3.43(*v*). Hence if $i < n$ we have

$$\begin{aligned} \beta(x, i) &= \text{rm}(\text{Exc } x, 1 + (i + 1)\text{L}x) \\ &= \text{rm}(v, m_i) \\ &= y_i. \end{aligned}$$

□

Definition 3.47. If f is a 1-place function with range ω, let $f^{(-1)}y = \mu x(fx = y)$ for all $y \in \omega$. We say that $f^{(-1)}$ is obtained from f by *inversion.*

Theorem 3.48 (Julia Robinson). *The class of recursive functions is the intersection of all classes A of functions such that $+$, $\mathfrak{s}$, Exc, $\text{U}_i^n \in A$ (for $0 \leq i < n$), and such that A is closed under the operations of composition, and of inversion (applied to functions with range ω).*

Proof. Clearly the indicated intersection is a subset of the class of recursive functions ($f^{(-1)}y = \mu x(|fx - y| = 0)$, so we have here a special case of minimalization). Now suppose that A is a class with the properties indicated in the statement of the theorem. We want to show that every recursive function is in A. This will take several steps.

The general idea of the proof is this: Inversion is a special case of minimalization, and the general case is obtained from inversion by using pairing functions. Primitive recursion is obtained by representing the computation of a function f as a finite sequence of the successive values of f, coding the sequence into one number using the β function, and selecting that number out by minimalization.

Our proof will begin with some preliminaries, giving a stock of members of A, which leads to the fact that the pairing functions are in A. First note that for any $x \in \omega$, $x^2 \leq x^2 + x < (x + 1)^2$, and hence $\text{Exc}(x^2 + x) = x$. Thus

(1) $$\text{Exc has range } \omega.$$

Next,

(2) $$\text{Exc}^{(-1)}(2x) = x^2 + 2x \qquad \text{for all } x \in \omega.$$

For, obviously $\text{Exc}(x^2 + 2x) = 2x$. If $\text{Exc}(y) = 2x$ with $y < x^2 + 2x$, we may write $y = z^2 + 2x < (z + 1)^2$ and so $z < x$ and hence $(z + 1)^2 = z^2 + 2z + 1 \leq z^2 + 2x < (z + 1)^2$, a contradiction. Thus (2) holds.

Again,

(3) $$\text{Exc}^{(-1)}(2x + 1) = x^2 + 4x + 2 \qquad \text{for all } x \in \omega.$$

For, $(x + 1)^2 = x^2 + 2x + 1 < x^2 + 4x + 2 < x^2 + 4x + 4 = (x + 2)^2$, and hence $\text{Exc}(x^2 + 4x + 2) = 2x + 1$. Now suppose $\text{Exc}(y) = 2x + 1$, with $y < x^2 + 4x + 2$. Choose z such that $y = z^2 + 2x + 1 < (z + 1)^2$. Then $z^2 + 2x + 1 = y < x^2 + 4x + 2 = (x + 1)^2 + 2x + 1$, and hence $z \leq x$. Hence $(z + 1)^2 = z^2 + 2z + 1 \leq z^2 + 2x + 1 = y < (z + 1)^2$, a contradiction. Thus (3) holds.

From (2) we see that $C_0^1 x = 0 = \text{Exc} \circ \text{Exc}^{(-1)}(x + x)$ for all $x \in \omega$; hence

(4) $$C_0^1 \in A.$$

Hence by composition with $\mathfrak{s}$,

(5) $$C_m^n \in A \qquad \text{for all } n > 0 \text{ and all } m \in \omega.$$

Now let $x \ominus y = \text{Exc}(\text{Exc}^{(-1)}(2x + 2y) + 3x + y + 4)$ for all $x, y \in \omega$. Thus

(6) $$\ominus \in A.$$

Now if $x \geq y$, then

$$\begin{aligned}(x + y + 2)^2 &= (x + y)^2 + 4x + 4y + 4 \\ &\leq (x + y)^2 + 2(x + y) + 3x + y + 4 \\ &= \text{Exc}^{(-1)}(2x + 2y) + 3x + y + 4 \qquad \text{by (2)} \\ &< (x + y)^2 + 6x + 6y + 9 \\ &= (x + y + 3)^2.\end{aligned}$$

Hence

(7) $$x \ominus y = x - y \qquad \text{if } y \leq x.$$

Let $fx = x^2$ for all $x \in \omega$. Then by (2), (7), $fx = \text{Exc}^{(-1)}(2x) - 2x$ for all $x \in \omega$, so

(8) $$f \in A.$$

Next note that $\text{sg}\, x = \text{Exc}\, \mathfrak{s}(x^2)$ and $\overline{\text{sg}}\, x = 1 \ominus \text{sg}\, x$ for all $x \in \omega$. Thus

(9) $$\text{sg}, \overline{\text{sg}} \in A.$$

Furthermore,

(10) $\quad\quad \mathrm{Exc} \circ \mathscr{s}$ has range ω.

For, $\mathrm{Exc}\, \mathscr{s}0 = 0$, and if $x \neq 0$, then $\mathrm{Exc}\, \mathscr{s}(x^2 + x - 1) = x$.

Now using 3.43(*iii*) we see that $\mathscr{p}x = \mathrm{Exc}\, (\mathrm{Exc} \circ \mathscr{s})^{(-1)}(x)$ for all $x \in \omega$. Hence

(11) $\quad\quad \mathscr{p} \in A.$

Recall that $\mathbb{E}$ is the set of even numbers. Next we show

(12) $\quad\quad \chi\mathbb{E}(x) = \mathrm{Exc}\, \mathscr{s}\mathscr{s}\, \mathrm{Exc}^{(-1)}x \quad$ for all $x \in \omega$.

For, if $x = 2y$, then

$$\begin{aligned} \mathrm{Exc}\, \mathscr{s}\mathscr{s}\, \mathrm{Exc}^{(-1)}\, x &= \mathrm{Exc}\, \mathscr{s}\mathscr{s}\, (y^2 + 2y) && \text{by (2)} \\ &= \mathrm{Exc}\, (y^2 + 2y + 2) \\ &= 1; \end{aligned}$$

if $x = 2y + 1$, then

$$\begin{aligned} \mathrm{Exc}\, \mathscr{s}\mathscr{s}\, \mathrm{Exc}^{(-1)}\, x &= \mathrm{Exc}\, \mathscr{s}\mathscr{s}\, (y^2 + 4y + 2) && \text{by (3)} \\ &= \mathrm{Exc}\, (y^2 + 4y + 4) \\ &= 0. \end{aligned}$$

From (12) we have:

(13) $\quad\quad \chi\mathbb{E} \in A.$

Now let $gx = 2\, \mathrm{Exc}\, x + \overline{\mathrm{sg}}\, \chi\mathbb{E}x$ for all $x \in \omega$. Thus

(14) $\quad\quad g \in A.$

We claim:

(15) $\quad\quad g$ has range ω.

For, if $x = 2y$ then, since $y^2 + y$ is even, $g(y^2 + y) = 2\, \mathrm{Exc}\, (y^2 + y) = 2y = x$. If $x = 2y + 1$ then, since $(y + 1)^2 + y$ is odd, $g((y + 1)^2 + y) = 2y + 1 = x$.

Let $hx = [x/2] =$ greatest integer $y \leq x/2$ for all $x \in \omega$. Then

(16) $\quad\quad hx = \mathrm{Exc}\, g^{(-1)}x \quad$ for all $x \in \omega$, and hence $h \in A$.

For, $2\, \mathrm{Exc}\, g^{(-1)}x + \overline{\mathrm{sg}}\, \chi\mathbb{E}g^{(-1)}x = x$ for any x; thus if x is even, then $2\, \mathrm{Exc}\, g^{(-1)}x = x$; while if x is odd, $2\, \mathrm{Exc}\, g^{(-1)}x + 1 = x$, as desired.

For any $x \in \omega$, let $kx = [(\mathrm{Exc}\, \mathscr{p}x)/2] + \mathrm{sg}\, x$. Thus

(17) $\quad\quad k \in A.$

Furthermore,

(18) $\quad\quad k(x^2) = x \quad$ for all x.

For, if $x = 0$ the result is obvious. If $x \neq 0$, then $\mathscr{p}x^2 = x^2 - 1 = (x - 1)^2 + 2x - 2$, $\mathrm{Exc}\, \mathscr{p}x^2 = 2x - 2$, and hence $k(x^2) = x$, as desired.

Let $lx = [\sqrt{x}]$ for all $x \in \omega$. Then by (18), $lx = k(x \ominus \text{Exc } x)$, so

(19) $$l \in A.$$

Hence by (8) and (19)

(20) $$\mathrm{J}, \mathrm{L} \in A.$$

(21) $$\text{if } x < y, \text{ then } x \ominus y = 3x + y + 3.$$

For,

$$\begin{aligned}(x + y + 1)^2 &= (x + y)^2 + 2x + 2y + 1 \\ &< (x + y)^2 + 5x + 3y + 4 \\ &< (x + y)^2 + 4x + 4y + 4 \\ &= (x + y + 2)^2.\end{aligned}$$

Since $x \ominus y = \text{Exc}\,((x + y)^2 + 2(x + y) + 3x + y + 4)$, (21) now follows.

(22) $\chi_{\geq}(x, y) = \text{sg}\,[(x \ominus y) \ominus (3x + y + 3)]$ for all $x, y \in \omega$, and hence $\chi_{\geq} \in A$.

For, if $x \geq y$ then

$$\begin{aligned}\text{sg}\,[(x \ominus y) \ominus (3x + y + 3)] &= \text{sg}\,[(x - y) \ominus (3x + y + 3)] && \text{by (7)} \\ &= \text{sg}\,(3x - 3y + 3x + y + 3 + 3) && \text{by (21)} \\ &= \text{sg}\,(6x - 2y + 6) \\ &= 1\end{aligned}$$

If $x < y$, then

$$\begin{aligned}\text{sg}\,[(x \ominus y) \ominus (3x + y + 3)] &= (3x + y + 3) \ominus (3x + y + 3) && \text{by (21)} \\ &= 0 && \text{(by (7)}\end{aligned}$$

(23) $$\cdot \in A.$$

For, $x \cdot y = [(((x + y)^2 \ominus x^2) \ominus y^2)/2]$ for all $x, y \in \omega$. Let $m(x, y) = |x - y|$ for all $x, y \in \omega$. Then

(24) $$m \in A,$$

for $m(x, y) = \chi_{\geq}(x, y) \cdot (x \ominus y) + \chi_{\geq}(y, x) \cdot (y \ominus x)$.

With the aid of the auxiliary functions which we have shown to be in A, we can now show how minimalization can be reduced to inversion.

(25) Suppose f is a 2-ary special function, $f \in A$. Let $gx = \mu y(f(x, y) = 0)$, for all $x \in \omega$. Then for all $x \in \omega$, $gx = \mathrm{L}\mu z(f(\text{Exc } z, \mathrm{L}z) = 0$, $\text{Exc } z + \mathrm{L}z = [\sqrt{[\sqrt{z}]}]$, and $\text{Exc } z = x)$ (and for each x there always is a z satisfying the conditions in parentheses).

To prove this, let $x \in \omega$ be given. Let $z = \mathrm{J}(x, gx)$. Then $f(\text{Exc } z, \mathrm{L}z) = f(x, gx) = 0$ by 3.43; $\text{Exc } z + \mathrm{L}z = x + gx = [\sqrt{[\sqrt{z}]}]$ by direct computation, and $\text{Exc } z = x$. Clearly also $\mathrm{L}z = gx$. It remains to show that our choice of z gives the least integer s satisfying the conditions of the μ – operator. Assume that $f(\text{Exc } s, \mathrm{L}s) = 0$, $\text{Exc } s + \mathrm{L}s = [\sqrt{[\sqrt{s}]}]$, and $\text{Exc } s = x$. Say

$s = p^2 + x < (p + 1)^2$. Then $[\sqrt{s}] = p$, and $\mathrm{L}s = \mathrm{Exc}\, p$. Say $p = q^2 + \mathrm{L}s < (q + 1)^2$. Then $[\sqrt{[\sqrt{s}]}] = q$. Thus $x + \mathrm{L}s = q$. Since $f(\mathrm{Exc}\, s, \mathrm{L}s) = 0$, we have $gx \leq \mathrm{L}s$. Hence $x + gx \leq q$, $(x + gx)^2 \leq q^2$, $(x + gx)^2 + gx \leq p$, $[(x + gx)^2 + gx]^2 \leq p^2$, and $z = \mathrm{J}(x, gx) \leq p^2 + x = s$, as desired. Thus (25) is established.

(26) Under the hypothesis of (25) we have $g \in A$.

For, let $nz = \overline{\mathrm{sg}}\, f(\mathrm{Exc}\, z, \mathrm{L}z) \cdot \overline{\mathrm{sg}}\, (|\mathrm{Exc}\, z + \mathrm{L}z - [\sqrt{[\sqrt{z}]}]|) \cdot \mathrm{Exc}\, z$. By (25), n has range ω, and clearly $n \in A$. Clearly for any $x \in \omega$ we have $gx = \overline{\mathrm{sg}}\, x \cdot g0 + \mathrm{L}n^{(-1)}x$, so $g \in A$.

Next,

(27) if f is special, $f \in A$, and g is obtained from f by minimalization, then $g \in A$.

For suppose f is m-ary, $m > 1$. We proceed by induction on m; the case $m = 2$ is given by (26). Inductively assume that $m > 2$. Define f' by

$$f'(x_0, \ldots, x_{m-2}) = f(\mathrm{Exc}\, x_0, \mathrm{L}x_0, x_1, \ldots, x_{m-2}),$$

for all $x_0, \ldots, x_{m-2} \in \omega$. Clearly f' is special, since f is. Let g' be obtained from f' by minimalization. By the induction hypothesis, $g' \in A$. Now if $x_0, \ldots, x_{m-2} \in \omega$, then

$$\begin{aligned} g(x_0, \ldots, x_{m-2}) &= \mu y(f(x_0, .., x_{m-2}, y) = 0) \\ &= \mu y(f(\mathrm{Exc}\, \mathrm{J}(x_0, x_1), \mathrm{LJ}(x_0, x_1), x_2, \ldots, x_{m-2}, y) = 0) \\ &= \mu y(f'(\mathrm{J}(x_0, x_1), x_2, \ldots, x_{m-2}, y) = 0) \\ &= g'(\mathrm{J}(x_0, x_1), x_2, \ldots, x_{m-2}); \end{aligned}$$

hence $g \in A$ by (20).

For all $x, y \in \omega$ let $g(x, y) = [x/y]$.

(28) $g \in A$.

For, if $x, y \in \omega$ then

$$\begin{aligned} [x/y] &= \mu z(y \cdot \sigma z > z \text{ or } y = 0) \\ &= \mu z(\chi_{\geq}(x, y \cdot \sigma z) \cdot y = 0), \end{aligned}$$

so $g \in A$ by (27) and (22).

Now since $\mathrm{rm}\,(x, y) = x \ominus ([x/y] \cdot y)$, we have

(29) $\mathrm{rm} \in A$.

Hence by (20),

(30) $\beta \in A$.

Now we can take care of primitive recursion.

(31) Suppose g is obtained from f and h by primitive recursion, f m-ary and h $(m + 2)$-ary, $m > 0$. Then for any $x_0, \ldots, x_{m-1}, y \in \omega$, $g(x_0, \ldots, x_{m-1}, y) = \beta(\mu z[\beta(z, 0) = f(x_0, \ldots, x_{m-1})$ and $\mu w(\beta(z, \sigma w) \neq h(x_0, \ldots, x_{m-1}, w, \beta(z, w))$ or $w = y) = y], y)$; such z and w always exist, for any $x_0, \ldots, x_{m-1}, y \in \omega$.

To prove (31), let $x_0, \ldots, x_{m-1}, y \in \omega$ be given. By Theorem 3.46 choose z such that $\beta(z, i) = g(x_0, \ldots, x_{m-1}, i)$ for each $i \leq y$. Thus if $\sigma w \leq y$ we have

$$\begin{aligned} \beta(z, \sigma w) &= g(x_0, \ldots, x_{m-1}, \sigma w) \\ &= h(x_0, \ldots, x_{m-1}, w, g(x_0, \ldots, x_{m-1}, w)) \\ &= h(x_0, \ldots, x_{m-1}, w, \beta(z, w)). \end{aligned}$$

Hence

$$\mu w(\beta(z, \sigma w) \neq h(x_0, \ldots, x_{m-1}, w, \beta(z, w)) \text{ or } w = y) = y.$$

Furthermore, $\beta(z, 0) = g(x_0, \ldots, x_{m-1}, 0) = f(x_0, \ldots, x_{m-1})$. Hence there is a z of the sort mentioned in (31). Let t be the least such z. By induction on i it is easily seen that for any $i \leq y$ we have $\beta(t, i) = g(x_0, \ldots, x_{m-1}, i)$. Hence $\beta(t, y) = g(x_0, \ldots, x_{m-1}, y)$, as desired.

(32) Under the hypothesis of (31), if in addition f and h are in A, then $g \in A$.

For, first let

$$k'(x_0, \ldots, x_{m-1}, y, z, w) = \overline{\mathrm{sg}}\, |\beta(z, \sigma w) - h(x_0, \ldots, x_{m-1}, w, \beta(z, w))| \cdot \mathrm{sg}\,(|w - y|)$$

for all $x_0, \ldots, x_{m-1}, y, z, w \in \omega$. Then $k' \in A$ by (9), (24), and (30). Furthermore, obviously k' is special and

$$\begin{aligned} g(x_0, \ldots, x_{m-1}, y) = \beta(\mu z[\beta(z, 0) &= f(x_0, \ldots, x_{m-1}) \quad \text{and} \\ &\mu w(k'(x_0, \ldots, x_{m-1}, y, z, w) = 0) = y], y) \end{aligned}$$

Let $k''(x_0, \ldots, x_{m-1}, y, z) = \mu w(k'(x_0, \ldots, x_{m-1}, y, z, w) = 0)$ for all $x_0, \ldots, x_{m-1}, y, z \in \omega$. Then $k'' \in A$ by (27). Let $k'''(x_0, \ldots, x_{m-1}, y, z) = \mathrm{sg}\,(|\beta(z, 0) - f(x_0, \ldots, x_{m-1})|) + \mathrm{sg}\,(|k''(x_0, \ldots, x_{m-1}, y, z) - y|)$. Then $k''' \in A$, and by (31) k''' is special; moreover,

$$g(x_0, \ldots, x_{m-1}, y) = \beta(\mu z(k'''(x_0, \ldots, x_{m-1}, y, z) = 0), y).$$

Hence $g \in A$, as desired.

(33) If g is obtained from a and h by primitive recursion, $a \in \omega$ and h binary and $h \in A$, then $g \in A$.

The proof is similar to that of (32).

Thus σ, $U_i^n \in A$, and A is closed under composition, primitive recursion, and minimalization (applied to special functions). Hence, every recursive function is in A, and the proof of 3.48 is complete. □

Definition 3.49. Let P be the two-place operation on one place functions such that

$$\mathrm{P}(f, g)(x) = fx + gx$$

for all one place functions f, g and all $x \in \omega$.

Theorem 3.50. *The class of* 1*-place recursive functions is the intersection of all sets A of* 1*-place functions such that* $\mathscr{s}$, Exc $\in A$ *and A is closed under* K_1^1, P, *and inversion (applied to functions with range* ω).

PROOF. Clearly the intersection indicated is included in the class of 1-place recursive functions. Now suppose that A satisfies the conditions of the theorem. Note that $U_0^1 \in A$, since $U_0^1 = K_1^1(\mathrm{Exc}, \mathrm{Exc}^{(-1)})$. If f is a 1-place recursive function, then $f = K_1^1(f, U_0^1)$. Hence in order to show that all 1-place recursive functions are in A (which is all that remains for the proof), it suffices to prove the statement

(*) if f is an m-ary general recursive function and $g_0, \ldots, g_{m-1} \in A$, then $K_1^m(f; g_0, \ldots, g_{m-1}) \in A$.

To prove (*), let B be the set of all f such that if f is m-ary and $g_0, \ldots, g_{m-1} \in A$ then $K_1^m(f; g_0, \ldots, g_{m-1}) \in A$. Note that for f unary we have $f \in A$ iff $f \in B$. Hence $+$, $\mathscr{s}$, Exc, $U_i^n \in B$ (for $0 \leq i < n < \omega$) and B is closed under inversion, applied to functions with range ω. To show that B is closed under composition, assume that f (m-ary) is in B, that $h_0, \ldots, h_{m-1}$ (all n-ary) are in B, and that $g_0, \ldots, g_{n-1} \in A$. Then by 2.2,

$$K_1^n(K_n^m(f; h_0, \ldots, h_{m-1}); g_0, \ldots, g_{n-1}) = K_1^m(f; K_1^n(h_0; g_0, \ldots, g_{n-1}), \ldots, K_1^n(h_{m-1}; g_0, \ldots, g_{n-1})).$$

Now $K_1^n(h_0; g_0, \ldots, g_{n-1}), \ldots, K_1^n(h_{m-1}; g_0, \ldots, g_{n-1}) \in A$, so

$$K_1^n(K_n^m(f; h_0, \ldots, h_{m-1}); g_0, \ldots, g_{n-1}) \in A.$$

Thus, $g_0, \ldots, g_{n-1}$ being arbitrary, $K_n^m(f; h_0, \ldots, h_{m-1}) \in B$. Hence by 3.48 the proof is complete. □

BIBLIOGRAPHY

1. Davis, M. *Computability and Unsolvability*. New York: McGraw-Hill (1958).
2. Hermes, H. *Enumerability, Decidability, Computability*. New York: Springer (1969).
3. Robinson, J. General recursive functions. *Proc. Amer. Math. Soc.*, *1* (1950), 703–718.

EXERCISES

3.51. Show that the set Γ in the proof of 3.5 is closed under primitive recursion with parameters.

3.52*. Let $f(0, y) = y + 1$, $f(x + 1, 0) = f(x, 1)$, and $f(x + 1, y + 1) = f(x, f(x + 1, y))$. Show that f is recursive.

3.53*. (continuing **3.52***). Show that f is not primitive recursive. *Hint:* prove the following in succession (for any $x, y \in \omega$):

(1) $y < f(x, y)$;
(2) $f(x, y) < f(x, y + 1)$;

(3) $f(x, y + 1) \leq f(x + 1, y)$;
(4) $f(x, y) < f(x + 1, y)$;
(5) $f(1, y) = y + 2$;
(6) $f(2, y) = 2y + 3$;
(7) for any $c_1, \ldots, c_r$ there is a d such that for all x, $\Sigma_{1 \leq j \leq r} f(c_j, x) \leq f(d, x)$ (prove first for $r = 2$, taking $d = \max(c_1, c_2) + 4$);
(8) for every primitive recursive function g (say with n places) there is a c such that for all $x_1, \ldots, x_n$, $g(x_1, \ldots, x_n) < f(c, x_1 + \cdots + x_n)$;
(9) f is not primitive recursive.

3.54. What difficulty would arise in deleting "primitive" from Lemma 3.5 [show that 3.5 would then be false, but also indicate how a proof roughly similar to that given for 3.5 would break down].

3.55. Express a Turing machine to compute + directly in terms of the machines of Chapter 1, i.e., don't use results of this section.

3.56. The set $\{\mathcal{g}M : M$ is a Turing machine with exactly five states$\}$ is elementary.

3.57′. Prove (33) in the proof of 3.48 in detail.

3.58. The class of recursive functions is the intersection of all classes A of functions such that $\mathfrak{s}$, $U_i^n \in A$ (each $n > 0$, each i with $i < n$), $+ \in A$, $\dot{-} \in A$, $\cdot \in A$, and A is closed under composition and under minimalization (applied to special functions). Thus we have here another equivalent definition of the notion of recursive function; this version, or slight variations of it, are frequently found in the literature.

3.59. Let $J_1(x, y) = 2^x . (2y + 1) - 1$ for all $x, y \in \omega$, let $K_1x = (x + 1)_0$, and let $L_1x = ([(x + 1)/\exp(2, K_1x)] - 1)/2$. Then show:

(1) J_1, K_1, L_1 are elementary
(2) $J_1(K_1x, L_1x) = x$
(3) $K_1J_1(x, y) = x$
(4) $L_1J_1(x, y) = y$
(5) J_1 maps $\omega \times \omega$ 1 − 1 onto ω

3.60. For any $x, y \in \omega$, let $J_2(x, y) = [(x + y)^2 + 3x + y]/2$, $Q_1x = [([\sqrt{(8x + 1)}] + 1)/2] \dot{-} 1$, $Q_2x = 2x - (Q_1x)^2$, $K_2x = (Q_2x - Q_1x)/2$, and $L_2x = Q_1x - K_2x$. Prove analogs of 3.59(1)–(5). *Hint*: Define $f: \omega \times \omega \rightarrow \omega \times \omega$ by putting

$$f(x, y) = \begin{cases} (x + 1, y - 1) & \text{if } y \neq 0, \\ (0, x + 1) & \text{if } y = 0. \end{cases}$$

(The function f describes a certain easily visualized procedure of going through all pairs (x, y).)

Prove that $J_2f(x, y) = J_2(x, y) + 1$ for all $x, y \in \omega$. Thus J_2 is the natural mapping $\omega \times \omega \rightarrow \omega$ associated with f. Then show successively that J_2 maps $\omega \times \omega$ onto ω and that for all $x, y \in \omega$, $Q_1J_2(x, y) = x + y$, $Q_2J_2(x, y) = 3x + y$, $K_2J_2(x, y) = x$, $L_2J_2(x, y) = y$. Then $J_2(K_2x, L_2x) = x$ follows easily since J_2 is onto.

3.61*. If f is a 1-place number-theoretic function, we define f^n (temporary notation) by induction:

$$f^0x = x \qquad \text{for all } x \in \omega,$$
$$f^{n+1}x = ff^nx \qquad \text{for all } x \in \omega.$$

The function g such that $gn = f^n0$ for all $n \in \omega$ is said to be obtained from f by *iteration*.

Prove the following theorem:

Theorem (R. M. Robinson). *The class of primitive recursive functions is the intersection of all classes A of functions such that σ, Exc, $+$, $U_i^n \in A$ whenever $i < n \in \omega$ and A is closed under composition and iteration.*

Hint: As in the proof of 3.48 the essential thing is to show that each primitive recursive function is in A, where A satisfies the conditions of the theorem. Proceed stepwise:

(1) C_q^n, sg, $\overline{\text{sg}} \in A$.
(2) Let $fx = x + 2\,\overline{\text{sg}}\,\text{Exc}\,(x + 4) + 1$. Then $f \in A$.
(3) Let $gx = x + 2[\sqrt{x}]$. Then $g \in A$.
(4) Let $hx = x^2$. Then $h \in A$.
(5) Let $x \ominus y = \text{Exc}\,[(x + y)^2 + 3x + y + 1]$. Then $\ominus \in A$.
(6) Let $\alpha x = \overline{\text{sg}}\,x + 2\,\overline{\text{sg}}\,(x \ominus 1)$. Then $\alpha \in A$.
(7) Let β be obtained from α by iteration, $\gamma x = x + 1 + \beta x$, ε obtained from γ by iteration, $kx = [x/2]$ for all x; then $3 > \beta x \equiv x \pmod 3$, and $k_x = \varepsilon x - x$, and $k \in A$.
(8) Let $ix = [\sqrt{x}]$. Then $i \in A$.
(9) $\cdot$, J, $L \in A$.
(10) Suppose $j \in A$, and k is defined from j as follows:

$$k0 = 0,$$
$$k(n + 1) = j(n, kn).$$

Then $k \in A$. *Hint:* define $k'n = J(n, kn)$ for all $n \in \omega$.
(11) Suppose $f_1 \in A$, and f_2 is defined from f_1 as follows:

$$f_2(a, 0) = a,$$
$$f_2(a, n + 1) = f_1(n, f_2(a, n)).$$

Then $f_2 \in A$. *Hint:* define $l0 = 0$, $l(n + 1) = f_2(\text{L}n, \text{Exc}n)$.
(12) Suppose $f_1, f_2 \in A$, and f_3 is defined as follows:

$$f_3(a, 0) = f_1a,$$
$$f_3(a, n + 1) = f_2(n, f_3(a, n)).$$

Then $f_3 \in A$. *Hint:* define $l(a, 0) = a$, $l(a, n + 1) = J(a, f_3(a, n))$
(13) If $f_4, f_5 \in A$, and f_6 is defined by:

$$f_6(a, 0) = f_4a,$$
$$f_6(a, n + 1) = f_5(a, n, f_6(a, n)),$$

then $f_6 \in A$.
(14) A is closed under primitive recursion.

3.62. Using 3.61, show that the class of all 1-ary primitive recursive functions is the intersection of all classes A such that $\mathscr{s}$, Exc $\in A$ and A is closed under iteration, K_1^1, and P.

3.63* (HERBRAND–GÖDEL–KLEENE CALCULUS). We outline another equivalent version of recursiveness. We need a small formal system:
Variables: $v_0, v_1, v_2, \ldots$;
Individual constant: $\mathbf{0}$;
Operation symbols: $\mathbf{f}_m$ (m-ary), $\mathbf{g}_{mn}$ (m-ary) for all $m \in \omega \sim 1$, $n \in \omega$; $\mathscr{s}$ (unary).

By induction we define $\triangle 0 = 0$, $\triangle(m + 1) = \mathscr{s}\, \triangle m$ for all $m \in \omega$; we denote $\triangle m$ sometimes by $\mathbf{m}$. Now we define terms:

(1) $\langle v_i \rangle$;
(2) $\langle \mathbf{0} \rangle$;
(3) if σ is a term, so is $\mathscr{s}\sigma$;
(4) if $m \in \omega \sim 1$ and $\sigma_0, \ldots, \sigma_{m-1}$ are terms, so are $\mathbf{f}_m\sigma_0 \cdots \sigma_{m-1}$ and $\mathbf{g}_{mn}\sigma_0 \cdots \sigma_{m-1}$ for each $n \in \omega$;
(5) terms are formed only in these ways.

An *equation* is an expression $\sigma \mathbf{=} \tau$ with σ, τ terms.

A *system of equations* is a finite sequence of equations. If E is a system of equations, say $E = \langle \varphi_0, \ldots, \varphi_{m-1} \rangle$, then an *E-derivation* is a finite sequence $\langle \psi_0, \ldots, \psi_{n-1} \rangle$ of equations such that for each $i < n$ one of the following holds:

(6) $\exists j < m\, \psi_i = \varphi_j$;
(7) $\exists j < i \exists$ variable $\alpha \exists\, m \in \omega$ (ψ_i is obtained from ψ_j by replacing each occurrence of α in ψ_j by $\mathbf{m}$);
(8) $\exists j, k < i\, \psi_k$ has the form $\sigma \mathbf{=} \tau$, ψ_j has the form $\mathbf{f}_p\mathbf{x}_0 \cdots \dot{\mathbf{x}}_{p-1} = \mathbf{y}$ or $\mathbf{g}_{pq}\mathbf{y}_0 \cdots \mathbf{x}_{p-1} = \mathbf{y}$, and ψ_k is obtained from ψ_k by replacing one occurrence of $\mathbf{f}_p\mathbf{x}_0 \cdots \mathbf{x}_{p-1}$ (or $\mathbf{g}_{pq}\mathbf{x}_0 \cdots \mathbf{x}_{p-1}$) in τ by $\mathbf{x}$.

We write $E \vdash \chi$ to mean that there is an E-derivation with last member χ.

Now an m-place number-theoretic function k is called *Herbrand–Gödel–Kleene* recursive if there is a system E of equations such that $\forall x_0 \cdots \forall x_{m-1} \forall y (E \vdash f_m\mathbf{x}_0 \cdots \mathbf{x}_{m-1} = \mathbf{y}$ iff $k(x_0, \ldots, x_{m-1}) = y)$.

Show that k is Herbrand–Gödel–Kleene recursive iff it is recursive.

Hint: To show that every recursive function is HGK recursive, let A be the collection of all functions k (say k is m-ary) such that there is a set E of equations and an assignment of n-ary operations to the n-ary operation symbols occurring in members of E (for all $n \in \omega$), k assigned to $\mathbf{f}_m$, under which all members of E become intuitively true for any values assigned to the variables and such that $\forall x_0 \cdots \forall x_{m-1} \exists y\, (E \vdash \mathbf{f}_m\mathbf{x}_0 \cdots \mathbf{x}_{m-1}$ $\mathbf{y})$. Show that A satisfies the conditions of Exercise 3.58 and hence that every recursive function is Herbrand–Gödel–Kleene recursive.

To show the converse, do a Gödel numbering. Let $T'_m = \{(e, x_0, \ldots, x_{m-1}, u)$: e is the Gödel number of a system E of equations and u is the Gödel number of an E-derivation with last term of the form $\mathbf{f}_m\mathbf{x}_0 \cdots \mathbf{x}_{m-1} = \mathbf{y}\}$. Given such a u, $\mathrm{V}'u$ is the y mentioned. Then see the proof of 3.38.

3.64* (INFINITE DIGITAL COMPUTER). Yet another equivalent form of recursiveness is obtained by generalizing a first-generation digital computer. We visualize our computer as an infinite array of storage boxes, labeled 0, 1, 2, ... Each storage box is allowed to hold any natural number. By convention we assume that all but finitely many of the boxes have 0 in them. Box 0 is the *instruction counter*. Box 1 is the *accumulator*. All other boxes are just fast memory cells. We supply only six instructions:

(1) add one to the contents of Box 1;
(2) subtract one from the contents of Box 1, or leave zero if already 0;
(3) replace the contents of storage n by the contents of storage 1 (for any n);
(4) replace the contents of storage 1 by the contents of storage n (for any n);
(5) (for each $n \in \omega$) if storage 1 has a zero in it, take the next instruction from storage n otherwise proceed as usual;
(6) stop.

For technical reasons there is no *start* instruction.

The machine works as follows. We set the storages initially to certain values (programming). Then the machine starts. It looks at box 0 and takes its instruction from the box specified there (each instruction will be assigned a number). After performing the instruction, the instruction counter advances one step (except possibly for instructions (5) and (6)), and then the next instruction is executed, etc. The machine continues until hitting the stop instruction, and then stops. It is possible that the machine will get in a "loop", and never stop.

An initial state of the machine is called a *program*. A program computes a 1-place function f as follows. We put x in storage 2 and press the start button. The machine grinds away, and finally stops; fx is then in the accumulator.

Now we express all of this rigorously. A storage description or *program* is a function F mapping ω into ω such that for some $m \in \omega$ we have $Fn = 0$ for all $n \geq m$.

An *instruction* is a number of the form $2^0 3^0$, $2^1 \cdot 3^0$, $2^2 \cdot 3^n$, $2^3 \cdot 3^n$, $2^4 \cdot 3^n$, 0, where $n \in \omega$. These instructions correspond to (1)–(6) above, respectively.

A *computation step* is a pair (F, G) such that F and G are storage descriptions and one of the following conditions holds:

(1) $FF0 = 2^0 \cdot 3^0$ and $G = (F^0_{F0+1})^1_{F1+1}$
(2) $FF0 = 2^1 \cdot 3^0$ and $G = (F^0_{F0+1})^1_{F1 \dot{-} 1}$
(3) $FF0 = 2^2 \cdot 3^n$ (for some n), and $G = (F^0_{F0+1})^n_{F1}$
(4) $FF0 = 2^3 \cdot 3^n$ (for some n), and $G = (F_{F0+1})^1_{Fn}$
(5) $FF0 = 2^4 \cdot 3^n$ (for some n), and

$$G = F^0_{F0+1} \quad \text{if } F1 \neq 0,$$
$$G = F^0_n \quad \text{if } F1 = 0.$$

A computation is a finite sequence $\langle F_0, \ldots, F_{m-1} \rangle$ of storage descriptions, with $m > 0$, such that (F_i, F_{i+1}) is a computation step for each $i < m - 1$ and $F_{m-1}F_{m-1}0 = 0$. We say that $\langle F_0, \ldots, F_{m-1} \rangle$ is a computation *beginning with* F_0 and *ending with* F_{m-1}. Now an m-ary function f is said to be *infinite-digital computed* by a program F provided that for all $x_0, \ldots, x_{m-1}$ there is a computation beginning with $F^{2, \ldots, m+1}_{x0, \ldots, x(m-1)}$ and ending with a program G such that $G1 = f(x_0, \ldots, x_{m-1})$.

Show that a function is infinite-digital computable iff it is recursive.

Markov Algorithms 4*

The present chapter is optional; it is devoted to another important and widely used version of effectiveness, Markov algorithms.

The theory of Markov algorithms is described carefully and in detail in Markov [3]. Here we shall only give enough of its development to prove equivalence with Turing computability and recursiveness. The equivalence was first proved in Detlovs [2]. For a brief outline of the theory see Curry [1].

Definition 4.1. Throughout this chapter, by a *word* we shall understand a finite sequence of 0's, 1's, and 2's. The empty word is admitted. A *Markov algorithm* is a matrix A of the form

$$\begin{matrix} a_0 & b_0 & c_0 \\ a_1 & b_1 & c_1 \\ \vdots & \vdots & \vdots \\ a_m & b_m & c_m \end{matrix}$$

such that $a_0, \ldots, a_m, b_0, \ldots, b_m$ are words and $c_0, \ldots, c_m \in \{0, 1\}$. A word a *occurs* in a word b if there are words c and d such that $b = cad$. Of course a may occur in b several times. An *occurrence* of a *in* b is a triple (c, a, d) such that $b = cad$. It is called the *first occurrence of* a *in* b if c has shortest length among all occurrences of a in b.

An *algorithmic step under* A is a pair (d, e) of words with the following properties:

(*i*) there is an $i \leq m$ such that a_i occurs in d;
(*ii*) if $i \leq m$ is minimum such that a_i occurs in d, and if (f, a_i, g) is the first occurrence of a_i in d, then $e = fb_ig$.

Such an algorithmic step is said to be *nonterminating*, if with i as in (*ii*), $c_i = 0$; otherwise (i.e., with $c_i = 1$), it is called *terminating*. A *computation under* A is a finite sequence $\langle d_0, \ldots, d_m \rangle$ of words such that for each $i < m - 1$, (d_i, d_{i+1}) is a nonterminating algorithmic step, while (d_{m-1}, d_m) is a terminating algorithmic step.

Now an m-ary function f is *algorithmic* if there is a Markov algorithm A as above such that for any $x_0, \ldots, x_{m-1} \in \omega$ there is a computation $\langle d_0, \ldots, d_n \rangle$ under A such that the following conditions hold:

(*iii*) $d_0 = 0\ 1^{(x0+1)}\ 0\ \cdots\ 0\ 1^{(x(m-1)+1)}\ 0\ 2$;
(*iv*) $\langle 2 \rangle$ occurs only once in d_n;
(*v*) $0\ 1^{(f(x0,\ldots,x(m-1))+1)}\ 0\ 2$ occurs in d_n.

We then say that A *computes* f.

A row $a_i\ b_i\ 0$ in a Markov algorithm will be indicated $a_i \to b_i$, while a row $a_i\ b_i\ 1$ will be indicated $a_i \to \cdot b_i$. A Markov algorithm lists out finitely many substitutions of one word for another, and an algorithmic computation consists in just mechanically applying these substitutions until reaching a substitution of the form $a_i \to \cdot b_i$. Clearly, then, an algorithmic function is effective in the intuitive sense. Markov algorithms are related to Post systems and to formal grammars. Now we shall give some examples of algorithms, which we shall not numerate since they are not needed later. The algorithm A_0:

$$\langle 0 \rangle \to \cdot \langle 0 \rangle$$

works as follows: any computation under A is of length 2 and simply repeats the word: $\langle a, a \rangle$, where $\langle 0 \rangle$ occurs in a. Consider the algorithm A_1:

$$\langle 0 \rangle \to \cdot \langle 01 \rangle.$$

Some examples of computations under A_1 are:

(1) $\langle \langle 0 \rangle, \langle 01 \rangle \rangle$
(2) $\langle \langle 00 \rangle, \langle 010 \rangle \rangle$
(3) $\langle \langle 11010 \rangle, \langle 110110 \rangle \rangle$.

Let A_2 be the following algorithm:

$$\langle 0 \rangle \to \langle 1 \rangle$$
$$\langle 1 \rangle \to \cdot \langle 1 \rangle.$$

The algorithm A_2 takes any word and replaces all 0's by 1's, then stops. Let A_3 be

$$\langle 1 \rangle \to \langle 11 \rangle.$$

Clearly no computation under A_3 exists. Starting with a word in which $\langle 1 \rangle$ occurs, A_3 manufactures more and more one's.

Lemma 4.2. *Every Turing computable function is algorithmic.*

PROOF. Let f (n-ary) be computed by a Turing machine M, with notation as in 1.1 and 3.9. With each row $ti = (c_{j(i)}, \varepsilon i, vi, di)$ of M ($1 \le i \le 2m$) we shall associate one or more rows $t'(i, 0), \ldots, t'(i, pi)$ of a Markov algorithm, depending on vi.

Case 1. $vi = 0$ or 1. We associate the row

$$\langle \varepsilon_i \quad 2 \quad 1^{(cji+1)} \quad 2\rangle \rightarrow \langle v_i \quad 2 \quad 1^{(di+1)} \quad 2\rangle.$$

Case 2. $vi = 2$. We associate the rows (in order)

$$\begin{aligned}
\langle 0 \quad \varepsilon_i \quad 2 \quad 1^{(cji+1)} \quad 2\rangle &\rightarrow \langle 0 \quad 2 \quad 1^{(di+1)} \quad 2 \quad \varepsilon_i\rangle \\
\langle 1 \quad \varepsilon_i \quad 2 \quad 1^{(cji+1)} \quad 2\rangle &\rightarrow \langle 1 \quad 2 \quad 1^{(di+1)} \quad 2 \quad \varepsilon_i\rangle \\
\langle \varepsilon_i \quad 2 \quad 1^{(cji+1)} \quad 2\rangle &\rightarrow \langle 0 \quad 2 \quad 1^{(di+1)} \quad 2 \quad \varepsilon_i\rangle
\end{aligned}$$

Case 3. $vi = 3$. We associate the rows (in order)

$$\begin{aligned}
\langle \varepsilon_i \quad 2 \quad 1^{(cji+1)} \quad 2 \quad 0\rangle &\rightarrow \langle \varepsilon_i \quad 0 \quad 2 \quad 1^{(di+1)} \quad 2\rangle \\
\langle \varepsilon_i \quad 2 \quad 1^{(cji+1)} \quad 2 \quad 1\rangle &\rightarrow \langle \varepsilon_i \quad 1 \quad 2 \quad 1^{(di+1)} \quad 2\rangle \\
\langle \varepsilon_i \quad 2 \quad 1^{(cji+1)} \quad 2\rangle &\rightarrow \langle \varepsilon_i \quad 0 \quad 2 \quad 1^{(di+1)} \quad 2\rangle.
\end{aligned}$$

Case 4. $vi = 4$. We associate the row

$$\langle \varepsilon_i \quad 2 \quad 1^{(cji+1)} \quad 2\rangle \rightarrow \cdot \langle \varepsilon_i \quad 2\rangle.$$

Now let A be the following Markov algorithm:

$$\begin{gathered}
t'(1, 0) \\
\vdots \\
t'(1, p1) \\
t'(2, 0) \\
\vdots \\
t'(2m, p(2m)) \\
\langle 2\rangle \rightarrow \langle 2 \quad 1^{(c1+1)} \quad 2\rangle.
\end{gathered}$$

We claim that A computes f. To see this, let $x_0, \ldots, x_{n-1} \in \omega$. Since M computes f, by 3.9 there is a computation $\langle (F, c_1, 0), (G_1, a_1, b_1), \ldots, (G_{q-1}, a_{q-1}, b_{q-1})\rangle$ of M with the following properties:

(1) $0 \; 1^{(x0+1)} \; 0 \; \cdots \; 0 \; 1^{(x(m-1)+1)}$ lies on F ending at -1, and F is 0 elsewhere;

(2) $1^{(x(m-1)+1)} \; 0 \; 1^{(f(x0,\ldots,x(m-1)+1)} \; 0$ lies on G_{q-1} ending at b_{q-1}.

Now let $G_0 = F$, $a_0 = c_1$, $b_0 = 0$. Let Q_{-1} be the word

$$0 \quad 1^{(x0+1)} \quad 0 \quad \cdots \quad 0 \quad 1^{(x(m-1)+1)} \quad 0 \quad 2.$$

Now we define N_i, P_i, Q_i for $i < q$ by induction. Let N_0 be $0 \; 1^{(x0+1)} \; 0 \; \cdots \; 0 \; 1^{(x(m-1)+1)} \; 0$, $P_0 = 0$ (the empty sequence), and $Q_0 = 0 \; 1^{(x0+1)} \; 0 \; \cdots \; 0 \; 1^{(x(m-1)+1)} \; 0 \; 2 \; 1^{(c1+1)} \; 2$. Suppose now that $i + 1 < q$ and that N_i, P_i, Q_i have been defined so that the following conditions hold:

(3) $N_i \neq 0$;

(4) N_i lies on G_i ending at b_i;

(5) P_i lies on G_i beginning at $b_i + 1$;

(6) G_i is 0 except for N_iP_i;
(7) exactly two 2's occur in Q_i;
(8) $N_i\ 2\ 1^{(ai+1)}\ 2\ P_i = Q_i$;
(9) if $i \neq 0$, then (Q_{i-1}, Q_i) is a nonterminating algorithmic step under A.

Clearly (3)–(9) hold for $i = 0$. We now define $N_{i+1}, P_{i+1}, Q_{i+1}$. Let the row of M beginning with $a_i\ G_ib_i$ be

$$a_i \quad G_ib_i \quad v \quad w.$$

We now distinguish cases depending on v. Note that, since $i < q - 1$, $v \neq 4$. In each case we define $N_{i+1}, P_{i+1}, Q_{i+1}$, and it will then be evident that (3)–(9) hold for $i + 1$ in that case. In each case, let Q_{i+1} be defined by (8) for $i + 1$.

Case 1. $v = 0$. Let N_{i+1} be N_i with its last entry replaced by 0, and let $P_{i+1} = P_i$.

Case 2. $v = 1$. Similarly.

Case 3. $v = 2$. Here we take two subcases:

Subcase 1. N_i has length at least 2. Write $N_i = N_{i+1}\varepsilon$, where $\varepsilon = 0$ or 1, and set $P_{i+1} = \varepsilon P_i$.

Subcase 2. N_i has length 1. Let $N_{i+1} = \langle 0\rangle$, $P_{i+1} = N_iP_i$.

Case 4. $v = 3$. Again we take two subcases:

Subcase 1. $P_i \neq 0$. Write $P_i = \varepsilon P_{i+1}$ with $\varepsilon = 0$ or 1, and set $N_{i+1} = N_i\varepsilon$.

Subcase 2. $P_i = 0$. Let $P_{i+1} = 0$, $N_{i+1} = N_i0$.

This completes the definition of N_i, P_i, Q_i for all $i < q$, so that (3)–(9) hold. Let Q_q be the word $N_{q-1}\ 2$. Then by (9) it follows that $\langle Q_{-1}, Q_0, \ldots, Q_q\rangle$ is a computation under A. Now by (2), (6), and (4), we can write

$$N_{q-1} = N'_{q-1} \quad 0 \quad 1^{(f(x0,\ldots,x(m-1)+1)} \quad 0;$$

hence $0\ 1^{(f(x0,\ldots,x(m-1)+1)}\ 0\ 2$ occurs in Q_q and 2 occurs only once in Q_q.

It follows that A computes f. □

We now turn to the problem of showing that every algorithmic function is recursive. This is done by the now familiar device of Gödel numbering.

Definition 4.3. If $a = \langle a_0, \ldots, a_{m-1}\rangle$ is a word, its Gödel number, $\mathcal{g}a$, is

$$\prod_{i<m} p_i^{ai+1}.$$

Thus the empty word has Gödel number 1.

Lemma 4.4. *The set of Gödel numbers of words is elementary.*

PROOF. m is the Gödel number of a word iff $m = 1$ or $m > 1$ and $\forall i \leqslant \mathrm{l}m$ $[(m)_i \leq 3$ and $1 \leq (m)_i]$. □

Definition 4.5. If A is a Markov algorithm as in 4.1, its Gödel number, $\mathcal{g}A$, is the number

$$\prod_{i<m} p_i^{ti},$$

where $ti = 2^{\mathcal{g}ai} \cdot 3^{\mathcal{g}bi} \cdot 5^{ci}$ for each $i \leq m$.

Lemma 4.6. *The set of Gödel numbers of Markov algorithms is elementary.*

PROOF. n is the Gödel number of a Markov algorithm iff $n \geq 2$ and $\forall i \leq \mathrm{l}n$ $[((n)_i)_0$ and $((n)_i)_1$ are Gödel numbers of words, $((n)_i)_2 \leq 1$, and $\mathrm{l}((n)_i) \leq 2]$. □

Definition 4.7. Let $R_0 = \{(m, n)$: m and n are Gödel numbers of words a and b respectively, and a occurs in $b\}$.

Lemma 4.8. *R_0 is elementary.*

PROOF $(m, n) \in R_0$ iff m is the Gödel number of a word, n is the Gödel number of a word, and $\exists x \leq n\, \exists y \leq n[\mathrm{Cat}\,(\mathrm{Cat}\,(x, m), y) = n]$. (Recall from 3.30 the definition of Cat.) □

Definition 4.9. $R_1 = \{(m, n, p, q)$: m, n, p, q are Gödel numbers of words a, b, c, d respectively, and (a, b, c) is the first occurrence of b in $d\}$.

Lemma 4.10. *R_1 is elementary.*

PROOF. $(m, n, p, q) \in R_1$ if m, n, p, q are Gödel numbers of words and $\mathrm{Cat}\,(\mathrm{Cat}\,(m, n), p) = q$ and $\forall x \leq q\, \forall y \leq q[\mathrm{l}x < \mathrm{l}m\ \&\ x$ and y are Gödel numbers of words $\Rightarrow \mathrm{Cat}\,(\mathrm{Cat}\,(x, n), y) \neq q]$. □

Definition 4.11. $R_2 = \{(p, m, n)$: p is the Gödel number of a Markov algorithm A, m, n are Gödel numbers of words a, b respectively, and (a, b) is a nonterminating computation step under $A\}$.

Lemma 4.12. *R_2 is elementary.*

PROOF. $(p, m, n) \in R_2$ iff p is the Gödel number of a Markov algorithm, m and n are Gödel numbers of words, $\exists i \leq \mathrm{l}p$ such that $(((p)_i)_0, m) \in R_0$, and $\forall i \leq \mathrm{l}p\, \forall x \leq m\, \forall y \leq m[(((p)_i)_0, m) \in R_0\ \&\ \forall j < i[(((p)_j)_0, m) \notin R_0]\ \&\ (x, ((p)_i)_0, y, m) \in R_1 \Rightarrow \mathrm{Cat}\,(\mathrm{Cat}\,(x, ((p)_i)_1, y) = n\ \&\ ((p)_i)_2 = 0]$. □

Definition 4.13. R_3 is like R_2 except with "terminating" instead of "nonterminating".

Lemma 4.14. *R_3 is elementary.*

Definition 4.15. If $\langle d_0, \ldots, d_m \rangle$ is a finite sequence of words, its Gödel number is

$$\prod_{i<m} p_i^{\mathcal{G}d_i}.$$

Also let $R_4 = \{(m, n) : m$ is the Gödel number of a Markov algorithm A, and n is the Gödel number of a computation under $A\}$.

Lemma 4.16. *R_4 is elementary.*

PROOF. $(m, n) \in R_4$ iff m is a Gödel number of a Markov algorithm, $\mathrm{l}n \geq 1$, and $\forall i < \mathrm{l}n \dot{-} 1[(m, (n)_i, (n)_{i+1}) \in R_2]$ and $(m, (n)_{\mathrm{l}n \dot{-} 1}, (n)_{\mathrm{l}n}) \in R_3$. □

Definition 4.17. $f_1 x = \prod_{i \leq x} p_i^2$.

Lemma 4.18. *f_1 is elementary.*

Definition 4.19. $f_2^1 x = \mathrm{Cat}\,(2, f_1 x) \cdot f_2^{m+1}(x_0, \ldots, x_m) = \mathrm{Cat}\,(f_2^m(x_0, \ldots, x_{m-1}),$ $f_2^1 x_m)$.

Lemma 4.20. *f_2^m is elementary, for each $m \in \omega \sim \{0\}$.*

Lemma 4.21. *$f_2^m(x_0, \ldots, x_{m-1})$ is the Gödel number of*

$$0 \quad 1^{(x0+1)} \quad 0 \quad \cdots \quad 0 \quad 1^{(x(m-1)+1)}.$$

The notations R_1, R_2, R_3, R_4, f_1, f_2^m will not be used beyond the present section.

Definition 4.22. $T'_m = \{(e, x_0, \ldots, x_{m-1}, c) : e$ is the Gödel number of a Markov algorithm A, and c is the Gödel number of a computation $\langle d_0, \ldots, d_n \rangle$ under A, $(c)_0 = \mathrm{Cat}\,(f_2^m(x_0, \ldots, x_{m-1}), 2 \cdot 3^3)$, and 2 occurs only once in $d_n\}$.

Lemma 4.23. *T'_m is elementary.*

Definition 4.24. $V'y = \mu x \leq y[(\mathrm{Cat}\,(f_2^1 x, 2 \cdot 3^3), (y)_{\mathrm{l}y}) \in R_0]$.

Lemma 4.25. *V' is elementary.*

Lemma 4.26. *Every algorithmic function is recursive.*

PROOF. Say f is m-ary and is computed by a Markov algorithm A. Let e be the Gödel number of A. Then for any $x_0, \ldots, x_{m-1} \in \omega$ we have

$$f(x_0, \ldots, x_{m-1}) = V'\mu z(\langle e, x_0, \ldots, x_{m-1}, z \rangle \in T'_m).$$

Thus f is recursive, as desired. □

Theorem 4.27. *Turing computable = recursive = algorithmic.*

BIBLIOGRAPHY

1. Curry, H. *Foundations of Mathematical Logic.* New York: McGraw-Hill (1963).
2. Detlovs, V. *The equivalence of normal algorithms and recursive functions.* A.M.S. Translations Ser. 2, Vol. 23, pp. 15–81.
3. Markov, A. *Theory of Algorithms.* Jerusalem: Israel Program for Scientific Translations (1961).

EXERCISES

4.28. Let A be the algorithm

$$\begin{aligned} 2\ 0 &\to 0\ 2 \\ 2\ 1 &\to 1\ 2 \\ 2 &\to \cdot\ 1^{(3)} \\ &\to 2 \end{aligned}$$

Show that A converts any word a on 0, 1 (i.e., involving only 0 and 1) into $a\ 1^{(3)}$.

4.29. Construct an algorithm which converts every word into a fixed word a.

4.30. Construct an algorithm which converts every word a into $1^{(n+1)}$, where n is the length of a.

4.31. Let a be a fixed word. Construct an algorithm which converts any word $\neq a$ into the empty word, but leaves a alone.

4.32. There is no algorithm which converts any word a into aa.

4.33. Construct an algorithm which converts any word a on 0, 1 into aa.

4.34*. Show directly that any algorithmic function is Turing computable.

5 Recursion Theory

We have been concerned so far with just the definitions of mathematical notions of effectiveness. We now want to give an introduction to the theory of effectiveness based on these definitions. Most of the technical details of the proofs of the results of this chapter are implicit in our earlier work. We wish to look at the proofs and results so far stated and try to see their significance.

In order to formulate some of the results in their proper degree of generality we need to discuss the notion of *partial functions.* An m-ary partial function on ω is a function f mapping some *subset* of ${}^m\omega$ into ω. The domain of f may be empty—then f itself is the empty set. The domain of f may be finite; it may also be infinite but not consist of all of ${}^m\omega$. Finally, it may be all of ${}^m\omega$, in which case f is an ordinary m-ary function on ω. When talking about partial functions, we shall sometimes refer to those f with $\mathrm{Dmn}\, f = {}^m\omega$ as *total.*

Intuitively speaking, a partial function f (say m-ary) is *effective* if there is an automatic procedure P such that for any $x_0, \ldots, x_{m-1} \in \omega$, if P is presented with the m-tuple $\langle x_0, \ldots, x_{m-1}\rangle$ then it proceeds to calculate, and if $\langle x_0, \ldots, x_{m-1}\rangle \in \mathrm{Dmn}\, f$, then after finitely many steps P produces the answer $f(x_0, \ldots, x_{m-1})$ and stops. In case $(x_0, \ldots, x_{m-1}) \notin \mathrm{Dmn}\, f$ the procedure P never stops. We do *not* require that there be an automatic method for recognizing membership in $\mathrm{Dmn}\, f$. Clearly if f is total then this notion of effectiveness coincides with our original intuitive notion (see p. 12). Now we want to give mathematical equivalents for the notion of an effective partial function.

Definition 5.1. Let f be an m-ary partial function. We say that f is *partial Turing computable* iff there is a Turing machine M as in 1.1 such that for every tape description F, all $q, n \in \mathbb{Z}$, and all $x_0, \ldots, x_{m-1} \in \omega$, if $0\ 1^{(x0+1)}\ 0 \cdots 0\ 1^{(x(m-1)+1)}$ lies on F beginning at q and ending at n, and if $Fi = 0$ for all $i > n$, then the two conditions

(*i*) $(x_0, \ldots, x_{m-1}) \in \mathrm{Dmn}\, F$,
(*ii*) there is a computation of M beginning with $(F, c_1, n + 1)$

are equivalent; and if one of them holds, and $\langle F, c_1, n + 1), (G_1, a_1, b_1), \ldots, (G_{p-1}, a_{p-1}, b_{p-1})\rangle$ is a computation of M, then (1)–(3) of 3.9(*ii*) hold.

Clearly any partial Turing computable function is effectively calculable.

Corollary 5.2. *Every Turing computable function is partial Turing computable. Every total partial Turing computable function is Turing computable.*

Next, we want to generalize our Definition 3.1 of recursive functions. To shorten some of our following exposition we shall use the informal notation

$$\cdots \simeq ---$$

to mean that $\cdots$ is defined iff --- is defined, and if $\cdots$ is defined, then $\cdots = ---$. For example, if f is the function with domain $\{2, 3\}$ then when we say

$$gx + hx \simeq f(x + 2) \qquad \text{for all } x \in \omega,$$

we mean that $\mathrm{Dmn}\, g \cap \mathrm{Dmn}\, h = \{0,1\}$ and for any $x \in \{0,1\}$, $gx + hx = f(x + 2)$.

Definition 5.3

(*i*) *Composition.* We extend the operator K_n^m of 2.1 to act upon partial functions. Let f be an m-ary partial function, and $g_0, \ldots, g_{m-1}$ n-ary partial functions. Then K_n^m is the n-ary partial function h such that for any $x_0, \ldots, x_{n-1} \in \omega$,

$$h(x_0, \ldots, x_{n-1}) \simeq f(g_0(x_0, \ldots, x_{n-1}), \ldots, g_{m-1}(x_0, \ldots, x_{n-1})).$$

(*ii*) *Primitive recursion with parameters.* If f is an m-ary partial function and h is an $(m + 2)$-ary partial function, then $\mathrm{R}^m(f, h)$ is the $(m + 1)$-ary partial function defined recursively by:

$$g(x_0, \ldots, x_{m-1}, 0) \simeq f(x_0, \ldots, x_{m-1})$$
$$g(x_0, \ldots, x_{m-1}, \sigma y) \simeq h(x_0, \ldots, x_{m-1}, y, g(x_0, \ldots, x_{m-1}, y))$$

for all $x_0, \ldots, x_{m-1}, y \in \omega$.

(*iii*) *Primitive recursion without parameters.* If $a \in \omega$ and h is a 2-ary partial function, then $\mathrm{R}^0(a, h)$ is the unary partial function g defined recursively by

$$g0 = a$$
$$g\sigma y \simeq h(y, gy)$$

for all $y \in \omega$.

(*iv*) *Minimalization*. Let f be an $(m + 1)$-ary partial function. An m-ary partial function g is obtained from f by *minimalization* provided that for all $x_0, \ldots, x_{m-1} \in \omega$,

$$g(x_0, \ldots, x_{m-1}) \simeq \text{least } y \text{ such that } \forall z \leq y((x_0, \ldots, x_{m-1}, z) \in \operatorname{Dmn} f) \text{ and } f(x_0, \ldots, x_{m-1}, y) = 0.$$

We then write $g(x_0, \ldots, x_{m-1}) \simeq \mu y(f(x_0, \ldots, x_{m-1}, y) = 0)$.

(*v*) The class of *partial recursive functions* is the intersection of all classes C of partial functions such that $\mathfrak{s} \in C$, $U_i^n \in C$ whenever $i < n \in \omega$, and C is closed under composition, primitive recursion, and minimalization.

Clearly every partial recursive function is effectively calculable. Note that it is not appropriate to simplify the definition of minimalization to

$$g(x_0, \ldots, x_{m-1}) \simeq \text{least } y \text{ such that } (x_0, \ldots, x_{m-1}, y) \in \operatorname{Dmn} f \text{ and } f(x_0, \ldots, x_{m-1}, y) = 0,$$

for all $x_0, \ldots, x_{m-1} \in \omega$. For, even if f is calculable there may be no clear way to calculate g. For example, suppose that $(x_0, \ldots, x_{m-1}, 0) \notin \operatorname{Dmn} f$, while $(x_0, \ldots, x_{m-1}, 1) \in \operatorname{Dmn} f$ and $f(x_0, \ldots, x_{m-1}, 1) = 0$. Without knowing that $(x_0, \ldots, x_{m-1}, 0) \notin \operatorname{Dmn} f$ it is unclear at what point in a computation of $g(x_0, \ldots, x_{m-1})$ one would be justified in setting $g(x_0, \ldots, x_{m-1}) = 1$. The above definition of minimalization clearly avoids this difficulty. One can give explicit examples where f is partial recursive but g, defined in this new way, is not. (See Exercise 5.38.)

Note that there *are* nontotal partial recursive functions. For example, clearly C_1^2 is partial recursive, and hence by 5.3(*iv*) so is the function g such that $gx \simeq \mu y(C_1^2(x, y) = 0)$. Obviously, however, g is the empty function.

Corollary 5.4. *Every general recursive function is partial recursive.*

In contrast to the situation for Turing computability, it is not at all immediately clear that every total partial recursive function is general recursive; this is, however, true, as our next theorem shows. The proof of this theorem is rather long when carried out from the beginning.

Theorem 5.5. *Partial Turing computable = partial recursive.*

PROOF. PARTIAL RECURSIVE ⇒ PARTIAL TURING COMPUTABLE. Here it is only necessary to read again the proofs of Lemmas 3.10–3.16 and check that they adapt to the situation of partial functions and the new Definitions 5.1 and 5.3.

PARTIAL TURING COMPUTABLE ⇒ PARTIAL RECURSIVE. Again one needs only to reread 3.17–3.38. □

Corollary 5.6. *Any total partial recursive function is recursive.*

A natural question occurs as to whether every partial recursive function can be extended to a recursive function; the answer is no:

Theorem 5.7. *There is a partial recursive function f such that f cannot be extended to a recursive function.*

PROOF. The rule for computing f is as follows. For a given $x \in \omega$, determine whether or not x is the Gödel number of a Turing machine. If it is not, set $fx = 0$. If it is, test in succession whether or not $(x, x, 0) \in T_1$, $(x, x, 1) \in T_1$, $(x, x, 2) \in T_1$, etc. The first time we find a u such that $(x, x, u) \in T_1$, set $fx = Vu + 1$. If we never find such a u, the computation never ends. Clearly f is intuitively a calculable partial function, and it has the following property: for any $x \in \omega$,

(1)
$fx = 0$ if x is not the Gödel number of a Turing machine,
$V\mu u((x, x, u) \in T_1) + 1$ if x is the Gödel number of a Turing machine and there is such a u,
fx is undefined, otherwise.

It is routine to show that f is partial recursive; we will prove this formally in this case, but usually not in the future. We can *define* f by (1). Clearly then, for any $x \in \omega$

$$fx \simeq [V\mu u(\chi_{T_1}(x, x, u) = 1 \text{ or } \chi_T x = 0) + 1]\cdot\chi_T x,$$

so f is partial recursive.

Now f cannot be extended to a general recursive function. For, suppose $f \subseteq h$ with h general recursive. By the proof of 3.38 there is an $e \in \omega$ such that, for all $x \in \omega$,

$$hx = V\mu u(e, x, u) \in T_1).$$

In particular, $he = V\mu u((e, e, u) \in T_1)$ and (by the definition of T_1, 3.35) e is the Gödel number of a Turing machine. Thus fe is defined, and

$$fe = V\mu u((e, e, u) \in T_1) + 1 = he + 1 = fe + 1,$$

contradiction. □

We now turn to the formulation of some basic results called the *normal form, iteration*, and *recursion theorems.*

Definition 5.8. For any $e \in \omega$ and $m \in \omega$ let φ_e^m be the m-ary partial recursive function such that for all $x_0, \ldots, x_{m-1} \in \omega$,

$$\varphi_e^m(x_0, \ldots, x_{m-1}) \simeq V\mu u((e, x_0, \ldots, x_{m-1}, u) \in T_m).$$

Note also that the $(m + 1)$-ary partial function φ' defined by

$$\varphi'(x_0, \ldots, x_{m-1}, e) \simeq V\mu u((e, x_0, \ldots, x_{m-1}, u) \in T_m)$$

for all $x_0, \ldots, x_{m-1}, e \in \omega$, is also partial recursive. This remark will be frequently useful in what follows.

Theorem 5.9 (Normal form theorem). *For any partial recursive function f (say m-ary) there is an $e \in \omega$ such that $f = \boldsymbol{\varphi}_e^m$.*

PROOF. By the proof of 5.5, second part. □

This theorem, which was implicitly used already in the proof of 5.7, has many important corollaries, which we shall now explore. First of all, its normal form nature is made a little more explicit in the following corollary.

Corollary 5.10. *For each $m \in \omega \sim 1$ there exist a 1-place elementary function f and an $(m + 2)$-place elementary function g such that for any m-ary partial recursive function h there is an $e \in \omega$ such that for all $x_0, \ldots, x_{m-1} \in \omega$,*

$$h(x_0, \ldots, x_{m-1}) \simeq f\mu u[g(e, x_0, \ldots, x_{m-1}, u) = 0].$$

PROOF. Let $f = \mathrm{V}$ and $g = \overline{\mathrm{sg}} \circ \chi_{\mathrm{T}m}$. □

This formulation suggests the possibility of improving the result by dropping f. (Another possibility, dropping μ, is impossible since there are recursive functions which are not primitive recursive.) As to this possibility, see Exercise 5.43; the answer is no.

Theorem 5.9 and its proof give rise to a certain *universal phenomenon* as follows.

Corollary 5.11 (Universal Turing machines). *There is a Turing machine M with the following property. If f is any unary partial Turing computable function and a Turing machine N computes it, and if e is the Gödel number of N, then if $0\ 1^{(e+1)}\ 0\ 1^{(x+1)}\ 0$ is placed upon an otherwise blank tape ending at -1 and if M is started at 0, then M will stop iff $x \in \mathrm{Dmn}\, f$, and if $x \in \mathrm{Dmn}\, f$, then after the machine stops $1^{(fx+1)}\ 0$ will lie on the tape beginning at 1.*

PROOF. Let g be the partial recursive function defined by

$$g(e, x) \simeq \mathrm{V}\mu u[(e, x, u) \in \mathrm{T}_1]$$

for all $e, x \in \omega$. Let M compute g. Clearly M is as desired. □

In more intuitive terms we can describe the way M is to act as follows: M is presented with two numbers e and x. First M checks if e is the Gödel number of some Turing machine. If it is, say $e = \mathcal{g}N$, then M begins checking one after the other whether 0 or 1 or $\cdots$ is the Gödel number of a computation under N with input x. If there is such a number, M takes the first such and reads off the result of the computation. It may be that e is not the Gödel number of a Turing machine or that there is no computation with input x; then M does not give an answer.

Corollary 5.12 (Universal partial recursive function). *There is a partial recursive function g of two variables such that for any partial recursive function f of one variable there is an $e \in \omega$ such that for all $x \in \omega$, $g(e, x) \simeq fx$.*

PROOF. Let g be as in the proof of 5.11. □

In view of the proof of 3.4, the reader might view 5.12 with some suspicion. Let us see what happens if we try the diagonal method on the g of 5.12. For any $x \in \omega$, let $fx \simeq g(x, x) + 1$. Then f is partial recursive, so by 5.12 there is an $e \in \omega$ such that for all $x \in \omega$, $g(e, x) \simeq fx$. Now if $g(e, e)$ is defined, then $g(e, e) = fe = g(e, e) + 1$. Conclusion: $g(e, e)$ is not defined. We are saved by g being a partial function. No contradiction arises.

We now turn to the iteration theorem. This basic result, although of a rather technical nature, is basic for most of the deeper results in recursion theory. See, e.g., the proofs of 5.15, 6.19, and 6.25.

Theorem 5.13 (Iteration theorem). *For any $m, n \in \omega \sim 1$ there is an $(m + 1)$-ary recursive function s_n^m such that for all $e, y_1, \ldots, y_m, x_1, \ldots, x_n \in \omega$,*

$$\boldsymbol{\varphi}_e^{m+n}(x_1, \ldots, x_n, y_1, \ldots, y_m) \simeq (\boldsymbol{\varphi}^n(\mathrm{s}_n^m(e, y_1, \ldots, y_m)))(x_1, \ldots, x_n).$$

PROOF. If M is any Turing machine and $y_1, \ldots, y_m \in \omega$, let $M^*_{y1,\ldots,ym}$ be the following Turing machine:

$$\text{Start} \to (T_{\text{left}} \to T_1)^{y1} \to (T_{\text{left}} \to T_1)^{y1} \to \cdots$$
$$\to (T_{\text{left}} \to T_1)^{ym} \to T_{\text{left}} \to M \to T^m_{1shift} \to \text{Stop}$$

Clearly there is an $(m + 1)$-ary recursive function s_n^m such that for any $e, y_1, \ldots, y_m \in \omega$, if e is the Gödel number of a Turing machine M, then $\mathrm{s}_n^m(e, y_1, \ldots, y_m)$ is the Gödel number of $M^*_{y1,\ldots,ym}$. Obviously s_n^m is as desired in the theorem. □

Actually a more detailed analysis would show that s_n^m in 5.13 can be taken to be elementary recursive, but we shall not use this fact. As a first application of the iteration theorem we give

Corollary 5.14. *There is no binary function f such that for all $x, y \in \omega$,*

$$f(x, y) = 1 \quad \text{if } y \in \text{Dmn } \varphi_x^1,$$
$$f(x, y) = 0 \quad \text{if } y \notin \text{Dmn } \varphi_x^1.$$

PROOF. Suppose there is such an f; say $f = \boldsymbol{\varphi}_e^2$. Now for any $x, y \in \omega$ let

$$g(x, y) \simeq \mu z[\mathrm{V}\mu u((y, x, x, u) \in \mathrm{T}_2) = 0].$$

Hence for any $x, y \in \omega$,

$g(x, y) = 0$ if y is the Gödel number of a Turing machine, $(x, x) \in \text{Dmn } \boldsymbol{\varphi}_y^2$, and $\boldsymbol{\varphi}_y^2(x, x) = 0$;
$g(x, y)$ is undefined, otherwise.

Say $g = \varphi_r^2$. Then by the iteration theorem,

$$
\begin{aligned}
s_1^1(r, e) \in \mathrm{Dmn}\, \varphi'(s_1^1(r, e)) \quad & \text{iff } (s_1^1(r, e), e) \in \mathrm{Dmn}\, \varphi_r^2 \\
& \text{iff } f(s_1^1(r, e), s_1^1(r, e)) = 0 \\
& \\
& \text{iff } s_1^1(r, e) \notin \mathrm{Dmn}\, \varphi'(s_1^1(r, e)),
\end{aligned}
$$

a contradiction. □

Thus there is no automatic method for determining of a pair (x, y) whether $y \in \mathrm{Dmn}\, \varphi_x^1$. Otherwise stated, there is no automatic method of determining of a Turing machine M and a number y whether M will eventually stop with an output when presented with input y. Thus Corollary 5.14 shows the *recursive unsolvability of the Halting problem for Turing machines.* We can give a more intuitive, informal proof of this result. Suppose we have an automatic method telling us whether a Turing machine M will stop with input y. Then we can construct a machine N such that for any $y \in \omega$ the following conditions are equivalent:

(1) N stops when given input y;
(2) y is not the Gödel number of a Turing machine, or it is the number of a machine T such that T does not stop when given input y.

Let N have Gödel number e. By (1) and (2) we reach a contradiction in trying to decide whether N stops, given input e.

Theorem 5.15 (Recursion theorem). *If $m > 1$ and f is an m-ary partial recursive function, then there is an $e \in \omega$ such that for all $x_0, \ldots, x_{m-2} \in \omega$,*

$$f(x_0, \ldots, x_{m-2}, e) \simeq \varphi_e^{m-1}(x_0, \ldots, x_{m-2}).$$

Proof. For any $x_0, \ldots, x_{m-1} \in \omega$ let

$$g(x_0, \ldots, x_{m-1}) \simeq f(x_0, \ldots, x_{m-2}, s_{n-1}^1(x_{m-1}, x_{m-1})).$$

Thus g is partial recursive; say $g = \varphi_r^m$. Let $e = s_{m-1}^1(r, r)$. Then by the iteration theorem, for all $x_0, \ldots, x_{m-2} \in \omega$,

$$
\begin{aligned}
\varphi_e^{m-1}(x_0, \ldots, x_{m-2}) & \simeq \varphi_r^m(x_0, \ldots, x_{m-2}, r) \\
& \simeq g(x_0, \ldots, x_{m-2}, r) \\
& \simeq f(x_0, \ldots, x_{m-2}, e).
\end{aligned}
$$
□

The recursion theorem is extremely useful in checking that functions defined by rather complicated recursive conditions are, in fact, general recursive. We shall illustrate its use by verifying again that the functions of 3.5 and 3.52 are recursive.

In the case of 3.5, we first define an auxiliary three-place function h' that is obviously partial recursive from the form of its definition, which goes by cases as in the definition of h, as follows. Let $x, y, e \in \omega$.

Case 1. $x = 2$. Let $h'(x, y, e) = (y)_0 + 1$ for all y.
Case 2. $x = 2^n \cdot 3^{i+1}$, where $i < n$. Let $h'(x, y, e) = (y)_i$ for all y.
Case 3. $x = 2^n \cdot 5^n \cdot p_3^q \cdot p_4^{r0} \cdot \ldots \cdot p_{m+3}^{r(m-1)}$, with $n, m > 0$.
Let

$$h'(x, y, e) \simeq \varphi_e^2(q, p_0^{t0} \cdot \ldots \cdot p_{m-1}^{t(m-1)}),$$

where $ti \simeq \varphi_e^2(ri, y)$ for each $i < m$.
Case 4. $x = 2 \cdot 7^q \cdot 11^a$ with $q > 0$. Let

$$h'(x, 1, e) = a,$$
$$h'(x, 2^{y+1}, e) \simeq \varphi_e^2(q, 2^y \cdot 3^v),$$

where $v \simeq \varphi_e^2(x, 2^y)$,

$$h'(x, z, e) = 0$$

for z not of the form 2^u.
Case 5. $x = 2^{m+1} \cdot 11^q \cdot 13^r$ with $m > 0$ and $q > 0$. Let y be given with $(y)_m = 0$. We set

$$h'(x, y, e) \simeq \varphi_e^2(q, y),$$
$$h'(x, y \cdot p_m^{z+1}, e) \simeq \varphi_e^2(r, y \cdot p_m^z \cdot p_{m+1}^v),$$

where $v \simeq \varphi_e^2(x, y \cdot p_m^z)$.
Case 6. For x not of one of the above forms, let $h'(x, y, e) = 0$ for all y, e.
Now we apply the recursion theorem and obtain an $e \in \omega$ such that for all $x, y \in \omega$,

$$h'(x, y, e) \simeq \varphi_e^2(x, y).$$

Now it is straightforward to check by complete induction on x that for all $x, y \in \omega$, $h(x, y) \simeq \varphi_e^2(x, y)$. Thus $h = \varphi_e^2$ is recursive.

It is similarly shown that the function in 3.52 is recursive. Namely, we define a partial recursive function f' as follows:

$$f'(0, y, e) = y + 1,$$
$$f'(x + 1, 0, e) \simeq \varphi_e^2(x, 1),$$
$$f'(x + 1, y + 1, e) \simeq \varphi_e^2(x, \varphi_e^2(x + 1, y)).$$

Let $e \in \omega$ be such that $f'(x, y, e) \simeq \varphi_e^2(x, y)$ for all $x, y \in \omega$. Then it is easily proved by induction on x, with induction on y in the induction step, that $f(x, y) \simeq \varphi_e^2(x, y)$ for all $x, y \in \omega$. Thus $f = \varphi_e^2$.

Theorem 5.16 (Fixed point theorem). *If f is a unary recursive function then there is an $e \in \omega$ such that $\varphi_e^1 = \varphi_{fe}^1$.*

Proof. For any $x, y \in \omega$ let

$$g(x, y) \simeq \mathrm{V}\mu u((fy, x, u) \in \mathrm{T}_1).$$

Thus g is partial recursive, and $g(x, y) \simeq \varphi_{fy}^1 x$ for all $x, y \in \omega$. Now we apply the recursion theorem to obtain on $e \in \omega$ such that $g(x, e) \simeq \varphi_e^1 x$ for all $x \in \omega$. Thus $\varphi_e^1 = \varphi_{fe}^1$, as desired. □

An important consequence of the fixed-point theorem is given in

Theorem 5.17 (Rice). *Let F be a set of one-place partial recursive functions such that $0 \neq F$ and F does not consist of all one-place partial recursive functions. Then $A = \{e : \varphi_e^1 \in F\}$ is not recursive.*

PROOF. Suppose it is. Let $a \in A$ and $b \notin A$. Now define

$$\begin{aligned} gx &= a \qquad x \notin A, \\ gx &= b \qquad x \in A. \end{aligned}$$

Then g is recursive. By 5.16 choose e such that $\varphi_e^1 = \varphi_{ge}^1$. Then if $e \in A$ we see that $\varphi_e^1 \in F$ (by the definition of A), hence $\varphi_{ge}^1 \in F$, so $ge \in A$; but $e \in A$ implies also $ge = b \notin A$, contradiction. Also, $e \notin A$ implies on the one hand $\varphi_e^1 \notin F$, $\varphi_{ge}^1 \notin F$, $ge \notin A$, and on the other hand implies $ge = a \in A$, contradiction. □

Rice's theorem has many important corollaries; we shall mention a few.

Corollary 5.18. *For any unary partial recursive function f, $\{e : \varphi_e^1 = f\}$ is not recursive.*

Corollary 5.19. *$\{x : \varphi_x^1$ is a constant function$\}$ is not recursive.*

Corollary 5.20. *$\{(x, y) : y$ is in the range of $\varphi_x^1\}$ is not recursive.*

PROOF. If the given set is recursive, then clearly so is

$$\{x : 0 \text{ is in the range of } \varphi_x^1\},$$

contradicting 5.17. □

Corollary 5.21. *$\{(x, y) : \varphi_x^1 = \varphi_y^1\}$ is not recursive.*

PROOF. If the given set is recursive, and $e \in \omega$, then

$$\{x : \varphi_x^1 = \varphi_e^1\}$$

is recursive, contradicting 5.17. □

Thus there is no automatic procedure for determining whether or not φ_e^1 is a given unary partial recursive function; or whether or not φ_x^1 is a constant function; or whether or not y is in the range of φ_x^1; or whether or not $\varphi_x^1 = \varphi_y^1$. Clearly 5.14 is also a consequence of Rice's theorem.

We can use 5.14 to establish the following result concerning the length of computations.

Theorem 5.22. *There is no binary recursive function f such that for all e, $x \in \omega$, $\exists u((e, x, u) \in \mathrm{T}_1)$ iff $\exists u \leq f(e, x)((e, x, u) \in \mathrm{T}_1)$.*

PROOF. Suppose there is such an f. Let

$$g(e, x) = 1 \quad \text{if } \exists u \le f(e, x)((e, x, u) \in \mathrm{T}_1),$$
$$g(e, x) = 0 \quad \text{otherwise.}$$

Thus g is recursive and

$$g(e, x) = 1 \quad \text{if } x \in \mathrm{Dmn}\ \boldsymbol{\varphi}_e^1,$$
$$g(e, x) = 0 \quad \text{if } x \notin \mathrm{Dmn}\ \boldsymbol{\varphi}_e^1,$$

contradicting 5.14. □

Thus there is no automatic procedure P such that, given a Turing machine M and a number x, P determines the maximum number of steps in an M-computation starting with x.

Our final topic of this section is the *arithmetical hierarchy*. The recursive relations, as we have argued, coincide with the effective number-theoretic relations. Certain other relations, namely those obtained by using the quantifiers $\exists$ or $\forall$ on the recursive relations, are also very natural relations to consider in many contexts. They can be arranged in the so-called arithmetical hierarchy, according to the depth of quantifiers used in defining them. We shall describe this hierarchy and its most important properties.

Our main result, 5.36, depends on the following normal form theorem.

Theorem 5.23. *Let $m > 1$. If R is an m-ary recursive relation, then there exist $e, e' \in \omega$ such that for all $x_0, \ldots, x_{m-2} \in \omega$,*

(i) $\exists y((x_0, \ldots, x_{m-2}, y) \in R)$ *iff* $\exists y((e, x_0, \ldots, x_{m-2}, y) \in \mathrm{T}_{m-1})$.

(ii) $\forall y((x_0, \ldots, x_{m-2}, y) \in R)$ *iff* $\forall y((e', x_0, \ldots, x_{m-2}, y) \notin \mathrm{T}_{m-1})$.

PROOF. For any $x_0, \ldots, x_{m-2} \in \omega$ let

$$f(x_0, \ldots, x_{m-2}) \simeq \mu y((x_0, \ldots, x_{m-2}, y) \in R).$$

Thus f is partial recursive, so by 5.9 there is an $e \in \omega$ such that for all $x_0, \ldots, x_{m-2} \in \omega$,

$$f(x_0, \ldots, x_{m-2}) \simeq \mathrm{V}\mu u((e, x_0, \ldots, x_{m-2}, u) \in \mathrm{T}_{m-1}).$$

Thus

$$\exists y((x_0, \ldots, x_{m-2}, y) \in R) \quad \text{iff } (x_0, \ldots, x_{m-2}) \in \mathrm{Dmn}\, f,$$
$$\text{iff } \exists y((e, x_0, \ldots, x_{m-2}, y) \in \mathrm{T}_{m-1}).$$

Thus *(i)* holds. Condition *(ii)* is easily obtained from *(i)*. □

Definition 5.24. Let $\Sigma_0 = \Pi_0 =$ set of all recursive relations. If $n, m > 0$, then an n-ary relation R is in Σ_m (respectively Π_m) provided there is an $(m + n)$-ary recursive relation S such that, if m is odd,

$$R = \{(x_0, \ldots, x_{n-1}) \in {}^n\omega : \exists y_0 \in \omega\ \forall y_1 \in \omega\ \exists y_2 \in \omega \cdots \forall y_{m-2} \in \omega\ \exists y_{m-1} \in \omega[(x_0, \ldots, x_{n-1}, y_0, \ldots, y_{m-1}) \in S]\}$$

(respectively

$$R = \{(x_0, \ldots, x_{n-1}) \in {}^n\omega : \forall y_0 \in \omega\ \exists y_1 \in \omega\ \forall y_2 \in \omega \cdots \exists y_{m-2} \in \omega\ \forall y_{m-1} \in \omega[(x_0, \ldots, x_{n-1}, y_0, \ldots, y_{m-1}) \in S]\}),$$

while, if m is even,

$$R = \{(x_0, \ldots, x_{m-1}) \in {}^n\omega : \exists y_0 \in \omega\, \forall y_1 \in \omega\, \exists y_2 \in \omega \cdots \\ \exists y_{m-2} \in \omega\, \forall y_{m-1} \in \omega[(x_0, \ldots, x_{n-1}, y_0, \ldots, y_{m-1}) \in S]\}$$

(respectively

$$R = \{(x_0, \ldots, x_{n-1}) \in {}^n\omega : \forall y_0 \in \omega\, \exists y_1 \in \omega\, \forall y_2 \in \omega \cdots \\ \forall y_{m-2} \in \omega\, \exists y_{m-1} \in \omega[(x_0, \ldots, x_{n-1}, y_0, \ldots, y_{m-1}) \in S]\}).$$

Members of Σ_m (respectively Π_m) are called *Σ_m-relations* (respectively *Π_m-relations*). Also let $\Delta_m = \Sigma_m \cap \Pi_m$. Any member of $\bigcup_{m \in \omega \sim 1} (\Sigma_m \cup \Pi_m)$ is said to be *arithmetical.*

Note that there are only $\aleph_0$ arithmetic relations, and hence most number-theoretic relations are not arithmetical. Now we want to describe the relationships between the various classes Σ_m, Π_m, and indicate some operations under which these classes are closed. The following obvious proposition indicates how these classes can be inductively defined, and furnishes a basis for inductive proofs of our further results.

Proposition 5.25

(*i*) *An n-ary relation R is in* Σ_{m+1} *iff there is an* $(n + 1)$*-ary relation S in* Π_m *such that for all* $x_0, \ldots, x_{n-1} \in \omega$, $(x_0, \ldots, x_{n-1}) \in R$ *iff* $\exists y \in \omega$ $((x_0, \ldots, x_{n-1}, y) \in S)$.

(*ii*) *An n-ary relation R is in* Π_{m+1} *iff there is an* $(n + 1)$*-ary relation S in* Σ_m *such that for all* $x_0, \ldots, x_{n-1} \in \omega$, $(x_0, \ldots, x_{n-1}) \in R$ *iff* $\forall y \in \omega$ $((x_0, \ldots, x_{n-1}, y) \in S)$.

The following three propositions are now easily established by induction, using 5.25:

Proposition 5.26. *If R is an n-ary* Σ_m*-relation,* $f_0, \ldots, f_{n-1}$ *are n-ary recursive functions, and*

$$S = \{(x_0, \ldots, x_{n-1}) : (f_0(x_0, \ldots, x_{n-1}), \ldots, f_{n-1}(x_0, \ldots, x_{n-1})) \in R\},$$

then $S \in \Sigma_m$. *Similarly for* Π_m *and* Δ_m.

Proposition 5.27 (Adjunction of apparent variables). *If R is an n-ary* Σ_m*-relation and* $S = \{(x_0, \ldots, x_n) : (x_1, \ldots, x_n) \in R\}$, *then* $S \in \Sigma_m$. *Similarly for* Π_m *and* Δ_m.

Proposition 5.28 (Identification of variables). *If R is an n-ary* Σ_m*-relation,* $n > 1$, *and* $S = \{(x_0, \ldots, x_{n-2}) : (x_0, x_0, x_1, x_2, \ldots, x_{n-2}) \in R\}$, *then* $R \in \Sigma_m$. *Similarly for* Π_m *and* Δ_m.

Proposition 5.29. *If R and S are n-ary* Σ_m*-relations, then so are* $R \cup S$ *and* $R \cap S$. *Similarly for* Π_m *and* Δ_m.

PROOF. The assertions for Δ_m follow from those for Σ_m and Π_m. The assertions for Σ_m and Π_m are proved simultaneously by induction on m. The case $m = 0$ is trivial. Now assume the assertions for m. We take just one typical assertion for $m + 1$:

Assume $R, S \in \Sigma_{m+1}$; we show that $R \cap S \in \Sigma_{m+1}$. By 5.25, choose $(n + 1)$-ary Π_m-relations R', S' such that for all $x_0, \ldots, x_{n-1} \in \omega$,

$$(x_0, \ldots, x_{n-1}) \in R \quad \text{iff } \exists y \in \omega[(x_0, \ldots, x_{n-1}, y) \in R'],$$
$$(x_0, \ldots, x_{n-1}) \in S \quad \text{iff } \exists y \in \omega[(x_0, \ldots, x_{n-1}, y) \in S'].$$

Then for any $x_0, \ldots, x_{n-1} \in \omega$,

$$(x_0, \ldots, x_{n-1}) \in R \cap S \quad \text{iff } \exists y \in \omega\ \exists z \in \omega[(x_0, \ldots, x_{n-1}, y) \in R' \text{ and } (x_0, \ldots, x_{n-1}, z) \in S'].$$

Now let $R'' = \{(x_0, \ldots, x_{n+1}) : (x_0, \ldots, x_n) \in R'\}$ and $S'' = \{(x_0, \ldots, x_{n+1}) : (x_0, \ldots, x_{n-1}, x_{n+1}) \in S'\}$. Using 5.26 and 5.27 it is easy to see that R'', $S'' \in \Pi_m$. Now, continuing from above, for any $x_0, \ldots, x_{n-1} \in \omega$,

$$(x_0, \ldots, x_{n-1}) \in R \cap S \quad \text{iff } \exists y \in \omega\ \exists z \in \omega[(x_0, \ldots, x_{n-1}, y, z) \in R'' \cap S'']$$
$$\text{iff } \exists y \in \omega[(x_0, \ldots, x_{n-1}, (y)_0, (y)_1) \in R'' \cap S''].$$

Since $R'' \cap S'' \in \Pi_m$ by the induction hypothesis, we get $R \cap S \in \Sigma_{m+1}$ by 5.25. □

For $m > 0$, neither Σ_m nor Π_m nor Δ_m is closed under complementation; see 5.36. The following proposition is evident.

Proposition 5.30. *If R is an n-ary Σ_m-relation with $n > 1$ and $m > 0$, and if $S = \{(x_0, \ldots, x_{n-2}) : \exists y \in \omega(x_0, \ldots, x_{n-2}, y) \in R\}$, then $S \in \Sigma_m$. Similarly with Π_m and $\forall$.*

Proposition 5.31. *If R is an n-ary Σ_m-relation, then so are the two relations*

$$S = \{(x_0, \ldots, x_{n-1}) : \exists y < x_{n-1}[(x_0, \ldots, x_{n-2}, y) \in R],$$
$$T = \{(x_0, \ldots, x_{n-1}) : \forall y < x_{n-1}[(x_0, \ldots, x_{n-2}, y) \in R].$$

Similarly for Π_m and Δ_m.

PROOF. Again we prove all cases simultaneously by induction on m. The case $m = 0$ is trivial. Assume that all of the statements are true for m. We take one typical case for $m + 1$:

Let R be an n-ary Σ_{m+1}-relation, and let T be as above. By 5.25, let R' be a Π_m-relation such that for all $x_0, \ldots, x_{n-1} \in \omega$,

$$(x_0, \ldots, x_{n-1}) \in R \quad \text{iff } \exists z \in \omega[(x_0, \ldots, x_{n-1}, z) \in R'].$$

Clearly, then, it suffices to show that for all $x_0, \ldots, x_{n-1} \in \omega$,

(1) $\forall y < x_{n-1}\ \exists z \in \omega[(x_0, \ldots, x_{n-2}, y, z) \in R']$
$$\text{iff } \exists z \in \omega\ \forall y < x_{n-1}[(x_0, \ldots, x_{n-2}, y, (z)_y) \in R'].$$

Clearly the right side of (1) implies the left side. If the left side holds, choose for each $y < x_{n-1}$ an integer $w_y \in \omega$ such that $(x_0, \ldots, x_{n-2}, y, w_y) \in R'$, and let $z = \prod_{y < x(n-1)} \mathrm{p}_y^{wy}$; clearly then z is as desired in the right side of (1). $\square$

The following proposition is obvious:

Proposition 5.32. *If R is an n-ary relation, then $R \in \Sigma_m$ iff ${}^n\omega \sim R \in \Pi_m$.*

Proposition 5.33. $\Sigma_m \cup \Pi_m \subseteq \Delta_{m+1}$.

PROOF. Let $R \in \Sigma_m$, say R is n-ary. Let $S = \{(x_0, \ldots, x_n) : (x_0, \ldots, x_{n-1}) \in R\}$. Then $S \in \Sigma_m$ by 5.26 and 5.27. Clearly $R = \{(x_0, \ldots, x_{n-1}) : \forall y \in \omega[(x_0, \ldots, x_{n-1}, y) \in S]\}$, so $R \in \Pi_{m+1}$. Thus $\Sigma_m \subseteq \Pi_{m+1}$, and similarly $\Pi_m \subseteq \Sigma_{m+1}$. An easy inductive argument shows that $\Sigma_m \subseteq \Sigma_{m+1}$ and $\Pi_m \subseteq \Pi_{m+1}$. $\square$

We will return to the following important result several times later on:

Theorem 5.34. $\Delta_1 = \Delta_0$.

PROOF. We know that $\Delta_0 \subseteq \Delta_1$. Suppose $R \in \Delta_1$, say R is n-ary. Then there are recursive S, $T((n + 1)$-ary) such that for all $x_0, \ldots, x_{n-1} \in \omega$,

$$\begin{aligned}(x_0, \ldots, x_{n-1}) \in R \quad & \text{iff } \exists y[(x_0, \ldots, x_{n-1}, y) \in S] \\ & \text{iff } \forall y[(x_0, \ldots, x_{n-1}, y) \in T].\end{aligned}$$

Hence, as is easily seen,

$$\chi_R(x_0, \ldots, x_{n-1}) = \chi_S(x_0, \ldots, x_{n-1}, \mu y[(x_0, \ldots, x_{n-1}, y) \in S \text{ or } (x_0, \ldots, x_{n-1}, y) \notin T]),$$

so R is recursive. $\square$

Intuitively, to determine whether or not $(x_0, \ldots, x_{n-1}) \in R$ we check in succession $(x_0, \ldots, x_{n-1}, 0)$, $(x_0, \ldots, x_{n-1}, 1), \ldots$ for membership in S and T. Eventually one of these is in S (hence $(x_0, \ldots, x_{n-1}) \in R$), or else one of them fails to be in T (hence $(x_0, \ldots, x_{n-1}) \notin R$).

Now we extend our normal form results up into the arithmetical hierarchy:

Theorem 5.35. *For $m, n > 0$ there is an $(n + 1)$-ary Σ_m-relation R_m^n with the following properties:*

(i) for every n-ary Σ_m-relation S there is an $e \in \omega$ such that $S = \{(x_0, \ldots, x_{n-1}) : (e, x_0, \ldots, x_{n-1}) \in R_m^n\}$;
(ii) for every n-ary Π_m-relation S there is an $e \in \omega$ such that $S = \{(x_0, \ldots, x_{n-1}) : (e, x_0, \ldots, x_{n-1}) \notin R_m^n\}$.

PROOF. We construct R_m^n by recursion on m. Let

$$R_1^n = \{(x_0, \ldots, x_n) : \exists y \in \omega[(x_0, \ldots, x_n, y) \in \mathrm{T}_n]\}.$$

If R^n_m has been defined for all n, let

$$R^n_{m+1} = \{(x_0, \dots, x_n) : \exists y \in \omega[(x_0, \dots, x_n, y) \notin R^{n+1}_m]\}.$$

It is easily seen by induction on m, using 5.23, that the desired conditions hold. □

Theorem 5.36 (Hierarchy theorem). *For any $m, n > 0$ there exists an n-ary relation $T \in \Sigma_m \sim \Pi_m$. Hence ${}^n\omega \sim T \in \Pi_m \sim \Sigma_m$. Furthermore, there is an n-ary relation $W \in \Delta_{m+1} \sim (\Sigma_m \cup \Pi_m)$.*

PROOF. Let R^n_m be as in 5.35. Let

$$T = \{(x_0, \dots, x_{n-1}) : (x_0, x_0, x_1, x_2, \dots, x_{n-1}) \in R^n_m\}.$$

Thus $T \in \Sigma_m$. If $T \in \Pi_m$, by 5.35 choose $e \in \omega$ so that $T = \{(x_0, \dots, x_{n-1}) : (e, x_0, \dots, x_{n-1}) \notin R^n_m\}$. Then

$$e^{(n+1)} \in R^n_m \quad \text{iff } e^{(n)} \in T \quad \text{iff } (e)^{n+1} \notin R^n_m,$$

a contradiction. Thus $T \notin \Pi_m$.

For the second part of the theorem, let T be as in the first part. Set

$$W = \{(x_0, \dots, x_{n-1}) : ((x_0)_0, (x_1)_0, \dots, (x_{n-1})_0) \notin T \quad \text{and} \quad ((x_0)_1, (x_1)_1, \dots, (x_{n-1})_1) \in T\}.$$

Now T, ${}^n\omega \sim T \in \Delta_{m+1}$ by 5.33, so $W \in \Delta_{m+1}$. Suppose $W \in \Sigma_m$. Choose $(t_0, \dots, t_{n-1}) \in T$ (T is obviously nonempty since $0 \in \Pi_m$). For any $x_0, \dots, x_{n-1} \in \omega$ we have

$$(x_0, \dots, x_{n-1}) \notin T \quad \text{iff } (2^{x0} \cdot 3^{t0}, \dots, 2^{x(n-1)} \cdot 3^{t(n-1)}) \in W,$$

so ${}^n\omega \sim T \in \Sigma_m$, contradiction. Similarly, $W \in \Pi_m$ leads to a contradiction. □

Thus the arithmetical hierarchy appears as in the following diagram, where the lines indicate proper inclusions:

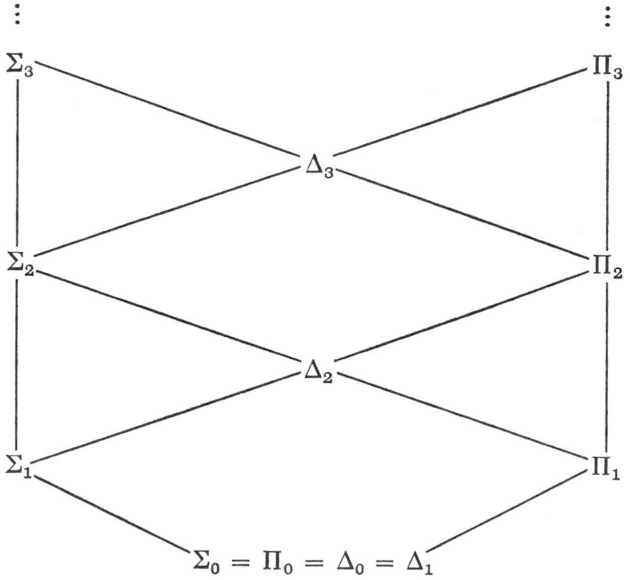

BIBLIOGRAPHY

1. Kleene, S. C. *Introduction to Metamathematics.* Princeton: van Nostrand (1952).
2. Malcev, A. I. *Algorithms and Recursive Functions.* Groningen: Wolters-Noordhoff (1970).
3. Rogers, H. *Theory of Recursive Functions and Effective Computability.* New York: McGraw-Hill (1967).

EXERCISES

5.37. If f is a finite function (i.e., it is a finite set and is a function), then f is partial recursive.

5.38. Give an example of a binary partial recursive function f such that if g is defined by

$$\begin{aligned} gx &= \text{least } y \text{ such that } f(x, y) \text{ is defined and } f(x, y) = 0, \\ gx &= \text{undefined if no such } y, \end{aligned}$$

then g is *not* partial recursive (cf. 5.3 and following remarks).

5.39. Give an example of a binary recursive function f such that if g is defined by

$$\begin{aligned} gx &= \text{least } y \text{ such that } f(x, y) = 0, \\ gx &= 0 \qquad \text{if no such } y, \end{aligned}$$

then g is *not* recursive.

5.40′. The class of partial recursive functions is the intersection of all classes C of partial functions such that $\mathscr{s} \in C$, $U_i^n \in C$ whenever $i < n$, and C is closed under composition, primitive recursion, and minimalization, all except composition applied only to total functions.

5.41. If f is an m-ary partial recursive function and Dmn f is recursive, then f can be extended to a general recursive function.

5.42. Give an example of an m-ary partial recursive function f which can be extended to a general recursive function, but has the property that Dmn f is not recursive.

5.43. There is a unary partial recursive function f such that for no binary recursive function g is it true that for all x, $fx \simeq \mu y[g(x, y) = 0]$. *Hint:* let $fx \simeq \varphi_x^1 x \cdot 0 + x$ for all x. If g works as above, let $hx = \varphi_x^1 x + 1$ if $g(x, x) = 0$, $hx = 0$ otherwise. Show h is recursive and obtain a contradiction.

5.44. For any total function f of one variable the following conditions are equivalent:

(1) there is a recursive function g of two variables such that for all $x \in \omega$, $fx = \mu y[g(x, y) = 0]$.
(2) $\{(x, fx) : x \in \omega\}$ is a recursive relation.

The conditions remain equivalent if in both (1) and (2) "recursive" is replaced by "primitive recursive" or by "elementary."

5.45. If f is a unary recursive function, then $\{(x, fx) : x \in \omega\}$ is a recursive relation. Similarly if we replace both words "recursive" by "primitive recursive" or by "elementary."

5.46. Give an example of a unary partial recursive function f such that $\{(x, fx) : x \in \mathrm{Dmn}\, f\}$ is not recursive.

5.47. There is a recursive set which is not elementary.

5.48. There is a unary recursive function f for which there is no binary elementary function g such that for all $x \in \omega$, $fx = \mu y[g(x, y) = 0]$. *Hint:* take $f = \chi_A$, where A is as in 5.47.

5.49. There is a total unary function f such that $\{(x, fx) : x \in \omega\}$ is elementary but f is not elementary.

5.50. There is no recursive procedure for deciding for an arbitrary e whether or not φ_e^1 has infinite range.

5.51. Assume $m > 1$. Let $A = \{e : \varphi_e^m$ is a special recursive function$\}$. Show that A is not recursive.

5.52. Show that the function f defined as follows is recursive.

$$\begin{aligned} f(0, y) &= y + 1, \\ f(1, y) &= y + 2, \\ f(x + 2, 0) &= f(x + 1, 1), \\ f(x + 2, y + 1) &= f(x, f(x + 1, f(x + 2, y))). \end{aligned}$$

5.53. Show that there is no recursive function f satisfying the following conditions:

$$\begin{aligned} f(0, y) &= y + 2, \\ f(x + 1, 0) &= f(x, 1), \\ f(x + 1, y + 1) &= f(x + 1, f(x, y)) + 1. \end{aligned}$$

6 Recursively Enumerable Sets

In this chapter we shall deal in some detail with the set Σ_1 of relations (see 5.24). Such relations are called *recursively enumerable* for reasons which will shortly become clear. The study of recursively enumerable relations is one of the main branches of recursive function theory. They play a large role in logic. In fact, for most theories the set of Gödel numbers of theorems is recursively enumerable. Thus many of the concepts introduced in this section will have applications in our discussion of decidable and undecidable theories in Part III. Unless otherwise stated, the functions in this chapter are unary.

A nonempty set is *effectively enumerable* provided there is an automatic method for listing out its members, one after the other. This does *not* imply that there is a decision method for determining membership in the set. The formal version of this notion is given in

Definition 6.1. A set $A \subseteq \omega$ is *recursively enumerable* (for brevity *r.e.*) if $A = 0$ or A is the range of a recursive function.

This definition can be given several equivalent forms, each having its own intuitive appeal:

Theorem 6.2. *For $A \subseteq \omega$ the following are equivalent;*

(*i*) *$A = 0$ or A is the range of an elementary function;*
(*ii*) *$A = 0$ or A is the range of a primitive recursive function;*
(*iii*) *A is recursively enumerable;*
(*iv*) *A is the range of a partial recursive function;*
(*v*) *A is the domain of a partial recursive function;*
(*vi*) *$A \in \Sigma_1$.*

PROOF. Obviously $(i) \Rightarrow (ii) \Rightarrow (iii)$. To show that $(iii) \Rightarrow (iv)$ we just need to show that 0 (the empty set) is the range of some partial recursive function; and obviously the only possibility for such a function is 0 (which is also the empty function). 0 is partial recursive by the argument following 5.3.

$(iv) \Rightarrow (v)$. Let $A = \text{Rng}\, \varphi_e^1$. For any $x \in \omega$ let

$$fx \simeq \mu y((e, (y)_0, (y)_1) \in \mathrm{T}_1 \text{ and } \mathrm{V}(y)_1 = x).$$

Clearly then $\text{Dmn}\, f = \text{Rng}\, \varphi_e^1 = A$, and f is partial recursive.

$(v) \Rightarrow (vi)$. Suppose $A = \text{Dmn}\, \varphi_e^1$. Then for all $x \in \omega$, $x \in A$ iff $\exists y((e, x, y) \in \mathrm{T}_1)$, so $A \in \Sigma_1$.

$(vi) \Rightarrow (i)$. Suppose $A \in \Sigma_1$. By 5.23 choose $e \in \omega$ such that $A = \{x : \exists y((e, x, y) \in \mathrm{T}_1)\}$. We may assume that $A \neq 0$; say $a \in A$. Now for any $x \in \omega$ let

$$\begin{aligned} fx &= (x)_0 \quad \text{if } (e, (x)_0, (x)_1) \in \mathrm{T}_1, \\ fx &= a \quad \text{otherwise.} \end{aligned}$$

Clearly f is an elementary function and $\text{Rng}\, f = A$, as desired. □

An intuitive proof of the equivalence of 6.2(iii) and 6.2(v) is instructive. First assume that A is recursively enumerable, $A \neq 0$. Say $A = \text{Rng}\, f$, recursive. We define a function g with domain A as follows. To calculate gx, we look along the list $f0, f1, \ldots$ for x. If we find it, we set $gx = 0$. If x is never found, gx is never computed. Clearly g is effectively calculable (see introduction to Chapter 5), and $\text{Dmn}\, g = A$.

Conversely, suppose $A = \text{Dmn}\, g$, g partial recursive, and assume that $A \neq 0$. Now we make the following calculations:

two steps in the calculation of $g0$
one step in the calculation of $g1$
three steps in the calculation of $g0$
two steps in the calculation of $g1$
one step in the calculation of $g2$
four steps in the calculation of $g0$
three steps in the calculation of $g1$
two steps in the calculation of $g2$
one step in the calculation of $g3$
⋮ ⋮ ⋮ ⋮ ⋮ ⋮ ⋮

During this process we will occasionally obtain answers. At regular intervals we list out all the x for which we have so far calculated gx. Since $A \neq 0$, eventually we will list at least one x, and then at regular intervals we put more on our list (with many repetitions). Calling the list $f0, f1, \ldots$, clearly f is an effectively calculable total function with range A.

Now we want to investigate the relationship between recursive and recursively enumerable sets. By 5.33 and 5.36 we have

Theorem 6.3. *Every recursive set is recursively enumerable. There is a recursively enumerable set which is not recursive.*

The second part of 6.3 is one of the most important results of recursion theory, so we give its proof here in a more direct form:

Definition 6.4. $K = \{x : \exists y((x, x, y) \in T_1)\}$.

Theorem 6.5. *K is recursively enumerable but not recursive.*

PROOF. Obviously $K \in \Sigma_1$ so K is recursively enumerable. Suppose K is recursive. Then so is $\omega \sim K$, so by 6.2(v) there is an $e \in \omega$ such that $\omega \sim K = \text{Dmn}\, \varphi_e^1$. Then

$$\begin{array}{ll} e \in K & \text{iff } e \in \text{Dmn}\, \varphi_e^1 \text{ by the definition of K,} \\ e \notin K & \text{iff } e \in \text{Dmn}\, \varphi_e^1 \text{ by the choice of } e, \end{array}$$

contradiction. □

The set K will be discussed further later on. Another important relationship between recursive and recursively enumerable sets is given in 5.34, which can be reformulated as follows:

Theorem 6.6. *Let $A \subseteq \omega$. The following conditions are equivalent*:

(i) *A is recursive;*
(ii) *A and $\omega \sim A$ are recursively enumerable.*

This theorem can be seen in the following fashion, working directly from Definition 6.1: Of course (ii) ⇒ (i) is the main part of 6.6. Assume (ii). We may suppose $0 \neq A \neq \omega$. Then let f and g be recursive functions with $\text{Rng}\, f = A$, $\text{Rng}\, g = \omega \sim A$. To determine whether $x \in A$ or not, list out $f0, g0, f1, g1, \ldots$. Eventually x will appear in the list; if $x = fn$ for some n, then $x \in A$, while if $x = gn$ for some n, then $x \in A$. Formally, for any $x \in \omega$,

$$\begin{array}{ll} \chi_A x = 1 & \text{if } f\mu y(fy = x \text{ or } gy = x) = x, \\ \chi_A x = 0 & \text{otherwise.} \end{array}$$

Theorem 6.7. *Let $A \subseteq \omega$. The following are equivalent*:

(i) *A is infinite and recursive*:
(ii) *there is a recursive function f with* $\text{Rng} f = A$ *and* $\forall x \in \omega(fx < f(x + 1))$.

PROOF. (i) ⇒ (ii). Let a be the least member of A. Define

$$\begin{aligned} f0 &= a \\ f(x + 1) &= \mu y(y \in A \text{ and } y > fx). \end{aligned}$$

Clearly f is as desired.

(ii) ⇒ (i). Assume f as in (ii). Then by induction on x,

(1) $$\forall x \in \omega \quad (x \leq fx).$$

Thus

(2) $$\forall y \in \text{Rng}\, f\, \exists x \leq y \quad (fx = y).$$

Hence for all $y \in \omega$,

$$\begin{aligned} \chi_A y &= 1 && \text{if } \exists x \leq y(fx = y) \\ \chi_A y &= 0 && \text{otherwise,} \end{aligned}$$

as desired. □

Theorem 6.8. *Any infinite recursively enumerable set has an infinite recursive subset.*

PROOF. Let A be infinite r.e., say $A = \text{Rng} f$, f recursive. We define g by induction:

$$\begin{aligned} g0 &= f0 \\ g(x+1) &= f\mu y(fy > gx). \end{aligned}$$

Thus $gx < g(x+1)$ for all $x \in \omega$, and hence, by 6.7, $\text{Rng}\, g$ is infinite and recursive. Obviously $\text{Rng}\, g \subseteq A$. □

Next, we want to investigate closure properties of the class of r.e. sets. Which operations on sets lead out of the class, and under which operation is the class closed? By 5.29, the class of r.e. sets is closed under union and intersection. We can give intuitive proofs of these facts directly from the definition. Let A and B be r.e. sets, and ignore the case when one of them is empty. Let f and g be recursive functions enumerating A and B respectively. One enumerates $A \cup B$ by: $f0, g0, f1, g1, \ldots$. One can enumerate $A \cap B$ by looking along this list and putting a number on a separate list as soon as it appears at both an odd and even step. Both of these procedures can be given a rigorous formulation.

By 6.3 and 6.6, the class of r.e. sets is not closed under complementation. Some further closure properties:

Theorem 6.9. *If A is r.e. and f is partial recursive, then f^*A is r.e.*

PROOF. We may assume that $A \neq 0$. Say $A = \text{Rng}\, g$, g recursive. Clearly $f^*A = \text{Rng}\,(f \circ g)$ and $f \circ g$ is partial recursive. □

Theorem 6.10. *If A is r.e. and f is partial recursive, then $f^{-1*}A$ is r.e.*

PROOF. Say $A = \text{Dmn}\, \boldsymbol{\varphi}_e^1$. Then $f^{-1*}A = \text{Dmn}\,(\boldsymbol{\varphi}_e^1 \circ f)$ as desired. □

Theorem 6.11. *If A is r.e., then $\bigcup_{x \in A} \text{Rng}\, \boldsymbol{\varphi}_e^1$ is r.e.*

PROOF. For any $y \in \omega$,

$$y \in \bigcup_{x \in A} \text{Rng}\, \boldsymbol{\varphi}_x^1 \quad \text{iff } \exists x \in A(y \in \text{Rng}\, \boldsymbol{\varphi}_x^1).$$

Since both A and each $\text{Rng}\, \varphi_x^1$ are in Σ_1, it follows easily that $\bigcup_{x \in A} \text{Rng}\, \varphi_x^1$ is in Σ_1. □

Before carrying the theory of r.e. sets further we wish to back up and extend our results obtained so far to relations.

Definition 6.12. A relation $R \subseteq {}^m\omega$ is *recursively enumerable* (for brevity *r.e.*) if $A = 0$ or there exist m recursive functions $f_0, \ldots, f_{m-1}$ such that

$$R = \{(f_0x, \ldots, f_{m-1}x) : x \in \omega\}.$$

Theorem 6.13. *For $R \subseteq {}^m\omega$ the following are equivalent:*

(*i*) *$R = 0$ or there exist elementary functions $f_0, \ldots, f_{m-1}$ with $R = \{(f_0x, \ldots, f_{m-1}x) : x \in \omega\}$;*

(*ii*) *like (i) with "elementary" replaced by "primitive recursive";*

(*iii*) *R is recursively enumerable;*

(*iv*) *there exist partial recursive functions $f_0, \ldots, f_{m-1}$ with $R = \{(f_0x, \ldots, f_{m-1}x) : x \in \text{Dmn } f_0 \cap \cdots \cap \text{Dmn } f_{m-1}\}$;*

(*v*) *$R = 0$ or there is an elementary function f with $R = \{((fx)_0, \ldots, (fx)_{m-1}) : x \in \omega\}$;*

(*vi*) *like (v), with "elementary" replaced by "primitive recursive";*

(*vii*) *like (v), with "elementary" replaced by "recursive";*

(*viii*) *there is a partial recursive function f such that $R = \{((fx)_0, \ldots, (fx)_{m-1}) : x \in \text{Dmn } f\}$;*

(*ix*) *there is an m-ary partial recursive function f such that $R = \text{Dmn } f$;*

(*x*) *$R \in \Sigma_1$.*

PROOF. Clearly (*i*) $\Rightarrow$ (*ii*) $\Rightarrow$ (*iii*) $\Rightarrow$ (*iv*).

(*iv*) $\Rightarrow$ (*v*). Assume (*iv*), with $f_0, \ldots, f_{m-1}$ partial recursive and $R = \{(f_0x, \ldots, f_{m-1}x) : x \in \text{Dmn } f_0 \cap \cdots \cap \text{Dmn } f_{m-1}\}$. We may assume that $R \neq 0$, say $(a_0, \ldots, a_{m-1}) \in R$. Say $f_0 = \varphi^1_{e0}, \ldots, f_{m-1} = \varphi^1_{e(m-1)}$. For any $x \in \omega$, let

$$gx = \prod_{i<m} \mathrm{p}_i^{\mathrm{V}((x)(i+1))} \qquad \text{if } (e_i, (x)_0, (x)_{i+1}) \in \mathrm{T}_1 \text{ for all } i < m,$$

$$gx = \prod_{i<m} \mathrm{p}_i^{ai} \qquad \text{otherwise.}$$

Clearly g is elementary and $R = \{((gx)_0, \ldots, (gx)_{m-1}) : x \in \omega\}$.

Obviously (*v*) $\Rightarrow$ (*vi*) $\Rightarrow$ (*vii*) $\Rightarrow$ (*viii*).

(*viii*) $\Rightarrow$ (*ix*). Suppose f is as in (*viii*). Say $f = \varphi^1_e$. For any $x_0, \ldots, x_{m-1} \in \omega$ let

$$g(x_0, \ldots, x_{m-1}) = \mu y((e, (y)_0, (y)_1) \in \mathrm{T}_1 \text{ and } \mathrm{V}(y)_1 = \prod_{i<m} \mathrm{p}_i^{xi}).$$

Clearly g is partial recursive and $R = \text{Dmn } g$.

(*ix*) $\Rightarrow$ (*x*). Suppose $R = \text{Dmn } \varphi^m_e$. Then

$$R = \{(x_0, \ldots, x_{m-1}) : \exists y((e, x_0, \ldots, x_{m-1}, y) \in \mathrm{T}_m)\},$$

so $R \in \Sigma_1$.

(*x*) $\Rightarrow$ (*i*). Suppose $R \in \Sigma_1$. By 5.23, choose $e \in \omega$ such that

$$R = \{(x_0, \ldots, x_{m-1}) : \exists y((e, x_0, \ldots, x_{m-1}, y) \in \mathrm{T}_m)\}.$$

We may assume that $R \neq 0$; say $(a_0, \ldots, a_{m-1}) \in R$. Now for $i < m$ and any $x \in \omega$ let

$$\begin{aligned} f_i x &= (x)_i && \text{if } (e, (x)_0, \ldots, (x)_m) \in \mathrm{T}_m \\ f_i x &= a_i && \text{otherwise.} \end{aligned}$$

Clearly each f_i is elementary and $R = \{(f_0 x, \ldots, f_{m-1} x) : x \in \omega\}$. □

Theorem 6.14. *Every recursive relation is recursively enumerable. For each positive m there is a recursively enumerable m-ary relation which is not recursive.*

PROOF. The first part is true by 6.13(x) and 5.33; for the second part, use 5.36. □

The following result is proved just as for sets.

Theorem 6.15. *Let $R \subseteq {}^m\omega$. The following conditions are equivalent:*

(*i*) *R is recursive;*
(*ii*) *R and ${}^m\omega \sim R$ are recursively enumerable.*

The following important theorem shows that the notion of a partial recursive function can be defined without resorting to the rather complicated notions discussed at the beginning of Chapter 5.

Theorem 6.16. *Let f be a unary partial function. Then the following conditions are equivalent:*

(*i*) *f is partial recursive;*
(*ii*) $\{(x, fx) : x \in \mathrm{Dmn}\, f\}$ *is r.e.*

PROOF. (*i*) ⇒ (*ii*). Assume (*i*). For any $x, y \in \omega$ let

$$g(x, y) \simeq \mu z(|y - fx| = 0).$$

Clearly g is partial recursive and $\mathrm{Dmn}\, g = \{(x, fx) : x \in \mathrm{Dmn}\, f\}$.

(*ii*) ⇒ (*i*). Assume (*ii*), and by 6.12 let g and h be recursive functions such that

$$\{(x, fx) : x \in \mathrm{Dmn}\, f\} = \{(gx, hx) : x \in \omega\}.$$

Then for any $x \in \omega$,

$$fx \simeq h\mu y(gy = x),$$

so f is partial recursive. □

We now turn to the study of some special r.e. sets.

Definition 6.17

(*i*) A set $A \subseteq \omega$ is *productive* if there is a recursive function f (called a *productive function for A*) such that for all $e \in \omega$, if $\mathrm{Dmn}\, \varphi_e^1 \subseteq A$ then $fe \in A \sim \mathrm{Dmn}\, \varphi_e^1$.

(*ii*) A set $A \subseteq \omega$ is *creative* if A is r.e. and $\omega \sim A$ is productive.

Thus a productive set A is strongly not recursively enumerable: there is an effective procedure for finding members of $A \sim B$ for any r.e. subset B of A. A creative set, while r.e., is strongly nonrecursive. The sets of Gödel numbers of theorems of many theories studied in Part III are creative, as we shall see.

Recall Definition 6.4.

Theorem 6.18. K *is creative.*

Proof. By 6.5, K is r.e. Now U_0^1 is a productive function for $\omega \sim K$. For if $e \in \omega$ and $\operatorname{Dmn} \varphi_e^1 \subseteq \omega \sim K$, then $e \in (\omega \sim K) \sim \operatorname{Dmn} \varphi_e^1$; for

$$\begin{aligned} e \in K &\Rightarrow e \in \operatorname{Dmn} \varphi_e^1 && \text{by definition of K, 6.4} \\ &\Rightarrow e \in \omega \sim K && \text{by assumption } \operatorname{Dmn} \varphi_e^1 \subseteq \omega \sim K \end{aligned}$$

so $e \in \omega \sim K$, and hence by definition of K, $e \notin \operatorname{Dmn} \varphi_e^1$. □

The next theorem shows that, in a sense, any r.e. set can be obtained from a creative set; cf. 6.10 and the initial section of Chapter 7.

Theorem 6.19. *If A is r.e. and C is creative, then there is a recursive function f such that $A = f^{-1*}C$.*

Proof. Say $A = \operatorname{Dmn} \varphi_d^1$, and let g be a productive function for $\omega \sim C$. For any $x, y, z \in \omega$ let

$$l(z, y, x) \simeq \mu u[z = gs_1^1(x, y)] + \varphi_d^1 y.$$

Thus l is partial recursive. By the recursion theorem (5.15) choose $e \in \omega$ such that for all $y, z \in \omega$,

$$l(z, y, e) \simeq \varphi_e^2(z, y).$$

Let $fy = gs_1^1(e, y)$ for all $y \in \omega$. We claim that $A = f^{-1*}C$. Since f is obviously recursive, this will complete the proof.

First suppose that $y \in A$. Then

(1) $$\operatorname{Dmn} \varphi^1(s_1^1(e, y)) = \{gs_1^1(e, y)\}.$$

In fact, by 5.13 we have

$$\begin{aligned} z \in \operatorname{Dmn} \varphi^1(s_1^1(e, y)) \quad &\text{iff } (z, y) \in \operatorname{Dmn} \varphi_e^2 \\ &\text{iff } (z, y, e) \in \operatorname{Dmn} l \text{ (by choice of } e) \\ &\text{iff } z = gs_1^1(e, y) \text{ and } y \in \operatorname{Dmn} \varphi_d^1 \\ &\text{iff } z = gs_1^1(e, y). \end{aligned}$$

Thus (1) holds. Now if $fy \notin C$, this means that $gs_1^1(e, y) \notin C$ and so by (1) $\operatorname{Dmn} \varphi^1(s_1^1(e, y)) \subseteq \omega \sim C$. Since g is a productive function for $\omega \sim C$ we would get

$$gs_1^1(e, y) \in (\omega \sim C) \sim \operatorname{Dmn} \varphi^1(s_1^1(e, y)),$$

contradicting (1). Thus $fy \in C$.

Second, suppose that $y \notin A$. Then $y \notin \text{Dmn } \boldsymbol{\varphi}_d^1$, so $\forall z\,(z, y, e) \notin \text{Dmn } l$, hence $\forall z((z, y) \notin \text{Dmn } \boldsymbol{\varphi}_e^2)$, so by 5.13 $\text{Dmn } \boldsymbol{\varphi}^1(\mathrm{s}_1^1(e, y)) = 0$. Thus, since g is productive,

$$fy = g\mathrm{s}_1^1(e, y) \in (\omega \sim C) \sim \text{Dmn } \boldsymbol{\varphi}^1(\mathrm{s}_1^1(e, y)),$$

in particular $fy \notin C$, as desired. □

The following result will not play a role in our logical discussion, but is important in the general theory of r.e. sets. See also the definition and results concerning simple sets below.

Theorem 6.20. *If A is productive, then A has an infinite recursive subset.*

Proof. By 6.8 it suffices to show that A has an infinite r.e. subset. Let f be a productive function for A. For any x, y, let

$$k(y, x) \simeq \mu i(i \leq \text{l}x \text{ and } y = (x)_i \dot{-} 1).$$

Clearly k is partial recursive; say $k = \boldsymbol{\varphi}_e^2$. Now for any $x \in \omega$,

(1) $$\text{Dmn } \boldsymbol{\varphi}^1(\mathrm{s}_1^1(e, x)) = \{(x)_i \dot{-} 1 : i \leq \text{l}x\}.$$

In fact, for any $y \in \omega$,

$$\begin{aligned} y \in \text{Dmn } \boldsymbol{\varphi}^1(\mathrm{s}_1^1(e, x)) \quad &\text{iff } (y, x) \in \text{Dmn } \boldsymbol{\varphi}_e^2 \quad \text{iff } (y, x) \in \text{Dmn } k \\ &\text{iff } \exists i \leq \text{l}x(y = (x)_i \dot{-} 1). \end{aligned}$$

Now let r be such that $\boldsymbol{\varphi}_r^1 = 0$, and define

$$\begin{aligned} g(x, y) &= fr \qquad \text{if } y = 0 \text{ or } 1, \\ g(x, y) &= f\mathrm{s}_1^1(e, y) \qquad \text{if } y \neq 0 \text{ and } y \neq 1. \end{aligned}$$

Thus g is recursive. Now define $t: \omega \to \omega$ by setting, for any $x \in \omega$, $tx = g(x, \bar{t}x)$. Here $\bar{t}x$ is defined in 2.31, and by 2.33, t is recursive. Now we claim for all $x \in \omega$,

(2) $$tx \in A \sim \{ty : y < x\}.$$

We establish (2) by induction on x. For $x = 0$,

$$t0 = g(0, \bar{t}0) = g(0, 1) = fr \in A$$

(since $\boldsymbol{\varphi}_r^1 = 0 \subseteq A$ and f is a productive function for A). Thus (2) holds for $x = 0$. Suppose (2) holds for all $x' < x$, where $x \neq 0$. Then $tx = g(x, \bar{t}x)$, and $\bar{t}x \neq 0, 1$, so $tx = f\mathrm{s}_1^1(e, \bar{t}x)$. Also

$$\text{Dmn } \boldsymbol{\varphi}^1(\mathrm{s}_1^1(e, \bar{t}x)) = \{ty : y < x\} \subseteq A$$

by (1) and the induction hypothesis. Since f is a productive function for A, $tx = f\mathrm{s}_1^1(e, \bar{t}x) \in A \sim \{ty : y < x\}$, as desired. Thus (2) holds. Hence Rng t is an infinite r.e. subset of A, and the proof is complete. □

We now give a method to arrive at creative sets.

Definition 6.21

(*i*) Two sets A and B are *recursively separable* if there is a recursive set C such that $A \subseteq C$ and $B \subseteq \omega \sim C$.

(*ii*) A and B are *recursively inseparable* if they are disjoint but not recursively separable.

(*iii*) A and B are *effectively inseparable* if they are disjoint and there is a 2-ary recursive function f such that for all e and r, if $A \subseteq \text{Dmn}\, \varphi_e^1$, $B \subseteq \text{Dmn}\, \varphi_r^1$, and $\text{Dmn}\, \varphi_e^1 \cap \text{Dmn}\, \varphi_r^1 = 0$, then $f(e, r) \in \omega \sim (\text{Dmn}\, \varphi_e^1 \cup \text{Dmn}\, \varphi_r^1)$.

Effectively inseparable sets will be constructed in abundance in Part III; most undecidability results actually yield such sets.

Obviously we have:

Theorem 6.22. *If A and B are effectively inseparable then they are recursively inseparable.*

The converse of 6.22 fails; see Exercises 6.47, 6.48.

Theorem 6.23. *If A and B are recursively enumerable and effectively inseparable, then both A and B are creative.*

Proof. By symmetry it suffices to show that A is creative, i.e. that $\omega \sim A$ is productive. Let f be as in 6.21(*iii*). Say $A = \text{Dmn}\, \varphi_u^1$ and $B = \text{Dmn}\, \varphi_s^1$. For any e, $x \in \omega$ let

$$g(x, e) \simeq \mu y((e, x, y) \in \text{T}_1 \text{ or } (s, x, y) \in T_1).$$

Thus $\text{Dmn}\, g = \{(x, e) : x \in \text{Dmn}\, \varphi_e^1 \cup B\}$. Clearly g is partial recursive; say $g = \varphi_r^2$. Now for any $e \in \omega$ we have, by 5.13,

$$\text{(1)} \qquad \text{Dmn}\, \varphi^1(\text{s}_1^1(r, e)) = \{x : (x, e) \in \text{Dmn}\, \varphi_r^2\} = \text{Dmn}\, \varphi_e^1 \cup B.$$

Let for any $e \in \omega$ $he = f(u, \text{s}_1^1(r, e))$. Thus h is recursive; we claim it is a productive function for $\omega \sim A$. In fact, suppose $\text{Dmn}\, \varphi_e^1 \subseteq \omega \sim A$. Then, using (1), $A = \text{Dmn}\, \varphi_u^1$, $B \subseteq \text{Dmn}\, \varphi^1(\text{s}_1^1(r, e))$, and $\text{Dmn}\, \varphi_u^1 \cap \text{Dmn}\, \varphi^1(\text{s}_1^1(r, e)) = 0$. Hence, by 6.21(*iii*), $he = f(u, \text{s}_1^1(r, e)) \in \omega \sim (\text{Dmn}\, \varphi_u^1 \cup \text{Dmn}\, \varphi^1(\text{s}_1^1(r, e)))$, i.e., $he \in \omega \sim (A \cup \text{Dmn}\, \varphi_e^1 \cup B)$, so $he \in (\omega \sim A) \sim \text{Dmn}\, \varphi_e^1$, as desired. □

Theorem 6.24. *There exist two recursively enumerable effectively inseparable sets.*

Proof. Let

$$K_1 = \{x : \exists y[((x)_0, x, y)\, \text{T}_1 \text{ and } \forall z \leq y(((x)_1, x, z) \notin \text{T}_1)]\},$$
$$K_2 = \{x : \exists y[((x)_1, x, y) \in \text{T}_1 \text{ and } \forall z \leq y(((x)_0, x, z) \notin \text{T}_1)]\}.$$

Clearly K_1 and K_2 are r.e. and $K_1 \cap K_2 = 0$. For any $e, r \in \omega$ let $f(e, r) = 2^r \cdot 3^e$. To verify 6.21(*iii*), assume that $K_1 \subseteq \text{Dmn}\ \varphi^1_e$ and $K_2 \subseteq \text{Dmn}\ \varphi^1_r$ with $\text{Dmn}\ \varphi^1_e \cap \text{Dmn}\ \varphi^1_r = 0$. Suppose $f(e, r) \in \text{Dmn}\ \varphi^1_e \cup \text{Dmn}\ \varphi^1_r$. By symmetry, say $f(e, r) \in \text{Dmn}\ \varphi^1_e$. Thus $\exists y((e, 2^r \cdot 3^e, y) \in \text{T}_1)$, and since $\text{Dmn}\ \varphi^1_e \cap \text{Dmn}\ \varphi^1_r = 0$, obviously $\forall z((r, 2^r \cdot 3^e, z) \notin \text{T}_1)$. Thus $2^r \cdot 3^e \in K_2$, so $2^r \cdot 3^e \in \text{Dmn}\ \varphi^1_r$, contradiction. □

The next theorem gives an important method of producing new effectively inseparable sets from old ones:

Theorem 6.25. *Suppose that A and B are effectively inseparable, f is a unary recursive function,* $C, D \subseteq \omega$, $C \cap D = 0$, $A \subseteq f^{-1*}C$, *and* $B \subseteq f^{-1*}D$. *Then C and D are effectively inseparable.*

Proof. Let h be a function given by 6.21(*iii*) because A and B are effectively inseparable. For any $e, x \in \omega$, let $g(x, e) \simeq \mu y((e, fx, y) \in \text{T}_1)$. Thus g is partial recursive; say $g = \varphi^2_r$. Now we can define a function k intended to satisfy 6.21(*iii*) for C and D: for any $e, u \in \omega$, let $k(e, u) = fh(\text{s}^1_1(r, e), \text{s}^1_1(r, u))$. Thus k is recursive. In order to verify 6.21(*iii*), assume that $C \subseteq \text{Dmn}\ \varphi^1_e$ and $D \subseteq \text{Dmn}\ \varphi^1_u$, where $\text{Dmn}\ \varphi^1_e \cap \text{Dmn}\ \varphi^1_u = 0$. It follows that $A \subseteq f^{-1*}\ \text{Dmn}\ \varphi^1_e$, $B \subseteq f^{-1*}\ \text{Dmn}\ \varphi^1_u$, and $f^{-1*}\ \text{Dmn}\ \varphi^1_e \cap f^{-1*}\ \text{Dmn}\ \varphi^1_u = 0$. Now for any $x \in \omega$,

$$\begin{aligned} x \in f^{-1*}\ \text{Dmn}\ \varphi^1_e \quad & \text{iff } fx \in \text{Dmn}\ \varphi^1_e \\ & \text{iff } \exists y((e, fx, y) \in \text{T}_1) \\ & \text{iff } (x, e) \in \text{Dmn}\ g = \text{Dmn}\ \varphi^2_r \\ & \text{iff } x \in \text{Dmn}\ \varphi^1 \text{s}^1_1(r, e). \end{aligned}$$

Similarly, $f^{-1*}\ \text{Dmn}\ \varphi^1_u = \text{Dmn}\ \varphi^1\text{s}^1_1(r, u)$. Thus $A \subseteq \text{Dmn}\ \varphi^1(\text{s}^1_1(r, e)$, $B \subseteq \text{Dmn}\ \varphi^1\text{s}^1_1(r, u)$, and $\text{Dmn}\ \varphi^1\text{s}^1_1(r, e) \cap \text{Dmn}\ \varphi^1\text{s}^1_1(r, u) = 0$. Hence by choice of h, $h(\text{s}^1_1(r, e), \text{s}^1_1(r, u)) \in \omega \sim (\text{Dmn}\ \varphi^1\text{s}^1_1(r, e) \cup \text{Dmn}\ \varphi^1\text{s}^1_1(r, u))$, and hence $k(e, u) \in \omega \sim (\text{Dmn}\ \varphi^1_e \cup \text{Dmn}\ \varphi^1_u)$, as desired. □

As our final topic in this chapter we briefly consider a kind of r.e. set much different from creative sets. We introduce them partly to give a class of sets which are not creative, and partly because there is a big literature concerning them.

Definition 6.26. A set $A \subseteq \omega$ is *simple* if A is r.e., $\omega \sim A$ is infinite, and $B \cap A \neq 0$ whenever B is an infinite r.e. set.

Theorem 6.27. *A simple set is neither recursive nor creative.*

Proof. If A is simple and recursive, then $\omega \sim A$ is an infinite r.e. set and $A \cap (\omega \sim A) = 0$, contradiction. If A is simple and creative, by 6.20 choose B infinite recursive such that $B \subseteq \omega \sim A$. Contradiction. □

Theorem 6.28. *Simple sets exist.*

PROOF. Let g be a recursive function universal for unary primitive recursive functions (see Lemma 3.5). For any $e \in \omega$ let

$$fe \simeq (\mu y[g(e, (y)_0) = (y)_1 \text{ and } (y)_1 > 2e])_1.$$

Thus f is partial recursive. For each $e \in \omega$ let $\psi_e x = g(e, x)$ for all $x \in \omega$. Clearly for any $e \in \omega$,

(1) if $e \in \operatorname{Dmn} f$, then $fe \in \operatorname{Rng} \psi_e$ and $fe > 2e$;
(2) if $\operatorname{Rng} \psi_e$ is infinite then $e \in \operatorname{Dmn} f$.

Now $\operatorname{Rng} f$ is simple. For, it is obviously r.e. Suppose B is any infinite r.e. set. By choice of g, choose $e \in \omega$ so that $\operatorname{Rng} \psi_e = B$. By (2) and (1), $fe \in \operatorname{Rng} \psi_e$. Thus $B \cap \operatorname{Rng} f \neq 0$. Finally, to show that $\omega \sim \operatorname{Rng} f$ is infinite, note

(3) if $n \in \omega$, then $2n \cap \operatorname{Rng} f \subseteq f^*n$.

For, let $i \in 2n \cap \operatorname{Rng} f$. Say $i = fj$. By (1), $2j < fj$, so $2j < i < 2n$. Thus $j < n$, so $i \in f^*n$.

Since (3) holds, $|2n \cap \operatorname{Rng} f| \leq n$, hence $|2n \sim \operatorname{Rng} f| \geq n$, for any $n \in \omega$. Thus $\omega \sim \operatorname{Rng} f$ is infinite. □

BIBLIOGRAPHY

1. Malcev, A. I. *Algorithms and Recursive Functions.* Groningen: Wolters-Noordhoff (1970).
2. Rogers, H. *Theory of Recursive Functions and Effective Computability.* New York: McGraw-Hill (1967).
3. Smullyan, R. M. *Theory of Formal Systems.* Princeton: Princeton University Press (1961).

EXERCISES

6.29. Let $f: \omega \to \omega$. Then the following conditions are equivalent:

(1) f is recursive;
(2) $\{(x, fx) : x \in \omega\}$ is an r.e. relation;
(3) $\{(x, fx) : x \in \omega\}$ is a recursive relation.

6.30. Prove that the class of r.e. sets is closed under union and intersection using the argument following 6.8, but rigorously.

6.31. Show that if A is a Σ_n-set, $n > 0$, and f is partial recursive, then f^*A is Σ_n.

6.32. If A and B are r.e. sets, then there exist r.e. sets C, D such that $C \subseteq A$, $D \subseteq B$, $C \cup D = A \cup B$, and $C \cap D = 0$.

6.33. Suppose that f and g are unary recursive functions, g is one-one, $\operatorname{Rng} g$ is recursive, and $\forall x(fx \geq gx)$. Show that $\operatorname{Rng} f$ is recursive.

6.34. For each of the following determine if the set in question is recursive, r.e., or has an r.e. complement:

(1) $\{x$: there are at least x consecutive 7's in the decimal representation of $\pi\}$;
(2) $\{x$: there is a run of exactly x consecutive 7's in the decimal representation of $\pi\}$;
(3) $\{x: \varphi_x^1$ is total$\}$;
(4) $\{x$: Dmn φ_x^1 is recursive$\}$.

6.35. There are $\aleph_0$ r.e. sets which are not recursive.

6.36. There is a recursive set A such that $\bigcap_{x \in A}$ Dmn φ_x^1 is not r.e.

6.37. If A is productive, then so is $\{e$: Dmn $\varphi_e^1 \subseteq A\}$.

6.38. There are $2^{\aleph_0}$ productive sets. *Hint:* Let $A = \{e$: Dmn $\varphi_e^1 \subseteq \omega \sim \mathrm{K}\}$. Show that $A \subseteq \omega \sim \mathrm{K}$, $(\omega \sim \mathrm{K}) \sim A$ is infinite, and any set P with $A \subseteq P \subseteq \omega \sim \mathrm{K}$ is productive.

6.39. Any infinite r.e. set is the disjoint union of a creative set and a productive set. *Hint:* say Rng $f = A$. Let $gn = f\mu i(fi \neq gj$ for all $j < n)$. Show that $g^*\mathrm{K}$ is creative and $A \sim g^*\mathrm{K}$ is productive.

6.40. If B is r.e. and $A \cap B$ is productive, then A is productive.

6.41. There is an r.e. set which is neither recursive, simple, nor creative. *Hint:* let A be simple and set $B = \{x : (x)_0 \in A\}$.

6.42. For $A \subseteq \omega$ the following are equivalent:

(1) A is recursive and $A \neq 0$;
(2) there is a recursive function f with Rng $f = A$ and $\forall x \in \omega (fx \leq f(x+1))$.

6.43. For $A \subseteq \omega$ the following are equivalent:

(1) A is productive;
(2) there is a partial recursive function f such that $\forall e \in \omega$ (if Dmn $\varphi_e^1 \subseteq A$ then fe is defined and $fe \in A \sim$ Dmn φ_e^1).

6.44. If A is creative, B is r.e., and $A \cap B = 0$, then $A \cup B$ is creative.

6.45. There is a set A such that both A and $\omega \sim A$ are productive.

6.46. If A is productive and B is simple, then $A \cap B$ is productive.

6.47. Two sets A and B are *strongly recursively inseparable* if $A \cap B = 0$, $\omega \sim (A \cup B)$ is infinite, and for every r.e. set C, $C \sim A$ infinite $\Rightarrow C \cap B \neq 0$, $C \sim B$ infinite $\Rightarrow C \cap A \neq 0$. Show that if A and B are r.e. but strongly recursively inseparable, then:

(1) A and B are recursively inseparable.
(2) $A \cup B$ is simple.
(3) neither A nor B is creative.
(4) A and B are not effectively inseparable.

6.48. Show that there exist two r.e. strongly recursively inseparable sets. *Hint:* let $E = \{(e, x) : \exists y((e, x, y) \in \mathrm{T}_1)\}$. Show that there exist recursive functions f, g such that

$$E = \{(fi, gi) : i < \omega\}.$$

Show that there exist recursive functions h, k such that

$$\begin{aligned} h0 &= \mu i(gi > 3fi); \\ k0 &= \mu i(gi > 3fi \text{ and } gi \neq gh0); \\ h(n+1) &= \mu i(gi > 3fi \mathbin{\&} \forall j \leq n(gi \neq gkj) \mathbin{\&} \\ &\qquad \forall j \leq n(fi \neq fhj) \mathbin{\&} \forall j \leq n(gi \neq ghj)); \\ k(n+1) &= \mu i(gi > 3fi \mathbin{\&} \forall j \leq n+1(gi \neq ghj) \mathbin{\&} \\ &\qquad \forall j \leq n(fi \neq fkj) \mathbin{\&} \forall j \leq n(gi \neq gkj)). \end{aligned}$$

Let $A = \operatorname{Rng}(g \circ h)$, $B = \operatorname{Rng}(g \circ k)$.

Survey of Recursion Theory 7

We have developed recursion theory as much as we need for our later purposes in logic. But in this chapter we want to survey, without proofs, some further topics. Most of these topics are also frequently useful in logical investigations.

Turing Degrees

Let g be a function mapping ω into ω. Imagine a Turing machine equipped with an oracle—an inpenetrable black box—which gives the answer gx when presented with x. The function g may be nonrecursive, so that the oracle is not an effective device. Rigorously, one defines a g-Turing machine just like Turing machines were defined in 1.1, except that $v_1, \ldots, v_{2m}$ are arbitrary members of $\{0, 1, 2, 3, 4, 5\}$. And one adds one more stipulation in 1.2:

> If $w = 5$, and $F(e-1) = 0$ or $Fe = 1$, then $F' = F$, $d' = f$, $e' = e$, while if $w = 5$ and $0\ 1^{(x+1)}\ 0$ lies on F ending at e, then $0\ 1^{(x+1)}\ 0\ 1^{(gx+1)}\ 0$ lies on F' ending at e', $e' = e + gx + 2$, F' is otherwise like F and $d' = f$.

Then the notion of *g-Turing computable* function is easily defined.

One can also define *g-recursive function*: in 3.1, each class A is required to have g as a member. These two notions, g-Turing computable and g-recursive function, are shown equivalent just as in Chapter 3. In fact, most considerations of Chapters 1 through 6 carry over to this situation. If h is g-recursive, we also say that h is *recursive in g*. One can extend the notion in an obvious way to a set of F of functions, arriving at the notion of a function being recursive in F. At present we restrict ourselves to the simpler notion. We say that h and g are *Turing equivalent* if each is recursive in the other.

This establishes an equivalence relation on the set of all functions mapping ω into ω. The equivalence classes are called *Turing degrees of unsolvability.* Each equivalence class has at most $\aleph_0$ members (actually exactly $\aleph_0$, as is easily seen), since there are only $\aleph_0$ possible Turing machines with oracles. Clearly then there are $\exp \aleph_0$ degrees. Let D be the set of degrees. For α, $\beta \in D$ we write $\alpha \leq \beta$ provided there exist $f \in \alpha$ and $g \in \beta$ with f recursive in g. This relation $\leq$ makes D into a partially ordered set. Clearly the degree of recursive functions, denoted by 0, is the least element of D.

Many of the important results about D are concerned with trying to describe the partial ordering $\leq$. A complete description is far from being known. The rather scattered results which we now want to mention are among the strongest facts known. Some of their proofs are quite complicated, involving *priority* arguments, a kind of argument seemingly unique to this area.

Proposition 7.1. *Any two elements of D have a least upper bound.*

Theorem 7.2. *There exist two elements of D without a greatest lower bound.*

Theorem 7.3. *In D, no ascending sequence* $\alpha_0 < \alpha_1 < \cdots$ *has a least upper bound.*

Proposition 7.4. *Every element of D has only countably many predecessors.*

An element α of D is *minimal,* if $0 < \alpha$ and there is no β with $0 < \beta < \alpha$.

Theorem 7.5. *There are* $\exp \aleph_0$ *minimal degrees.*

A subset E of D is an *initial segment of D* provided that for all $\alpha, \beta \in D$, if $\alpha < \beta \in E$, then $\alpha \in E$.

Theorem 7.6. *Any finite distributive lattice can be embedded as an initial segment of D; likewise any countable Boolean algebra and any countable ordinal.*

One of the main open problems in the theory of degrees is the conjecture that every finite lattice can be embedded as an initial segment in D.

There are some special degrees of particular importance for applications to logic. A degree α is *recursively enumerable* (r.e.) provided that $\chi_A \in \alpha$ for some r.e. set A. Note that there are only $\aleph_0$ r.e. degrees. We let $0'$ be the degree of χ_K. We know from Theorem 6.19, p. 98 that $\chi_B \in 0'$ for any creative set B; and $0'$ is the largest r.e. degree.

Theorem 7.7. *No r.e. degree is minimal.*

Theorem 7.8. *There are two minimal degrees with join* $0'$.

Corollary 7.9. *There are degrees* $\leq 0'$ *which are not r.e.*

Theorem 7.10. *For every nonzero r.e. degree* α *there is a minimal degree* $\leq \alpha$.

Partial Recursive Functionals

A *partial functional* is a function F such that for some $m, n \in \omega$, the domain of F is a subset of ${}^m({}^\omega\omega) \times {}^n\omega$, while the range of F is a subset of ω; additionally we assume $m + n > 0$. In case $m = 0$ we are dealing with the partial functions of Chapter 5. In case the domain of F is all of ${}^m({}^\omega\omega) \times {}^n\omega$, we call F *total.* An (m, n)*-relation* R is any subset of ${}^m({}^\omega\omega) \times {}^n\omega$. We now wish to give a reasonable meaning to F and R being recursive, and to R being recursively enumerable. Since a function cannot be presented in its entirety to a machine it is natural to seek a definition of these notions in which only initial segments of functions are given. If $\mathfrak{A} = (f_0, \ldots, f_{m-1}, x_0, \ldots, x_{n-1}) \in {}^m({}^\omega\omega) \times {}^n\omega$, we let for any $y \in \omega$

$$\tilde{\mathfrak{A}}y = (\tilde{f}_0 y, \ldots, \tilde{f}_{n-1} y, x_0, \ldots, x_{n-1}) \in {}^{m+n}\omega.$$

Now we say that an (m, n)-relation R is *recursively enumerable* (r.e.) provided that there is an $(m + n + 1)$-ary recursive relation $S \subseteq {}^{m+n+1}\omega$ such that for all $\mathfrak{A} \in {}^m({}^\omega\omega) \times {}^n\omega$,

$$\mathfrak{A} \in R \qquad \text{iff } \exists x \in \omega[(\tilde{\mathfrak{A}}x, x) \in S].$$

Obviously this definition coincides with the definition of r.e. relation if $m = 0$. The definition is motivated as follows. We generate the members of S one after the other. Having generated a member $(y_0, \ldots, y_{m-1}, z_0, \ldots, z_{n-1}, x)$ of S, we have implicitly generated each member $\mathfrak{A}$ of R such that $\tilde{\mathfrak{A}}x = (y_0, \ldots, y_{m-1}, z_0, \ldots, z_{n-1})$. Eventually each member of R is generated in this fashion. A partial functional F is *partial recursive* provided that its *graph*

$$R = \{(\mathfrak{A}, x) : \mathfrak{A} \in \text{Dmn } F, F\mathfrak{A} = x\}$$

is r.e. Again this notion coincides with the old definition for $m = 0$. Given $\mathfrak{A} \in \text{Dmn } F$, clearly the above generation of R constitutes an effective calculation of $F\mathfrak{A}$ (provided there is some way to recognize effectively that $\tilde{\mathfrak{A}}x = (y_0, \ldots, y_{m-1}, z_0, \ldots, z_{n-1})$ for given $(y_0, \ldots, y_{m-1}, z_0, \ldots, z_{n-1})$). An (m, n)-relation R is *recursive* provided χ_R is recursive. These definitions form the basis for a generalized recursion theory. This generalization, expounded at length in Shoenfield [9], has many of the properties of ordinary recursion theory; the enumeration, iteration, and recursion theorems carry over, as well as the considerations concerning the arithmetical hierarchy. As is suggested above, there is a strong connection between generalized recursion theory and relative recursiveness:

Theorem 7.11. *A function* $f\colon \omega \to \omega$ *is recursive in a function* $g\colon \omega \to \omega$ *iff there is a total recursive functional* $F\colon {}^\omega\omega \times \omega \to \omega$ *such that for all* $x \in \omega$, $fx = F(g, x)$.

The notion of a functional also enables one to clarify the role of the sets Δ_n in the arithmetical hierarchy:

Theorem 7.12. *A relation is Δ_{n+1} iff it is recursive in $\{\chi_A: A \text{ is } \Pi_n\}$ iff it is recursive in $\{\chi_A: A \text{ is } \Sigma_n\}$.*

The notion of recursive functionals also makes possible the construction of a new hierarchy. An (m, n)-relation R is Σ_m^1 (resp. Π_m^1) where $m \geq 1$ provided there is a recursive relation S so that for all $\mathfrak{A} \in {}^m({}^\omega\omega) \times {}^n\omega$ we have

$$\mathfrak{A} \in R \qquad \text{iff } Q_1 \cdots Q_{m+1}[(\mathfrak{A}, \mathfrak{B}) \in S],$$

where $Q_1, \ldots, Q_m$ are quantifiers $\forall$ or $\exists$ on functions (members of ${}^\omega\omega$), alternately $\forall$ and $\exists$ (with $Q_1 = \exists$ (resp. $Q_1 = \forall$), while Q_{m+1} is a quantifier $\forall x$ or $\exists x$ on numbers. By collapsing quantifiers, it is easy to see that any second-order prefix can be put in this form (see Chapter 30). The classification of relations in the sets Σ_m^1 and Π_m^1 forms the *analytical hierarchy*. Again we set $\Delta_m^1 = \Sigma_m^1 \cap \Pi_m^1$. The theory of this hierarchy shows considerable similarity, in results and proofs, to the classical topological theory of analytic sets. For example, we have

Theorem 7.13 *If P and Q are disjoint Σ_1^1 relations, then there is a Δ_1^1 relation R such that $P \subseteq R$ and $Q \subseteq \sim R$.*

Isols

Two sets $A, B \subseteq \omega$ are said to be *recursively equivalent* if there is a one-one partial recursive function f such that $A \subseteq \text{Dmn} f$ and $f^*A = B$. This establishes an equivalence relation on the set of all subsets of ω; the equivalence classes are called *recursive equivalence types* (RET's). They are the effective version of cardinal numbers.

Proposition 7.14. *If α and β are* RET's, *then there exist $A \in \alpha$ and $B \in \beta$ such that A and B are recursively separable.*

Proposition 7.15. *If α and β are* RET's, *A, $A' \in \alpha$, B, $B' \in \beta$, A and B are recursively separable, and A' and B' are recursively separable, then $A \cup B$ is recursively equivalent to $A' \cup B'$.*

By 7.14 and 7.15, we can define a binary operation $+$ on RET by setting $\alpha + \beta =$ recursive equivalence type of $A \cup B$, where $A \in \alpha$, $B \in \beta$, and A and B are recursively separable.

Recall the function J_2 from 3.60. It is a one-one function mapping $\omega \times \omega$ onto ω.

Proposition 7.16. *If α, $\beta \in$ RET, A, $A' \in \alpha$, and B, $B' \in \beta$, then $J_2^*(A \times B)$ is recursively equivalent to $J_2^*(A' \times B')$.*

It follows that we can define $\cdot$ on RET by setting $\alpha \cdot \beta$ = recursive equivalence type of $J_2^*(A \times B)$, where $A \in \alpha$ and $B \in \beta$.

Proposition 7.17. *Addition and multiplication of* RET's *are commutative and associative. Multiplication is distributive over addition.*

The structure (RET, $+$, $\cdot$) is not, however, a ring, and it cannot be embedded in a ring. This can be seen for example, from the fact that $\alpha + \beta = \alpha$ where α and β are respectively the recursive equivalence types of ω and of 1. Since $\beta + \beta \neq \beta$, β is not the additive zero of (RET, $+$, $\cdot$), so this structure cannot even be embedded in a ring.

For each $n \in \omega$, let $\bar{n}$ be the recursive equivalence type of n. Then $^-$ is an isomorphic embedding of (ω, $+$, $\cdot$) into (RET, $+$, $\cdot$).

The structure (RET, $+$, $\cdot$) has a simple substructure which is much more closely related to (ω, $+$, $\cdot$). To define it, let us call a set $A \subseteq \omega$ *isolated* if it is not recursively equivalent to any proper subset $B \subset A$. An RET α is an *isol* if it has an isolated member. We denote by ISOL the collection of all isol's.

Theorem 7.18. ISOL *is closed under* $+$ *and* $\cdot$. *For any* α, β, $\gamma \in$ ISOL *we have*:

(*i*) $\alpha + \beta = \alpha + \gamma$ *implies* $\beta = \gamma$;
(*ii*) $\alpha \cdot \beta = \alpha \cdot \gamma$ *and* $\alpha \neq \bar{0}$ *imply* $\beta = \gamma$;
(*iii*) (ω, $+$, $\cdot$) *is a substructure of* (ISOL, $+$, $\cdot$).

The structure (ISOL, $+$, $\cdot$) can be embedded in a ring ISOL*, which has the ordinary ring of integers as a substructure. It has many interesting properties. Since it has zero divisors, it cannot be embedded in a field.

Recursive Real Numbers

It is natural to try to effectivize common notions of mathematics, such as the notion of a real number. We give here a few of the relevant definitions and results.

Let $\mathbb{Q}$ be the set of rational numbers. A sequence $r \in {}^{\omega}\mathbb{Q}$ is *recursive* iff there exist unary recursive functions f, g, h such that for all $n \in \omega$,

$$rn = (fn - gn)/(1 + hn).$$

Thus if $\varepsilon n = 1/2^n$ for all $n \in \omega$, then ε is recursive. In fact we may take $fn = 1$ for all n, $gn = 0$ for all n, and $hn = 2^n - 1$ for all n. Now a recursive sequence $r \in {}^{\omega}\mathbb{Q}$ *recursively converges* to a real number α provided there is a unary recursive function k such that for all $n \in \omega$ and all $n \geq km$ we have $|rn - \alpha| < \varepsilon m$. A real number α is *recursive* if there is a recursive sequence of rationals which recursively converges to α.

Theorem 7.19. *The set of recursive real numbers forms a subfield F of the field of real numbers. Every rational number is recursive. F is countable.*

There is a Cauchy recursive sequence of rationals which does not converge recursively.

Theorem 7.20. *If $r \in {}^{\omega}\mathbb{Q}$ is recursive, strictly monotone, converges to a recursive real number α, then r recursively converges to α.*

A sequence $r \in {}^{\omega}F$ is *recursive* provided there are binary recursive functions f, g, h, k such that for all $m, n \in \omega$ and all $p \geq k(m, n)$ we have

$$|r_n - \{[f(p, n) - g(p, n)]/[1 + h(p, n)]\}| < \varepsilon m.$$

Many other concepts of ordinary mathematics can be given effective formulations in a similar way.

Word Problem for Groups

There is a classical problem in group theory which has been given a negative solution using notions of recursive function theory. We shall give a precise formulation of it. Let X be a nonempty set. We form the *free group generated by* X as follows. For each $x \in X$ let $x' = (X, x)$. Note that $'$ is a one-one function whose range is disjoint from X. A finite sequence (perhaps 0) of elements of $X \cup \text{Rng}'$ is called a *word* on X; we let W_X be the set of all words on X. Let $\equiv$ be the smallest equivalence relation on W_X containing all pairs $(0, aa')$ and $(0, a'a)$ with $a \in X$. It is easily seen that if $a, b, c, d \in W_X$, $a \equiv b$, and $c \equiv d$, then $ac \equiv bd$. Hence there is a binary operation $\cdot$ on the set F_X of equivalence classes under $\equiv$ such that $[a]\cdot[b] = [ab]$ for all a, $b \in W_X$. Under this operation F_X becomes a group, called the *free group generated by* X. A *defining relation over* X is a pair (a, b) of words over X. If R is a set of defining relations over X, we let R^* be the normal subgroup of F_X generated by all elements $[a]\cdot[b]^{-1}$ with $(a, b) \in R$. Let $F_{X,R} = F_X/R^*$. A group G is *determined* by generators X and defining relations R if it is isomorphic to $F_{X,R}$; then (X, R) is a *presentation* of G. It is easily seen that every group has a presentation. If X and R are finite, then (X, R) is a *finite presentation* and G is *finitely presentable*. If f is a one-one map of $X \cup X'$ into ω, then any word x of W_X can be given a Gödel number $\mathcal{g}_f x$ by

$$\mathcal{g}_f x = \prod_{i<m} p_i^{fxi+1}$$

where x is of length m. We say that the word problem for (X, R) is *recursively solvable* provided that for some such f,

$$\{(\mathcal{g}_f a, g_f b) : h[a] = h[b]\}$$

is recursive, where h is the natural homomorphism of F_X onto $F_{X,R}$. For X and R finite, this definition does not depend on the choice of f.

Theorem 7.21 (Novikov). *There is a group G with a finite presentation* (X, R) *which is recursively unsolvable. Thus there is no automatic procedure for determining of a pair of words* (a, b) *whether they become equal upon applying the relations in R.*

Solvability of Diophantine Equations.

A *diophantine equation* is an equation of the form $P(x_0, \ldots, x_{m-1}) = 0$, where $P(x_0, \ldots, x_{m-1})$ is a polynomial in indeterminants $x_0, \ldots, x_{m-1}$ with integer coefficients. A classical problem of number theory, called *Hilbert's tenth problem* (see Davis [2]) is whether there is an automatic method for determining whether an arbitrary diophantine equation has an integral solution. By means of Gödel numbering this question can be given a rigorous form. The answer (Theorem 7.24) is negative, and follows from an even stronger result which we now want to formulate. An n-ary relation $R \subseteq {}^n\omega$ is called *diophantine* if there is a polynomial $P(x_0, \ldots, x_{n-1}, y_0, \ldots, y_{m-1})$ with integral coefficients such that

$$R = \{x \in {}^n\omega : \text{there exist } y_0, \ldots, y_{m-1} \in \omega \text{ such that } P(x_0, \ldots, x_{n-1}, y_0, \ldots, y_{m-1}) = 0\}.$$

Theorem 7.22 (J. Robinson, M. Davis, Y. Matiyasevic). *A relation is r.e. iff it is diophantine.*

As an interesting corollary we have

Theorem 7.23. *There is a polynomial with integral coefficients such that its positive values, when members of* ω *are substituted, are exactly all positive primes.*

It is also easy to derive the solution to Hilbert's tenth problem from 7.22:

Theorem 7.24. *There is no automatic method which, presented with a diophantine equation* ε, *will decide whether* ε *has a solution.*

BIBLIOGRAPHY

1. Boone, W. W. The word problem. *Ann. Math.*, *70* (1959), 207–265.
2. Davis, M. Hilbert's tenth problem is unsolvable. *Amer. Math. Monthly*, *80* (1973), 233–269.
3. Dekker, J. C. E. *Les Fonctions Combinatoires et les Isols.* Paris: Gauthier-Villars (1966).
4. Dekker, J. C. E. and Myhill, J. Recursive equivalence types. *Univ. Calif. Publ. Math.*, *3* (1960), 67–214.
5. Hermes, H. *Enumerability, Decidability, Computability.* New York: Springer (1969).

PART II

Elements of Logic

We now begin the study of logic proper. In this part we shall introduce the basic notions of a first-order language and the structures appropriate for it. We shall only present here the most basic definitions and results, up to and including the completeness theorem—which expresses the most important relationship between the syntactical and semantic notions. In Parts III and IV many deeper results in these two domains will be found. The emphasis throughout this book, as well as in the mathematical world, is on first-order languages. There are many other languages, some of which will be briefly discussed in Part V. First-order languages are, roughly speaking, the simplest languages with which one can conveniently formulate any mathematical theory. Moreover, one can prove about these languages and their structures many interesting and useful results which cannot be extended to more involved languages.

8* Sentential Logic

Before considering first-order languages we consider some simpler languages, whose study will be a simplified model of the more involved study of first-order languages themselves. These sentential languages enable one to express only such primitive logical notions as *not*, *implication*, *and*, *or*, etc. These connectives between sentences are such that the truth of a complicated sentence can be inferred just from the truth or falsity of its components.

We shall choose two of these connectives, negation and implication, as primitive, and show at a later stage that all other connectives can be expressed in terms of them. We show how the notion of a sentence can be rigorously defined, how axioms for valid sentences can be given, and the connection between the axioms and the notion of truth. In the exercises we outline the theories of some nonclassical versions of logic, such as three-valued and intuitionistic logic.

Definition 8.1. A sentential language is a triple (n, c, P) such that $n \neq c$, $\{n, c\} \cap P = 0$, and $P \neq 0$.

We think of n and c as a negation symbol and implication symbol respectively, while P is a set of atomic, indivisible sentences. With these intuitive meanings it is possible to define in a rigorous way additional intuitive notions, such as an expression—any sequence of symbols—and a sentence—an expression which has a meaningful form:

Definition 8.2. Let $\mathscr{P} = (n, c, P)$ be a sentential language. The set $\{n, c\} \cup P$ is the set of *symbols* of $\mathscr{P}$. An *expression* of $\mathscr{P}$ is any finite sequence of symbols of $\mathscr{P}$; the empty set (=empty sequence) is admitted. We define one-place and two-place operations $\neg$ and $\rightarrow$ on the set of expressions by setting

$$\neg\varphi = \langle n\rangle\varphi,$$
$$\varphi \rightarrow \psi = \langle c\rangle\varphi\psi,$$

for all expressions φ and ψ. Recall from 1.11 that $\varphi\psi$ is just the juxtaposition of the two finite sequences φ and ψ. $\mathrm{Sent}_{\mathscr{P}}$, the set of *sentences* of $\mathscr{P}$, is the intersection of all classes Γ of expressions of $\mathscr{P}$ such that $\langle s\rangle \in \Gamma$ for each $s \in P$ and Γ is closed under $\neg$ and $\rightarrow$.

Note that we have not used parentheses in our language; the notation used here and later on is *parenthesis-free*, or *Polish* notation. Of course we shall use parentheses freely in our informal language (metalanguage). For example, $\varphi \rightarrow (\psi \rightarrow \chi)$ has a clear meaning for expressions φ, ψ, χ; in fact, it is the same as the expression $\langle c\rangle\varphi\langle c\rangle\psi\chi$. The most important feature of the parenthesis-free notation, as well as the more usual notation with parentheses, is the *unique readability of sentences* (Theorem 8.5).

Throughout this section, unless otherwise indicated, $\mathscr{P} = (n, c, P)$ will be a fixed but arbitrary sentential language; also, we shall use small Greek letters to denote sentences, and large ones to denote sets of sentences. The following two obvious results will be used frequently:

Proposition 8.3 (Induction principle for sentences). *If $\langle s\rangle \in \Gamma$ for each $s \in P$ and Γ is closed under $\neg$ and $\rightarrow$, then* $\mathrm{Sent}_{\mathscr{P}} \subseteq \Gamma$.

Proposition 8.4 (Construction sequence for sentences). $\varphi \in \mathrm{Sent}_{\mathscr{P}}$ *iff there is a finite sequence $\langle \psi_0, \ldots, \psi_{n-1}\rangle$ of expressions of $\mathscr{P}$ such that $n > 0$, $\psi_{n-1} = \varphi$, and for each $i < n$ one of the following condition holds:*
(i) $\psi_i = \langle s\rangle$ for some $s \in P$;
(ii) $\psi_i = \neg\psi_j$ for some $j < i$;
(iii) $\psi_i = \psi_j \rightarrow \psi_k$ for some $j, k < i$.

With regard to 8.4, cf. Proposition 2.3.

Theorem 8.5 (Unique readability)
(i) Every sentence is of positive length.
(ii) If φ is a sentence, then either $\varphi = \langle s\rangle$ for some $s \in P$, $\varphi = \neg\psi$ for some sentence ψ, or $\varphi = \psi \rightarrow \chi$ for some sentences ψ, χ.
(iii) If $\varphi = \langle \varphi_0, \ldots, \varphi_{m-1}\rangle$ is a sentence and $i < m - 1$, then $\langle \varphi_0, \ldots, \varphi_i\rangle$ is not a sentence.
(iv) In (ii), the three possibilities are mutually exclusive, and the sentences ψ and χ are uniquely determined by φ.

Proof. Conditions (*i*) and (*ii*) are easily established using 8.3; for example, to prove (*i*) we let Γ be the set of all expressions of positive length. We establish (*iii*) by induction on m. If $m = 1$, then by (*ii*) and (*i*), $\varphi = \langle s\rangle$ for some

$s \in P$. Then (*iii*) holds, by (*i*). Now assume, inductively, that $m > 1$. By (*ii*) we have two cases. First suppose $\varphi = \neg\psi$ for some sentence ψ. Thus $\varphi_0 = n$, and $\psi = \langle\varphi_1, \ldots, \varphi_{m-1}\rangle$. If $\langle\varphi_0, \ldots, \varphi_i\rangle$ is a sentence, then by (*ii*) we easily infer that $\langle\varphi_1, \ldots, \varphi_i\rangle$ is a sentence, contradicting the induction hypothesis, since ψ is a sentence. Second, suppose $\varphi = \psi \to \chi$ for sentences ψ and χ, and suppose the hypothesis of (*iii*) holds. Then $\varphi_0 = c$, and there is a j with $1 < j < m - 1$ and $\psi = \langle\varphi_1, \ldots, \varphi_j\rangle$, $\chi = \langle\varphi_{j+1}, \ldots, \varphi_{m-1}\rangle$. Suppose that $\langle\varphi_0, \ldots, \varphi_i\rangle$ is a sentence. Then by (*ii*) there are sentences θ, σ such that $\langle\varphi_0, \ldots, \varphi_i\rangle = \theta \to \sigma$; hence there is a k with $1 < k < i$ such that $\theta = \langle\varphi_1, \ldots, \varphi_k\rangle$ and $\sigma = \langle\varphi_{k+1}, \ldots, \varphi_i\rangle$. Since $k, j < m - 1$ and ψ and θ are sentences, the induction hypothesis gives $k = j$. But then σ is a proper initial segment of χ, contradicting the induction hypothesis.

Part (*iv*) is easily established using (*iii*). □

From 8.5 the following result of a purely set-theoretical nature follows.

Theorem 8.6 (Recursion principle for sentences). *Let A be any set, f a function mapping P into A, g a function mapping $A \times \mathrm{Sent}_{\mathscr{P}}$ into A, and h a function mapping $A \times A \times \mathrm{Sent}_{\mathscr{P}} \times \mathrm{Sent}_{\mathscr{P}}$ into A. Then there is a unique function k mapping $\mathrm{Sent}_{\mathscr{P}}$ into A such that the following conditions hold:*

(*i*) *$k\langle s\rangle = fs$ for all $s \in P$,*
(*ii*) *$k \neg\varphi = g(k\varphi, \varphi)$ for every sentence φ,*
(*iii*) *$k(\varphi \to \psi) = h(k\varphi, k\psi, \varphi, \psi)$ for all sentences φ, ψ.*

We shall now define the syntactical notion of *theorem* in a sentential logic. Here we meet in its most primitive form the very important process of axiomatization in mathematics. In this case we wish to axiomatize the notion of logical truth appropriate for the primitive logic we are now dealing with.

Definition 8.7. A sentence is a *logical axiom* of $\mathscr{P}$ iff it has one of the following forms (where φ, ψ, χ are arbitrary sentences):
(A1) $\varphi \to (\psi \to \varphi)$,
(A2) $[\varphi \to (\psi \to \chi)] \to [(\varphi \to \psi) \to (\varphi \to \chi)]$,
(A3) $(\neg\varphi \to \neg\psi) \to (\psi \to \varphi)$.

Now let Γ be a set of sentences. The set of *Γ-theorems$_{\mathscr{P}}$* is the intersection of all sets Δ of sentences such that $\Gamma \subseteq \Delta$, each logical axiom is in Δ, and $\psi \in \Delta$ whenever $\varphi \in \Delta$ and $\varphi \to \psi \in \Delta$. We say that ψ is obtained from φ and $\varphi \to \psi$ by *detachment* or *modus ponens*. We write $\Gamma \vdash_{\mathscr{P}} \varphi$ to abbreviate "φ is a Γ-theorem$_{\mathscr{P}}$," and we write $\vdash_{\mathscr{P}}\varphi$ for $0 \vdash_{\mathscr{P}} \varphi$. The subscript $\mathscr{P}$ is frequently omitted.

We give one of the usual consequences of a definition like 8.7:

Theorem 8.8. *$\Gamma \vdash \varphi$ iff there is a finite sequence $\langle\psi_0, \ldots, \psi_{m-1}\rangle$, $m > 0$, of sentences of $\mathscr{P}$ such that $\psi_{m-1} = \varphi$ and for each $i < m$ one of the following holds:*

(*i*) ψ_i *is a logical axiom of* $\mathscr{P}$,
(*ii*) $\psi_i \in \Gamma$,
(*iii*) $\psi_k = \psi_j \rightarrow \psi_i$ *for some* $j, k < i$.

A sequence $\langle \psi_0, \ldots, \psi_{m-1} \rangle$ as in 8.8 is called a *formal proof of* φ *from the hypotheses* Γ.

Some simple but frequently useful properties of $\vdash$ are given in the following theorem.

Theorem 8.9. *Let* $\Gamma, \Delta \subseteq \mathrm{Sent}_{\mathscr{P}}$ *and* $\varphi, \psi \in \mathrm{Sent}_{\mathscr{P}}$. *Then:*

(*i*) *if* $\Delta \subseteq \Gamma$ *and* $\Delta \vdash \varphi$, *then* $\Gamma \vdash \varphi$;
(*ii*) *if* $\Gamma \vdash \varphi$, *then* $\Theta \vdash \varphi$ *for some finite subset* Θ *of* Γ;
(*iii*) *if* $\Gamma \vdash \chi$ *for each* $\chi \in \Delta$ *and* $\Delta \vdash \varphi$, *then* $\Gamma \vdash \varphi$;
(*iv*) *if* $\Gamma \vdash \varphi$ *and* $\Gamma \vdash \varphi \rightarrow \psi$, *then* $\Gamma \vdash \psi$.

Lemma 8.10. $\vdash \varphi \rightarrow \varphi$.

PROOF. Just this once we give a formal proof, with justifications listed in the column to the right.

(1) $\{\varphi \rightarrow [(\varphi \rightarrow \varphi) \rightarrow \varphi]\} \rightarrow \{[\varphi \rightarrow (\varphi \rightarrow \varphi)] \rightarrow (\varphi \rightarrow \varphi)\}$	A2
(2) $\varphi \rightarrow [(\varphi \rightarrow \varphi) \rightarrow \varphi]$	A1
(3) $[\varphi \rightarrow (\varphi \rightarrow \varphi)] \rightarrow (\varphi \rightarrow \varphi)$	(1), (2), detachment
(4) $\varphi \rightarrow (\varphi \rightarrow \varphi)$	A1
(5) $\varphi \rightarrow \varphi$	(3), (4), detachment □

Theorem 8.11 (Deduction theorem). *If* $\Gamma \cup \{\varphi\} \vdash \psi$, *then* $\Gamma \vdash \varphi \rightarrow \psi$.

PROOF. By induction on m we show that for every nonzero $m \in \omega$, if $\langle \chi_0, \ldots, \chi_{m-1} \rangle$ is a formal proof of ψ from $\Gamma \cup \{\varphi\}$, then $\Gamma \vdash \varphi \rightarrow \psi$. Suppose this is true for all $n < m$, and suppose $\langle \chi_0, \ldots, \chi_{m-1} \rangle$ is a formal proof of ψ from $\Gamma \cup \{\varphi\}$. By 8.8 we have four cases:

Case 1. ψ is a logical axiom. Now $\vdash \psi \rightarrow (\varphi \rightarrow \psi)$ by A1, and obviously $\vdash \psi$, so $\vdash \varphi \rightarrow \psi$ and hence $\Gamma \vdash \varphi \rightarrow \psi$.

Case 2. $\psi \in \Gamma$. This is treated similarly to Case 1.

Case 3. $\psi = \varphi$. By 8.10 we have $\vdash \varphi \rightarrow \psi$; hence $\Gamma \vdash \varphi \rightarrow \psi$.

Case 4. $\exists j, k < m - 1$ such that $\chi_k = \chi_j \rightarrow \psi$. Now by the induction assumption $\Gamma \vdash \varphi \rightarrow \chi_k$ and $\Gamma \vdash \varphi \rightarrow \chi_j$. The following Γ-proof shows that $\Gamma \vdash \varphi \rightarrow \psi$:

$$\left.\begin{matrix} \text{—} \\ \vdots \\ \text{—} \end{matrix}\right\} \text{ a } \Gamma\text{-proof of } \varphi \rightarrow \chi_k$$

$$\left.\begin{matrix} \text{—} \\ \vdots \\ \text{—} \end{matrix}\right\} \text{ a } \Gamma\text{-proof of } \varphi \rightarrow \chi_j$$

$[\varphi \rightarrow (\chi_j \rightarrow \psi)] \rightarrow [(\varphi \rightarrow \chi_j) \rightarrow (\varphi \rightarrow \psi)]$	A2
$(\varphi \rightarrow \chi_j) \rightarrow (\varphi \rightarrow \psi)$	detachment
$\varphi \rightarrow \psi$	detachment □

The deduction theorem is a formalization of a common method of proof in mathematics. Frequently when one wants to establish an implication $\varphi \rightarrow \psi$ in common mathematical reasoning, one first adjoins φ to the mathematical assumptions Γ. After arguing that ψ holds, one concludes that $\varphi \rightarrow \psi$ follows from the original assumptions.

Now we show $\vdash \theta$ for various sentences θ. These facts will serve as lemmas for the completeness theorem, which enables us to check for $\vdash \theta$ in a routine, mechanical way.

Lemma 8.12. $\vdash (\varphi \rightarrow \psi) \rightarrow [(\psi \rightarrow \chi) \rightarrow (\varphi \rightarrow \chi)]$.

Proof

$$\{\varphi \rightarrow \psi, \psi \rightarrow \chi, \varphi\} \vdash \varphi$$
$$\{\varphi \rightarrow \psi, \psi \rightarrow \chi, \varphi\} \vdash \varphi \rightarrow \psi$$
$$\{\varphi \rightarrow \psi, \psi \rightarrow \chi, \varphi\} \vdash \psi$$
$$\{\varphi \rightarrow \psi, \psi \rightarrow \chi, \varphi\} \vdash \psi \rightarrow \chi$$
$$\{\varphi \rightarrow \psi, \psi \rightarrow \chi, \varphi\} \vdash \chi$$
$$\{\varphi \rightarrow \psi, \psi \rightarrow \chi\} \vdash \varphi \rightarrow \chi$$
$$\{\varphi \rightarrow \psi\} \vdash (\psi \rightarrow \chi) \rightarrow (\varphi \rightarrow \chi)$$
$$\vdash (\varphi \rightarrow \psi) \rightarrow [(\psi \rightarrow \chi) \rightarrow (\varphi \rightarrow \chi)]$$

□

Easy applications of the deduction theorem give:

Lemma 8.13. $\vdash [\varphi \rightarrow (\psi \rightarrow \chi)] \rightarrow [\psi \rightarrow (\varphi \rightarrow \chi)]$.

Lemma 8.14. $\vdash \varphi \rightarrow (\neg\varphi \rightarrow \psi)$.

Proof

$$\{\varphi, \neg\varphi\} \vdash \neg\varphi \rightarrow (\neg\psi \rightarrow \neg\varphi) \qquad \text{A1}$$
$$\{\varphi, \neg\varphi\} \vdash \neg\varphi$$
$$\{\varphi, \neg\varphi\} \vdash \neg\psi \rightarrow \neg\varphi$$
$$\{\varphi, \neg\varphi\} \vdash (\neg\psi \rightarrow \neg\varphi) \rightarrow (\varphi \rightarrow \psi) \qquad \text{A3}$$
$$\{\varphi, \neg\varphi\} \vdash \varphi \rightarrow \psi$$
$$\{\varphi, \neg\varphi\} \vdash \varphi$$
$$\{\varphi, \neg\varphi\} \vdash \psi$$

Two applications of the deduction theorem finish the proof. □

By a similar proof, or using 8.13, we get:

Lemma 8.15. $\vdash \neg\varphi \rightarrow (\varphi \rightarrow \psi)$.

Lemma 8.16. $\vdash \neg\neg\varphi \rightarrow \varphi$.

Proof

$$\{\neg\neg\varphi\} \vdash \neg\neg\neg\neg\varphi \rightarrow \neg\neg\varphi \qquad \text{using A1}$$
$$\{\neg\neg\varphi\} \vdash \neg\varphi \rightarrow \neg\neg\neg\varphi \qquad \text{A3}$$
$$\{\neg\neg\varphi\} \vdash \neg\neg\varphi \rightarrow \varphi \qquad \text{A3}$$
$$\{\neg\neg\varphi\} \vdash \varphi$$
$$\vdash \neg\neg\varphi \rightarrow \varphi$$

□

Lemma 8.17. $\vdash(\varphi \to \psi) \to (\neg\psi \to \neg\varphi)$.

PROOF

$\{\varphi \to \psi, \neg\psi, \neg\neg\varphi\} \vdash \varphi$	8.16
$\{\varphi \to \psi, \neg\psi, \neg\neg\varphi\} \vdash \psi$	
$\{\varphi \to \psi, \neg\psi, \neg\neg\varphi\} \vdash \neg\psi$	
$\{\varphi \to \psi, \neg\psi, \neg\neg\varphi\} \vdash \neg\neg\psi$	8.15
$\{\varphi \to \psi, \neg\psi\} \vdash \neg\neg\varphi \to \neg\neg\psi$	
$\{\varphi \to \psi, \neg\psi\} \vdash \neg\psi \to \neg\varphi$	A3
$\{\varphi \to \psi, \neg\psi\} \vdash \neg\varphi$	
$\vdash (\varphi \to \psi) \to (\neg\psi \to \neg\varphi)$	□

Lemma 8.18. $\vdash\varphi \to \neg\neg\varphi$.

PROOF

$\{\varphi, \neg\neg\neg\varphi\} \vdash \neg\varphi$	8.16
$\{\varphi\} \vdash \neg\neg\neg\varphi \to \neg\varphi$	
$\{\varphi\} \vdash \varphi \to \neg\neg\varphi$	A3
$\{\varphi\} \vdash \neg\neg\varphi$	
$\vdash \varphi \to \neg\neg\varphi$	□

Lemma 8.19. $\vdash(\varphi \to \neg\varphi) \to \neg\varphi$.

PROOF

$\{\varphi \to \neg\varphi, \neg\neg\varphi\} \vdash \varphi$	8.16
$\{\varphi \to \neg\varphi, \neg\neg\varphi\} \vdash \neg\varphi$	
$\{\varphi \to \neg\varphi, \neg\neg\varphi\} \vdash \neg(\varphi \to \varphi)$	8.14
$\{\varphi \to \neg\varphi\} \vdash \neg\neg\varphi \to \neg(\varphi \to \varphi)$	
$\{\varphi \to \neg\varphi\} \vdash (\varphi \to \varphi) \to \neg\varphi$	
$\{\varphi \to \neg\varphi\} \vdash \neg\varphi$	8.10
$\vdash (\varphi \to \neg\varphi) \to \neg\varphi$	□

Our last two lemmas are easily obtained using the methods in the proofs above.

Lemma 8.20. $\vdash(\neg\varphi \to \varphi) \to \varphi$.

Lemma 8.21. $\vdash\varphi \to (\neg\psi \to \neg(\varphi \to \psi))$.

Now we introduce a semantical consequence relation $\Gamma \models \varphi$. Roughly speaking, the difference between syntactical notions like $\vdash$ and semantical ones like $\models$ is this: Syntactical notions are defined purely in terms of the formal symbols, with only the mathematical notions being used which are essential for the definition. In semantical notions, however, mathematical ideas of a very different sort from formal notions play an essential role; almost always some version of the idea of a *model*, or *mathematical realization*, of the formal notions plays a role, along with a rigorous notion of *truth*.

Definition 8.22. Let $\mathscr{P} = (n, c, P)$ be a sentential language. Members of ${}^P2$ are called *models* of $\mathscr{P}$. (Intuitively, 0 means *falsity*, 1 means *truth*, and a function $f \in {}^P2$ is just an assignment of a truth value to each sentence of P.) Using the recursion principle for sentences, we can associate with each $f \in {}^P2$ a function $f^+ : \mathrm{Sent}_{\mathscr{P}} \to 2$ such that for any $s \in P$ and any $\varphi, \psi \in \mathrm{Sent}_{\mathscr{P}}$

$$\begin{aligned} f^+\langle s\rangle &= fs, & \\ f^+ \neg\varphi &= 1 & &\text{if } f^+\varphi = 0, \\ f^+ \neg\varphi &= 0 & &\text{if } f^+\varphi = 1, \\ f^+(\varphi \to \psi) &= 0 & &\text{iff } f^+\varphi = 1 \text{ and } f^+\varphi = 0. \end{aligned}$$

(f^+ intuitively tells us about the truth or falsity of any sentence of $\mathscr{P}$, given the truth or falsity of members of P.) We say that f is a *model* of φ if $f^+\varphi = 1$; f is a *model* of a set Γ of sentences iff $f^+\varphi = 1$ for all $\varphi \in \Gamma$. We write $\Gamma \models_{\mathscr{P}} \varphi$ iff every model of Γ is a model of φ, and we write $\models_{\mathscr{P}}\varphi$ instead of $0 \models_{\mathscr{P}} \varphi$. Sentences φ with $\models_{\mathscr{P}}\varphi$ are called *tautologies*.

Whether or not a sentence φ is a tautology can be decided by the familiar truth table method: one writes in rows all possible $f \in {}^P2$ and for each such f calculates $f^+\varphi$ from inside out. Of course instead of all $f \in {}^P2$ it suffices to list only the $f \in {}^Q2$, where Q is the set of $s \in P$ which occur in φ. For example, the following table shows that $\langle s_1\rangle \to (\langle s_2\rangle \to \langle s_1\rangle)$ is a tautology:

s_1	s_2	$\langle s_2\rangle \to \langle s_1\rangle$	$\langle s_1\rangle \to (\langle s_2\rangle \to \langle s_1\rangle)$
1	1	1	1
1	0	1	1
0	1	0	1
0	0	1	1

The following table shows that $\neg\langle s_1\rangle \to (\neg\langle s_1\rangle \to \langle s_1\rangle)$ is not a tautology:

s_1	$\neg\langle s_1\rangle$	$\neg\langle s_1\rangle \to \langle s_1\rangle$	$\neg\langle s_1\rangle \to (\neg\langle s_1\rangle \to \langle s_1\rangle)$
1	0	1	1
0	1	0	0

Clearly this truth table procedure provides an effective procedure for determining whether or not a sentence is a tautology. This statement could be made precise for sentential languages $\mathscr{P} = (n, c, P)$ with P countable by the usual procedure of Gödel numbering. (See 10.19–10.22, where this is done in detail for first-order languages.) In practice, to check that a statement is or is not a tautology it is frequently better to argue informally, assuming the given sentence is not true and trying to infer a contradiction from this. For example, if $\langle s_1\rangle \to (\langle s_2\rangle \to \langle s_1\rangle)$ is false, then s_1 is true and $\langle s_2\rangle \to \langle s_1\rangle$ is false; but this is impossible; $\langle s_2\rangle \to \langle s_1\rangle$ is true since s_1 is true. Thus $\langle s_1\rangle \to (\langle s_2\rangle \to \langle s_1\rangle)$ is a tautology.

We are going to show shortly that the relations $\vdash$ and $\models$ are identical. To do this we need some preliminary statements.

Lemma 8.23. *If* $\Gamma \vdash \varphi$, *then* $\Gamma \models \varphi$.

PROOF. Let $\Delta = \{\varphi : \text{every model of } \Gamma \text{ is a model of } \varphi\}$. It is easy to check, using truth tables for the logical axioms, that $\Gamma \subseteq \Delta$, every logical axiom is in Δ, and Δ is closed under detachment. Hence all Γ-theorems are in Δ. The lemma follows. □

Definition 8.24. Γ is *consistent* iff $\Gamma \nvdash \varphi$ for some φ.

Theorem 8.25. *The following conditions are equivalent:*

(*i*) Γ *is inconsistent.*
(*ii*) $\Gamma \vdash \neg(\varphi \rightarrow \varphi)$ *for every sentence* φ.
(*iii*) $\Gamma \vdash \neg(\varphi \rightarrow \varphi)$ *for some sentence* φ.

PROOF. Obviously (*i*) $\Rightarrow$ (*ii*) $\Rightarrow$ (*iii*). Now suppose $\Gamma \vdash \neg(\varphi \rightarrow \varphi)$ for a certain sentence φ. Let ψ be any sentence. By A1, $\Gamma \vdash (\varphi \rightarrow \varphi) \rightarrow [\neg\psi \rightarrow (\varphi \rightarrow \varphi)]$; from 8.10 we infer that $\Gamma \vdash \neg\psi \rightarrow (\varphi \rightarrow \varphi)$, and then 8.17 yields $\Gamma \vdash \neg(\varphi \rightarrow \varphi) \rightarrow \neg\neg\psi$. Hence $\Gamma \vdash \neg\neg\psi$. So by 8.16, $\Gamma \vdash \psi$: ψ being any sentence, Γ is inconsistent. □

Theorem 8.26. $\Gamma \cup \{\varphi\}$ *is inconsistent iff* $\Gamma \vdash \neg\varphi$.

PROOF. $\Rightarrow$: Since $\Gamma \cup \{\varphi\} \vdash \psi$ for any sentence ψ, we have $\Gamma \cup \{\varphi\} \vdash \neg\varphi$, so by the deduction theorem $\Gamma \vdash \varphi \rightarrow \neg\varphi$. By 8.19, $\Gamma \vdash \neg\varphi$. $\Leftarrow$: $\Gamma \cup \{\varphi\} \vdash \neg\varphi$ and $\Gamma \cup \{\varphi\} \vdash \varphi$, so by 8.14, $\Gamma \cup \{\varphi\} \vdash \psi$ for any sentence ψ. □

Theorem 8.27. 0 *is consistent.*

PROOF. Since $\neg(\varphi \rightarrow \varphi)$ always receives the value 0 under any model, for any sentence φ, by 8.23 we have not $(\vdash \neg(\varphi \rightarrow \varphi))$. □

Theorem 8.28 (Extended completeness theorem). *Every consistent set of sentences has a model.*

PROOF. Let Γ be a consistent set of sentences. Let $\mathscr{A} = \{\Delta : \Gamma \subseteq \Delta, \Delta \text{ is consistent}\}$. Since $\Gamma \in \mathscr{A}$, $\mathscr{A}$ is nonempty. Suppose $\mathscr{B}$ is a subset of $\mathscr{A}$ simply ordered by inclusion, $\mathscr{B} \neq 0$. Then $\Gamma \subseteq \bigcup \mathscr{B}$. Also, $\bigcup \mathscr{B}$ is consistent, for, if not, there would be, by 8.25, a sentence φ such that $\bigcup \mathscr{B} \vdash \neg(\varphi \rightarrow \varphi)$. Then by 8.9, $\{\psi_0, \ldots, \psi_{m-1}\} \vdash \neg(\varphi \rightarrow \varphi)$ for some finite subset $\{\psi_0, \ldots, \psi_{m-1}\}$ of $\bigcup \mathscr{B}$. Say $\psi_0 \in \Delta_0 \in \mathscr{B}, \ldots, \psi_{m-1} \in \Delta_{m-1} \in \mathscr{B}$. Since $\mathscr{B}$ is simply ordered, there is an $i < m$ such that $\Delta_j \subseteq \Delta_i$ for all $j < m$. Thus $\psi_0 \in \Delta_i, \ldots, \psi_{m-1} \in \Delta_i$, so $\Delta_i \vdash \neg(\varphi \rightarrow \varphi)$. Thus Δ_i is inconsistent by 8.25, contradicting $\Delta_i \in \mathscr{B}$. Thus $\bigcup \mathscr{B}$ *is* consistent.

Hence we may apply Zorn's lemma to obtain a member Δ of $\mathscr{A}$ maximal under inclusion. Now we establish some important properties of Δ.

(1) $$\Delta \vdash \varphi \text{ implies that } \varphi \in \Delta.$$

For, if $\Delta \vdash \varphi$ and $\varphi \notin \Delta$, then $\Delta \cup \{\varphi\}$ is inconsistent, so by 8.26, $\Delta \vdash \neg\varphi$. Then by 8.14, Δ is inconsistent, contradiction.

(2) $$\text{if } \varphi \in \text{Sent, then } \varphi \in \Delta \text{ or } \neg\varphi \in \Delta.$$

For, suppose $\varphi \notin \Delta$. Then $\Delta \cup \{\varphi\}$ is inconsistent, so by 8.26 $\Delta \vdash \neg\varphi$, and (1) yields $\neg\varphi \in \Delta$.

(3) $$\varphi \rightarrow \psi \in \Delta \text{ iff } \neg\varphi \in \Delta \text{ or } \psi \in \Delta.$$

To prove this, first suppose $\neg\varphi \in \Delta$. By 8.15 and (1), $\varphi \rightarrow \psi \in \Delta$. If $\psi \in \Delta$, then $\varphi \rightarrow \psi \in \Delta$ by A1 and (1). Thus $\Leftarrow$ in (3) holds. Now suppose $\neg\varphi \notin \Delta$ and $\psi \notin \Delta$. By (2) we have $\varphi \in \Delta$ and $\neg\psi \in \Delta$. Hence by 8.21 and (1), $\neg(\varphi \rightarrow \psi) \in \Delta$, so $\varphi \rightarrow \psi \notin \Delta$ since Δ is consistent. Thus (3) holds.

We now define the desired model f. For any $s \in P$ let

$$fs = 1 \quad \text{if } \langle s \rangle \in \Delta,$$
$$fs = 0 \quad \text{otherwise.}$$

We claim:

(4) $$\text{for any sentence } \varphi, f^+\varphi = 1 \text{ iff } \varphi \in \Delta.$$

For, let $\Theta = \{\varphi : f^+\varphi = 1 \text{ iff } \varphi \in \Delta\}$. By the definition of f, $\langle s \rangle \in \Theta$ for each $s \in P$. Now suppose $\varphi \in \Theta$. Then

$$\begin{aligned} f^+(\neg\varphi) = 1 \quad & \text{iff } f^+\varphi = 0 && \text{definition of } {}^+ \\ & \text{iff } f^+\varphi \neq 1 && \\ & \text{iff } \varphi \notin \Delta && \text{since } \varphi \in \Theta \\ & \text{iff } \neg\varphi \in \Delta && \text{(2), consistency of } \Delta \end{aligned}$$

Thus $\neg\varphi \in \Theta$. Finally, suppose $\varphi, \psi \in \Theta$. Then

$$\begin{aligned} f^+(\varphi \rightarrow \psi) = 1 \quad & \text{iff } f^+\varphi = 0 \text{ or } f^+\psi = 1 && \text{definition of } {}^+ \\ & \text{iff } f^+\varphi \neq 1 \text{ or } f^+\psi = 1 && \\ & \text{iff } \varphi \notin \Delta \text{ or } \psi \in \Delta && \text{since } \varphi, \psi \in \Theta \\ & \text{iff } \neg\varphi \in \Delta \text{ or } \psi \in \Delta && \text{(2), consistency of } \Delta \\ & \text{iff } \varphi \rightarrow \psi \in \Delta && \text{(3)} \end{aligned}$$

Thus $\varphi \rightarrow \psi \in \Theta$. Hence $\text{Sent}_{\mathscr{P}} \subseteq \Theta$, and (4) holds. If $\varphi \in \Gamma$, then $\varphi \in \Delta$ and hence $f^+\varphi = 1$ by (4). Thus f is a model of Γ, as desired. □

Theorem 8.29 (Completeness theorem). $\Gamma \vdash \varphi$ *iff* $\Gamma \models \varphi$.

PROOF. $\Rightarrow$: by 8.23. $\Leftarrow$: Suppose not ($\Gamma \vdash \varphi$). Then not ($\Gamma \vdash \neg\neg\varphi$) by 8.16, so $\Gamma \cup \{\neg\varphi\}$ is consistent, by 8.26. By 8.28, let f be a model of $\Gamma \cup \{\neg\varphi\}$. Thus f is a model of Γ but not of φ, so not ($\Gamma \models \varphi$). □

This completeness theorem is our first result of a modern model-theoretical character. It shows the complete equivalence of the syntactical notion $\vdash$ and the semantical notion $\models$. Hence in considering various problems in sentential logic we can use syntactical or semantical methods with equal right. Another interesting consequence of the equivalence, $\vdash\varphi$ iff $\models\varphi$, is that there is an automatic method for determining whether $\vdash\varphi$ or not (the truth-table method above). From just the definition of $\vdash\varphi$ it is difficult to infer that such a decision method exists. This does not mean that for any Γ there is an effective procedure to determine whether $\Gamma \vdash \varphi$ or not. For example, P may be infinite, we may have $\Gamma \subseteq P$, and under a natural Gödel numbering Γ may be nonrecursive; in this case there is no such decision method.

We have now seen that our axioms for sentential logic formulated for negation and implication are sound and complete. Next, we indicate how various other notions of sentential logic can be expressed in terms of negation and implication. Some important results, like duality and normal form theorems, can be formulated using the new notions.

Definition 8.30. For any sentences φ and ψ let

$\varphi \vee \psi = \neg\varphi \rightarrow \psi$ (*disjunction* of φ and ψ);
$\varphi \wedge \psi = \neg(\varphi \rightarrow \neg\psi)$ (*conjunction* of φ and ψ);
$\varphi \leftrightarrow \psi = (\varphi \rightarrow \psi) \wedge (\psi \rightarrow \varphi)$ (*biimplication* between φ and ψ).

The following obvious proposition enables one to tell easily whether a sentence formulated using these defined symbols is a tautology or not:

Proposition 8.31. *If f is a model of $\mathscr{P}$ and φ and ψ are sentences of $\mathscr{P}$, then*

(*i*) $f^+(\varphi \vee \psi) = 1$ *iff* $f^+\varphi = 1$ *or* $f^+\psi = 1$;
(*ii*) $f^+(\varphi \wedge \psi) = 1$ *iff* $f^+\varphi = 1$ *and* $f^+\psi = 1$;
(*iii*) $f^+(\varphi \leftrightarrow \psi) = 1$ *iff* $f^+\varphi = f^+\psi$.

Having all of these sentential notions available, we can discuss the notion of the dual of a sentence:

Definition 8.32. For any sentence φ we define the dual of φ, denoted by φ^{d}, as follows (using 8.6):

(*i*) $\langle s\rangle^{\mathrm{d}} = \langle s\rangle$ for each $s \in P$;
(*ii*) $(\neg\varphi)^{\mathrm{d}} = \neg\varphi^{\mathrm{d}}$;
(*iii*) $(\varphi \rightarrow \psi)^{\mathrm{d}} = \neg\varphi^{\mathrm{d}} \wedge \psi^{\mathrm{d}}$.

The following proposition shows that $\wedge$ and $\vee$ are dual notions. The most useful part of the proposition is probably (*viii*), by which one can conclude that $\varphi^{\mathrm{d}} \leftrightarrow \psi^{\mathrm{d}}$ is a tautology after proving that $\varphi \leftrightarrow \psi$ is.

Proposition 8.33

(*i*) $\vdash (\varphi \wedge \psi)^{\mathrm{d}} \leftrightarrow \varphi^{\mathrm{d}} \vee \psi^{\mathrm{d}}$;
(*ii*) $\vdash (\varphi \vee \psi)^{\mathrm{d}} \leftrightarrow \varphi^{\mathrm{d}} \wedge \psi^{\mathrm{d}}$;
(*iii*) $\vdash \varphi \leftrightarrow \varphi^{\mathrm{dd}}$;
(*iv*) *if f and g are models of $\mathscr{P}$ and $fs = g^+ \neg \langle s \rangle$ for all $s \in P$, then $f^+\varphi = g^+(\neg\varphi^{\mathrm{d}})$ for every sentence φ*;
(*v*) $\Gamma \vdash \varphi$ *iff* $\{\neg\psi^{\mathrm{d}} : \psi \in \Gamma\} \vdash \neg\varphi^{\mathrm{d}}$;
(*vi*) $\vdash \varphi$ *iff* $\vdash \neg \varphi^{\mathrm{d}}$;
(*vii*) $\vdash \varphi \rightarrow \psi$ *iff* $\vdash \psi^{\mathrm{d}} \rightarrow \varphi^{\mathrm{d}}$;
(*viii*) $\vdash \varphi \leftrightarrow \psi$ *iff* $\vdash \varphi^{\mathrm{d}} \leftrightarrow \psi^{\mathrm{d}}$.

PROOF. In the case of (*i*) and (*ii*) it is just necessary to apply Definition 8.32 to the sentences $(\varphi \wedge \psi)^{\mathrm{d}}$ and $(\varphi \vee \psi)^{\mathrm{d}}$ in order to see what combination of φ^{d} and ψ^{d} they are and then check that then (*i*) and (*ii*) are tautologies. For example,

$$\begin{aligned}(\varphi \wedge \psi)^{\mathrm{d}} &= (\neg(\varphi \rightarrow \neg\psi))^{\mathrm{d}} = \neg(\varphi \rightarrow \neg\psi)^{\mathrm{d}} \\ &= \neg(\neg\varphi^{\mathrm{d}} \wedge (\neg\psi)^{\mathrm{d}}) = \neg(\neg\varphi^{\mathrm{d}} \wedge \neg\psi^{\mathrm{d}}),\end{aligned}$$

and $\neg(\neg\varphi^{\mathrm{d}} \wedge \neg\psi^{\mathrm{d}}) \leftrightarrow \varphi^{\mathrm{d}} \vee \psi^{\mathrm{d}}$ is a tautology (by, say, the truth-table method).

We prove (*iii*) by induction on φ. For φ of the form $\langle s \rangle$ with $s \in P$, we have $\varphi^{\mathrm{dd}} = \varphi$, and (*iii*) is clear. Now we assume (*iii*) for φ and prove it for $\neg\varphi$. Now $(\neg\varphi)^{\mathrm{dd}} = \neg\varphi^{\mathrm{dd}}$, and $(\varphi \leftrightarrow \varphi^{\mathrm{dd}}) \rightarrow (\neg\varphi \leftrightarrow \neg\varphi^{\mathrm{dd}})$ is a tautology, so (*iii*) for φ clearly implies (*iii*) for $\neg\varphi$. Finally, assume (*iii*) for φ and ψ; we establish it for $\varphi \rightarrow \psi$. We have

$$\begin{aligned}(\varphi \rightarrow \psi)^{\mathrm{dd}} &= (\neg\varphi^{\mathrm{d}} \wedge \psi^{\mathrm{d}})^{\mathrm{d}} = (\neg(\neg\varphi^{\mathrm{d}} \rightarrow \neg\psi^{\mathrm{d}}))^{\mathrm{d}} \\ &= \neg(\neg\varphi^{\mathrm{d}} \rightarrow \neg\psi^{\mathrm{d}})^{\mathrm{d}} = \neg(\neg(\neg\varphi^{\mathrm{d}})^{\mathrm{d}} \wedge (\neg\psi^{\mathrm{d}})^{\mathrm{d}}) \\ &= \neg(\neg\neg\varphi^{\mathrm{dd}} \wedge \neg\psi^{\mathrm{dd}}) = \neg\neg(\neg\neg\varphi^{\mathrm{dd}} \rightarrow \neg\neg\psi^{\mathrm{dd}});\end{aligned}$$

thus $(\varphi \leftrightarrow \varphi^{\mathrm{dd}}) \rightarrow ((\psi \leftrightarrow \psi^{\mathrm{dd}}) \rightarrow [(\varphi \rightarrow \psi) \leftrightarrow (\varphi \rightarrow \psi)^{\mathrm{dd}}])$ is a tautology, so (*iii*) follows for $\varphi \rightarrow \psi$. Therefore (*iii*) holds in general.

Assume the hypothesis of (*iv*). We prove its conclusion also by induction on φ. The case $\varphi = \langle s \rangle$ with $s \in P$ is trivial from the hypothesis of (*iv*). Assuming $f^+\varphi = g^+ \neg \varphi^{\mathrm{d}}$, we have

$$\begin{aligned}f^+ \neg \varphi = 0 \quad &\text{iff } f^+\varphi = 1 \\ &\text{iff } g^+ \neg \varphi^{\mathrm{d}} = 1 \\ &\text{iff } g^+ \neg \neg\varphi^{\mathrm{d}} = 0 \\ &\text{iff } g^+ \neg(\neg\varphi)^{\mathrm{d}} = 0.\end{aligned}$$

Thus if the conclusion of (*iv*) holds for φ, then it also holds for $\neg\varphi$. Finally, assume that $f^+\varphi = g^+(\neg\varphi^{\mathrm{d}})$ and $f^+\psi = g^+(\neg\psi^{\mathrm{d}})$. Then

$$\begin{aligned}f^+(\varphi \rightarrow \psi) = 0 \quad &\text{iff } f^+\varphi = 1 \text{ and } f^+\psi = 0 \\ &\text{iff } g^+ \neg \varphi^{\mathrm{d}} = 1 \text{ and } g^+ \neg \psi^{\mathrm{d}} = 0 \\ &\text{iff } g^+ \neg \varphi^{\mathrm{d}} = 1 \text{ and } g^+\psi^{\mathrm{d}} = 1 \\ &\text{iff } g^+(\neg\varphi^{\mathrm{d}} \wedge \psi^{\mathrm{d}}) = 1 \\ &\text{iff } g^+(\varphi \rightarrow \psi)^{\mathrm{d}} = 1 \\ &\text{iff } g^+ \neg (\varphi \rightarrow \psi)^{\mathrm{d}} = 0.\end{aligned}$$

Hence the conclusion of (*iv*) holds for $\varphi \to \psi$ if it holds for φ and for ψ. Hence (*iv*) holds.

We can derive (*v*) from (*iv*) easily using the completeness theorem. First assume $\Gamma \vdash \varphi$. Thus by the completeness theorem $\Gamma \models \varphi$. We shall now establish that $\{\neg\psi^{\mathrm{d}} : \psi \in \Gamma\} \models \neg\varphi^{\mathrm{d}}$. To this end, let g be any model such that $g^{+} \neg \psi^{\mathrm{d}} = 1$ for all $\psi \in \Gamma$. Define $fs = g^{+} \neg \langle s \rangle$ for all $s \in P$. By (*iv*), $f^{+}\psi = g^{+} \neg \psi^{\mathrm{d}}$ for every sentence ψ. Hence f is a model of Γ, and so $f^{+}\varphi = 1$. Therefore $g^{+} \neg \varphi^{\mathrm{d}} = 1$. Since g is arbitrary, $\{\neg\psi^{\mathrm{d}} : \psi \in \Gamma\} \models \neg\varphi^{\mathrm{d}}$. By the completeness theorem, $\{\neg\psi^{\mathrm{d}} : \psi \in \Gamma\} \vdash \neg\varphi^{\mathrm{d}}$. The converse is proved in exactly the same way; (*v*) follows.

The condition (*vi*) is a special case of (*v*). Concerning (*vii*), recall that $(\varphi \to \psi)^{\mathrm{d}} = \neg\varphi^{\mathrm{d}} \wedge \psi^{\mathrm{d}}$. Hence $\vdash \neg(\varphi \to \psi)^{\mathrm{d}} \leftrightarrow (\psi^{\mathrm{d}} \to \varphi^{\mathrm{d}})$, so (*vii*) follows from (*vi*). Finally, (*viii*) follows from (*vii*) since $\vdash \chi \leftrightarrow \theta$ is equivalent to the conjunction of $\vdash \chi \to \theta$ and $\vdash \theta \to \chi$, for any sentences χ and θ. □

Some other important sentential connectives are the finite generalizations of conjunction and disjunction:

Definition 8.34. Let $\langle \varphi_0, \ldots, \varphi_{m-1} \rangle$ be a finite sequence of sentences, with $m > 0$. We define the general conjunction $\bigwedge_{i \le j} \varphi_i$ and the general disjunction $\bigvee_{i \le j} \varphi_i$ for $j < m$ by recursion:

$$\bigwedge_{i \le 0} \varphi_i = \bigvee_{i \le 0} \varphi_i = \varphi_0;$$

$$\bigwedge_{i \le j+1} \varphi_i = \left(\bigwedge_{i \le j} \varphi_i\right) \wedge \varphi_{j+1};$$

$$\bigvee_{i \le i+1} \varphi_i = \left(\bigvee_{i \le j} \varphi_i\right) \vee \varphi_{j+1}.$$

We sometimes write $\varphi_0 \wedge \cdots \wedge \varphi_j$ instead of $\bigwedge_{i \le j} \varphi_i$, and $\varphi_0 \vee \cdots \vee \varphi_j$ instead of $\bigvee_{i \le j} \varphi_i$.

The following proposition shows how to calculate truth values with these new connectives; it is easily established by induction on j.

Proposition 8.35. *Under the assumptions of Definition* 8.34, *for any* $j < m$ *and any model* f *we have*

(*i*) $f^{+} \bigwedge_{i \le j} \varphi_i = 1$ *iff* $\forall i \le j\, (f^{+}\varphi_i = 1)$;
(*ii*) $f^{+} \bigvee_{i \le j} \varphi_i = 1$ *iff* $\exists i \le j\, (f^{+}\varphi_i = 1)$.

With the aid of generalized conjunction we can formulate a generalized form of the deduction theorem 8.11 which is frequently useful in the further study of sentential logic:

Theorem 8.36. *If* $\Gamma \cup \Delta \vdash \psi$ *and* $\Delta \neq 0$, *then there is an* $m \in \omega$ *and a* $\varphi \in {}^{m+1}\Delta$ *such that* $\Gamma \vdash \varphi_0 \wedge \cdots \wedge \varphi_m \to \psi$.

PROOF. By 8.9(*ii*) we may assume that Δ is finite. Hence it is enough to prove by induction on m that for all $m \in \omega$ and all $\varphi \in {}^{m+1}\mathrm{Sent}_{\mathscr{P}}$, if $\Gamma \cup \{\varphi_i : i \leq m\} \vdash \psi$ then $\Gamma \vdash \bigwedge_{i \leq m} \varphi_i \rightarrow \psi$. This statement for $m = 1$ is just the deduction theorem. Now we assume our statement for m (and for all sentences ψ), and we assume that $\varphi \in {}^{m+2}\mathrm{Sent}_{\mathscr{P}}$ and $\Gamma \cup \{\varphi_i : i \leq m + 1\} \vdash \psi$. By the deduction theorem, $\Gamma \cup \{\varphi_i : i \leq m\} \vdash \varphi_{m+1} \rightarrow \psi$. Hence by the induction assumption, $\Gamma \vdash \bigwedge_{i \leq m} \varphi_i \rightarrow (\varphi_{m+1} \rightarrow \psi)$. Now using Proposition 8.35 it is easily checked that the following sentence is a tautology, and hence is a Γ-theorem:

$$\left[\bigwedge_{i \leq m} \varphi_i \rightarrow (\varphi_{m+1} \rightarrow \psi)\right] \rightarrow \left(\bigwedge_{i \leq m+1} \varphi_i \rightarrow \psi\right).$$

It follows that $\Gamma \vdash \bigwedge_{i \leq m+1} \varphi_i \rightarrow \psi$, as desired. □

We can also give a useful criterion for inconsistency:

Theorem 8.37. *For any set Γ of sentences the following conditions are equivalent:*

(*i*) *Γ is inconsistent;*

(*ii*) *there is an $m \in \omega$ and a $\varphi \in {}^{m+1}\Gamma$ such that $\vdash \bigvee_{i \leq m} \neg \varphi_i$.*

PROOF. (*i*) $\Rightarrow$ (*ii*). Assuming (*i*), we have $\Gamma \vdash \psi \wedge \neg\psi$ for any sentence ψ; we fix ψ. From 8.27 we know that $\Gamma \neq 0$. Hence by Theorem 8.36 there is an $m \in \omega$ and a $\varphi \in {}^{m+1}\Gamma$ such that

$$\vdash \bigwedge_{i \leq m} \varphi_i \rightarrow \psi \wedge \neg\psi.$$

It remains only to notice that the sentence

$$\left(\bigwedge_{i \leq m} \varphi_i \rightarrow \psi \wedge \neg\psi\right) \rightarrow \bigvee_{i \leq m} \neg\varphi_i$$

is a tautology.

(*ii*) $\Rightarrow$ (*i*). Assume (*ii*). Then for any sentence ψ we have $\Gamma \models \psi$. In fact, let f be a model of Γ. Then in particular $f^+\varphi_i = 1$ for all $i \leq m$. But, by virtue of Proposition 8.35, this contradicts the fact that $\bigvee_{i \leq m} \neg \varphi_i$ is a tautology. Conclusion: Γ has no models. Thus, vacuously, every model of Γ is a model of ψ. Hence, indeed, $\Gamma \models \psi$, so $\Gamma \vdash \psi$ by the completeness theorem. Thus Γ is inconsistent. □

The following theorem is perhaps the most important result in the classic theory of sentential logic:

Theorem 8.38 (Distinguished disjunctive normal form). *Let $\mathscr{P} = \langle n, c, \{s_0, \ldots, s_m\}\rangle$ be a sentential language, with $m \in \omega$. Let φ be a sentence which has at least one model. Then there is a function ψ such that:*

(*i*) *the domain of ψ is $(p + 1) \times (m + 1)$ for some $p \in \omega$;*

(*ii*) *for each $i \leq p$ and $j \leq m$, $\psi_{ij} = \langle s_j\rangle$ or $\psi_{ij} = \neg\langle s_j\rangle$;*

(*iii*) $\vdash \varphi \leftrightarrow \bigvee_{i \leq p} \bigwedge_{j \leq m} \psi_{ij}$.

Proof. Let M be the set of all models of φ; $M \neq 0$ by assumption. Let g be a one-one function mapping some integer $p + 1$ onto M. For each $i \leq p$ and $j \leq m$ let

$$\psi_{ij} = \begin{cases} \langle s_j \rangle & \text{if } g_i s_j = 1, \\ \neg\langle s_j \rangle & \text{if } g_i s_j = 0. \end{cases}$$

Thus $g_i^+ \bigwedge_{j \leq m} \psi_{ij} = 1$, while $f^+ \bigwedge_{j \leq m} \psi_{ij} = 0$ if $f \neq g_i$, for each $i \leq p$. From this it follows easily that $f^+\varphi = f^+ \bigvee_{i \leq p} \bigwedge_{j \leq m} \psi_{ij}$ for every model f, as desired. □

By duality we obtain

Theorem 8.39 (Distinguished conjunctive normal form). *Let $\mathscr{P} = \langle n, c, \{s_0, \ldots, s_{m-1}\}\rangle$ be a sentential logic, with $m \in \omega$. Let φ be a sentence which is not a tautology. Then there is a function ψ such that:*

(*i*) *the domain of ψ is $(p + 1) \times (m + 1)$ for some $p \in \omega$;*
(*ii*) *for each $i \leq p$ and each $j \leq m$, $\psi_{ij} = \langle s_j \rangle$ or $\psi_{ij} = \neg\langle s_j \rangle$;*
(*iii*) $\vdash \varphi \leftrightarrow \bigwedge_{i \leq p} \bigvee_{j \leq m} \psi_{ij}$.

Proof. Assume the hypothesis. Thus not($\vdash\varphi$), so by 8.33(*vi*), φ^{d} has a model. Hence by 8.38 we can choose ψ so that (*i*) and (*ii*) hold and

$$\vdash \varphi^{\mathrm{d}} \leftrightarrow \bigvee_{i \leq p} \bigwedge_{j \leq m} \psi_{ij}.$$

Now an application of 8.33(*viii*) gives the desired result (8.33(*i*) and 8.33(*ii*) must first be generalized in the obvious way).

As the final topic of this chapter we want to systematically consider possible variants of our choice of connectives negation and implication, and show that all other connectives can be defined from these two. The full meaning of our sentential logic is given by a sentential language (n, c, P) together with the definition of f^+ for $f \in {}^P2$, which amounts to assigning the usual truth tables for the meanings of n and c. We shall see that any connective, with its meaning given by some truth table, can be expressed in terms of n and c. There are also other connectives which can serve, like n and c, to express any connective. To prove all of this rigorously, we need a notion of *general sentential logic*:

Definition 8.40. For each set P, a *P-truth function* is any function mapping ${}^P2$ into 2. [Thus, for example, the 2-truth function f corresponding to c is given by the stipulations

$$f\langle 1, 1\rangle = 1,$$
$$f\langle 1, 0\rangle = 0,$$
$$f\langle 0, 1\rangle = 1,$$
$$f\langle 0, 0\rangle = 1.]$$

A *general sentential logic* is a quadruple $\mathscr{P} = (K, P, f, g)$ such that $P \neq 0$, $K \cap P = 0$, f is a function mapping K into $\omega \sim 1$, and g is a function assigning to each $k \in K$ an f_k-truth function g_k. *Symbols* of $\mathscr{P}$ are the members of $K \cup P$; *expressions* of $\mathscr{P}$ are finite sequences of symbols. With each $k \in K$ we associate an f_k-ary operation $\hat{k}$ on expressions:

$$\hat{k}(\varphi_0, \ldots, \varphi_{m-1}) = \langle k \rangle \varphi_0 \cdots \varphi_{m-1}$$

where $m = f_k$, for any expressions $\varphi_0, \ldots, \varphi_{m-1}$. The set of *sentences* of $\mathscr{P}$ is the intersection of all classes Γ of expressions of $\mathscr{P}$ such that $\{\langle s \rangle : s \in P\} \subseteq \Gamma$ and Γ is closed under all the operations $\hat{k}$ for $k \in K$. On the basis of these definitions, analogs of 8.3–8.6 are easily established. This enables us to extend the notion $^+$ of 8.22. A model of $\mathscr{P}$ is again any mapping of P into 2. Given a model h, a mapping h^+ of the set of sentences of $\mathscr{P}$ into 2 is defined by recursion as follows:

$$h^+\langle s \rangle = hs \qquad \text{for } s \in P,$$
$$h^+\hat{k}\varphi_0 \cdots \varphi_{m-1} = g_k(h^+\varphi_0, \ldots, h^+\varphi_{m-1}),$$

where $m = f_k$ and $\varphi_0, \ldots, \varphi_{m-1}$ are sentences of $\mathscr{P}$. Then with any sentence φ we can associate a P-truth function $\mathscr{T}_\varphi$ such that $\mathscr{T}_\varphi f = f^+\varphi$ for any $f \in {}^P2$. The logic $\mathscr{P}$ is *functionally complete* provided that $\{\mathscr{T}_\varphi : \varphi \text{ a sentence of } \mathscr{P}\}$ is the set of all P-truth functions.

We can identify sentential logics in our earlier sense (language together with semantic notions) with certain general sentential logics, namely those of the form $(\{n, c\}, P, f, g)$, where $n \neq c$, $\{n, c\} \cap P = 0$, $fn = 1$, $fc = 2$, g_n is the function given by $g\langle 1 \rangle = 0$, $g\langle 0 \rangle = 1$, and g_c is the function mentioned above in 8.40. Then we have

Theorem 8.41. *Let $\mathscr{P} = \langle \{n, c\}, P, f, g \rangle$ be a sentential logic in the original sense, with P finite. Then $\mathscr{P}$ is functionally complete.*

PROOF. Note that there are exp exp $|P|$ P-truth functions. Hence it suffices to show that there are at least that many truth functions of the form $\mathscr{T}_\varphi$. Let $|P| = m + 1$, say $P = \{s_i : i \leq m\}$. For each $h \in {}^P2$ we define χ_h, an $(m + 1)$-termed sequence of sentences of $\mathscr{P}$: for $j \leq m$, $\chi_{hj} = \langle s_j \rangle$ if $hs_j = 1$, $\chi_{hj} = \neg\langle s_j \rangle$ if $hs_j = 0$. Now for any nonempty subset Γ of ${}^P2$, let

$$\varphi_\Gamma = \bigvee_{h \in \Gamma} \bigwedge_{j \leq m} \chi_{hj}.$$

Then it is easily verified that Γ is the set of all models of φ_Γ (cf. the proof of 8.38). Hence $\mathscr{T}_{\varphi\Gamma} \neq \mathscr{T}_{\varphi\Delta}$ if $\Gamma \neq \Delta$. Furthermore, clearly $\mathscr{T}_{\varphi\Gamma} \neq \mathscr{T}_\psi$ for all nonempty $\Gamma \subseteq {}^P2$, where ψ is a sentence with no models. Thus we have exhibited exp exp $|P|$ different functions $\mathscr{T}_\theta$, as desired. □

This theorem is a rigorous expression of the above mentioned fact that any connective can be defined in terms of n and c. As also stated, other choices of connectives can be used for this purpose. To show this, the following lemma is useful.

Lemma 8.42. *Let neg be the truth function corresponding to n in any sentential logic, and imp that corresponding to c. Let* $\mathscr{P} = (K, P, f, g)$ *be a general sentential logic such that* $K \cap 2 = 0$*, P finite,* $neg \in \{\mathscr{T}_\varphi : \varphi$ *a sentence of* $(K, 1, f, g)\}$*, and* $imp \in \{\mathscr{T}_\varphi : \varphi$ *a sentence of* $(K, 2, f, g\}$*. Then* $\mathscr{P}$ *is functionally complete.*

Proof. Let $\mathscr{P}' = (K, 1, f, g)$, $\mathscr{P}'' = (K, 2, f, g)$, and let $\mathscr{P}''' = (\{n', c'\}, P, f', g')$ be a sentential logic. For φ a sentence of $\mathscr{P}'$ and ψ a sentence of $\mathscr{P}$, let $\varphi[\psi]$ be the sentence of $\mathscr{P}$ obtained from φ by substituting ψ for $\langle 0 \rangle$ throughout φ. If h is a model of $\mathscr{P}$ and ψ is a sentence of $\mathscr{P}$, then h'_ψ is the model of $\mathscr{P}'$ such that $h'_\psi 0 = h^+\psi$. Now

(1) if φ is a sentence of $\mathscr{P}'$, ψ a sentence of $\mathscr{P}$, and h is a model of $\mathscr{P}$, then $h^+\varphi[\psi] = h'^+_\psi \varphi$.

This is easily proved by induction on φ. We work similarly with $\mathscr{P}''$. For φ a sentence of $\mathscr{P}''$ and ψ, χ sentences of $\mathscr{P}$, let $\varphi[\psi, \chi]$ be the sentence of $\mathscr{P}$ obtained from φ by substituting ψ for $\langle 0 \rangle$ and χ for $\langle 1 \rangle$ throughout φ. If h is a model of $\mathscr{P}$ and ψ and χ are sentences of $\mathscr{P}$, then $h''_{\psi\chi}$ is the model of $\mathscr{P}''$ such that $h''_{\psi\chi}0 = h^+\psi$ and $h''_{\psi\chi}1 = h^+\chi$. By induction, one easily establishes

(2) if φ is a sentence of $\mathscr{P}''$, ψ and χ sentences of $\mathscr{P}$, and h a model of $\mathscr{P}$, then $h^+\varphi[\psi, \chi] = h''^+_\psi \varphi$.

Now by hypothesis choose a sentence φ of $\mathscr{P}'$ and a sentence ψ of $\mathscr{P}''$ so that $neg = \mathscr{T}_\varphi$ and $imp = \mathscr{T}_\psi$. With each sentence χ of $\mathscr{P}'''$ we now associate a sentence χ^* of $\mathscr{P}$, defined by recursion:

$$\begin{aligned} \langle s \rangle^* &= \langle s \rangle; \\ (\neg\chi)^* &= \varphi[\chi^*]; \\ (\chi_0 \rightarrow \chi_1)^* &= \psi[\chi_0^*, \chi_1^*]; \end{aligned}$$

for any sentences χ, χ_0, χ_1 of $\mathscr{P}'''$ and any $s \in P$. Now we prove the following statement by induction on χ:

(3) for any sentence χ of $\mathscr{P}'''$ and any model h of $\mathscr{P}'''$, $h^+\chi^* = h^+\chi$.

The statement is clear for $\chi = \langle s \rangle$, $s \in P$. Now assume it true for χ. Then

$$\begin{aligned} h^+(\neg\chi)^* &= h^+\varphi[\chi^*] = h'^+_{\chi^*}\varphi && \text{by (1)} \\ &= \mathscr{T}_\varphi h'_{\chi^*} = neg\ h'_{\chi^*} \end{aligned}$$

and $neg\ h'_{\chi^*} = 0$ iff $h'_{\chi^*}\langle 0 \rangle = 1$ iff $h^+\chi^* = 1$ iff $h^+\chi = 1$ iff $h^+(\neg\chi) = 0$, i.e., $neg\ h'_{\chi^*} = h^+\neg\chi$. Similarly, $h^+(\chi_0 \rightarrow \chi_1)^* = imp\ h''_{\chi_0^*\chi_1^*}$, and $imp\ h''_{\chi_0^*\chi_1^*} = 0$ iff $h''_{\chi_0^*\chi_1^*}0 = 1$ and $h''_{\chi_0^*\chi_1^*}1 = 0$, i.e., iff $h^+\chi_0^* = 1$ and $h^+\chi_1^* = 0$, i.e., iff $h^+\chi_0 = 1$ and $h^+\chi_1 = 0$, that is, iff $h^+(\chi_0 \rightarrow \chi_1) = 0$. Hence $imp\ h''_{\chi_0^*\chi_1^*} = h^+(\chi_0 \rightarrow \chi_1)$. Thus (3) holds. It follows directly from (3) that, since $\mathscr{P}'''$ is functionally complete, so is $\mathscr{P}$: for, if l is any truth function, write $l = \mathscr{T}_\chi$ with χ a sentence of $\mathscr{P}'''$. By (3), $\mathscr{T}_\chi = \mathscr{T}_{\chi^*}$, as desired. □

Theorem 8.43. *Negation and conjunction give rise to a functionally complete sentential logic. That is, if $\mathscr{P} = \langle\{n, k\}, P, f, g\rangle$ is a general sentential logic with $fn = 1, fk = 2$, and g given by:*

$$g_n\langle 0\rangle = 1, \qquad g_n\langle 1\rangle = 0;$$
$$g_k\langle 1, 1\rangle = 1, \qquad g_k\langle 1, 0\rangle = g_k\langle 0, 1\rangle = g_k\langle 0, 0\rangle = 0;$$

and if P is finite, then $\mathscr{P}$ is functionally complete.

PROOF. We may assume that $P \cap 2 = 0$. Let $\neg\varphi = \langle n\rangle\varphi$ and $\varphi \wedge \psi = \langle k\rangle\varphi\psi$ for all sentences φ, ψ. Clearly $neg = \mathscr{T}_{\langle 0\rangle}$ in $\langle\{n, k\}, 1, f, g\rangle$, and $imp = \mathscr{T}_\varphi$ in $\langle\{n, k\}, 2, f, g\rangle$, where φ is the sentence

$$\neg(\langle 0\rangle \wedge \neg\langle 1\rangle).$$

Thus the theorem follows from Lemma 8.42. □

Similarly:

Theorem 8.44. *Negation and disjunction give rise to a functionally complete sentential logic.*

It is of some interest that a single connective can be used in sentential logic:

Theorem 8.45. *Let $\mathscr{P} = \langle\{s\}, P, f, g\rangle$ be a general sentential logic such that $fs = 1$ and g is given by*

$$g_s\langle 1, 1\rangle = 0,$$
$$g_s\langle 0, 1\rangle = g_s\langle 1, 0\rangle = g_s\langle 0, 0\rangle = 1,$$

and with P finite. Then $\mathscr{P}$ is functionally complete.

PROOF. $\varphi|\psi = \langle s\rangle\varphi\psi$ for any sentences φ, ψ. (The connective here is called the *Sheffer stroke*.) Then $neg = \mathscr{T}_\varphi$ and $imp = \mathscr{T}_\psi$ where φ is $\langle 0\rangle|\langle 0\rangle$ and ψ is $\langle 0\rangle|(\langle 1\rangle|\langle 1\rangle)$. Thus Lemma 8.43 implies. □

BIBLIOGRAPHY

1. Schmidt, H. A. *Mathematische Gesetze der Logik*. Berlin: Springer (1960).
2. Lukasiewicz, J. and Tarski, A. Investigations into the sentential calculus. In *Logic, Semantics, Metamathematics; Collected Papers of A. Tarski*. Oxford: Oxford Univ. Press (1956).
3. Post, E. *The Two-Valued Iterative Systems of Mathematical Logic*. Princeton: Princeton Univ. Press (1941).

EXERCISES

8.46. Let $\mathscr{P} = (n, c, \{s, t\})$ be a sentential language, with $s \neq t$. Determine which of the following expressions are sentences: $\langle s\rangle$, $\langle c, c, n, s, n, t, c, t, s\rangle$, $\langle c, s, c, s, t, s\rangle$, $\langle n, n, n, s, s\rangle$, $\langle c, n, p, c, p, q\rangle$.

8.47. By a *sentential language with parentheses* we mean a system $\mathscr{P} = (n, c, l, r, P)$ such that n, c, l, r are all distinct, $P \neq 0$, and $\{n, c, l, r\} \cap P = 0$. We introduce operations $\neg$, $\rightarrow$ on expressions as follows:

$$\neg\varphi = \langle n\rangle\varphi$$
$$\varphi \rightarrow \psi = \langle l\rangle\varphi\langle c\rangle\psi\langle r\rangle,$$

for all expressions φ, ψ. $\text{Sent}_{\mathscr{P}}$ is the intersection of all sets Γ of expressions such that $\langle s\rangle \in \Gamma$ for each $s \in P$ and $\neg\varphi$, $\varphi \rightarrow \psi$ for all $\varphi, \psi \in \Gamma$. Prove an analog of 8.5.

8.48. Modify 8.2 by changing the definition of $\rightarrow$:

$$\varphi \rightarrow \psi = \varphi\langle c\rangle\psi$$

for all expressions φ, ψ. Show then that parts (*iii*) and (*iv*) of 8.5 fail.

8.49. Show that the following schemas, due to Lukasiewicz, can replace (A1)–(A3) of 8.7:

(a) $(\psi \rightarrow \chi) \rightarrow [(\chi \rightarrow \theta) \rightarrow (\psi \rightarrow \theta)]$,
(b) $(\neg\psi \rightarrow \psi) \rightarrow \psi$
(c) $\psi \rightarrow (\neg\psi \rightarrow \chi)$

Hint: prove the following schemas in succession from the above schemas:

(1) $[(\neg\chi \rightarrow \psi) \rightarrow (\neg\psi \rightarrow \psi)] \rightarrow [\chi \rightarrow (\neg\psi \rightarrow \psi)]$
(2) $(\neg\psi \rightarrow \neg\chi) \rightarrow [\chi \rightarrow (\neg\psi \rightarrow \psi)]$
(3) $(\neg\psi \rightarrow \neg\chi) \rightarrow ([\neg\psi \rightarrow \psi) \rightarrow \psi] \rightarrow (\chi \rightarrow \psi))$
(4) $\psi \rightarrow ([\neg\psi \rightarrow \psi) \rightarrow \psi] \rightarrow (\chi \rightarrow \psi))$
(5) $\chi \rightarrow [(\neg\psi \rightarrow \psi) \rightarrow \psi]$
(6) $([(\neg\psi \rightarrow \psi) \rightarrow \psi] \rightarrow (\chi \rightarrow \psi)) \rightarrow [\neg(\chi \rightarrow \psi) \rightarrow (\chi \rightarrow \psi)]$
(7) $([(\neg\psi \rightarrow \psi) \rightarrow \psi] \rightarrow (\chi \rightarrow \psi)) \rightarrow (\chi \rightarrow \psi)$
(8) $\psi \rightarrow (\chi \rightarrow \psi)$
(9) $(\neg\psi \rightarrow \neg\chi) \rightarrow (\chi \rightarrow \psi)$
(10) $\neg\chi \rightarrow (\chi \rightarrow \psi)$
(11) $[(\chi \rightarrow \psi) \rightarrow \chi] \rightarrow \chi$
(12) $[\chi \rightarrow (\chi \rightarrow \psi)] \rightarrow ([(\chi \rightarrow \psi) \rightarrow \chi] \rightarrow (\chi \rightarrow \psi))$
(13) $[\chi \rightarrow (\chi \rightarrow \psi)] \rightarrow (\chi \rightarrow \psi)$
(14) $\chi \rightarrow ((\chi \rightarrow \psi) \rightarrow [(\chi \rightarrow \psi) \rightarrow \psi])$
(15) $\chi \rightarrow [(\chi \rightarrow \psi) \rightarrow \psi]$
(16) $([(\psi \rightarrow \theta) \rightarrow \theta] \rightarrow (\chi \rightarrow \theta)) \rightarrow [\psi \rightarrow (\chi \rightarrow \theta)]$
(17) $[\chi \rightarrow (\psi \rightarrow \theta)] \rightarrow [\psi \rightarrow (\chi \rightarrow \theta)]$
(18) $(\psi \rightarrow \theta) \rightarrow [(\chi \rightarrow \psi) \rightarrow (\chi \rightarrow \theta)]$
(19) $((\psi \rightarrow \chi) \rightarrow [\psi \rightarrow (\psi \rightarrow \theta)]) \rightarrow [(\psi \rightarrow \chi) \rightarrow (\psi \rightarrow \theta)]$
(20) $[\psi \rightarrow (\chi \rightarrow \theta)] \rightarrow [(\psi \rightarrow \chi) \rightarrow (\psi \rightarrow \theta)]$

8.50. Show that the following single schema, due to Meredith, can replace the schemas (A1)–(A3):

(a) $(\{[(\varphi \rightarrow \psi) \rightarrow (\neg\chi \rightarrow \neg\theta)] \rightarrow \chi\} \rightarrow \tau) \rightarrow [(\tau \rightarrow \varphi) \rightarrow (\theta \rightarrow \varphi)]$
Hint: prove the following formulas from this schema:

(1) $\{[(\tau \to \varphi) \to (\theta \to \varphi)] \to (\varphi \to \psi)\} \to [\chi \to (\varphi \to \psi)]$
(2) $\{[\chi \to (\neg\varphi \to \psi)] \to \tau\} \to (\varphi \to \tau)$
(3) $[(\varphi \to \varphi) \to \chi] \to (\theta \to \chi)$
(4) $\psi \to [\theta \to (\varphi \to \varphi)]$
(5) $\{[\theta \to (\varphi \to \varphi)] \to \chi\} \to (\psi \to \chi)$
(6) $[(\psi \to \chi) \to \varphi] \to (\chi \to \varphi)$
(7) $\tau \to [(\tau \to \varphi) \to (\theta \to \varphi)]$
(8) $[(\{[\chi \to (\neg\varphi \to \psi)] \to \tau\} \to (\varphi \to \tau)] \to \mu) \to (\theta \to \mu)$
(9) $\{(\theta \to \varphi) \to [\chi \to (\neg\neg\varphi \to \psi)]\} \to \{\tau \to [\chi \to (\neg\neg\varphi \to \psi)]\}$
(10) $(\varphi \to \psi) \to (\neg\neg\varphi \to \psi)$
(11) $\chi \to \{[(\psi \to \chi) \to \varphi] \to (\theta \to \varphi)\}$
(12) $(\{([\tau \to \{[(\varphi \to \psi) \to (\neg\chi \to \neg\theta)] \to \chi\}] \to \mu) \to (\nu \to \mu)\} \to \varphi) \to (\theta \to \varphi)$
(13) $((\theta \to \varphi) \to [\tau \to \{[(\varphi \to \psi) \to (\neg\chi \to \neg\theta)] \to \chi\}])$
$\to (\mu \to [\tau \to \{[(\varphi \to \psi) \to (\neg\chi \to \neg\theta)] \to \chi\}])$
(14) $[(\theta \to \varphi) \to \chi] \to \{[(\varphi \to \psi) \to (\neg\chi \to \neg\theta)] \to \chi\}$
(15) $[(\varphi \to \psi) \to \{\neg[\{[\tau \to (\theta \to \varphi)] \to \chi\} \to (\mu \to \chi)] \to \neg\theta\}]$
$\to [\{[\tau \to (\theta \to \varphi)] \to \chi\} \to (\mu \to \chi)]$
(16) $\{[([\tau \to \{\theta \to [\theta \to \psi]\}] \to \chi) \to (\mu \to \chi)] \to \varphi\} \to (\nu \to \varphi)$
(17) $[(\nu \to \varphi) \to \{\tau \to [\theta \to (\varphi \to \psi)]\}] \to [\chi \to \{\tau \to [\theta \to (\varphi \to \psi)]\}]$
(18) $[\varphi \to (\varphi \to \psi)] \to [\theta \to (\varphi \to \psi)]$
(19) $[\varphi \to (\varphi \to \psi)] \to (\varphi \to \psi)$
(20) $\varphi \to (\neg\varphi \to \psi)$
(21) $\{[(\varphi \to \psi) \to (\neg\neg\varphi \to \psi)] \to \chi\} \to (\theta \to \chi)$
(22) $\{[(\varphi \to \psi) \to (\neg\neg\varphi \to \psi)] \to \chi\} \to \chi$
(23) $(\neg\varphi \to \varphi) \to (\theta \to \varphi)$
(24) $(\neg\varphi \to \varphi) \to \varphi$
(25) $\theta \to (\neg\neg\chi \to \chi)$
(26) $\{[\theta \to (\neg\neg\chi \to \chi)] \to \varphi\} \to (\psi \to \varphi)$
(27) $\{[\theta \to (\neg\neg\chi \to \chi)] \to \varphi\} \to \varphi$
(28) $\neg\neg\{[\theta \to (\neg\neg\chi \to \chi)] \to \varphi\} \to \varphi$
(29) $[(\psi \to [\neg\neg\{[\theta \to (\neg\neg\chi \to \chi)] \to \varphi\} \to \varphi]) \to \tau] \to (\mu \to \tau)$
(30) $([\mu \to \neg\{[\theta \to (\neg\neg\chi \to \chi)] \to \neg\varphi\}] \to \psi) \to (\varphi \to \psi)$
(31) $\varphi \to \neg\{[\theta \to (\neg\neg\chi \to \chi)] \to \neg\varphi\}$
(32) $\{[\psi \to (\varphi \to \neg\{[\theta \to (\neg\neg\chi \to \chi)] \to \neg\varphi\})] \to \tau\} \to (\mu \to \tau)$
(33) $[(\psi \to [\varphi \to \neg\{[\theta \to (\neg\neg\chi \to \chi)] \to \neg\varphi\}]) \to \tau] \to \tau$
(34) $(\varphi \to \psi) \to [\{[\theta \to (\neg\neg\chi \to \chi)] \to \neg\neg\varphi\} \to \psi]$
(35) $\{[\tau \to (\neg\neg\mu \to \mu)] \to \neg\neg(\varphi \to \psi)\} \to [\{[\theta \to (\neg\neg\chi \to \chi)] \to \neg\neg\varphi\} \to \psi]$
(36) $([\{[\theta \to (\neg\neg\chi \to \chi)] \to \neg\neg\varphi\} \to \psi] \to \tau) \to [(\varphi \to \psi) \to \tau]$
(37) $(\varphi \to \psi) \to [(\psi \to \chi) \to (\varphi \to \chi)]$

8.51. Let $\mathscr{P} = (n, c, P)$ and $\mathscr{P}' = (n', c', P')$ be two sentential languages. We write $\mathscr{P} \leq \mathscr{P}'$ iff $n = n'$, $c = c'$, and $P \subseteq P'$. Assume that $\mathscr{P} \leq \mathscr{P}'$ and prove the following:

(1) $\mathrm{Sent}_{\mathscr{P}} \subseteq \mathrm{Sent}_{\mathscr{P}'}$;
(2) if $f: P' \to 2$, then $f^+ \restriction \mathrm{Sent}_{\mathscr{P}} = (f \restriction P)^+$;
(3) if $\Gamma \cup \{\varphi\} \subseteq \mathrm{Sent}_{\mathscr{P}}$, then $\Gamma \vdash_{\mathscr{P}} \varphi$ iff $\Gamma \vdash_{\mathscr{P}'} \varphi$.

8.52. Let ISent be the set of all sentences (of a fixed but arbitrary sentential language) in which the negation symbol does not occur. Our aim is to outline an independent development of sentential languages without negation. Let IAxm be the set of all sentences of the following kinds (where φ, ψ, $\chi \in$ ISent):

$$\varphi \to (\psi \to \varphi)$$
$$[\varphi \to (\psi \to \chi)] \to [(\varphi \to \psi) \to (\varphi \to \chi)]$$
$$[(\varphi \to \psi) \to \varphi] \to \varphi.$$

Let Γ – IThm be the intersection of all sets of I-sentences including $\Gamma \cup$ IAxm and closed under detachment. We write $\Gamma \vdash_{\mathrm{I}} \varphi$ for $\varphi \in \Gamma$ – IThm. Establish the following, where all sentences and sets of sentences are taken from ISent:

(a) $0 \vdash_{\mathrm{I}} \varphi \to \varphi$.
(b) the deduction theorem for $\vdash_{\mathrm{I}}$.
(c) $0 \vdash_{\mathrm{I}} (\varphi \to \psi) \to [(\psi \to \chi) \to (\varphi \to \chi)]$.
(d) If $\Gamma \vdash_{\mathrm{I}} \langle s \rangle$ for each $s \in P$, then $\Gamma \vdash_{\mathrm{I}} \varphi$ for each $\varphi \in$ ISent.
(e) $0 \vdash_{\mathrm{I}} (\varphi \to \psi) \to ([(\varphi \to \chi) \to \psi] \to \psi)$.
(f) Call a set Γ of I-Sentences I-*consistent* if not ($\Gamma \vdash_{\mathrm{I}} \varphi$) for some I-sentence φ. If Γ is I-consistent, then Γ has a model which is not identically 1.

Hint: use these steps:

(1) there is an $s \in P$ such that not ($\Gamma \vdash_{\mathrm{I}} \langle s \rangle$);
(2) there is a maximal $\Delta \supseteq \Gamma$ such that not ($\Delta \vdash_{\mathrm{I}} \langle s \rangle$);
(3) $\Delta \vdash_{\mathrm{I}} \varphi$ implies that $\varphi \in \Delta$;
(4) $\varphi \to \psi \in \Delta$ iff $\varphi \notin \Delta$ or $\psi \in \Delta$;
(5) continue as in the proof of 8.28.
(g) $\Gamma \vdash_{\mathrm{I}} \varphi$ iff $\Gamma \models \varphi$ iff $\Gamma \vdash \varphi$. *Hint:* for the hard direction, assume not ($\Gamma \vdash_{\mathrm{I}} \varphi$). Then not ($\Gamma \cup \{\varphi \to \psi : \psi \in \text{ISent}\} \vdash \varphi$), so by (f) let f be a model of $\Gamma \cup \{\varphi \to \psi : \psi \in \text{ISent}\}$ which is not identically 1. Show that $f^+\varphi = 0$.

8.53. We work in a sentential logic in which negation and disjunction are taken as primitive. For any sentences φ and ψ let $\varphi \to \psi = \neg\varphi \vee \psi$. Take as axioms all sentences of the following forms:

(a) $(\varphi \vee \varphi) \to \varphi$;
(b) $\varphi \to (\varphi \vee \psi)$;
(c) $(\varphi \vee \psi) \to (\psi \vee \varphi)$;
(d) $(\varphi \to \psi) \to [(\chi \vee \varphi) \to (\chi \vee \psi)]$.

Prove analogs of 8.28 and 8.29. *Hint:* apply the result of 8.49. In order to do this it suffices (proof!) to derive the following:

(e) $(\varphi \to \psi) \to [(\chi \to \varphi) \to (\chi \to \psi)]$;
(f) $\varphi \to \varphi$;
(g) $\varphi \to \neg\neg\varphi$;
(h) $\neg\neg\varphi \to \varphi$;
(i) $\vdash \varphi \to \psi$ and $\vdash \varphi' \to \psi'$ imply $\vdash \neg\neg\varphi \vee \varphi' \to \psi \vee \psi'$;
(j) $\vdash \varphi \to \psi$ and $\vdash \varphi' \to \psi'$ imply $\vdash \varphi \vee \varphi' \to \neg\neg\psi \vee \psi'$;
(k) $\vdash (\varphi \to \psi) \to (\neg\psi \to \neg\varphi)$;
(l) $\vdash (\neg\varphi \to \neg\psi) \to (\psi \to \varphi)$;

(m) $\vdash(\varphi \to \psi) \to [(\psi \to \chi) \to (\varphi \to \chi)]$;
(n) $\vdash(\neg\varphi \to \varphi) \to \varphi$;
(o) $\vdash\varphi \to (\neg\varphi \to \psi)$;
(p) $\vdash\psi \to (\varphi \to \varphi)$;
(q) $\vdash \neg(\varphi \to \varphi) \to \psi$.

8.54. Again we work with a sentential logic in which negation and disjunction are taken as primitive.

We want to describe now a definition of "$\vdash\varphi$" in which rules of inference rather than axioms take precedence. We define four rules of inference.

(1) *Association.* This rule is a relation $R_0 \subseteq \text{Sent} \times \text{Sent}$. If φ is a sentence, $(\psi \vee \chi) \vee \theta$ is a consecutive part of φ, with $\psi, \chi, \theta \in \text{Sent}$ (i.e., $\varphi = \theta'((\psi \vee \chi) \vee \theta)\theta''$ for some expressions θ', θ''), and if σ is obtained from φ by replacing a consecutive part of φ of this sort by $\psi \vee (\chi \vee \theta)$ (i.e., if $\sigma = \theta'(\psi \vee (\chi \vee \theta))\theta''$ with notation as before), then $(\varphi, \sigma) \in R_0$.

(2) *Commutation.* We define $R_1 \subseteq \text{Sent} \times \text{Sent}$. If φ is a sentence, $\psi \vee \chi$ is a consecutive part of φ, with ψ, χ Sent, and σ is obtained from φ by replacing a consecutive part of φ of this sort by $\chi \vee \psi$, then $(\varphi, \sigma) \in R_1$.

(3) *Double negation.* R_2 is to consist of all pairs of the following sort, where $\psi, \chi, \theta \in \text{Sent}$: $((\psi \vee \chi) \vee \theta, (\psi \vee \neg\neg\chi) \vee \theta)$; $(\chi \vee \theta, \neg\neg\chi \vee \theta)$; $(\psi \vee \chi, \psi \vee \neg\neg\chi)$, $(\chi, \neg\neg\chi)$.

(4) *Conjunction.* R_3 consists of all triples of the following sort, where $\psi, \chi, \theta, \sigma \in \text{Sent}$:

$((\psi \vee \neg\chi) \vee \theta, (\psi \vee \neg\sigma) \vee \theta, (\psi \vee \neg(\chi \vee \sigma)) \vee \theta)$;
$(\neg\chi \vee \theta, \neg\sigma \vee \theta, \neg(\chi \vee \sigma) \vee \theta)$;
$(\psi \vee \neg\chi, \psi \vee \neg\sigma, \psi \vee \neg(\chi \vee \sigma))$; $(\neg\chi, \neg\sigma, \neg(\chi \vee \sigma))$.

A set $\Gamma \subseteq \text{Sent}$ is closed under $R_0, \ldots, R_3$ provided that $\varphi \in \Gamma$ whenever ψ and $\chi \in R_0$ and $(\psi, \varphi) \in R_0$, $(\psi, \varphi) \in R_1$, $(\psi, \varphi) \in R_2$, or $(\psi, \chi, \varphi) \in R_3$. We call a sentence φ *basic* if it has one of the following forms:

$$\begin{aligned} \langle s\rangle \vee \neg\langle s\rangle &\quad \text{with } s \in P,\\ \psi \vee \chi &\quad \text{with } \chi \text{ basic},\\ \chi \vee \psi &\quad \text{with } \chi \text{ basic},\\ (\theta \vee \chi) \vee \psi &\quad \text{with } \chi \text{ basic}, \end{aligned}$$

(formulate this as a recursive definition). Thm is the intersection of all classes of sentences including all the basic sentences and closed under R_0, R_1, R_2, and R_3. We write $\vdash\varphi$ for $\varphi \in \text{Thm}$. Prove that $\vdash\varphi$ iff $\models\varphi$. *Hint:* the following facts lead to an easy solution:

(1) If $\models[\psi \vee \neg(\chi \vee \theta)] \vee \sigma$, then $\models(\psi \vee \neg\chi) \vee \sigma$ and $\models(\psi \vee \neg\theta) \vee \sigma$.
(2) If $\models(\psi \vee \neg\neg\chi) \vee \theta$, then $\models(\psi \vee \chi) \vee \theta$.
(3) If $\models\varphi$ and φ cannot be obtained from some other sentence by f_2 or f_3 (after applying association and commutation), then φ is basic.

8.55. Show that a logic with negation alone is not functionally complete. Similarly for implication, disjunction, conjunction.

8.56 (*Three-valued logic*). It is natural to consider sentential logics in which three or more truth values are allowed. We describe here a version of three-valued logic. The truth values are 0 (falsity), 1 (truth), and 2 (indeterminate). The truth tables for $\neg$ and $\rightarrow$ are as follows:

φ	$\neg\varphi$
0	1
1	0
2	2

φ	ψ	$\varphi \rightarrow \psi$
1	1	1
1	0	0
1	2	2
0	1	1
0	0	1
0	2	1
2	1	1
2	0	2
2	2	1

(a) Show that the following sentence is not always true in three-valued logic, but is never false:

$$[\varphi \rightarrow (\psi \rightarrow \chi)] \rightarrow [(\varphi \rightarrow \psi) \rightarrow (\varphi \rightarrow \chi)].$$

Now we shall describe an axiom system for three-valued logic. The axioms are all sentences of the following forms:

(A1) $\varphi \rightarrow (\psi \rightarrow \varphi)$
(A2) $(\varphi \rightarrow \psi) \rightarrow [(\psi \rightarrow \chi) \rightarrow (\varphi \rightarrow \chi)]$
(A3) $(\neg\varphi \rightarrow \neg\psi) \rightarrow (\psi \rightarrow \varphi)$
(A4) $((\varphi \rightarrow \neg\varphi) \rightarrow \varphi) \rightarrow \varphi$
(A5) $\{\varphi \rightarrow [\varphi \rightarrow (\psi \rightarrow \chi)]\} \rightarrow \{[\varphi \rightarrow (\varphi \rightarrow \psi)] \rightarrow [\varphi \rightarrow (\varphi \rightarrow \chi)]\}$
(A6) $[\varphi \rightarrow (\varphi \rightarrow \neg\varphi)] \rightarrow \{[\neg\psi \rightarrow (\neg\psi \rightarrow \psi)] \rightarrow (\varphi \rightarrow \psi)\}$
(A7) $\varphi \rightarrow [(\varphi \rightarrow \psi) \rightarrow \psi]$
(A8) $\{[\varphi \rightarrow (\varphi \rightarrow \neg\varphi)] \rightarrow \neg[\varphi \rightarrow (\varphi \rightarrow \neg\varphi)]\} \rightarrow \varphi$

Let detachment be the only rule of inference. This gives a new notion $\vdash$. Let $\models$ be the semantic notion for three-valued logic.

(b) Show that $\Gamma \vdash \varphi$ implies $\Gamma \models \varphi$.

(c) For further purposes, prove the following:

(1) $\vdash \neg\varphi \rightarrow (\varphi \rightarrow \psi)$;
(2) $\vdash \psi \rightarrow [\varphi \rightarrow (\varphi \rightarrow \psi)]$;
(3) $\vdash \neg\neg\varphi \rightarrow \varphi$;
(4) $\vdash \varphi \rightarrow \neg\neg\varphi$;
(5) $\vdash \neg(\varphi \rightarrow \psi) \rightarrow \neg\psi$;
(6) $\Delta \vdash \varphi \rightarrow \psi$ and $\Delta \vdash \neg\psi$ imply $\Delta \vdash \neg\varphi$.

(d) Prove analogs of the completeness theorems 8.28 and 8.29. *Hint:* to prove the analog of 8.28, begin as in the proof of 8.28, obtaining the set Δ. Show:

$$\begin{aligned} \Delta \cup \{\varphi\} \vdash \psi &\quad \text{implies } \Delta \vdash \varphi \rightarrow (\varphi \rightarrow \psi), \\ \varphi \notin \Delta &\quad \text{implies } \Delta \vdash \varphi \rightarrow (\varphi \rightarrow \neg\varphi), \\ \neg\varphi \notin \Delta &\quad \text{implies } \Delta \vdash \neg\varphi \rightarrow (\neg\varphi \rightarrow \varphi), \\ \Delta \vdash \varphi &\quad \text{implies } \varphi \in \Delta. \end{aligned}$$

Now define for each basic sentential symbol s

$$\begin{aligned} fs &= 1 && \text{if } \langle s\rangle \in \Delta, \\ fs &= 0 && \text{if } \neg\langle s\rangle \in \Delta, \\ fs &= 2 && \text{otherwise.} \end{aligned}$$

Prove by induction on φ that

$$\begin{aligned} f^+\varphi &= 1 && \text{implies } \varphi \in \Delta \\ f^+\varphi &= 0 && \text{implies } \neg\varphi \in \Delta, \\ f^+\varphi &= 2 && \text{implies } \varphi \notin \Delta \text{ and } \neg\varphi \notin \Delta. \end{aligned}$$

To prove the analog of 8.29, note that if $\Gamma \vdash \varphi$ then $\Gamma \cup \{\varphi \rightarrow (\varphi \rightarrow \neg\varphi)\}$ is consistent.

(e) Define the appropriate notion of functional completeness for three-valued logic. Show that our connectives are not functionally complete.

8.57 (*Intuitionistic logic*). We outline here a short introduction to intuitionistic sentential logic (see the introduction to the book). According to intuitionism, every mathematical assertion, to be called mathematical and not theological, should be expressible in the form "I have effected a certain construction A in my mind." If φ and ψ are assertions, then:

(1) We can assert $\varphi \wedge \psi$ iff both φ and ψ can be asserted.
(2) We can assert $\varphi \vee \psi$ iff at least one of the sentences φ and ψ can be asserted.
(3) We can assert $\neg\varphi$ iff we possess a construction which leads from the supposition of φ to a contradiction.
(4) We can assert $\varphi \rightarrow \psi$ iff we possess a construction which, when joined to the construction mentioned in φ (which is not, however assumed to exist), leads to a construction of the sort mentioned in ψ).

Examples:

(a) $\varphi \vee \neg\varphi$ cannot be asserted for every assertion φ, for this would imply the existence of a general method such that, given a construction A, the method tells us how to carry out A, or else how to infer a contradiction from the assumption that A has been carried out. For example, let φ be: "I have constructed integers x, y, z, m such that $m > 2$, and $x^m + y^m = z^m$." I cannot assert φ. But also I cannot obtain a contradiction from the assumption that the construction mentioned in φ has been carried out, i.e., I cannot assert $\neg\varphi$. Thus I cannot assert $\varphi \vee \neg\varphi$.

(b) $\neg\neg\varphi \rightarrow \varphi$ cannot be asserted for every assertion φ. Indeed, take the following example of Brouwer. Write the decimal expansion of π:

$$\begin{aligned} \pi &= 3.14159\ldots \\ \rho &= .33333\ldots \end{aligned}$$

Underneath write $.3333\ldots$, breaking this off as soon as the sequence of digits 0123456789 has occurred in the expansion of π (classically it is unknown whether this will ever occur). Let φ be the assertion: ρ is rational. Classically φ is obviously true. Intuitionistically φ is to be interpreted as "I possess a construction of integers m, n such that $\rho = m/n$." Suppose $\neg\varphi$; then ρ cannot have the form $.33\ldots300\ldots$, so $\rho = \frac{1}{3}$, i.e., φ holds, contradiction. Assuming $\neg\varphi$ we get a contradiction. Hence (intuitionistically)

$\neg\neg\varphi$. But we cannot assert φ intuitionistically, and hence we cannot assert $\neg\neg\varphi \rightarrow \varphi$.

Now we seek to axiomatize those assertions which are intuitionistically valid. We take negation, conjunction and implication as primitive. As axioms we take all sentences of the following kinds:

(I1) $\varphi \rightarrow (\psi \rightarrow \varphi)$;
(I2) $[\varphi \rightarrow (\psi \rightarrow \chi)] \rightarrow [(\varphi \rightarrow \psi) \rightarrow (\varphi \rightarrow \chi)]$;
(I3) $(\varphi \rightarrow \neg\varphi) \rightarrow \neg\varphi$;
(I4) $\neg\varphi \rightarrow (\varphi \rightarrow \psi)$;
(I5) $\varphi \rightarrow (\psi \rightarrow \varphi \wedge \psi)$;
(I6) $\varphi \wedge \psi \rightarrow \varphi$;
(I7) $\varphi \wedge \psi \rightarrow \psi$.

Detachment is the only rule of inference. "$\vdash_i \varphi$" is defined analogously to the classical case.

(α) Check that the axioms are intuitionistically valid in the intuitive sense.

(β) Check that the deduction theorem goes through.

Our notion of intuitionistic validity is informal, and hence we cannot hope to prove a completeness theorem until we find a mathematical version of this validity. Without trying to justify it, we introduce the following mathematical version of intuitionistic validity:

Given a sentential language $\mathscr{P}$, an *intuitionistic model for* $\mathscr{P}$ is a function f mapping P into the collection of open sets in some topological space X. Given such a function f, we define

$$\begin{aligned} f^+\langle s\rangle &= fs \text{ for all } s \in P; \\ f^+\neg\varphi &= \text{interior of } (X \sim f^+\varphi); \\ f^+(\varphi \wedge \psi) &= f^+\varphi \cap f^+\psi; \\ f^+(\varphi \rightarrow \psi) &= \text{interior of } [(X \sim f^+\varphi) \cup f^+\psi]. \end{aligned}$$

This gives rise to a notion $\Gamma \models_i \varphi$.

(γ) Prove that $\Gamma \vdash_i \varphi$ implies $\Gamma \models_i \varphi$.

(δ) Show that if Γ is consistent, then there is a model f with associated topological space X such that for any sentence φ, $\Gamma \vdash \varphi$ iff $f^+\varphi = X$. *Hint:* let $X = \{\Delta : \Gamma \subseteq \Delta, \Delta \text{ consistent}\}$. For each sentence φ, let

$$U_\varphi = \{\Delta \in X : \Delta \vdash \varphi\}.$$

Show that the U_φ's form a base for a topology on X. Let $fs = U_{\langle s\rangle}$ for each $s \in P$, and show that $f^+\varphi = U_\varphi$ for each sentence φ.

(ε) $\Gamma \vdash_i \varphi$ iff $\Gamma \models_i \varphi$.

(ζ) An interesting connection between intuitionistic and ordinary logic is given by the following result: $\Gamma \vdash \varphi$ iff $\{\neg\neg\psi : \psi \in \Gamma\} \vdash_i \neg\neg\varphi$; in particular, $\vdash\varphi$ iff $\vdash_i \neg\neg\varphi$. *Hint:* pick a suitable set of axioms Δ for classical logic and show that

$$\Gamma \cup \Delta \subseteq \{\chi : \{\neg\neg\psi : \psi \in \Gamma\} \vdash_i \neg\neg\chi\},$$

and that the latter set is closed under detachment. Then use (ε).

8.58 (*Modal logic*). We now briefly discuss the addition of new connectives to sentential logic which express such intuitive notions as *possibility* or *necessity*. A *modal language* is a quadruple $\mathscr{P} = (n, c, d, P)$ such that $|\{n, c, d\}| = 3$, $P \neq 0$, and $\{n, c, d\} \cap P = 0$. The symbol d is the *necessity* symbol. For any expression φ we write $\square\varphi = \langle d\rangle\varphi$. Other notation is carried over from the sentential case. If φ is a sentence, then $\square\varphi$ expresses, intuitively speaking, that φ is not only true, but is true in all possible worlds. Thus if φ is the sentence "It is sunny now," then φ may be true at a given place and time, but it is not necessarily true. On the other hand, if φ is "A rose is a rose," then φ is necessarily true. The notion of necessity is, however, somewhat unclear and indefinite in its intuitive meaning, much more indefinite than the notions in three-valued logic or intuitionism. For example, is the sentence $\square\varphi \rightarrow \square\square\varphi$ to be accepted, or not? Debate over these matters has led to a large number of systems of modal logic, of which we present only one.

Our axioms are as follows:

(M1) $\varphi \rightarrow (\psi \rightarrow \varphi)$,
(M2) $[\varphi \rightarrow (\psi \rightarrow \chi)] \rightarrow [(\varphi \rightarrow \psi) \rightarrow (\varphi \rightarrow \chi)]$,
(M3) $(\neg\varphi \rightarrow \neg\psi) \rightarrow (\psi \rightarrow \varphi)$,
(M4) $\square\varphi \rightarrow \varphi$,
(M5) $\square\varphi \rightarrow \square\square\varphi$,
(M6) $\square(\varphi \rightarrow \psi) \rightarrow (\square\varphi \rightarrow \square\psi)$

For rules of inference we allow detachment, and the inference of $\square\varphi$ from φ. This gives a notion $\Gamma \vdash_{\mathrm{m}} \varphi$. If X is a topological space and $f: P \rightarrow \mathbf{S}X$, we define $f^+ : \mathrm{Sent}_{\mathscr{P}} \rightarrow \mathbf{S}X$ as follows:

$$
\begin{aligned}
f^+\langle s\rangle &= fs \\
f^+(\neg\varphi) &= X \sim f^+\varphi \\
f^+(\varphi \rightarrow \psi) &= (X \sim f^+\varphi) \cup f^+\varphi \\
f^+\square\varphi &= \text{interior of } f^+\varphi.
\end{aligned}
$$

Of course this gives a notion of $\Gamma \vDash_{\mathrm{m}} \varphi$.

(α) Show that $\Gamma \vdash_{\mathrm{m}} \varphi$ implies $\Gamma \vDash_{\mathrm{m}} \varphi$.

(β) If Γ is consistent, then Γ has a model f such that for every sentence φ, $\Gamma \vdash_{\mathrm{m}} \varphi$ iff $f^+\varphi = X$. *Hint:* let $X = \{\Delta : \Delta$ is maximal consistent in ordinary sentential logic and for every sentence φ, $\Gamma \vdash_{\mathrm{m}} \varphi$ implies $\varphi \in \Delta\}$. For each sentence φ, let

$$U_\varphi = \{\Delta \in X : \Delta \vdash \square\varphi\}.$$

Show that these sets form a base for a topology on X. Let $fs = \{\Delta \in X : \Delta \vdash \langle s\rangle\}$. Show that $f^+\varphi = \{\Delta \in X : \Delta \vdash \varphi\}$ for every sentence φ.

(γ) $\Gamma \vdash_{\mathrm{m}} \varphi$ iff $\Gamma \vDash_{\mathrm{m}} \varphi$, and in particular $\vdash_{\mathrm{m}} \varphi$ iff $\vDash_{\mathrm{m}} \varphi$.

Now for any sentence φ not including $\square$ we associate a new sentence φ^{m}:

$$
\begin{aligned}
\langle s\rangle^{\mathrm{m}} &= \square\langle s\rangle, \\
(\neg\varphi)^{\mathrm{m}} &= \square \neg \varphi^{\mathrm{m}}, \\
(\varphi \rightarrow \psi)^{\mathrm{m}} &= \square(\varphi^{\mathrm{m}} \rightarrow \psi^{\mathrm{m}}).
\end{aligned}
$$

(δ) $\vdash_{\mathrm{i}} \varphi$ iff $\vdash_{\mathrm{m}} \varphi^{\mathrm{m}}$. *Hint:* Assume that $\vdash_{\mathrm{i}} \varphi$, and let $f: P \rightarrow \mathbf{S}X$, X a topological space. For each $s \in P$ let $gs =$ interior of fs. Show that $g^{+\mathrm{i}}\psi = f^{+\mathrm{m}}\psi^{\mathrm{m}}$

for any sentence ψ (g^{+i} is the extension of g defined in Exercise 8.57, while f^{+m} is the extension of f defined in this exercise). This easily leads to $\models_m \varphi^m$, hence $\models \varphi^m$. A similar argument works for the converse.

8.59. We have considered several different systems of sentential logic: two-valued, three-valued, and intuitionistic logic. How many different systems are there altogether? Here is a "best possible" answer: Let $\mathscr{P} = (n, c, \{s\})$ be a sentential language. Given any set $\Gamma \subseteq \mathrm{Sent}_{\mathscr{P}}$, let S_Γ be the intersection of all classes including Γ and closed under detachment. Let $\mathscr{S} = \{S_\Gamma : \Gamma \subseteq \mathrm{Sent}_{\mathscr{P}}\}$. Show that $|\mathscr{S}| = \exp \aleph_0$. *Hint:* define φ_n recursively:

$$\varphi_0 = \langle s \rangle \rightarrow \langle s \rangle;$$
$$\varphi_{n+1} = \varphi_n \rightarrow \langle s \rangle.$$

For each subset M of ω let $\Gamma M = \{\varphi_{3n} : n \in M\} \cup \{\varphi_{3n+1} : n \notin M\}$. Show that $M \neq N \Rightarrow S_{\Gamma M} \neq S_{\Gamma N}$.

Boolean Algebra 9*

In this chapter we give a brief introduction to the theory of Boolean algebras. As we shall see, these are algebraic structures which stand in an intimate relationship to sentential logics. They will also form a source for some of the applications of logic which we shall give later. Boolean algebras are actually most simply motivated by a reference to elementary set theory, however, as follows:

Definition 9.1. A *field of sets* is a set $\mathscr{A}$ such that $\bigcup \mathscr{A} \in \mathscr{A}$, $\bigcup \mathscr{A} \sim X \in \mathscr{A}$ whenever $X \in \mathscr{A}$, and $X \cup Y \in \mathscr{A}$ whenever $X, Y \in \mathscr{A}$. We also say that $\mathscr{A}$ is a *field of subsets of* $\bigcup \mathscr{A}$.

Proposition 9.2. *Let $\mathscr{A}$ be a field of sets. Then*

(*i*) $0 \in \mathscr{A}$;
(*ii*) *if $X, Y \in \mathscr{A}$ then $X \cap Y \in \mathscr{A}$.*

We now want to consider an abstraction from the concrete notion of a field of sets. We consider algebraic structures $\langle A, +, \cdot, -, 0, 1\rangle$ where A is intended to correspond to a field of sets $\mathscr{A}$, $+$ to $\cup$, $\cdot$ to $\cap$, $-$ to $\sim$ (relative to $\bigcup \mathscr{A}$), 0 to the empty set 0, and 1 to $\bigcup \mathscr{A}$. We write down axioms on $+, \cdot, -, 0, 1$ intended to axiomatize the concrete notions fully; later we show that in fact our axiomatization captures all the true identities among the concrete notions. Note that we treat $+, \cdot, -, 0, 1$ as variables ranging over the corresponding notions in all Boolean algebras. Of course 0 and 1 are also used in an entirely different sense for the integers 0 and 1.

Definition 9.3. A *Boolean algebra* (BA) is a system $\mathfrak{A} = \langle A, +, \cdot, -, 0, 1\rangle$ such that $+$ and $\cdot$ are binary operations on A, $-$ is a unary operation on A, $0, 1 \in A$, and the following conditions hold for all $x, y, z \in A$:

(*i*) $x + y = y + x$ and $x \cdot y = y \cdot x$;
(*ii*) $x + (y + z) = (x + y) + z$ and $x \cdot (y \cdot z) = (x \cdot y) \cdot z$;
(*iii*) $x \cdot y + y = y$ and $(x + y) \cdot y = y$;
(*iv*) $x \cdot (y + z) = x \cdot y + x \cdot z$ and $x + y \cdot z = (x + y) \cdot (x + z)$;
(*v*) $x \cdot -x = 0$ and $x + -x = 1$.

We define $x \leq y$ iff $x + y = y$. A is called the *universe*, or *underlying set*, of $\mathfrak{A}$.

By routine checking we have

Corollary 9.4. *If $\mathscr{A}$ is a field of sets, then $\langle \mathscr{A}, \cup, \cap, \sim, 0, \bigcup \mathscr{A} \rangle$ is a* BA, *called a* Boolean set algebra *of subsets of $\bigcup \mathscr{A}$.*

Throughout this chapter, unless otherwise indicated, we deal with an arbitrary BA $\mathfrak{A}$ as above, and with arbitrary elements x, y, z, etc. of A. Note the following proposition, which is obvious from the form of the axioms in 9.3:

Proposition 9.5. *If $\mathfrak{A} = \langle A, +, \cdot, -, 0, 1 \rangle$ is a* BA, *then so is $\langle A, \cdot, +, -, 1, 0 \rangle$.*

A duality principle follows from 9.5: if we have proved a statement about all BA's, then a second dual statement, obtained from the first one by interchanging $+$ and $\cdot$ and also 0 and 1, is true. We shall not bother to make this principle more precise, since our applications of it can be justified by 9.5 directly.

The following proposition summarizes the elementary arithmetic of BA's. It will be used later without specific reference to it.

Proposition 9.6

(*i*) $x + x = x$;
(*ii*) $x \cdot x = x$;
(*iii*) $x \leq x$;
(*iv*) *if* $x \leq y$ *and* $y \leq x$, *then* $x = y$;
(*v*) *if* $x \leq y$ *and* $y \leq z$, *then* $x \leq z$;
(*vi*) $x \leq y$ *iff* $x \cdot y = x$;
(*vii*) $x \cdot 0 = 0$;
(*viii*) $x \cdot 1 = x$;
(*ix*) $x + 0 = x$;
(*x*) $x + 1 = 1$;
(*xi*) $x \leq y$ *iff* $x \cdot -y = 0$;
(*xii*) $x = -y$ *iff* $x + y = 1$ *and* $x \cdot y = 0$;
(*xiii*) $--x = x$;
(*xiv*) $-(x + y) = -x \cdot -y$;
(*xv*) $-(x \cdot y) = -x + -y$;
(*xvi*) $0 \leq x \leq 1$;

(*xvii*) *if* $x \leq z$ *and* $y \leq z$, *then* $x + y \leq z$;
(*xviii*) *if* $x \leq y$ *and* $x \leq z$, *then* $x \leq y \cdot z$.

PROOF. For (*i*), we calculate:

$$\begin{aligned} x &= x \cdot x + x && 9.3(iii) \\ &= x + x \cdot x && 9.3(i) \\ &= (x + x) \cdot (x + x) && 9.3(iv) \\ &= x \cdot (x + x) + x \cdot (x + x) && 9.3(iv) \\ &= (x + x) \cdot x + (x + x) \cdot x && 9.3(i) \\ &= x + x && 9.3(iii) \end{aligned}$$

By duality we have (*ii*), and (*iii*) is immediate from (*i*) and the definition of $\leq$. To prove (*iv*), assume that $x \leq y$ and $y \leq x$. Then, using the definition of $\leq$ and 9.3(*i*), $y = x + y = y + x = x$. Condition (*v*) is proved similarly: assume that $x \leq y$ and $y \leq z$. Thus $x + y = y$ and $y + z = z$, so

$$\begin{aligned} x + z &= x + (y + z) = (x + y) + z && 9.3(ii) \\ &= y + z = z. \end{aligned}$$

For (*vi*), assume first that $x \leq y$. Thus $x + y = y$, so $x \cdot y = x \cdot (x + y) = x$ by 9.3(*iii*). Thus $x + y = y$ implies $x \cdot y = x$. By duality the converse holds, and this establishes (*vi*). Condition (*vii*) is proved as follows:

$$\begin{aligned} x \cdot 0 &= x \cdot (x \cdot -x) && 9.3(v) \\ &= (x \cdot x) \cdot -x && 9.3(ii) \\ &= x \cdot -x = 0 && (ii), 9.3(v) \end{aligned}$$

Condition (*viii*) is easy, using 9.3(*v*), 9.3(*i*), 9.3(*iii*):

$$x \cdot 1 = x \cdot (x + -x) = (-x + x) \cdot x = x.$$

By duality, we obtain (*ix*) and (*x*) from (*viii*) and (*vii*) respectively. To prove (*xi*), first assume $x \leq y$. Then by (*vi*), $x \cdot -y = (x \cdot y) \cdot -y = x \cdot (y \cdot -y) = x \cdot 0 = 0$ (using also 9.3(*ii*), 9.3(*v*), and (*vii*)). Now assume $x \cdot -y = 0$. Then

$$\begin{aligned} x \cdot y &= x \cdot y + 0 = x \cdot y + x \cdot -y && (ix) \\ &= x \cdot (y + -y) = x \cdot 1 = x && 9.3(iv), 9.3(v), (viii) \end{aligned}$$

Hence $x \leq y$ by (*vi*). The direction $\Rightarrow$ in (*xii*) is clear from 9.3(*i*) and 9.3(*v*). Now assume $x + y = 1$ and $x \cdot y = 0$. Then

$$\begin{aligned} x &= x \cdot 1 = x \cdot (y + -y) = x \cdot y + x \cdot -y && (viii), 9.3(v), 9.3(iv) \\ &= 0 + x \cdot -y = x \cdot -y + 0 = x \cdot -y + y \cdot -y && 9.3(i), 9.3(v) \\ &= (x + y) \cdot -y = 1 \cdot -y = -y && 9.3(iv), 9.3(i), (viii) \end{aligned}$$

Now (*xiii*) is immediate from (*xii*). To prove (*xiv*) we also use (*xii*):

$$\begin{aligned} x + y + -x \cdot -y &= x + (x + -x) \cdot y + -x \cdot -y \\ &= x + x \cdot y + -x \cdot y + -x \cdot -y \\ &= x + -x \cdot (y + -y) = x + -x = 1; \\ (x + y) \cdot -x \cdot -y &= x \cdot -x \cdot -y + y \cdot -x \cdot -y = 0. \end{aligned}$$

Thus $-(x + y) = -x \cdot -y$ by (*xii*). By duality we have (*xv*). (*xvi*) is immediate from (*vii*), (*viii*), (*vi*). To prove (*xvii*), assume $x \leq z$ and $y \leq z$. Thus $x + z = z$ and $y + z = z$. Hence $x + y + z = x + z = z$. (*xviii*) follows by duality (note that the dual of $x \leq y$ is $y \leq x$). □

Now we shall consider several common algebraic notions in their special form for Boolean algebras. Many of these notions and the theorems we prove about them apply to much more general situations; see for example Part IV.

Definition 9.7. Let $\mathfrak{A} = \langle A, +, \cdot, -, 0, 1 \rangle$ and $\mathfrak{B} = \langle B, +', \cdot', -', 0', 1' \rangle$ be BA's. We say that $\mathfrak{A}$ is a *subalgebra* of $\mathfrak{B}$ if $A \subseteq B$, $0 = 0'$, $1 = 1'$, and for all $x, y \in A$, $x + y = x +' y$, $x \cdot y = x \cdot' y$, and $-x = -'x$. For any $X \subseteq B$, X is a *subuniverse* of $\mathfrak{B}$ if $0', 1' \in X$ and X is closed under $+', \cdot', -'$.

Thus if $\mathfrak{A}$ is a subalgebra of $\mathfrak{B}$, then A is a subuniverse of $\mathfrak{B}$. Any subuniverse of $\mathfrak{B}$ is the underlying set of some subalgebra of $\mathfrak{B}$. A subuniverse is just a subalgebra with the structure ignored. Some useful equivalent definitions of subuniverse are given in

Proposition 9.8. *Let $\mathfrak{A}$ be a* BA, $X \subseteq A$. *Then the following conditions are equivalent:*

(*i*) *X is a subuniverse of $\mathfrak{A}$;*
(*ii*) *$X \neq 0$, and X is closed under $+$ and $-$;*
(*iii*) *$X \neq 0$, and X is closed under $\cdot$ and $-$.*

Proof. Obviously (*i*) $\Rightarrow$ (*ii*). Now assume (*ii*). Then X is closed under $\cdot$, since $x \cdot y = -(-x + -y)$; hence (*iii*) holds. Assume (*iii*). Then X is closed under $+$ by the dual of the argument just given. Choose $x_0 \in X$. Then $0 = x_0 \cdot -x_0 \in X$, and $1 = x_0 + -x_0 \in X$. Thus X is a subuniverse of $\mathfrak{A}$. □

Proposition 9.9. *If $\mathscr{A}$ is a nonempty collection of subuniverses of a* BA $\mathfrak{A}$, *then $\bigcap \mathscr{A}$ is a subuniverse of $\mathfrak{A}$.*

The proof is trivial. Proposition 9.9 justifies the following definition (observe that A is always a subuniverse of $\mathfrak{A}$):

Definition 9.10. If $X \subseteq A$, $\mathfrak{A}$ a BA, then the set $\bigcap \{Y : X \subseteq Y, Y$ a subuniverse of $\mathfrak{A}\}$ is called the *subuniverse of $\mathfrak{A}$ generated by X.*

The following theorem expresses this notion in a simpler fashion. Note the similarity to 8.38.

Theorem 9.11. *If X is a subset of A, then the subuniverse of $\mathfrak{A}$ generated by X consists of* 0, 1, *and all elements of A of the form*

$$\sum_{i<m} \prod_{j<n_i} y_{ij}, \tag{1}$$

where for each i, j, either $y_{ij} \in X$ or $-y_{ij} \in X$.

PROOF. Clearly each such element is in the subuniverse of $\mathfrak{A}$ generated by X. Thus it suffices to show that the set S of all such elements is a subuniverse of $\mathfrak{A}$ containing X. Obviously $X \subseteq S$ and S is closed under $+$. By 9.8(i) it remains only to show that S is closed under $-$. Let z be any element of S, as in (1). Then

$$-z = \prod_{i<m} \sum_{j<ni} -y_{ij} = \sum_{f\in F} \prod_{i<m} -y_{i,fi},$$

where $F = \mathrm{P}_{i<m}\, n_i$. Here we have used a generalization of the distributive law 9.3(iv) which is easy to establish by induction on m. Thus $-z \in S$, as desired. □

The next general algebraic notions we shall consider for Boolean algebras are homomorphisms and isomorphisms:

Definition 9.12. Let $\mathfrak{A} = \langle A, +, \cdot, -, 0, 1\rangle$ and $\mathfrak{A}' = \langle A', +', \cdot', -', 0', 1'\rangle$ be BA's. A function f mapping A into A' is a *homomorphism* of $\mathfrak{A}$ into $\mathfrak{A}'$ provided that $f0 = 0'$, $f1 = 1'$, and for all $x, y \in A$, $f(x + y) = fx +' fy$, $f(x\cdot y) = fx \cdot' fy$, and $f(-x) = -'fx$. If f maps onto A' we say that f is a homomorphism of $\mathfrak{A}$ *onto* $\mathfrak{A}'$. An *isomorphism into* is a one-one homomorphism. The BA's $\mathfrak{A}$ and $\mathfrak{A}'$ are *isomorphic* provided there is an isomorphism from $\mathfrak{A}$ onto $\mathfrak{A}'$. A homomorphism of $\mathfrak{A}$ into itself is called an *endomorphism* of $\mathfrak{A}$.

As is usual in algebra, one is only interested in the properties of BA's up to isomorphism.

Proposition 9.13. *Let $\mathfrak{A}$ and $\mathfrak{A}'$ be* BA's, *as in* 9.12, *and let f be a map of A into A'. Then the following conditions are equivalent;*

(i) *f is a homomorphism;*
(ii) *$f(-x) = -'fx$ and $f(x + y) = fx +' fy$ for all $x, y \in A$;*
(iii) *$f(-x) = -'fx$ and $f(x\cdot y) = fx \cdot' fy$ for all $x, y \in A$;*
(iv) *$f0 = 0'$, $f1 = 1'$, $f(x + y) = fx +' fy$ for all $x, y \in A$, and $fx\cdot' fy = 0'$ whenever $x\cdot y = 0$.*

PROOF. Obviously (i) ⇒ (ii). Now assume (ii), and let $x, y \in A$. Then

$$f(x\cdot y) = f(-(-x + -y)) = -'(-'fx +' -'fy) = fx\cdot' fy,$$

so (iii) holds. Analogously, (iii) ⇒ (ii). If (ii) holds, then f preserves $\cdot$, as already proved. Furthermore, $f0 = f(0\cdot -0) = f0 \cdot' -'f0 = 0'$ and $f1 = f(1 + -1) = f1 +' -'f1 = 1'$. Hence ($i$) holds. Thus ($i$)–($iii$) are equivalent. Obviously (i) ⇒ (iv). Now assume (iv); to show (ii) it suffices to show that f preserves $-$. We have for any $x \in A$

$$1 = f(x + -x) = fx +' f(-x);$$
$$fx\cdot' f(-x) = 0' \qquad \text{by hypothesis of } (iv).$$

Thus $f(-x) = -'fx$, as desired. □

Proposition 9.14. *Let* $\mathfrak{A}$ *and* $\mathfrak{A}'$ *be* BA's, *as in* 9.12, *and let* f *be a map of* A *into* A'. *Then the following conditions are equivalent:*

(*i*) f *is an isomorphism of* $\mathfrak{A}$ *onto* $\mathfrak{A}'$;
(*ii*) f *maps* A *onto* A', *and for all* $x, y \in A$, $x \leq y$ *iff* $fx \leq' fy$.

Proof. Obviously (*i*) $\Rightarrow$ (*ii*). Now assume (*ii*). Now if $fx = fy$, then $fx \leq' fy$ and $fy \leq' fx$, so $x \leq y$ and $y \leq x$, hence $x = y$. Thus f is one-one. Now we shall apply 9.13(*iv*). For any $x, y \in A$ we have $fx \leq' f(x + y)$ and $fy \leq' f(x + y)$, since $x \leq x + y$ and $y \leq x + y$. Thus $fx +' fy \leq' f(x + y)$. Let $fx +' fy = fz$ (since f is onto A'). Then $fx \leq' fz$, so $x \leq z$. Similarly $y \leq z$, so $x + y \leq z$. But $fz \leq' f(x + y)$ implies $z \leq x + y$. Thus $z = x + y$, and hence $fx +' fy = fz = f(x + y)$. Next, choose x so that $fx = 0'$. Then $0 \leq x$, so $f0 \leq' fx = 0'$ and hence $f0 = 0'$. Similarly $f1 = 1'$. Now assume $x \cdot y = 0$. Write $fx \cdot' fy = fz$. Then $fz \leq' fx$ and $fz \leq' fy$, so $z \leq x$ and $z \leq y$, hence $z \leq x \cdot y = 0$, and $z = 0$. Thus $fx \cdot' fy = f0 = 0'$, as desired. □

Now we introduce the important notions of ideals and filters, which as we shall see, enable one to study homomorphisms within a given BA rather than between two BA's.

Definition 9.15. A subset I of $\mathfrak{A}$ is an *ideal* of $\mathfrak{A}$ if $I \neq 0$, $a + b \in I$ whenever $a, b \in I$, and $a \in I$ whenever $a \leq b \in I$. A subset F of $\mathfrak{A}$ is a *filter* of $\mathfrak{A}$ iff $F \neq 0$, $a \cdot b \in F$ whenever $a, b \in F$, and $a \in F$ whenever $a \geq b \in F$.

Note that 0 is a member of any ideal I. In fact, since $I \neq 0$ choose $b \in I$. Then $0 \leq b \in I$, so $0 \in I$. Similarly, 1 is a member of any filter. The notions of ideal and filter are dual. Later we give a more exact description of the relationship of the notions.

Proposition 9.16. *Let* I *be an ideal of* $\mathfrak{A}$. *Let* $R = \{(x, y) : x \cdot -y + y \cdot -x \in I\}$. *Then* R *is an equivalence relation on* A. *Moreover,*

(*i*) *if* xRy, *then* $-xR-y$;
(*ii*) *if* xRy *and* $x'Ry'$, *then* $(x + x')R(y + y')$;
(*iii*) *if* xRy *and* $x'Ry'$, *then* $(x \cdot x')R(y \cdot y')$.

Proof. R is reflexive on A: if $x \in A$, then $x \cdot -x + -x \cdot x = 0 \in I$. Obviously R is symmetric. R is transitive: assume that $xRyRz$. Thus $x \cdot -y + y \cdot -x \in I$ and $y \cdot -z + z \cdot -y \in I$. Now

$$x \cdot -z = x \cdot (y + -y) \cdot -z = x \cdot y \cdot -z + x \cdot -y \cdot -z.$$

But $x \cdot y \cdot -z \leq y \cdot -z \leq y \cdot -z + z \cdot -y \in I$, so $x \cdot y \cdot -z \in I$. And $x \cdot -y \cdot -z \leq x \cdot -y \leq x \cdot -y + y \cdot -x \in I$, so $x \cdot -y \cdot -z \in I$. Hence $x \cdot -z = x \cdot y \cdot -z + x \cdot -y \cdot -z \in I$. Similarly, $z \cdot -x \in I$, so $x \cdot -z + z \cdot -x \in I$, i.e., xRz. Thus R is an equivalence relation on A.

Now we turn to (*i*). Assume that xRy. Thus $x\cdot -y + y\cdot -x \in I$. Hence

$$-x\cdot --y + -y\cdot --x = -x\cdot y + -y\cdot x \in I,$$

so $-xR-y$. To prove (*ii*), assume that xRy and $x'Ry'$. Thus $x\cdot -y + y\cdot -x \in I$ and $x'\cdot -y' + y'\cdot -x' \in I$. Hence

$$(x + x')\cdot -(y + y') = (x + x')\cdot -y\cdot -y' = x\cdot -y\cdot -y' + x'\cdot -y\cdot -y'$$
$$\leq x\cdot -y + x'\cdot -y' \in I$$

and hence $(x + x')\cdot -(y + y') \in I$. Similarly $(y + y')\cdot -(x + x') \in I$, so $(x + x')R(y + y')$. Finally, (*iii*) clearly follows from (*i*) and (*ii*). □

Proposition 9.16 justifies the following definition.

Definition 9.17. Let I be an ideal of a BA $\mathfrak{A} = \langle A, +, \cdot, -, 0, 1\rangle$. Let A/I be the set of all equivalence classes under the equivalence relation R of 9.16; the equivalence class of an element $a \in I$ is denoted by $[a]_I$, or simply $[a]$. Operations $+'$, $\cdot'$, and $-'$ on A/I are uniquely determined by the stipulation that, for all $a, b \in A$,

$$[a] +' [b] = [a + b];$$
$$[a]\cdot'[b] = [a\cdot b];$$
$$-'[a] = [-a].$$

We let $\mathfrak{A}/I = \langle A/I, +', \cdot', -', [0], [1]\rangle$. Let I^* be the function mapping A onto A/I defined by $I^*a = [a]_I$ for all $a \in A$. $\mathfrak{A}/I$ is called a *quotient algebra* of $\mathfrak{A}$.

The following proposition is easily established:

Proposition 9.18. *If I is an ideal of a* BA $\mathfrak{A}$, *then $\mathfrak{A}/I$ is a* BA, *and I^* is a homomorphism from $\mathfrak{A}$ onto $\mathfrak{A}/I$.*

The most important facts about the relationship between ideals and homomorphisms are expressed in the following theorem:

Theorem 9.19 (The homomorphism theorem). *Let f be a homomorphism from a* BA $\mathfrak{A}$ *onto a* BA $\mathfrak{B}$. *Let $I = \{x : x \in A$ and $fx = 0\}$. Then I is an ideal of $\mathfrak{A}$, and $\mathfrak{A}/I$ is isomorphic to $\mathfrak{B}$. I is called the* kernel *of f. Furthermore, f is one-one iff $I = \{0\}$.*

Proof. Since $f0 = 0$, we have $0 \in I$ and hence $I \neq 0$. Suppose $x, y \in I$. Then $f(x + y) = fx + fy = 0 + 0 = 0$. Thus $x + y \in I$. Suppose $x \leq y \in I$. Then $fx \leq fy = 0$, so $fx = 0$ and hence $x \in I$. Thus I is an ideal. Now let $g = \{([a], fa) : a \in A\}$. Then g is a function: if $[a] = [b]$, then $a\cdot -b + b\cdot -a \in I$, hence $f(a\cdot -b + b\cdot -a) = 0$; since f is a homomorphism, it follows that

$fa \cdot -fb + fb \cdot -fa = 0$, so $fa \cdot -fb = 0 = fb \cdot -fa$, hence $fa \leq fb \leq fa$, hence $fa = fb$. Thus g is a function mapping A/I onto B. g preserves $-$:

$$g(-[a]) = g[-a] = f(-a) = -fa = -g[a].$$

And g preserves $+$:

$$g([a] + [b]) = g[a + b] = f(a + b) = fa + fb = g[a] + g[b].$$

This completes the proof that g is an isomorphism of $\mathfrak{A}/I$ onto $\mathfrak{B}$. If f is one-one, obviously $I = \{0\}$. Conversely, if $I = \{0\}$ and $x \neq y$, then $x \cdot -y + -x \cdot y \neq 0$, so

$$0 \neq f(x \cdot -y + -x \cdot y) = fx \cdot -fy + -fx \cdot fy$$

and $fx \neq fy$. □

It is time to tie up the notions of ideal and filter; the following proposition is easily proved:

Proposition 9.20. *There is a one-one correspondence between ideals and filters on a given* BA $\mathfrak{A}$. *In more detail, if I is an ideal of $\mathfrak{A}$ and F is a filter of $\mathfrak{A}$, then:*

(i) $I^f = \{x : -x \in I\}$ *is a filter of* $\mathfrak{A}$;
(ii) $F^i = \{x : -x \in F\}$ *is an ideal of* $\mathfrak{A}$;
(iii) $I^{fi} = I$;
(iv) $F^{if} = F$.

The following proposition is also clear:

Proposition 9.21. *The intersection of any nonempty family of ideals (filters) is an ideal (filter).*

This justifies the following definition:

Definition 9.22. For any $X \subseteq A$, the ideal (filter) generated by X is the ideal (filter)

$$\bigcap \{Y : X \subseteq Y, \ Y \text{ is an ideal (filter) of } \mathfrak{A}\}.$$

Corollary 9.23. *For any $X \subseteq A$, the ideal generated by X is the set*

$$\{y : \text{for some finite sequence } x_0, \ldots, x_{m-1} \in X, \ y \leq x_0 + \cdots + x_{m-1}\},$$

while the filter generated by X is the set

$$\{y : \text{for some finite sequence } x_0, \ldots, x_{m-1} \in X, \ x_0 \cdot \ldots \cdot x_{m-1} \leq y\}.$$

Now we turn to the study of some special ideals and filters.

Definition 9.24. An ideal I of $\mathfrak{A}$ is *maximal* if $I \neq A$ and there is no ideal J such that $I \subset J \subset A$. Maximal filters, usually called ultrafilters, are similarly defined.

We let $\mathbf{2} = \langle 2, +, \cdot, -, 0, 1\rangle$, where $0, 1, 2 \in \omega$, and $+, \cdot, -$ are operations on 2 given by the following table:

x	y	$x + y$	$x \cdot y$	$-x$
1	1	1	1	0
1	0	1	0	0
0	1	1	0	1
0	0	0	0	1

The algebra **2** is a BA, as is easily checked. Maximal ideals are sometimes called prime ideals. It is also easy to check that maximal ideals and ultrafilters correspond to each other under the correspondence described in 9.20. The following proposition gives some useful equivalent definitions for maximal ideals.

Proposition 9.25. *For any ideal I in a* BA $\mathfrak{A}$ *the following conditions are equivalent:*

(*i*) *I is maximal;*
(*ii*) *$I \neq A$, and for any $x, y \in A$, if $x \cdot y \in I$ then $x \in I$ or $y \in I$;*
(*iii*) *$I \neq A$, and for any $x \in A$, either $x \in I$ or $-x \in I$;*
(*iv*) *the algebra $\mathfrak{A}/I$ is isomorphic to* **2**.

Proof

(*i*) $\Rightarrow$ (*ii*). Assume that I is maximal, but that there are elements $x, y \in A$ with $x \cdot y \in I$ while $x \notin I$ and $y \notin I$. Then the ideal generated by $I \cup \{x\}$ is A, so by 9.23 there is a $u \in I$ such that $1 = u + x$. Similarly, there is a $v \in I$ such that $1 = v + y$. Hence

$$1 = (u + x)\cdot(v + y) = u\cdot v + u\cdot y + x\cdot v + x\cdot y$$
$$\leq u + v + x\cdot y \in I,$$

so $1 \in I$. But then, since $y \leq 1$ for any $y \in A$, $A = I$. This contradicts I being maximal.

(*ii*) $\Rightarrow$ (*iii*). Assume (*ii*). Now $0 = x\cdot -x \in I$, so by (*ii*), $x \in I$ or $-x \in I$.

(*iii*) $\Rightarrow$ (*iv*). Assume (*iii*). Then $[0] \neq [1]$ since $I \neq A$. For any $x \in I$, $[x] = [0]$ and $[-x] = [1]$, so $[0]$ and $[1]$ are the only elements of $\mathfrak{A}/I$, by (*iii*). Clearly the tables for $+, \cdot, -$ are just like in 9.24, so $\mathfrak{A}/I$ is isomorphic to **2**.

(*iv*) $\Rightarrow$ (*i*). Since $|A/I| > 1$, I is proper. If J is an ideal such that $I \subset J \subset A$, choose $x \in J \sim I$. Then $[0]_I \neq [x]_I$ since $I \neq J$, and $[x]_I \neq [1]_I$ since $J \neq A$. Thus $|A/I| > 2$, contradiction. □

There is a theorem dual to 9.25 concerning filters; we will apply it when needed. The same applies to our next theorem. Of course the dual of 9.25(*iv*) requires a careful formulation of the notion $\mathfrak{A}/F$, F a filter. This is straightforward using 9.20. For example, the associated equivalence relation is $\{(x, y) : x \cdot y + -x \cdot -y \in F\}$.

Now we prove a fundamental existence theorem for maximal ideals.

Theorem 9.26 (Boolean prime ideal theorem). *If I is a proper ideal of $\mathfrak{A}$ (i.e., I is an ideal of $\mathfrak{A}$, and $I \neq A$), then there is a maximal ideal J of $\mathfrak{A}$ such that $I \subseteq J$.*

PROOF. Let $\mathscr{A} = \{J : I \subseteq J, J \text{ is a proper ideal of } \mathfrak{A}\}$. Then $\mathscr{A} \neq 0$, since obviously $I \in \mathscr{A}$. Let $\mathscr{B}$ be a nonempty subset of $\mathscr{A}$ simply ordered by inclusion. We now show that $\bigcup \mathscr{B} \in \mathscr{A}$. Since $\mathscr{B}$ is a nonempty collection of ideals $\supseteq I$, clearly $\bigcup \mathscr{B} \neq 0$, and $I \subseteq \bigcup \mathscr{B}$. Let $x, y \in \bigcup \mathscr{B}$. Say $x \in J \in \mathscr{B}$ and $y \in K \in \mathscr{B}$. By symmetry we may assume that $J \subseteq K$. Then $x, y \in K$, and since K is an ideal it follows that $x + y \in K$ and hence $x + y \in \bigcup \mathscr{B}$. If $x \leq y \in \bigcup \mathscr{B}$, say $y \in J \in \mathscr{B}$. Then $x \in J$ since J is an ideal; thus $x \in \bigcup \mathscr{B}$. If $1 \in \bigcup \mathscr{B}$, then $1 \in J \in \mathscr{B}$ for some J, contradicting J proper. Thus $\bigcup \mathscr{B}$ is proper, and we have established that $\bigcup \mathscr{B} \in \mathscr{A}$.

We can now apply Zorn's lemma to obtain a member J of $\mathscr{A}$ maximal under inclusion. Clearly J is as desired. □

The Boolean prime ideal theorem has been studied extensively from the point of view of axiomatic set theory. It is weaker than the axiom of choice (which is equivalent to Zorn's lemma), but it cannot be established on the basis of the usual axioms of set theory without choice, both statements assuming those axioms are consistent.

The prime ideal theorem can be used to establish the following important result. In its proof we make a switch from working with ideals to working with filters. Ideals seem more appropriate in purely algebraic contexts, as above, while filters are natural when considering matters, like here, which verge on topology; see Exercise 9.69.

Theorem 9.27 (Boolean representation theorem). *Every* BA *is isomorphic to a Boolean set algebra.*

PROOF. Let $\mathfrak{A}$ be a BA. Let $X = \{F : F \text{ is an ultrafilter of } \mathfrak{A}\}$. It suffices to find an isomorphism f of $\mathfrak{A}$ into the BA $\langle \mathbf{S}X, \cup, \cap, \sim, 0, X\rangle$. For each $a \in A$, let $fa = \{F : a \in F \in X\}$. Then for any $F \in X$ and any $a, b \in A$,

$$F \in f(-a) \quad \text{iff } -a \in F \text{ iff } a \notin F \qquad \text{(by 9.25}(iii))$$
$$\text{iff } F \notin fa \text{ iff } F \in X \sim fa;$$

so $f(-a) = X \sim fa$;

$$F \in f(a + b) \quad \text{iff } a + b \in F \text{ iff } a \in F \text{ or } b \in F \qquad \text{(by 9.25}(ii))$$
$$\text{iff } F \in fa \text{ or } F \in fb$$
$$\text{iff } F \in fa \cup fb;$$

thus $f(a + b) = fa \cup fb$. It remains to show that f is one-one. Suppose $x \in A$, $x \neq 0$. Let $F = \{z : z \in A, x \leq z\}$. Clearly F is a proper filter. By 9.26, let G be an ultrafilter such that $F \subseteq G$. Now $x \in G$, so $G \in fx$, as desired. □

The Boolean representation theorem shows that our axioms for BA's are complete, in the sense that the only models of these axioms are the Boolean set algebras, up to isomorphism. By this theorem, we can establish properties of arbitrary BA's by establishing them for the more special Boolean set algebras. On the other hand, we can apply ordinary algebraic constructions, some of which, like homomorphisms and quotient algebras, lead out of the class of Boolean set algebras, to establish properties of Boolean set algebras themselves.

We now turn to some important notions special for Boolean algebras.

Definition 9.28. An element $x \in A$ is an *atom* of $\mathfrak{A}$ if $x \neq 0$, while for all y, $0 \leq y \leq x$ implies $0 = y$ or $y = x$. $\mathfrak{A}$ is *atomic* if for every nonzero $x \in A$ there is an atom $a \leq x$. $\mathfrak{A}$ is *atomless* if $\mathfrak{A}$ has no atoms.

In the BA of all subsets of a set X, the singletons $\{x\}$ for $x \in X$ are the atoms, and this BA is atomic. An easy example of an atomless BA is the BA of subsets of the set of rational numbers generated by all half-open intervals $[r, s)$. There are BA's which are neither atomic nor atomless. A fundamental property of atomless BA's is given in 9.48 below. With the aid of the notion of an atom we can determine all finite BA's; see 9.30.

Proposition 9.29. *Every finite* BA *is atomic.*

Proof. Suppose $\mathfrak{A}$ is nonatomic. Choose a nonzero $a_0 \in A$ such that there is no atom $\leq a_0$. Having found $a_0, \ldots, a_n \in A$ with $a_0 > a_1 > \cdots > a_n > 0$, since there is no atom $\leq a_n$ choose $a_{n+1} \in A$ with $a_n > a_{n+1} > 0$. The so constructed sequence $a \in {}^{\omega}A$ is one-one, so $|A| \geq \aleph_0$. □

Proposition 9.30. *If $\mathfrak{A}$ is a finite* BA *with m atoms, then $\mathfrak{A}$ is isomorphic to the Boolean set algebra of all subsets of m.*

Proof. Let $a \in {}^{m}A$ enumerate all atoms of $\mathfrak{A}$. For each $x \in A$, let

$$fx = \{i : a_i \leq x\}.$$

Thus f maps A into the Boolean set algebra $\mathfrak{B}$ of all subsets of m. Suppose $i \in f(x + y)$, while $i \notin fx$. Thus $a_i \leq x + y$ but $a_i \nleq x$. Hence $a_i \cdot x < a_i$, so $a_i \cdot x = 0$ since a_i is an atom. Therefore $a_i = a_i \cdot (x + y) = a_i \cdot y \leq y$, i.e., $i \in fy$. This proves that $f(x + y) \subseteq fx \cup fy$; the converse inclusion is obvious, so $f(x + y) = fx \cup fy$. Clearly $fx \cap f - x = 0$. For any $i \in m$ we have $a_i = a_i \cdot (x + -x) = a_i \cdot x + a_i \cdot -x$, so $a_i \cdot x \neq 0$ or $a_i \cdot -x \neq 0$; these two alternatives imply (since a_i is an atom) $a_i \leq x$ or $a_i \leq -x$ respectively. Thus $i \in fx$

or $i \in f(-x)$. Hence $fx \cup f(-x) = m$, so $m \sim fx = f(-x)$. Thus f is a homomorphism of $\mathfrak{A}$ into $\mathfrak{B}$. If $x \neq 0$, then, since $\mathfrak{A}$ is atomic by 9.29, there is an $i \in m$ with $a_i \leq x$; thus $fx \neq 0$. It follows by 9.19 that f is one-one. Finally, suppose $T \subseteq m$. Let $x = \sum_{i \in T} a_i$; we claim that $fx = T$. If $i \in T$, obviously $a_i \leq x$ and $i \in fx$. Now suppose $i \in fx$. Thus $a_i \leq x$, so $a_i = \sum_{j \in T} a_i \cdot a_j$. Hence there is a $j \in T$ with $a_i \cdot a_j \neq 0$. Since a_i and a_j are both atoms, $a_i = a_i \cdot a_j = a_j$ and so $i = j \in T$, as desired. □

Corollary 9.31. *For any $n \in \omega$ the following conditions are equivalent:*

(*i*) *there is a* BA *of power n;*
(*ii*) $n = 2^m$ *for some m.*

Corollary 9.32. *If $\mathfrak{A}$ and $\mathfrak{B}$ are finite* BA's, *then $\mathfrak{A} \cong \mathfrak{B}$ iff $|A| = |B|$.*

Definition 9.33. Let $\mathfrak{A}$ be a BA, $X \subseteq A$. The *least upper bound* of X, if it exists, is the unique element ΣX such that $\forall x \in X\ (x \leq \Sigma X)$ (i.e., ΣX is an upper bound of X) and $\forall y \in A\ [\forall x \in X\ (x \leq y) \Rightarrow \Sigma X \leq y]$. Dually, the *greatest lower bound* of X, if it exists, is the unique element ΠX such that $\forall x \in X\ (\Pi X \leq x)$ (i.e., ΠX is a lower bound of X) and $\forall y \in A\ [\forall x \in X\ (y \leq x) \Rightarrow y \leq \Pi X]$.

Proposition 9.34. *The following conditions are equivalent:*

(*i*) $\mathfrak{A}$ *is atomic;*
(*ii*) *for every $a \in A$, $\Sigma\{x : x \leq a$, x is an atom of $\mathfrak{A}\}$ exists and equals a.*

Proof

(*i*) $\Rightarrow$ (*ii*). Assume (*i*), and let $a \in A$. Obviously a is an upper bound for $X = \{x : x \leq a, x \text{ is an atom of } \mathfrak{A}\}$. Suppose that b is another upper bound for X. If $a \nleq b$, then $a \cdot -b \neq 0$, and there is an atom x with $x \leq a \cdot -b$. Thus $x \leq a$, so $x \in X$ and $x \leq b$. Thus $x \leq b \cdot -b = 0$, contradiction. Thus $a \leq b$, as desired.

(*ii*) $\Rightarrow$ (*i*): obvious. □

Corollary 9.35. *If $\mathfrak{A}$ is atomic and has only finitely many atoms, then $\mathfrak{A}$ is finite.*

Proposition 9.36. *If ΣX exists, then $\Pi\{-x : x \in X\}$ exists and equals $-\Sigma X$.*

Proof. If $x \in X$, then $x \leq \Sigma X$ and hence $-\Sigma X \leq -x$. Thus $-\Sigma X$ is a lower bound for $Y = \{-x : x \in X\}$. Let a be any lower bound for Y. Thus $\forall x \in X\ (a \leq -x)$, so $\forall x \in X\ (x \leq -a)$. Hence $\Sigma X \leq -a$ and $a \leq -\Sigma X$.

Proposition 9.37. *If ΣX exists and $a \in A$, then $\Sigma\{x \cdot a : x \in X\}$ exists and equals $a \cdot \Sigma X$.*

PROOF. For any $x \in X$, $x \leq \Sigma X$ and hence $a \cdot x \leq a \cdot \Sigma X$. Now suppose that b is any upper bound of $\{x \cdot a : x \in X\}$. Then for any $x \in X$, $x \cdot a \leq b$ and so $x \leq -a + b$. Thus $\Sigma X \leq -a + b$, so $a \cdot \Sigma X \leq b$. □

The next general algebraic notion we shall consider for BA's is that of direct or Cartesian products.

Definition 9.38. Let $\langle \mathfrak{A}_i : i \in I \rangle$ be a system of BA's,

$$\mathfrak{A}_i = \langle A_i, +_i, \cdot_i, -_i, 0_i, 1_i \rangle$$

for each $i \in I$. By the *product* of the system $\langle \mathfrak{A}_i : i \in I \rangle$ we mean the structure

$$\mathsf{P}_{i \in I} \mathfrak{A}_i = \langle B, +, \cdot, -, 0, 1 \rangle,$$

where $B = \mathsf{P}_{i \in I} A_i = \{f : f$ is a function with domain I and $f_i \in A_i$ for all $i \in I\}$ and $+$, $\cdot$, $-$ are defined as follows (for arbitrary $f, g \in B$, $i \in I$):

$$\begin{aligned} (f + g)_i &= f_i +_i g_i; \\ (f \cdot g)_i &= f_i \cdot_i g_i; \\ (-f)_i &= -_i f_i; \end{aligned}$$

The members 0, 1 of B are of course given by hypothesis. If each $\mathfrak{A}_i = \mathfrak{C}$, we denote $\mathsf{P}_{i \in I} \mathfrak{A}_i$ by ${}^I\mathfrak{C}$. If $I = 2$, we denote $\mathsf{P}_{i \in I} \mathfrak{A}_i$ by $\mathfrak{A}_0 \times \mathfrak{A}_1$. Thus $\mathfrak{A}_0 \times \mathfrak{A}_1$ consists of all pairs $\langle x, y \rangle$ with $x \in A_0$, $y \in A_1$, with the operations given by

$$\begin{aligned} \langle x_0, y_0 \rangle + \langle x_1, y_1 \rangle &= \langle x_0 +_0 x_1, y_0 +_1 y_1 \rangle, \\ \langle x_0, y_0 \rangle \cdot \langle x_1, y_1 \rangle &= \langle x_0 \cdot_0 x_1, y_0 \cdot_1 y_1 \rangle, \\ -\langle x_0, y_0 \rangle &= \langle -_0 x_0, -_1 y_0 \rangle, \\ 0 &= \langle 0_0, 0_1 \rangle, \\ 1 &= \langle 1_0, 1_1 \rangle. \end{aligned}$$

The following proposition is straightforward:

Proposition 9.39. *If* $\langle \mathfrak{A}_i : i \in I \rangle$ *is a system of* BA's, *then* $\mathsf{P}_{i \in I} \mathfrak{A}_i$ *is a* BA.

Proposition 9.40. *The* BA $\langle \mathbf{S}X, \cup, \cap, \sim, 0, X \rangle$ *is isomorphic to* ${}^X\mathbf{2}$.

PROOF. For each $Y \subseteq X$, let χ_Y be the characteristic function Y, i.e., for all $x \in X$ let

$$\begin{aligned} \chi_Y x &= 1 \quad \text{if } x \in Y, \\ \chi_Y x &= 0 \quad \text{if } x \in Y. \end{aligned}$$

It is easily verified that χ is the desired isomorphism. □

If we combine this proposition with the representation theorem, we obtain

Corollary 9.41. *Any* BA *can be isomorphically embedded in* ${}^X\mathbf{2}$ *for some set* X.

This corollary shows another completeness property of our axioms for BA's. Namely, if we wish to check that an equation holds in all BA's, it is enough to check that it holds in the two-element BA **2**. For, then it will also clearly hold in ${}^X\mathbf{2}$, and any subalgebra of ${}^X\mathbf{2}$, and hence by 9.41 in any BA. This gives a decision procedure for checking when equations hold in all BA's; in fact, under a natural Gödel numbering the set of Gödel numbers of equations holding in all BA's is recursive, indeed even elementary recursive. The decision procedure really coincides with the truth table method described for sentential logic.

Another important fact about products of BA's can be expressed using the notion of relativized BA:

Definition 9.42. For any BA $\mathfrak{A} = \langle A, +, \cdot, -, 0, 1\rangle$ and any $a \in A$ we let $\mathfrak{A} \upharpoonright a = \langle A \upharpoonright a, +', \cdot', -', 0, a\rangle$, where

$$
\begin{aligned}
A \upharpoonright a &= \{x \in A : x \leq a\},\\
x +' y &= x + y,\\
x \cdot' y &= x \cdot y,\\
-'x &= a \cdot -x,
\end{aligned}
$$

for any $x, y \in A$. $\mathfrak{A} \upharpoonright a$ is the *relativization* of $\mathfrak{A}$ to a.

It is easy to check the following proposition:

Proposition 9.43. $\mathfrak{A} \upharpoonright a$ *is a* BA.

Proposition 9.44. *For any* $a \in A$, $\mathfrak{A}$ *is isomorphic to* $(\mathfrak{A} \upharpoonright a) \times (\mathfrak{A} \upharpoonright -a)$.

PROOF. For any $x \in A$, let $fx = \langle x \cdot a, x \cdot -a\rangle$. It is easily checked that f is the desired isomorphism. □

Definition 9.45. Given a product $\mathsf{P}_{i\in I}\, \mathfrak{A}_i$ of BA's, and any $i \in I$, pr_i is the function mapping $\mathsf{P}_{i\in I}\, A_i$ into A_i given by $\mathrm{pr}_i x = x_i$ for all $x \in \mathsf{P}_{i\in I}\, A_i$.

Corollary 9.46. pr_i *is a homomorphism from* $\mathsf{P}_{i\in I}\, \mathfrak{A}_i$ *onto* $\mathfrak{A}_i$.

The following lemma expresses an important property of the notion of relativization.

Lemma 9.47 (Vaught). *Let R be a binary relation between countable* BA's *such that the following conditions hold:*

(*i*) *if* $\mathfrak{A}R\mathfrak{B}$, *then* $\mathfrak{B}R\mathfrak{A}$;
(*ii*) *if* $\mathfrak{A}R\mathfrak{B}$, *then* $|A| = |B|$;
(*iii*) *if* $\mathfrak{A}R\mathfrak{B}$ *and* $a \in A$, *then there is a* $b \in B$ *such that* $(\mathfrak{A} \upharpoonright a)R(\mathfrak{B} \upharpoonright b)$ *and* $(\mathfrak{A} \upharpoonright -a)R(\mathfrak{B} \upharpoonright -b)$.

Then $\mathfrak{A} \cong \mathfrak{B}$ *whenever* $\mathfrak{A}R\mathfrak{B}$.

PROOF. Assume $\mathfrak{A}R\mathfrak{B}$. If A is finite, then $\mathfrak{A} \cong \mathfrak{B}$ by (*ii*) and Corollary 9.32. Hence assume A is infinite. By (*ii*), B is also infinite. Let $a: \omega \rightarrowtail\!\!\!\twoheadrightarrow A$ and $b: \omega \rightarrowtail\!\!\!\twoheadrightarrow B$. We now define two sequences $\langle x_i : i \in \omega \rangle \in {}^{\omega}A$ and $\langle y_i : i \in \omega \rangle \in {}^{\omega}B$ by recursion so that for every $m \in \omega$ and every $t \subseteq m$,

$$(1) \qquad \mathfrak{A} \upharpoonright \left(\prod_{i \in t} x_i \cdot \prod_{i \in m \sim t} - x_i \right) R\, \mathfrak{B} \upharpoonright \left(\prod_{i \in t} y_i \cdot \prod_{i \in m \sim t} - y_i \right).$$

Let $x_0 = a_0$, and by (*iii*) choose $y_0 \in B$ so that $(\mathfrak{A} \upharpoonright x_0)R(\mathfrak{B} \upharpoonright y_0)$ and $(\mathfrak{A} \upharpoonright -x_0)R(\mathfrak{B} \upharpoonright -y_0)$. Thus (1) holds for $m = 1$. Now suppose x_i and y_i defined for all $i < n$, where (1) holds for all $m \leq n$, and $n \geq 1$. For each $t \subseteq n$ let

$$u_t = \prod_{i \in t} x_i \cdot \prod_{i \in n \sim t} - x_i,$$

$$v_t = \prod_{i \in t} y_i \cdot \prod_{i \in n \sim t} - y_i.$$

For n odd we proceed as follows. Let j be minimum such that $b_j \notin \{y_i : i < n\}$, and set $y_n = b_j$. Now let $t \subseteq n$. We have $(\mathfrak{A} \upharpoonright u_t)R(\mathfrak{B} \upharpoonright v_t)$ by (1), so by (*i*) and (*iii*) choose $w_t \in A \upharpoonright u_t$ so that $[(\mathfrak{A} \upharpoonright u_t) \upharpoonright w_t]R[(\mathfrak{B} \upharpoonright v_t) \upharpoonright (v_t \cdot y_n)]$ and $[(\mathfrak{A} \upharpoonright u_t) \upharpoonright -' w_t]R[(\mathfrak{B} \upharpoonright v_t) \upharpoonright -''(v_t \cdot y_n)]$, where $-'$ and $-''$ are the minus operations in $\mathfrak{A} \upharpoonright u_t$ and $\mathfrak{B} \upharpoonright v_t$ respectively. Thus

$$(2) \qquad [\mathfrak{A} \upharpoonright w_t]R[\mathfrak{B} \upharpoonright (v_t \cdot y_n)] \text{ and } [\mathfrak{A} \upharpoonright (u_t \cdot - w_t)]R[\mathfrak{B} \upharpoonright (v_t \cdot - y_n)],$$

where $w_t \leq u_t$. Set $x_n = \sum_{t \subseteq n} w_t$. We now check (1) for $m = n + 1$. Given any subset s of $n + 1$, we take two cases.

Case 1. $n \in s$. Then we write $s = t \cup \{n\}$, where $t \subseteq n$. Thus $\prod_{i \in s} x_i \cdot \prod_{i \in (n+1) \sim s} - x_i = u_t \cdot x_n$ and $\prod_{i \in s} y_i \cdot \prod_{i \in (n+1) \sim s} - y_i = v_t \cdot y_n$. Now if $t' \subseteq n$ and $t \neq t'$, clearly $u_t \cdot u_{t'} = 0$. Hence $u_t \cdot x_n = w_t$. Hence the desired result is immediate from (2).

Case 2. $n \notin s$. Thus $s \subseteq n$, and $\prod_{i \in s} x_i \cdot \prod_{i \in (n+1) \sim s} - x_i = u_s \cdot - x_n$, while $\prod_{i \in s} y_i \cdot \prod_{i \in (n+1) \sim s} - y_i = v_s \cdot - y_n$. Clearly $u_s \cdot - x_n = u_s \cdot - w_s$, so again the desired conclusion follows from (2).

If n is even, we proceed as above except with the roles of $\mathfrak{A}$ and $\mathfrak{B}$, a and b, x and y interchanged. Thus our sequences are defined, and (1) holds for every $m \in \omega$ and $t \subseteq m$. In particular, for any $m \in \omega$ and $t \subseteq m$, using (*ii*),

$$(3) \qquad \prod_{i \in t} x_i \cdot \prod_{i \in m \sim t} - x_i = 0 \text{ iff } \prod_{i \in t} y_i \cdot \prod_{i \in m \sim t} - y_i = 0.$$

Now we claim

$$(4) \qquad \text{if } i, j < \omega, \text{ then } x_i \leq x_j \text{ iff } y_i \leq y_j.$$

For, let $m = \max\{i, j\} + 1$. By induction on n one easily shows that for every $n \in \omega$,

$$1 = \sum_{t \subseteq n} \left(\prod_{i \in t} x_i \cdot \prod_{i \in n \sim t} - x_i \right)$$

and hence

$$x_i \cdot -x_j = \sum \left\{ \prod_{k \in t} x_k \cdot \prod_{k \in m \sim t} - x_k : t \subseteq m, i \in t, j \notin t \right\},$$

and, similarly,

$$y_i \cdot -y_j = \sum \left\{ \prod_{k \in t} y_k \cdot \prod_{k \in m \sim t} - y_k : t \subseteq m, i \in t, j \in t \right\}.$$

Thus (4) follows from (3). By induction, the following statement is easily established:

(5) for all $i \in \omega$, $a_i \in \{x_0, \ldots, x_{2i}\}$ and $b_i \in \{y_0, \ldots, y_{2i+1}\}$.

Combining (4) and (5), we see from 9.14 that $\{(x_i, y_i) : i \in \omega\}$ is the desired isomorphism. □

Theorem 9.48. *Any two denumerable atomless* BA's *are isomorphic.*

PROOF. Let $R = \{(\mathfrak{A}, \mathfrak{B}) : |A| = |B| = 1$ or $\mathfrak{A}$ and $\mathfrak{B}$ are denumerable and atomless$\}$. The hypothesis of 9.47 is easily verified. □

The final algebraic notion which we shall consider for BA's is that of a free algebra:

Definition 9.49. $\mathfrak{A}$ is *freely generated by X* provided:

(*i*) X generates $\mathfrak{A}$;
(*ii*) if $\mathfrak{B}$ is any BA and f is any function mapping X into B, then f can be extended to a homomorphism of $\mathfrak{A}$ into $\mathfrak{B}$.

If $\mathfrak{A}$ is freely generated by X, we say that $\mathfrak{A}$ is a *free* BA *with free generators X.*

We first note that free BA's are determined up to isomorphism by their sets of free generators:

Proposition 9.50. *If $\mathfrak{A}$ and $\mathfrak{B}$ are each freely generated by X, then they are isomorphic.*

PROOF. Let f be a homomorphism from $\mathfrak{A}$ into $\mathfrak{B}$ extending Id $\restriction X$ (identity on X), and let g be a homomorphism from $\mathfrak{B}$ into $\mathfrak{A}$ extending Id $\restriction X$. Then $\{a \in A : gfa = a\}$ is a subuniverse of $\mathfrak{A}$ including X; since $\mathfrak{A}$ is generated by X, this set is equal to A. Thus $g \circ f = \text{Id} \restriction A$. Similarly, $f \circ g = \text{Id} \restriction B$. Hence f is the desired isomorphism. □

To prove the existence of free algebras we need a couple of lemmas.

Lemma 9.51. *If X generates $\mathfrak{A}$, then $|A| \leq |X| + \aleph_0$.*

Proof. Let $Y_0 = X$ and, for $m \in \omega$,

$$Y_{m+1} = Y_m \cup \{-x : x \in Y_m\} \cup \{x + y : x, y \in Y_m\} \cup \{0, 1\}.$$

It is easily checked that $\bigcup_{m\in\omega} Y_m = A$. Furthermore, by induction on m one easily shows that $|Y_m| \leq |X| + \aleph_0$ for each $m \in \omega$. Hence

$$\begin{aligned} |A| = \left|\bigcup_{m\in\omega} Y_m\right| &\leq \sum_{n\in\omega} |Y_m| \leq \sum_{m\in\omega} (|X| + \aleph_0) \\ &= (|X| + \aleph_0)\cdot\aleph_0 = |X| + \aleph_0. \end{aligned}$$

□

Lemma 9.52. *If $\mathfrak{A}$ is a* BA *and f is a one-one function from A onto B, then there is a* BA *$\mathfrak{B}$ with universe B such that f is an isomorphism from $\mathfrak{A}$ onto $\mathfrak{B}$.*

Proof. Let $\mathfrak{A} = \langle A, +, \cdot, -, 0, 1\rangle$. We define the operations of $\mathfrak{B}$ as follows. For any $b, c \in B$,

$$\begin{aligned} b +' c &= f(f^{-1}b + f^{-1}c), \\ b \cdot' c &= f(f^{-1}b \cdot f^{-1}c), \\ -'b &= f(-f^{-1}b), \\ 0' &= f0, \\ 1' &= f1. \end{aligned}$$

Then we set $\mathfrak{B} = \langle B, +', \cdot', 0', 1'\rangle$. The conclusions of the lemma are then routine. □

Theorem 9.53. *For any nonempty set X there is a* BA *freely generated by X.*

Proof. Let A be any set such that $|A| = |X| + \aleph_0$. Let $I = \{(\mathfrak{B}, f) : \mathfrak{B}$ is a BA, $B \subseteq A$, and f is a function mapping X into $B\}$. Let $\mathfrak{C} = \mathsf{P}_{(\mathfrak{B},f)\in I}\, \mathfrak{B}$. We define a function g mapping X into C by setting $(gx)_{\mathfrak{B}f} = fx$ for all $x \in X$ and $(\mathfrak{B}, f) \in I$. Then

(1) $\qquad g$ is a one-one function.

For, assume that $x, y \in X$ and $x \neq y$. Now there is a BA $\mathfrak{B}$ with $|B| > 1$ and $B \subseteq A$; to see this, consider the BA **2**, note that $|A| \geq 2$, and apply 9.52. Let f be any function mapping X into B such that $fx \neq fy$. Thus $(\mathfrak{B}, f) \in I$, and

$$(gx)_{\mathfrak{B}f} = fx \neq fy = (gy)_{\mathfrak{B}f}.$$

Thus (1) holds. Now let $\mathfrak{D}$ be the subalgebra of $\mathfrak{C}$ generated by g^*X. Let h be any one-one function with domain D such that $g^{-1} \subseteq h$. Finally, by 9.52 let $\mathfrak{F}$ be a BA such that h is an isomorphism from $\mathfrak{D}$ onto $\mathfrak{F}$. We claim that $\mathfrak{F}$ is the desired algebra freely generated by X. Since g^*X generates $\mathfrak{D}$ and the isomorphism h maps g^*X onto X, clearly X generates $\mathfrak{F}$. Now let k be any function mapping X into some BA $\mathfrak{E}$. Let $\mathfrak{H}$ be the subalgebra of $\mathfrak{E}$ generated by k^*X. Then 9.51, $|H| \leq |X| + \aleph_0 = |A|$. Let l be any one-one function mapping H into A, and by 9.52 let $\mathfrak{K}$ be a BA such that l is an isomorphism of $\mathfrak{H}$ onto $\mathfrak{K}$. Note that $(\mathfrak{K}, l \circ k) \in I$. Now set $k^+ = l^{-1} \circ \mathrm{pr}_{(\mathfrak{K}, l \circ k)} \circ h^{-1}$; we claim that k^+ is the desired extension of k. Obviously k^+ is a homomorphism of $\mathfrak{F}$ into $\mathfrak{C}$. If $x \in X$, then

$$\begin{aligned} k^+x &= l^{-1}\, \mathrm{pr}_{(\mathfrak{K}, l \circ k)} h^{-1}x = l^{-1}(h^{-1}x)_{(\mathfrak{K}, l \circ k)} = l^{-1}(gx)_{(\mathfrak{K}, l \circ k)} \\ &= l^{-1}lkx = kx, \end{aligned}$$

as desired. □

We shall give another proof of 9.53 below; see Theorem 9.58 and the remark after it. This new proof will be based on the correspondence between Boolean algebras and sentential logics, to which we now turn. We shall see that there is a full correspondence between these two kinds of mathematical objects.

Definition 9.54. Let $\mathscr{P}$ be a sentential language and $\Gamma \subseteq \mathrm{Sent}_{\mathscr{P}}$. We let $\equiv_\Gamma^{\mathscr{P}} = \{(\varphi, \psi) : \varphi, \psi \in \mathrm{Sent}_{\mathscr{P}} \text{ and } \Gamma \vdash_{\mathscr{P}} \varphi \leftrightarrow \psi\}$.

We shall sometimes write $\equiv_\Gamma$, or even $\equiv$, when no ambiguity is likely. The following proposition can be routinely checked.

Proposition 9.55. *Let $\mathscr{P}$ be a sentential language and $\Gamma \subseteq \mathrm{Sent}_{\mathscr{P}}$. Then for any $\varphi, \psi, \varphi', \psi' \in \mathrm{Sent}_{\mathscr{P}}$,*

(*i*) $\equiv$ *is an equivalent relation on* $\mathrm{Sent}_{\mathscr{P}}$;
(*ii*) *if* $\varphi \equiv \psi$, *then* $\neg\varphi \equiv \neg\psi$;
(*iii*) *if* $\varphi \equiv \psi$, *and* $\varphi' \equiv \psi'$, *then* $(\varphi \vee \varphi') \equiv (\psi \vee \psi')$ *and* $(\varphi \wedge \varphi') \equiv (\psi \wedge \psi')$;
(*iv*) $(\varphi \wedge \neg\varphi) \equiv (\psi \wedge \neg\psi)$;
(*v*) $(\varphi \vee \neg\varphi) \equiv (\psi \vee \neg\psi)$.

This proposition justifies the following definition:

Definition 9.56. Let $\mathscr{P}$ be a sentential language and $\Gamma \subseteq \mathrm{Sent}_{\mathscr{P}}$. We let $\mathfrak{M}_\Gamma^{\mathscr{P}}$ be the algebra $(\mathrm{Sent}_{\mathscr{P}}/\equiv_\Gamma^{\mathscr{P}}, +, \cdot, -, 0, 1)$ with the operations determined by the following stipulations, where $\varphi, \psi \in \mathrm{Sent}_{\mathscr{P}}$:

$$\begin{aligned} [\varphi] + [\psi] &= [\varphi \vee \psi], \\ [\varphi] \cdot [\psi] &= [\varphi \wedge \psi], \\ -[\varphi] &= [\neg\varphi], \\ 0 &= [\varphi \wedge \neg\varphi], \\ 1 &= [\varphi \vee \neg\varphi]. \end{aligned}$$

The following proposition is easy to check:

Proposition 9.57. *If $\mathscr{P}$ is a sentential language and $\Gamma \subseteq \mathrm{Sent}_{\mathscr{P}}$, then $\mathfrak{M}_\Gamma^{\mathscr{P}}$ is a* BA.

Theorem 9.58. *If $\mathscr{P} = (n, c, P)$ is a sentential language, then $\mathfrak{M}_0^{\mathscr{P}}$ is freely generated by $\{[\langle s\rangle] : s \in P\}$, and $\langle[\langle s\rangle] : s \in P\rangle$ is a one-one function.*

PROOF. We check the second statement first. If $s, t \in P$ and $s \neq t$, clearly $\nvDash \langle s\rangle \leftrightarrow \langle t\rangle$ and hence $\nvdash \langle s\rangle \leftrightarrow \langle t\rangle$, so $[\langle s\rangle] \neq [\langle t\rangle]$. By induction on sentences it is easily seen that $[\varphi]$ is in the subalgebra generated by $X = \{[\langle s\rangle] : s \in P\}$, for each sentence φ. Thus X generates $\mathfrak{M}_0^{\mathscr{P}}$. Now let f be any function mapping X into a BA $\mathfrak{A}$. By recursion (8.6), we define a function g mapping $\text{Sent}_{\mathscr{P}}$ into A:

$$\begin{aligned} g\langle s\rangle &= f[\langle s\rangle], \\ g \neg \varphi &= -g\varphi, \\ g(\varphi \rightarrow \psi) &= -g\varphi + g\psi, \end{aligned}$$

for any $s \in P$ and any $\varphi, \psi \in \text{Sent}_{\mathscr{P}}$. We now claim

(1) $$\text{if } \vdash\varphi, \text{ then } g\varphi = 1.$$

To prove (1), let $\Gamma = \{\varphi : \varphi \in \text{Sent}_{\mathscr{P}}, g\varphi = 1\}$. It is easily seen that each logical axiom is in Γ. For example, to check (A3) we note

$$\begin{aligned} g((\neg\varphi \rightarrow \neg\psi) \rightarrow (\psi \rightarrow \varphi)) &= -g(\neg\varphi \rightarrow \neg\psi) + g(\psi \rightarrow \varphi) \\ &= -(-g(\neg\varphi) + g(\neg\psi)) + -g\psi + g\varphi \\ &= -(--g\varphi + -g\psi) + -g\psi + g\varphi \\ &= -g\varphi \cdot g\psi + -g\psi + g\varphi \\ &\geq -g\varphi \cdot g\psi + -g\varphi \cdot -g\psi + g\varphi \\ &= -g\varphi + g\varphi = 1. \end{aligned}$$

Also, clearly Γ is closed under detachment. Thus each logical theorem is in Γ, and (1) follows.

Now if $\vdash \varphi \leftrightarrow \psi$, then $g(\varphi \leftrightarrow \psi) = 1$, and hence it follows easily that $g\varphi = g\psi$. Thus there is a function f^+ mapping $M_\Gamma^{\mathscr{P}}$ into A such that $f^+[\varphi] = g\varphi$ for any sentence φ. It is easily checked that f^+ is the desired extension of f. □

From 9.58 one easily obtains a new proof of the existence of free algebras. In fact, given any set X, there is a sentential language $\mathscr{P}$ of the form (n, c, X). By 9.58, $\mathfrak{M}_0^{\mathscr{P}}$ is freely generated by $\{[\langle s\rangle] : s \in X\}$, and $|\{[\langle s\rangle] : s \in X\}| = |X|$. Thus by 9.52 we can easily infer that there is a BA freely generated by X.

There is a natural correspondence between notions in sentential logic and notions in Boolean algebra. We give one instance of this correspondence next, and state some others in the exercises.

Proposition 9.59. *Let F be a filter in a BA $\mathfrak{M}_\Gamma^{\mathscr{P}}$, and set $\Delta = \bigcup F$. Then $\Gamma \subseteq \Delta$, and $\mathfrak{M}_\Gamma^{\mathscr{P}}/F$ is isomorphic to $\mathfrak{M}_\Delta^{\mathscr{P}}$.*

PROOF. Note first that Γ is a subset of the unit element 1 of $\mathfrak{M}_\Gamma^{\mathscr{P}}$; since $1 \in F$, clearly $\Gamma \subseteq \Delta$. We write below $[\varphi]_\Gamma$, $[x]_F$, $[\varphi]_\Delta$ for the equivalence classes under the equivalence relations associated with Γ, F, Δ respectively. Assume that $[[\varphi]_\Gamma]_F = [[\psi]_\Gamma]_F$. Then $[\varphi]_\Gamma \cdot [\psi]_\Gamma + -[\varphi]_\Gamma \cdot -[\psi]_\Gamma \in F$. Recalling the definitions of the operations in $\mathfrak{M}_\Gamma^{\mathscr{P}}$, we see that $[(\varphi \wedge \psi) \vee (\neg\varphi \wedge \neg\psi)]_\Gamma \in F$ and hence, using an easy tautology, $[\varphi \leftrightarrow \psi]_\Gamma \in F$. Thus $\varphi \leftrightarrow \psi \in \Delta$ and so $[\varphi]_\Delta = [\psi]_\Delta$. Similarly $[\varphi]_\Delta = [\psi]_\Delta$ implies that $[[\varphi]_\Gamma]_F = [[\psi]_\Gamma]_F$. Hence there is a one-one function f mapping $\mathfrak{M}_\Gamma^\mathscr{P}/F$ onto $\mathfrak{M}_\Delta^\mathscr{P}$ such that $f[[\varphi]_\Gamma]_F = [\varphi]_\Delta$ for every sentence φ. Clearly f is the desired isomorphism. □

The preceding two results lead to the following theorem, which is another kind of completeness theorem for Boolean algebras. It shows that any BA is one of the algebras $\mathfrak{M}_\Gamma^\mathscr{P}$, up to isomorphism. Hence we may say that the theories of Boolean algebras and of sentential logics are equivalent, in some sense.

Theorem 9.60. *Any* BA *is isomorphic to* $\mathfrak{M}_\Gamma^\mathscr{P}$ *for some sentential language* $\mathscr{P}$ *and some* $\Gamma \subseteq \mathrm{Sent}_\mathscr{P}$.

PROOF. Let $\mathfrak{A}$ be any BA. Let $\mathscr{P} = (n, c, A)$ be a sentential language, and let f be the function such that $f[\langle a\rangle] = a$ for $a \in A$. By 9.58, we can extend f to a homomorphism f^+ of $\mathfrak{M}_0^\mathscr{P}$ onto $\mathfrak{A}$. Let F be the filter-kernel of f^+, i.e., $F = \{x : x \in M_0^\mathscr{P}, f^+x = 1\}$. By the homomorphism theorem we have $\mathfrak{A}$ isomorphic to $\mathfrak{M}_0^\mathscr{P}/F$. Let $\Gamma = \bigcup F$. Then by 9.59, $\mathfrak{M}_0^\mathscr{P}/F$ is isomorphic to $\mathfrak{M}_\Gamma^\mathscr{P}$. Thus $\mathfrak{A}$ is isomorphic to $\mathfrak{M}_\Gamma^\mathscr{P}$, as desired.

BIBLIOGRAPHY

1. Halmos, P. R. *Lectures on Boolean Algebras*. Princeton: van Nostrand (1963).
2. Sikorski, R. *Boolean Algebras*, 3rd ed. New York, Berlin: Springer (1969).

EXERCISES

9.61. Show that in any BA we have

$$-(-x + -y + z) + -(-x + y) + -x + z = 1.$$

9.62. Show that there is a natural one-one correspondence between Boolean algebras and rings with identity such that $x \cdot x = x$ for all x. (Only the most elementary facts about rings are needed to solve this problem. Given a BA $\mathfrak{A} = \langle A, +, \cdot, -, 0, 1\rangle$, let $x +' y = x \cdot -y + y \cdot -x$ for any $x, y \in A$, and show that under $+'$ and $\cdot$, A forms a ring of the above sort. Given a ring of this sort, expand $(x + y)\cdot(x + y)$ to show that the ring is commutative and $x + x = 0$ for all x in the ring; then let $-'x = 1 + x$, $x +' y = x + y + x \cdot y$ and show that A with these operations, $\cdot$, and the 0 and 1 of the ring is a BA.)

9.63. There is a BA $\mathfrak{A}$ and a nonempty subset X of A such that $0, 1 \in X$, and X is closed under $+$ and $\cdot$, but X is not a subuniverse of $\mathfrak{A}$.

9.64. For each infinite cardinal $\mathfrak{m}$, there is a BA with exactly $\mathfrak{m}$ elements.

9.65. If $\mathfrak{A}$ is a finitely generated BA, say generated by X, X finite, then $\mathfrak{A}$ is finite, in fact has at most $\exp\exp|X|$ elements. This bound can be attained.

9.66. There is an infinite BA with exactly 3 atoms.

9.67. Every infinite BA has an infinite subset X such that if $x, y \in X$, and $x \neq y$ then $x \cdot y = 0$, and an infinite subset Y such that if $x, y \in Y$, then $x \leq y$ or $y \leq x$.

9.68. Without using any principle beyond ZF (set theory without the axiom of choice) show that the following two statements are equivalent:
(1) the Boolean prime ideal theorem,
(2) every BA with at least two elements has a maximal ideal.

9.69. For any BA $\mathfrak{A}$, let $F\mathfrak{A}$ be the set $X = \{F : F \text{ is an ultrafilter on } \mathfrak{A}\}$, with the topology on X given by the base $\{\{F : x \in F \in X\} : x \in A\}$. Then $F\mathfrak{A}$ is a compact Hausdorff space in which the simultaneously closed and open sets form a base. $F\mathfrak{A}$ is called the *Stone space* of $\mathfrak{A}$. In fact, $\{\{F : x \in F \in X\} : x \in A\}$ is the set of all closed and open sets in $F\mathfrak{A}$. The space $F\mathfrak{A}$ is homomorphic to the closed subspace of ${}^{A}\mathbf{2}$ consisting of all homomorphisms from $\mathfrak{A}$ into $\mathbf{2}$, with the discrete topology.

9.70. A Stone space is a compact Hausdorff space in which the closed-open sets form a base. For any topological space X, the set of closed-open sets forms a field of sets; $\mathfrak{S}X$ is the associated BA. Then for X a Stone space, $\mathfrak{S}X$ is isomorphic to the BA, a subalgebra of the direct power ${}^{X}\mathbf{2}$, whose universe consists of all continuous functions mapping X into $\mathbf{2}$ (latter with discrete topology). Furthermore, $F\mathfrak{S}X$ is homeomorphic to X. Finally, if $\mathfrak{A}$ is any BA, then $\mathfrak{S}F\mathfrak{A}$ is isomorphic to $\mathfrak{A}$.

9.71. There is a BA $\mathfrak{A}$ such that ${}^{2}\mathfrak{A}$ is isomorphic to $\mathfrak{A}$.

9.72. For any set X, the BA associated with the product space ${}^{X}\mathbf{2}$ ($\mathbf{2}$ discrete) is freely generated by $\{\{f : f_i = 1\} : i \in X\}$.

9.73. If X generates $\mathfrak{A}$, then X freely generates $\mathfrak{A}$ iff for every $m \in \omega \sim 1$, every one-one $x \in {}^{m}X$, and every $\varepsilon \in {}^{m}2$, $\prod_{i<m} x_i^{\varepsilon i} \neq 0$, where for any $y \in A$, $y^0 = -y$ and $y^1 = y$.

9.74. Without using any principle beyond ZF, show that the following statements are equivalent:
(1) extended completeness theorem for sentential logic,
(2) Boolean representation theorem.
Hint: For (1) $\Rightarrow$ (2) it suffices by 9.68 and the proof of the Boolean representation theorem, along with 9.60 and its proof, to show that (1) $\Rightarrow$ (in any BA $\mathscr{M}_{\Gamma}^{\mathscr{P}}$ with at least two elements there is an ultrafilter). Let $\Delta = \{\varphi : \varphi \text{ a sentence of } \mathscr{P}, [\varphi] = 1\}$. Assuming $|M_{\Gamma}^{\mathscr{P}}| > 1$, show that Δ is consistent, and apply (1) to get a model f. Show that $F = \{[\varphi] : f^{+}\varphi = 1\}$ is an ultrafilter in $\mathscr{M}_{\Gamma}^{\mathscr{P}}$. For (2) $\Rightarrow$ (1), assume that Γ is a consistent set of sentences in a sentential language $\mathscr{P}$. Let h be an isomorphism from $\mathscr{M}_{\Gamma}^{\mathscr{P}}$ onto a Boolean set algebra of subsets of a set X. Since Γ is consistent, $X \neq 0$. Choose $x_0 \in X$, and let f be the model of $\mathscr{P}$ such that $fs = 1$ iff $x_0 \in h[\langle s \rangle]$. Show that f is a model of Γ.

9.75 (*Tarski*). Let $\mathfrak{A}$ be the Boolean set algebra of recursive subsets of ω. Show that a BA $\mathfrak{B}$ is isomorphic to $\mathfrak{A}$ iff $\mathfrak{B}$ is denumerable, atomic, and for every $b \in B$, if $\mathfrak{B} \restriction b$ is infinite then there are $c, d \leq b$ with $c \cdot d = 0$, $c + d = b$, and both $\mathfrak{B} \restriction c$ and $\mathfrak{B} \restriction d$ infinite. *Hint:* use Vaught's Lemma 9.47.

10 Syntactics of First-order Languages

In this chapter we give the basic definitions and results concerning the syntax of first-order languages: terms, formulas, proofs, etc. As we proceed we shall also check the effectiveness of many of the notions, although at a later stage we shall just appeal to the weak Church's thesis (see the comments preceding 3.3). This long section contains only very elementary facts, which will be used later mainly without citation. The basic definitions of syntactical notions occupy 10.1–10.18. The remainder of the chapter is concerned with elements of proof theory; this plays an important role in our discussion of decidable and undecidable theories in Part III, but will not be used in the discussion of model theory (Part IV). The basic notion of a first-order language, with which we shall be working for most of the remainder of this book, is as follows:

Definition 10.1. A first-order language is a quadruple $\mathscr{L} = (L, v, \mathscr{O}, \mathscr{R})$ with the following properties:

(*i*) L, v, $\mathscr{O}$ and $\mathscr{R}$ are functions such that $\operatorname{Rng} L$, $\operatorname{Rng} v$, $\operatorname{Dmn} \mathscr{O}$, and $\operatorname{Dmn} \mathscr{R}$ are pairwise disjoint.

(*ii*) $\operatorname{Dmn} L = 5$, and L is one-one. L_0 is the *negation symbol* of $\mathscr{L}$, L_1 the *disjunction symbol* of $\mathscr{L}$, L_2 the *conjunction symbol* of $\mathscr{L}$, L_3 the *universal quantifier* of $\mathscr{L}$, and L_4 the *equality symbol* of $\mathscr{L}$.

(*iii*) $\operatorname{Dmn} v = \omega$, and v is one-one. v_i is called the *ith individual variable.*

(*iv*) $\operatorname{Rng} \mathscr{O} \subseteq \omega$. For $\mathbf{O} \in \operatorname{Dmn} \mathscr{O}$, $\mathbf{O}$ is called an *operation symbol of rank* $\mathscr{O}\mathbf{O}$; in case $\mathscr{O}\mathbf{O} = 0$, we also refer to $\mathbf{O}$ as an *individual constant.*

(*v*) $\operatorname{Rng} \mathscr{R} \subseteq \omega \sim 1$. For $\mathbf{R} \in \operatorname{Dmn} \mathscr{R}$, $\mathbf{R}$ is called a *relation symbol of rank* $\mathscr{R}\mathbf{R}$.

In an intended interpretation (model) of a first-order language, a certain set A is chosen as the domain or universe over which the variables v_i are to range. Also, corresponding to each $\mathbf{O} \in \text{Dmn}\,\mathcal{O}$ an $\mathcal{O}\mathbf{O}$-ary operation on A is selected as the "meaning" of $\mathbf{O}$. Finally, corresponding to each $\mathbf{R} \in \text{Dmn}\,\mathcal{R}$ an $\mathcal{R}\mathbf{R}$-ary relation on A is chosen as the meaning of $\mathbf{R}$. The logical symbols, i.e., members of $\text{Rng}\,L$, are interpreted in their intuitive senses. We shall make all of this precise in the next chapter.

We will shortly define precisely the notions of term, formula (grammatical expression), etc. First let us give some examples of first-order languages.

Language of equality. We let $L = \langle \omega + i : i \in 5\rangle$, $v = \langle m : m \in \omega\rangle$, $\mathcal{O} = \mathcal{R} = 0$. This language is suitable just for expressing statements about equality alone. For example, the statement that equality is a symmetric relation is expressed by the formula

$$\forall v_0 \forall v_1 (v_0 = v_1 \rightarrow v_1 = v_0)$$

and the statement that the universe has at least three elements is expressed by the formula

$$\exists v_0 \exists v_1 \exists v_2 (\neg v_0 = v_1 \wedge \neg v_0 = v_2 \wedge \neg v_1 = v_2).$$

(For the meaning of some of the symbols here, see below.) Note that there are other languages which are clearly equivalent in all respects to this language. For example, we might let $L = \langle i : i \in 5\rangle$, $v = \langle m + 5 : m \in \omega\rangle$, $\mathcal{O} = \mathcal{R} = 0$. In general, if we have two languages $\mathcal{L} = (L, v, \mathcal{O}, \mathcal{R})$ and $\mathcal{L}' = (L', v', \mathcal{O}', \mathcal{R}')$ and there is a one-one function f from $\text{Rng}\,L \cup \text{Rng}\,v \cup \text{Dmn}\,\mathcal{O} \cup \text{Dmn}\,\mathcal{R}$ onto $\text{Rng}\,L' \cup \text{Rng}\,v' \cup \text{Dmn}\,\mathcal{O}' \cup \text{Dmn}\,\mathcal{R}'$ such that $f \circ L = L'$, $f \circ v = v'$, $\mathcal{O}' \circ f = \mathcal{O}$, and $\mathcal{R}' \circ f = \mathcal{R}$, then $\mathcal{L}$ and $\mathcal{L}'$ are equivalent in all senses of interest to us. For this reason, we usually do not specify a language in full detail; any exact specification the reader imagines should be all right for our purposes. In particular, there is no need to specify L or v. It is enough just to indicate the number and rank of the operation and relation symbols.

Language of set theory. There are no operation symbols and only one relation symbol, $\in$, which is to be binary. The statement that the union of two sets exists is:

$$\forall v_0 \forall v_1 \exists v_2 \forall v_3 (v_3 \in v_2 \leftrightarrow v_3 \in v_0 \vee v_3 \in v_1).$$

Language of rings. There are no relation symbols. The operation symbols are: $+$, $\cdot$ (binary); $-$ (unary); $\mathbf{0}$, $\mathbf{1}$ (0-ary). In this language, specific polynomials with integer coefficients can be expressed. For example, the polynomial $x^3 + 3x - 1$ is expressed by

$$((v_0 \cdot v_0) \cdot v_0 + ((\mathbf{1} + \mathbf{1}) + \mathbf{1}) \cdot v_0) + (-\mathbf{1}).$$

Language of ordered fields. This is like the language of rings, except that we add a binary relation symbol $<$.

Full language of a nonempty set A. Here we have for each operation (say m-ary) f on A an m-ary operation symbol $\mathbf{O}_f$, and for each relation (say n-ary) R on A an m-ary relation symbol $\mathbf{R}_R$ on A. More precisely, let B and C be two disjoint sets with $\alpha : \bigcup_{m\in\omega} {}^{m A}A \rightarrowtail\!\!\!\twoheadrightarrow B$ and $\beta : \bigcup_{m\in\omega\sim 1} \mathrm{S}({}^m A) \rightarrowtail\!\!\!\twoheadrightarrow C$. Then $\mathscr{L}$ is a language $(L, v, \mathscr{O}, \mathscr{R})$ such that $\mathrm{Dmn}\ \mathscr{O} = B$, $\mathrm{Dmn}\ \mathscr{R} = C$, $\mathscr{O}\alpha f =$ arity of f for each finitary operation f on A, and $\mathscr{R}\beta R =$ arity of R for each finitary relation R on A.

As we see from the last example, some of the languages we consider can have uncountably many symbols. For many of the countable languages it is of interest to consider the effectiveness of various notions to be introduced. For this purpose we make the following definition:

Definition 10.2. If $\mathscr{L} = (L, v, \mathscr{O}, \mathscr{R})$ is a first-order language, then members of $\mathrm{Dmn}\ \mathscr{O} \cup \mathrm{Dmn}\ \mathscr{R}$ are called *nonlogical constants*. The set $\mathrm{Sym}_{\mathscr{L}} = \mathrm{Rng}\ L \cup \mathrm{Rng}\ v \cup \mathrm{Dmn}\ \mathscr{O} \cup \mathrm{Dmn}\ \mathscr{R}$ is the set of *symbols* of $\mathscr{L}$.

An *effectivized first-order language* is a quintuple $\mathscr{L} = (L, v, \mathscr{O}, \mathscr{R}, \mathscr{g})$ such that $(L, v, \mathscr{O}, \mathscr{R})$ is a first-order language and in addition the following conditions hold:

(*i*) $\mathscr{g}$ is a one-one function mapping $\mathrm{Sym}_{\mathscr{L}}$ into ω;
(*ii*) $\mathscr{g} \circ v$ is recursive, $\mathrm{Rng}\,(\mathscr{g} \circ v)$ is recursive, and $v^{-1} \circ \mathscr{g}^{-1}$ is partial recursive;
(*iii*) $\mathscr{g}^*\ \mathrm{Dmn}\ \mathscr{O}$ is recursive and $\mathscr{O} \circ \mathscr{g}^{-1}$ is partial recursive;
(*iv*) $\mathscr{g}^*\ \mathrm{Dmn}\ \mathscr{R}$ is recursive and $\mathscr{R} \circ \mathscr{g}^{-1}$ is partial recursive.

Most of the languages appropriate for ordinary mathematical theories can be effectivized. For example, the language of equality above which was explicitly described as a certain quadruple $\mathscr{L} = (L, v, \mathscr{O}, \mathscr{R})$ can be effectivized as $\mathscr{L}' = (L, v, \mathscr{O}, \mathscr{R}, \mathscr{g})$, where $\mathscr{g}(\omega + i) = i$ for each $i \in 5$ and $\mathscr{g}m = m + 5$ for each $m \in \omega$.

The Gödel-numbering function $\mathscr{g}$ of 10.2 will be extended to various concepts introduced later, and the extension will always be denoted by $\mathscr{g}^+$. Unless otherwise indicated, the statements which follow refer to an arbitrary first-order language $\mathscr{L}$; when appropriate, to an arbitrary effectivized first-order language $\mathscr{L}$. The following proposition is obvious.

Proposition 10.3. *$\mathscr{g}^*\ \mathrm{Sym}_{\mathscr{L}}$ is recursive.*

Intuitively, of course, 10.3 says that there is an automatic method for recognizing a symbol of a given effectivized first-order language.

Definition 10.4. $\mathrm{Expr}_{\mathscr{L}}$, the set of *expressions* of $\mathscr{L}$, is the set of all finite sequences of symbols of $\mathscr{L}$. The empty sequence is admitted. If $\sigma = \langle \sigma_0, \ldots, \sigma_{m-1} \rangle$ is an expression of $\mathscr{L}$, we define

$$\mathscr{g}^+\sigma = \prod_{i<m} \mathrm{p}_i^{\mathscr{g}\sigma i + 1}.$$

Proposition 10.5. *$\mathscr{g}^{+*}\ \mathrm{Expr}_{\mathscr{L}}$ is recursive.*

PROOF. For any $x \in \omega$, we have $x \in \mathscr{g}^{+*}\ \mathrm{Expr}_{\mathscr{L}}$ iff $x = 1$ or else $x > 1$ and $\forall i \le \mathrm{l}x((x)_i - 1 \in \mathscr{g}^*\ \mathrm{Sym}_{\mathscr{L}})$. □

Thus there is an effective method for recognizing expressions in a given effective language.

Definition 10.6. We introduce some operations on expressions φ, ψ:

(*i*) $\neg\varphi = \langle L_0\rangle\varphi$,
(*ii*) $\varphi \vee \psi = \langle L_1\rangle\varphi\psi$,
(*iii*) $\varphi \wedge \psi = \langle L_2\rangle\varphi\psi$,
(*iv*) $\forall\alpha\varphi = \langle L_3\rangle\langle\alpha\rangle\varphi$, where α is any individual variable,
(*v*) $\varphi \equiv \psi = \langle L_4\rangle\varphi\psi$,
(*vi*) $\varphi \rightarrow \psi = \neg\varphi \vee \psi$,
(*vii*) $\varphi \leftrightarrow \psi = (\varphi \rightarrow \psi) \wedge (\psi \rightarrow \varphi)$,
(*viii*) $\exists\alpha\varphi = \neg\forall\alpha \neg \varphi$, where α is any individual variable.

Now we introduce analogous operations on ω. For the definition of Cat, see 3.30. For $m, n \in \omega$, let

(*i*)′ $\neg' m = \text{Cat}\,(2^{\mathscr{g}L0+1}, m)$,
(*ii*)′ $m \vee' n = \text{Cat}\,(\text{Cat}\,(2^{\mathscr{g}L1+1}, m), n)$,
(*iii*)′ $m \wedge' n = \text{Cat}\,(\text{Cat}\,(2^{\mathscr{g}L2+1}, m), n)$,
(*iv*)′ $\forall' mn = \text{Cat}\,(\text{Cat}\,(2^{\mathscr{g}L3+1}, m), n)$,
(*v*)′ $m \equiv' n = \text{Cat}\,(\text{Cat}\,(2^{\mathscr{g}L4+1}, m), n)$,
(*vi*)′ $m \rightarrow' n = \neg' m \vee' n$,
(*vii*)′ $m \leftrightarrow' n = (m \rightarrow' n) \wedge' (n \rightarrow' m)$,
(*viii*)′ $\exists' mn = \neg'\forall' m \neg' n$.

In addition we use $\bigvee_{i<n} \varphi_i$ and $\bigwedge_{i<n} \varphi_i$ for the finite iterations of $\vee$ and $\wedge$:

$$\bigvee_{i<1} \varphi_i = \varphi_0, \qquad \bigvee_{i<m+1} \varphi_i = \bigvee_{i<m} \varphi_i \vee \varphi_m \text{ for } m > 0;$$

$$\bigwedge_{i<1} \varphi_i = \varphi_0, \qquad \bigwedge_{i<m+1} \varphi_i = \bigwedge_{i<m} \varphi_i \wedge \varphi_m \text{ for } m > 0.$$

Strictly speaking, all of the operations in 10.6 are relative to $\mathscr{L}$. We might indicate this sometimes with a subscript, e.g., $\neg_{\mathscr{L}}$, $\rightarrow'_{\mathscr{L}}$. Note that, as for sentential languages, our languages do not have parentheses but yet we can use ordinary notation as in 10.6. The actual expressions of a language will rarely be written. That is, we will usually prefer to write an expression in the form $\forall v_0(v_0 \equiv v_1 \rightarrow v_1 \equiv v_0)$ for example rather than in the equal form

$$\langle L_3, v_0, L_1, L_0, L_4, v_0, v_1, L_4, v_1, v_0\rangle.$$

We use the boldface symbols for operations $\rightarrow$, $\equiv$, etc. on expressions to distinguish from the intuitive symbols. Note that the operations on ω given in 10.6 act on Gödel numbers of expressions just like the corresponding operations act on expressions. Thus, for example, $\mathscr{g}^+ \neg\varphi = \neg'\mathscr{g}^+\varphi, g^+\forall\alpha\varphi = \forall' 2^{\mathscr{g}\alpha+1}\mathscr{g}^+\varphi$, etc. The following proposition is obvious.

Proposition 10.7. *The operations* $\neg'$, $\vee'$, $\wedge'$, $\forall'$, $\equiv'$, $\rightarrow'$, $\leftrightarrow'$, $\exists'$ *are recursive.*

Definition 10.8

(*i*) For $m \in \omega \sim 1$ we define an m-ary operation Con_m on ω by induction on m. For any $x, y_0, \ldots, y_m \in \omega$,

$$\mathrm{Con}_1 \, x = x,$$
$$\mathrm{Con}_{m+1} \, (y_0, \ldots, y_m) = \mathrm{Cat} \, (\mathrm{Con}_m \, (y_0, \ldots, y_{m-1}), y_m).$$

Also, we define Con′ by primitive recursion. For any $x, y \in \omega$,

$$\mathrm{Con}' \, (x, 0) = (x)_0,$$
$$\mathrm{Con}' \, (x, y + 1) = \mathrm{Cat} \, (\mathrm{Con}' \, (x, y), (x)_{y+1}).$$

Finally, we set $\mathrm{Con}'' \, y = \mathrm{Con}' \, (x, \mathrm{l}x)$ for any $x \in \omega$.

(*ii*) $\mathrm{Trm}_{\mathscr{L}}$, the collection of *terms* of $\mathscr{L}$, is the intersection of all classes Γ of expressions such that the following conditions hold:

(1) $\langle v_m \rangle \in \Gamma$ for each $m \in \omega$;

(2) if $\mathbf{O} \in \mathrm{Dmn}\, \mathscr{O}$, say with $\mathscr{O}\mathbf{O} = m$, and if $\psi_0, \ldots, \psi_{m-1} \in \Gamma$, then $\langle \mathbf{O} \rangle \psi_0 \cdots \psi_{m-1} \in \Gamma$.

Note, with regard to 10.8(*ii*), that if $\mathscr{O}\mathbf{O} = \mathbf{O}$, then the condition merely says that $\langle \mathbf{O} \rangle \in \Gamma$. The number-theoretic functions in (*i*) are introduced so that the following properties of Gödel numbers will hold. If $\varphi_0, \ldots, \varphi_{m-1}$ are expressions, $m > 0$, then

$$\mathrm{Con}_m \, (\mathscr{g}^+\varphi_0, \ldots, \mathscr{g}^+\varphi_{m-1}) = \mathscr{g}^+(\varphi_0 \cdots \varphi_{m-1});$$

and if $x = \prod_{i<m} \mathrm{p}_i^{\mathscr{g}^+\varphi i}$, then

$$\mathrm{Con}'' \, x = \mathscr{g}^+(\varphi_0 \cdots \varphi_{m-1}).$$

The notion of term in 10.8 is the generalization to arbitrary first-order languages of the common mathematical notion of a polynomial. In case the first-order language is a language for rings (see above after 10.1), then we obtain exactly the ordinary notion of a polynomial with integer coefficients.

The following construction property of terms is easily established.

Proposition 10.9. *An expression σ is a term iff there is a finite sequence $\langle \tau_0, \ldots, \tau_{n-1} \rangle$ of expressions with $\tau_{n-1} = \sigma$ such that for each $i < n$ one of the following conditions holds:*

(*i*) $\tau_i = \langle v_m \rangle$ *for some* $m \in \omega$,

(*ii*) *there is an* $\mathbf{O} \in \mathrm{Dmn}\, \mathscr{O}$, *say with* $\mathscr{O}\mathbf{O} = m$, *and there are* $j_0, \ldots, j_{p-1} < i$ *such that* $\tau_i = \langle \mathbf{O} \rangle \tau_{j0} \cdots \tau_{j(p-1)}$.

There is an effective procedure for recognizing when an expression is a term:

Proposition 10.10. *The functions* Con_m, Con′, *and* Con″ *are recursive. The set* $\mathscr{g}^{+*}\,\mathrm{Trm}$ *is recursive.*

PROOF. The first statement is obvious. To prove the second, the reader should check the following statement, using 10.9. For any $x \in \omega$, $x \in g^{+*}$ Trm iff $x > 1$ and there is a $y \leq \mathrm{p}_{1x}^{x \cdot 1x}$ such that $(y)_{1y} = x$ and for each $i \leq 1y$ one of the following conditions holds:

(1) there is an $m < (y)_i$ such that $(y)_i = 2^{gvm+1}$;
(2) there is a $k \leq x$ such that $k \in g^*$ Dmn $\mathcal{O}$, $\mathcal{O}g^{-1}k = 0$, and $(y)_i = 2^{k+1}$;
(3) there exist $k \leq x$ and $z \leq \mathrm{p}_{1x}^{x \cdot 1x}$ such that $k \in g^*$ Dmn $\mathcal{O}$, $\mathcal{O}g^{-1}k > 0$, $1z = \mathcal{O}g^{-1}k - 1$, for each $j \leq 1z$ there is an $s < i$ such that $(z)_j = (y)_s$, and $(y)_i = \mathrm{Cat}\,(2^{k+1}, \mathrm{Con}''\, z)$. □

The following proposition follows directly from the definition of terms and is frequently of use.

Proposition 10.11 (Induction on terms). *If* $\langle v_m \rangle \in \Gamma$ *for all* $m \in \omega$, *and* $\langle 0 \rangle \sigma_0 \cdots \sigma_{m-1} \in \Gamma$ *whenever* $0 \in \mathrm{Dmn}\, \mathcal{O}$, $\mathcal{O}0 = m$, *and* $\sigma_0, \ldots, \sigma_{m-1} \in \Gamma$, *then* $\mathrm{Trm} \subseteq \Gamma$.

Proposition 10.12 (Unique readability)

(*i*) *Every term is nonempty.*

(*ii*) *If* σ *is a term, then either* $\sigma = \langle v_m \rangle$ *for some* $m \in \omega$, *or else there exist* $\mathbf{O} \in \mathrm{Dmn}\, \mathcal{O}$, *say with* $\mathcal{O}\mathbf{O} = m$, *and* $\tau_0, \ldots, \tau_{m-1} \in \mathrm{Trm}$ *such that* $\sigma = \langle \mathbf{O} \rangle \tau_0 \cdots \tau_{m-1}$.

(*iii*) *If* σ *is a term and* $i <$ Dmn σ, *then* $\langle \sigma_0, \ldots, \sigma_{i-1} \rangle$ *is not a term.*

(*iv*) *If* $\mathbf{O}, \mathbf{P} \in \mathrm{Dmn}\, \mathcal{O}$, *say* $\mathcal{O}\mathbf{O} = m$, $\mathcal{O}\mathbf{P} = n$, *and* $\sigma \in {}^m\mathrm{Trm}$, $\tau \in {}^n\mathrm{Trm}$, *and* $\langle \mathbf{O} \rangle \sigma_0 \cdots \sigma_{m-1} = \langle \mathbf{P} \rangle \tau_0 \cdots \tau_{n-1}$, *then* $m = n$, $\mathbf{O} = \mathbf{P}$, *and* $\sigma = \tau$.

PROOF. In each of the cases (*i*) and (*ii*), let Γ be the collection of terms σ for which the desired condition holds, and apply 10.11. We prove (*iii*) by induction on Dmn σ. The case Dmn $\sigma = 1$ is clear by (*i*). Now assume that Dmn $\sigma > 1$ and that (*iii*) is true for all terms τ such that Dmn $\tau <$ Dmn σ. By (*ii*) we may write $\sigma = \langle \mathbf{O} \rangle \tau_0 \cdots \tau_{m-1}$, where $\mathbf{O} \in \mathrm{Dmn}\, \mathcal{O}$, $\mathcal{O}\mathbf{O} = m$, and $\tau \in {}^m\mathrm{Trm}$. If $i = 0$, then $0 = \langle \sigma_0, \ldots, \sigma_{i-1} \rangle$ is not a term, by (*i*). If $i = 1$, then $\langle \sigma_0, \ldots, \sigma_{i-1} \rangle = \langle \mathbf{O} \rangle$, which fails to be a term because (1) $i <$ Dmn σ implies $m > 0$ and (2) by (*ii*), for $\langle \mathbf{O} \rangle$ to be a term we would have to have $\mathcal{O}\mathbf{O} = 0$. So, assume that $i > 1$. Then $\langle \sigma_0, \ldots, \sigma_{i-1} \rangle = \langle \mathbf{O} \rangle \tau_0 \cdots \tau_{j-1} \rho$ for some $j < m$ and some expression ρ which is an initial segment of τ_j. Assume that $\langle \sigma_0, \ldots, \sigma_{i-1} \rangle$ is a term. By (*ii*) there exist terms $\xi_0, \ldots, \xi_{m-1}$ such that $\langle \sigma_0, \ldots, \sigma_{i-1} \rangle = \langle \mathbf{O} \rangle \xi_0 \cdots \xi_{m-1}$. By the induction hypothesis we easily infer that $\xi_0 = \tau_0, \ldots, \xi_{j-1} = \tau_{j-1}$. Hence $\rho = \xi_j \cdots \xi_{m-1}$. Hence ξ_j is an initial segment of τ_j, so by the induction hypothesis $\xi_j = \tau_j$ and $j = m - 1$. But then $\langle \sigma_0, \ldots, \sigma_{i-1} \rangle = \sigma$, contradiction.

Condition (*iv*) follows from (*iii*) in an obvious fashion. □

The following useful proposition follows from 10.12 purely set theoretically:

Proposition 10.13 (Recursion on terms). *Let f map* Rng v *into a set* A, *and for each* $\mathbf{O} \in \text{Dmn}\, \mathcal{O}$ *(say with* $\mathcal{O}\mathbf{O} = m$*), let* $g_{\mathbf{O}}$ *map* ${}^mA \times {}^m\text{Trm}$ *into* A. *Then there is an* $h\colon \text{Trm} \to A$ *such that:*

(*i*) $h\langle v_n \rangle = fv_n$ *for all* $n \in \omega$;

(*ii*) $h\langle \mathbf{O} \rangle \sigma_0 \cdots \sigma_{m-1} = g_{\mathbf{O}}(h\sigma_0, \ldots, h\sigma_{m-1}, \sigma_0, \ldots, \sigma_{m-1})$ *whenever* $\mathbf{O} \in \text{Dmn}\, \mathcal{O}$, $\mathcal{O}\mathbf{O} = m$, *and* $\sigma \in {}^m\text{Trm}$.

Now we can define the most important syntactical notion, that of a formula (grammatically built expression).

Definition 10.14. An expression of the form $\sigma \equiv \tau$, where σ and τ are terms, is called an *atomic equality formula*. If $\mathbf{R} \in \text{Dmn}\, \mathcal{R}$, say $\mathcal{R}\mathbf{R} = m$, and $\sigma_0, \ldots, \sigma_{m-1}$ are terms, then $\langle \mathbf{R} \rangle \sigma_0 \cdots \sigma_{m-1}$ is an *atomic nonequality formula*. These two kinds of expressions together constitute the *atomic formulas*. The set of *formulas*, Fmla, is the intersection of all sets Γ of expressions such that each atomic formula is in Γ and Γ is closed under all the operations $\neg$, $\vee$, $\wedge$, $\forall\alpha$ (for each individual variable α).

There is an automatic method for recognizing when an expression is a formula:

Proposition 10.15. g^{+*} Fmla *is recursive.*

PROOF. The proposition becomes obvious after checking the following statement whose proof is based on a proposition for formulas which is entirely similar to 10.9. For any $x \in \omega$, $x \in g^{+*}$ Fmla iff there is a $y \leq \text{p}_{1x}^{x \cdot 1x}$ such that $(y)_{1y} = x$ and for each $i \leq 1y$ one of the following conditions holds:

(1) There are $u, v \in g^{+*}$ Trm such that $(y)_i = u \equiv' v$.
(2) There exist a $k \leq x$ and $z \leq \text{p}_{1x}^{x \cdot 1x}$ such that $k \in g^*$ Dmn $\mathcal{R}$, $1z = \mathcal{O}g^{-1}k - 1$, $(z)_j \in$ Trm for each $j \leq 1z$, and $(y)_i = \text{Cat}\,(2^{h+1}, \text{Con}''\, z)$.
(3) There is a $j < i$ such that $(y)_i = \neg'(y)_j$.
(4) There exist $j, k < i$ such that $(y)_i = (y)_j \vee' (y)_k$.
(5) There exist $j, k < i$ such that $(y)_i = (y)_j \wedge' (y)_k$.
(6) There exist $m < (y)_i$ and $j < i$ such that $(y)_i = \forall' 2^{gvm+1}(y)_j$. □

The following results are established in a fashion analogous to the corresponding results for terms:

Proposition 10.16 (Induction on formulas). *If* Γ *is a set of formulas containing each atomic formula and closed under all operations* $\neg$, $\vee$, $\wedge$, $\forall\alpha$ *(for each individual variable* α*), then* Fmla $\subseteq \Gamma$.

Proposition 10.17 (Unique readability)

(*i*) *Every formula is nonempty.*

(*ii*) *If* φ *is a formula, then one of the following holds:*

(1) *φ is an atomic equality formula, i.e., there are terms σ and τ with $\varphi = \sigma \equiv \tau$;*
(2) *φ is an atomic nonequality formula, i.e., there is an $\mathbf{R} \in \text{Dmn}\,\mathscr{R}$, say with $\mathscr{R}\mathbf{R} = m$, and there are $\sigma_0, \ldots, \sigma_{m-1} \in \text{Trm}$ such that $\varphi = \langle\mathbf{R}\rangle\sigma_0 \cdots \sigma_{m-1}$;*
(3) *$\varphi = \neg\psi$ for some formula ψ;*
(4) *$\varphi = \psi \vee \chi$ for some formulas ψ, χ;*
(5) *$\varphi = \psi \wedge \chi$ for some formulas ψ, χ;*
(6) *$\varphi = \forall\alpha\psi$ for some individual variable α and some formula ψ.*

(*iii*) *If φ is a formula and $i < \text{Dmn}\,\varphi$, then $\langle\varphi_0, \ldots, \varphi_{i-1}\rangle$ is not a formula.*

(*iv*) *for each formula φ, exactly one of* (1)–(6) *above holds, and the terms, formulas, etc. $\sigma, \tau, \psi, \ldots$ asserted to exist are uniquely determined by φ.*

Proposition 10.18 (Recursion on formulas). *Let f map the set of atomic formulas into a set A, let g map $A \times \text{Fmla}$ into A, h and k map ${}^2A \times {}^2\text{Fmla}$ into A, and let l map $A \times \text{Rng}\,v \times \text{Fmla}$ into A. Then there is an $s\colon \text{Fmla} \to A$ such that for any formulas φ, ψ and any variable α,*

(*i*) *$s\varphi = f\varphi$ if φ is atomic;*
(*ii*) *$s(\neg\varphi) = g(s\varphi, \varphi)$;*
(*iii*) *$s(\varphi \vee \psi) = h(s\varphi, s\psi, \varphi, \psi)$;*
(*iv*) *$s(\varphi \wedge \psi) = k(s\varphi, s\psi, \varphi, \psi)$;*
(*v*) *$s(\forall\alpha\varphi) = l(s\varphi, \alpha, \varphi)$.*

This takes care of the basic facts about grammar in first-order languages. Now we want to formulate precisely the basic proof-theoretic notions of logical axioms and theorems. Essentially we want to give a mathematical analysis of the intuitive notion of a proof. To begin with, we need to define the notion of *tautology* in a first-order language. Intuitively speaking, a tautology is a formula which is logically valid purely on the basis of the intended meanings of $\neg$, $\vee$, $\wedge$ alone. They were discussed in detail in Chapter 8, but our treatment here is self-contained.

Definition 10.19

(*i*) A *truth valuation* is a function f mapping Fmla into 2 with the following properties, for any formulas φ, ψ:

(1) $f(\neg\varphi) = 1$ iff $f\varphi = 0$;
(2) $f(\varphi \vee \psi) = 1$ iff $f\varphi = 1$ or $f\psi = 1$;
(3) $f(\varphi \wedge \psi) = 1$ iff $f\varphi = 1$ and $f\chi = 1$.

(*ii*) A formula φ is a *tautology* iff $f\varphi = 1$ for every truth valuation f.

(*iii*) An expression ψ *occurs* in an expression φ iff $\varphi = \chi\psi\theta$ for some expressions χ, θ.

(*iv*) ψ is a *subformula* of φ iff both ψ and φ are formulas and ψ occurs in φ.

The following lemma gives a simplified definition for the notion of a tautology:

Lemma 10.20. *For any formula φ the following conditions are equivalent:*

(i) φ is a tautology;

(ii) for any f, if f maps the set S of subformulas of φ into 2 and satisfies the following conditions;

(a) for any ψ, if $\psi \in S$ and $\neg\psi \in S$, then $f(\neg\psi) = 1$ iff $f\psi = 0$;

(b) for any ψ, χ, if $\psi, \chi, \psi \vee \chi \in S$, then $f(\psi \vee \chi) = 1$ iff ($f\psi = 1$ or $f\chi = 1$);

(c) for any ψ, χ, if $\psi, \chi, \psi \wedge \chi \in S$, then $f(\psi \wedge \chi) = 1$ iff ($f\psi = 1$ and $f\chi = 1$);

then $f\varphi = 1$.

PROOF

(*i*) ⇒ (*ii*). Assume that φ is a tautology, and let f satisfy the hypothesis of (*ii*). By recursion on formulas we define a function h: Fmla → 2 as follows:

$$
\begin{aligned}
h\psi &= f\psi \quad &&\text{if } \psi \text{ is an atomic formula} \in S,\\
h\psi &= 0 &&\text{if } \psi \text{ is an atomic formula} \notin S,\\
h(\neg\psi) &= 1 - h\psi &&\text{for any formula } \psi,\\
h(\psi \vee \chi) &= 1 &&\text{if } h\psi = 1 \text{ or } h\chi = 1,\\
h(\psi \vee \chi) &= 0 &&\text{if } h\psi = h\chi = 0,\\
h(\psi \wedge \chi) &= h\psi \cdot h\chi,\\
h\forall\alpha\psi &= f\forall\alpha\psi &&\text{if } \forall\alpha\psi \in S,\\
h\forall\alpha\psi &= 0 &&\text{if } \forall\alpha\psi \notin S.
\end{aligned}
$$

Clearly h is a truth valuation, so $h\varphi = 1$. Now it is easily shown by induction on formulas that

$$\forall\psi \in \text{Fmla}\ (\psi \in S \Rightarrow h\psi = f\psi).$$

Hence $f\varphi = 1$, as desired.

(*ii*) ⇒ (*i*). Assume (*ii*), and, in order to show that φ is a tautology, let h be any truth valuation. Let $f = h \upharpoonright S$. Clearly f satisfies the hypothesis of (*ii*), so $h\varphi = f\varphi = 1$. □

Clearly 10.20 gives an automatic method for checking whether a given formula φ is a tautology. We simply list out all members of S, then check for each of the finitely many f: $S \to 2$ whether the condition (*ii*) holds. Now we want to make this argument rigorous.

Proposition 10.21. $\{(\mathscr{g}^+\psi, \mathscr{g}^+\varphi) : \psi$ *is a subformula of* $\varphi\}$ *is recursive.*

PROOF. Let R be the set in question. Then for any $m, n \in \omega$, $(m, n) \in R$ iff $m, n \in \mathscr{g}^{+*}$ Fmla and $\exists x, y \leq n(n = \text{Cat}(\text{Cat}(x, m), y))$. □

Proposition 10.22. *The set of Gödel numbers of tautologies is recursive.*

PROOF. Let R be the relation of 10.21. Using 10.20, it is easy to check that for any $m \in \omega$, m is the Gödel number of a tautology iff:

$$\exists n \leq \mathrm{p}_{\mathrm{l}m}^{m \cdot \mathrm{l}m} \{n > 1 \;\&\; \forall p \leq m[(p, m) \in R \Rightarrow \exists i \leq \mathrm{l}n(p = (n)_i)] \;\&\; \forall i \leq \mathrm{l}n[((n)_i, m) \in R] \;\&\; (n)_{\mathrm{l}n} = m \;\&\; \forall f \leq \mathrm{p}_{\mathrm{l}m}^{m} [\mathrm{l}f = \mathrm{l}n \;\&\; \forall i \leq \mathrm{l}f [(f)_i = 1 \text{ or } (f)_i = 2] \;\&\; \forall i \leq \mathrm{l}f \forall j \leq \mathrm{l}f [\neg'(n)_j = (n)_i \Rightarrow ((f)_i = 2 \Leftrightarrow (f)_j = 1)] \;\&\; \forall i \leq \mathrm{l}f \forall j \leq \mathrm{l}f \forall k \leq \mathrm{l}f [(n)_j \vee' (n)_k = (n)_i \Rightarrow ((f)_i = 2 \Leftrightarrow (f)_j = 2 \text{ or } (f)_k = 2)] \;\&\; \forall i \leq \mathrm{l}f \forall j \leq \mathrm{l}f \forall k \leq \mathrm{l}f [(n)_j \wedge' (n)_k = (n)_i \Rightarrow ((f)_i = 2 \text{ iff } (f)_j = 2 \text{ and } (f)_k = 2)] \Rightarrow (f)_{\mathrm{l}f} = 2]\}.$$

Note here that n is used to code the set S of all subformulas of φ, if $m = g^+\varphi$, while f is used to code any function as in 10.20(*ii*). Of course, the form of the above displayed expression shows that $\{m : m$ is the Gödel number of a tautology$\}$ is recursive. □

Tautologies are by no means the only logically valid formulas in a first order language. Some additional ones are, for example, the formulas $v_0 = v_0$, $\forall v_0(v_0 = v_1) \rightarrow v_2 = v_3$, $\forall v_0[v_0 = v_1 \vee \neg(v_0 = v_1)]$. We now want to formulate the notion of a logically valid formula, or *theorem*, in a general way, but still syntactically. We defer to the next section the proof of the completeness theorem, which states that the theorems are exactly the logically valid formulas. See also the next section for the rigorous definition of a logically valid formula.

Definition 10.23

(*i*) $\mathrm{Axm}_{\mathscr{L}}$, the set of *logical axioms* of $\mathscr{L}$, is the set of all formulas of the following forms (where φ and ψ are formulas, $i < \omega$, and σ and τ are terms):

(1) φ, φ any tautology;
(2) $\forall v_i(\varphi \rightarrow \psi) \rightarrow (\forall v_i\varphi \rightarrow \forall v_i\psi)$;
(3) $\varphi \rightarrow \forall v_i\varphi$, if v_i does not occur in φ;
(4) $\exists v_i(v_i = \sigma)$, if v_i does not occur in σ;
(5) $\sigma = \tau \rightarrow (\varphi \rightarrow \psi)$, if φ and ψ are atomic formulas and ψ is obtained from φ by replacing an occurrence of σ in φ by τ.

(*ii*) Let Γ be a set of formulas. Then Γ-$\mathrm{Thm}_{\mathscr{L}}$, the set of all Γ-*theorems* of $\mathscr{L}$, is the intersection of all subsets Δ of $\mathrm{Fmla}_{\mathscr{L}}$ such that

(1) $\Gamma \cup \mathrm{Axm}_{\mathscr{L}} \subseteq \Delta$;
(2) $\psi \in \Delta$ whenever φ, $\varphi \rightarrow \psi \in \Delta$ (closure under *detachment* or *modus ponens*);
(3) $\forall v_i\varphi \in \Delta$ whenever $\varphi \in \Delta$ and $i < \omega$ (closure under *generalization*).

We write $\mathrm{Thm}_{\mathscr{L}}$ instead of 0-$\mathrm{Thm}_{\mathscr{L}}$, $\Gamma \vdash_{\mathscr{L}} \varphi$ instead of $\varphi \in \Gamma$-$\mathrm{Thm}_{\mathscr{L}}$, and $\vdash_{\mathscr{L}}\varphi$ instead of $0 \vdash_{\mathscr{L}} \varphi$. Formulas φ with $\vdash\varphi$ are called *logical theorems*. If $\Gamma \vdash_{\mathscr{L}} \varphi$, we call φ a (syntactical) *consequence* of Γ. Of course when no confusion is likely we shall omit the subscript $\mathscr{L}$ in all these cases.

As usual, we give an equivalent version of this closure definition of theorems:

Proposition 10.24. *For* $\Gamma \cup \{\varphi\} \subseteq \text{Fmla}$, $\Gamma \vdash \varphi$ *iff there is a finite sequence* $\langle \psi_0, \ldots, \psi_{m-1} \rangle$ *of formulas such that* $\psi_{m-1} = \varphi$ *and for each* $i < m$ *one of the following conditions holds:*

(*i*) $\psi_i \in \text{Axm}$,
(*ii*) $\psi_i \in \Gamma$,
(*iii*) $\exists j, k < i$ *such that* $\psi_j = \psi_k \to \psi_i$,
(*iv*) $\exists j < i \, \exists k \in \omega \, (\psi_i = \forall v_k \psi_j)$.

A sequence of the sort described in 10.24 is called a *formal proof of* φ *from* Γ, or a Γ-*formal proof of* φ. We denote by Γ-Prf the set of all Γ-proofs. This is our rigorous formulation of the intuitive notion of a proof. In fact, as stated in the introduction, we consider mathematics itself to be formalized on the basis of set theory. More precisely, mathematical language can be identified with a certain definitional expansion of the language of set theory describing following 10.1. The axioms Γ of mathematics are just the usual axioms for set theory together with definitions of all the defined symbols. It is our conviction that any mathematical proof can be expanded, somewhat routinely, to eventually reach the form of a formal proof from Γ in the above sense. Of course, this conviction is another instance, like the weak Church's thesis (see the comments following 3.1), of a judgement of applied mathematics that is not subject to a rigorous proof. We are just stating that our rigorous notion of proof is a fully adequate mathematical version of the mathematical proofs actually found in articles and books.

It is clear that there is an effective method for recognizing when an expression is a logical axiom. More rigorously, we have the following theorem:

Proposition 10.25. $\mathcal{g}^{+*}$ Axm *is recursive.*

Proof. For any $x \in \omega$, $x \in \mathcal{g}^{+*}$ Axm iff one of the following conditions holds:

(1) x is the Gödel number of a tautology;
(2) $\exists y, z, w \le x[y, z \in \mathcal{g}^{+*}$ Fmla and $w \in \text{Rng}(\mathcal{g} \circ v)$ and $x = \forall' 2^{w+1}(y \to' z) \to' (\forall' 2^{w+1} y \to' \forall' 2^{w+1} z)]$;
(3) $\exists y, z \le x[y \in \mathcal{g}^{+*}$ Fmla and $z \in \text{Rng}(\mathcal{g} \circ v)$ and $\neg \exists u, w \le y$ $y = \text{Cat}(\text{Cat}(u, 2^{z+1}), w)$ and $x = y \to' \forall' 2^{z+1} y]$;
(4) $\exists y, z \le x[y \in \mathcal{g}^{+*}$ Trm and $z \in \text{Rng}(\mathcal{g} \circ v)$ and $\neg \exists u, w \le y$ $y = \text{Cat}(\text{Cat}(u, 2^{z+1}), w)$ and $x = \exists' 2^{z+1}(2^{z+1} =' y)]$;
(5) $\exists s, t, y, z \le x[s, t \in \mathcal{g}^{+*}$ Trm and y and z are Gödel numbers of atomic formulas and $\exists m, n \le y[y = \text{Cat}(\text{Cat}(m, s), n)$ and $z = \text{Cat}(\text{Cat}(m, t), n)]$ and $x = (s =' t) \to' (y \to' z)]$. □

Since there is an automatic method for recognizing logical axioms, it is clear from 10.24 that if there is an automatic method for recognizing members of Γ, then there is an automatic method for checking when a sequence of formulas is a Γ-formal proof. We now proceed to prove this rigorously.

Definition 10.26. We extend $\mathcal{g}$ and $\mathcal{g}^+$ further to finite sequences of expressions. If $\varphi = \langle \varphi_0, \ldots, \varphi_{m-1} \rangle$ is a finite sequence of expressions, let

$$\mathcal{g}^{++}\varphi = \prod_{i<m} \mathrm{p}_i^{\mathcal{g}^+\varphi_i}.$$

Proposition 10.27. *Let Γ be a set of formulas such that $\mathcal{g}^{+*}\Gamma$ is recursive. Then $\mathcal{g}^{++}(\Gamma\text{-Prf})$ is recursive.*

Proof. For any $x \in \omega$ we have: $x \in \mathcal{g}^{++*}(\Gamma\text{-Prf})$ iff $x > 1$ and for every $i \leq \mathrm{l}x$ one of the following conditions holds:

(1) $(x)_i \in \mathcal{g}^{+*}\,\mathrm{Axm}$,
(2) $(x)_i \in \mathcal{g}^{+*}\Gamma$,
(3) $\exists j, k < i\ (x)_j = (x)_k \rightarrow' (x)_i$,
(4) $\exists j < i\ \exists k \leq x[k \in \mathrm{Rng}\,(\mathcal{g} \circ v)$ and $(x)_i = \forall' 2^{k+1}(x)_j]$. □

From this proof we also obviously obtain:

Proposition 10.28. *If Γ is a set of formulas such that $\mathcal{g}^{+*}\Gamma$ is recursively enumerable, then $\mathcal{g}^{++*}(\Gamma\text{-Prf})$ is recursively enumerable.*

The following easy consequence of this proposition is one of the most important results of elementary logic.

Theorem 10.29. *Let Γ be a set of formulas such that $\mathcal{g}^{+*}\Gamma$ is recursively enumerable. Then $\mathcal{g}^{+*}(\Gamma\text{-Thm})$ is recursively enumerable.*

Proof. For any $x \in \omega$, we have $x \in \mathcal{g}^{+*}(\Gamma\text{-Thm})$ iff $\exists y[y \in \mathcal{g}^{++*}(\Gamma\text{-Prf})$ and $(y)_{\mathrm{l}y} = x]$. □

An intuitive proof of Theorem 10.29 runs as follows. We assume that Γ can be listed by some effective procedure A. Let B be an effective procedure which lists all formulas. Now we describe an effective procedure C listing all Γ-theorems. We start the procedures A and B going simultaneously. At the kth stage of the procedure C, having accomplished the kth stage of both A and B, we list out all sequences of length $\leq k$ of formulas already produced by B. For each such sequence φ, we check whether it is a Γ-proof using for members of Γ only the formulas already produced by A. For each sequence φ for which the answer is affirmative, procedure C produces as an output the last term of φ. Clearly the procedure C so described generates precisely the Γ-theorems.

It is quite possible in 10.29 to have $\mathcal{g}^{+*}(\Gamma\text{-Thm})$ nonrecursive; this is in fact the defining characteristic of the undecidable theories which will be discussed extensively in Part III.

Most of the remainder of this chapter is devoted to establishing elementary results about the relation $\vdash$ defined in 10.23. Some high points of this development, which is rather tedious because of our economical system of axioms, are as follows: a formal expression of the principle of substitution of equals for equals, 10.49; the principle for changing bound variables, 10.59; universal specification—dropping a universal quantifier, 10.61; substitutivity of equivalent formulas, 10.71; prenex normal form theorem, 10.81; provability for sentences alone, 10.85; and the notion of consistency, 10.89–10.93.

Proposition 10.30. *For any $m \in \omega$, if $\alpha \in {}^m\mathrm{Rng}\, v$ and φ and ψ are formulas, then*

$$\vdash \forall\alpha_0 \cdots \forall\alpha_{m-1}(\varphi \to \psi) \to (\forall\alpha_0 \cdots \forall\alpha_{m-1}\varphi \to \forall\alpha_0 \cdots \forall\alpha_{m-1}\psi).$$

PROOF. We proceed by induction on m. The case $m = 0$ is trivial, since $(\varphi \to \psi) \to (\varphi \to \psi)$ is a tautology. Now assume the result for m, and suppose that $\varphi \in {}^{m+1}\,\mathrm{Rng}\, v$. Then

$$\vdash \forall\alpha_1 \cdots \forall\alpha_m(\varphi \to \psi) \to (\forall\alpha_1 \cdots \forall\alpha_m\varphi \to \forall\alpha_1 \cdots \forall\alpha_m\psi)$$

by the induction hypothesis. Generalizing on α_0, noticing an instance of the axioms 10.23(2), and applying detachment, we get

$$\vdash \forall\alpha_0 \cdots \forall\alpha_m(\varphi \to \psi) \to \forall\alpha_0(\forall\alpha_1 \cdots \forall\alpha_m\varphi \to \forall\alpha_1 \cdots \forall\alpha_m\psi).$$

Now by another instance of 10.23(2), a tautology, and detachment, we get our desired result for $m + 1$. □

Related to 10.30 is the following important theorem, to the effect that there is a version of $\vdash$ not involving the rule of generalization:

Definition 10.31. Let $\Gamma \subseteq \mathrm{Fmla}_{\mathscr{L}}$. We let Γ-Thm$'$ be the intersection of all sets $\Delta \subseteq \mathrm{Fmla}_{\mathscr{L}}$ satisfying the following conditions:

(*i*) if $\varphi \in \mathrm{Axm}_{\mathscr{L}} \cup \Gamma$, $m \in \omega$, and $\alpha \in {}^m\mathrm{Rng}\, v$, then $\forall\alpha_0 \cdots \forall\alpha_{m-1}\varphi \in \Delta$;

(*ii*) $\psi \in \Delta$ whenever $\varphi, \varphi \to \psi \in \Delta$.

We write $\mathrm{Thm}'_{\mathscr{L}}$ instead of $0\text{-}\mathrm{Thm}'_{\mathscr{L}}$, $\Gamma \vdash'_{\mathscr{L}} \varphi$ instead of $\varphi \in \Gamma\text{-}\mathrm{Thm}'_{\mathscr{L}}$, and $\vdash'_{\mathscr{L}}\varphi$ instead of $0 \vdash'_{\mathscr{L}} \varphi$.

Theorem 10.32. $\Gamma \vdash \varphi$ *iff* $\Gamma \vdash' \varphi$.

PROOF. Γ-Thm clearly satisfies the conditions 10.31(*i*) and 10.31(*ii*). Hence Γ-Thm$' \subseteq \Gamma$-Thm, i.e., $\Gamma \vdash' \varphi \Rightarrow \Gamma \vdash \varphi$. To prove the converse, let

$$\Delta = \{\varphi \in \mathrm{Fmla} : \text{for all } m \in \omega \text{ and all } \alpha \in {}^m\mathrm{Rng}\, v,\ \Gamma \vdash' \forall\alpha_0 \cdots \forall\alpha_{m-1}\varphi\}.$$

Obviously $\mathrm{Axm} \cup \Gamma \subseteq \Delta$ and Δ is closed under generalization. To show that Δ is closed under detachment it is clearly sufficient to establish

(1) if $n \in \omega$, $\alpha \in {}^n\mathrm{Rng}\, v$, $\Gamma \vdash' \forall\alpha_0 \cdots \forall\alpha_{n-1}\varphi$, and $\Gamma \vdash' \forall\alpha_0 \cdots \forall\alpha_{n-1}(\varphi \to \psi)$, then $\Gamma \vdash' \forall\alpha_0 \cdots \forall\alpha_{n-1}\psi$.

We prove (1) by induction on n. The case $n = 0$ is trivial. Suppose true for n, $\Gamma \vdash' \forall\alpha_0 \cdots \forall\alpha_n\varphi$, $\Gamma \vdash' \forall\alpha_0 \cdots \forall\alpha_n(\varphi \rightarrow \psi)$. Now

$$\Gamma \vdash' \forall\alpha_0 \cdots \forall\alpha_{n-1}(\forall\alpha_n(\varphi \rightarrow \psi) \rightarrow (\forall\alpha_n\varphi \rightarrow \forall\alpha_n\psi))$$

and $\Gamma \vdash' \forall\alpha_0 \cdots \forall\alpha_{n-1}\, \forall\alpha_n(\varphi \rightarrow \psi)$, so by the induction hypothesis, $\Gamma \vdash' \forall\alpha_0 \cdots \forall\alpha_{n-1}(\forall\alpha_n\varphi \rightarrow \forall\alpha_n\psi)$. By the induction hypothesis again, $\Gamma \vdash' \forall\alpha_0 \cdots \forall\alpha_n\psi$, as desired, hence Γ-Thm $\subseteq \Delta$. In particular (taking $m = 0$ in the definition of Δ), $\Gamma \vdash \varphi \Rightarrow \Gamma \vdash' \varphi$. □

In the following proposition we summarize some simple properties of $\vdash$ which will be used frequently without citation.

Proposition 10.33. *Let* $\Gamma, \Delta \subseteq$ Fmla *and* $\varphi, \psi \in$ Fmla. *Then*
(*i*) *if* $\Gamma \subseteq \Delta$ *and* $\Gamma \vdash \varphi$, *then* $\Delta \vdash \varphi$;
(*ii*) *if* $\Gamma \vdash \varphi$, *then* $\Theta \vdash \varphi$ *for some finite subset* Θ *of* Γ;
(*iii*) *if* $\Gamma \vdash \chi$ *for each* $\chi \in \Delta$ *and* $\Delta \vdash \varphi$, *then* $\Gamma \vdash \varphi$;
(*iv*) *if* $\Gamma \vdash \varphi$ *and* $\Gamma \vdash \varphi \rightarrow \psi$, *then* $\Gamma \vdash \psi$.

In what follows, α, β will denote arbitrary variables, σ, τ, ρ will denote arbitrary terms, and φ, ψ, χ arbitrary formulas.

Proposition 10.34. $\vdash \sigma \equiv \sigma$.

PROOF. Let α be a variable not occurring in σ. Then

$\vdash \alpha \equiv \sigma \rightarrow (\alpha \equiv \sigma \rightarrow \sigma \equiv \sigma)$	by 10.23(5)
$\vdash \neg(\sigma \equiv \sigma) \rightarrow \neg(\alpha \equiv \sigma)$	by a tautology, detachment
$\vdash \forall\alpha[\neg(\sigma \equiv \sigma) \rightarrow \neg(\alpha \equiv \sigma)]$	generalization
$\vdash \forall\alpha\ \neg(\sigma \equiv \sigma) \rightarrow \forall\alpha\ \neg(\alpha \equiv \sigma)$	10.23(2), detachment
$\vdash \neg(\sigma \equiv \sigma) \rightarrow \forall\alpha\ \neg(\sigma \equiv \sigma)$	10.23(3)
$\vdash \neg(\sigma \equiv \sigma) \rightarrow \forall\alpha\ \neg(\alpha \equiv \sigma)$	tautology, etc.
$\vdash \exists\alpha(\alpha \equiv \sigma) \rightarrow \sigma \equiv \sigma$	tautology, etc.
$\vdash \sigma \equiv \sigma$	using 10.23(4) □

Proposition 10.35. $\vdash \sigma \equiv \tau \rightarrow \tau \equiv \sigma$.

PROOF.

$\vdash \sigma \equiv \tau \rightarrow (\sigma \equiv \sigma \rightarrow \tau \equiv \sigma)$	10.23(5)
$\vdash \sigma \equiv \tau \rightarrow \tau \equiv \sigma$	using a tautology □

Proposition 10.36. $\vdash \sigma \equiv \tau \rightarrow (\tau \equiv \rho \rightarrow \sigma \equiv \rho)$.

PROOF.

$\vdash \tau \equiv \sigma \rightarrow (\tau \equiv \rho \rightarrow \sigma \equiv \rho)$	10.23(5)
$\vdash \sigma \equiv \tau \rightarrow (\tau \equiv \rho \rightarrow \sigma \equiv \rho)$	using 10.35 and a tautology □

Lemma 10.37. *If* **O** *is an operation symbol of rank m, $i < m$, $\sigma_0, \ldots, \sigma_{m-1}$ are terms, and τ is a term, then*

$$\vdash \sigma_i \equiv \tau \rightarrow \mathbf{O}\sigma_0 \cdots \sigma_{m-1} = \mathbf{O}\sigma_0 \cdots \sigma_{i-1}\tau\sigma_{i+1} \cdots \sigma_{m-1}.$$

PROOF. The following formula is an instance of 10.23:

$$\sigma_i \equiv \tau \rightarrow (\mathbf{O}\sigma_0 \cdots \sigma_{m-1} = \mathbf{O}\sigma_0 \cdots \sigma_{m-1} \rightarrow \mathbf{O}\sigma_0 \cdots \sigma_{m-1} \equiv \mathbf{O}\sigma_0 \cdots \sigma_{i-1}\tau\sigma_{i+1} \cdots \sigma_{m-1}).$$

Hence 10.34 and a suitable tautology give the desired result. □

Using this lemma, we give our first form of the principle of substitution of equals for equals:

Theorem 10.38. *If* **O** *is an operation symbol of rank m, and $\sigma_0, \ldots, \sigma_{m-1}$, $\tau_0, \ldots, \tau_{m-1}$ are terms, then*

$$\vdash \sigma_0 \equiv \tau_0 \wedge \cdots \wedge \sigma_{m-1} \equiv \tau_{m-1} \rightarrow \mathbf{O}\sigma_0 \cdots \sigma_{m-1} \equiv \mathbf{O}\tau_0 \cdots \tau_{m-1}.$$

PROOF. By 10.37 we have, for each $i < m$,

$$\vdash \sigma_i \equiv \tau_i \rightarrow \mathbf{O}\tau_0 \cdots \tau_{i-1}\sigma_i\sigma_{i+1} \cdots \sigma_{m-1} \equiv \mathbf{O}\tau_0 \cdots \tau_i\sigma_{i+1} \cdots \sigma_{m-1}.$$

Hence 10.36 and an easy but long tautology give the desired result. □

Lemma 10.39. *If $\vdash \sigma \equiv \tau \rightarrow (\varphi \rightarrow \psi)$ and $\alpha \in \operatorname{Rng} v$ does not occur in σ or in τ, then $\vdash \sigma \equiv \tau \rightarrow (\forall\alpha\varphi \rightarrow \forall\alpha\psi)$.*

PROOF.

$\vdash \sigma \equiv \tau \rightarrow (\varphi \rightarrow \psi)$	hypothesis
$\vdash \forall\alpha(\sigma \equiv \tau) \rightarrow \forall\alpha(\varphi \rightarrow \psi)$	generalization, 10.23(2)
$\vdash \sigma \equiv \tau \rightarrow \forall\alpha(\sigma \equiv \tau)$	10.23(3)
$\vdash \sigma \equiv \tau \rightarrow (\forall\alpha\varphi \rightarrow \forall\alpha\psi)$	using 10.23(2) and a tautology □

To proceed further, we must introduce the basic notions of free and bound occurrences of variables in formulas. The definitions are based on the following Proposition, which is easily proved by induction on φ, using 10.17.

Proposition 10.40. *Let $\varphi = \langle\varphi_0, \ldots, \varphi_{m-1}\rangle$ be a formula, and suppose that $i < m$. If $\varphi_i \in \operatorname{Rng} L$, then there is a unique j such that $i < j \leq m - 1$ and $\langle\varphi_i, \ldots, \varphi_j\rangle$ is a formula.*

Definition 10.41. Let $\varphi = \langle\varphi_0, \ldots, \varphi_{m-1}\rangle$ be a formula, and let $i < m$. We say that $\forall$ *is a quantifier on α at the* ith *place in φ with scope ψ* if $\varphi_i = L_3$, $\varphi_{i+1} = \alpha$, and $\psi = \langle\varphi_i, \ldots, \varphi_j\rangle$ is the unique formula given by 10.40. A variable σ *occurs bound at the* kth *place of φ* if $\varphi_k = \alpha$ and there exist i, $j < m$ such that $\forall$ is a quantifier on α at the ith place in φ with scope $\langle\varphi_i, \ldots, \varphi_j\rangle$ and $i < k \leq j$. If $\varphi_k = \alpha$ but α does not occur bound at the kth place of φ, we say that α *occurs free at the* kth *place of φ.*

Before returning to proof-theoretic matters concerning these notions, we as usual want to briefly discuss the effectivity of the notions. It is, in fact, obvious that there is an effective procedure for recognizing if a certain variable occurs bound or free at a given place in a formula. The following three propositions give a rigorous formulation of this fact.

Proposition 10.42. *Let $R = \{(m, i, x, y) : m = \mathscr{g}\alpha$ for some variable α, $x = \mathscr{g}^+\varphi$ and $y = \mathscr{g}^+\psi$ for some formulas φ, ψ, $i <$ Dmn φ, and $\forall$ is a quantifier on α at the ith place in φ with scope $\psi\}$. Then R is recursive.*

PROOF. $(m, i, x, y) \in R$ iff $m \in \mathrm{Rng}(\mathscr{g} \circ v)$ and $x, y \in \mathscr{g}^{+*}$ Fmla and $i \leq \mathrm{l}x$ and $(x)_i = \mathscr{g}L_3 + 1$ and $(x)_{i+1} = m + 1$ and $i + \mathrm{l}y \leq \mathrm{l}x$ and $\forall j \leq \mathrm{l}y[(y)_j = (x)_{i+j}]$. □

Proposition 10.43. *Let $S = \{(m, k, x) : m = \mathscr{g}\alpha$ for some variable α, $x = \mathscr{g}^+\varphi$ for some formula φ, $k <$ Dmn φ, and α occurs bound at the kth place of $\varphi\}$. Then S is recursive.*

PROOF. $(m, k, x) \in S$ iff $m \in \mathrm{Rng}(\mathscr{g} \circ v)$, $x \in \mathscr{g}^{+*}$ Fmla, $k \leq \mathrm{l}x$, $(x)_k = m + 1$ and $\exists i \leq \mathrm{l}x\, \exists y \leq x[y \in \mathscr{g}^{+*}$ Fmla, $i < k \leq i + \mathrm{l}y$, and $(m, i, x, y) \in R]$, where R is as in 10.42. □

Proposition 10.44. *Let $T = \{(m, k, x) : m = \mathscr{g}\alpha$ for some variable α, $x = \mathscr{g}^+\varphi$ for some formula φ, $k <$ Dmn φ, and α occurs free at the kth place of $\varphi\}$. Then T is recursive.*

PROOF. $(m, k, x) \in T$ iff $m \in \mathrm{Rng}(\mathscr{g} \circ v)$, $x \in g^{+*}$ Fmla, $k \leq \mathrm{l}x$, $(x)_k = m + 1$, and $(m, k, x) \notin S$, where S is as in 10.43. □

The following proposition is analogous to 10.40, and is proved similarly:

Proposition 10.45. *Let $\varphi = \langle \varphi_0, \ldots, \varphi_{m-1} \rangle$ be a formula, and suppose that $i < m$. If $\varphi_i \in$ Dmn $\mathcal{O} \cup$ Rng v, then there is a unique j such that $i \leq j \leq m - 1$ and $\langle \varphi_i, \ldots, \varphi_j \rangle$ is a term.*

We now extend the notion of free variable to terms:

Definition 10.46. A term σ *occurs free at the* ith *place in* φ provided there is a j with $i \leq j <$ Dmn φ such that $\sigma = \langle \varphi_i, \ldots, \varphi_j \rangle$ and for any k, if $i \leq k \leq j$ and a variable α occurs at the kth place in φ, then α occurs free at that place in φ.

This notion, too, is effective:

Proposition 10.47. *Let $U = \{(x, i, y) : x = \mathscr{g}^+\sigma$ for some term σ, $y = \mathscr{g}^+\varphi$ for some formula φ, $i <$ Dmn φ, and σ occurs free at the ith place in $\varphi\}$. Then U is recursive.*

PROOF. $(x, i, y) \in U$ iff $x \in \mathscr{g}^{+*}$ Trm, $y \in \mathscr{g}^{+*}$ Fmla, $i \leq ly$, and $\exists j \leq ly[i \leq j$ and $\forall k \leq j[i \leq k \Rightarrow (y)_k = (x)_{k-i}]$ and $\forall k \leq j(i \leq k$ and $(x)_k \in \text{Rng}(\mathscr{g} \circ v) \Rightarrow ((x)_k, k, y) \in T)]$, where T is as in 10.44. □

Now we return to proof-theoretic matters, still aiming toward our main theorem on substitution of equals for equals.

Lemma 10.48. *If φ and ψ are formulas, and ψ is obtained from φ by replacing one free occurrence of σ in φ by a free occurrence of τ in ψ, then $\vdash \sigma \equiv \tau \rightarrow (\varphi \leftrightarrow \psi)$.*

PROOF. We proceed by induction on φ. For φ atomic,

$$\vdash \sigma \equiv \tau \rightarrow (\varphi \rightarrow \psi) \qquad 10.23(5)$$
$$\vdash \tau \equiv \sigma \rightarrow (\psi \rightarrow \varphi) \qquad 10.23(5)$$

Hence by 10.35 we have

$$\vdash \sigma \equiv \tau \rightarrow (\varphi \leftrightarrow \psi).$$

The induction steps from φ to $\neg\varphi$, φ and ψ to $\varphi \vee \psi$, and φ and ψ to $\varphi \wedge \psi$ are all trivial, using suitable tautologies. Now suppose that our result holds for φ, α is a variable, and $\forall\alpha\psi$ is obtained from $\forall\alpha\varphi$ by replacing one free occurrence of σ in $\forall\alpha\varphi$ by a free occurrence of τ in $\forall\alpha\psi$. Then α does not occur in σ or in τ, and ψ is obtained from φ by replacing one free occurrence of σ in φ by a free occurrence of τ in ψ. Hence

$$\vdash \sigma \equiv \tau \rightarrow (\varphi \leftrightarrow \psi) \qquad \text{induction hypothesis}$$
$$\vdash \sigma \equiv \tau \rightarrow (\forall\alpha\varphi \leftrightarrow \forall\alpha\psi) \qquad \text{10.39, suitable tautologies}$$

This completes the proof. □

Theorem 10.49 (Substitution of equals for equals). *If φ and ψ are formulas, and ψ is obtained from φ by replacing zero or more free occurrences of σ in φ by free occurrences of τ in ψ, then $\vdash \sigma \equiv \tau \rightarrow (\varphi \leftrightarrow \psi)$.*

This theorem is obtained from 10.48 by induction on the number of free occurrences of σ in φ which are replaced to obtain ψ.

Corollary 10.50. *If ρ and ξ are terms, and ξ is obtained from ρ by replacing zero or more occurrences of σ in ρ by τ, then $\vdash \sigma \equiv \tau \rightarrow \rho \equiv \xi$.*

PROOF. By 10.49, $\vdash \sigma \equiv \tau \rightarrow (\rho \equiv \rho \leftrightarrow \rho \equiv \xi)$. Hence the desired result follows by 10.34. □

We now introduce another important notion connected with the notion of free occurrence of a variable.

Definition 10.51. For any variable α, term σ, and formula φ, let $\text{Subf}^{\alpha}_{\sigma}\varphi$ be the formula obtained from φ by replacing every free occurrence of α in φ by σ.

This definition contains a hidden assumption, easily established, that the indicated substitutions always convert a formula into another formula. Note that we do not insist that the substitutions be what is sometimes called *proper*: it may be that some of the substituted occurrences of σ in $\mathrm{Subf}^{\alpha}_{\sigma}\varphi$ do not occur free in $\mathrm{Subf}^{\alpha}_{\sigma}\varphi$. For example, $\mathrm{Subf}^{\alpha}_{\beta}\,\forall\beta(\alpha = \beta) = \forall\beta(\beta = \beta)$. Clearly the formation of $\mathrm{Subf}^{\alpha}_{\sigma}\varphi$ is an effective procedure starting from α, σ, and φ. Formally:

Proposition 10.52. *If m is the Gödel number of a variable α, x is the Gödel number of a term σ, and y is the Gödel number of a formula φ, let $f(m, x, y)$ be the Gödel number of* $\mathrm{Subf}^{\alpha}_{\sigma}\varphi$; *otherwise, let $f(m, x, y) = 0$. Then f is recursive.*

PROOF. We will indicate a simple procedure for obtaining $\mathrm{Subf}^{\alpha}_{\sigma}\varphi$ which makes the effectiveness very clear. First, it is clearly an effective matter to take the first free occurrence of α in φ (if any) and replace it by σ. Formally, for any $m, x, y \in \omega$ we define a function f' by considering two cases:

Case 1: $m = \mathcal{g}\alpha$ for some variable α, $x = \mathcal{g}^{+}\sigma$ for some term σ, $y = \mathcal{g}^{+}\varphi$ for some formula φ, and α occurs freely in φ. Let ψ be obtained from φ by replacing the first free occurrence of α in φ by σ, and set $f'(m, x, y) = \mathcal{g}^{+}\psi$.

Case 2: Case 1 does not hold. Let $f'(m, x, y) = y$.

This function f' is recursive, since for any $m, x, y \in \omega$ we have

$$\begin{aligned} f'(m, x, y) = \mu z\{(&m \in \mathrm{Rng}\,(\mathcal{g} \circ v) \;\&\; x \in \mathcal{g}^{+*}\,\mathrm{Trm} \;\&\; y \in \mathcal{g}^{+*}\,\mathrm{Fmla} \\ &\& \;\exists i \le \mathrm{l}y\{(m, i, y) \in T \;\&\; \forall j < i[(m, j, y) \notin T] \;\& \\ &\exists u, w \le y[\mathrm{l}u = i - 1 \;\&\; y = \mathrm{Cat}\,(\mathrm{Cat}\,(u, 2^{m}), w) \;\&\; z = \\ &\mathrm{Cat}\,(\mathrm{Cat}\,(u, x), w]\}) \text{ or } [(m \notin \mathrm{Rng}\,(\mathcal{g} \circ v) \text{ or } x \notin \mathcal{g}^{+*}\,\mathrm{Trm} \\ &\text{or } y \notin \mathcal{g}^{+*}\,\mathrm{Fmla} \text{ or } \forall i \le \mathrm{l}y[(m, i, y) \notin T]) \text{ and } z = y]\}. \end{aligned}$$

where T is as in 10.44. Thus f' is recursive. Second, we obtain f from f' in a certain way. Unfortunately, we cannot obtain f just by iterating f', in general. The reason is that α may occur in σ; then iterating f' would not only not give the desired result but would give longer and longer formulas. We can circumvent this difficulty by first replacing all free occurrences of α by a new variable β, and then replacing each free occurrence of β by σ; both of these procedures can be obtained by iterating f'. We obtain a suitable β by a function f'':

$$\begin{aligned} f''(m, x, y) = \mu z\{z \in \mathrm{Rng}\,(\mathcal{g} \circ v) \;\&\; z \neq m \;\&\; \forall i \le \mathrm{l}x \\ [z \neq (x)_i] \;\&\; \forall i \le \mathrm{l}y[z \neq (y)_i]\}. \end{aligned}$$

Now we iterate f' to replace α by β:

$$\begin{aligned} f'''(m, x, y, 0) &= y, \\ f'''(m, x, y, n + 1) &= f'(m, f''(m, x, y), f'''(m, x, y, n)); \end{aligned}$$

and set $f^{\mathrm{iv}}(m, x, y) = f'''(m, x, y, \mathrm{l}y + 1)$. Finally, we again iterate f' to replace β by σ:

$$\begin{aligned} f^{\mathrm{v}}(m, x, y, 0) &= f^{\mathrm{iv}}(m, x, y), \\ f^{\mathrm{v}}(m, x, y, n + 1) &= f'(f''(m, x, y), x, f^{\mathrm{v}}(m, x, y, n)). \end{aligned}$$

Clearly then $f(m, x, y) = f^{\mathrm{v}}(m, x, y, \mathrm{l}f^{\mathrm{iv}}(m, x, y) + 1)$. □

Important properties of $\mathrm{Subf}^{\alpha}_{\sigma}\varphi$ will be given shortly. At the moment we need this notion only to establish an important theorem allowing change of bound variables, for which we need several lemmas. The first two are special cases of important results which will be established later.

Lemma 10.53. *If α is a variable which does not occur bound in φ and does not occur in σ, and if no free occurrence of α in φ is within the scope of a quantifier on a variable occurring in σ, then* $\vdash \forall\alpha\varphi \rightarrow \mathrm{Subf}^{\alpha}_{\sigma}\varphi$.

Proof

$\vdash \alpha = \sigma \rightarrow (\varphi \rightarrow \mathrm{Subf}^{\alpha}_{\sigma}\varphi)$	10.49
$\vdash \varphi \rightarrow (\neg \mathrm{Subf}^{\alpha}_{\sigma}\varphi \rightarrow \neg\alpha = \sigma)$	suitable tautology
$\vdash \forall\alpha\varphi \rightarrow (\forall\alpha\ \neg\ \mathrm{Subf}^{\alpha}_{\sigma}\varphi \rightarrow \forall\alpha\ \neg\ \alpha = \sigma)$	using 10.23(2)
$\vdash \neg\forall\alpha\ \neg\ \alpha = \sigma \rightarrow (\forall\alpha\varphi \rightarrow \neg\forall\alpha\ \neg\ \mathrm{Subf}^{\alpha}_{\sigma}\varphi)$	suitable tautology
$\vdash \neg\forall\alpha\ \neg\ \alpha = \sigma$	10.23(4)
$\vdash \forall\alpha\varphi \rightarrow \neg\forall\alpha\ \neg\ \mathrm{Subf}^{\alpha}_{\sigma}\varphi$	
$\vdash \neg\mathrm{Subf}^{\alpha}\varphi \rightarrow \forall\alpha\ \neg\ \mathrm{Subf}^{\alpha}_{\sigma}\varphi$	10.23(3)
$\vdash \forall\alpha\varphi \rightarrow \mathrm{Subf}^{\alpha}_{\sigma}\varphi$	suitable tautology □

Lemma 10.54. *If ψ is obtained from φ by replacing an occurrence of χ in φ by θ, and if $\vdash \chi \leftrightarrow \theta$, then $\vdash \varphi \leftrightarrow \psi$.*

Proof. Induction on φ. □

Lemma 10.55. *If α and β are distinct variables, α does not occur bound in φ, and β does not occur in φ at all, then* $\vdash \forall\alpha\varphi \leftrightarrow \forall\beta\ \mathrm{Subf}^{\alpha}_{\beta}\varphi$.

Proof

$\vdash \forall\alpha\varphi \rightarrow \mathrm{Subf}^{\alpha}_{\beta}\varphi$	10.53
$\vdash \forall\beta\ \forall\alpha\varphi \rightarrow \forall\beta\ \mathrm{Subf}^{\alpha}_{\beta}\varphi$	using 10.23(2)
$\vdash \forall\alpha\varphi \rightarrow \forall\alpha\ \forall\alpha\varphi$	10.23(3)
$\vdash \forall\alpha\varphi \rightarrow \forall\beta\ \mathrm{Subf}^{\alpha}_{\beta}\varphi$	

Now we can apply the result just obtained to β, α, $\mathrm{Subf}^{\alpha}_{\beta}\varphi$ instead of α, β, φ and obtain $\vdash \forall\beta\ \mathrm{Subf}^{\alpha}_{\beta}\varphi \rightarrow \forall\alpha\ \mathrm{Subf}^{\alpha}_{\beta}\ \mathrm{Subf}^{\beta}_{\alpha}\varphi$. Since $\mathrm{Subf}^{\beta}_{\alpha}\ \mathrm{Subf}^{\alpha}_{\beta}\varphi = \varphi$, the desired theorem follows. □

Definition 10.56. By $\mathrm{Subb}^{\alpha}_{\beta}\varphi$ we mean the formula obtained from φ by replacing each bound occurrence of α in φ by β.

Again one should check that $\mathrm{Subb}^{\alpha}_{\beta}\varphi$ is really always again a formula. The formation of $\mathrm{Subb}^{\alpha}_{\beta}\varphi$ from φ is clearly effective. The following proposition is established analogously to 10.52.

Proposition 10.57. *If m is the Gödel number of a variable α, x is the Gödel number of a variable β, and y is the Gödel number of a formula φ, let $h(m, x, y)$ be the Gödel number of* $\mathrm{Subb}^{\alpha}_{\beta}\varphi$; *otherwise, let $h(m, x, y) = 0$. Then h is recursive.*

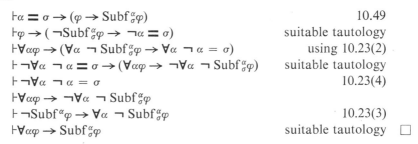

Lemma 10.58. *If α occurs bound in φ, then there is a formula $\forall\alpha\psi$ which occurs in φ such that α does not occur bound in ψ.*

PROOF. Induction on φ. □

Theorem 10.59 (Change of bound variables). *If β does not occur in φ, then* $\vdash\varphi\leftrightarrow \mathrm{Subb}^{\alpha}_{\beta}\varphi$.

PROOF. We proceed by induction on the number m of bound occurrences of α in φ. If $m = 0$, the desired conclusion is trivial. We now assume that $m > 0$, and that our result is known for all formulas with fewer than m bound occurrences of α. By Lemma 10.58, let $\forall\alpha\psi$ be a formula occurring in φ such that α does not occur bound in ψ. Let γ be a variable not occurring in φ (hence not in ψ) with $\gamma \neq \alpha, \beta$. Then by 10.55 we have

(1) $$\vdash\forall\alpha\psi\leftrightarrow\forall\gamma\, \mathrm{Subf}^{\alpha}_{\gamma}\psi.$$

Now let χ be obtained from φ by replacing an occurrence of $\forall\alpha\psi$ in φ by $\forall\gamma\, \mathrm{Subf}^{\alpha}_{\gamma}\psi$. Then by (1) and 10.54 we have

(2) $$\vdash\varphi\leftrightarrow\chi.$$

Now β does not occur in χ, and χ has fewer than m bound occurrences of α. Hence by the induction assumption,

(3) $$\vdash\chi\leftrightarrow \mathrm{Subb}^{\alpha}_{\beta}\chi.$$

Now clearly $\mathrm{Subb}^{\alpha}_{\beta}\varphi$ can be obtained from $\mathrm{Subb}^{\alpha}_{\beta}\chi$ by replacing an occurrence of $\forall\gamma\, \mathrm{Subf}^{\alpha}_{\gamma}\psi$ in $\mathrm{Subb}^{\alpha}_{\beta}\chi$ by $\forall\beta\, \mathrm{Subf}^{\alpha}_{\beta}\psi$. Thus by 10.54 it suffices to show

(4) $$\vdash\forall\gamma\, \mathrm{Subf}^{\alpha}_{\gamma}\psi\leftrightarrow\forall\beta\, \mathrm{Subf}^{\alpha}_{\beta}\psi$$

In fact, $\gamma \neq \beta$, γ does not occur bound in $\mathrm{Subf}^{\alpha}_{\gamma}\psi$, and β does not occur in $\mathrm{Subf}^{\alpha}_{\gamma}\psi$ at all; furthermore, $\mathrm{Subf}^{\gamma}_{\beta}\, \mathrm{Subf}^{\alpha}_{\gamma}\psi = \mathrm{Subf}^{\alpha}_{\beta}\psi$. Thus 10.55 yields (4), and the proof is complete. □

The restriction on β in 10.59 is necessary, as one sees intuitively by the example $\varphi = \exists\alpha\ \neg\,\alpha = \beta$ with $\alpha \neq \beta$. Here $\mathrm{Subb}^{\alpha}_{\beta}\varphi = \exists\beta\ \neg\,\beta = \beta$, and $\varphi\rightarrow \mathrm{Subb}^{\alpha}_{\beta}\varphi$ is obviously not a valid formula, so $\nvdash\varphi\rightarrow \mathrm{Subb}^{\alpha}_{\beta}\varphi$. This can be rigorously established after we have introduced the notion of truth (see 11.50).

We now turn to properties of $\mathrm{Subf}^{\alpha}_{\sigma}\varphi$. The main result, 10.61, removes unnecessary hypotheses from 10.53.

Lemma 10.60. *If the variable α does not occur in σ, and if no free occurrence of α in φ is within the scope of a quantifier on a variable occurring in σ, then* $\vdash\forall\alpha\varphi\rightarrow \mathrm{Subf}^{\alpha}_{\sigma}\varphi$.

PROOF. Let β be a variable not occurring in φ, not occurring in σ, and different from α. Then by change of bound variables,

$$\vdash\varphi\leftrightarrow \mathrm{Subb}^{\alpha}_{\beta}\varphi.$$

Hence using 10.23(2) we infer that

(1) $$\vdash \forall\alpha\varphi \leftrightarrow \forall\alpha\, \mathrm{Subb}^{\alpha}_{\beta}\varphi.$$

Now α does not occur bound in $\mathrm{Subb}^{\alpha}_{\beta}\varphi$, so by 10.53 we see that

(2) $$\vdash \forall\alpha\, \mathrm{Subb}^{\alpha}_{\beta}\varphi \rightarrow \mathrm{Subf}^{\alpha}_{\sigma}\, \mathrm{Subb}^{\alpha}_{\beta}\varphi.$$

Now α does not occur at all in $\mathrm{Subf}^{\alpha}_{\sigma}\, \mathrm{Subb}^{\alpha}_{\beta}\varphi$, and clearly

$$\mathrm{Subb}^{\beta}_{\alpha}\, \mathrm{Subf}^{\alpha}_{\sigma}\, \mathrm{Subb}^{\alpha}_{\beta}\varphi = \mathrm{Subf}^{\alpha}_{\sigma}\varphi,$$

so by change of bound variable,

(3) $$\vdash \mathrm{Subf}^{\alpha}_{\sigma}\, \mathrm{Subb}^{\alpha}_{\beta}\varphi \leftrightarrow \mathrm{Subf}^{\alpha}_{\sigma}\varphi.$$

Conditions (1), (2), (3) immediately yield the desired result. □

Theorem 10.61 (Universal specification). *If no free occurrence in φ of the variable α is within the scope of a quantifier on a variable occurring in σ, then $\vdash \forall\alpha\varphi \rightarrow \mathrm{Subf}^{\alpha}_{\sigma}\varphi$.*

PROOF. Let β be a variable not occurring in φ or in σ, and distinct from α. Then

	$\vdash \forall\alpha\varphi \rightarrow \mathrm{Subf}^{\alpha}_{\beta}\varphi$	10.60
(1)	$\vdash \forall\beta\,\forall\alpha\varphi \rightarrow \forall\beta\, \mathrm{Subf}^{\alpha}_{\beta}\varphi$	using 10.23(2)
(2)	$\vdash \forall\alpha\varphi \rightarrow \forall\beta\,\forall\alpha\varphi$	10.23(3)

Now no free occurrence of β in $\mathrm{Subf}^{\alpha}_{\beta}\varphi$ is within the scope of a quantifier on a variable occurring in σ. Clearly also $\mathrm{Subf}^{\beta}_{\sigma}\, \mathrm{Subf}^{\alpha}_{\beta}\varphi = \mathrm{Subf}^{\alpha}_{\sigma}\varphi$. Hence by 10.60

(3) $$\vdash \forall\beta\, \mathrm{Subf}^{\alpha}_{\beta}\varphi \rightarrow \mathrm{Subf}^{\alpha}_{\sigma}\varphi.$$

Conditions (1), (2), (3) immediately yield the desired result. □

Theorem 10.61 gives the most important property of $\mathrm{Subf}^{\alpha}_{\sigma}\varphi$. This property is frequently taken as one of the axiom schemas for derivability. Again, the hypothesis on α is necessary, as is seen by the example $\varphi = \exists\beta\ \neg(\alpha \equiv \beta)$; $\forall\alpha\varphi \rightarrow \mathrm{Subf}^{\alpha}_{\beta}\varphi$ is not logically valid. We say here that a *clash* of bound variables has occurred.

We now give some important corollaries of 10.61.

Corollary 10.62. $\vdash \forall\alpha\varphi \rightarrow \varphi$.

Corollary 10.63. *If the variable α does not occur free in φ, then $\vdash \varphi \leftrightarrow \forall\alpha\varphi$.*

PROOF. By 10.62,

(1) $$\vdash \forall\alpha\varphi \rightarrow \varphi.$$

For the other direction, let β be a variable not occurring in φ. Then by change of bound variable,

(2) $$\vdash \varphi \leftrightarrow \mathrm{Subb}^{\alpha}_{\beta}\varphi.$$

Hence using 10.23(2) we obtain

(3) $$\vdash \forall\alpha\ \mathrm{Subb}^{\alpha}_{\beta}\varphi \rightarrow \forall\alpha\varphi,$$

and by 10.23(3) we obtain

(4) $$\vdash \mathrm{Subb}^{\alpha}_{\beta}\varphi \rightarrow \forall\alpha\ \mathrm{Subb}^{\alpha}_{\beta}\varphi.$$

From (1)–(4) the desired conclusion easily follows. □

Proposition 10.64. *If the variable α does not occur free in φ, then* $\vdash \forall\alpha(\varphi \rightarrow \psi) \rightarrow (\varphi \rightarrow \forall\alpha\psi)$.

PROOF. $\vdash \varphi \leftrightarrow \forall\alpha\varphi$ by 10.63; hence use 10.23(2). □

Proposition 10.65. $\vdash \forall\alpha\forall\beta\varphi \rightarrow \forall\beta\forall\alpha\varphi$.

PROOF

$\vdash \forall\alpha\forall\beta\varphi \rightarrow \varphi$	10.62 (twice)
$\vdash \forall\alpha\forall\alpha\forall\beta\varphi \rightarrow \forall\alpha\varphi$	10.23(2)
$\vdash \forall\alpha\forall\beta\varphi \rightarrow \forall\alpha\forall\alpha\forall\beta\varphi$	10.63
$\vdash \forall\alpha\forall\beta\varphi \rightarrow \forall\alpha\varphi$	
$\vdash \forall\alpha\forall\beta\varphi \rightarrow \forall\beta\forall\alpha\varphi$	similarly

The desired result now follows easily by symmetry. □

The following proposition is easily established.

Proposition 10.66

(*i*) $\vdash \neg\forall\alpha\varphi \leftrightarrow \exists\alpha \neg \varphi$

(*ii*) $\vdash \forall\alpha \neg \varphi \leftrightarrow \neg\exists\alpha\varphi$

(*iii*) $\vdash \forall\alpha\varphi \leftrightarrow \neg\exists\alpha \neg \varphi$

Corollary 10.67. *If no free occurrence of α in φ is within the scope of a quantifier on a variable occurring in σ, then* $\vdash \mathrm{Subf}^{\alpha}_{\sigma}\varphi \rightarrow \exists\alpha\varphi$.

PROOF

$\vdash \forall\alpha \neg \varphi \rightarrow \mathrm{Subf}^{\alpha}_{\sigma} \neg \varphi$	universal specification
$\vdash \mathrm{Subf}^{\alpha}_{\sigma}\varphi \rightarrow \exists\alpha\varphi$	suitable tautology □

Corollary 10.68. $\vdash \varphi \rightarrow \exists\alpha\varphi$.

Corollary 10.69. $\vdash \forall\alpha\varphi \rightarrow \exists\alpha\varphi$.

Proposition 10.70. $\vdash \exists\alpha\forall\beta\varphi \rightarrow \forall\beta\exists\alpha\varphi$.

PROOF

$\vdash \varphi \rightarrow \exists\alpha\varphi$	10.68
$\vdash \forall\beta\varphi \rightarrow \forall\beta\exists\alpha\varphi$	10.23(2)
$\vdash \neg\forall\beta\exists\alpha\varphi \rightarrow \neg\forall\beta\varphi$	tautology
$\vdash \neg\forall\beta\exists\alpha\varphi \rightarrow \forall\alpha\ \neg\ \forall\beta\varphi$	10.23(2), 10.64
$\vdash \exists\alpha\forall\beta\varphi \rightarrow \forall\beta\exists\alpha\varphi$	tautology □

The formula $\forall\beta\exists\alpha\varphi \rightarrow \exists\alpha\forall\beta\varphi$ is not, in general, valid.

The following important syntactical theorem is a version for formulas of the principle of substitution of equals for equals:

Theorem 10.71 (Substitutivity of equivalence). *Let φ, ψ, χ be formulas and $\alpha \in {}^m\text{Rng}\, v$. Suppose that if β occurs free in φ or in ψ but bound in χ then $\beta \in \{\alpha_i : i < m\}$. Let θ be obtained from χ by replacing zero or more occurrences of φ in χ by ψ. Then*

$$\vdash \forall\alpha_0 \cdots \forall\alpha_{m-1}(\varphi \leftrightarrow \psi) \rightarrow (\chi \leftrightarrow \theta).$$

PROOF. We proceed by induction on χ. We may assume that $\theta \neq \chi$. If χ is atomic, then $\chi = \varphi$ and $\psi = \theta$; this case is trivial. Suppose χ is $\neg\chi'$. Then θ is of the form $\neg\theta'$, and the induction hypothesis easily gives the desired result. The induction steps involving $\vee$ and $\wedge$ are similar. Now suppose $\chi = \forall\beta\chi'$. Then by the induction hypothesis,

$$\vdash \forall\alpha_0 \cdots \forall\alpha_{m-1}(\varphi \leftrightarrow \psi) \rightarrow (\chi' \leftrightarrow \theta').$$

Note that β does not occur free in $\forall\alpha_0 \cdots \forall\alpha_{m-1}(\varphi \leftrightarrow \psi)$. Hence, using 10.61, we easily obtain

$$\vdash \forall\alpha_0 \cdots \forall\alpha_{m-1}(\varphi \leftrightarrow \psi) \rightarrow (\forall\beta\chi' \leftrightarrow \forall\beta\theta'),$$

as desired. □

Again note that implicit in 10.71 is the assertion that the expression θ formed from χ is again a formula; this is easily established.

Now we introduce a notation which will be frequently used in the remainder of this book.

Definition 10.72. Let φ be a formula, $m \in \omega$, $\sigma \in {}^m\text{Trm}$. Choose k maximum such that v_k occurs in φ or in σ_j for some $j < m$, $k = 0$ if no variable occurs in φ or in any σ_j. Let $\alpha_0, \ldots, \alpha_{n-1}$ be a list of all variables which occur bound in φ but also occur in some σ_j, with $\alpha_0 < \cdots < \alpha_{n-1}$ in the natural order $v_0, v_1, \ldots$ of the variables. Let ψ be the formula

$$\text{Subb}^{\alpha 0}_{v(k+1)}\, \text{Subb}^{\alpha 1}_{v(k+2)} \cdots \text{Subb}^{\alpha(n-1)}_{v(k+n)}\, \varphi,$$

and let $\varphi(\sigma_0, \ldots, \sigma_{m-1})$ be the formula obtained from ψ by simultaneously replacing all free occurrences of $v_0, \ldots, v_{m-1}$ by $\sigma_0, \ldots, \sigma_{m-1}$ respectively.

The purpose of first forming ψ is to eliminate any possible clash of bound variables. The following two corollaries give the essential properties of this notation.

Corollary 10.73. $\vdash \forall v_0 \cdots \forall v_{m-1} \varphi \rightarrow \varphi(\sigma_0, \ldots, \sigma_{m-1})$.

Corollary 10.74. $\vdash \varphi(\sigma_0, \ldots, \sigma_{m-1}) \rightarrow \exists v_0 \cdots \exists v_{m-1} \varphi$.

Both of these corollaries are immediate consequences of earlier results, upon noticing that simultaneous substitution can be obtained by iterated ordinary substitution; in the notation of 10.72,

$$\begin{aligned}\varphi(\sigma_0, \ldots, \sigma_{m-1}) = {} & \mathrm{Subf}^{v(k+n+1)}_{\sigma 0}\, \mathrm{Subf}^{v(k+n+2)}_{\sigma 0} \cdots \mathrm{Subf}^{v(k+n+m)}_{\sigma(m-1)} \\ & \mathrm{Subf}^{v0}_{v(k+n+1)}\, \mathrm{Subf}^{v1}_{v(k+n+2)} \cdots \mathrm{Subf}^{v(m-1)}_{v(k+n+m)}\, \psi.\end{aligned}$$

This fact is also useful in checking formally that this substitution notion is effective:

Proposition 10.75. *If x is the Gödel number of a formula φ, $m \in \omega$, $y_0, \ldots, y_{m-1}$ are Gödel numbers of terms $\sigma_0, \ldots, \sigma_{m-1}$ respectively, let $s(x, y_0, \ldots, y_{m-1}) = \mathcal{g}^+\varphi(\sigma_0, \ldots, \sigma_{m-1})$; if x and $y_0, \ldots, y_{m-1}$ do not satisfy these conditions, let $s(x, y_0, \ldots, y_{m-1}) = 0$. Then s is recursive.*

PROOF. First we need a function picking out the integer k described in 10.72. For any $x, y_0, \ldots, y_{m-1} \in \omega$, let

$$\begin{aligned}& g(x, y_0, \ldots, y_{m-1}) = \mu k[(\exists i \leq \mathrm{l}x\{((x)_i = \mathcal{g}vk + 1) \text{ or} \\ & \textstyle\bigvee_{j<m} \exists i \leq \mathrm{l}y_j[(y_j)_i = \mathcal{g}vk + 1]\} \text{ and } \forall i \leq \mathrm{l}x[(x)_i \dot{-} \\ & 1 \in \mathrm{Rng}\,(\mathcal{g} \circ v) \Rightarrow v^{-1}\mathcal{g}^{-1}((x)_i - 1) \leq k] \text{ and} \\ & \textstyle\bigwedge_{j<m} \forall i \leq \mathrm{l}y_j[(y_j)_i \dot{-} 1 \in \mathrm{Rng}\,(\mathcal{g} \circ v) \Rightarrow v^{-1}\mathcal{g}^{-1}((y_j)_i - 1) \leq k]) \\ & \text{or } (\forall i \leq \mathrm{l}x[(x)_i \dot{-} 1 \notin \mathrm{Rng}\,(\mathcal{g} \circ v) \text{ and } \textstyle\bigwedge_{j<m} \forall i \leq \\ & \mathrm{l}y_j[(y_j)_i \dot{-} 1 \notin \mathrm{Rng}\,(\mathcal{g} \circ v)]])].\end{aligned}$$

Now let f be the function of 10.52, let f' be the function of 10.57, and let S be the relation of 10.43. We now define a function f'' which codes the set of variables which occur bound in φ but also occur in some σ_j:

$$\begin{aligned}& f''(x, z_0, \ldots, z_{m-1}) = \mu y((y)_{\mathrm{l}y} = 1 \text{ and } \forall i \leq \mathrm{l}y(\exists j \leq \\ & \mathrm{l}x[((y)_i, j, x) \in S] \wedge \textstyle\bigvee_{j<m} \exists u \leq z_j \exists w \leq z_j[z_j = \mathrm{Cat}\,(u, \mathrm{Cat} \\ & (2^{(y)i+1}, w))]) \text{ and } \forall j \leq \mathrm{l}x\, \forall k \leq x\{(k, j, x) \in S \text{ and} \\ & \textstyle\bigvee_{n<m} \exists u \leq z_n \exists w \leq z_n[z_n = \mathrm{Cat}\,(u, \mathrm{Cat}\,(2^{k+1}, w))] \Rightarrow \exists i < \\ & \mathrm{l}y[(y)_i = k]\} \text{ and } \forall i, j < \mathrm{l}y[i < j \Rightarrow v^{-1}\mathcal{g}^{-1}(y)_i < v^{-1}\mathcal{g}^{-1}(y)_j]).\end{aligned}$$

Next we define a function f^{iv} yielding the formula ψ of 10.72, via an auxiliary function f''':

$$\begin{aligned} f'''(x, y_0, \ldots, y_{m-1}, 0) &= x, \\ f'''(x, y_0, \ldots, y_{m-1}, i + 1) &= f'((f''(x, y_0, \ldots, y_{m-1}))_i, \\ & \qquad \mathcal{g}v(h(x, y_0, \ldots, y_{m-1}) + i + 1), \\ & \qquad f'''(x, y_0, \ldots, y_{m-1}, i))\end{aligned}$$

and

$$f^{iv}(x, y_0, \ldots, y_{m-1}) = f'''(x_0, y_0, \ldots, y_{m-1}, lf''(x, y_0, \ldots, y_{m-1})).$$

We then define a function s' using the fact stated after 10.74:

$$s'(x, y_0, \ldots, y_{m-1}) = f(gv(h(x, y_0, \ldots, y_{m-1}) + lf''(x, y_0, \ldots, y_{m-1}) + 1), y_0, f(gv(h(x, y_0, \ldots, y_{m-1}) + lf''(x, y_0, \ldots, y_{m-1}) + 2), y_1, \ldots f(gv(h(x, y_0, \ldots, y_{m-1}) + lf''(x, y_0, \ldots, y_{m-1}) + m, y_{m-1}, f(gv0, gv(h(x, y_0, \ldots, y_{m-1}) + lf''(x, y_0, \ldots, y_{m-1}) + 1), f(gv1, gv(h(x, y_0, \ldots, y_{m-1}) + lf''(x, y_0, \ldots, y_{m-1}) + 2), \ldots, f(gv(m-1), gv(h(x, y_0, \ldots, y_{m-1}) + lf''(x, y_0, \ldots, y_{m-1}) + m), f^{iv}(x, y_0, \ldots, y_{m-1}))\cdots).$$

The desired function s is obtained from s' by a simple and obvious definition by cases. □

Our next main result 10.81 concerns prenex normal form.

Lemma 10.76. *If α does not occur free in ψ, then* $\vdash \forall\alpha\varphi \vee \psi \leftrightarrow \forall\alpha(\varphi \vee \psi)$.

PROOF

	$\vdash \forall\alpha\varphi \rightarrow \varphi$	10.62
	$\vdash \forall\alpha\varphi \vee \psi \rightarrow \varphi \vee \psi$	
	$\vdash \forall\alpha(\forall\alpha\varphi \vee \psi \rightarrow \varphi \vee \psi)$	
(1)	$\vdash \forall\alpha\varphi \vee \psi \rightarrow \forall\alpha(\varphi \vee \psi)$	using 10.64
	$\vdash \forall\alpha(\varphi \vee \psi) \rightarrow \varphi \vee \psi$	10.62
	$\vdash \forall\alpha(\varphi \vee \psi) \wedge \neg\psi \rightarrow \varphi$	
	$\vdash \forall\alpha[\forall\alpha(\varphi \vee \psi) \wedge \neg\psi \rightarrow \varphi]$	
	$\vdash \forall\alpha(\varphi \vee \psi) \wedge \neg\psi \rightarrow \forall\alpha\varphi$	10.64
(2)	$\vdash \forall\alpha(\varphi \vee \psi) \rightarrow \forall\alpha\varphi \vee \psi$	

Now (1) and (2) give the desired conclusion. □

The proof of the following lemma is just like that for 10.76.

Lemma 10.77. *If α does not occur free in ψ, then* $\vdash \forall\alpha\varphi \wedge \psi \leftrightarrow \forall\alpha(\varphi \wedge \psi)$.

Lemma 10.78. *If α does not occur free in ψ, then* $\vdash \exists\alpha\varphi \vee \psi \leftrightarrow \exists\alpha(\varphi \vee \psi)$.

PROOF

$\vdash \exists\alpha(\varphi \vee \psi) \leftrightarrow \neg\forall\alpha\neg(\varphi \vee \psi)$	
$\vdash \neg\forall\alpha\neg(\varphi \vee \psi) \leftrightarrow \neg\forall\alpha(\neg\varphi \wedge \neg\psi)$	
$\vdash \neg\forall\alpha(\neg\varphi \wedge \neg\psi) \leftrightarrow \neg(\forall\alpha\neg\varphi \wedge \neg\psi)$	10.77
$\vdash \neg(\forall\alpha\neg\varphi \wedge \neg\psi) \leftrightarrow \exists\alpha\varphi \vee \psi.$	□

Similarly:

Lemma 10.79. *If α does not occur free in ψ, then* $\vdash \exists\alpha\varphi \wedge \psi \leftrightarrow \exists\alpha(\varphi \wedge \psi)$.

Definition 10.80. A formula φ is *quantifier-free* if L_3 does not occur in it. A formula ψ is in *prenex normal form* provided there is an $m \in \omega$, a $\mathbf{Q} \in {}^m\{\forall, \exists\}$, an $\alpha \in {}^m\mathrm{Rng}\, v$, and a quantifier-free formula φ such that

$$\psi = \mathbf{Q}_0\alpha_0 \cdots \mathbf{Q}_{m-1}\alpha_{m-1}\varphi.$$

The formula φ is called the *matrix* of ψ; $\mathbf{Q}_0\alpha_0 \cdots \mathbf{Q}_{m-1}\alpha_{m-1}$ is the *prefix* of ψ.

Theorem 10.81 (Prenex normal form theorem). *For any formula φ there is a formula ψ in prenex normal form such that $\vdash \varphi \leftrightarrow \psi$, and such that a variable occurs free in φ iff it occurs free in ψ.*

Proof. By induction on φ. The only difficult step is the passage from φ_1 and φ_2 to $\varphi_1 \vee \varphi_2$ (or to $\varphi_1 \wedge \varphi_2$). Since these cases are symmetric, we deal only with the first. Thus assume, with obvious notation,

$$\vdash \varphi_1 \leftrightarrow \mathbf{Q}_0\alpha_0 \cdots \mathbf{Q}_{m-1}\alpha_{m-1}\psi_1$$
$$\vdash \varphi_2 \leftrightarrow \mathbf{Q}'_0\beta_0 \cdots \mathbf{Q}'_{n-1}\beta_{n-1}\psi_2.$$

By change of bound variable we may assume that none of $\alpha_0, \ldots, \alpha_{m-1}$ occur in $\mathbf{Q}'_0\beta_0 \cdots \mathbf{Q}'_{n-1}\beta_{n-1}\psi_2$, and that none of $\beta_0, \ldots, \beta_{n-1}$ occur in $\mathbf{Q}_0\alpha_0 \cdots \mathbf{Q}_{m-1}\alpha_{m-1}\psi_1$. Thus by 10.76 and 10.78,

$$\vdash \varphi_1 \vee \varphi_2 \leftrightarrow \mathbf{Q}_0\alpha_0 \cdots \mathbf{Q}_{m-1}\alpha_{m-1}\mathbf{Q}'_0\beta_0 \cdots \mathbf{Q}'_{n-1}\beta_{n-1}(\psi_1 \vee \psi_2),$$

as desired. □

Definition 10.82. φ is a *sentence*, in symbols $\varphi \in \mathrm{Sent}_{\mathscr{L}}$, if no variable occurs free in φ.

Intuitively speaking, only sentences express complete statements. A formula with free variables has no definite meaning until values are assigned to the free variables. Obviously the notion of sentence is effective:

Proposition 10.83. g^{+*} Sent *is recursive.*

Because one is generally only interested in sentences, and not formulas in general, it is convenient to have notions of theorem and proof where only sentences are mentioned:

Definition 10.84. Let $\Gamma \subseteq \mathrm{Sent}_{\mathscr{L}}$. We let Γ-Thm$''$ be the intersection of all sets $\Delta \subseteq \mathrm{Sent}_{\mathscr{L}}$ satisfying the following conditions:

(*i*) if $\varphi \in \mathrm{Axm}_{\mathscr{L}} \cup \Gamma$, $m \in \omega$, $\alpha \in {}^m\mathrm{Rng}\, v$, and $\forall\alpha_0 \cdots \forall\alpha_{m-1}\varphi$ is a sentence, then $\forall\alpha_0 \cdots \forall\alpha_{m-1}\varphi \in \Delta$; also, $\forall\alpha\varphi \rightarrow \varphi \in \Delta$ for any $\alpha \in \mathrm{Rng}\, v$ and any sentence φ;

(*ii*) $\psi \in \Delta$ whenever $\varphi, \varphi \rightarrow \psi \in \Delta$.

We write $\mathrm{Thm}''_{\mathscr{L}}$ instead of 0-$\mathrm{Thm}''_{\mathscr{L}}$, $\Gamma \vdash''_{\mathscr{L}} \varphi$ instead of $\varphi \in \Gamma$-$\mathrm{Thm}''_{\mathscr{L}}$, and $\vdash''_{\mathscr{L}}\varphi$ instead of $0 \vdash''_{\mathscr{L}} \varphi$.

Theorem 10.85. *For* $\Gamma \cup \{\varphi\} \subseteq \text{Sent}_{\mathscr{L}}$, $\Gamma \vdash \varphi$ *iff* $\Gamma \vdash'' \varphi$.

PROOF. (Γ-Thm) $\cap$ Sent in place of Δ clearly satisfies 10.84 (*i*)–(*ii*). Hence $\Gamma\text{-Thm}'' \subseteq (\Gamma\text{-Thm}) \cap \text{Sent}$, so $\Gamma \vdash'' \varphi \Rightarrow \Gamma \vdash \varphi$.

To prove the converse, let

$$\Delta = \{\varphi \in \text{Fmla}: \text{for all } m \in \omega \text{ and all } \alpha \in {}^m\text{Rng } v, \text{ if } \forall\alpha_0 \cdots \forall\alpha_{m-1}\varphi \text{ is a sentence then } \Gamma \vdash'' \forall\alpha_0 \cdots \forall\alpha_{m-1}\varphi\}.$$

Clearly Axm $\cup\, \Gamma \subseteq \Delta$ and Δ is closed under generalization. To show that Δ is closed under detachment it clearly suffices to show

(1) if $n \in \omega$, $\alpha \in {}^n\text{Rng } v$, $\forall\alpha_0 \cdots \forall\alpha_{n-1}(\varphi \to \psi)$ is a sentence, $\Gamma \vdash'' \forall\alpha_0 \cdots \forall\alpha_{n-1}\varphi$, and $\Gamma \vdash'' \forall\alpha_0 \cdots \forall\alpha_{n-1}(\varphi \to \psi)$, then $\Gamma \vdash'' \forall\alpha_0 \cdots \forall\alpha_{n-1}\psi$.

This is established just like (1) in the proof of 10.32.

Hence $\Gamma\text{-Thm} \subseteq \Delta$, so $\Gamma \vdash \varphi \Rightarrow \Gamma \vdash'' \varphi$. □

Theorem 10.86 (Deduction theorem). *If φ is a sentence and* $\Gamma \cup \{\varphi\} \vdash \psi$, *then* $\Gamma \vdash \varphi \to \psi$.

PROOF. Let $\Delta = \{\chi : \Gamma \vdash \varphi \to \psi\}$. If χ is an axiom or a member of Γ, then

$$\begin{gathered}\chi\\ \chi \to (\varphi \to \chi)\\ \varphi \to \chi\end{gathered}$$

is a Γ-formal proof of $\varphi \to \chi$, so $\chi \in \Delta$. Obviously $\varphi \in \Delta$. It is easily seen that Δ is closed under detachment. Now suppose that $\chi \in \Delta$ and $\alpha \in \text{Rng } v$. Then

$$\begin{aligned}&\Gamma \vdash \varphi \to \chi && \text{since } \chi \in \Delta\\ &\Gamma \vdash \forall\alpha(\varphi \to \chi) &&\\ &\Gamma \vdash \varphi \to \forall\alpha\chi && \text{using 10.64}\end{aligned}$$

Thus $\forall\alpha\chi \in \Delta$. Hence $(\Gamma \cup \{\varphi\})$-Thm $\subseteq \Delta$, as desired. □

The deduction theorem formalizes an important procedure in intuitive proof theory. Frequently when one wishes to prove an implication $\varphi \to \psi$ one will add φ to the standing assumptions Γ and then derive ψ. The deduction theorem then says that $\varphi \to \psi$ is derivable from Γ alone.

The deduction theorem does not hold if φ is merely assumed to be a formula. For example, $\{v_0 = v_1\} \vdash v_0 = v_2$ (as is easily seen), but $\nvdash v_0 = v_1 \to v_0 = v_2$ (see Exercise 11.52). This comes about because the free variables in formulas of Γ, where $\Gamma \vdash \varphi$, are treated as though they were universally quantified for most practical purposes.

Corollary 10.87. *Assume* $\Gamma \subseteq$ Sent, $m \in \omega$, *and* $\varphi \in {}^m$Sent. *Then the following conditions are equivalent:*

(*i*) $\Gamma \cup \{\varphi_0, \ldots, \varphi_{m-1}\} \vdash \psi$;
(*ii*) $\Gamma \vdash \varphi_0 \wedge \cdots \wedge \varphi_{m-1} \to \psi$.

PROOF. (*i*) ⇒ (*ii*). We proceed by induction on m, the case $m = 0$ being trivial (condition (*ii*) is then to be interpreted as $\Gamma \vdash \psi$, just like (*i*)). Assume that the implication holds for m, and suppose that $\Gamma \cup \{\varphi_0, \ldots, \varphi_m\} \vdash \psi$. Then by 10.86, $\Gamma \cup \{\varphi_0, \ldots, \varphi_{m-1}\} \vdash \varphi_m \rightarrow \psi$, so by the induction assumption, $\Gamma \vdash \varphi_0 \wedge \cdots \wedge \varphi_{m-1} \rightarrow (\varphi_m \rightarrow \psi)$. A suitable tautology then gives $\Gamma \vdash \varphi_0 \wedge \cdots \wedge \varphi_m \rightarrow \psi$, as desired.

The implication (*ii*) ⇒ (*i*) is trivial. □

Recalling 10.33(*ii*), we easily obtain

Corollary 10.88. *Assume* $\Gamma \cup \Delta \cup \{\varphi\} \subseteq$ Sent. *Then the following conditions are equivalent:*

(*i*) $\Gamma \cup \Delta \vdash \varphi$;
(*ii*) $\Gamma \vdash \psi_0 \wedge \cdots \wedge \psi_{m-1} \rightarrow \varphi$ *for some* $m \in \omega$ *and some* $\psi \in {}^m\Delta$.

The notion of a consistent set of formulas is one of the most important notions of elementary logic:

Definition 10.89. A set $\Gamma \subseteq$ Fmla is *consistent* if $\Gamma \nvdash \varphi$ for some φ.

Some obvious equivalents of this notion are as follows:

Proposition 10.90. *For any* $\Gamma \subseteq$ Fmla *the following conditions are equivalent:*

(*i*) Γ *is inconsistent;*
(*ii*) $\Gamma \vdash \varphi$ *and* $\Gamma \vdash \neg\varphi$ *for some formula* φ;
(*iii*) $\Gamma \vdash \varphi \wedge \neg\varphi$ *for some formula* φ.

Theorem 10.91. 0 *is consistent.*

PROOF. For any formula φ, let φ' be obtained from φ by first simultaneously replacing all atomic formulas in φ by $v_0 \equiv v_0 \rightarrow v_0 \equiv v_0$ and then deleting all quantifiers. Let $\Delta = \{\varphi : \varphi' \text{ is a tautology}\}$. It is easily checked that Axm $\subseteq \Delta$ and Δ is closed under detachment and generalization. Hence Thm $\subseteq \Delta$ by Definition 10.23. Since, for example, $\neg(v_0 \equiv v_0) \notin \Delta$, it follows that 0 is consistent. □

Of course 10.91 is an important result, showing that all of our work concerning the notion $\vdash$ is not completely trivial. It is obvious on intuitive grounds, since $\vdash\varphi$ implies that φ is intuitively true, while there are certainly formulas that are not intuitively true. This can be made precise using the model-theoretic notions of the next chapter.

The following two facts concerning consistency are frequently useful:

Proposition 10.92. *Assume* $\Gamma \cup \Delta \subseteq$ Sent *and* $\Delta \neq 0$. *Then the following conditions are equivalent:*

(*i*) $\Gamma \cup \Delta$ *is inconsistent;*
(*ii*) *there exist an* $m \in \omega \sim 1$ *and a* $\varphi \in {}^m\Delta$ *such that* $\Gamma \vdash \neg\varphi_0 \vee \cdots \vee \neg\varphi_{m-1}$.

PROOF
(*i*) $\Rightarrow$ (*ii*). By 10.88 we obtain $m \in \omega$ and $\varphi \in {}^m\Delta$ such that

$$\Gamma \vdash \varphi_0 \wedge \cdots \wedge \varphi_{m-1} \rightarrow \neg\forall v_0(v_0 = v_0).$$

We may assume $m \neq 0$. Since $\Gamma \vdash \forall v_0(v_0 = v_0)$, we easily obtain using a suitable tautology $\Gamma \vdash \neg\varphi_0 \vee \cdots \vee \neg\varphi_{m-1}$.

(*ii*) $\Rightarrow$ (*i*). Since $\Gamma \vdash \neg(\varphi_0 \wedge \cdots \wedge \varphi_{m-1})$ and $\Gamma \cup \Delta \vdash \varphi_0 \wedge \cdots \wedge \varphi_{m-1}$, $\Gamma \cup \Delta$ is inconsistent. □

Proposition 10.93. *If* $\Gamma \subseteq$ Sent, *then the following conditions are equivalent:*

(*i*) Γ *is inconsistent;*
(*ii*) *there exist* $m \in \omega \sim 1$ *and* $\varphi \in {}^m\Gamma$ *so that* $\vdash \neg\varphi_0 \vee \cdots \vee \neg\varphi_{m-1}$.

To derive 10.93 from 10.92 it is necessary to use 10.91.

Since, as mentioned earlier, we are mainly interested in sentences, as opposed to formulas in general, it is useful to have a standard method for forming a sentence from a formula:

Definition 10.94. Let φ be a formula. Then there is a unique $m \in \omega$ and $\alpha \in {}^m\text{Rng}\, v$ such that $\{\alpha_0, \ldots, \alpha_{m-1}\}$ is the collection of all variables occurring free in φ and $v^{-1}\alpha_0 < \cdots < v^{-1}\alpha_{m-1}$. We let $[[\varphi]]$, the *closure* of φ, be the sentence $\forall\alpha_0 \cdots \forall\alpha_{m-1}\varphi$.

The following proposition summarizes some usual facts about this notion:

Proposition 10.95
(*i*) $\vdash [[\varphi]] \rightarrow \varphi$;
(*ii*) *if* $\Gamma \vdash \varphi$, *then* $\Gamma \vdash [[\varphi]]$;
(*iii*) *if* $\Gamma \subseteq$ Fmla *and* $\Gamma' = \{[[\varphi]] : \varphi \in \Gamma\}$, *then* Γ-Thm $= \Gamma'$-Thm.

In conclusion of this section, we wish to indicate a modified notion of effectiveness for first-order languages which is frequently useful. An *elementary effectivized first-order language* is a quintuple as in 10.2, except with (*ii*), (*iii*), (*iv*) replaced by the following conditions:

(*ii'*) $g \circ v$ is elementary, $\text{Rng}(g \circ v)$ is elementary, and $(v^{-1} \circ g^{-1}) \cup \langle 0 : m \in \omega \sim \text{Rng}(g \circ v)\rangle$ is elementary;
(*iii'*) $g^*\, \text{Dmn}\, \mathcal{O}$ is elementary and $(\mathcal{O} \circ g^{-1}) \cup \langle 0 : m \in \omega \sim g^*\, \text{Dmn}\, \mathcal{O}\rangle$ is elementary;
(*iv'*) $g^*\, \text{Dmn}\, \mathcal{R}$ is elementary and $(\mathcal{R} \circ g^{-1}) \cup \langle 0 : m \in \omega \sim g^*\, \text{Dmn}\, \mathcal{R}\rangle$ is elementary.

Most of the languages naturally occurring in mathematics can be considered as elementary in this sense. Furthermore, the various effectiveness results of this chapter extend with minor modifications of proofs to prove elementariness of the various notions. Some exercises below indicate where some of these modifications must occur.

BIBLIOGRAPHY

1. Church, A. *Introduction to Mathematical Logic*, Vol. I. Princeton: Princeton Univ. Press (1956).
2. Shoenfield, J. *Mathematical Logic*. Reading: Addison-Wesley (1967).
3. Tarski, A. A simplified formalization of predicate logic with identity. *Arch. f. Math. Logik u. Grundl.*, 7 (1965), 61–79.

EXERCISES

10.96. Show that in 10.2(*ii*) the requirement that $v^{-1} \circ \mathscr{g}^{-1}$ is partial recursive can be replaced by the requirement that $(v^{-1} \circ \mathscr{g}^{-1}) \cup \langle 0 : m \notin \operatorname{Rng}(\mathscr{g} \circ v)\rangle$ is recursive.

10.97. Let $\mathscr{L}$ be a first-order language, as in 10.1. Let $P = \{\varphi : \varphi$ is a formula of $\mathscr{L}$ whose first symbol is either in Dmn $\mathscr{R}$, or is L_3 or $L_4\}$. Let n, c be distinct sets not in P. Then $\mathscr{P} = (n, c, P)$ is a sentential language. With each sentence φ of $\mathscr{P}$ we associate an expression φ' of $\mathscr{L}$:

$$\begin{aligned} \langle\psi\rangle' &= \psi \qquad \text{for } \psi \in P; \\ (\neg\varphi)' &= \neg\varphi'; \\ (\varphi \rightarrow \psi)' &= \varphi' \rightarrow \psi'. \end{aligned}$$

Also, with each formula φ of $\mathscr{L}$ we associate an expression $\varphi^\dagger$ of $\mathscr{P}$;

$$\begin{aligned} \varphi^\dagger &= \langle\varphi\rangle \qquad \text{if } \varphi \text{ is atomic}; \\ (\neg\varphi)^\dagger &= \neg\varphi^\dagger, \qquad (\varphi \vee \psi)^\dagger = \varphi^\dagger \vee \psi^\dagger, \qquad (\varphi \wedge \psi)^\dagger = \varphi^\dagger \wedge \psi^\dagger; \\ (\forall\alpha\varphi)^\dagger &= \langle\forall\alpha\varphi\rangle. \end{aligned}$$

Show that $'$ maps $\mathrm{Sent}_{\mathscr{P}}$ into $\mathrm{Fmla}_{\mathscr{L}}$ (but is not onto), and † maps $\mathrm{Fmla}_{\mathscr{L}}$ into $\mathrm{Sent}_{\mathscr{P}}$ (but is not onto). For any sentence φ of $\mathscr{P}$, $\vdash_{\mathscr{P}}\varphi \leftrightarrow \varphi'^\dagger$. For any formula φ of $\mathscr{L}$, $\varphi \leftrightarrow \varphi^{\dagger\prime}$ is a tautology. For any sentence φ of $\mathscr{L}$, φ is a tautology iff φ' is a tautology. And for any formula φ of $\mathscr{L}$, φ is a tautology iff $\varphi^\dagger$ is a tautology.

10.98. Prove a version of 10.52 for elementary effective languages.

10.99. Prove a version of 10.75 for elementary effective languages.

10.100. Let α be a variable not occurring in σ and such that no free occurrence of α in φ is within the scope of a quantifier on a variable occurring in σ. Prove:

(a) $\vdash \mathrm{Subf}_\sigma^\alpha\varphi \leftrightarrow \forall\alpha(\alpha \equiv \sigma \rightarrow \varphi)$;
(b) $\vdash \mathrm{Subf}_\sigma^\alpha\varphi \leftrightarrow \exists\alpha(\alpha \equiv \sigma \wedge \varphi)$.

10.101. Let α, β, γ be distinct variables. Prove:

(a) $\vdash \exists\alpha(\neg\alpha \equiv \beta \wedge \neg\alpha \equiv \gamma) \leftrightarrow [\beta \equiv \gamma \wedge \exists\alpha\exists\beta(\neg\alpha \equiv \beta)]$
$\vee [\neg\beta \equiv \gamma \wedge \exists\alpha\exists\beta\exists\gamma(\neg\alpha \equiv \beta \wedge \neg\beta \equiv \gamma \wedge \neg\alpha \equiv \gamma)]$;
(b) $\vdash \exists\alpha[\varphi \wedge \psi \wedge \exists\beta(\varphi \wedge \neg\psi)] \rightarrow \exists\beta(\exists\alpha\varphi \wedge \neg\alpha \equiv \beta)$;
(c) $\vdash \exists\gamma\{\exists\alpha[\exists\gamma(\exists\alpha\varphi \wedge \exists\beta\psi) \wedge \alpha \equiv \gamma] \wedge \exists\beta\chi\} \leftrightarrow$
$\exists\gamma[\exists\beta(\exists\gamma\{\exists\alpha[\exists\gamma(\exists\beta\psi \wedge \beta \equiv \gamma) \wedge \alpha = \gamma] \wedge \exists\beta\chi\} \wedge \beta \equiv \gamma) \wedge \exists\alpha\varphi]$;
(d) $\vdash \exists\alpha\varphi \wedge \exists\beta\psi \wedge \exists\gamma\chi \rightarrow \exists\alpha\exists\beta\exists\gamma[\exists\alpha(\exists\beta\chi \wedge \exists\gamma\psi)$
$\wedge \exists\beta(\exists\gamma\varphi \wedge \exists\alpha\chi) \wedge \exists\gamma(\exists\alpha\psi \wedge \exists\beta\varphi)]$.

10.102. Prove the following generalization of the deduction theorem: if φ is a formula and $\Gamma \cup \{\varphi\} \vdash \psi$, then $\Gamma \vdash [[\varphi]] \rightarrow \psi$.

10.103. For any $\Gamma \subseteq$ Fmla, let Γ-Thm$'''$ be the intersection of all sets $\Delta \subseteq$ Fmla such that:

(1) $\Gamma \cup \text{Thm} \subseteq \Delta$;
(2) Δ is closed under detachment.
We write $\Gamma \vdash''' \varphi$ for $\varphi \in \Gamma$-Thm$'''$. Prove:
(3) if $\Gamma \subseteq$ Sent, then $\Gamma \vdash \varphi$ iff $\Gamma \vdash''' \varphi$;
(4) if $\Gamma \cup \{\varphi\} \vdash''' \psi$, then $\Gamma \vdash''' \varphi \rightarrow \psi$;
(5) if $\Gamma \vdash''' \varphi$ and α is a variable which does not occur free in any formula of Γ, then $\Gamma \vdash''' \forall\alpha\varphi$.

Exercise 10.103 gives a common definition of Γ-theorem, which differs from ours in the case of Γ having formulas with free variables.

The next exercise gives the most common selection of logical axioms:

10.104. Let Axm^{iv} consist of all formulas of the forms in 10.23(1), 10.34, 10.48, 10.61, and 10.64. Let Γ-Thm^{iv} be the intersection of all $\Delta \subseteq$ Fmla such that $\Gamma \cup \text{Axm}^{iv} \subseteq \Delta$ and Δ is closed under detachment and generalization. Show Γ-Thm $= \Gamma$-Thm^{iv}.

Axioms and rules widely used by Hilbert and his followers are given in the next exercise.

10.105. Let Axm^{v} consist of all formulas of the forms in 10.23(1), 10.34, 10.48. 10.61, and 10.67. Let Γ-Thm^{v} be the intersection of all $\Delta \subseteq$ Fmla such that $\Gamma \cup \text{Axm}^{v} \subseteq \Delta$ and:

(1) Δ is closed under detachment;
(2) if $\varphi \rightarrow \psi \in \Delta$ and α does not occur free in φ, then $\varphi \rightarrow \forall\alpha\psi \in \Delta$;
(3) if $\varphi \rightarrow \psi \in \Delta$ and α does not occur free in ψ, then $\exists\alpha\varphi \rightarrow \psi \in \Delta$.
Show that Γ-Thm $= \Gamma$-Thm^{v}.

The following axiom system is due to Quine:

10.106. Let Axm^{vi} consist of all sentences of the following forms:

(a) $[[\varphi]]$, φ a tautology;
(b) $[[\forall\alpha\forall\beta\varphi \rightarrow \forall\beta\forall\alpha\varphi]]$;
(c) $[[\forall\alpha(\varphi \rightarrow \psi) \rightarrow (\forall\alpha\varphi \rightarrow \forall\alpha\psi)]]$;
(d) $[[\forall\alpha\varphi \rightarrow \varphi]]$;

(e) $[[\varphi \rightarrow \forall\alpha\varphi]]$ if α does not occur free in φ;
(f) $[[\varphi]]$, φ as in 10.23(4);
(g) $[[\varphi]]$, φ as in 10.23(5).

For $\Gamma \subseteq$ Sent, let Γ-Thm^{vi} be the intersection of all sets $\Delta \subseteq$ Sent such that $\Gamma \cup$ Axm$^{vi} \subseteq \Delta$ and Δ is closed under detachment. Show that Γ-Thm^{vi} = Sent $\cap$ Γ-Thm.

10.107. Let $\mathscr{L}$ be a first-order language in which $\mathcal{O} = 0$. An atomic formula $\mathbf{R}v_{i0} \cdots v_{i(m-1)}$ is *standard* if $i0 = 0, \ldots, i(m-1) = m-1$; any atomic equality formula is *standard*. A formula is *standard* if all of its atomic parts are standard. Let Fmla^s be the collection of all standard formulas. Prove that for any formula φ there is a standard formula ψ with the same free variables as φ such that $\vdash \varphi \leftrightarrow \psi$.

10.108. (Continuing 10.107.) Let Axm^s consist of all standard formulas of the following kinds:

(a) φ, φ a tautology;
(b) $\forall\alpha(\varphi \rightarrow \psi) \rightarrow (\forall\alpha\varphi \rightarrow \forall\alpha\psi)$;
(c) $\varphi \rightarrow \forall\alpha\varphi$, if α does not occur in φ;
(d) $\neg\forall\alpha\varphi \rightarrow \forall\alpha \neg\forall\alpha\varphi$;
(e) $\forall\alpha\forall\beta\varphi \rightarrow \forall\beta\forall\alpha\varphi$;
(f) $\exists\alpha(\alpha \equiv \beta)$;
(g) $\alpha \equiv \beta \rightarrow (\alpha \equiv \gamma \rightarrow \beta \equiv \gamma)$;
(h) $\alpha \equiv \beta \rightarrow (\varphi \rightarrow \forall\alpha(\alpha \equiv \beta \rightarrow \varphi))$ if $\alpha \neq \beta$.

For $\Gamma \subseteq$ Fmla^s, let Γ-Thm^s be the intersection of all sets $\Delta \subseteq$ Fmla^s such that $\Gamma \cup$ Axm$^s \subseteq \Delta$ and Δ is closed under detachment and generalization. Prove that if $\Gamma \subseteq$ Fmla^s, then Γ-Thm^s = Γ-Thm $\cap$ Fmla^s.

11 Some Basic Results of First-order Logic

We now introduce the basic notions of model and truth for first-order languages. Then we prove the completeness theorem, which shows the equivalence between the proof-theoretic notion $\vdash$ and the corresponding semantic notion. Following this we give a series of simple but basic results concerning first-order logic. Namely, we will discuss compactness, the elimination of operation symbols, extensions by definitions, Skolem functions, Herbrand's theorem, and interpretations from one language to another. These results will be useful in discussing decidable and undecidable theories as well as in the model-theoretic portion of the book.

Definition 11.1. Let $\mathscr{L} = (L, v, \mathscr{O}, \mathscr{R})$ be a first-order language. An $\mathscr{L}$-*structure* is a triple $\mathfrak{A} = (A, f, R)$ such that:

(*i*) $A \neq 0$;
(*ii*) $\text{Dmn } f = \text{Dmn } \mathscr{O}$, and for each $\mathbf{O} \in \text{Dmn } \mathscr{O}$, f_0 is an $\mathscr{O}\mathbf{O}$-ary operation on A (with simply $f_0 \in A$ if $\mathscr{O}\mathbf{O} = 0$);
(*iii*) $\text{Dmn } R = \text{Dmn } \mathscr{R}$, and for each $\mathbf{R} \in \text{Dmn } \mathscr{R}$, $R_{\mathbf{R}}$ is an $\mathscr{R}\mathbf{R}$-ary relation on A.

Thus an $\mathscr{L}$-structure gives a domain A over which the variables of $\mathscr{L}$ can range, and assigns a meaning to each nonlogical symbol of $\mathscr{L}$. Frequently $\mathscr{L}$-structures will be written in other forms, e.g. $\mathfrak{A} = (A, f_i, R_j)_{i\in I, j\in J}$, where $I = \text{Dmn } \mathscr{O}$, $J = \text{Dmn } \mathscr{R}$, $\mathfrak{A} = (A, f)$ if $\mathscr{R} = 0$, etc. Or we may write $\mathfrak{A} = (A, \mathbf{R}^{\mathfrak{A}}, \mathbf{O}^{\mathfrak{A}})_{\mathbf{R}\in \text{Dmn } \mathscr{R}, \mathbf{O}\in \text{Dmn } \mathscr{O}}$. For $\mathbf{R} \in \text{Dmn } \mathscr{R}$, $R_{\mathbf{R}}$ is called the *denotation* of $\mathbf{R}$ in $\mathfrak{A}$; and f_0 is the *denotation* of $\mathbf{O}$ in $\mathfrak{A}$, for $\mathbf{O} \in \text{Dmn } \mathscr{O}$. Note that if $\mathfrak{A} = (A, f, R)$ is any mathematical object with $A \neq 0$, f a function with range a collection of (finitary) operations on A, and R a function with range a set of (finitary) relations on A, and if $\text{Dmn } f \cap \text{Dmn } R = 0$, then there is a first-order language $\mathscr{L}$ such that $\mathfrak{A}$ is an $\mathscr{L}$-structure. In this case we call $\mathscr{L}$ an $\mathfrak{A}$-*language*. Let us now consider $\mathscr{L}$-structures for the languages $\mathscr{L}$ described following 10.1.

Language of equality. The $\mathscr{L}$-structures are (essentially) just the nonempty sets.

Language of set theory. The $\mathscr{L}$-structures are the pairs (A, R), where A is a nonempty set and R is a binary relation on A. If $R = \{(x, y) : x, y \in A$ and $x \in y\}$, then we obtain a particularly natural $\mathscr{L}$-structure.

Language of rings. Any $\mathscr{L}$-structure has the form $\mathfrak{A} = (A, +, \cdot, -, 0, 1)$ where A is a nonempty set, $+$ and $\cdot$ are binary operations on A, $-$ is a unary operation on A, and $0, 1 \in A$. In particular, every ring is an $\mathscr{L}$-structure. Of course, there are also many $\mathscr{L}$-structures which are not rings.

Full language of a nonempty set A. An $\mathscr{L}$-structure has the form $\mathscr{D} = (D, f, R)$ of the general sort; D in general has no relationship to A, and may even have smaller or larger cardinality than A. One particular $\mathscr{L}$-structure is $\mathfrak{A} = (A, f, R)$, where f and R list all of the operations and relations on A; thus if $f \in \bigcup_{m\in\omega} {}^{mA}A$ then $f\alpha f = f$ and if $R \in \bigcup_{m\in\omega\sim 1} \mathbf{S}({}^mA)$ then $R\beta R = R$.

Our next objective is to describe precisely how an $\mathscr{L}$-structure determines the meaning of terms and formulas.

Definition 11.2. For any term σ of $\mathscr{L}$ and any $\mathscr{L}$-structure $\mathfrak{A}$ we define $\sigma^{\mathfrak{A}}: {}^{\omega}A \to A$: for any $x \in {}^{\omega}A$,

$$v_i^{\mathfrak{A}}x = x_i;$$
$$(\mathbf{O}\sigma_0\cdots\sigma_{m-1})^{\mathfrak{A}}x = \mathbf{O}^{\mathfrak{A}}(\sigma_0^{\mathfrak{A}}x, \ldots, \sigma_{m-1}^{\mathfrak{A}}x),$$

where $\mathcal{O}\mathbf{O} = m$. We say that σ *term-defines* the function $\sigma^{\mathfrak{A}}$ in $\mathfrak{A}$.

Note that if $\mathbf{O}$ is 0-ary, then $\mathbf{O}^{\mathfrak{A}}x = \mathbf{O}^{\mathfrak{A}}$ for any $x \in {}^{\omega}A$.

Proposition 11.3. *If every variable occurring in σ is in the set $\{v_i : i \in \Gamma\}$, where $\Gamma \subseteq \omega$, then $\sigma^{\mathfrak{A}}x = \sigma^{\mathfrak{A}}y$ whenever $x \upharpoonright \Gamma = y \upharpoonright \Gamma$.*

This proposition is easily established by induction on σ. It justifies the following definition:

Definition 11.4. If σ is a term with variables among $v_0, \ldots, v_{n-1}$, let ${}^n\sigma^{\mathfrak{A}}$ be the n-ary operation on A such that ${}^n\sigma^{\mathfrak{A}}x = \sigma^{\mathfrak{A}}y$ whenever $x \in {}^nA, y \in {}^{\omega}A$, and $x \subseteq y$. We say that σ *term-defines* ${}^n\sigma^{\mathfrak{A}}$ in $\mathfrak{A}$.

Note that $n = 0$ is possible in 11.4, this being the case if σ is built up from individual constants alone, with no variables occurring in it. Then ${}^0\sigma^{\mathfrak{A}}$ is just a certain element of $\mathfrak{A}$. Now we turn to the meaning of formulas; naturally this also depends on the values assigned to the variables, as follows:

Definition 11.5. For any formula φ of $\mathscr{L}$ and any $\mathscr{L}$-structure $\mathfrak{A}$ we define $\varphi^{\mathfrak{A}} \subseteq {}^{\omega}A$: for any $x \in {}^{\omega}A$,

$$x \in (\sigma = \tau)^{\mathfrak{A}} \text{ iff } \sigma^{\mathfrak{A}}x = \tau^{\mathfrak{A}}y;$$
$$x \in (\mathbf{R}\sigma_0 \cdots \sigma_{m-1})^{\mathfrak{A}} \text{ iff } (\sigma_0^{\mathfrak{A}}x, \ldots, \sigma_{m-1}^{\mathfrak{A}}x) \in \mathbf{R}^{\mathfrak{A}},$$
$$\text{where } \mathscr{R}\mathbf{R} = m; \ (\neg\varphi)^{\mathfrak{A}} = {}^{\omega}A \sim \varphi^{\mathfrak{A}};$$
$$(\varphi \vee \psi)^{\mathfrak{A}} = \varphi^{\mathfrak{A}} \cup \psi^{\mathfrak{A}};$$
$$(\varphi \wedge \psi)^{\mathfrak{A}} = \varphi^{\mathfrak{A}} \cap \psi^{\mathfrak{A}};$$
$$x \in (\forall v_i\varphi)^{\mathfrak{A}} \text{ iff for all } a \in A,\ x_a^i \in \varphi^{\mathfrak{A}}.$$

We say that φ *elementarily defines* $\varphi^{\mathfrak{A}}$ in $\mathfrak{A}$. If $x \in \varphi^{\mathfrak{A}}$ we say that x *satisfies* φ in $\mathfrak{A}$, and write $\mathfrak{A} \models \varphi[x]$.

It is easy to give rules for the effect of our defined connectives on the truth of $x \in \varphi^{\mathfrak{A}}$. For example,

$$x \in (\varphi \rightarrow \psi)^{\mathfrak{A}} \text{ iff } (x \in \varphi^{\mathfrak{A}} \text{ implies } x \in \psi^{\mathfrak{A}});$$
$$x \in (\exists v_i\varphi)^{\mathfrak{A}} \text{ iff there is an } a \in A \text{ with } x_a^i \in \varphi^{\mathfrak{A}}.$$

Proposition 11.6. *If every variable occurring free in a formula φ is in the set $\{v_i : i \in \Gamma\}$, where $\Gamma \subseteq \omega$, and if $x \restriction \Gamma = y \restriction \Gamma$, then $x \in \varphi^{\mathfrak{A}}$ iff $y \in \varphi^{\mathfrak{A}}$. In particular, if φ is a sentence, then $\varphi^{\mathfrak{A}} = {}^{\omega}A$ or $\varphi^{\mathfrak{A}} = 0$.*

This proposition is easily established by induction on φ, and justifies the following definition.

Definition 11.7. If φ is a formula with free variables among $v_0, \ldots, v_{n-1}$, we let

$${}^{n}\varphi^{\mathfrak{A}} = \{x \in {}^{n}A : \exists y \in {}^{\omega}A(x \subseteq y \in \varphi^{\mathfrak{A}})\}.$$

We say that φ *elementarily defines* ${}^{n}\varphi^{\mathfrak{A}}$ *in* $\mathfrak{A}$.

If $x \in {}^{n}\varphi^{\mathfrak{A}}$ we say that x *satisfies* φ in $\mathfrak{A}$, and again write $\mathfrak{A} \models \varphi[x]$. In general, a formula φ *holds* or is *true* in $\mathfrak{A}$ if $\varphi^{\mathfrak{A}} = {}^{\omega}A$; we also say that $\mathfrak{A}$ *satisfies* φ or is a *model* of φ and we write $\mathfrak{A} \models \varphi$. We write $\mathfrak{A} \models \Delta$ if each $\varphi \in \Delta$ holds in $\mathfrak{A}$; $\mathbf{K} \models \varphi$ if each $\mathfrak{A} \in \mathbf{K}$ is a model of φ, and $\mathbf{K} \models \Delta$ if $\mathbf{K} \models \varphi$ for each $\varphi \in \Delta$. Furthermore, we write $\Gamma \models \varphi$ if each model of Γ is a model of φ, and $\Gamma \models \Delta$ if $\Gamma \models \varphi$ for each $\varphi \in \Delta$. We say that φ is *universally valid* if φ holds in every $\mathscr{L}$-structure, and we then write $\models\varphi$.

In this definition we have given the main semantic notions that we will work with in the model-theoretic portion of the book. We now turn to the discussion of the relationships of these notions with the proof-theoretic notions. The main result we want to establish is that $\Gamma \vdash \varphi$ iff $\Gamma \models \varphi$. This will enable us in the second part of this section to establish some important proof-theoretic results by model-theoretic means. One half of this important equivalence is rather easy to establish:

Lemma 11.8. $\Gamma \vdash \varphi$ *implies* $\Gamma \models \varphi$.

PROOF. Using a simple inductive argument based on 10.24, it obviously suffices to show that each logical axiom is universally valid. We consider the five kinds of logical axioms given in 10.23:

(1) φ is a tautology. Let $x \in {}^{\omega}A$; we wish to show that $x \in \varphi^{\mathfrak{A}}$. For any formula ψ, let

$$\begin{aligned} f\psi &= 1 \qquad \text{if } x \in \psi^{\mathfrak{A}}, \\ f\psi &= 0 \qquad \text{if } x \notin \psi^{\mathfrak{A}}. \end{aligned}$$

Clearly f is a truth valuation in the sense of 10.19. Hence $f\varphi = 1$, i.e., $x \in \varphi^{\mathfrak{A}}$, as desired.

(2) φ is the formula $\forall v_i(\psi \rightarrow \chi) \rightarrow (\forall v_i\psi \rightarrow \forall v_i\chi)$. Again let $x \in {}^{\omega}A$. To establish that $x \in \varphi^{\mathfrak{A}}$, we use the remark following 11.5: we assume that $x \in (\forall v_i(\psi \rightarrow \chi))^{\mathfrak{A}}$, $x \in (\forall v_i\psi)^{\mathfrak{A}}$, and prove that $x \in (\forall v_i\chi)^{\mathfrak{A}}$. Let $a \in A$ be arbitrary. Then $x^i_a \in (\psi \rightarrow \chi)^{\mathfrak{A}}$ and $x^i_a \in \psi^{\mathfrak{A}}$, so $x^i_a \in \chi^{\mathfrak{A}}$, as desired.

(3) φ is $\psi \rightarrow \forall v_i\psi$, where v_i does not occur in ψ. Obviously $\varphi^{\mathfrak{A}} = {}^{\omega}A$, by Proposition 11.6.

(4) φ is $\exists v_i(v_i \equiv \sigma)$, where v_i does not occur in σ. Let $x \in {}^{\omega}A$, and set $a = \sigma^{\mathfrak{A}}x$. Then by 11.3, $\sigma^{\mathfrak{A}}x = \sigma^{\mathfrak{A}}x^i_a$. Hence

$$v_i^{\mathfrak{A}}(x^i_a) = a = \sigma^{\mathfrak{A}}x = \sigma^{\mathfrak{A}}x^i_a,$$

so $\mathfrak{A} \models (v_i \equiv \sigma)[x^i_a]$. Hence $\mathfrak{A} \models \exists v_i(v_i \equiv \sigma)[x]$, as desired.

(5) φ is $\sigma \equiv \tau \rightarrow (\psi \rightarrow \chi)$, where ψ and χ are atomic formulas and χ is obtained from ψ by replacing an occurrence of σ in ψ by τ. The desired result clearly follows from the following lemma:

(5′) if ξ is obtained from the term ρ by replacing one occurrence of σ in ρ by τ, and if $x \in {}^{\omega}A$ and $\sigma^{\mathfrak{A}}x = \tau^{\mathfrak{A}}x$, then $\rho^{\mathfrak{A}}x = \xi^{\mathfrak{A}}x$.

The lemma is easily proved by induction on ρ. □

Lemma 11.8 is already of some interest. In particular, using it we can establish that a formula φ is not derivable by exhibiting a model of $\neg\varphi$. Thus if φ has the form $\psi \wedge \neg\psi$, clearly any $\mathscr{L}$-structure is a model of $\neg\varphi$, and so $\nvdash\varphi$ by 11.8. This provides another proof that the empty set is consistent (cf. 10.91).

We can now establish rigorously that various restrictions on axioms and theorems in Chapter 10 are essential in order to preserve the property given in 11.8. For example, consider the restrictions in 10.23(3) and 10.23(4). Formulas of the form $\varphi \rightarrow \forall v_i\varphi$ are not universally valid, in general, if v_i occurs in φ. For example, let $i = 0$ and consider the formula $\varphi = (v_0 \equiv v_1)$. Let $\mathfrak{A}$ be any $\mathscr{L}$-structure with at least two elements, and let $x \in {}^{\omega}A$ be such that $x_0 = x_1$. Clearly then $\mathfrak{A} \models \varphi[x]$, while if $a \in A \sim \{x_0\}$, then $\mathfrak{A} \models \neg\varphi[x^0_a]$ and hence $\mathfrak{A} \models \neg\forall v_i\varphi[x]$; thus x does not satisfy $\varphi \rightarrow \forall v_i\varphi$ in $\mathfrak{A}$. For an example concerning 10.23(4), let $\mathscr{L}$ be a language with a unary operation symbol f and let φ be the formula $v_0 \equiv \mathbf{f}v_0$. Let $\mathfrak{A}$ be an $\mathscr{L}$-structure in which $A = \omega$ and $\mathbf{f}$ is interpreted by $\mathbf{s}$. Clearly then $\mathfrak{A} \models \neg\exists v_0\varphi$. Thus the restriction in 10.23(4) is also necessary.

To prove the converse of 11.8, it is enough to restrict attention to a sentence φ and a set Γ of sentences. The contrapositive of the converse of 11.8 then says that if $\Gamma \nvdash \varphi$, then $\Gamma \not\models \varphi$; by 10.89, this means that: if $\Gamma \cup \{\neg\varphi\}$ is consistent, then $\Gamma \cup \{\neg\varphi\}$ has a model. Thus our task is really to show that any consistent set of sentences has a model. We shall first show this for special sets of sentences where, in a sense, quantifiers can be eliminated:

Definition 11.9. Let $\mathscr{L}$ be a first-order language. A set Γ of sentences of $\mathscr{L}$ is *rich* if for every sentence of $\mathscr{L}$ of the form $\exists\alpha\varphi$ there is an individual constant $\mathbf{c}$ of $\mathscr{L}$ such that $\Gamma \vdash \exists\alpha\varphi \rightarrow \mathrm{Subf}^{\alpha}_{\mathbf{c}}\varphi$. A set Γ of sentences of $\mathscr{L}$ is *complete* provided that for any sentence φ of $\mathscr{L}$, $\Gamma \vdash \varphi$ or $\Gamma \vdash \neg\varphi$.

Note that $\vdash \mathrm{Subf}^{\alpha}_{\mathbf{c}}\varphi \rightarrow \exists\alpha\varphi$; hence the requirement in 11.9 essentially ensures that any existential quantifier can be eliminated. In fact, the following proposition is easily established by induction on φ:

Proposition 11.10. *Let Γ be a rich set of sentences. Then for any sentence φ there is a sentence ψ in which no quantifier occurs such that $\Gamma \vdash \varphi \leftrightarrow \psi$.*

The following proposition expresses some simple but useful properties of complete sets of sentences:

Proposition 11.11. *For any $\Gamma \subseteq \mathrm{Sent}_{\mathscr{L}}$ the following conditions are equivalent:*

(*i*) *Γ is complete and consistent;*
(*ii*) *$\{\varphi : \varphi \in \mathrm{Sent}_{\mathscr{L}}, \Gamma \vdash \varphi\}$ is a maximal consistent set of sentences;*
(*iii*) *for any sentences φ, ψ, if $\Gamma \vdash \varphi \vee \psi$, then $\Gamma \vdash \varphi$ or $\Gamma \vdash \psi$, and Γ is consistent.*

PROOF

(*i*) $\Rightarrow$ (*ii*). Obviously $\Delta = \{\varphi : \varphi \in \mathrm{Sent}_{\mathscr{L}}, \Gamma \vdash \varphi\}$ is consistent. Suppose $\Delta \subset \Theta$; say $\varphi \in \Theta \sim \Delta$. Thus $\Gamma \nvdash \varphi$, so by (*i*) $\Gamma \vdash \neg\varphi$. Hence $\neg\varphi \in \Delta \subset \Theta$, so both φ and $\neg\varphi$ are in Θ. Hence Θ is inconsistent.

(*ii*) $\Rightarrow$ (*iii*). Assume that $\Gamma \nvdash \varphi$ and $\Gamma \nvdash \psi$, and let Δ be as above. Then $\Delta \subset \Delta \cup \{\varphi\}$, so $\Delta \cup \{\varphi\}$ is inconsistent. Hence by 10.92, $\Delta \vdash \neg\varphi$, so $\Gamma \vdash \neg\varphi$. Similarly $\Gamma \vdash \neg\psi$ and so by a suitable tautology, $\Gamma \vdash \neg(\varphi \vee \psi)$. Thus $\Gamma \nvdash \varphi \vee \psi$ since Γ is consistent.

(*iii*) $\Rightarrow$ (*i*). Given a sentence φ, we have $\Gamma \vdash \varphi \vee \neg\varphi$, so by (*iii*), $\Gamma \vdash \varphi$ or $\Gamma \vdash \neg\varphi$. Thus Γ is complete. □

We now show, by a direct construction, that any complete, rich, consistent set of sentences has a model:

Theorem 11.12. *If Γ is a complete, rich, consistent set of sentences in a language $\mathscr{L}$, then Γ has a model $\mathfrak{A}$ such that the cardinality of A is at most*

$$|\{\sigma : \sigma \text{ is a term of } \mathscr{L} \text{ in which no variable occurs}\}|.$$

PROOF. Let $B = \{\sigma : \sigma$ is a term of $\mathscr{L}$, and no variable occurs in $\sigma\}$. We define $\equiv$ to be the set

$$\{(\sigma, \tau) : \sigma, \tau \in B \text{ and } \Gamma \vdash \sigma \mathbf{=} \tau\}.$$

By 10.34–10.36, $\equiv$ is an equivalence relation on B. Recall that $[\,]$ is the function assigning to each $\sigma \in B$ its equivalence class under $\equiv$. Let $A = B/\!\equiv$. By 10.38, for each $\mathbf{O} \in \mathrm{Dmn}\, \mathscr{O}$, say of rank m, there is an m-ary operation $\mathbf{O}^{\mathfrak{A}}$ on A such that for any $\sigma_0, \ldots, \sigma_{m-1} \in B$,

$$\mathbf{O}^{\mathfrak{A}}([\sigma_0], \ldots, [\sigma_{m-1}]) = [\mathbf{O}\sigma_0 \cdots \sigma_{m-1}].$$

For any $\mathbf{R} \in \mathrm{Dmn}\, \mathscr{R}$, say $\mathbf{R}$ m-ary, we let $\mathbf{R}^{\mathfrak{A}}$ be the collection of all m-tuples of the form $([\sigma_0], \ldots, [\sigma_{m-1}])$ such that $\Gamma \vdash \mathbf{R}\sigma_0 \cdots \sigma_{m-1}$. This defines $\mathfrak{A}$. We claim that $\mathfrak{A}$ is the desired model of Γ. The proof of this is based on the following auxiliary considerations. First, the following statement is easily established by induction on τ:

(1) $$\text{for any } \tau \in B,\ {}^0\tau^{\mathfrak{A}} = [\tau].$$

Next, let $\sigma \in {}^{\omega}B$. If τ is any term, we let $S_\sigma\tau$ be the term obtained from τ by simultaneously replacing each variable v_i in τ by σ_i. Then by induction on τ it is easily shown that

(2) $$\text{for any term } \tau,\ \tau^{\mathfrak{A}}([\,] \circ \sigma) = [S_\sigma\tau].$$

Similarly, for any formula φ, let $S_\sigma\varphi$ be obtained from φ by simultaneously replacing each free occurrence of v_i in φ by σ_i for each $i < \omega$. Then:

(3) $$\text{For any formula } \varphi,\ \mathfrak{A} \models \varphi[[\,] \circ \sigma] \text{ iff } \mathfrak{A} \models S_\sigma\varphi.$$

Of course (3) is also established by induction on φ; we consider two steps as illustration. If φ is $\tau \mathbf{=} \rho$, then

$$\begin{aligned} \mathfrak{A} \models \varphi[[\,] \circ \sigma] \quad & \text{iff } \tau^{\mathfrak{A}}([\,]) \circ \sigma) = \rho^{\mathfrak{A}}([\,] \circ \sigma) \\ & \text{iff } {}^0(S_\sigma\tau)^{\mathfrak{A}} = {}^0(S_\sigma\rho)^{\mathfrak{A}} \text{ by (1), (2)} \\ & \text{iff } \mathfrak{A} \models S_\sigma\varphi. \end{aligned}$$

Now, suppose φ is $\forall v_i\psi$. Assume first $\mathfrak{A} \models \varphi[[\,] \circ \sigma]$. Then $S_\sigma\varphi$ has the form $\forall v_i\psi'$, where ψ' is obtained from ψ by simultaneously replacing each free occurrence of v_j in ψ by σ_j, for each $j \in \omega \sim \{i\}$. To show that $S_\sigma\varphi$ holds in $\mathfrak{A}$, let $a = [\tau]$ be arbitrary. Then $\mathfrak{A} \models \psi[([\,] \circ \sigma)^i_a]$, i.e., $\mathfrak{A} \models \psi[[\,] \circ \sigma^i_\tau]$. By the induction hypothesis, $\mathfrak{A} \models S(\sigma^i_\tau)\psi$. Clearly $S(\sigma^i_\tau)\psi = S(\sigma^i_\tau)\psi'$, so again by the induction assumption, $\mathfrak{A} \models \psi'[[\,] \circ \sigma^i_\tau]$, i.e., $\mathfrak{A} \models \psi'[([\,] \circ \sigma)^i_a]$. Thus $S_\sigma\varphi$ holds in $\mathfrak{A}$, as desired. The converse is similar.

Now we return to the proof that $\mathfrak{A}$ is a model of Γ. This is immediate from the following stronger statement:

(4) for any sentence φ, $\Gamma \vdash \varphi$ iff φ holds in $\mathfrak{A}$.

We prove (4) by induction on the *sentence* φ. If φ is $\sigma \equiv \tau$, then

$$\begin{aligned} \Gamma \vdash \sigma \equiv \tau \quad & \text{iff } [\sigma] = [\tau] \\ & \text{iff } {}^0\sigma^{\mathfrak{A}} = {}^0\tau^{\mathfrak{A}} \qquad \text{by (1)} \\ & \text{iff } \mathfrak{A} \models \sigma \equiv \tau. \end{aligned}$$

Next, let φ be $\mathbf{R}\sigma_0 \cdots \sigma_{m-1}$. First suppose $\Gamma \vdash \varphi$. Then by the definition of $\mathbf{R}^{\mathfrak{A}}$, $([\sigma_0], \ldots, [\sigma_{m-1}]) \in \mathbf{R}^{\mathfrak{A}}$. By (1) it follows that $\mathfrak{A} \models \mathbf{R}\sigma_0 \cdots \sigma_{m-1}$. Conversely, suppose $\mathfrak{A} \models \mathbf{R}\sigma_0 \cdots \sigma_{m-1}$. Thus $({}^0\sigma_0^{\mathfrak{A}}, \ldots, {}^0\sigma_{m-1}^{\mathfrak{A}}) \in \mathbf{R}^{\mathfrak{A}}$, i.e., by (1), $([\sigma_0], \ldots, [\sigma_{m-1}]) \in \mathbf{R}^{\mathfrak{A}}$. Hence by the definition of $\mathbf{R}^{\mathfrak{A}}$, there exist $\tau_0, \ldots, \tau_{m-1} \in B$ with $[\sigma_0] = [\tau_0], \ldots, [\sigma_{m-1}] = [\tau_{m-1}]$, and $\Gamma \vdash \mathbf{R}\tau_0 \cdots \tau_{m-1}$. By an easy argument using 10.23(5), $\Gamma \vdash \varphi$.

For $\varphi = \neg\psi$ we have

$$\begin{aligned} \Gamma \vdash \varphi \quad & \text{iff } \Gamma \nvdash \psi \qquad \text{(by consistency and completeness)} \\ & \text{iff } \mathfrak{A} \nvDash \psi \qquad \text{(induction hypothesis)} \\ & \text{iff } \mathfrak{A} \models \varphi. \end{aligned}$$

The cases $\varphi = \psi \vee \chi$ and $\varphi = \psi \wedge \chi$ are similar. Finally, suppose φ is $\forall v_i\psi$. First suppose $\Gamma \vdash \varphi$. To show that φ holds in $\mathfrak{A}$, let $a = [\tau]$ be arbitrary. Since $\vdash \varphi \rightarrow \mathrm{Subf}_\tau^{vi}\psi$ by universal specification, we have $\Gamma \vdash \mathrm{Subf}_\tau^{vi}\psi$. Hence $\mathfrak{A} \models \mathrm{Subf}_\tau^{vi}\psi$, by the induction assumption. By (3) we have $\mathfrak{A} \models \psi[[\ ] \circ \sigma]$, where $\sigma \in {}^\omega B$ is any sequence with $\sigma_i = \tau$. Thus φ holds in $\mathfrak{A}$. Conversely, suppose φ holds in $\mathfrak{A}$. Since Γ is rich, let $\mathbf{c}$ be an individual constant of $\mathscr{L}$ such that $\Gamma \vdash \exists v_i \ \neg \psi \rightarrow \mathrm{Subf}_{\mathbf{c}}^{vi} \ \neg \psi$. Thus $\Gamma \vdash \mathrm{Subf}_{\mathbf{c}}^{vi}\psi \rightarrow \forall v_i\psi$. Let $\sigma \in {}^\omega B$ be any sequence such that $\sigma_i = \mathbf{c}$. Since φ holds in $\mathfrak{A}$, $\mathfrak{A} \models \psi[[\ ] \circ \sigma]$. By (3), $\mathrm{Subf}_{\mathbf{c}}^{vi}\psi$ holds in $\mathfrak{A}$, so $\Gamma \vdash \mathrm{Subf}_{\mathbf{c}}^{vi}\psi$ by the induction assumption. Hence $\Gamma \vdash \varphi$, as desired. □

To obtain the completeness theorem, we still need to see how any consistent set Γ of sentences can be extended to a complete, rich, consistent set. First we deal with the case of extension to a complete consistent set. The theorem in question has many applications in addition to our immediate concern with the completeness theorem:

Theorem 11.13 (Lindenbaum). *If Γ is a consistent set of sentences of $\mathscr{L}$, then there is a complete consistent set Δ of sentences in $\mathscr{L}$ such that $\Gamma \subseteq \Delta$.*

Proof. Let $\mathscr{A} = \{\Delta : \Gamma \subseteq \Delta \subseteq \mathrm{Sent}_{\mathscr{L}}, \Delta \text{ consistent}\}$. If $\mathscr{B}$ is a nonempty subset of $\mathscr{A}$ simply ordered by $\subseteq$, then $\Gamma \subseteq \bigcup\mathscr{B} \subseteq \mathrm{Sent}_{\mathscr{L}}$. Also, $\bigcup\mathscr{B}$ is consistent. For, otherwise, by 10.93 there is an $m \in \omega \sim 1$ and a $\varphi \in {}^m\bigcup\mathscr{B}$ such that $\vdash \neg\varphi_0 \vee \cdots \vee \neg\varphi_{m-1}$. Since $\mathscr{B}$ is simply ordered, $\varphi \in {}^m\Delta$ for some $\Delta \in \mathscr{B}$. Thus by 10.93, Δ is inconsistent, contradicting our assumption that $\mathscr{B} \subseteq \mathscr{A}$. Now we can conclude by Zorn's lemma that $\mathscr{A}$ has a maximal member Δ. By 11.11, Δ is complete. □

To extend a consistent set Γ to a rich consistent set Γ is technically somewhat difficult. If $\mathscr{L}$ has no individual constants, then it is obviously impossible to do this extension within $\mathscr{L}$ itself. In general we must expand the language. We now briefly discuss the idea of expansion of languages.

Definition 11.14. A language $\mathscr{L}' = (L', v', \mathscr{O}', \mathscr{R}')$ is an *expansion* of a language $\mathscr{L} = (L, v, \mathscr{O}, \mathscr{R})$ provided that $L = L'$, $v = v'$, $\mathscr{O} \subseteq \mathscr{O}'$ and $\mathscr{R} \subseteq \mathscr{R}'$; we also say then that $\mathscr{L}$ is a *reduct* of $\mathscr{L}'$. In case $\mathscr{L}' = (L', v', \mathscr{O}', \mathscr{R}', g')$ and $\mathscr{L} = (L, v, \mathscr{O}, \mathscr{R}, g)$ are effectivized first-order languages, then we insist additionally that $g \subseteq g'$, and we call $\mathscr{L}'$ an *effective expansion* of $\mathscr{L}$. Assume that $\mathscr{L}'$ is an expansion of $\mathscr{L}$. If $\mathfrak{A} = (A, f, R)$ is an $\mathscr{L}$-structure and $\mathfrak{B} = (B, f', R')$ is an $\mathscr{L}'$-structure, we say that $\mathfrak{A}$ is the *$\mathscr{L}$-reduct* of $\mathfrak{B}$, or that $\mathfrak{B}$ is an *$\mathscr{L}'$-expansion* of $\mathfrak{A}$ provided that $A = B$, $f \subseteq f'$, and $R \subseteq R'$. We then write $\mathfrak{A} = \mathfrak{B} \upharpoonright \mathscr{L}$.

In discussing expansions we will usually not be as rigorous as in 11.14. Thus we might say "let $\mathscr{L}'$ be obtained from $\mathscr{L}$ by adjoining a new binary relation symbol" or "let $\mathscr{L}'$ be obtained from $\mathscr{L}$ by adjoining m individual constants." Note that the ways of making such statements precise are all equivalent, in some sense; see the comments following 10.1.

In the following proposition we summarize the most important elementary facts about expansions.

Proposition 11.15. *Let* $\mathscr{L}$, $\mathscr{L}'$, $\mathfrak{A}$, $\mathfrak{B}$ *be as in* 11.14. *Then:*

(*i*) $\mathrm{Trm}_{\mathscr{L}} \subseteq \mathrm{Trm}_{\mathscr{L}'}$, $\mathrm{Fmla}_{\mathscr{L}} \subseteq \mathrm{Fmla}_{\mathscr{L}'}$, $\mathrm{Axm}_{\mathscr{L}} \subseteq \mathrm{Axm}_{\mathscr{L}'}$, *and* $\mathrm{Sent}_{\mathscr{L}} \subseteq \mathrm{Sent}_{\mathscr{L}'}$;

(*ii*) *if* $\sigma \in \mathrm{Trm}_{\mathscr{L}}$, *then* $\sigma^{\mathfrak{A}} = \sigma^{\mathfrak{B}}$;

(*iii*) *if* $\varphi \in \mathrm{Fmla}_{\mathscr{L}}$, *then* $\varphi^{\mathfrak{A}} = \varphi^{\mathfrak{B}}$;

(*iv*) *if* $\Gamma \subseteq \mathrm{Fmla}_{\mathscr{L}}$, *then* $\mathfrak{A}$ *is a model of* Γ *iff* $\mathfrak{B}$ *is a model of* Γ;

(*v*) *if* $\Gamma \cup \{\varphi\} \subseteq \mathrm{Fmla}_{\mathscr{L}}$, *then* $\Gamma \models_{\mathscr{L}} \varphi$ *iff* $\Gamma \models_{\mathscr{L}'} \varphi$.

This proposition is easy to prove; (*i*)–(*iii*) are proved by induction, and (*iv*) and (*v*) follow from (*i*)–(*iii*). By the completeness theorem which we will shortly prove, 11.15(*v*) also holds for the notion $\vdash$. This statement is needed in our proof of the completeness theorem, however, so we will give a proof-theoretic proof of it:

Proposition 11.16. *Let* $\mathscr{L}'$ *be an expansion of* $\mathscr{L}$, *and assume that* $\Gamma \cup \{\varphi\} \subseteq \mathrm{Fmla}_{\mathscr{L}}$. *Then* $\Gamma \vdash_{\mathscr{L}} \varphi$ *iff* $\Gamma \vdash_{\mathscr{L}'} \varphi$.

PROOF. Obviously $\Gamma \vdash_{\mathscr{L}} \varphi \Rightarrow \Gamma \vdash_{\mathscr{L}'} \varphi$. Now assume that $\Gamma \vdash_{\mathscr{L}'} \varphi$. Say that $\psi_0, \ldots, \psi_{m-1}$ is a Γ-formal proof in $\mathscr{L}'$ with $\psi_{m-1} = \varphi$. Nonlogical constants of $\mathscr{L}'$ which do not appear in $\mathscr{L}$ will be called *new* constants. Let α be a variable not occurring in any of the formulas $\psi_0, \ldots, \psi_{m-1}$. We now associate with any formula or term χ of $\mathscr{L}'$ a formula or term χ^- of $\mathscr{L}$, defined by recursion:

$$
\begin{aligned}
&(v_i)^- = v_i; \\
&(\mathbf{O}\sigma_0\cdots\sigma_{m-1})^- = \mathbf{O}\sigma_0^-\cdots\sigma_{m-1}^- && \text{if } \mathbf{O} \text{ is an old operation symbol;} \\
&(\mathbf{O}\sigma_0\cdots\sigma_{m-1})^- = \alpha && \text{if } \mathbf{O} \text{ is a new operation symbol;} \\
&(\sigma = \tau)^- = (\sigma^- = \tau^-); \\
&(\mathbf{R}\sigma_0\cdots\sigma_{m-1})^- = \mathbf{R}\sigma_0^-\cdots\sigma_{m-1}^- && \text{if } \mathbf{R} \text{ is an old relation symbol;} \\
&(\mathbf{R}\sigma_0\cdots\sigma_{m-1})^- = (\alpha = \alpha) && \text{if } R \text{ is a new relation symbol;} \\
&(\neg\varphi)^- = \neg\varphi^-; \\
&(\varphi \vee \psi)^- = \varphi^- \vee \psi^-; \\
&(\varphi \wedge \psi)^- = \varphi^- \wedge \psi^-; \\
&(\forall v_i\varphi)^- = \forall v_i\varphi^-.
\end{aligned}
$$

We claim:

(1) if $i < m$ and ψ_i is a logical axiom, then ψ_i^- is a logical theorem in $\mathscr{L}$.

To prove (1) we take up the five possibilities for ψ_i according to 10.23. First suppose that ψ_i is a tautology. Then ψ_i^- is a tautology. For, let f be any truth valuation in $\mathscr{L}$. Define $f^+\varphi = f\varphi^-$ for any formula φ of $\mathscr{L}'$. It is clear that f^+ is a truth valuation, so $f^+\psi_i = 1$. Thus $f\psi_i^- = 1$, as desired. Second, suppose ψ_i has the form $\forall v_i(\varphi \rightarrow \chi) \rightarrow (\forall v_i\varphi \rightarrow \forall v_i\chi)$. Then ψ_i^- is

$$\forall v_i(\varphi^- \rightarrow \chi^-) \rightarrow (\forall v_i\varphi^- \rightarrow \forall v_i\chi^-),$$

so ψ_i^- is still a logical axiom. Third, suppose ψ_i is the formula $\varphi \rightarrow \forall v_i\varphi$, where v_i does not occur in φ. Then ψ_i^- is the formula $\varphi^- \rightarrow \forall v_i\varphi^-$, and, by choice of α, v_i does not occur in φ^-. Thus ψ_i^- is a logical axiom. For ψ_i of the form $\exists v_i(v_i = \sigma)$ where v_i does not occur in σ, a similar argument works. Finally, suppose ψ_i has the form $\sigma = \tau \rightarrow (\varphi \rightarrow \chi)$, as in 10.23(5). If φ has the form $\mathbf{R}\rho_0\cdots\rho_{m-1}$ where $\mathbf{R}$ is a new relation symbol, then ψ_i is $\sigma^- = \tau^- \rightarrow (\alpha = \alpha \rightarrow \alpha = \alpha)$, which is a logical theorem. If φ does not have this form, then the desired result follows easily from the following statement:

(2) if $\rho = \mu\sigma\nu$ and $\rho' = \mu\tau\nu$ are terms, then either $\rho^- = \rho'^-$ or else ρ^- and ρ'^- have the respective forms $\mu'\sigma^-\nu'$, $\mu'\tau^-\nu'$.

This statement is easily established by induction on ρ. Thus (1) holds. Clearly:

(3) for σ a term of $\mathscr{L}$, $\sigma^- = \sigma$; for φ a formula of $\mathscr{L}$, $\varphi^- = \varphi$.

Now using (1) and (3) it is easily checked that ψ_i^- is a Γ-theorem in $\mathscr{L}$ for each $i < m$. Hence $\Gamma \vdash_{\mathscr{L}} \varphi$, as desired. □

The same method of proof gives the following proposition which is also needed in our construction of rich extensions:

Proposition 11.17 *Let* **c** *be an individual constant not occurring in any formula of* $\Gamma \cup \{\varphi\}$. *Assume that* $\Gamma \vdash \mathrm{Subf}_{\mathbf{c}}^{\alpha}\varphi$. *Then* $\Gamma \vdash \varphi$.

Proof. Let $\langle\psi_0, \ldots, \psi_{m-1}\rangle$ be a Γ-formal proof with $\psi_{m-1} = \mathrm{Subf}_{\mathbf{c}}^{\alpha}\varphi$. Let β be a variable not occurring in any of the formulas $\psi_0, \ldots, \psi_{m-1}$. For any $i < m$ let ψ_i^- be obtained from ψ_i by replacing $\mathbf{c}$ throughout ψ_i by β. As in the proof of 11.16 it is seen that $\langle\psi_0^-, \ldots, \psi_{m-1}^-\rangle$ is a Γ-formal proof. Since $(\mathrm{Subf}_{\mathbf{c}}^{\alpha}\varphi)^- = \mathrm{Subf}_{\beta}^{\alpha}\varphi$, we have $\Gamma \vdash \mathrm{Subf}_{\beta}^{\alpha}\varphi$. Hence $\Gamma \vdash \forall\beta\, \mathrm{Subf}_{\beta}^{\alpha}\varphi$ and $\Gamma \vdash \varphi$ by universal specification. □

Finally we are in a position to construct rich extensions:

Lemma 11.18. *Let $\mathscr{L}$ be any first-order language and let Γ be a consistent set of sentences in $\mathscr{L}$. Let $\mathscr{L}'$ be an expansion of $\mathscr{L}$ obtained by adjoining $|\mathrm{Fmla}_{\mathscr{L}}|$ new individual constants. Then there is a consistent rich set Δ of sentences of $\mathscr{L}'$ such that $\Gamma \subseteq \Delta$.*

Proof. Let $\langle\varphi_\alpha : \alpha < \mathfrak{m}\rangle$ be a list of all sentences of $\mathscr{L}'$ of the form $\exists\beta\psi$, where $\mathfrak{m}$ is an infinite cardinal number. Note that $\mathfrak{m} = |\mathrm{Fmla}_{\mathscr{L}}|$. We now define a sequence $\langle\mathbf{d}_\alpha : \alpha < \mathfrak{m}\rangle$ of new individual constants. Suppose we have already defined $\mathbf{d}\beta$ for all $\beta < \alpha$, where $\alpha < \mathfrak{m}$. Then

$$\{\mathbf{d}\beta : \beta < \alpha\} \cup \{\mathbf{c} : \mathbf{c} \text{ is a new individual constant occurring in some } \varphi_\beta \text{ with } \beta \le \alpha\}$$

has cardinality $<\mathfrak{m}$, so we can let $\mathbf{d}_\alpha$ be some new individual constant not in this set, say the first one under some given well-ordering of all the new individual constants. This completes the definition of the sequence $\langle\mathbf{d}_\alpha : \alpha < \mathfrak{m}\rangle$.

Now for each $\alpha < \mathfrak{m}$ let φ_α be the sentence $\exists\beta_\alpha\psi_\alpha$. For each $\alpha \le \mathfrak{m}$ let

$$\Theta_\alpha = \Gamma \cup \{\exists\beta_\gamma\psi_\gamma \rightarrow \mathrm{Subf}_{\mathbf{d}\gamma}^{\beta\alpha} : \gamma < \alpha\}.$$

We claim that each set Θ_α is consistent in $\mathscr{L}'$, and we shall establish this by induction on α. We have $\Theta_0 = \Gamma$, so Θ_0 is consistent in $\mathscr{L}'$ by 11.16. The induction step to a limit ordinal is clear from 10.92. So now assume that Θ_α is consistent; we prove that $\Theta_{\alpha+1}$ is consistent. If $\Theta_{\alpha+1}$ is inconsistent, then by 10.92 we have

$$\Theta_\alpha \vdash \neg(\exists\beta_\alpha\psi_\alpha \rightarrow \mathrm{Subf}_{\mathbf{d}\alpha}^{\beta\alpha}\psi_\alpha).$$

Hence

(1) $$\Theta_\alpha \vdash \exists\beta_\alpha\psi_\alpha$$

and also

$$\Theta_\alpha \vdash \neg\mathrm{Subf}_{\mathbf{d}\alpha}^{\beta\alpha}\psi_\alpha.$$

We may apply 11.17 to this, by choice of $\mathbf{d}\alpha$. Thus $\Theta_\alpha \vdash \neg\psi_\alpha$, and hence

(2) $$\Theta_\alpha \vdash \neg\exists\beta_\alpha\psi_\alpha.$$

By (1) and (2), Θ_α is inconsistent, contradicting the induction assumption.

Thus each set Θ_α is consistent. Clearly $\Theta_{\mathfrak{m}}$ is rich and contains Γ, as desired. □

Now we can prove the completeness theorem.

Theorem 11.19 (Completeness theorem, first form). *Any consistent set of sentences has a model. The model can be taken to have power* $\leq |\mathrm{Fmla}_{\mathscr{L}}|$.

PROOF. Let Γ be a consistent set of sentences in the language $\mathscr{L}$. Let $\mathscr{L}'$ be obtained from $\mathscr{L}$ by adjoining $|\mathrm{Fmla}_{\mathscr{L}}|$ new individual constants. By 11.18, let Δ be a consistent rich set of sentences in $\mathscr{L}'$ with $\Gamma \subseteq \Delta$. By 11.13, let Θ be a complete consistent set of sentences in $\mathscr{L}'$ such that $\Delta \subseteq \Theta$. Obviously Θ is still rich. By 11.12, Θ has a model $\mathfrak{A}$. Clearly (by 11.15(*iv*)), $\mathfrak{A} \upharpoonright \mathscr{L}$ is a model of Γ. From all of these results it also follows that $|A| \leq |\mathrm{Fmla}_{\mathscr{L}}|$. □

As already observed informally, 11.19 leads to the equivalence of $\vdash$ and $\models$:

Theorem 11.20 (Completeness theorem, second form). $\Gamma \vdash \varphi$ *iff* $\Gamma \models \varphi$.

PROOF. $\Rightarrow$: by 11.8. $\Leftarrow$: Assume $\Gamma \nvdash \varphi$. Thus by 10.95(*iii*), $\{[[\psi]] : \psi \in \Gamma\} \nvdash [[\varphi]]$ so by 10.92, $\Delta = \{[[\psi]] : \psi \in \Gamma\} \cup \{\neg[[\varphi]]\}$ is consistent. Hence Δ has a model $\mathfrak{A}$, by 11.20. Clearly $\mathfrak{A}$ is a model of Γ but not of φ, as desired. □

Because of this theorem, we can formulate results from now on using either $\vdash$ or $\models$. Since we shall use model-theoretic methods almost exclusively, it seems more appropriate to formulate them using $\models$, as we shall do. In the future our only use for the notion $\vdash$ will be in proofs of model-theoretic results where we use the completeness theorem. The following theorem expresses one of the main model-theoretic results, proved very essentially using $\vdash$:

Theorem 11.21 (Weak completeness theorem). *If* $\mathscr{L}$ *is an effectivized first-order language, then* $\{\mathscr{g}^+\varphi : \models \varphi\}$ *is recursively enumerable.*

PROOF. By the completeness theorem, the given set is identical with $\{\mathscr{g}^+\varphi : \vdash \varphi\}$. This set is r.e. by Theorem 10.29. □

The following important theorem follows immediately from the first form of the completeness theorem.

Theorem 11.22 (Compactness theorem). *If* Γ *is a set of sentences such that every finite subset of* Γ *has a model, then* Γ *has a model.*

PROOF. By 11.19 it suffices to show that Γ is consistent. Suppose not: say $\Gamma \vdash \varphi \wedge \neg\varphi$. By 10.33, $\Delta \vdash \varphi \wedge \neg\varphi$ for some finite subset Δ of Γ. By 11.20 (in fact, the easy part of 11.20 given in 11.8), $\Delta \models \varphi \wedge \neg\varphi$. By hypothesis, Δ has a model $\mathfrak{A}$. Hence $\varphi \wedge \neg\varphi$ holds in $\mathfrak{A}$, which is impossible. □

The compactness theorem lies at the start of model theory, and it will play a very important role in Part IV. For some motivation for the name compactness theorem, see Exercise 11.59. We give here just a few corollaries to indicate the flavor of applications of the compactness theorem (see also the exercises).

Corollary 11.23. *Let $\mathscr{L}$ be any first-order language. Then there is no set Γ of sentences of $\mathscr{L}$ such that for any $\mathscr{L}$-structure $\mathfrak{A}$, $\mathfrak{A}$ is a model of Γ iff $\mathfrak{A}$ is finite.*

PROOF. Assume the contrary. Expand $\mathscr{L}$ to $\mathscr{L}'$ by adjoining new individual constants $\mathbf{c}_m$ for $m \in \omega$. Let Δ be Γ together with all sentences $\neg(\mathbf{c}_i = \mathbf{c}_j)$ for $i \neq j$. Every finite subset Δ' of Δ has a model. For, choose $T \subseteq \omega$ finite so that $\Delta' \subseteq \Gamma \cup \{\neg(\mathbf{c}_i = \mathbf{c}_j) : i, j \in T, i \neq j\}$. Let $\mathfrak{A}$ be any finite $\mathscr{L}'$-structure in which $\mathbf{c}_i^{\mathfrak{A}} \neq \mathbf{c}_j^{\mathfrak{A}}$ for all distinct $i, j \in T$, where $\mathbf{c}_i^{\mathfrak{A}}$ is the denotation of $\mathbf{c}_i$ in $\mathfrak{A}$ for each $i \in T$. Thus $\mathfrak{A} \restriction \mathscr{L}$ is a finite $\mathscr{L}$-structure, so by assumption on Γ, $\mathfrak{A} \restriction \mathscr{L}$ is a model of Γ. Thus $\mathfrak{A}$ is a model of Δ'.

Hence by the compactness theorem, Δ has a model $\mathfrak{B}$. Since $\mathbf{c}_i^{\mathfrak{B}} \neq \mathbf{c}_j^{\mathfrak{B}}$ for all distinct $i, j \in \omega$ (where, now, $\mathbf{c}_i^{\mathfrak{B}}$ is the denotation of $\mathbf{c}_i$ in $\mathfrak{B}$ for all $i \in \omega$), $\mathfrak{B}$ is infinite. But $\mathfrak{B} \restriction \mathscr{L}$ is a model of Γ, contradiction. □

Corollary 11.24. *Let $\mathscr{L}$ be a first-order language with just one nonlogical constant, a binary relation symbol $\leq$. Let $\mathbf{K}$ be the class of all $\mathscr{L}$-structures $\mathfrak{A} = (A, \leq)$ such that $\leq$ is a well-ordering of A. Then there is no set Γ of sentences of $\mathscr{L}$ such that $\mathbf{K} = \{\mathfrak{A} : \mathfrak{A}$ is a model of $\Gamma\}$.*

PROOF. Assume the contrary. Again adjoin new individual constants $\mathbf{c}_m$ for $m \in \omega$. Let Δ be Γ together with all sentences $\mathbf{c}_{m+1} < \mathbf{c}_m$, for $m \in \omega$. Every finite subset of Δ clearly has a model. Hence by the compactness theorem, Δ has a model $\mathfrak{A}$. The $\mathscr{L}$-reduct $(A, \leq)$ of $\mathfrak{A}$ is such that $\leq$ is not a well-ordering, but $(A, \leq)$ is a model of Γ, contradiction. □

Corollary 11.25. *If a sentence φ holds in every infinite $\mathscr{L}$-structure, then there is an $m \in \omega$ such that φ holds in every finite $\mathscr{L}$-structure of power $>m$.*

PROOF. Suppose the conclusion fails. Thus for every $m \in \omega$ there is a finite $\mathscr{L}$-structure of power $>m$ in which φ fails to hold. Adjoin new individual constants $\mathbf{c}_m$ for $m \in \omega$. Let

$$\Gamma = \{\neg\varphi\} \cup \{\neg(\mathbf{c}_i = \mathbf{c}_j) : i, j \in \omega, i \neq j\}.$$

Then every finite subset of Γ has a model, so Γ has a model $\mathfrak{A}$. Thus $\mathfrak{A} \restriction \mathscr{L}$ is an infinite model of $\neg\varphi$, so the hypothesis of 11.25 fails. □

We now turn to some basic results about first-order languages which we prove in a model-theoretic way. These results have to do with introducing new symbols or eliminating symbols. First we prove a theorem which shows, roughly speaking, that any first-order language is equivalent to a language in which there are no operation symbols.

Definition 11.26

(*i*) Let $\mathscr{L} = (L, v, \mathscr{O}, \mathscr{R})$ be a first-order language. A *relational version* of $\mathscr{L}$ is a first-order language of the form $\mathscr{L}' = (L, v, \mathscr{O}', \mathscr{R}')$ such that $\mathscr{O}' = 0$, $\mathscr{R} \subseteq \mathscr{R}'$, and such that there is a one-one function T (called a *translation* of $\mathscr{L}$ into $\mathscr{L}'$) mapping Dmn $\mathscr{O}$ onto Dmn $\mathscr{R}' \sim$ Dmn $\mathscr{R}$ such that $\mathscr{R}'T_{\mathbf{O}} = \mathscr{O}\mathbf{O} + 1$ for all $\mathbf{O} \in \text{Dmn } \mathscr{O}$.

(*ii*) Let $\mathscr{L}$, $\mathscr{L}'$, T be as in (*i*). With each formula φ of $\mathscr{L}$ we associate its $\mathscr{L}'$, *T-translate* φ' as follows. First we define φ' for formulas φ of the form $\sigma \equiv v_i$, σ a term:

$$(v_j \equiv v_i)' = v_j \equiv v_i;$$
$$(\mathbf{O} \equiv v_i)' = T_{\mathbf{O}}v_i \qquad \text{for } \mathbf{O} \text{ an individual constant.}$$

If $\mathbf{O}$ is an operation symbol of rank $m > 1$ and φ is $\mathbf{O}\sigma_0 \cdots \sigma_{m-1} \equiv v_i$, let $\alpha_0, \ldots, \alpha_{m-1}$ be the first m variables not occurring in φ, and set

$$\varphi' = \exists\alpha_0 \cdots \exists\alpha_{m-1}[(\sigma_0 \equiv \alpha_0)' \wedge \cdots \wedge (\sigma_{m-1} \equiv \alpha_{m-1})' \wedge T_{\mathbf{O}}\alpha_0 \cdots \alpha_{m-1}v_i].$$

Next we define φ' for φ of the form $\sigma \equiv \tau$, τ not a variable. Let α be the first variable not occurring in φ, and let

$$\varphi' = \exists\alpha((\sigma \equiv \alpha)' \wedge (\tau \equiv \alpha)').$$

If $\varphi = \mathbf{R}\sigma_0 \cdots \sigma_{m-1}$ with each σ_i a variable, let $\varphi' = \varphi$. If at least one σ_i is not a variable, let $\alpha_0, \ldots, \alpha_{m-1}$ be the first m variables not occurring in φ, and set

$$\varphi' = \exists\alpha_0 \cdots \exists\alpha_{m-1}[\mathbf{R}\alpha_0 \cdots \alpha_{m-1} \wedge (\sigma_0 \equiv \alpha_0)' \wedge \cdots \wedge (\sigma_{m-1} \equiv \alpha_{m-1})'].$$

Finally we set $(\neg\varphi)' = \neg\varphi'$, $(\varphi \vee \psi)' = \varphi' \vee \psi'$, $(\varphi \wedge \psi)' = \varphi' \wedge \psi'$, and $(\forall\alpha\varphi)' = \forall\alpha\varphi'$.

(*iii*) Let $\mathscr{L}$, $\mathscr{L}'$, T be as in (*i*). For $\mathbf{O} \in \text{Dmn } \mathscr{O}$, the *existence* condition for $\mathbf{O}$ is the following sentence of $\mathscr{L}'$ (where $\mathbf{O}$ is m-ary):

$$\forall v_0 \cdots \forall v_{m-1}\exists v_m(T_{\mathbf{O}}v_0 \cdots v_m);$$

the *uniqueness condition* for $\mathbf{O}$ is the sentence

$$\forall v_0 \cdots \forall v_{m+1}(T_{\mathbf{O}}v_0 \cdots v_m \wedge T_{\mathbf{O}}v_0 \cdots v_{m-1}v_{m+1} \rightarrow v_m \equiv v_{m+1}).$$

The set of *translation conditions* is the set of all existence and uniqueness conditions for all $\mathbf{O} \in \text{Dmn } \mathscr{O}$.

(*iv*) If $\mathfrak{A}$ is any $\mathscr{L}$-structure, the *relational version* $\mathfrak{A}'$ of $\mathfrak{A}$ is obtained from $\mathfrak{A}$ by replacing each operation of $\mathfrak{A}$ by the associated relation. That is, if $\mathfrak{A} = (A, f, R)$, then $\mathfrak{A}' = (A, R')$, where $\mathfrak{A}'$ is the $\mathscr{L}'$-structure such that $R \subseteq R'$, while if $\mathbf{O} \in \text{Dmn } \mathscr{O}$, with $\mathbf{O}$ m-ary, then

$$R'T_{\mathbf{O}} = \{(x_0, \ldots, x_m) : f_{\mathbf{O}}x_0 \cdots x_{m-1} = x_m\}.$$

(*v*) If $\mathscr{L} = (L, v, \mathscr{O}, \mathscr{R}, g)$ is an effectivized first-order language, then an *effectivized relational version* of $\mathscr{L}$ is an effectivized first-order language

$\mathscr{L}' = (L, v, \mathcal{O}', \mathscr{R}', g')$ such that $(L, v, \mathcal{O}', \mathscr{R}')$ is a relational version of $\mathscr{L}$, $g \restriction (\text{Rng } L \cup \text{Rng } v \cup \text{Dmn } \mathscr{R}) = g' \restriction (\text{Rng } L \cup \text{Rng } v \cup \text{Dmn } \mathscr{R})$, and there is a translation T of $(L, v, \mathcal{O}, \mathscr{R})$ into $(L, v, \mathcal{O}', \mathscr{R}')$ such that $g' \circ T \circ g^{-1} \restriction g^* \text{ Dmn } \mathcal{O}$ is partial recursive.

The following proposition is easily (weak Church's thesis!) established.

Proposition 11.27. *Let $\mathscr{L}$, $\mathscr{L}'$, T be as in 11.26(v). For any $m \in \omega$, define*

$$fm = g'^{+}\varphi' \quad \text{if } m = g^{+}\varphi \text{ for some } \varphi \in \text{Fmla}_{\mathscr{L}'}, \text{ where } \varphi' \text{ is the } \mathscr{L}', T\text{-translate of } \varphi,$$
$$fm = 0 \quad \text{otherwise.}$$

Then f is a recursive function.

In the next theorem we give the main properties of relational versions of languages.

Theorem 11.28. *Let $\mathscr{L}'$ be a relational version of $\mathscr{L}$ with translation, T, notation as in 11.26(i).*

(i) If φ is a formula of $\mathscr{L}$ in which no operation symbol occurs, then $\varphi' = \varphi$;

(ii) Let $\mathfrak{A}$ be an $\mathscr{L}$-structure and $\mathfrak{A}'$ the relational version of $\mathfrak{A}$, with notation as in 11.26(iv). Then $\mathfrak{A}'$ is a model of the translation conditions. Furthermore, if φ is any formula of $\mathscr{L}$, and $x \in {}^{\omega}A$, then $\mathfrak{A} \models \varphi[x]$ iff $\mathfrak{A}' \models \varphi'[x]$. For any formula φ of $\mathscr{L}'$ there is a formula φ^* of $\mathscr{L}$ such that for all $x \in {}^{\omega}A$, $\mathfrak{A} \models \varphi^*[x]$ iff $\mathfrak{A}' \models \varphi[x]$. If $\mathfrak{B}$ is an $\mathscr{L}'$-structure which is a model of the translation conditions, then $\mathfrak{B} = \mathfrak{A}'$ for some $\mathscr{L}$-structure $\mathfrak{A}$.

(iii) Let $\Gamma \cup \{\varphi\}$ be a set of formulas of $\mathscr{L}$. Then $\Gamma \models \varphi$ iff $\{\psi' : \psi \in \Gamma\} \cup \{\psi : \psi \text{ is a translational condition}\} \models \varphi'$.

Proof. (*i*) and the first part of (*ii*) are obvious. We prove the second part of (*ii*) by induction on φ, following 11.26(*ii*). If φ is $v_j \equiv v_i$, obviously $\mathfrak{A} \models \varphi[x]$ iff $\mathfrak{A}' \models \varphi'[x]$. Suppose φ is $\mathbf{O} \equiv v_i$, where $\mathbf{O}$ is an individual constant. If $\mathfrak{A} \models \varphi[x]$, then $x_i = \mathbf{O}^{\mathfrak{A}}$, and hence $x_i \in T_{\mathbf{O}}^{\mathfrak{A}'}$; thus $\mathfrak{A}' \models \varphi'[x]$. The converse is similar. Now suppose, inductively as in 11.26(*ii*), that φ is $\mathbf{O}\sigma_0 \cdots \sigma_{m-1} \equiv v_i$. Assume that $\mathfrak{A} \models \varphi[x]$. Thus $\mathbf{O}^{\mathfrak{A}}\sigma_0^{\mathfrak{A}}x \cdots \sigma_{m-1}^{\mathfrak{A}}x = x_i$. Let y be like x except that $y\alpha_0 = \sigma_0^{\mathfrak{A}}x, \ldots, y\alpha_{m-1} = \sigma_{m-1}^{\mathfrak{A}}x$. Thus by 11.3, $\mathfrak{A} \models \sigma_0 \equiv \alpha_0[y], \ldots, \mathfrak{A} \models \sigma_{m-1} \equiv \alpha_{m-1}[y]$. So, by the induction hypothesis, $\mathfrak{A}' \models (\sigma_0 \equiv \alpha_0)'[y], \ldots, \mathfrak{A}' \models (\sigma_{m-1} \equiv \alpha_{m-1})'[y]$. It follows easily that $\mathfrak{A}' \models \varphi'[x]$. The converse is similar. The remaining steps in this inductive proof are similar. Next, given a formula φ of $\mathscr{L}'$ we construct φ^* by replacing all atomic subformulas of φ of the form $T_{\mathbf{O}}\alpha_0 \cdots \alpha_m$ by $\mathbf{O}\alpha_0 \cdots \alpha_{m-1} = \alpha_m$. The desired property of φ^* is easily established by induction on φ. The final condition of (*ii*) is clear.

To prove (*iii*), first assume $\Gamma \models \varphi$. Let $\mathfrak{B}$ be any model (an $\mathscr{L}'$-structure) of $\{\psi' : \psi \in \Gamma\} \cup \{\psi : \psi \text{ is a translation condition}\}$. By (*ii*) we may write $\mathfrak{B} = \mathfrak{A}'$ with $\mathfrak{A}$ an $\mathscr{L}$-structure. Since $\mathfrak{A}' \models \psi'$ for each $\psi \in \Gamma$ it follows by (*ii*) that $\mathfrak{A}$ is a model of Γ. Hence $\mathfrak{A} \models \varphi$, so by (*ii*), $\mathfrak{A}' \models \varphi'$ as desired. Second, assume $\{\psi' : \psi \in \Gamma\} \cup \{\psi : \psi \text{ is a translation condition}\} \models \psi'$. Let $\mathfrak{A}$ be a model of Γ. By (*ii*), $\mathfrak{A}'$ is a model of $\{\psi' : \psi \in \Gamma\} \cup \{\psi : \psi \text{ is a translation condition}\}$. So $\mathfrak{A}' \models \varphi'$, and by (*ii*) again, $\mathfrak{A} \models \varphi$, as desired. □

We now want to give the main facts concerning the role of definitions in first-order languages.

Definition 11.29. A *theory* is a pair $(\Gamma, \mathscr{L})$ such that Γ is a set of sentences of $\mathscr{L}$ and $\varphi \in \Gamma$ whenever $\Gamma \models \varphi$, for each sentence φ in $\mathscr{L}$. We call Γ itself a theory when $\mathscr{L}$ is implicitly understood, or we say that Γ is a theory *over* or *in* $\mathscr{L}$. If $\mathfrak{A}$ is an $\mathscr{L}$-structure, the *$\mathscr{L}$-theory of* $\mathfrak{A}$ is the pair $(\Gamma, \mathscr{L})$, where $\Gamma = \{\varphi \in \mathrm{Sent}_{\mathscr{L}} : \mathfrak{A} \models \varphi\}$; clearly $(\Gamma, \mathscr{L})$ *is* a theory. A theory $(\Gamma', \mathscr{L}')$ is an *extension* of a theory $(\Gamma, \mathscr{L})$ provided that $\mathscr{L}'$ is an expansion of $\mathscr{L}$ and $\Gamma \subseteq \Gamma'$. We say that $(\Gamma', \mathscr{L}')$ is a *conservative* extension of $(\Gamma, \mathscr{L})$ provided that, in addition, $\Gamma = \Gamma' \cap \mathrm{Sent}_{\mathscr{L}}$. If Γ is a theory in $\mathscr{L}$, a set $\Delta \subseteq \mathrm{Sent}_{\mathscr{L}}$ is a set of *axioms* for Γ provided that $\Gamma = \{\varphi \in \mathrm{Sent}_{\mathscr{L}} : \Delta \models \varphi\}$. Now let $\mathscr{L}'$ be an expansion of $\mathscr{L}$ and let Γ and Γ' be theories over $\mathscr{L}$ and $\mathscr{L}'$ respectively.

(*i*) If $\mathbf{R}$ is a relation symbol of $\mathscr{L}'$ but not of $\mathscr{L}$, then a *possible definition of* $\mathbf{R}$ *over* Γ is any formula φ of $\mathscr{L}$ with free variables among $\{v_0, \ldots, v_{m-1}\}$, where m is the rank of $\mathbf{R}$.

(*ii*) If $\mathbf{O}$ is an operation symbol of $\mathscr{L}'$ but not of $\mathscr{L}$, then a *possible definition of* $\mathbf{O}$ *over* Γ is a formula φ of $\mathscr{L}$ with free variables among $\{v_0, \ldots, v_m\}$, where m is the rank of $\mathbf{O}$, such that the following existence and uniqueness conditions are in Γ:

$$\forall v_0 \cdots \forall v_{m-1} \exists v_m \varphi;$$
$$\forall v_0 \cdots \forall v_{m+1}[\varphi(v_0, \ldots, v_m) \wedge \varphi(v_0, \ldots, v_{m-1}, v_{m+1}) \rightarrow v_m = v_{m+1}].$$

(*iii*) We say that $(\Gamma', \mathscr{L}')$ is a *definitional extension of* $(\Gamma, \mathscr{L})$ provided that for every nonlogical constant $\mathbf{C}$ of $\mathscr{L}'$ but not of $\mathscr{L}$ there is a possible definition $\varphi_{\mathbf{C}}$ of $\mathbf{C}$ over Γ such that

$$\Gamma' = \{\varphi : \varphi \in \mathrm{Sent}_{\mathscr{L}'} \text{ and } \Gamma \cup \{\varphi'_{\mathbf{C}} : \mathbf{C} \text{ a nonlogical constant of } \mathscr{L}' \text{ but not of } \mathscr{L}\} \models \varphi\},$$

where $\varphi'_{\mathbf{C}}$ is the sentence

$$\forall v_0 \cdots \forall v_m(\mathbf{C} v_0 \cdots v_{m-1} \leftrightarrow \varphi_{\mathbf{C}})$$

if $\mathbf{C}$ is a relation symbol of rank m, while $\varphi'_{\mathbf{C}}$ is

$$\forall v_0 \cdots \forall v_m(\mathbf{C} v_0 \cdots v_{m-1} = v_m \leftrightarrow \varphi_{\mathbf{C}})$$

if $\mathbf{C}$ is an operation symbol of rank m.

(*iv*) If $\mathscr{L} = (L, v, \mathcal{O}, \mathscr{R}, g)$ and $\mathscr{L}' = (L', v', \mathcal{O}', \mathscr{R}', g')$ are effectivized first-order languages, the above notation applies to them also. We assume that $\mathscr{L}'$ is an effective expansion of $\mathscr{L}$. We say that $(\Gamma', \mathscr{L}')$ is an *effective* definitional extension of $(\Gamma, \mathscr{L})$ if $(\Gamma', \mathscr{L}')$ is a definitional expansion of $(\Gamma, \mathscr{L})$ for which there is a function φ as in *(iii)* such that $\mathscr{g}^+ \circ \varphi \circ \mathscr{g}'^{-1} \upharpoonright \{\mathscr{g}'\mathbf{C} : \mathbf{C}$ is a nonlogical constant of $\mathscr{L}'$ but not of $\mathscr{L}\}$ is partial recursive.

There are two central results concerning definitions. The first is that defined symbols can always be eliminated in favor of old symbols:

Theorem 11.30. *Let* $(\Gamma', \mathscr{L}')$ *be a definitional extension of* $(\Gamma, \mathscr{L})$, *with notation as in* 11.29. *Then for any formula* ψ *of* $\mathscr{L}'$ *there is a formula* ψ' *of* $\mathscr{L}$ *with the same free variables as* ψ *such that* $\Gamma' \models \psi \leftrightarrow \psi'$. *If* Γ' *is an effective definitional extension of* Γ, *then such a formula* ψ' *can be effectively obtained from* ψ.

PROOF. We construct ψ' by induction on ψ. In each step the desired properties of ψ' are easy to prove, and we prove the desired result in only one step as an illustration. To begin with, we construct ψ' for ψ of the form $\sigma = v_i$ by induction on σ. If σ is a variable, we set $\psi' = \psi$. Now suppose inductively that σ is $\mathbf{O}\tau_0 \cdots \tau_{n-1}$. Let $\alpha_0, \ldots, \alpha_{n-1}$ be the first m variables not occurring in ψ. If $\mathbf{O}$ is a symbol of $\mathscr{L}$, then we let ψ' be the formula

$$\exists\alpha_0 \cdots \exists\alpha_{n-1}[\mathbf{O}\alpha_0 \cdots \alpha_{n-1} = v_i \wedge \bigwedge_{j<n} (\tau_j = \alpha_j)'.$$

If $\mathbf{O}$ is a symbol of $\mathscr{L}'$ but not of $\mathscr{L}$, we let ψ' be the formula

$$\exists\alpha_0 \cdots \exists\alpha_{n-1}[\varphi_{\mathbf{O}}(\alpha_0, \ldots, \alpha_{n-1}, v_i) \wedge \bigwedge_{j<n} (\tau_j = \alpha_j)'].$$

In this case we check explicitly that $\Gamma' \models \psi \leftrightarrow \psi'$. Let $\mathfrak{A}$ be any model of Γ'. First suppose that $\mathfrak{A} \models \psi[x]$ with $x \in {}^\omega A$. Thus $\sigma^{\mathfrak{A}}x = x_i$, so $\mathbf{O}^{\mathfrak{A}}(\tau_0^{\mathfrak{A}}x, \ldots, \tau_{n-1}^{\mathfrak{A}}x) = x_i$. Hence $\mathfrak{A} \models (\mathbf{O}v_0 \cdots v_{m-1} = v_m)[\tau_0^{\mathfrak{A}}x, \ldots, \tau_{n-1}^{\mathfrak{A}}x, x_i]$. Since $\Gamma' \models \varphi_{\mathbf{O}}'$, it follows that $\mathfrak{A} \models \varphi_{\mathbf{O}}[\tau_0^{\mathfrak{A}}x, \ldots, \tau_{n-1}^{\mathfrak{A}}x, x_i]$. Now let y be like x except that $y\alpha_j = \tau_j^{\mathfrak{A}}x$ for each $j < n$. Clearly then $\mathfrak{A} \models \varphi_{\mathbf{O}}(\alpha_0, \ldots, \alpha_{n-1}, v_i)[y]$ and $\mathfrak{A} \models (\tau_j = \alpha_j)[y]$ for each $j < n$. By the induction assumption, $\mathfrak{A} \models (\tau_j = \alpha_j)'[y]$ for each $j < n$. Thus $\mathfrak{A} \models \psi'[x]$. The converse is similar.

Now we continue the inductive definition of ψ'; it is complete for ψ of the form $\sigma = v_i$.

If ψ is $\sigma = \tau$, where τ is not a variable, let α be the first variable not occurring in ψ, and set $\psi' = \exists\alpha((\sigma = \alpha)' \wedge (\tau = \alpha)')$. Next, suppose ψ is $\mathbf{R}\sigma_0 \cdots \sigma_{n-1}$. Let $\alpha_0, \ldots, \alpha_{n-1}$ be the first n variables not occurring in ψ. If $\mathbf{R}$ is a symbol of $\mathscr{L}$, set

$$\psi' = \exists\alpha_0 \cdots \exists\alpha_{n-1}\left[\mathbf{R}\alpha_0 \cdots \alpha_{n-1} \wedge \bigwedge_{j<n} (\sigma_j = \alpha_j)'\right].$$

If $\mathbf{R}$ is a symbol of $\mathscr{L}'$ but not of $\mathscr{L}$, set

$$\psi' = \exists\alpha_0 \cdots \exists\alpha_{n-1}\left[\varphi_{\mathbf{R}}(\alpha_0, \ldots, \alpha_{n-1}) \wedge \bigwedge_{j<n} (\sigma_j = \alpha_j)'\right].$$

Finally, let $(\neg\psi)' = \neg\psi'$, $(\psi \vee \chi)' = \psi' \vee \chi'$, $(\psi \wedge \chi)' = \psi' \wedge \chi'$, and $(\forall v_i\psi)' = \forall v_i\psi'$.

The final assertion of the theorem concerning effectiveness is clear from the above. □

The second important fact concerning definitions is that nothing new can be proved:

Theorem 11.31. *Let* $(\Gamma', \mathscr{L}')$ *be a definitional expansion of* $(\Gamma, \mathscr{L})$. *Then* $(\Gamma', \mathscr{L}')$ *is a conservative extension of* $(\Gamma, \mathscr{L})$.

Proof. Again we take all notation as in 11.29. Suppose ψ is a formula of $\mathscr{L}$ and $\Gamma' \models \psi$; we must show that $\Gamma \models \psi$. Let $\mathfrak{A}$ be any model of Γ ($\mathfrak{A}$ is an $\mathscr{L}$-structure). For $\mathbf{C}$ a nonlogical constant of $\mathscr{L}$, let $\mathbf{C}^{\mathfrak{A}'} = \mathbf{C}^{\mathfrak{A}}$. Let $\mathbf{R}$ be a relation symbol of $\mathscr{L}'$ but not of $\mathscr{L}$, say $\mathbf{R}$ is m-ary. We define $\mathbf{R}^{\mathfrak{A}'} = {}^m\varphi_{\mathbf{R}}^{\mathfrak{A}}$. Now suppose that $\mathbf{O}$ is an operation symbol of $\mathscr{L}'$ but not of $\mathscr{L}$; say $\mathbf{O}$ is m-ary. Since $\varphi_{\mathbf{O}}$ is a possible definition of $\mathbf{O}$ over Γ and $\mathfrak{A}$ is a model of Γ, we may define $\mathbf{O}^{\mathfrak{A}'}$ as follows. For any $x_0, \ldots, x_{m-1} \in A$, let $\mathbf{O}^{\mathfrak{A}'}(x_0, \ldots, x_{m-1})$ be the unique $y \in A$ such that $\mathfrak{A} \models \varphi^{\mathfrak{A}}[x_0, \ldots, x_{m-1}, y]$. This defines the $\mathscr{L}'$-structure $\mathfrak{A}'$. Clearly $\mathfrak{A}'$ is a model of Γ', so $\mathfrak{A}'$ is a model of ψ. Hence $\mathfrak{A}$ is also a model of ψ, since ψ is a formula of $\mathscr{L}$. Thus $\Gamma \models \psi$, as desired. □

Closely related to this fact about definitions is the following result.

Theorem 11.32. *Let* Γ *be a theory in a language* $\mathscr{L}$, *and* φ *a formula of* $\mathscr{L}$ *with free variables among* $v_0, \ldots, v_m$. *Assume that* $\Gamma \models \forall v_0 \cdots \forall v_{m-1} \exists v_m \varphi$. *Let* $\mathscr{L}'$ *be an expansion of* $\mathscr{L}$ *by adding a new m-ary operation symbol* $\mathbf{O}$. *Let* Γ' *be a theory in* $\mathscr{L}'$ *with axioms* Γ *together with the sentence* $\forall v_0 \cdots \forall v_{m-1}\varphi(v_0, \ldots, v_{m-1}, \mathbf{O}v_0 \cdots v_{m-1})$. *Then* Γ' *is a conservative extension of* Γ.

Proof. Assume that $\Gamma' \models \psi$, where ψ is a formula of $\mathscr{L}$. To prove $\Gamma \models \psi$, let $\mathfrak{A}$ be any model of Γ. By the axiom of choice, there is an m-ary function g on A such that for all $x_0, \ldots, x_{m-1} \in A$, $\mathfrak{A} \models \varphi[x_0, \ldots, x_{m-1}, g(x_0, \ldots, x_{m-1})]$. Let $\mathfrak{A}'$ be the expansion of $\mathfrak{A}$ to an $\mathscr{L}'$-structure in which $\mathbf{O}$ is interpreted as g. Then $\mathfrak{A}'$ is a model of Γ', so $\mathfrak{A}'$ is a model of ψ. Since ψ is a formula of $\mathscr{L}$, $\mathfrak{A}$ is a model of ψ. □

Theorem 11.32 justifies the common intuitive practice in mathematics of introducing notation for objects proved to exist. For example, after proving that an algebraic equation of a specified type has a solution, one introduces a name c for such a solution. By 11.32, nothing new not involving c can be proved after this that was not provable before. The procedure can be used even if one cannot pick out a unique such c.

We now want to generalize the process described in Theorem 11.32. The construction we give is essential for our present purpose of proving the Skolem normal form theorem and Herbrand's theorem, and also plays an important role in model theory.

Definition 11.33. For any formula φ, let Fv φ be the set of all variables occurring freely in φ. For any first-order language $\mathscr{L}$, a *primitive Skolem expansion* of $\mathscr{L}$ is an expansion $\mathscr{L}'$ of $\mathscr{L}$ such that there is a function S mapping $\{\exists v_i\varphi : \varphi \in \text{Fmla}_{\mathscr{L}}\}$ one-one onto the set of all nonlogical constants of $\mathscr{L}'$ which are not constants of $\mathscr{L}$, such that for each $\varphi \in \text{Fmla}_{\mathscr{L}}$ and each variable α, $S_{\exists\alpha\varphi}$ is an operation symbol of rank $|\text{Fv}\ \exists\alpha\varphi|$. In case $\mathscr{L}$ and $\mathscr{L}'$ are effectivized with Gödel numbering functions $\mathscr{g}$, $\mathscr{g}'$, we call $\mathscr{L}'$ an effective primitive Skolem expansion provided that $\mathscr{L}'$ is an effective expansion of $\mathscr{L}$ and the function

$$\mathscr{g}' \circ S \circ \mathscr{g}^{+-1} \upharpoonright \mathscr{g}^{+*}\{\exists v_i\varphi : \varphi \in \text{Fmla}_{\mathscr{L}}\}$$

is partial recursive.

Given first-order languages $\mathscr{L} = (L, v, \mathscr{O}, \mathscr{R})$, and $\mathscr{L}' = (L, v, \mathscr{O}', \mathscr{R})$, we say that $\mathscr{L}'$ is a *Skolem expansion* of $\mathscr{L}$ provided there is a sequence $\langle \mathscr{L}_i : i \in \omega \rangle$ of first-order languages $\mathscr{L}_i = (L, v, \mathscr{O}_i, \mathscr{R})$ such that $\mathscr{L}_0 = \mathscr{L}$, for each $i \in \omega$ $\mathscr{L}_{i+1}$ is a primitive Skolem expansion of $\mathscr{L}_i$ with associated function S^i, and $\mathscr{O}' = \bigcup_{i\in\omega} \mathscr{O}_i$. If all of these languages are effectivized, $\mathscr{L}'$ is an *effective* Skolem expansion of $\mathscr{L}$ provided that each $\mathscr{L}_{i+1}$ is an effective primitive Skolem expansion of $\mathscr{L}_i$, and $\mathscr{L}'$ is an effective expansion of each $\mathscr{L}_i$.

Assuming that $\mathscr{L}'$ is a Skolem expansion of $\mathscr{L}$, with notation as above, the *Skolem set of* $\mathscr{L}'$ *over* $\mathscr{L}$ is the set of all sentences

$$[[\exists v_i\varphi(v_0, \ldots, v_i) \rightarrow \varphi(v_0, \ldots, v_{i-1}, \sigma)]]$$

where $\sigma = S^j_{\exists vi\varphi}\alpha_0 \cdots \alpha_{m-1}$, $m = |\text{Fv}\ \exists v_i\varphi|$, $\text{Fv}\ \exists v_i\varphi = \{\alpha_0, \ldots, \alpha_{m-1}\}$ with $v^{-1}\alpha_0 < \cdots < v^{-1}\alpha_{m-1}$, and j is minimal such that $\exists v_i\varphi$ is a formula of $\mathscr{L}_j$. (Recall that $[[\chi]]$ denotes the universal closure of χ; see 10.94.)

The following two propositions are obvious.

Proposition 11.34. *Any first-order language has a Skolem expansion.*

Proposition 11.35. *If $\mathscr{L}'$ is a Skolem expansion of $\mathscr{L}$, then* $|\text{Fmla}_{\mathscr{L}}| = |\text{Fmla}_{\mathscr{L}'}|$.

The following lemma is fundamental for our main results.

Lemma 11.36. *If $\mathscr{L}'$ is a Skolem expansion of $\mathscr{L}$, then any $\mathscr{L}$-structure can be expanded to a model of the Skolem set of $\mathscr{L}'$ over $\mathscr{L}$.*

PROOF. From the definition of Skolem expansions, we see that it is enough to prove the following statement:

Statement. Let $\mathscr{L}'$ be a primitive Skolem expansion of $\mathscr{L}$, and $\mathfrak{A}$ an $\mathscr{L}$-structure. Then $\mathfrak{A}$ can be expanded to an $\mathscr{L}'$-structure which is a model of all sentences $[[\exists v_i\varphi(v_0, \ldots, v_i) \rightarrow \varphi(v_0, \ldots, v_{i-1}, \sigma)]]$, where φ is a formula of $\mathscr{L}$, $\sigma = S_{\exists vi\varphi}\alpha_0 \cdots \alpha_{m-1}$, $m = |\mathrm{Fv}\, \exists v_i\varphi|$, and $\mathrm{Fv}\, \exists v_i\varphi = \{\alpha_0, \ldots, \alpha_{m-1}\}$ with $v^{-1}\alpha_0 < \cdots < v^{-1}\alpha_{m-1}$.

To prove this statement, let C be a choice function for nonempty subsets of A. Let $\psi = \exists v_i\varphi$ be a formula of $\mathscr{L}$, with $\mathrm{Fv}\, \exists v_i\varphi = \{v_{j0}, \ldots, v_{j(m-1)}\}$, where $j0 < \cdots < j(m-1)$. We define an m-ary operation t_ψ on A as follows. For any $x_0, \ldots, x_{m-1} \in A$, set

$$\begin{aligned} t_\psi(x_0, \ldots, x_{m-1}) &= C\{a : \text{there is a } y \in \varphi^{\mathfrak{A}} \text{ such that } y_i = a \\ &\qquad \text{and } y_{jk} = x_k \text{ for each } k < m\}, \\ t_\psi(x_0, \ldots, x_{m-1}) &= CA \text{ if the above set is empty.} \end{aligned}$$

Let $\mathfrak{A}'$ be the expansion of $\mathfrak{A}$ to an $\mathscr{L}'$-structure in which each symbol S_ψ is interpreted as t_ψ. To show that $\mathfrak{A}'$ is as desired, consider any formula $\psi = \exists v_i\varphi$ of $\mathscr{L}$ with notation as above. Suppose $x \in {}^\omega A$ and $\mathfrak{A}' \models \exists v_i\varphi[x]$. Then $\mathfrak{A} \models \exists v_i\varphi[x]$ since $\exists v_i\varphi$ is a formula of $\mathscr{L}$. Thus there is an $a \in A$ such that $\mathfrak{A} \models \varphi[x^i_a]$. Hence the first clause in the definition of $t_\psi(x_{j0}, \ldots, x_{j(m-1)})$ gives an element $b \in A$ such that $\mathfrak{A} \models \varphi[x^i_b]$. Let σ be the term $S_{\exists vi\varphi}v_{j0} \cdots v_{j(m-1)}$. Recall that the formula $\varphi(v_0, \ldots, v_{i-1}\sigma)$ has the form $\mathrm{Subf}^{vi}_\sigma\varphi'$, where φ' is obtained from φ by replacing bound variables suitably. Clearly $\mathfrak{A} \models \varphi'[x^i_b]$, where $b = t_\psi(x_{j0}, \ldots, x_{j(m-1)})$. Hence we easily infer that $\mathfrak{A} \models \mathrm{Subf}^{vi}_\sigma\varphi'[x]$. Thus $\mathfrak{A}'$ is a model of $[[\exists v_i\varphi(v_0, \ldots, v_i) \rightarrow \varphi(v_0, \ldots, v_{i-1}, \sigma)]]$, as desired. □

The functions introduced in the expansion of $\mathfrak{A}$ to $\mathfrak{A}'$ in the proof of 11.36 are called *Skolem functions*. The entire method associated with Skolem expansions is sometimes called the method of Skolem functions.

One of the main properties of Skolem expansions is that every formula becomes equivalent, in a certain sense, to a prenex formula having only universal quantifiers:

Definition 11.37. A formula is *universal* if it is in prenex normal form with only universal quantifiers. Let $\mathscr{L}'$ be a Skolem expansion of $\mathscr{L}$, with notation as in 11.33. With each prenex formula φ of $\mathscr{L}'$ we associate a formula φ^{S}:

$$\begin{aligned} \varphi^{\mathrm{S}} &= \varphi \text{ if } \varphi \text{ is quantifier free;} \\ (\forall\alpha\varphi)^{\mathrm{S}} &= \forall\alpha\varphi^{\mathrm{S}}; \\ (\exists v_i\varphi)^{\mathrm{S}} &= \varphi^{\mathrm{S}}(v_0, \ldots, v_{i-1}, \sigma), \end{aligned}$$

where $\sigma = S^j_{\exists vi\varphi}\beta_0 \cdots \beta_{m-1}$, $\mathrm{Fv}\, \exists v_i\varphi = \{\beta_0, \ldots, \beta_{m-1}\}$ with $v^{-1}\beta_0 < \cdots < v^{-1}\beta_{m-1}$, and j is chosen minimal so that $\exists v_i\varphi$ is a formula of $\mathscr{L}_j$.

Theorem 11.38 (Skolem normal form theorem). *Let $\mathscr{L}'$ be a Skolem expansion of $\mathscr{L}$. For every prenex formula φ of $\mathscr{L}'$, the formula φ^{S} is universal and the same variables occur free in φ as do in φ^{S}. Furthermore, for any prenex formula φ of $\mathscr{L}'$ we have:*

(*i*) $\models \varphi^{S} \rightarrow \varphi$;

(*ii*) *if* $\mathfrak{A}$ *is a model of the Skolem set of* $\mathscr{L}'$ *over* $\mathscr{L}$, *then* $\mathfrak{A} \models \varphi \rightarrow \varphi^{S}$;

(*iii*) *if* Γ *is a theory in* $\mathscr{L}$ *and* $\Gamma' = \{\varphi^{S} : \varphi \in \Gamma\}\text{-Thm}_{\mathscr{L}'}$, *then the following conditions hold:*

(*a*) *if* $\mathfrak{A}$ *is a model of* Γ *and* $\mathfrak{A}'$ *is an expansion of* $\mathfrak{A}$ *to a model of the Skolem set of* $\mathscr{L}'$ *over* $\mathscr{L}$, *then* $\mathfrak{A}'$ *is a model of* Γ';

(*b*) *if* $\mathfrak{A}'$ *is a model of* Γ', *then* $\mathfrak{A}' \upharpoonright \mathscr{L}$ *is a model of* Γ;

(*c*) $(\Gamma', \mathscr{L}')$ *is a conservative extension of* $(\Gamma, \mathscr{L})$;

(*iv*) φ *has a model iff* φ^{S} *has a model.*

PROOF. Clearly φ^{S} is universal and $\text{Fv}\,\varphi = \text{Fv}\,\varphi^{S}$. Condition (*i*) is easily proved by induction on φ, as is (*ii*). To prove (*iii*)(*a*), assume its hypothesis and let $\varphi \in \Gamma$. Then $\mathfrak{A} \models \varphi$ by hypothesis, so by (*ii*), $\mathfrak{A}' \models \varphi^{S}$. Thus $\mathfrak{A}'$ is a model of Γ', as desired. Condition (*iii*)(*b*) follows from (*i*). To prove (*iii*)(*c*), assume first that $\Gamma \models \varphi$, φ a sentence of $\mathscr{L}$. If $\mathfrak{A}'$ is any model of Γ', then by (*iii*)(*b*), $\mathfrak{A}' \upharpoonright \mathscr{L}$ is a model of Γ and so $\mathfrak{A}' \upharpoonright \mathscr{L} \models \varphi$ and $\mathfrak{A}' \models \varphi$. Thus $\Gamma' \models \varphi$. Now assume that $\Gamma' \models \varphi$. Let $\mathfrak{A}$ be any model of Γ. By Lemma 11.36, let $\mathfrak{A}'$ be an expansion of $\mathfrak{A}$ to a model of the Skolem set of $\mathscr{L}'$ over $\mathscr{L}$. Then by (*iii*)(*a*), $\mathfrak{A}'$ is a model of Γ', so $\mathfrak{A}' \models \varphi$ by assumption. Thus $\mathfrak{A} \models \varphi$. Hence $\Gamma \models \varphi$, as desired. Condition (*iv*) is immediate from 11.36, (*iii*)(*a*), and (*iii*)(*b*) (with Γ axiomatized by $\{\varphi\}$). □

The starting point of our considerations concerning Herbrand's theorem is 11.38(*iv*). From it one can easily obtain an equivalent condition for a prenex sentence φ to be universally valid. Indeed, this is true iff $\neg\varphi$ has no model, and the sentence $\neg\varphi$ is equivalent in a natural way to a prenex sentence ψ. Thus $\models\varphi$ iff ψ has no model iff ψ^{S} has no model (by 11.38(*iv*)) iff $\models \neg\psi^{S}$. The sentence $\neg\psi^{S}$ is equivalent to a certain prenex sentence with only existential quantifiers. We define it explicitly as follows.

Definition 11.39. A formula φ is *existential* if it is in prenex normal form with only existential quantifiers. Let $\mathscr{L}'$ be a Skolem expansion of $\mathscr{L}$, with notation as in 11.33. With each prenex formula φ of $\mathscr{L}'$ we associate the prenex formula φ^{n} obtained from φ by interchanging $\exists$ and $\forall$ and replacing the matrix ψ of φ by $\neg\psi$. Now with each prenex formula φ of $\mathscr{L}'$ we associate a formula φ^{H}:

$$\begin{aligned} \varphi^{H} &= \varphi \text{ if } \varphi \text{ is quantifier free;} \\ (\exists\alpha\varphi)^{H} &= \exists\alpha\varphi^{H}; \\ (\forall v_i\varphi)^{H} &= \varphi^{H}(v_0, \ldots, v_{i-1}, \sigma), \end{aligned}$$

where $\sigma = S^{j}_{\psi}\beta_0 \cdots \beta_{m-1}$, $\psi = \exists v_i\varphi^{n}$, $\text{Fv}\,\psi = \{\beta_0, \ldots, \beta_{m-1}\}$ with $v^{-1}\beta_0 < \cdots < v^{-1}\beta_{m-1}$, and j is chosen minimal so that ψ is a formula of $\mathscr{L}_j$.

Theorem 11.40. *Let* $\mathscr{L}'$ *be a Skolem expansion of* $\mathscr{L}$, *and let* φ *be a prenex formula of* $\mathscr{L}'$. *Then:*

(*i*) φ^{H} *is an existential formula;*

(*ii*) $\models \varphi^{\mathrm{H}} \leftrightarrow \neg \varphi^{\mathrm{nS}}$;

(*iii*) *if* φ *is a sentence, then* $\models \varphi$ *iff* $\models \varphi^{\mathrm{H}}$.

PROOF. (*i*) is obvious from the definitions. To prove (*ii*), first note that $\vdash \neg \varphi \leftrightarrow \varphi^{\mathrm{n}}$. Next we prove (*ii*) by induction on the length of φ; it is clear if φ is quantifier free. Now $(\exists \alpha \varphi)^{\mathrm{H}} = \exists \alpha \varphi^{\mathrm{H}}$, and we assume inductively that $\models \varphi^{\mathrm{H}} \Leftrightarrow \neg \varphi^{\mathrm{nS}}$. Thus $\models (\exists \alpha \varphi)^{\mathrm{H}} \leftrightarrow \exists \alpha \neg \varphi^{\mathrm{nS}}$. Also $\models \exists \alpha \neg \varphi^{\mathrm{nS}} \leftrightarrow \neg \forall \alpha \varphi^{\mathrm{nS}}$, and $(\forall \alpha \varphi^{\mathrm{n}})^{\mathrm{S}} = \forall \alpha \varphi^{\mathrm{nS}}$. Since $(\exists \alpha \varphi)^{\mathrm{n}} = \forall \alpha \varphi^{\mathrm{n}}$, it follows that $\models (\exists \alpha \varphi)^{\mathrm{H}} \leftrightarrow \neg (\exists \alpha \varphi)^{\mathrm{nS}}$, as desired. Next, let $(\forall v_i \varphi)^{H} = \varphi^{H}(v_0, \ldots, v_{n-1}, \sigma)$, with notation as in 11.39. By the induction hypothesis, $\models \varphi^{\mathrm{H}} \leftrightarrow \neg \varphi^{\mathrm{nS}}$, so $\models \varphi^{\mathrm{H}}(v_0, \ldots, v_{i-1}, \sigma) \leftrightarrow \neg \varphi^{\mathrm{nS}}(v_0, \ldots, v_{i-1}, \sigma)$. Recalling 11.37, we see that $\varphi^{\mathrm{nS}}(v_0, \ldots, v_{i-1}, \sigma) = (\exists v_i \varphi^{\mathrm{n}})^{\mathrm{S}}$. Note that $(\forall v_i \varphi)^{\mathrm{n}} = \exists v_i \varphi^{\mathrm{n}}$. Thus $\vdash (\forall v_i \varphi)^{\mathrm{H}} \leftrightarrow \neg (\forall v_i \varphi)^{\mathrm{nS}}$, as desired.

Finally, we prove (*iii*):

$\models \varphi$ iff $\neg \varphi$ does not have a model
iff φ^{n} does not have a model
iff φ^{nS} does not have a model (by 11.38(*iv*))
iff $\models \neg \varphi^{\mathrm{nS}}$ iff $\models \varphi^{\mathrm{H}}$ (by (*ii*)). □

Herbrand's theorem in a sense reduces provability to checking tautologies. The following result, interesting in itself, is one of the main lemmas for the theorem.

Theorem 11.41. *If* φ *is a quantifier-free formula not involving equality and* $\models \varphi$, *then* φ *is a tautology.*

PROOF. Assume that φ is not a tautology; let f be a truth valuation (10.19) such that $f\varphi = 0$. Let $A = \mathrm{Trm}_{\mathscr{L}}$. For any relation symbol $\mathbf{R}$, say $\mathbf{R}$ of rank m, let

$$\mathbf{R}^{\mathfrak{A}} = \{(\tau_0, \ldots, \tau_{m-1}) : f\mathbf{R}\tau_0 \cdots \tau_{m-1} = 1\}.$$

Also, for any operation symbol $\mathbf{O}$, say of rank m, let

$$\mathbf{O}^{\mathfrak{A}}(\tau_0, \ldots, \tau_{m-1}) = \mathbf{O}\tau_0 \cdots \tau_{m-1}.$$

With these denotations for operation symbols and relation symbols we obtain an $\mathscr{L}$-structure $\mathfrak{A}$. Note that $\sigma^{\mathfrak{A}}x = \sigma$ for every term σ where $x_i = \langle v_i \rangle$ for each $i \in \omega$. Hence by induction on ψ we easily obtain:

for any quantifier-free formula ψ not involving equality, $\mathfrak{A} \models \psi[x]$ iff $f\psi = 1$.

Since $f\varphi = 0$, it follows that $\mathfrak{A} \not\models \varphi[x]$, so $\not\models \varphi$, as desired. □

Now we can give our version of Herbrand's theorem. Several versions of this theorem can be found in the literature. It has found considerable use, especially in finitary proofs of the consistency of theories.

Theorem 11.42 (Herbrand). *Let* $\mathscr{L}'$ *be a Skolem expansion of* $\mathscr{L}$, *and let* φ *be a prenex sentence of* $\mathscr{L}'$ *not involving equality. Say* $\varphi^{\mathrm{H}} = \exists \alpha_0 \cdots \exists \alpha_{m-1} \psi$

with ψ quantifier-free. Then $\models\varphi$ *iff some disjunction of instances* $\mathrm{Subf}^{\alpha 0}_{\sigma 0}\cdots\mathrm{Subf}^{\alpha(m-1)}_{\sigma(m-1)}\psi$ *of* ψ *is a tautology, where* $\sigma_0, \ldots, \sigma_{m-1}$ *are variable-free terms.*

PROOF. $\Rightarrow$. Assume $\models\varphi$. Thus by 11.40, $\models\varphi^{\mathrm{H}}$. Now let Γ be the set of all sentences

$$\neg\mathrm{Subf}^{\alpha 0}_{\sigma 0}\cdots\mathrm{Subf}^{\alpha(m-1)}_{\sigma(m-1)}\psi$$

where $\sigma_0, \ldots, \sigma_{m-1}$ are variable-free terms of $\mathscr{L}'$. We claim that Γ is inconsistent. For, if it is consistent, then it has a model $\mathfrak{A}$. Let B be the set of all elements ${}^0\sigma^{\mathfrak{A}}$ of A (see 11.4), where σ is a variable-free term. If $\mathbf{R}$ is a relation symbol of rank n, denoted by $\mathbf{R}^{\mathfrak{A}}$ in $\mathfrak{A}$, let

$$\mathbf{R}^{\mathfrak{B}} = \{(x_0, \ldots, x_{n-1}) : x_0, \ldots, x_{n-1} \in B \text{ and } (x_0, \ldots, x_{n-1}) \in \mathbf{R}^{\mathfrak{A}}\}.$$

If $\mathbf{O}$ is an operation symbol of rank n, denoted by $\mathbf{O}^{\mathfrak{A}}$ in $\mathfrak{A}$, let for variable free terms $\sigma_0, \ldots, \sigma_{n-1}$

$$\mathbf{O}^{\mathfrak{B}}{}^0\sigma_0^{\mathfrak{A}}\cdots{}^0\sigma_{n-1}^{\mathfrak{A}} = {}^0(\mathbf{O}\sigma_0\cdots\sigma_{n-1})^{\mathfrak{A}}.$$

This definition is easily justified. Thus we obtain a structure $\mathfrak{B}$. Since each member of Γ is quantifier-free, $\mathfrak{B}$ is a model of Γ. Since $\models\varphi^{\mathrm{H}}$, choose $x_0, \ldots, x_{m-1} \in B$ so that $\mathfrak{B} \models \psi[x_0, \ldots, x_{m-1}]$. Say $x_i = {}^0\sigma_i^{\mathfrak{A}}$ for each $i < m$. Thus $\mathrm{Subf}^{v0}_{\sigma 0}\cdots\mathrm{Subf}^{v(m-1)}_{\sigma(m-1)}\psi$ holds in $\mathfrak{B}$, which is a contradiction since $\mathfrak{B}$ is a model of Γ.

Thus Γ is inconsistent. Hence by 10.92, some disjunction of negations of members of Γ is valid, so by 11.41 it is a tautology, as desired.

$\Leftarrow$. Obvious. □

We give one simple application of Herbrand's theorem: Consider the formula $\psi = \exists\alpha\forall\beta\varphi \rightarrow \forall\beta\exists\alpha\varphi$, where α and β are distinct variables and φ is any formula (possibly involving equality). We want to prove that $\models\psi$ by using Herbrand's theorem. It is of course easy to prove $\models\psi$ by a direct semantic argument, but as we shall see, an application of Herbrand's theorem is more routine. Let $\mathscr{L}$ be our original language. Choose $n > 0$ so that $\mathrm{Fv}\,\varphi \subseteq \{v_0, \ldots, v_{n-1}\}$. Expand the language to $\mathscr{L}'$ by adding a new n-ary relation symbol $\mathbf{R}$. Now let $\varphi' = \mathbf{R}v_0\cdots v_{n-1}$, $\psi' = \exists\alpha\forall\beta\varphi' \rightarrow \forall\beta\exists\alpha\varphi'$. We first show that $\models\psi'$. Let γ and δ be new variables. Then a prenex formula equivalent to ψ' is

$$\chi = \forall\gamma\forall\beta\exists\delta\exists\alpha(\neg\mathrm{Subf}^{\alpha}_{\gamma}\,\mathrm{Subf}^{\beta}_{\delta}\varphi' \vee \varphi'),$$

and χ^{H} has the form

$$\exists\delta\exists\alpha(\neg\mathrm{Subf}^{\alpha}_{\mathbf{c}}\,\mathrm{Subf}^{\beta}_{\delta}\varphi' \vee \mathrm{Subf}^{\beta}_{\mathbf{d}}\varphi'),$$

where $\mathbf{c}$ and $\mathbf{d}$ are new individual constants, in a Skolem expansion $\mathscr{L}''$ of $\mathscr{L}'$. An instance of the matrix of this prenex formula is the tautology

$$\neg\mathrm{Subf}^{\alpha}_{\mathbf{c}}\,\mathrm{Subf}^{\beta}_{\mathbf{d}}\varphi' \vee \mathrm{Subf}^{\alpha}_{\mathbf{c}}\,\mathrm{Subf}^{\beta}_{\mathbf{d}}\varphi'.$$

Thus $\models\psi'$ by Herbrand's theorem. Now let Γ be the theory consisting of all consequences of the sentence

$$\forall v_0 \cdots \forall v_{n-1}(\mathbf{R}v_0 \cdots v_{n-1} \leftrightarrow \varphi).$$

Clearly $(\Gamma, \mathscr{L}')$ is a definitional expansion of $(\{\varphi \in \mathrm{Sent}_{\mathscr{L}} : \models\varphi\}, \mathscr{L})$.

It is also easy to see that $\Gamma \models \psi \leftrightarrow \psi'$. Hence $\Gamma \models \psi$ since we know that $\models\psi'$. By Theorem 11.31, $(\Gamma, \mathscr{L}')$ is a conservative extension of $(\{\varphi \in \mathrm{Sent}_{\mathscr{L}} : \models\varphi\}, \mathscr{L})$, so $\models\psi$, as desired.

The final topic of this chapter is the notion of an interpretation of one theory in another. This can be viewed as a syntactical counterpart of the notion of a model. All of the definitions relevant to this new notion are given in the following

Definition 11.43. Let $\mathscr{L} = (L, v, \mathcal{O}, \mathscr{R})$ and $\mathscr{L}' = (L', v', \mathcal{O}', \mathscr{R}')$ be two first-order languages. A *syntactical $\mathscr{L}$-structure in $\mathscr{L}'$* is a quadruple $\mathfrak{A} = (\chi, f, R, \Gamma')$ such that χ is a formula of $\mathscr{L}'$ with $\mathrm{Fv}\,\chi \subseteq \{v_0\}$, $f : \mathrm{Dmn}\,\mathcal{O} \to \mathrm{Dmn}\,\mathcal{O}'$ and $\mathcal{O}'f\mathbf{O} = \mathcal{O}\mathbf{O}$ for each $\mathbf{O} \in \mathrm{Dmn}\,\mathcal{O}$, $R : \mathrm{Dmn}\,\mathscr{R} \to \mathrm{Fmla}_{\mathscr{L}'}$, and $\mathrm{Fv}\,R_{\mathbf{R}} \subseteq \{v_0, \ldots, v_{m-1}\}$, with $m = \mathscr{R}\mathbf{R}$, for each $\mathbf{R} \in \mathrm{Dmn}\,\mathscr{R}$, while Γ' is a set of sentences of $\mathscr{L}'$ satisfying the following conditions:

(*i*) $\Gamma' \models \exists v_0\chi$;

(*ii*) $\Gamma' \models \forall v_0 \cdots \forall v_{m-1}(\bigwedge_{i<m} \chi(v_i) \to \chi(f_{\mathbf{O}}v_0 \cdots v_{m-1}))$

for each operation symbol $\mathbf{O}$ of $\mathscr{L}$ (say of rank m).

Given a syntactical $\mathscr{L}$-structure $\mathfrak{A}$ with the notation above, with each term σ and formula φ of $\mathscr{L}$ we associate a term $\sigma^{\mathfrak{A}}$ and formula $\varphi^{\mathfrak{A}}$ of $\mathscr{L}'$ as follows (this is the syntactical counterpart of Definitions 11.2 and 11.5):

$$\begin{aligned}
v_i^{\mathfrak{A}} &= v_i';\\
(\mathbf{O}\sigma_0 \cdots \sigma_{m-1})^{\mathfrak{A}} &= f_{\mathbf{O}}\sigma_0^{\mathfrak{A}} \cdots \sigma_{m-1}^{\mathfrak{A}};\\
(\sigma \equiv \tau)^{\mathfrak{A}} &= \sigma^{\mathfrak{A}} \equiv \tau^{\mathfrak{A}};\\
(\mathbf{R}\sigma_0 \cdots \sigma_{m-1})^{\mathfrak{A}} &= R_{\mathbf{R}}(\sigma_0^{\mathfrak{A}} \cdots \sigma_{m-1}^{\mathfrak{A}});\\
(\varphi \vee \psi)^{\mathfrak{A}} &= \varphi^{\mathfrak{A}} \vee \psi^{\mathfrak{A}};\\
(\varphi \wedge \psi)^{\mathfrak{A}} &= \varphi^{\mathfrak{A}} \wedge \psi^{\mathfrak{A}};\\
(\neg\varphi)^{\mathfrak{A}} &= \neg\varphi^{\mathfrak{A}};\\
(\forall\alpha\varphi)^{\mathfrak{A}} &= \forall\alpha(\chi(\alpha) \to \varphi^{\mathfrak{A}}).
\end{aligned}$$

If Γ is a theory in $\mathscr{L}$, we say that $\mathfrak{A}$ is an *interpretation of* Γ *in* Γ' provided that $\Gamma' \models \varphi^{\mathfrak{A}}$ for each $\varphi \in \Gamma$.

With each model $\mathfrak{B}$ of Γ' we associate an $\mathscr{L}$-structure $\mathfrak{B}^{\Gamma'\mathfrak{A}} = \mathfrak{B}(\Gamma', \mathfrak{A})$ as follows: $B^{\Gamma'\mathfrak{A}} = {}^1\chi^{\mathfrak{B}}$; for $\mathbf{O}$ an operation symbol of $\mathscr{L}$, say m-ary, and for any $b_0, \ldots, b_{m-1} \in B^{\Gamma'\mathfrak{A}}$, let $\mathbf{O}^{\mathfrak{B}(\Gamma',\mathfrak{A})}(b_0, \ldots, b_{m-1}) = f^{\mathfrak{B}}(b_0, \ldots, b_{m-1})$ ((*ii*) assures that $B^{\Gamma'\mathfrak{A}}$ is closed under $\mathbf{O}^{\mathfrak{B}(\Gamma',\mathfrak{A})}$); for $\mathbf{R}$ a relation symbol of $\mathscr{L}$, say $\mathbf{R}$ m-ary, let $\mathbf{R}^{\mathfrak{B}(\Gamma',\mathfrak{A})} = {}^m\mathbf{R}_{\mathbf{R}}^{\mathfrak{B}} \cap {}^1\chi^{\mathscr{L}}$.

In case, to start with, $\mathscr{L}$ and $\mathscr{L}'$ are effectivized first-order languages with corresponding Gödel numbering functions g and g', a syntactical $\mathscr{L}$-structure $\mathfrak{A}$ in $\mathscr{L}'$, with notation as above, is *effective* provided that the functions $g' \circ f \circ g^{-1} \restriction g^*\,\mathrm{Dmn}\,\mathcal{O}$ and $g'^{+} \circ R \circ g^{-1} \restriction g^*\,\mathrm{Dmn}\,\mathscr{R}$ are partial recursive (no further restrictions on Γ').

We give a few simple facts about these notions. First, concerning effective interpretations, the following proposition is clear.

Proposition 11.44. *Let $\mathfrak{A}$ be an effective syntactical $\mathscr{L}$-structure in $\mathscr{L}'$. Then the formation of $\sigma^{\mathfrak{A}}$ and $\varphi^{\mathfrak{A}}$ is effective. That is,* $\mathscr{g}'^{+} \circ f \circ \mathscr{g}^{+-1} \upharpoonright \mathscr{g}^{+*}\,\mathrm{Trm}_{\mathscr{L}}$ *and* $\mathscr{g}'^{+} \circ f \circ \mathscr{g}^{+-1} \upharpoonright \mathscr{g}^{+*}\,\mathrm{Fmla}_{\mathscr{L}}$ *are partial recursive, where f is the function assigning $\sigma^{\mathfrak{A}}$ to σ and $\varphi^{\mathfrak{A}}$ to φ.*

The fundamental model-theoretic relationships concerning interpretations are expressed in the following proposition, which is easily proved by induction on the formula φ.

Proposition 11.45. *Let $\mathfrak{A} = (\chi, f, R, \Gamma')$ be a syntactical $\mathscr{L}$-structure in $\mathscr{L}'$, and let $\mathfrak{B}$ be a model of Γ'. Then for any formula φ of $\mathscr{L}$ and any $x \in {}^{\omega}B^{\Gamma'\mathfrak{A}}$ we have $\mathfrak{B}^{\Gamma'\mathfrak{A}} \models \varphi[x]$ iff $\mathfrak{B} \models \varphi^{\mathfrak{A}}[x]$.*

Corollary 11.46. *Let Γ be a theory in $\mathscr{L}$, and let $\mathfrak{A} = (\chi, f, R, \Gamma')$ be an interpretation of Γ in $\mathscr{L}'$. Then for any formula φ of $\mathscr{L}$, the condition $\Gamma \models \varphi$ implies that $\Gamma' \models \varphi^{\mathfrak{A}}$.*

PROOF. Assume that $\Gamma \models \varphi$, and let $\mathfrak{B}$ be any model of Γ' (thus $\mathfrak{B}$ is an $\mathscr{L}'$-structure). Since $\Gamma' \models \psi^{\mathfrak{A}}$ for each $\psi \in \Gamma$, it follows that $\mathfrak{B} \models \psi^{\mathfrak{A}}$ for each $\psi \in \Gamma$, and hence by 11.45 $\mathfrak{B}^{\Gamma'\mathfrak{A}}$ is a model of Γ. Thus $\mathfrak{B}^{\Gamma'\mathfrak{A}} \models \varphi$ by assumption, so by 11.45 again, $\mathfrak{B} \models \varphi^{\mathfrak{A}}$. Thus $\Gamma' \models \varphi^{\mathfrak{A}}$, as desired. □

BIBLIOGRAPHY

1. Chang, C. C., Keisler, H. J. *Model Theory*. Amsterdam: North-Holland Publ. Co. (1974).
2. Shoenfield, J. *Mathematical Logic*. Reading: Addison-Wesley (1967).
3. Tarski, A., Mostowski, A., Robinson, R. M. *Undecidable Theories*. Amsterdam: North-Holland Publ. Co. (1953).

EXERCISES

11.47. Show that the restriction that α not occur in σ or in τ is essential in Lemma 10.39.

11.48. Show that the restrictions in 10.49, that only free occurrences of σ are replaced and that the new occurrences of τ are free, are essential.

11.49. Show that in 10.59 it is essential that β not occur in φ.

11.50. Show that the restriction of 10.61 is essential.

11.51. Show that in 10.71 the string of quantifiers $\forall\alpha_0 \cdots \forall\alpha_{m-1}$ cannot be deleted.

11.52. Show that $\not\models v_0 \equiv v_1 \rightarrow v_0 \equiv v_2$ (cf. the remarks after 10.86).

11.53. Carry out a modification of the proof of 11.12 in which B is the set of all individual constants of $\mathscr{L}$.

11.54. Show that any consistent set of formulas in a language $\mathscr{L}$ can be extended to a maximal consistent set of formulas of $\mathscr{L}$.

11.55. Let **K** be the class of all groups in which every element is of finite order, and let $\mathscr{L}$ be an appropriate first-order language. Show that there is no set Γ of sentences of $\mathscr{L}$ such that $\mathbf{K} = \{\mathfrak{A} : \mathfrak{A}$ is a model of $\Gamma\}$.

11.56. Let **K** be the class of all fields of prime characteristic, and let $\mathscr{L}$ be an appropriate first-order language. Show that there is no set Γ of sentences of $\mathscr{L}$ such that $\mathbf{K} = \{\mathfrak{A} : \mathfrak{A}$ is a model of $\Gamma\}$.

11.57. If a sentence φ holds in all non-Archimedean ordered fields, then φ holds in all ordered fields.

11.58. Let Γ be a theory in a language $\mathscr{L}$. Suppose that for every finite subset Δ of Γ there is a model of Δ which is not a model of Γ. Then there is no finite set $\mathscr{L}$ of sentences which has exactly the same models as Γ.

11.59. Let **K** be a nonempty set of $\mathscr{L}$-structures. For each $\mathbf{L} \subseteq \mathbf{K}$ let $C\mathbf{L} = \{\mathfrak{A} \in \mathbf{K} : \mathfrak{A}$ is a model of every sentence which holds in all members of $\mathbf{L}\}$. Show that with respect to C as a closure operator, **K** is a compact topological space.

11.60. Establish using Herbrand's theorem that any formula of the last type in 10.101 is universally valid.

Cylindric Algebras 12*

Cylindric algebras stand in the same relationship to first-order logic as Boolean algebras stand to sentential logic. We present in this chapter an introduction to the theory of these algebras paralleling our treatment of Boolean algebras. Again, the simplest motivation for the study of these algebras is from elementary set theory:

Definition 12.1. Let U be a nonempty set and α an ordinal. For each $\kappa < \alpha$ we define a one place operation C_κ on subsets of ${}^\alpha U$ by setting, for any $X \subseteq {}^\alpha U$,

$$C_\kappa X = \{x \in {}^\alpha U : x \restriction \alpha \sim \{\kappa\} = y \restriction \alpha \sim \{\kappa\} \text{ for some } y \in X\}.$$

For κ, $\lambda < \alpha$ we set

$$D_{\kappa\lambda} = \{x \in {}^\alpha U : x_\kappa = x_\lambda\}.$$

Thus $C_\kappa X$ is the generalized cylinder obtained by moving X parallel to the κ-axis in α-space, while $D_{\kappa\lambda}$ is a diagonal hyperplane. An *α-dimensional cylindric field of sets* is a set $\mathscr{A}$ of subsets of ${}^\alpha U$ (for some U) which is a field of sets in the Boolean sense and which is closed under all operations C_κ and contains all sets $D_{\kappa\lambda}$ as elements. A *cylindric set algebra of dimension* α is a structure $\langle \mathscr{A}, \cup, \cap, \sim, 0, {}^\alpha U, C_\kappa, D_{\kappa\lambda}\rangle_{\kappa,\lambda<\alpha}$ in which $\mathscr{A}$ is an α-dimensional cylindric field of sets and all of the operations are natural.

The general notion of a cylindric algebra is obtained by abstraction from this notion:

Definition 12.2. By a *cylindric algebra of dimension* α (where α is an ordinal number), for brevity a CA_α, we mean a structure

$$\mathfrak{A} = \langle A, +, \cdot, -, 0, 1, \mathrm{c}_\kappa, \mathrm{d}_{\kappa\lambda}\rangle_{\kappa,\lambda<\alpha}$$

such that $\langle A, +, \cdot, -, 0, 1\rangle$ is a BA, c_κ is a unary operation on A for each $\kappa < \alpha$, $d_{\kappa\lambda} \in A$ for all $\kappa, \lambda < \alpha$, and the following axioms hold (for all κ, λ, $\mu < \alpha$ and all $x \in A$):

$$(C_1) \qquad c_\kappa 0 = 0;$$
$$(C_2) \qquad x \leq c_\kappa x;$$
$$(C_3) \qquad c_\kappa(x \cdot c_\kappa y) = c_\kappa x \cdot c_\kappa y;$$
$$(C_4) \qquad c_\kappa c_\lambda x = c_\lambda c_\kappa x;$$
$$(C_5) \qquad d_{\kappa\kappa} = 1;$$
$$(C_6) \qquad \text{if } \kappa \neq \lambda, \mu, \text{ then } d_{\lambda\mu} = c_\kappa(d_{\lambda\kappa} \cdot d_{\kappa\mu});$$
$$(C_7) \qquad \text{if } \kappa \neq \lambda, \text{ then } c_\kappa(d_{\kappa\lambda} \cdot x) \cdot c_\kappa(d_{\kappa\lambda} \cdot -x) = 0.$$

By routine checking one obtains

Corollary 12.3. *Every cylindric set algebra is a cylindric algebra.*

The following proposition summarizes elementary properties of CA_α's which will be used later without specific reference. In what follows results are implicitly relative to an arbitrary CA_α $\mathfrak{A}$, arbitrary κ, λ, $\mu < \alpha$, and arbitrary x, y, $z \in A$, unless otherwise stated.

Proposition 12.4

(*i*) *If* $c_\kappa x = 0$, *then* $x = 0$.
(*ii*) $c_\kappa 1 = 1$.
(*iii*) $c_\kappa c_\kappa x = c_\kappa x$.
(*iv*) $x \in \operatorname{Rng} c_\kappa$ *iff* $c_\kappa x = x$.
(*v*) $x \cdot c_\kappa y = 0$ *iff* $c_\kappa x \cdot y = 0$.
(*vi*) $c_\kappa(x + y) = c_\kappa x + c_\kappa y$.
(*vii*) *If* $x \leq y$, *then* $c_\kappa x \leq c_\kappa y$.
(*viii*) $c_\kappa - c_\kappa x = -c_\kappa x$.
(*ix*) $d_{\kappa\lambda} = d_{\lambda\kappa}$.
(*x*) $c_\kappa d_{\kappa\lambda} = 1$.
(*xi*) *If* $\kappa \neq \lambda, \mu$, *then* $c_\kappa d_{\lambda\mu} = d_{\lambda\mu}$.
(*xii*) *If* $\kappa \neq \lambda$, *then* $c_\kappa(d_{\kappa\lambda} \cdot -x) = -c_\kappa(d_{\kappa\lambda} \cdot x)$.
(*xiii*) *If* $\kappa \neq \lambda$, *then* $d_{\kappa\lambda} \cdot c_\kappa(d_{\kappa\lambda} \cdot x) = d_{\kappa\lambda} \cdot x$.

PROOF. (*i*), (*ii*): immediate by (C_2). (*iii*): Putting $x = 1$ in (C_3) we obtain $c_\kappa c_\kappa y = c_\kappa 1 \cdot c_\kappa y$, so (*iii*) follows from (*ii*). (*iv*): if $x \in \operatorname{Rng} c_\kappa$, say $x = c_\kappa y$. Then $c_\kappa x = c_\kappa c_\kappa y = c_\kappa y = x$ by (*iii*). The converse is trivial. (*v*): if $x \cdot c_\kappa y = 0$, then

$$\begin{aligned} 0 &= c_\kappa(x \cdot c_\kappa y) && \text{by } (C_1) \\ &= c_\kappa x \cdot c_\kappa y && \text{by } (C_3) \\ &= c_\kappa(c_\kappa x \cdot y) && \text{by } (C_3) \end{aligned}$$

so $c_\kappa x \cdot y = 0$ by (*i*). The converse is the same statement. (*vi*): Since $c_\kappa(x + y) \cdot -c_\kappa(x + y) = 0$, by (*v*) we have $(x + y) \cdot c_\kappa - c_\kappa(x + y) = 0$, so $x \cdot c_\kappa -$

$c_\kappa(x + y) = 0$ and (v) yields $c_\kappa x \cdot -c_\kappa(x + y) = 0$. Thus $c_\kappa x \leq c_\kappa(x + y)$, and similarly $c_\kappa y \leq c_\kappa(x + y)$. Next, $c_\kappa x \cdot (-c_\kappa x \cdot -c_\kappa y) = 0$, so $x \cdot c_\kappa(-c_\kappa x \cdot -c_\kappa y) = 0$ by (v). Similarly, $y \cdot c_\kappa(-c_\kappa x \cdot -c_\kappa y) = 0$, so $(x + y) \cdot c_\kappa(-c_\kappa x \cdot -c_\kappa y) = 0$, and an application of (v) yields $c_\kappa(x + y) \cdot -c_\kappa x \cdot -c_\kappa y = 0$, i.e., $c_\kappa(x + y) \leq c_\kappa x + c_\kappa y$. Thus (vi) holds. (vii): if $x \leq y$, then $x + y = y$, so $c_\kappa x \leq c_\kappa x + c_\kappa y = c_\kappa(x + y) = c_\kappa y$, using (vi). $(viii)$: The inequality $-c_\kappa x \leq c_\kappa - c_\kappa x$ is an instance of (C_2). By (iii), $-c_\kappa x \cdot c_\kappa c_\kappa x = 0$, so by (v), $c_\kappa - c_\kappa x \cdot c_\kappa x = 0$, i.e., $c_\kappa - c_\kappa x \leq -c_\kappa x$, as desired. (ix): by (C_5) we may assume that $\kappa \neq \lambda$. Then, since $c_\kappa(d_{\kappa\lambda} \cdot d_{\lambda\kappa}) = d_{\lambda\lambda} = 1$ by (C_6) and (C_5), we have

$$\begin{aligned} 0 &= c_\kappa(d_{\kappa\lambda} \cdot d_{\lambda\kappa}) \cdot c_\kappa(d_{\kappa\lambda} \cdot -d_{\lambda\kappa}) && \text{by } (C_7) \\ &= c_\kappa(d_{\kappa\lambda} \cdot -d_{\lambda\kappa}), \end{aligned}$$

so $d_{\kappa\lambda} \cdot -d_{\lambda\kappa} = 0$ by (i). Thus $d_{\kappa\lambda} \leq d_{\lambda\kappa}$, so (ix) follows by symmetry. (x): by (C_5) and (ii) we may assume that $\kappa \neq \lambda$. Then

$$\begin{aligned} c_\kappa d_{\kappa\lambda} &= c_\kappa(d_{\kappa\lambda} \cdot d_{\lambda\kappa}) && \text{by } (ix) \\ &= d_{\lambda\lambda} = 1 && \text{by } (C_6), (C_5) \end{aligned}$$

(xi): Assuming $\kappa \neq \lambda, \mu$, we have

$$\begin{aligned} c_\kappa d_{\lambda\mu} &= c_\kappa c_\kappa(d_{\lambda\kappa} \cdot d_{\kappa\mu}) && \text{by } (C_6) \\ &= c_\kappa(d_{\lambda\kappa} \cdot d_{\kappa\mu}) = d_{\lambda\mu} && \text{by } (iii), (C_6) \end{aligned}$$

(xii): by (C_7), $c_\kappa(d_{\kappa\lambda} \cdot -x) \cdot c_\kappa(d_{\kappa\lambda} \cdot x) = 0$. On the other hand,

$$\begin{aligned} c_\kappa(d_{\kappa\lambda} \cdot -x) + c_\kappa(d_{\kappa\lambda} \cdot x) &= c_\kappa(d_{\kappa\lambda} \cdot -x + d_{\kappa\lambda} \cdot x) && \text{by } (vi) \\ &= c_\kappa d_{\kappa\lambda} = 1 && \text{by } (x) \end{aligned}$$

Hence $c_\kappa(d_{\kappa\lambda} \cdot -x) = -c_\kappa(d_{\kappa\lambda} \cdot x)$ by 9.6(xii). $(xiii)$: we have

$$d_{\kappa\lambda} \cdot -x \leq c_\kappa(d_{\kappa\lambda} \cdot -x) = -c_\kappa(d_{\kappa\lambda} \cdot x) \qquad \text{by } (C_2), (xii)$$

so $d_{\kappa\lambda} \cdot -x \cdot c_\kappa(d_{\kappa\lambda} \cdot x) = 0$ and hence $d_{\kappa\lambda} \cdot c_\kappa(d_{\kappa\lambda} \cdot x) \leq x$, from which $(xiii)$ easily follows. □

An elementary operation which will play an important role in later considerations is as follows.

Definition 12.5. For $\kappa, \lambda < \alpha$ and $x \in A$ we set

$$\begin{aligned} s^\kappa_\lambda x &= x && \text{if } \kappa = \lambda, \\ s^\kappa_\lambda x &= c_\kappa(d_{\kappa\lambda} \cdot x) && \text{if } \kappa \neq \lambda. \end{aligned}$$

For later use we give some properties of this operation; they are easily established on the basis of 12.4:

Proposition 12.6

(*i*) s^κ_λ *is an endomorphism of* $\langle A, +, \cdot, -, 0, 1\rangle$.

(*ii*) *If* $\kappa \neq \mu$, *then* $s^\kappa_\lambda d_{\kappa\mu} = d_{\lambda\mu}$.

(*iii*) *If* $\kappa \neq \mu, \nu$, *then* $s^{\kappa}_{\lambda}d_{\mu\nu} = d_{\mu\nu}$.
(*iv*) $d_{\kappa\lambda} \cdot s^{\kappa}_{\lambda}x = d_{\kappa\lambda} \cdot x$.
(*v*) *If* $\mu \neq \kappa, \lambda$, *then* $d_{\kappa\lambda} \cdot s^{\mu}_{\kappa}x = d_{\kappa\lambda} \cdot s^{\mu}_{\lambda}x$.
(*vi*) $s^{\kappa}_{\lambda}c_{\kappa}x = c_{\kappa}x$.
(*vii*) *If* $\mu \neq \kappa, \lambda$, *then* $s^{\kappa}_{\lambda}c_{\mu}x = c_{\mu}s^{\kappa}_{\lambda}x$.
(*viii*) *If* $\kappa \neq \mu$, *then* $s^{\kappa}_{\lambda}s^{\kappa}_{\mu}x = s^{\kappa}_{\mu}x$.
(*ix*) $s^{\kappa}_{\lambda}s^{\mu}_{\kappa}x = s^{\kappa}_{\lambda}s^{\mu}_{\lambda}x$.
(*x*) *If* $\lambda \neq \mu \neq \kappa \neq \nu$, *then* $s^{\mu}_{\nu}s^{\kappa}_{\lambda}x = s^{\kappa}_{\lambda}s^{\mu}_{\nu}x$.
(*xi*) $s^{\kappa}_{\lambda}s^{\mu}_{\lambda}x = s^{\mu}_{\lambda}s^{\kappa}_{\lambda}x$.
(*xii*) $s^{\kappa}_{\lambda}s^{\mu}_{\kappa}c_{\kappa}x = s^{\mu}_{\lambda}c_{\kappa}x$.
(*xiii*) $s^{\kappa}_{\lambda}s^{\mu}_{\kappa}c_{\kappa}c_{\nu}x = s^{\nu}_{\lambda}s^{\mu}_{\nu}c_{\kappa}c_{\nu}x$.

For later purposes it is convenient to introduce the notion of the *dimension set* Δx of an element x, giving the coordinates in which x is not a cylinder:

Definition 12.7. $\Delta x = \{\kappa : c_{\kappa}x \neq x\}$.

The following proposition gives some easily established properties of this notion:

Proposition 12.8
(*i*) $\Delta 0 = \Delta 1 = 0$.
(*ii*) *If* $d_{\kappa\lambda} \neq 1$, *then* $\Delta d_{\kappa\lambda} = \{\kappa, \lambda\}$.
(*iii*) $\Delta(x + y) \subseteq \Delta x \cup \Delta y$.
(*iv*) $\Delta x = \Delta - x$.
(*v*) $\Delta(x \cdot y) \subseteq \Delta x \cup \Delta y$.
(*vi*) $\Delta c_{\kappa}x \subseteq \Delta x \sim \{\kappa\}$.
(*vii*) $\Delta s^{\kappa}_{\lambda}x \subseteq (\Delta x \sim \{\kappa\}) \cup \{\lambda\}$.

Now we consider some algebraic notions as applied to cylindric algebras.

Definition 12.9. Let $\mathfrak{A} = \langle A, +, \cdot, -, 0, 1, c_{\kappa}, d_{\kappa\lambda}\rangle_{\kappa,\lambda<\alpha}$ and $\mathfrak{B} = \langle B, +', \cdot', -', 0', 1', c'_{\kappa}, d'_{\kappa\lambda}\rangle_{\kappa,\lambda<\alpha}$ be CA_{α}'s. We say that $\mathfrak{A}$ is a *subalgebra* of $\mathfrak{B}$ if $A \subseteq B$, $0 = 0'$, $1 = 1'$, $d_{\kappa\lambda} = d'_{\kappa\lambda}$ for all $\kappa, \lambda < \alpha$, and for all $x, y \in A$, $x + y = x +' y$, $x \cdot y = x \cdot' y$, $-x = -'x$, and $c_{\kappa}x = c'_{\kappa}x$ for all $\kappa < \alpha$. For any $X \subseteq B$, X is a *subuniverse* of $\mathfrak{B}$ if $0', 1' \in X$, $d'_{\kappa\lambda} \in X$ for all κ, $\lambda < \alpha$, and X is closed under $+'$, $\cdot'$, $-'$ and c'_{κ} for each $\kappa < \alpha$.

Proposition 12.10. *If* $\mathscr{A}$ *is a nonempty collection of subuniverses of a* CA_{α} $\mathfrak{A}$, *then* $\bigcap\mathscr{A}$ *is a subuniverse of* $\mathfrak{A}$.

Definition 12.11. If $X \subseteq A$, $\mathfrak{A}$ a CA_{α}, then the set $\bigcap\{Y : X \subseteq Y, Y$ a subuniverse of $\mathfrak{A}\}$ is called the *subuniverse of* $\mathfrak{A}$ *generated by* X.

In general there is no simple expression for the elements of the subuniverse generated by X, unlike the situation for BA's (cf. 9.11).

Definition 12.12. Let $\mathfrak{A}$ and $\mathfrak{B}$ be CA_α's, with notation as in 12.9. A *homomorphism* from $\mathfrak{A}$ into $\mathfrak{B}$ is a function f mapping A into B such that for all $x, y \in A$ and $\kappa, \lambda < \alpha$,

(*i*) $f(x + y) = fx +' fy$;
(*ii*) $f(x \cdot y) = fx \cdot' fy$;
(*iii*) $f(-x) = -'fx$;
(*iv*) $f0 = 0'$;
(*v*) $f1 = 1'$;
(*vi*) $fc_\kappa x = c'_\kappa fx$;
(*vii*) $fd_{\kappa\lambda} = d'_{\kappa\lambda}$.

The terms *homomorphism onto*, *isomorphism into*, and *isomorphism onto* have the obvious meaning. We write $\mathfrak{A} \cong \mathfrak{B}$ if there is an isomorphism of $\mathfrak{A}$ into $\mathfrak{B}$.

Proposition 12.13. *If f is a homomorphism from $\mathfrak{A}$ into $\mathfrak{B}$ and $x \in A$, then* $\Delta fx \subseteq \Delta x$.

Definition 12.14. Let $\mathfrak{A} = \langle A, +, \cdot, -, 0, 1, c_\kappa, d_{\kappa\lambda}\rangle_{\kappa,\lambda<\alpha}$ be a CA_α. An *ideal* of $\mathfrak{A}$ is an ideal of $\langle A, +, \cdot, -, 0, 1\rangle$ such that for all $\kappa < \alpha$, if $x \in I$ then $c_\kappa x \in I$.

Proposition 12.15. *Let I be an ideal in a* CA_α $\mathfrak{A}$, *and let* $R = \{(x, y) : x \cdot -y + -x \cdot y \in I\}$ *(cf.* 9.16). *Then for any $\kappa < \alpha$ and $x, y \in A$, if xRy then* $c_\kappa x R c_\kappa y$.

Proof. Assume that xRy and $\kappa < \alpha$. Thus $x \cdot -y + -x \cdot y \in I$. Now

$$c_\kappa x = c_\kappa(x \cdot -y + x \cdot y) = c_\kappa(x \cdot -y) + c_\kappa(x \cdot y) \leq c_\kappa(x \cdot -y) + c_\kappa y;$$

hence

$$c_\kappa x \cdot -c_\kappa y \leq c_\kappa(x \cdot -y) \leq c_\kappa(x \cdot -y + y \cdot -x) \in I.$$

Thus $c_\kappa x \cdot -c_\kappa y \in I$, and by symmetry, $c_\kappa y \cdot -c_\kappa x \in I$, so $c_\kappa x \cdot -c_\kappa y + -c_\kappa x \cdot c_\kappa y \in I$ and $c_\kappa x R c_\kappa y$. □

Along with 9.16 and 9.17, Proposition 12.15 justifies the following definition:

Definition 12.16. Let I be an ideal in a CA_α $\mathfrak{A} = \langle A, +, \cdot, -, 0, 1, c_\kappa, d_{\kappa\lambda}\rangle_{\kappa,\lambda<\alpha}$. We define $\mathfrak{A}/I = \langle A/I, +', \cdot', -', 0', 1', c'_\kappa, d'_{\kappa\lambda}\rangle_{\kappa,\lambda<\alpha}$, where

$$\langle A, +, \cdot, -, 0, 1\rangle/I = \langle A/I, +', \cdot', -', 0', 1'\rangle$$

in accordance with 9.17, $c'_\kappa[a] = [c_\kappa a]$ for all $a \in A$ and $\kappa < \alpha$, and $d'_{\kappa\lambda} = [d_{\kappa\lambda}]$ for all $\kappa, \lambda < \alpha$.

Proposition 12.17. *If I is an ideal in a* CA_α $\mathfrak{A}$, then $\mathfrak{A}/I$ *is a* CA_α, *and I^* is a homomorphism from $\mathfrak{A}$ onto $\mathfrak{A}/I$.*

Proposition 12.18. *If f is a homomorphism from a* CA_α $\mathfrak{A}$ *onto a* CA_α $\mathfrak{B}$ *and $I = \{x \in A : fx = 0\}$, then I is an ideal of $\mathfrak{A}$, and $\mathfrak{B} \cong \mathfrak{A}/I$.*

Proposition 12.19. *The intersection of any nonempty family of ideals in a* CA_α *is an ideal.*

Definition 12.20. If $\mathfrak{A}$ is a CA_α and $x \subseteq A$, then the *ideal generated by X* is the set

$$\bigcap \{I : X \subseteq I, I \text{ an ideal of } \mathfrak{A}\}.$$

We can directly generalize 9.23 to give a simple expression for the members of the ideal generated by a set:

Proposition 12.21. *If $X \subseteq A$, $\mathfrak{A}$ a* CA_α, *then the ideal generated by X is the collection of all $y \in A$ such that there exist $m, n \in \omega$ and $x \in {}^mX$, $\kappa \in {}^n\alpha$ with $y \leq c_{\kappa 0} \cdots c_{\kappa(n-1)}(x_0 + \cdots + x_{m-1})$.*

PROOF. Let I be the collection of all $y \in A$ such that such m, n, x, κ exist. Clearly I is contained in the ideal generated by X. Thus it is enough to show that $X \subseteq I$ and I is an ideal. Taking $m = 1$ and $n = 0$ we easily see that $X \subseteq I$. Taking $m = n = 0$, we see that $0 \in I$ and hence $I \neq 0$. If $z \leq y \in I$, obviously also $z \in I$. If $y \in I$, with m, n, x, κ as above, and if $\lambda < \alpha$, then

$$c_\lambda y \leq c_\lambda c_{\kappa 0} \cdots c_{\kappa(n-1)}(x_0 + \cdots + x_{m-1}),$$

so $c_\lambda y \in I$. Finally, suppose $y, y' \in I$, with m, n, x, κ and m', n', x', κ' satisfying the corresponding conditions. Then

$$\begin{aligned} y + y' &\leq c_{\kappa 0} \cdots c_{\kappa(n-1)}(x_0 + \cdots + x_{m-1}) + c_{\kappa' 0} \cdots c_{\kappa'(n'-1)}(x'_0 + \cdots + x'_{m'-1}) \\ &\leq c_{\kappa 0} \cdots c_{\kappa(n-1)} c_{\kappa' 0} \cdots c_{\kappa'(n'-1)}(x_0 + \cdots + x_{m-1} + x'_0 + \cdots + x'_{m'-1}), \end{aligned}$$

so $y + y' \in I$ also. □

We shall not develop the algebraic theory of CA_α's any further. Instead, we now turn to the relationships between first-order logic and cylindric algebras. In this regard the following definition is fundamental.

Definition 12.22. For $\mathscr{L}$ a first-order language and Γ a set of sentences in $\mathscr{L}$ we set

$$\equiv_\Gamma^\mathscr{L} = \{(\varphi, \psi) : \varphi \text{ and } \psi \text{ are formulas of } \mathscr{L} \text{ and } \Gamma \models \varphi \leftrightarrow \psi\}.$$

Furthermore, we let $\mathfrak{M}_\Gamma^{\mathscr{L}} = \langle \mathrm{Fmla}_{\mathscr{L}}/{\equiv_\Gamma^{\mathscr{L}}}, +, \cdot, -, 0, 1, \mathrm{c}_\kappa, \mathrm{d}_{\kappa\lambda}\rangle_{\kappa,\lambda<\omega}$, where for any $\varphi, \psi \in \mathrm{Fmla}_{\mathscr{L}}$ and any $\kappa, \lambda \in \omega$,

$$\begin{aligned}
[\varphi] + [\psi] &= [\varphi \vee \psi];\\
[\varphi]\cdot[\psi] &= [\varphi \wedge \psi];\\
-[\varphi] &= [\neg\varphi];\\
0 &= [\neg v_0 = v_0];\\
1 &= [v_0 = v_0];\\
\mathrm{c}_\kappa[\varphi] &= [\exists \mathrm{v}_\kappa\varphi];\\
\mathrm{d}_{\kappa\lambda} &= [\mathrm{v}_\kappa = \mathrm{v}_\lambda].
\end{aligned}$$

This definition is easily justified (see 9.54–9.56). Routine checking gives:

Proposition 12.23. *$\mathfrak{M}_\Gamma^{\mathscr{L}}$ is a* CA_ω.

As in the case of sentential logic and Boolean algebras, there is a natural correspondence between notions of first-order logic and notions of cylindric algebras. We give two instances of this correspondence. The first one indicates the close relationship between set algebras and models of a theory:

Proposition 12.24. *Let Γ be a set of sentences in a first-order language $\mathscr{L}$, and let $\mathfrak{A}$ be a model of Γ. Then $\{\varphi^{\mathfrak{A}} : \varphi \in \mathrm{Fmla}_{\mathscr{L}}\}$ is an ω-dimensional field of sets. Let $\mathfrak{B}$ be the associated cylindric set algebra. Then the function f such that $f[\varphi] = \varphi^{\mathfrak{A}}$ for each $\varphi \in \mathrm{Fmla}_{\mathscr{L}}$ is a homomorphism of $\mathfrak{M}_\Gamma^{\mathscr{L}}$ onto $\mathfrak{B}$ (f is easily seen to be well defined).*

This proposition can be routinely checked. The following proposition is established just like 9.59.

Proposition 12.25. *Let I be an ideal in a* CA_ω *$\mathfrak{M}_\Gamma^{\mathscr{L}}$, and set $\Delta = \{\varphi \in \mathrm{Sent}_{\mathscr{L}} : -[\varphi] \in I\}$. Then $\Gamma \subseteq \Delta$, and $\mathfrak{M}_\Gamma^{\mathscr{L}}/I$ is isomorphic to $\mathfrak{M}_\Delta^{\mathscr{L}}$.*

The algebras $\mathfrak{M}_\Gamma^{\mathscr{L}}$ possess a special property not possessed by other CA_ω's; the definition of this property is given in

Definition 12.26. Let $\mathfrak{A}$ be a CA_α. We say that $\mathfrak{A}$ is *locally finite dimensional* provided that for all $a \in A$, Δa is finite.

In the case of an algebra $\mathfrak{M}_\Gamma^{\mathscr{L}}$, if φ is any formula and v_κ is a variable not occurring in φ, then $\models \exists v_\kappa\varphi \leftrightarrow \varphi$, and hence $\mathrm{c}_\kappa[\varphi] = [\exists \mathrm{v}_\kappa\varphi] = [\varphi]$. Thus $\Delta[\varphi] \subseteq \{\kappa : v_\kappa \text{ occurs in } \varphi\}$, and hence $\Delta[\varphi]$ is finite. Hence:

Proposition 12.27. *$\mathfrak{M}_\Gamma^{\mathscr{L}}$ is locally finite dimensional.*

The following result is an analog of 9.58; it shows that the algebras $\mathfrak{M}_0^{\mathscr{L}}$ have a certain freeness property.

Proposition 12.28. *Let $\mathscr{L} = (L, v, \mathcal{O}, \mathscr{R})$ be a first-order language with no operation symbols. Suppose $\mathfrak{A}$ is a CA_ω, and $f\colon \mathrm{Dmn}\, \mathscr{R} \to A$ is a function such that $\Delta f_{\mathbf{R}} \subseteq \mathscr{R}_{\mathbf{R}}$ for all $\mathbf{R} \in \mathrm{Dmn}\, \mathscr{R}$. Then there is a homomorphism g from $\mathfrak{M}_0^{\mathscr{L}}$ into $\mathfrak{A}$ such that $g[\mathbf{R}v_0 \cdots v_{m-1}] = f_{\mathbf{R}}$ for each $\mathbf{R} \in \mathrm{Dmn}\, \mathscr{R}$ (with, say, $\mathscr{R}_{\mathbf{R}} = m$).*

Proof. The proof is similar to that of 9.58, but is much more tedious. We first define a function $h\colon \mathrm{Fmla}_{\mathscr{L}} \to A$; $h\varphi$ is defined by induction on φ. The most complicated part of the definition is the case in which φ is an atomic formula of the form $\mathbf{R}v_{i0} \cdots v_{i(m-1)}$. In this case, let $j_0, \ldots, j_{m-1}$ be the first m integers in $\omega \sim \{i0, \ldots, i(m-1), 0, \ldots, m-1\}$, and set

$$h\mathbf{R}v_{i0} \cdots v_{i(m-1)} = \mathsf{s}_{i0}^{j0} \cdots \mathsf{s}_{i(m-1)}^{j(m-1)} \mathsf{s}_{j0}^{0} \cdots \mathsf{s}_{j(m-1)}^{m-1} f_{\mathbf{R}};$$

for motivation, cf. the comments following 10.74. The inductive definition of $h\varphi$ proceeds very simply: $h(v_\kappa = v_\lambda) = \mathsf{d}_{\kappa\lambda}$, $h(\varphi \vee \psi) = h\varphi + h\psi$, $h(\varphi \wedge \psi) = h\varphi \cdot h\psi$, $h(\neg\varphi) = -h\varphi$, $h\forall v_\kappa \varphi = -\mathsf{c}_\kappa - h\varphi$. Analogously to the sentential case, we now claim

$$\text{(1)} \qquad \text{if } \vdash\varphi, \text{ then } h\varphi = 1.$$

To prove (1), let $\Gamma = \{\varphi : \varphi \in \mathrm{Fmla}_{\mathscr{L}}, h\varphi = 1\}$. Now each logical axiom is in Γ. We prove this in detail for each kind of axiom. First suppose $h\varphi \neq 1$; we show that φ is not a tautology. Let I be a maximal ideal of the BA $\mathfrak{B} = \langle \mathfrak{M}_0^{\mathscr{L}}, +, \cdot, -, 0, 1 \rangle$ such that $h\varphi \in I$, and let π be the natural homomorphism of $\mathfrak{B}$ onto $\mathfrak{B}/I$. Then B/I is a two-element BA, and $\pi \circ h$ can be considered as a truth valuation. Since $\pi h\varphi = 0$, φ is not a tautology. Next, suppose φ has the form $\forall v_\kappa(\psi \to \chi) \to (\forall v_\kappa \psi \to \forall v_\kappa \chi)$. Then

$$\begin{aligned} h\varphi &= --\mathsf{c}_\kappa-(-h\psi + h\chi) + --\mathsf{c}_\kappa - h\psi + -\mathsf{c}_\kappa - h\chi \\ &= \mathsf{c}_\kappa(h\psi \cdot -h\chi) + \mathsf{c}_\kappa - h\psi + -\mathsf{c}_\kappa - h\chi \\ &\geq \mathsf{c}_\kappa(h\psi \cdot -h\chi) + \mathsf{c}_\kappa(-h\psi \cdot -h\chi) + -\mathsf{c}_\kappa - h\chi \\ &= \mathsf{c}_\kappa - h\chi + -\mathsf{c}_\kappa - h\chi = 1. \end{aligned}$$

For the next kind of axiom we need a lemma:

$$\text{(2)} \qquad \text{if } v_\kappa \text{ does not occur in } \psi, \text{ then } \kappa \notin \Delta h\psi.$$

To prove (2) we proceed by induction on ψ. First suppose ψ is $\mathbf{R}v_{i0} \cdots v_{i(m-1)}$. Then with notation as in the definition of $h\psi$, by 12.8(*vii*) we have $\Delta h\psi \subseteq \{i0, \ldots, i(m-1)\}$. Thus (2) holds in this case. All other steps in the inductive proof of (2) are easy using appropriate parts of 12.8. Thus (2) holds. Hence if φ is $\psi \to \forall v_\kappa \psi$ where v_κ does not occur in ψ, then

$$h\varphi = -h\psi + -\mathsf{c}_\kappa - h\psi = -h\psi + --h\psi = 1 \quad (\text{using } 12.8(iv)).$$

The desired conclusion for φ of the form $\exists v_\kappa(v_\kappa = v_\lambda)$ is clear. Finally, suppose φ is $v_\kappa = v_\lambda \to (\psi \to \chi)$, where χ and ψ are atomic and χ is obtained from ψ by replacing an occurrence of v_κ in ψ by v_λ. We may suppose that $\kappa \neq \lambda$, since if $\kappa = \lambda$ then $\psi = \chi$ and obviously $h\varphi = 1$. If ψ is $v_\kappa = v_\kappa$, then χ is

$v_\kappa \equiv v_\lambda$ or $v_\lambda \equiv v_\kappa$, and again obviously $h\varphi = 1$. If ψ is $v_\kappa \equiv v_\mu$ or $v_\mu \equiv v_\kappa$ with $\kappa \neq \mu$, then χ is $v_\lambda \equiv v_\mu$ or $v_\mu \equiv v_\lambda$, and

$$\mathrm{d}_{\kappa\lambda} \cdot \mathrm{d}_{\kappa\mu} \leq \mathrm{c}_\kappa(\mathrm{d}_{\kappa\lambda} \cdot \mathrm{d}_{\kappa\mu}) = \mathrm{d}_{\lambda\mu};$$

hence

$$h\varphi = -\mathrm{d}_{\kappa\lambda} + -\mathrm{d}_{\lambda\mu} = 1.$$

So we may assume that ψ has the form $\mathbf{R}v_{i0} \cdots v_{i(t-1)}v_\kappa v_{i(t+1)} \cdots v_{i(m-1)}$ and χ is $\mathbf{R}v_{i0} \cdots v_{i(t-1)}v_\lambda v_{i(t+1)} \cdots v_{i(m-1)}$. In this case we need several lemmas.

(3) Let $j_0, \ldots, j_{m-1} \in \omega$, and let $k_0, \ldots, k_{m-1}, l$ be distinct integers in $\omega \sim \{j_0, \ldots, j_{m-1}, 0, \ldots, m-1\}$, and assume that $u < m$. Then

$$\mathbf{s}^{k0}_{j0} \cdots \mathbf{s}^{k(m-1)}_{j(m-1)} \mathbf{s}^{0}_{k0} \cdots \mathbf{s}^{m-1}_{k(m-1)} f_{\mathbf{R}} = \mathbf{s}^{k0}_{j0} \cdots \mathbf{s}^{k(u-1)}_{j(u-1)} \mathbf{s}^{l}_{ju} \mathbf{s}^{k(u+1)}_{j(u+1)} \cdots \mathbf{s}^{k(m-1)}_{j(m-1)} \mathbf{s}^{0}_{k0} \cdots \mathbf{s}^{u-1}_{k(u-1)} \mathbf{s}^{u}_{l} \mathbf{s}^{u+1}_{k(u+1)} \cdots \mathbf{s}^{m-1}_{k(m-1)} f_{\mathbf{R}}.$$

The proof is straightforward:

$$\begin{aligned}
\mathbf{s}^{k0}_{j0} \cdots \mathbf{s}^{k(m-1)}_{j(m-1)} \mathbf{s}^{0}_{k0} \cdots \mathbf{s}^{m-1}_{k(m-1)} f_{\mathbf{R}} &= \mathbf{s}^{k0}_{j0} \cdots \mathbf{s}^{k(m-1)}_{j(m-1)} \mathbf{s}^{0}_{k0} \cdots \mathbf{s}^{m-1}_{k(m-1)} \mathrm{c}_{ku} \mathrm{c}_{l} f_{\mathbf{R}} \\
&= \mathbf{s}^{k0}_{j0} \cdots \mathbf{s}^{k(u-1)}_{j(u-1)} \mathbf{s}^{k(u+1)}_{j(u+1)} \cdots \mathbf{s}^{k(m-1)}_{j(m-1)} \mathbf{s}^{ku}_{ju} \mathbf{s}^{u}_{ku} \mathrm{c}_{ku} \mathrm{c}_{l} \mathbf{s}^{0}_{k0} \cdots \mathbf{s}^{u-1}_{k(u-1)} \\
&\qquad \mathbf{s}^{u+1}_{k(u+1)} \cdots \mathbf{s}^{m-1}_{k(m-1)} f_{\mathbf{R}} \qquad \text{(using 12.6}(x)\text{ and 12.6}(vii)\text{)} \\
&= \mathbf{s}^{k0}_{j0} \cdots \mathbf{s}^{k(u-1)}_{j(u-1)} \mathbf{s}^{k(u+1)}_{j(u+1)} \cdots \mathbf{s}^{k(m-1)}_{j(m-1)} \mathbf{s}^{l}_{ju} \mathbf{s}^{u}_{l} \mathrm{c}_{ku} \mathrm{c}_{l} \mathbf{s}^{0}_{k0} \cdots \mathbf{s}^{u-1}_{k(u-1)} \\
&\qquad \mathbf{s}^{u+1}_{k(u+1)} \cdots \mathbf{s}^{m-1}_{k(m-1)} f_{\mathbf{R}} \qquad \text{(by 12.6}(xiii)\text{)} \\
&= \mathbf{s}^{k0}_{j0} \cdots \mathbf{s}^{k(u-1)}_{j)u-1)} \mathbf{s}^{l}_{ju} \mathbf{s}^{k(u+1)}_{j(u+1)} \cdots \mathbf{s}^{k(m-1)}_{j(m-1)} \mathbf{s}^{0}_{k0} \cdots \mathbf{s}^{u-1}_{k(u-1)} \mathbf{s}^{u}_{l} \mathbf{s}^{u+1}_{k(u+1)} \cdots \\
&\qquad \mathbf{s}^{m-1}_{k(m-1)} f_{\mathbf{R}} \qquad \text{(by reversing above steps)}
\end{aligned}$$

Thus (3) holds. By induction from (3) we obtain

(4) Let $j0, \ldots, j(m-1) \in \omega$, and let $k0, \ldots, k(m-1), l0, \ldots, l(m-1)$ be distinct integers in $\omega \sim \{j0, \ldots, j(m-1), 0, \ldots, m-1\}$. Then

$$\mathbf{s}^{k0}_{j0} \cdots \mathbf{s}^{k(u-1)}_{j(m-1)} \mathbf{s}^{0}_{k0} \cdots \mathbf{s}^{m-1}_{k(m-1)} f_{\mathbf{R}} = \mathbf{s}^{l0}_{j0} \cdots \mathbf{s}^{l(m-1)}_{j(m-1)} \mathbf{s}^{0}_{l0} \cdots \mathbf{s}^{m-1}_{l(m-1)} f_{\mathbf{R}}.$$

We can also prove the following slightly stronger statement:

(5) Let $j0, \ldots, j(m-1) \in \omega$, let $k0, \ldots, k(m-1)$ be distinct integers in $\omega \sim \{j0, \ldots, j(m-1), 0, \ldots, m-1\}$, and let $l0, \ldots, l(m-1)$ also be distinct integers in $\omega \sim \{j0, \ldots, j(m-1), 0, \ldots, m-1\}$. Then the conclusion of (4) still holds.

In fact, let $p0, \ldots, p(m-1)$ be distinct integers in $\omega \sim \{j0, \ldots, j(m-1), 0, \ldots, m-1, k0, \ldots, k(m-1), l0, \ldots, l(m-1)\}$. Then, using (4) twice,

$$\begin{aligned}
\mathbf{s}^{k0}_{j0} \cdots \mathbf{s}^{k(m-1)}_{j(m-1)} \mathbf{s}^{0}_{k0} \cdots \mathbf{s}^{m-1}_{k(m-1)} f_{\mathbf{R}} &= \mathbf{s}^{p0}_{j0} \cdots \mathbf{s}^{p(m-1)}_{j(m-1)} \mathbf{s}^{0}_{p0} \cdots \mathbf{s}^{m-1}_{p(m-1)} f_{\mathbf{R}} \\
&= \mathbf{s}^{l0}_{j0} \cdots \mathbf{s}^{l(m-1)}_{j(m-1)} \mathbf{s}^{0}_{l0} \cdots \mathbf{s}^{m-1}_{l(m-1)} f_{\mathbf{R}}.
\end{aligned}$$

Now we return to checking that $h\varphi = 1$ (see above prior to (3)). Choose $j0, \ldots, j(m-1) \in \omega \sim \{i0, \ldots, i(t-1), i(t+1), \ldots, i(m-1), 0, \ldots, m-1, \kappa, \lambda\}$, and set

$$x = \mathbf{s}_{i0}^{j0} \cdots \mathbf{s}_{i(t-1)}^{j(t-1)} \mathbf{s}_{i(t+1)}^{j(t+1)} \cdots \mathbf{s}_{i(m-1)}^{j(m-1)} \mathbf{s}_{j0}^{0} \cdots \mathbf{s}_{j(m-1)}^{m-1} f_{\mathbf{R}}.$$

Then by (5) and 12.6(x), $h\psi = \mathbf{s}_\kappa^{jt} x$ and $h\chi = \mathbf{s}_\lambda^{jt} x$. Now

$$\mathrm{d}_{\kappa\lambda} \cdot \mathbf{s}_\kappa^{jt} x \leq \mathbf{s}_\lambda^{jt} x$$

by 12.6(v), so $h\varphi = -\mathrm{d}_{\kappa\lambda} + -\mathbf{s}_\kappa^{jt} x + \mathbf{s}_\lambda^{jt} x = 1$, as desired.

We have now shown that $h\varphi = 1$ for each logical axiom φ, i.e., each logical axiom is in the set Γ defined following (1). Obviously Γ is closed under detachment and generalization. Thus each logical theorem is in Γ, and (1) follows.

Now if $\vdash \varphi \leftrightarrow \psi$, then $h(\varphi \leftrightarrow \psi) = 1$, and hence it follows easily that $h\varphi = h\psi$. Hence (since $\vdash \, = \, \models$) there is a function g mapping $M_0^{\mathscr{L}}$ into A such that $g[\varphi] = h\varphi$ for any $\varphi \in \mathrm{Fmla}_{\mathscr{L}}$. It is easily checked that g is the desired homomorphism. □

The following kind of logical representation theorem is an obvious consequence of 12.18, 12.25, and 12.28.

Theorem 12.29. *If $\mathfrak{A}$ is a locally finite dimensional* CA_ω, *then* $\mathfrak{A} \cong \mathfrak{M}_\Gamma^{\mathscr{L}}$ *for some* $\mathscr{L}$, Γ.

As an easy consequence of 12.29 we obtain the following analog of the Boolean representation theorem.

Theorem 12.30. *If $\mathfrak{A}$ is a locally finite dimensional* CA_ω, $|A| > 1$, *then there is a homomorphism of $\mathfrak{A}$ onto a cylindric set algebra.*

PROOF. By 12.29 we may assume that $\mathfrak{A} = \mathfrak{M}_\Gamma^{\mathscr{L}}$ for some $\mathscr{L}$, Γ. Then Γ is consistent, since $|A| > 1$. Let $\mathscr{L}$ be a model of Γ. The desired conclusion now follows from 12.24.

BIBLIOGRAPHY

1. Halmos, P. R. *Algebraic Logic*. New York: Chelsea (1962).
2. Henkin, L., and Tarski, A. Cylindric algebras. In *Proc. Symp. Pure Math. (AMS)*, *2* (1961), 83–113.
3. Henkin, L., Monk, J. D., and Tarski, A. *Cylindric Algebras*, Part I. Amsterdam: North-Holland (1971).

EXERCISES

12.31. Prove $\mathrm{c}_\kappa - \mathrm{c}_\lambda x \cdot \mathrm{c}_\lambda - \mathrm{c}_\kappa - x = 0$.

12.32. Prove $\mathrm{c}_0 \mathrm{c}_1 - \mathrm{d}_{01} = \mathrm{c}_0 - \mathrm{d}_{01}$ (even if $\alpha = 2$).

12.33.* Prove $c_0c_1(-d_{01} \cdot -d_{02} \cdot -d_{12}) = c_0c_1c_2(-d_{01} \cdot -d_{02} \cdot -d_{12})$ (even if $\alpha = 3$).

12.34. If $\alpha > 1$ and c_0 is the identity mapping, then each c_κ is the identity mapping, and $d_{\kappa\lambda} = 1$ for all $\kappa, \lambda < \alpha$.

12.35. Every finitely generated CA_1 is finite.

12.36. For any $\alpha \geq 2$ there is an infinite CA_α generated by one element. *Hint:* consider the cylindric set algebra of subsets of ${}^\alpha\omega$ generated by $\{x \in {}^\alpha\omega : x_0 < x_1\}$.

12.37. If $\mathfrak{A}$ is a locally finite dimensional CA_ω and $0 \neq x \in A$, then there is a homomorphism f of $\mathfrak{A}$ onto a cylindric set algebra such that $fx \neq 0$.

12.38. If $\mathfrak{A}$ is a CA_α, $\alpha < \omega$, and $|A| > 1$, then $\mathfrak{A} \times \mathfrak{A}$ (understood in the natural sense) is not isomorphic to a cylindric set algebra. *Hint:* in a cylindric set algebra $\mathfrak{B}$ of dimension α, if $x \neq 0$ then $c_0 \cdots c_{\alpha-1}x = 1$.

PART III

Decidable and Undecidable Theories

The main topic of this part is the application of the definitions and results of recursive function theory to logic. The question that is central to our endeavors here is to determine for various particular mathematically interesting theories Γ whether or not there is an effective procedure to determine of any sentence φ the truth or falsity of $\Gamma \models \varphi$.

Some Decidable Theories 13

First we give the basic definitions with which we shall be working in this part.

Definition 13.1. Let Γ be a theory in an effectivized first-order language $\mathscr{L} = (L, v, \mathcal{O}, \mathscr{R}, g)$. We say that Γ is *decidable* if $g^{+*}\Gamma$ is recursive, and *undecidable* if $g^{+*}\Gamma$ is not recursive.

In this chapter we give a few examples of decidable theories. The methods for proving theories decidable are numerous. Some of the easiest methods are model-theoretic, so we shall give more examples of decidable theories in Part IV. For proving theories decidable, the extensive mechanism of recursive function theory is not really needed. Almost all of the work can be done on the intuitive level of recognizing that certain procedures are effective. Everything is made rigorous by applying the weak Church's thesis (see p. 46).

Many of the theories which have been proved to be decidable are rather simple. The table below may help the reader to get an idea of the complexity decidable theories can have, especially when compared with our list of undecidable theories on p. 279. We also list in each case a convenient method of proof for the decidability of the theory.

We shall give a detailed treatment in this chapter for the theories 1, 5, 7 in this table. The method we will use is that of *elimination of quantifiers*. This method can be described in rough terms as follows. In our given language we single out effectively certain formulas as *basic formulas*. These will usually *not* be quantifier free. Then we show (eliminating quantifiers) that any formula is effectively equivalent within our given theory to a sentential combination of basic formulas, i.e., a combination using only $\vee$, $\wedge$, $\neg$. Finally, we give an effective procedure for determining whether or not such a combination is

Some decidable theories

Theory of	Proved by	A method of proof	In this book
1. Equality (no nonlogical constants, no axioms)	Löwenheim 1915	Elimination of quantifiers	p. 241
2. Finitely many sets (m unary relation symbols, no axioms)		Elimination of quantifiers	p. 243
3. One equivalence relation	Janiczak 1953	m-elementary equivalence	p. 354
4. One unary function	Ehrenfeucht 1959	Tree automata	—
5. $(\omega, \mathbf{s})$	—	Elimination of quantifiers	p. 236
6. Two successor functions*	Rabin 1968	Tree automata	—
7. $(\mathbb{Z}, +)$	Presburger 1929	Elimination of quantifiers	p. 240
8. Simple ordering	Ehrenfeucht 1959	Tree automata	—
9. $(\mathbf{S}I, \cup, \cap, \sim, 0, I)$	Skolem 1917		—
10. Boolean algebras	Tarski 1949	Model completeness	—
11. Free groups	Malcev 1961		—
12. Absolutely free algebras	Malcev 1961		—
13. Abelian groups	Szmielew 1949	Model completeness	—
14. Ordered abelian groups	Gurevich 1964		—
15. Algebraically closed fields	Tarski 1949	Vaught's test	p. 351
16. Real-closed fields	Tarski 1949	Model completeness	p. 362
17. p-adic fields	Ax, Kochen; Ershov 1965		—
18. Euclidean geometry	Tarski 1949	Reduction to 16	—
19. Hyperbolic geometry	Schwabhäuser 1959	Reduction to 16	—

* This is the theory of $\mathfrak{A} = (A, \mathfrak{s}_0, \mathfrak{s}_1)$, where $A = \bigcup_{m\in\omega} {}^m2$, $\mathfrak{s}_0 w = w\langle 0\rangle$, $\mathfrak{s}_1 w = w\langle 1\rangle$, for all $w \in A$.

a consequence of the theory. This method yields much more information than just the decidability of the theory, as we shall see.

The Theory of $(\omega, \mathfrak{s})$

For technical reasons, instead of this theory we first consider the theory of $(\omega, \mathfrak{s}, 0)$. First we need some notions from sentential logic.

Definition 13.2. Let $\Gamma \subseteq \text{Fmla}_{\mathscr{L}}$ (where $\mathscr{L}$ is an arbitrary first-order language). The set $\text{Qf}\,\Gamma$ of *quantifier-free combinations* of members of Γ is the intersection of all sets $\Delta \subseteq \mathrm{Fmla}_{\mathscr{L}}$ such that $\Gamma \subseteq \Delta$ and Δ is closed under $\vee$, $\wedge$, and $\neg$.

Theorem 13.3 (Disjunctive normal form theorem). *Let $\Gamma \subseteq \mathrm{Fmla}_{\mathscr{L}}$. Then for any $\varphi \in \mathrm{Qf}\Gamma$ such that $\neg\varphi$ is not a tautology there exist $p, m \in \omega$ and a function ψ such that:*

(*i*) *the domain of ψ is $(p + 1) \times (m + 1)$;*
(*ii*) *for each $i \leq p$ and $j \leq m$, $\psi_{ij} \in \Gamma$ or $\psi_{ij} = \neg\chi$ with $\chi \in \Gamma$;*
(*iii*) $\models \varphi \leftrightarrow \bigvee_{i \leq p} \bigwedge_{j \leq m} \psi_{ij}$.

For the proof, see 8.38.

We now turn to the decidability proof for the theory of $(\omega, \mathscr{s}, 0)$. We work in an effectivized language with a unary operation symbol $\mathbf{s}$ and an individual constant $\mathbf{O}$. By induction we set

$$\Delta 0 = \mathbf{O}; \qquad \Delta(m + 1) = \mathbf{s}\Delta m; \qquad \mathbf{s}^1 = \mathbf{s}; \qquad \mathbf{s}^{m+1} = \mathbf{s}\mathbf{s}^m; \qquad \mathbf{m} = \Delta m.$$

The terms of this language are just of two kinds: $\mathbf{s}^m\alpha$ for some variable α, and $\mathbf{m}$ for some $m \in \omega$, where $\mathbf{s}^0\alpha$ is just α. A formula will be called *basic* if it has one of the following forms:

$$\mathbf{s}^m v_i = \sigma, \ \sigma \text{ a term not involving } v_i;$$
$$\mathbf{O} = \mathbf{O}.$$

Clearly there is an effective method for recognizing when a formula is basic. Let Γ_1 be the set of all sentences which hold in $(\omega, \mathscr{s}, 0)$. Obviously Γ_1 is a complete and consistent theory. Two formulas φ and ψ are *equivalent* (*under* Γ_1) provided that $\varphi \leftrightarrow \psi$ holds in $(\omega, \mathscr{s}, 0)$.

Lemma 13.4. *For any formula φ one can effectively find a formula ψ equivalent under Γ_1 to φ such that ψ is a quantifier-free combination of basic formulas and* $\mathrm{Fv}\psi \subseteq \mathrm{Fv}\varphi$.

PROOF. We proceed by induction on φ. First suppose φ is atomic; thus φ has the form $\sigma = \tau$. If no variable occurs in φ, then φ has the form $\mathbf{m} = \mathbf{n}$. This is equivalent to $\mathbf{O} = \mathbf{O}$ or to $\neg(\mathbf{O} = \mathbf{O})$ according as $m = n$ or $m \neq n$. Suppose a variable occurs in φ. If φ has the form $\mathbf{s}^m\alpha = \mathbf{s}^n\alpha$ for some variable α, then φ is equivalent to $\mathbf{O} = \mathbf{O}$ or to $\neg(\mathbf{O} = \mathbf{O})$ according as $m = n$ or $m \neq n$. The only forms left are $\mathbf{s}^m\alpha = \sigma$ or $\sigma = \mathbf{s}^m\alpha$ (for some variable α), where σ does not involve α. Both are equivalent to $\mathbf{s}^m\alpha = \sigma$. This takes care of the atomic case.

The induction steps using $\neg$, $\vee$, $\wedge$ are trivial. To make the induction step using $\forall\alpha$ it suffices to show that if φ is a quantifier free combination of basic formulas then $\exists\alpha\varphi$ is equivalent to a quantifier free combination of basic formulas determined effectively from $\exists\alpha\varphi$. Since

$$\exists\alpha(\psi \vee \chi) \leftrightarrow \exists\alpha\psi \vee \exists\alpha\chi$$

is logically valid, we may by 13.3 assume that φ is a conjunction of basic formulas and their negations. Now in general if α does not occur in ψ then

$$\exists\alpha(\psi \wedge \chi) \leftrightarrow \psi \wedge \exists\alpha\chi$$

is logically valid. Hence we may assume that each conjunct of φ actually involves α. Thus we may assume that $\exists\alpha\varphi$ is the formula

$$\exists\alpha[\mathbf{s}^{k0}\alpha = \sigma_0 \wedge \cdots \wedge \mathbf{s}^{k(m-1)}\alpha = \sigma_{m-1} \\ \wedge \neg(\mathbf{s}^{km}\alpha = \sigma_m) \wedge \cdots \wedge \neg(\mathbf{s}^{k(n-1)} = \sigma_{n-1})]$$

where $0 \leq m \leq n > 0$, and $\sigma_0, \ldots, \sigma_{n-1}$ do not involve α. Noting that $\sigma = \tau$ is equivalent under Γ_1 to $\mathbf{s}\sigma = \mathbf{s}\tau$, if we let l be the maximum of $k_0, \ldots, k_{n-1}$ we easily see that $\exists\alpha\varphi$ is equivalent to a formula of the form

$$\exists\alpha[\mathbf{s}^l\alpha = \tau_0 \wedge \cdots \wedge \mathbf{s}^l\alpha \\ = \tau_{m-1} \wedge \neg(\mathbf{s}^l\alpha = \tau_m) \wedge \cdots \wedge \neg(\mathbf{s}^l\alpha = \tau_{n-1})]$$

where $\tau_0, \ldots, \tau_{n-1}$ do not involve α. In turn, this formula is obviously equivalent to

$$\exists\alpha[\neg(\alpha = 0) \wedge \cdots \wedge \neg(\alpha = \Delta(l-1)) \wedge \alpha \\ = \tau_0 \wedge \cdots \wedge \alpha = \tau_{m-1} \wedge \neg(\alpha = \tau_m) \wedge \cdots \wedge \neg(\alpha = \tau_{n-1})].$$

Thus $\exists\alpha\varphi$ is equivalent to a formula of the form

$$\exists\alpha[\alpha = \rho_0 \wedge \cdots \wedge \alpha = \rho_{m-1} \wedge \neg(\alpha = \rho_m) \wedge \cdots \wedge \neg(\alpha = \rho_p)]$$

where $\rho_0, \ldots, \rho_p$ do not involve α. Now if $m = 0$, then $\exists\alpha\varphi$ is obviously equivalent to $\mathbf{O} = \mathbf{O}$. If $m \neq 0$, then $\exists v_i\varphi$ is clearly equivalent to

$$\bigwedge_{j<k<m} (\rho_j = \rho_k) \wedge \neg(\rho_0 = \rho_m) \wedge \cdots \wedge \neg(\rho_0 = \rho_p),$$

as desired. □

Theorem 13.5. *Γ_1 is decidable.*

PROOF. Let φ be any sentence in our language. Let ψ be found from φ by 13.4: ψ is a quantifier-free combination of basic formulas and $\mathrm{Fv}\psi \subseteq \mathrm{Fv}\varphi$, therefore ψ is a sentence. Now by 13.3, whose proof is obviously effective, there exist m and n and sentences χ_{ij} such that

$$\vdash \psi \leftrightarrow \bigvee_{i<m} \bigwedge_{j<n} \chi_{ij}$$

where each χ_{ij} is a basic sentence or the negation of one. But the only basic *sentence* is $\mathbf{O} = \mathbf{O}$, which is obviously a member of Γ_1. Therefore we have

$$\begin{aligned} \varphi \in \Gamma_1 \quad & \text{iff } \psi \in \Gamma_1 \\ & \text{iff } \exists i < m \forall j < n (\chi_{ij} \text{ is } \mathbf{O} = \mathbf{O}). \end{aligned}$$ □

Corollary 13.6. *The theory of $(\omega, \mathfrak{s})$ is decidable.*

PROOF. There is an effective method for recognizing when a sentence does not involve $\mathbf{O}$. Such a sentence holds in $(\omega, \mathfrak{s})$ iff it holds in $(\omega, \mathfrak{s}, 0)$. □

It is clearly possible to choose our original language $\mathscr{L}_1$ to be elementarily effective (see p. 190). By examining the above proofs it is clear that in 13.5 and 13.6 the decision method then is elementary, i.e., the set $\mathscr{g}^{+*}\Gamma$ is elementary for the Γ's of 13.5 and 13.6. Similar remarks apply to our other two decidability results of this chapter.

The proof of 13.4 gives the following important corollary:

Corollary 13.7. *A set $A \subseteq \omega$ is elementarily definable in $(\omega, \mathfrak{s})$ or in $(\omega, \mathfrak{s}, 0)$, iff it is finite or cofinite.*

Proof. First we treat $(\omega, \mathfrak{s}, 0)$. $\Rightarrow$. Suppose φ elementarily defines A. Thus $\mathrm{Fv}\varphi \subseteq \{v_0\}$ and ${}^1\varphi^{\mathfrak{A}} = A$, where $\mathfrak{A} = (\omega, \mathfrak{s}, 0)$. By 13.4 let ψ be a quantifier-free combination of basic formulas equivalent under Γ_1 to φ with $\mathrm{Fv}\psi \subseteq \{v_0\}$. Thus ψ also elementarily defines A. Now ψ is built up from formulas $\mathbf{O} \mathbf{=} \mathbf{O}$, $\mathbf{s}^m v_0 = \mathbf{n}$ using $\neg$, $\wedge$, $\vee$. Note that $\mathbf{s}^m v_0 = \mathbf{n}$ elementarily defines 0 if $m > n$, and $\{n - m\}$ if $m \leq n$. Thus A is built up from sets ω, 0, $\{p\}$ using $\sim$, $\cap$, $\cup$. Hence A is finite or cofinite. The converse is trivial. □

To treat the structure $(\omega, \mathfrak{s})$ it is enough to note that if φ elementarily defines A in $(\omega, \mathfrak{s})$, then it also does in $(\omega, \mathfrak{s}, 0)$ and hence A is finite or cofinite. Conversely, any finite or cofinite subset of ω is clearly definable in $(\omega, \mathfrak{s})$.

No really nice characterizations of the elementarily definable n-ary relations are known to the author for $n > 1$.

The Theory of $(\mathbb{Z}, +)$

We shall instead consider the theory of $\mathfrak{A} = (\mathbb{Z}, +, <, 0, 1, -)$; this will clearly give the desired result for $(\mathbb{Z}, +)$ as in the proof of 13.6. We work now in a language with binary operation symbols $\mathbf{+}$ and $\mathbf{-}$, individual constants $\mathbf{O}$ and $\mathbf{1}$, and a binary relation symbol $\mathbf{<}$. Γ_2 is the set of all sentences of this language holding in $\mathfrak{A}$. We shall write $\sigma \mathbf{+} \tau$ instead of $\langle \mathbf{+} \rangle \sigma\tau$. The expressions $\sigma \mathbf{-} \tau$ and $\sigma \mathbf{<} \tau$ are to be similarly understood. Now we let

$$\begin{aligned}
\Delta 0 &= \mathbf{O}, \qquad \Delta(m+1) = \Delta m \mathbf{+} \mathbf{1} \qquad \text{for } m \in \omega,\\
\Delta m &= \mathbf{O} \mathbf{-} \Delta(-m) \qquad \text{for } m \in \mathbb{Z}, \quad m < 0,\\
\mathbf{m} &= \Delta m \qquad \text{for } m \in \mathbb{Z},\\
0\sigma &= \mathbf{O}, \qquad (m+1)\sigma = m\sigma \mathbf{+} \sigma \qquad \text{for } m \in \omega,\\
m\sigma &= \mathbf{O} \mathbf{-} (-m)\sigma \qquad \text{for } m < 0.
\end{aligned}$$

For $m > 1$ and σ, τ terms we let $\sigma \equiv_m \tau$ be the formula

$$\exists \alpha(\sigma \mathbf{-} \tau \mathbf{=} m\alpha)$$

where α is the first variable (in the natural order $v_0, v_1, \ldots$) not appearing in σ or τ.

A formula φ is *basic* if it has one of the following three forms:

$$\sigma = \tau$$
$$\sigma < \tau$$
$$\sigma \equiv_m \tau.$$

Note that the third kind of basic formulas involves quantifiers. As indicated in the introduction to this section, this is typical of the elimination of quantifiers method when it is applied to theories of any complexity at all.

Lemma 13.8. *For any formula φ one can effectively find a formula ψ equivalent under Γ_2 to φ such that ψ is a quantifier-free combination of basic formulas and* $\mathrm{Fv}\psi \subseteq \mathrm{Fv}\varphi$.

PROOF. The assertion is obvious for atomic formulas, and the induction steps involving $\neg$, $\vee$, $\wedge$ are trivial. As in the proof of 13.4 it is hence sufficient to prove the lemma for φ of the form $\exists\alpha\psi$, ψ a conjunction of basic formulas and their negations. Now the following formulas hold in $\mathfrak{A}$:

$$\neg(\sigma = \tau) \leftrightarrow \sigma < \tau \vee \tau < \sigma$$
$$\neg(\sigma < \tau) \leftrightarrow \tau = \sigma \vee \tau < \sigma$$
$$\neg(\sigma \equiv_m \tau) \leftrightarrow \sigma + 1 \equiv_m \tau \vee \cdots \vee \sigma + \Delta(m - 1) \equiv_m \tau.$$

Therefore we may actually assume that ψ is a conjunction of basic formulas. As in the proof of 13.4, we may assume that each conjunct actually involves α. Now if σ is a term involving α, then there is a term τ not involving α and an $m \in \mathbb{Z}$ such that

$$\sigma = m\alpha + \tau$$

holds in $\mathfrak{A}$. It follows that formulas of the forms

$$\sigma = \tau$$
$$\sigma < \tau$$
$$\sigma \equiv_m \tau$$

are respectively equivalent to formulas of the forms

$$n\alpha = \rho$$
$$n\alpha < \rho \qquad \text{or } \rho < n\alpha$$
$$n\alpha \equiv_m \rho$$

where $n \geq 0$ and ρ does not involve α; and we may clearly assume $n > 0$. Hence we may assume that φ is equivalent to a formula of the form

$$(1)\quad \begin{aligned}&\exists\alpha[n_0\alpha = \rho_0 \wedge \cdots \wedge n_{i-1}\alpha = \rho_{i-1} \wedge n_i\alpha < \rho_i \wedge \cdots \wedge n_{j-1}\alpha \\ &< \rho_{j-1} \wedge \rho_j < n_j\alpha \wedge \cdots \wedge \rho_{k-1} < n_{k-1}\alpha \wedge n_k\alpha \equiv_{mk} \rho_k \wedge \cdots \\ &\wedge n_{l-1}\alpha \equiv_{m(l-1)} \rho_{l-1}]\end{aligned}$$

where $0 \leq i \leq j \leq k \leq l > 0$, $n_0, \ldots, n_{l-1} > 0$, $m_k, \ldots, m_{l-1} > 1$, and ρ_0,

$\ldots, \rho_{l-1}$ do not involve α. Now for $p > 0$ the following pairs of formulas are equivalent:

$$\begin{array}{ll} \sigma = \tau & \text{and } p\sigma = p\tau; \\ \sigma < \tau & \text{and } p\sigma < p\tau; \\ \sigma \equiv_m \tau & \text{and } p\sigma \equiv_{mp} p\tau. \end{array}$$

It follows that in (1) we may assume that $n_0 = \cdots = n_{l-1} > 1$. Thus (1) becomes of the form

$$\exists\alpha(p\alpha = \xi_0 \wedge \cdots \wedge p\alpha = \xi_{i-1} \wedge p\alpha < \xi_i \wedge \cdots \wedge p\alpha < \xi_{j-1} \wedge \xi_j < p\alpha \wedge \cdots \wedge \xi_{k-1} < p\alpha \wedge p\alpha \equiv_{qk} \xi_k \wedge \cdots \wedge p\alpha \equiv_{q(l-1)} \xi_{l-1})$$

where $0 \leq i \leq j \leq k \leq l > 0$, $p > 1$, $q_0, \ldots, q_{l-1} > 1$, and $\xi_0, \ldots, \xi_{l-1}$ do not involve α. This in turn is clearly equivalent to

$$\exists\alpha(\alpha = \xi_0 \wedge \cdots \wedge \alpha = \xi_{i-1} \wedge \alpha < \xi_i \wedge \cdots \wedge \alpha < \xi_{j-1} \wedge \xi_j < \alpha \wedge \cdots \wedge \xi_{k-1} < \alpha \wedge \alpha \equiv_p 0 \wedge \alpha \equiv_{qk} \xi_{mk} \wedge \cdots \wedge \alpha \equiv_{q(l-1)} \xi_{l-1}). \tag{2}$$

In case $i > 0$, (2) is equivalent to

$$\xi_0 = \xi_1 \wedge \cdots \wedge \xi_0 = \xi_{i-1} \wedge \xi_0 < \xi_i \wedge \cdots \wedge \xi_0 < \xi_{j-1} \wedge \xi_j < \xi_0 \wedge \cdots \wedge \xi_{k-1} < \xi_0 \wedge \xi_0 \equiv_p 0 \wedge \xi_0 \equiv_{qk} \xi_k \wedge \cdots \wedge \xi_0 \equiv_{q(l-1)} \xi_{l-1}.$$

Thus we may assume that $i = 0$; so (2) is equivalent to a formula of the form

$$\exists\alpha(\alpha < \eta_0 \wedge \cdots \wedge \alpha < \eta_{s-1} \wedge \eta_s < \alpha \wedge \cdots \wedge \eta_{t-1} < \alpha \wedge \alpha \equiv_{rt} \eta_t \wedge \cdots \wedge, \alpha \equiv_{r(u-1)} \eta_{u-1}) \tag{3}$$

where $0 \leq s \leq t < u$, $r_t, \ldots, r_{u-1} > 1$, and $\eta_0, \ldots, \eta_{u-1}$ do not involve α. Next, we claim that we may assume that $u = t + 1$. This clearly follows from the following number-theoretic fact:

(4) Let a, b, m, n be integers with $m, n > 1$. Let $p = \text{lcm}(m, n)$; then $\gcd(p/m, p/n) = 1$, so there exist integers c, d such that $c(p/m) + d(p/n) = 1$. It follows that for any integer y the following two conditions are equivalent:
(*i*) $y \equiv a \pmod m$ and $y \equiv b \pmod n$;
(*ii*) $a \equiv b \pmod{\gcd(m, n)}$ and $y \equiv c(p/m)a + d(p/n)b \pmod p$.

To prove (4), assume its hypothesis. Suppose (*i*) holds. Then $y - a = em$ and $y - b = fn$ for some e, f, so $a - b = fn - em$, which is divisible by $\gcd(m, n)$. Thus $a \equiv b \pmod{\gcd(m, n)}$. Also,

$$\begin{aligned} y - c(p/m)a - d(p/n)b &= c(p/m)y + d(p/n)y - c(p/m)a - d(p/n)b \\ &= c(p/m)(y - a) + d(p/n)(y - b) \\ &= c(p/m)em + d(p/n)fn \\ &\equiv 0 \pmod p. \end{aligned}$$

Thus (*ii*) holds. Conversely, suppose that (*ii*) holds. Write $a - b = v \gcd(m, n)$. Then

$$\begin{aligned} y - a &= y - c(p/m)a - d(p/n)a \\ &= y - c(p/m)a - d(p/n)b - d(p/n)\, v \gcd(m, n) \\ &\equiv 0 \pmod m \end{aligned}$$

since $y - c(p/m)a - d(p/n)b \equiv 0 \pmod{p}$ and $d(p/n)\gcd(m, n) \equiv 0 \pmod{m}$. Similarly, $y \equiv b \pmod{n}$. Hence (4) has been checked, and hence in (3) we may assume that $u = t + 1$.

If $s = 0$ or $t = s$, it is clear that (3) is equivalent to $\mathbf{O} = \mathbf{O}$. So, assume that $0 < s < t$. Then (3) is equivalent to the following formula:

$$\bigvee_{0 \leq i < s,\, s \leq j < t} \ \bigwedge_{0 \leq c < s,\, s \leq d < t} \Big[\eta_d < \eta_j + \mathbf{1} \wedge \eta_i < \eta_c + \mathbf{1} \wedge \bigwedge_{e < rt} (\eta_j + \mathbf{e} + \mathbf{1} < \eta_i \wedge \eta_j + \mathbf{e} + \mathbf{1} \equiv_{rt} \eta_i)\Big] \quad \square$$

Theorem 13.9. *Γ_2 is decidable.*

Proof. By the method of proof of 13.5 we see that it suffices to describe a method for determining the truth in $\mathfrak{A}$ of basic *sentences*. The basic sentences are easily seen to be effectively equivalent to sentences of the forms

$$\mathbf{m} = \mathbf{n},$$
$$\mathbf{m} < \mathbf{n},$$
$$\mathbf{m} \equiv_n \mathbf{p},$$

which are true in $\mathfrak{A}$ iff, respectively, $m = n$, $m < n$, or $m \equiv_n p$. Obviously there is a decision method for determining these latter three questions.

Theorem 13.10. *The theory of $(\mathbb{Z}, +)$ is decidable.*

The Pure Theory of Equality

Our language here has no nonlogical constants. Γ_3 consists of all sentences φ such that $\vdash \varphi$. Thus in this case, unlike the preceding two cases, the theory we investigate is not complete. For example, $\forall v_0 \forall v_1 (v_0 = v_1)$ holds in one-element structures but not in any others. The general procedure is still the same, however. First we distinguish some basic formulas, which again are not special atomic formulas but some of which are rather complicated sentences. We let

$$\varepsilon_1 = \exists v_0 (v_0 = v_0)$$

and, for $m > 1$,

$$\varepsilon_m = \exists v_0 \cdots \exists v_{m-1} \bigwedge_{i < j < m} \neg (v_i = v_j).$$

By a *basic formula* we mean a formula $v_i = v_j$ or a formula ε_m.

Note that the $\mathscr{L}$-structures here are just sets. The basic lemma, as usual in this chapter, is:

Lemma 13.11. *For any formula φ one can effectively find a formula ψ equivalent under Γ_3 to φ such that ψ is a quantifier-free combination of basic formulas and* $\mathrm{Fv}\psi \subseteq \mathrm{Fv}\varphi$.

PROOF. The atomic case is trivial, and the induction steps using $\neg$, $\vee$, $\wedge$ are trivial. As in previous proofs it now suffices to assume that φ has the form $\exists\alpha\psi$, ψ a quantifier-free combination of basic formulas. Since the formulas ε_m are sentences, and since $\alpha = \alpha$ is equivalent to ε_1, we may assume that φ has the form

$$\exists\alpha[\alpha \equiv \beta_0 \wedge \cdots \wedge \alpha \equiv \beta_{m-1} \wedge \neg(\alpha \equiv \beta_m) \wedge \cdots \wedge \neg(\alpha \equiv \beta_{n-1})]$$

where $\beta_0, \ldots, \beta_{n-1}$ are distinct variables $\neq \alpha$. If $m > 0$, φ is clearly equivalent to

$$\bigwedge_{i<j<m} \beta_i \equiv \beta_j \wedge \neg(\beta_0 \equiv \beta_m) \wedge \cdots \wedge \neg(\beta_0 \equiv \beta_{n-1})$$

(or to ε_1 if $m = 1$ and $n = m$). Hence assume that $m = 0$. Let $I = \{i : m \leq i < n\}$. We claim

(1) $$\models \varphi \leftrightarrow \bigwedge_{J \subseteq I} (\bigwedge_{i,\, j \in J,\, i \neq j} \neg (\beta_i \equiv \beta_j) \rightarrow \varepsilon_{|J|+1}).$$

To prove (1), first let A be any set and let $x \in {}^{\omega}A$. Suppose x satisfies φ in $\mathfrak{A}$, $J \subseteq I$, and x satisfies $\bigwedge_{i,j \in J, i \neq j} \neg (\beta_i \equiv \beta_j)$ in A. Say $\beta_i = vk_i$ for all $i \in I$, and $\alpha = v_l$. Thus $xk_i \neq xk_j$ whenever $i, j \in J$ and $i \neq j$. Since x satisfies φ in $\mathfrak{A}$, choose $a \in A$ so that x^l_a satisfies $\bigwedge_{i \in J} \neg (\alpha \equiv \beta_i)$. Thus $a \neq xk_i$ for each $i \in I$, in particular for each $i \in J$, so $|A| \geq |J| + 1$. Thus $\varepsilon_{|J|+1}$ holds in A.

Conversely, suppose x satisfies the right side of (1) in A. Define $i \equiv j$ iff $i, j \in I$ and $xk_i = xk_j$. Clearly $\equiv$ is an equivalence relation on I. Let J be a subset of I which has exactly one element in common with each $\equiv$ equivalence class. If $i, j \in J$ and $i \neq j$, then $xk_i \neq xk_j$. Thus x satisfies $\bigwedge_{i,j \in J, i \neq j} \neg (\beta_i \equiv \beta_j)$ in A. Since x satisfies the right side of (1) in A, it follows that $\varepsilon_{|J|+1}$ holds in A, i.e., A has at least $|J| + 1$ elements. Hence we may choose

$$a \in A \sim \{xk_i : i \in J\}.$$

Clearly then x^l_a satisfies $\bigwedge_{i \in I} \neg (\alpha \equiv \beta_i)$, so x satisfies φ. Thus (1) holds, and this completes the proof of 13.11. □

Theorem 13.12. Γ_3 *is decidable.*

PROOF. The only basic sentences are ε_m, $m \in \omega \sim 1$. Clearly $\varepsilon_1 \in \Gamma_3$ while $\varepsilon_m \notin \Gamma_3$ if $m \neq 1$. Let Δ be the set of all sentential combinations of basic sentences. Note:

(1) $m \leq |A|$ iff ε_m holds in the set A.

For any $\varphi \in \Delta$ let m_φ be the maximum m such that ε_m is a part of φ. Using (1) it is easy to check that

(2) if $\varphi \in \Delta$ and $m_\varphi \leq |A|, |B|$, then φ holds in A iff φ holds in B.

Now by (2) we have

(3) if $\varphi \in \Delta$, then φ holds in every A with $m_\varphi \leq |A|$ iff φ holds in the set m_φ.

Thus

(4) if $\varphi \in \Delta$, then $\models \varphi$ iff φ holds in the set n for every nonzero $n \leq m_\varphi$.

Condition (4) clearly provides an effective procedure for determining whether $\varphi \in \Gamma_3$ for $\varphi \in \Delta$. Namely, for each $n \leq m_\varphi$, $n \neq 0$ one checks whether φ holds in n, this being essentially just a matter of checking tautologies by virtue of (1). □

Corollary 13.13. *If A is any nonempty set, then a subset B of A is elementarily definable iff $B = 0$ or $B = A$.*

As a further consequence of our decision method for Γ_3 we can find all complete extensions of Γ_3; i.e., all theories $\Delta \supseteq \Gamma_3$ which are complete. For each $m \in \omega \sim 1$ let Γ_3^m be the set of all sentences φ such that

$$\{\varepsilon_m, \neg\varepsilon_{m+1}\} \models \varphi$$

and let Γ_3^∞ be the set of all sentences φ such that

$$\{\varepsilon_m : m \in \omega \sim 1\} \models \varphi.$$

Theorem 13.14

(*i*) *The theories Γ_3^m and Γ_3^∞ are each complete, consistent, and decidable.*

(*ii*) *The theories Γ_3^m and Γ_3^∞ constitute all the complete and consistent extensions of Γ_3.*

PROOF

(*i*) By (1) in the proof of 13.12, $\varepsilon_m \wedge \neg\varepsilon_{m+1}$ holds in a set A iff it holds in the set m. Hence $\Gamma_3^m = \{\varphi : \varphi \text{ holds in } m\}$, so Γ_3^m is complete and consistent. Similarly, $\Gamma_3^\infty = \{\varphi : \varphi \text{ holds in } \omega\}$, so Γ_3^∞ is complete and consistent. Obviously then each theory Γ_3^m is decidable. Γ_3^∞ is easily seen to be decidable using (1) in the proof of 13.12, and 13.11.

(*ii*) Obviously Γ_3 is a subset of Γ_3^∞ and of each set Γ_3^m. Now suppose that Δ is any complete and consistent extension of Γ_3. By the completeness theorem, let A be a model for Δ. If $|A| = m < \omega$, clearly then $\Gamma_3^m = \Delta$. If $|A| \geq \omega$, by (2) in the proof of 13.12 we easily see that $\Gamma_3^\infty \subseteq \Delta$, so $\Gamma_3^\infty = \Delta$. □

BIBLIOGRAPHY

1. Ershov, Yu., Lavrov, I., Taimanov, A., Taitslin, M. Elementary theories. *Russian Mathematical Surveys, 20* (1965), 35–105.

2. Rabin, M. Decidability of second-order theories and automata on infinite trees. *Trans. Amer. Math. Soc., 141* (1969), 1–35.

3. Tarski, A., Mostowski, A., Robinson, R. M. *Undecidable Theories.* Amsterdam: North-Holland (1953).

EXERCISES

13.15. Determine all elementarily definable subsets of $\mathbb{Z}$ with respect to the structure $(\mathbb{Z}, +, <)$.

13.16. Show that the theory of $(\omega, +)$ is decidable.

13.17. Show that the theory of one set is decidable.

13.18. Using the method of elimination of quantifiers, show that the theory of $(\mathbb{Q}, <)$ is decidable.

13.19. Determine the elementarily definable binary relations in any infinite set (i.e., using the language with no nonlogical symbols and the theory Γ_3^∞).

14 Implicit Definability in Number Theories

In this chapter we consider several ways in which number-theoretic functions and relations can be implicitly defined in number theories. We do not mean elementarily definable as in Chapter 11; the present notions of definability are expressed in terms of theories and not of structures. As we shall see, the notions lead to new equivalents of the notion of recursiveness; see 14.12, 14.20, and 14.26. They also form the basis for diagonalization procedures which produce many undecidable theories (see the next chapter). We shall be concerned with two types of implicit definability. The first, syntactic definability, follows; the second, spectral representability, is given in 14.22.

Definition 14.1. $\mathscr{L}_{\mathrm{nos}}$ is a fixed effective first-order language with the following nonlogical constants:

$\mathbf{O}$	an individual constant,
$\mathbf{s}$	a unary operation symbol,
$\mathbf{+}, \cdot$	binary operation symbols.

For each $n \in w$ we let $\Delta(n + 1) = \mathbf{s}\Delta n$ (inductively), with $\Delta 0 = \mathbf{O}$; frequently we shall write $\mathbf{n}$ in place of Δn. For terms σ, τ, the expression $\sigma \leq \tau$ is used as an abbreviation for the formula $\exists v_i(v_i \mathbf{+} \sigma \mathbf{=} \tau)$, where i is the least integer such that v_i does not occur in σ or in τ.

Now let Γ be a theory in $\mathscr{L}_{\mathrm{nos}}$. We say that an m-ary function $f\colon {}^m\omega \to \omega$ is *syntactically defined in* Γ *by* a formula φ iff $Fv\varphi \subseteq \{v_0, \ldots, v_m\}$ and for all $x_0, \ldots, x_{m-1} \in \omega$, the following sentences are in Γ, where $y = f(x_0, \ldots, x_{m-1})$:

$$\varphi(\mathbf{x}_0, \ldots, \mathbf{x}_{m-1}, \mathbf{y})$$
$$\forall v_m[\varphi(\mathbf{x}_0, \ldots, \mathbf{x}_{m-1}, v_m) \to v_m \mathbf{=} \mathbf{y}].$$

On the other hand, an m-ary relation $R \subseteq {}^m\omega$ is *syntactically defined in* Γ *by* φ iff $Fv\varphi \subseteq \{v_0, \ldots, v_{m-1}\}$ and for all $x_0, \ldots, x_{m-1} \in \omega$,

$$\langle x_0, \ldots, x_{m-1}\rangle \in R \Rightarrow \varphi(\mathbf{x}_0, \ldots, \mathbf{x}_{m-1}) \in \Gamma;$$
$$\langle x_0, \ldots, x_{m-1}\rangle \notin R \Rightarrow \neg\varphi(\mathbf{x}_0, \ldots, \mathbf{x}_{m-1}) \in \Gamma.$$

Finally, an m-ary relation $R \subseteq {}^m\omega$ is *weakly syntactically defined in* Γ *by* φ iff $Fv\varphi \subseteq \{v_0, \ldots, v_{m-1}\}$ and for all $x_0, \ldots, x_{m-1} \in \omega$,

$$\langle x_0, \ldots, x_{m-1}\rangle \in R \Leftrightarrow \varphi(\mathbf{x}_0, \ldots, \mathbf{x}_{m-1}) \in \Gamma.$$

We first want to indicate some relationships between these notions and the notion of recursiveness. Then we shall turn starting with 14.9 to specific theories Γ in which many functions and relations are syntactically definable.

Proposition 14.2. *Let $R \subseteq {}^m\omega$. Let Γ be a theory in $\mathscr{L}_{\text{nos}}$, and assume that $\neg(\mathbf{0} = \mathbf{1}) \in \Gamma$. Then the following conditions are equivalent:*

(i) R is syntactically definable in Γ;
(ii) χ_R is syntactically definable in Γ.

Proof. $(i) \Rightarrow (ii)$. Let φ syntactically define R in Γ. Let ψ be the formula

$$(\varphi \wedge v_m = \mathbf{1}) \vee (\neg\varphi \wedge v_m = \mathbf{0}).$$

Thus for any $x_0, \ldots, x_{m-1} \in \omega$,

$$\langle x_0, \ldots, x_{m-1}\rangle \in R \Rightarrow \varphi(\mathbf{x}_0, \ldots, \mathbf{x}_{m-1}) \in \Gamma$$
$$\Rightarrow \psi(\mathbf{x}_0, \ldots, \mathbf{x}_{m-1}, \mathbf{1}) \in \Gamma.$$

and similarly $\langle x_0, \ldots, x_{m-1}\rangle \notin R \Rightarrow \psi(\mathbf{x}_0, \ldots, \mathbf{x}_{m-1}, \mathbf{0}) \in \Gamma$. Thus $\psi(\mathbf{x}_0, \ldots, \mathbf{x}_{m-1}, \mathbf{y}) \in \Gamma$, where $y = \chi_R(x_0, \ldots, x_{m-1})$. On the other hand, if $\langle x_0, \ldots, x_{m-1}\rangle \in R$, then $\varphi(\mathbf{x}_0, \ldots, \mathbf{x}_{m-1}) \in \Gamma$ and hence by a suitable tautology

$$\Gamma \models \psi(\mathbf{x}_0, \ldots, \mathbf{x}_{m-1}, v_m) \rightarrow v_m = \mathbf{1}.$$

Similarly if $\langle x_0, \ldots, x_{m-1}\rangle \notin R$ then

$$\Gamma \models \psi(\mathbf{x}_0, \ldots, \mathbf{x}_{m-1}, v_m) \rightarrow v_m = \mathbf{0}.$$

Hence χ_R is syntactically defined by ψ.

$(ii) \Rightarrow (i)$. Let ψ syntactically define χ_R in Γ. Let φ be the formula $\psi(v_0, \ldots, v_{m-1}, \mathbf{1})$. Suppose that $\langle x_0, \ldots, x_{m-1}\rangle \in R$. Thus $\chi_R(x_0, \ldots, x_{m-1}) = 1$, so $\psi(\mathbf{x}_0, \ldots, \mathbf{x}_{m-1}, \mathbf{1}) \in \Gamma$, i.e., $\varphi(\mathbf{x}_0, \ldots, \mathbf{x}_{m-1}) \in \Gamma$. On the other hand, suppose that $\langle x_0, \ldots, x_{m-1}\rangle \notin R$. Thus $\chi_R(x_0, \ldots, x_{m-1}) = 0$, so by the second requirement on syntactic definability for functions,

$$\Gamma \models \psi(\mathbf{x}_0, \ldots, \mathbf{x}_{m-1}, \mathbf{1}) \rightarrow \mathbf{1} = \mathbf{0}.$$

Hence, since $\neg(\mathbf{0} = \mathbf{1}) \in \Gamma$, we have $\neg\psi(\mathbf{x}_0, \ldots, \mathbf{x}_{m-1}, 1) \in \Gamma$, i.e., $\neg\varphi(\mathbf{x}_0, \ldots, \mathbf{x}_{m-1}) \in \Gamma$. □

The following proposition is obvious:

Proposition 14.3. *Let $R \subseteq {}^n\omega$ and let Γ be a consistent theory in $\mathscr{L}_{\text{nos}}$. If R is syntactically definable in Γ then R is weakly syntactically definable in Γ.*

Recall from 11.29 the notion of a set of axioms for a theory. We effectivize this notion:

Definition 14.4. If $\mathscr{L}$ is an effectivized first-order language, a theory Γ in $\mathscr{L}$ is *recursively axiomatizable* if there is a set Δ of sentences of $\mathscr{L}$ with $\mathscr{g}^{+*}\Delta$ recursive such that Δ axiomatizes Γ.

If Γ is finitely axiomatizable, i.e., it is axiomatized by a finite set, then Γ is recursively axiomatizable. The connections of the notion of 14.4 with the notions of decidability and undecidability will be explored in the next chapter. For now we relate the notion to our notions of syntactic definability.

Theorem 14.5. *Let $f\colon {}^m\omega \to \omega$ and let Γ be a consistent recursively axiomatizable theory in $\mathscr{L}_{\text{nos}}$ such that $\neg(\mathbf{n} \equiv \mathbf{p}) \in \Gamma$ whenever $n \neq p$. If f is syntactically definable in Γ then f is recursive. A similar statement holds for relations $R \subseteq {}^m\omega$.*

PROOF. Say Γ is recursively axiomatized by Δ. Let

$\mathbf{T}_m^\Delta = \{(e, x_0, \ldots, x_{m-1}, z, y)$: e is the Gödel number of a formula φ with free variables among $v_0, \ldots, v_m$, and y is the Gödel number of a Δ-formal proof of $\varphi(\mathbf{x}_0, \ldots, \mathbf{x}_{m-1}, \mathbf{z})\}$.

Clearly $\mathbf{T}_m^\Delta$ is recursive. Now let f be syntactically defined in Γ by φ. Clearly then for any $x_0, \ldots, x_{m-1} \in \omega$ we have, with $e = \mathscr{g}^+\varphi$,

$$f(x_0, \ldots, x_{m-1}) = (\mu y((e, x_0, \ldots, x_{m-1}, (y)_0, (y)_1) \in \mathbf{T}_m^\Delta))_0.$$

Thus f is recursive. The proof for relations follows from 14.2. □

An intuitive account of 14.5 may be helpful. Suppose that f (unary for simplicity) is syntactically defined by φ in Γ, where Γ has a recursive set Δ of axioms. We compute fx as follows. Start generating all Δ-theorems, one after the other (see the remarks following 10.29). As the theorems are generated, we look for one of the form $\varphi(\mathbf{x}, \mathbf{y})$. As soon as one such appears, we can say that $y = fx$. That eventually a formula $\varphi(\mathbf{x}, \mathbf{y})$ will appear in our list of Γ-theorems is assured by the stipulation in 14.1 that $\varphi(\mathbf{x}, \mathbf{y}) \in \Gamma$; the uniqueness of y follows from the fact that $\forall v_1(\varphi(\mathbf{x}, v_1) \to v_1 \equiv \mathbf{z}) \in \Gamma$, where $z = fx$.

Theorem 14.6. *Let $R \subseteq {}^m\omega$ and let Γ be a recursively axiomatizable theory in $\mathscr{L}_{\text{nos}}$. If R is weakly syntactically definable in Γ, then R is recursively enumerable.*

PROOF. Say Γ is recursively axiomatized by Δ. Let $\mathbf{T}_m^\Delta$ be as in the above proof. Let R be weakly syntactically defined in Γ by φ, and let $e = \mathscr{g}^+\varphi$. Clearly for all $x_0, \ldots, x_{m-1} \in \omega$ we have

$$\langle x_0, \ldots, x_{m-1}\rangle \in R \qquad \text{iff } \exists y((e, x_0, \ldots, x_{m-1}, y) \in \mathbf{T}_m^\Delta),$$

so R is r.e. □

The intuitive basis of 14.6 is like that for 14.5. Given R, unary, weakly syntactically defined by φ in Γ, where Γ is axiomatized by an effective set Δ, first start listing out all Δ theorems. Whenever $\varphi(\mathbf{x})$ appears on the list, give output x. In this way R will be effectively enumerated.

The preceding two theorems give a method for exhibiting logical equivalents of recursiveness. Thus if we can find a consistent recursively axiomatizable theory Γ in $\mathscr{L}_{\text{nos}}$ in which every recursive function is syntactically definable and with $\neg(\mathbf{m} \equiv \mathbf{n}) \in \Gamma$ whenever $m \neq n$, then by 14.5, a function will be recursive iff it is syntactically definable in Γ. By 14.2, the same will be true of relations. Shortly we shall, indeed, produce such theories.

Definition 14.7. A theory Γ in $\mathscr{L}_{\text{nos}}$ is *ω-consistent* iff for every formula φ with free variables among $\{v_0\}$, if $\varphi(\mathbf{x}) \in \Gamma$ for each $x \in \omega$, then $\exists v_0 \neg \varphi(v_0) \notin \Gamma$.

Proposition 14.8. *If Γ is an ω-consistent theory in $\mathscr{L}_{\text{nos}}$, $\neg(\mathbf{m} \equiv \mathbf{n}) \in \Gamma$ whenever $m \neq n$, and every unary recursive function is syntactically definable in Γ, then every r.e. set is weakly syntactically definable in Γ.*

PROOF. First note that every ω-consistent theory is also consistent. Hence the formula $\neg(v_0 \equiv v_0)$ clearly weakly syntactically defines the empty set. Now let A be any nonempty r.e. set, say $A = \text{Rng} f$ where f is recursive. By hypothesis, there is a formula φ which syntactically defines f in Γ. Let ψ be the formula $\exists v_1 \varphi(v_1, v_0)$; we claim that ψ weakly defines A. To prove this, first suppose $x \in A$. Say $fy = x$. Then $\varphi(\mathbf{y}, \mathbf{x}) \in \Gamma$, so $\psi(\mathbf{x}) \in \Gamma$. Second, suppose that $x \notin A$. Thus for each $y \in \omega$ we have $fy \neq x$. Now $\varphi(\mathbf{y}, \mathbf{x}) \rightarrow \mathbf{x} \equiv \mathbf{z}$ is a member of Γ, where $z = fy$, and $\neg(\mathbf{x} \equiv \mathbf{z})$ is a member of Γ, so $\neg\varphi(\mathbf{y}, \mathbf{x})$ is a member of Γ, for each $\mathbf{y} \in \omega$. By ω-consistency $\psi(\mathbf{x}) \notin \Gamma$, as desired. □

Now we are ready to discuss our weakest form of number theory:

Definition 14.9. The theory R in $\mathscr{L}_{\text{nos}}$ is the collection of all sentences φ of $\mathscr{L}_{\text{nos}}$ such that $\Delta \models \varphi$, where Δ consists of the following sentences (for all $m, n \in \omega$):

(*i*) $\mathbf{m} + \mathbf{n} \equiv \mathbf{p}$, where $m + n = p$;
(*ii*) $\mathbf{m} \cdot \mathbf{n} \equiv \mathbf{p}$, where $m \cdot n = p$;
(*iii*) $\neg(\mathbf{m} \equiv \mathbf{n})$ for $m \neq n$;
(*iv*) $\forall v_0(v_0 \leq n \rightarrow v_0 \equiv \mathbf{O} \vee \cdots \vee v_0 \equiv \mathbf{n})$;
(*v*) $\forall v_0(v_0 \leq \mathbf{n} \vee \mathbf{n} \leq v_0)$.

Proposition 14.10

(*i*) *R is recursively axiomatizable.*

(*ii*) *R is ω-consistent.*

PROOF. Condition (*i*) is obvious. To prove (*ii*), first note that $\mathfrak{A} = \langle \omega, +, \cdot, \mathfrak{s}, 0 \rangle$ is a model of R. Now if $Fv\varphi \subseteq \{v_0\}$ and $\varphi(\mathbf{x}) \in R$ for each $x \in \omega$, then $\mathfrak{A} \models \varphi[\mathbf{x}]$ for each $x \in \omega$ and hence $\mathfrak{A} \models \forall v_0 \varphi(v_0)$. Since $\mathfrak{A}$ is a model of R it follows that $\exists v_0 \neg \varphi(v_0) \notin \Gamma$. □

The next theorem constitutes one of the two main theorems of this section. It forms the main underpinning of all of our results which use the notion of syntactic definability.

Theorem 14.11. *Every recursive function is syntactically definable in R.*

PROOF. Let $A = \{f : f$ is syntactically definable in $R\}$. We shall use Julia Robinson's theorem 3.48 to show that every recursive function is in A.

I. $+$ *is in A*. For, let φ be the formula $v_0 \mathbf{+} v_1 \mathbf{=} v_2$; obviously the desired conditions of 14.1 hold.

II. $\mathfrak{s}$ *is in A*. In this case let φ be $\mathbf{s}v_0 \mathbf{=} v_1$.

III. $U_i^n \in A$ $(0 \le i < n)$. Let φ be $v_i \mathbf{=} v_n$.

IV. $\mathrm{Exc} \in A$. Let φ be the following formula:

$$\exists v_2[v_2 \mathbf{\le} v_0 \wedge v_1 \mathbf{\le} v_2 \mathbf{+} v_2 \wedge v_0 \mathbf{=} (v_2 \mathbf{\cdot} v_2) \mathbf{+} v_1].$$

We show that φ syntactically defines Exc in R. To check the first requirement of 14.1, assume that $x \in \omega$. By the definition of Exc, choose $y \in \omega$ such that $x = y^2 + \mathrm{Exc}\, x < (y + 1)^2$. Since $(y + 1)^2 = y^2 + 2y + 1$, it follows that $\mathrm{Exc}\, x \le 2y$; and clearly $y \le x$. Thus by 14.9(*i*), $R \models \mathbf{z} + \mathbf{y} = \mathbf{x}$, where $z = x - y$, so

(1) $$R \models \mathbf{y} \mathbf{\le} \mathbf{x}.$$

Similarly, $R \models \Delta(2y - \mathrm{Exc}\, x) \mathbf{+} \Delta\, \mathrm{Exc}\, x \mathbf{=} \Delta 2y$ by 14.9(*i*) and $R \models \mathbf{y} \mathbf{+} \mathbf{y} \mathbf{=} \Delta(2y)$, so

(2) $$R \models \Delta\, \mathrm{Exc}\, x \mathbf{\le} \mathbf{y} \mathbf{+} \mathbf{y}.$$

Next, by 14.9(*ii*) $R \models \mathbf{y} \mathbf{\cdot} \mathbf{y} \mathbf{=} \Delta(y^2)$ and by 14.9(*i*) $R \models \Delta(y^2) \mathbf{+} \Delta\, \mathrm{Exc}\, \mathbf{x} \mathbf{=} x$, so

(3) $$R \models x \mathbf{=} \mathbf{y} \mathbf{\cdot} \mathbf{y} \mathbf{+} \Delta\, \mathrm{Exc}\, x.$$

From (1)–(3) it follows that $R \models \varphi(\mathbf{x}, \Delta\, \mathrm{Exc}\, x)$, which is the first of the two desired conditions for $\mathrm{Exc}\, x$. To check the second condition, let ψ be the formula

$$v_2 \mathbf{\le} \mathbf{x} \wedge v_1 \mathbf{\le} v_2 \mathbf{+} v_2 \wedge \mathbf{x} \mathbf{=} (v_2 \mathbf{\cdot} v_2) \mathbf{+} v_1.$$

Thus $\varphi(\mathbf{x}, v_1)$ is the same formula as $\exists v_2 \psi$. We shall prove that

(4) $$R \models \psi \rightarrow v_1 \mathbf{=} \Delta\, \mathrm{Exc}\, x,$$

which obviously implies that $R \models \varphi(\mathbf{x}, v_1) \rightarrow v_1 = \Delta \operatorname{Exc} x$, as desired. First note by 14.9(*iv*) that

(5) $$R \models \psi \rightarrow v_2 = \mathbf{O} \vee \cdots \vee v_2 = \mathbf{x}.$$

Now let $i \leq x$. Using 14.9(*i*) and 14.9(*iv*) we easily obtain

(6) $$R \models \psi \wedge v_2 = \mathbf{i} \rightarrow v_1 = \mathbf{O} \vee \cdots \vee v_1 = \Delta(2i).$$

Let $j \leq 2i$. We now note the following little number-theoretic fact:

(7) $$\text{if } i \neq y \quad \text{or } j \neq \operatorname{Exc} x, \quad \text{then } x \neq i^2 + j.$$

In fact, if $i < y$, then $i^2 + j \leq i^2 + 2i < i^2 + 2i + 1 = (i + 1)^2 \leq y^2 \leq x$, so $i^2 + j \neq x$, while if $y < i$, then $x < (y + 1)^2 \leq i^2 \leq i^2 + j$; thus $x \neq i^2 + j$. If $i = y$ but $j \neq \operatorname{Exc} x$, obviously $x \neq i^2 + j$. Hence (7) holds.

Now we clearly have, using 14.9(*i*) and 14.9(*ii*),

(8) $$R \models \psi \wedge v_2 = \mathbf{i} \wedge v_1 = \mathbf{j} \rightarrow \mathbf{x} = \Delta(i^2 + j),$$

while from (7) and 14.9(*iii*) we get

(9) $$\text{if } i \neq y \quad \text{or } j \neq \operatorname{Exc} x, \quad \text{then } R \vdash \neg(\mathbf{x} = \Delta(i^2 + j)).$$

Thus if $i \neq y$ but $j \leq 2i$ is arbitrary, (8) and (9) give

$$R \models \psi \wedge v_2 = \mathbf{i} \rightarrow \neg(v_1 = \mathbf{j}).$$

Thus, since $j \leq 2i$ is arbitrary, (6) gives us

$$R \models \psi \rightarrow \neg(v_2 = \mathbf{i}).$$

Hence by (5) we infer that

(10) $$R \models \psi \rightarrow v_2 = \mathbf{y}.$$

Now if $i = y$ but $j \neq \operatorname{Exc} x$, (8) and (9) give

$$R \models \psi \wedge v_2 = \mathbf{i} \rightarrow \neg(v_1 = \mathbf{j}).$$

Combining this with (10) and (6) we obtain the desired conclusion (4).

V. *A is closed under composition.* To prove this, assume that $f, g_0, \ldots, g_{m-1} \in A$, where f is m-ary and each g_i is n-ary; let $h = K_n^m(f, g_0, \ldots, g_{m-1})$. Say $f, g_0, \ldots, g_{m-1}$ are syntactically defined by $\varphi, \psi_0, \ldots, \psi_{m-1}$ in R, respectively. Let χ be the formula

$$\exists v_{n+1} \cdots \exists v_{n+m} \left[\bigwedge_{i<m} \psi_i(v_0, \ldots, v_{n-1}, v_{n+i+1}) \wedge \varphi(v_{n+1}, \ldots, v_{n+m}, v_n) \right].$$

We show that χ syntactically defines h in R. For the first condition of 14.1, assume that $x_0, \ldots, x_{n-1} \in \omega$. Then for any $i < m$, since ψ_i syntactically defines g_i in R,

(11) $$R \models \psi_i(\mathbf{x}_0, \ldots, \mathbf{x}_{n-1}, \mathbf{y}_i),$$

where $y_i = g_i(x_0, \ldots, x_{n-1})$. Since φ syntactically defines f in R,

(12) $$R \models \varphi(\mathbf{y}_0, \ldots, \mathbf{y}_{m-1}, \mathbf{z}),$$

where $z = f(y_0, \ldots, y_{m-1}) = h(x_0, \ldots, x_{n-1})$. By (11) and (12) we obviously obtain $R \models \chi(\mathbf{x}_0, \ldots, \mathbf{x}_{n-1}, \mathbf{z})$, as desired in the first part of 14.1.

Now let θ be the formula

$$\bigwedge_{i<m} \psi_i(\mathbf{x}_0, \ldots, \mathbf{x}_{n-1}, v_{n+i+1}) \wedge \varphi(v_{n+1}, \ldots, v_{n+m}, v_n).$$

Then for any $i < m$, since ψ_i syntactically defines g_i in R,

$$R \models \theta \rightarrow v_{n+i+1} \equiv \mathbf{y}_i.$$

It follows that $R \models \theta \rightarrow \varphi(\mathbf{y}_0, \ldots, \mathbf{y}_{m-1}, v_n)$ and hence, since φ syntactically defines f in R, $R \models \theta \rightarrow v_n \equiv \mathbf{z}$, as desired in the second part of 14.1.

VI. *A is closed under inversion (applied to functions with range ω).*

In fact, let $f \in A$, with f having range ω. Say φ syntactically defines f in R. Let ψ be the formula

$$\varphi(v_1, v_0) \wedge \forall v_2[\varphi(v_2, v_0) \rightarrow v_1 \leq v_2].$$

We show that ψ syntactically defines $f^{(-1)}$ in R. Let $x \in \omega$, and set $f^{(-1)}x = y$. Therefore, since φ syntactically defines f in R and $fy = x$,

(13) $$R \models \varphi(\mathbf{y}, \mathbf{x}).$$

Before continuing to check the two conditions of 14.1, we need a small lemma expressed in (14) below. Note by 14.9(*iv*) we have

$$R \models v_2 \leq \mathbf{y} \rightarrow \bigvee_{i \leq y} (v_2 \equiv \mathbf{i}).$$

If $i < y$, then $fi \neq x$ and hence by 14.9(*iii*),

$$R \models \neg(\Delta(fi) \equiv \mathbf{x}) \qquad \text{for } i < y.$$

On the other hand, since φ syntactically defines f in R we have

$$\mathrm{R} \models \varphi(\mathbf{i}, \mathbf{x}) \rightarrow \mathbf{x} \equiv \Delta(fi) \qquad \text{for any } i \in \omega.$$

Putting these three statements together, we have

(14) $$\mathrm{R} \models \varphi(v_2, \mathbf{x}) \wedge v_2 \leq \mathbf{y} \rightarrow v_2 \equiv \mathbf{y}.$$

Now we return to checking the conditions of 14.1. By 14.9(*v*) we have $\mathrm{R} \models v_2 \leq \mathbf{y} \vee \mathbf{y} \leq v_2$, so (14) yields

(15) $$\mathrm{R} \models \varphi(v_2, \mathbf{x}) \rightarrow v_2 \equiv \mathbf{y} \vee \mathbf{y} \leq v_2.$$

But $\mathrm{R} \models \mathbf{0} \dotplus \mathbf{y} \equiv \mathbf{y}$ by 14.9(*i*), so $\mathrm{R} \models \mathbf{y} \leq \mathbf{y}$ and hence (15) gives $\mathrm{R} \models \varphi(v_2, \mathbf{x}) \rightarrow \mathbf{y} \leq v_2$. Therefore by (13) we have $\mathrm{R} \models \psi(\mathbf{x}, \mathbf{y})$, as desired in the first part of 14.1. Now by (13) and the definition of ψ we easily obtain $\mathrm{R} \models \psi(\mathbf{x}, v_1) \rightarrow v_1 \leq \mathbf{y} \wedge \varphi(v_1, \mathbf{x})$, so by (14) we have $\mathrm{R} \models \psi(\mathbf{x}, v_1) \rightarrow v_1 \equiv \mathbf{y}$, as desired in the second part of 14.1. □

From 14.5 and 14.11 we obtain the following syntactical equivalent of recursiveness.

Corollary 14.12. *A function is recursive iff it is syntactically definable in* R.

By 14.3, 14.6, 14.8, 14.10, and 14.11, we have:

Corollary 14.13. *A relation is r.e. iff it is weakly syntactically definable in* R.

Another consequence of 14.11 is a result concerning elementary definability; it depends on the following simple lemma:

Lemma 14.14. *Let* $\mathfrak{A} = (\omega, +, \cdot, \mathfrak{s}, 0)$, *and let* Γ *be a theory in* $\mathscr{L}_{\text{nos}}$ *with* $\mathfrak{A}$ *as a model. Then if a function or relation is syntactically definable in* Γ *it is also elementarily definable in* $\mathfrak{A}$.

Proof. First suppose f, an m-ary function, is syntactically defined by a formula φ in Γ. Then ${}^{m+1}\varphi^{\mathfrak{A}} = f$, i.e., f is elementarily defined by φ (see 11.7). In fact, first suppose $fx_0 \cdots x_{m-1} = x_m$. Then by 14.1, $\Gamma \models \varphi(\mathbf{x}_0, \ldots, \mathbf{x}_m)$. Since $\mathfrak{A}$ is a model of Γ, $\mathfrak{A} \models \varphi(\mathbf{x}_0, \ldots, \mathbf{x}_m)$ and so $\mathfrak{A} \models \varphi[x_0, \ldots, x_m]$. Second, suppose $fx_0 \cdots x_{m-1} \neq x_m$, say $fx_0 \cdots x_{m-1} = y \neq x_m$. Then by the second requirement on syntactic definability, $\Gamma \models \varphi(\mathbf{x}_0, \ldots, \mathbf{x}_m) \rightarrow \mathbf{y} \equiv \mathbf{x}_m$ so, since $\mathfrak{A}$ is a model of Γ and $\mathfrak{A} \models \neg \mathbf{y} \equiv \mathbf{x}_m$, also $\mathfrak{A} \models \neg\varphi(\mathbf{x}_0, \ldots, \mathbf{x}_m)$ and so it is not the case that $\mathfrak{A} \models \varphi[x_0, \ldots, x_m]$.

The case of relations is even easier. □

From 14.11 we then obtain

Corollary 14.15. *Every recursive function and relation is elementarily definable in the structure* $(\omega, +, \cdot, \mathfrak{s}, 0)$.

Recalling the definition of the arithmetical hierarchy from Chapter 5, we then have:

Corollary 14.16. *For any number-theoretic relation* R *the following conditions are equivalent.*

(*i*) *R appears in the arithmetical hierarchy;*
(*ii*) *R is elementarily definable in* $(\omega, +, \cdot, \mathfrak{s}, 0)$.

There are three other number theories which we shall consider in the future:

Definition 14.17

(*i*) The theory Q in $\mathscr{L}_{\text{nos}}$ is the collection of all sentences φ such that $\Delta \vdash \varphi$, where Δ consists of the following sentences:

(*a*) $\forall v_0 \forall v_1 (\mathbf{s}v_0 \equiv \mathbf{s}v_1 \rightarrow v_0 \equiv v_1)$;
(*b*) $\forall v_0 \neg (\mathbf{s}v_0 \equiv \mathbf{O})$;

(*c*) $\forall v_0[\neg(v_0 = \mathbf{O}) \rightarrow \exists v_1(\mathbf{s}v_1 = v_0)]$;
(*d*) $\forall v_0(v_0 + \mathbf{O} = v_0)$;
(*e*) $\forall v_0 \forall v_1[v_0 + \mathbf{s}v_1 = \mathbf{s}(v_0 + v_1)]$;
(*f*) $\forall v_0(v_0 \cdot \mathbf{O} = \mathbf{O})$;
(*g*) $\forall v_0 \forall v_1(v_0 \cdot \mathbf{s}v_1 = v_0 \cdot v_1 + v_0)$.

(*ii*) The theory P in $\mathscr{L}_{\text{nos}}$ of *Peano arithmetic* is the collection of all sentences φ such that $\Delta \vdash \varphi$, where Δ consists of the sentences (*a*), (*b*), (*d*)–(*g*) above and also, for each formula ψ, the following sentence:

$$[[\psi(\mathbf{O}) \wedge \forall v_0[\psi(v_0) \rightarrow \psi(\mathbf{s}v_0)] \rightarrow \forall v_0 \psi(v_0)]].$$

(Recall from 10.94 that for any formula χ, $[[\chi]]$ is the universal closure of χ; also recall the notation $\psi(0)$, $\psi(\mathbf{s}v_0)$, etc., from 10.72 (note that ψ may have free variables other than v_0).)

(*iii*) The theory N in $\mathscr{L}_{\text{nos}}$ is the collection of all sentences φ which hold in $(\omega, +, \cdot, \mathscr{s}, 0)$.)

The following proposition is obvious (cf. the proof of 14.10).

Proposition 14.18

(*i*) Q *is finitely axiomatizable.*
(*ii*) P *is recursively axiomatizable.*
(*iii*) Q, P, *and* N *are* ω*-consistent.*

We shall see later that N is *not* recursively axiomatizable. The connection between our number theories is given in

Theorem 14.19. $\text{R} \subseteq \text{Q} \subseteq \text{P} \subseteq \text{N}$.

Proof. Obviously $\text{P} \subseteq \text{N}$. To prove $\text{Q} \subseteq \text{P}$ it is obviously enough to show that the sentence $\forall v_0[\neg(v_0 = \mathbf{0}) \rightarrow \exists v_1(\mathbf{s}v_1 = v_0)]$ is in P. To this end, let φ be the formula $\neg(v_0 = \mathbf{O}) \rightarrow \exists v_1(\mathbf{s}v_1 = v_0)$. Since $\models \mathbf{O} = \mathbf{O}$, we obviously have $\text{P} \models \varphi(\mathbf{O})$. It is equally obvious that $\models \varphi(\mathbf{s}v_0)$, so $\text{P} \models \forall v_0[\varphi(v_0) \rightarrow \varphi(\mathbf{s}v_0)]$. Hence $\text{P} \models \forall v_0 \varphi(v_0)$ by the special axiom for P, as desired.

Now to show that $\text{R} \subseteq \text{Q}$ it suffices to check that the sentences of 14.9 are in Q.

14.9(*i*) We show this by ordinary mathematical induction on n. The case $n = 0$ is clear by 14.17(*i*)(*d*). Now assume that $\text{Q} \models \mathbf{m} + \mathbf{n} = \mathbf{\Delta}(m + n)$. By 14.17(*i*)(*e*), $\text{Q} \models \mathbf{m} + \Delta(\mathscr{s}n) = \mathbf{s}(\mathbf{m} + \mathbf{n})$, and hence $\text{Q} \models \mathbf{m} + \mathbf{\Delta}\mathscr{s}n = \mathbf{\Delta}\mathscr{s}(m + n)$.

14.9(*ii*) This is similarly checked.

14.9(*iii*) It clearly suffices to show that $\text{Q} \models \neg(\mathbf{m} = \mathbf{n})$ whenever $n < m$. This we do by induction on n. For $n = 0$, we have by 14.17(*i*)(*b*) that $\text{Q} \models \neg[\mathbf{\Delta}(\mathscr{s}(m - 1)) = \mathbf{O}]$, as desired. Now assume that $\text{Q} \models \neg(\mathbf{m} = \mathbf{n})$ for all $m > n$, for a certain n. Now assume $\mathscr{s}n < m$. Now $\text{Q} \models \mathbf{\Delta}\mathscr{s}n = \mathbf{m} \rightarrow \mathbf{n} = \mathbf{\Delta}(m - 1)$, and $n < m - 1$ implies $\text{Q} \models \neg(\mathbf{n} = \mathbf{\Delta}(m - 1))$ by the induction hypothesis, so $\text{Q} \vdash \neg(\mathbf{\Delta}\mathscr{s}n = \mathbf{m})$.

14.9(*iv*). Again we proceed by ordinary induction on n. Now $v_0 \leq \mathbf{O}$ is the formula $\exists v_1(v_1 + v_0 = \mathbf{O})$. Note that

$$Q \models v_1 + v_0 = \mathbf{O} \wedge \neg(v_0 = \mathbf{O}) \rightarrow \exists v_2(\mathbf{s}v_2 = v_0) \qquad \text{by 14.17}(i)(c),$$

so, making use of 14.17(*i*)(*e*),

$$Q \models v_1 + v_0 = \mathbf{O} \wedge \neg(v_0 = \mathbf{O}) \rightarrow \exists v_2[\mathbf{s}(v_1 + v_2) = \mathbf{O}].$$

Hence 14.17(*i*)(*b*) yields $Q \vdash v_0 \leq \mathbf{O} \rightarrow v_0 = \mathbf{O}$, as desired. Now we assume that

$$(1) \qquad Q \models v_0 \leq \mathbf{n} \rightarrow v_0 = \mathbf{O} \vee \cdots \vee v_0 = \mathbf{n}.$$

Now

$$\begin{aligned} &Q \models v_1 + v_0 = \Delta(n+1) \wedge \neg(v_0 = \mathbf{O}) \rightarrow \exists v_2(\mathbf{s}v_2 = v_0) && 14.17(i)(c)\\ &Q \models v_1 + \mathbf{s}v_2 = \Delta(n+1) \rightarrow \mathbf{s}(v_1 + v_2) = \Delta(n+1) && 14.17(i)(e)\\ &Q \models \mathbf{s}(v_1 + v_2) = \Delta(n+1) \rightarrow v_1 + v_2 = \mathbf{n} && 14.17(i)(a) \end{aligned}$$

Thus $Q \models v_0 \leq \Delta(n+1) \rightarrow v_0 = 0 \vee \exists v_2(\mathbf{s}v_2 = v_0 \wedge v_2 \leq \mathbf{n})$, and $Q \models \mathbf{s}v_2 = v_0 \wedge v_2 \leq \mathbf{n} \rightarrow v_2 = \mathbf{O} \vee \cdots \vee v_2 = \mathbf{n}$ by (1), so $Q \models v_0 \leq \Delta(n+1) \rightarrow v_0 = \mathbf{O} \vee \cdots \vee v_0 = \Delta(n+1)$, as desired.

14.9(*v*). Again we proceed by induction on n. Note that $Q \models v_0 + \mathbf{O} = v_0$ by 14.17(*i*)(*d*), so $Q \models \mathbf{O} \leq v_0$ and $Q \models v_0 \leq \mathbf{O} \vee \mathbf{O} \leq v_0$. Now assume that $Q \models v_0 \leq \mathbf{n} \vee \mathbf{n} \leq v_0$. Now $Q \models v_0 \leq \mathbf{n} \rightarrow v_0 = \mathbf{O} \vee \cdots \vee v_0 = \mathbf{n}$ by 14.9(*iv*), which has already been established, and for each $i \leq n$ we have $Q \models \Delta(n - i + 1) + \mathbf{i} = \Delta(n+1)$ by 14.9(*i*), so $Q \models \mathbf{i} \leq \Delta(n+1)$. It follows that

$$(2) \qquad Q \models v_0 \leq \mathbf{n} \rightarrow v_0 \leq \Delta(n+1).$$

Now

$$(3) \qquad Q \models \mathbf{s}v_0 + \mathbf{m} = v_0 + \Delta(\mathfrak{s}m) \qquad \text{for every } m \in \omega.$$

In fact, $Q \models \mathbf{s}v_0 + \mathbf{O} = \mathbf{s}v_0$, $Q \models \mathbf{s}v_0 = \mathbf{s}(v_0 + \mathbf{O})$, and $Q \models \mathbf{s}(v_0 + \mathbf{O}) = v_0 + \mathbf{sO}$ using 14.17(*i*)(*d*) and 14.17(*i*)(*e*). Thus $Q \models \mathbf{s}v_0 + \mathbf{O} = v_0 + \mathbf{1}$. Assume that $Q \models \mathbf{s}v_0 + \mathbf{m} = v_0 + \Delta(\mathfrak{s}m)$. Then $Q \models \mathbf{s}v_0 + \Delta(m+1) = \mathbf{s}(\mathbf{s}v_0 + \mathbf{m})$ by 14.17(*i*)(*e*), $Q \models \mathbf{s}(\mathbf{s}v_0 + \mathbf{m}) = \mathbf{s}(v_0 + \Delta(\mathfrak{s}m))$ by our assumption, and $Q \models \mathbf{s}(v_0 + \Delta(\mathfrak{s}m)) = v_0 + \Delta(\mathfrak{s}\mathfrak{s}m)$ by 14.17(*i*)(*e*). Thus $Q \models \mathbf{s}v_0 + \Delta(m+1) = v_0 + \Delta(m+2)$, and (3) has been established by induction. Now

$$\begin{aligned} &Q \models v_1 + \mathbf{n} = v_0 \wedge \neg(v_1 = \mathbf{O}) \rightarrow \exists v_2(\mathbf{s}v_2 = v_1) && 14.14(i)(c)\\ &Q \models \mathbf{s}v_2 + \mathbf{n} = v_2 + \Delta(\mathfrak{s}n) && \text{by (3)} \end{aligned}$$

and hence $Q \models \mathbf{n} \leq v_0 \rightarrow \mathbf{n} = v_0 \vee \Delta(n+1) \leq v_0$. We easily infer from this, the inductive hypothesis and (2) that $Q \models v_0 \leq \Delta(n+1) \vee \Delta(n+1) \leq v_0$. □

Corollary 14.20. *The following conditions are equivalent*:

(*i*) *f is recursive*;
(*ii*) *f is syntactically definable in* R, Q, *or* P.

Corollary 14.21. *The following conditions are equivalent*:

(*i*) *R is r.e.*;
(*ii*) *R is weakly syntactically definable in* R, Q, *or* P.

Now we turn to our second notion of implicit definability in number theory.

Definition 14.22. $\mathscr{L}_{\text{un}}$ (the *universal language*) is an effective first-order language with no operation symbols but such that for each $m \in \omega \sim 1$ and each $n \in \omega$ there is an m-ary relation symbol $\mathbf{R}_n^m$ (with $\mathbf{R}_n^m \neq \mathbf{R}_{n'}^{m'}$ if $m \neq m'$ or $n \neq n'$).

An n-ary number-theoretic function f is *spectrally represented* by a sentence φ of $\mathscr{L}_{\text{un}}$ provided that for all $x_0, \ldots, x_{n-1} \in \omega$ the following two conditions hold:

(*i*) there is a finite model $\mathfrak{A}$ of φ such that $|\mathbf{R}_i^{1\mathfrak{A}}| = x_i$ for each $i < n$;
(*ii*) if $\mathfrak{A}$ is any model of φ such that $|\mathbf{R}_i^{1\mathfrak{A}}| = x_i$ for each $i < n$, then $|\mathbf{R}_n^{1\mathfrak{A}}| = f(x_0, \ldots, x_{n-1})$.

Lemma 14.23. *Any spectrally representable function is recursive.*

Proof. We give an intuitive proof for this, thus appealing to the weak Church's thesis (see p. 46). Let f, n-ary, be spectrally represented by φ. Let $\Gamma = \{\mathbf{R}_i^m : m \in \omega \sim 1, i \in \omega, \mathbf{R}_i^m$ occurs in φ or $m = 1$ and $i \leq n\}$. Thus Γ is finite. Let $\mathscr{L}'$ be the reduct of $\mathscr{L}_{\text{un}}$ to the symbols of Γ. We know that the truth of φ in an $\mathscr{L}_{\text{un}}$-structure $\mathfrak{A}$ is completely determined by the truth of φ in the $\mathscr{L}'$-reduct of $\mathfrak{A}$. Clearly we can effectively construct all $\mathscr{L}'$-structures with universe a fixed $m \in \omega \sim 1$, and any $\mathscr{L}'$-structure $\mathfrak{A}$ with $|A| = m$ is isomorphic to one of them. Our decision method for f is as follows. Given $x_0, \ldots, x_{n-1} \in \omega$, we begin constructing all $\mathscr{L}'$-structures $\mathfrak{A}$ with $A = 1$, all with $A = 2$, etc. For each structure $\mathfrak{A}$, as soon as it has been constructed, we check if φ holds in $\mathfrak{A}$ (surely an effective question) and if $|\mathbf{R}_i^{1\mathfrak{A}}| = x_i$ for each $i < n$. As soon as we get an affirmative answer to all of these questions, which must eventually happen by 14.22(*i*), we set $f(x_0, \ldots, x_{n-1}) = |\mathbf{R}_n^{1\mathfrak{A}}|$ (by 14.22(*ii*)). □

We shall shortly prove that every recursive function is spectrally representable. For this theorem and one in Chapter 16 we need the following rather technical lemma. It essentially shows how to spectrally represent not only x and fx but also x and all values $f0, f1, \ldots, fx$ simultaneously.

Lemma 14.24. *Let f be a unary function spectrally represented by a sentence φ. Then there is a sentence ψ which also spectrally represents f such that for any model $\mathfrak{A}$ of ψ the following conditions hold*:

(*i*) $|\mathbf{R}_2^{1\mathfrak{A}}| = |\mathbf{R}_0^{1\mathfrak{A}}| + 1$;
(*ii*) $\mathbf{R}_0^{2\mathfrak{A}}$ *is a simple ordering of* $\mathbf{R}_2^{1\mathfrak{A}}$ *in the* $<$*-sense, with respect to which*

$\mathbf{R}_2^{1\mathfrak{A}}$ has a smallest and a largest element and under which each element of $\mathbf{R}_2^{1\mathfrak{A}}$ different from the greatest element has an immediate successor;
(iii) for any $x \in \mathbf{R}_2^{1\mathfrak{A}}$, if $\{y : (y, x) \in \mathbf{R}_0^{2\mathfrak{A}}\}$ is finite, then $|\{y : (y, x) \in \mathbf{R}_1^{2\mathfrak{A}}\}| = f|\{y : (y, x) \in \mathbf{R}_0^{2\mathfrak{A}}\}|$.

PROOF. We first divide the symbols of $\mathscr{L}_{\text{un}}$ different from $\mathbf{R}_0^1$, $\mathbf{R}_1^1$, $\mathbf{R}_2^1$, $\mathbf{R}_0^2$, and $\mathbf{R}_1^2$ into two disjoint classes, consisting of distinct symbols ${}^0\mathbf{R}_r^q$ and ${}^1\mathbf{R}_r^q$ for $q \in \omega \sim 1$, $r \in \omega$, where ${}^i\mathbf{R}_r^q$ is q-ary. The symbols ${}^0\mathbf{R}_r^q$ will be used for a translation of φ, while the other symbols ${}^1\mathbf{R}_r^q$ will be used for auxiliary purposes. To define the translation of φ, first let v_l be a variable not occurring in φ, fixed for the rest of the proof. Let Δ be the set of all formulas of $\mathscr{L}_{\text{un}}$ whose relation symbols and variables occur in φ. For each formula $\chi \in \Delta$ we associate a formula χ':

$$(v_s = v_t)' = v_s = v_t;$$
$$(\mathbf{R}_r^q v_{s0} \cdots v_{s(q-1)})' = {}^0\mathbf{R}_r^{q+1} v_{s0} \cdots v_{s(q-1)} v_l;$$
$$(\neg\chi') = \neg\chi'; \qquad (\chi \vee \theta)' = \chi' \vee \theta'; \qquad (\chi \wedge \theta)' = \chi' \wedge \theta';$$
$$(\forall v_s \chi)' = \forall v_s({}^1\mathbf{R}_0^2 v_s v_l \rightarrow \chi').$$

Now we can describe our desired sentence ψ. Namely, there clearly is a sentence ψ such that for any $\mathscr{L}_{\text{un}}$-structure $\mathfrak{A}$, $\mathfrak{A} \models \psi$ iff the following conditions hold:

(1) ${}^1\mathbf{R}_2^{2\mathfrak{A}}$ is a one–one function mapping $\mathbf{R}_0^{1\mathfrak{A}}$ into $\mathbf{R}_2^{1\mathfrak{A}}$, there being only one element of $\mathbf{R}_2^{1\mathfrak{A}}$ not in the range of ${}^1\mathbf{R}_2^{2\mathfrak{A}}$;
(2) condition (*ii*) of 14.24;
(3) if $x \in {}^\omega A$ and $x_l \in \mathbf{R}_2^{1\mathfrak{A}}$, then:

(a) $\mathfrak{A} \models \varphi'[x]$;
(b) there is an $a \in A$ with $(a, x_l) \in {}^1\mathbf{R}_0^{2\mathfrak{A}}$;
(c) $\{(a, b) : (a, b, x_l) \in {}^1\mathbf{R}_0^{3\mathfrak{A}}\}$ is a one–one function mapping $\{a : (a, x_l) \in \mathbf{R}_0^{2\mathfrak{A}}\}$ onto $\{a : (a, x_l) \in {}^0\mathbf{R}_0^{2\mathfrak{A}}\}$, and $\{a : (a, x_l) \in {}^0\mathbf{R}_0^{2\mathfrak{A}}\} \subseteq \{a : (a, x_l) \in {}^1\mathbf{R}_0^{2\mathfrak{A}}\}$;
(d) $\{(a, b) : (a, b, x_l) \in {}^1\mathbf{R}_1^{3\mathfrak{A}}\}$ is a one–one function mapping $\{a : (a, x_l) \in {}^0\mathbf{R}_1^{2\mathfrak{A}}\}$ onto $\{a : (a, x_l) \in \mathbf{R}_1^{2\mathfrak{A}}\}$, and $\{a : (a, x_l) \in {}^0\mathbf{R}_1^{2\mathfrak{A}}\} \subseteq \{a : (a, x_l) \in {}^1\mathbf{R}_0^{2\mathfrak{A}}\}$;

(4) if y is the greatest element of $\mathbf{R}_2^{1\mathfrak{A}}$ under $\mathbf{R}_0^{2\mathfrak{A}}$, then ${}^1\mathbf{R}_1^{2\mathfrak{A}}$ is a one–one function mapping $\mathbf{R}_1^{1\mathfrak{A}}$ onto $\{a : (a, y) \in \mathbf{R}_1^{2\mathfrak{A}}\}$.

Thus conditions (*i*) and (*ii*) of the lemma are now trivial. We now check condition (*iii*), and then come back to the proof that ψ still spectrally represents f. So, let $\mathfrak{A}$ be a model of ψ. We now associate with each y such that $\{a : (a, y) \in {}^1\mathbf{R}_0^{2\mathfrak{A}}\} \neq 0$ a new $\mathscr{L}_{\text{un}}$-structure $\mathfrak{B}_y$. Namely we set

(5) $$B_y = \{a : (a, y) \in {}^1\mathbf{R}_0^{2\mathfrak{A}}\},$$

while for each symbol $\mathbf{R}_n^m$ we let

(6) $$\mathbf{R}_n^{m\mathfrak{B}_y} = \{(x_0, \ldots, x_{m-1}) : x_0, \ldots, x_{m-1} \in B_y \quad \text{and} \quad (x_0, \ldots, x_{m-1}, y) \in {}^0\mathbf{R}_n^{(m+1)\mathfrak{A}}\}.$$

Now we claim:

(7) if $B_y \neq 0$, $x \in {}^{\omega}B_y$, and χ is any member of Δ, then

$$\mathfrak{B}_y \models \chi[x] \quad \text{iff } \mathfrak{A} \models \chi'[x_y^l].$$

This condition is easily checked by induction on χ; it explains to some extent why χ' was constructed in the above fashion. From it our condition (*iii*) easily follows. In fact, let $x \in \mathbf{R}_2^{1\mathfrak{A}}$, and assume that $\{y : (y, x) \in \mathbf{R}_0^{2\mathfrak{A}}\}$ is finite. We know by (3)(b) that $\{a : (a, x) \in {}^1\mathbf{R}_0^{2\mathfrak{A}}\} \neq 0$, so $\mathfrak{B}_x$ is defined. Take any $z \in {}^{\omega}B_x$. Then by (3)(a) we have $\mathfrak{A} \models \varphi'[z_x^l]$, so (7) yields

(8) $$\mathfrak{B}_x \models \varphi.$$

Now by (3)(c), (5) and (6) we have

(9) $$|\mathbf{R}_0^{1\mathfrak{B}_x}| = |\{a : (a, x) \in \mathbf{R}_0^{2\mathfrak{A}}\}|,$$

and by (3)(d), (5) and (6) we have

(10) $$|\mathbf{R}_1^{1\mathfrak{B}_x}| = |\{a : (a, x) \in \mathbf{R}_1^{2\mathfrak{A}}\}|.$$

But since φ spectrally represents f, it follows from (8), (9), (10) that $f|\{a : (a, x) \in \mathbf{R}_0^{2\mathfrak{A}}\}| = |\{a : (a, x) \in \mathbf{R}_1^{2\mathfrak{A}}\}|$, as desired in (*iii*).

Now we turn to showing that ψ still spectrally represents f. To this end, let $x \in \omega$, fixed for the rest of this proof. To check 14.22(*ii*), suppose that $\mathfrak{A}$ is a model of ψ and $|\mathbf{R}_0^{1\mathfrak{A}}| = x$. Then, by what we have shown above, conditions (*i*)–(*iii*) of our lemma 14.24 hold. By (*ii*), let y be the greatest number of $\mathbf{R}_2^{1\mathfrak{A}}$ under $\mathbf{R}_0^{2\mathfrak{A}}$; then by (*i*),

(11) $$|\{a : (a, y) \in \mathbf{R}_0^{2\mathfrak{A}}\}| = x.$$

By (4) and (*iii*) it follows from (10) and (11) that $|\mathbf{R}_1^{1\mathfrak{A}}| = fx$, as desired in 14.22(*ii*).

It remains to check 14.22(*i*), i.e., to construct a finite model $\mathfrak{A}$ of ψ such that $|\mathbf{R}_0^{1\mathfrak{A}}| = x$. To do this, we first apply 14.22(*i*) for φ to produce for each $y \leq x$ a finite model $\mathfrak{B}_y$ of φ with $|\mathbf{R}_0^{1\mathfrak{B}_y}| = y$. For the universe of our model $\mathfrak{A}$ we take

(12) $$A = \bigcup_{y \leq x} B_y \cup (x + 1).$$

Now we define the relations of $\mathfrak{A}$ as follows:

(13) $\mathbf{R}_0^{1\mathfrak{A}} = x$ and $\mathbf{R}_2^{1\mathfrak{A}} = x + 1$;
(14) $\mathbf{R}_0^{2\mathfrak{A}} = \{(y, z) : y < z \leq x\}$;
(15) $\mathbf{R}_1^{1\mathfrak{A}} = \mathbf{R}_1^{1\mathfrak{B}_x}$;
(16) $\mathbf{R}_1^{2\mathfrak{A}} = \{(a, y) : y \leq x \text{ and } a \in \mathbf{R}_1^{1\mathfrak{B}_y}\}$;
(17) ${}^0\mathbf{R}_r^{(q+1)\mathfrak{A}} = \{(z_0, \ldots, z_{q-1}, y) : y \leq x \text{ and } (z_0, \ldots, z_{q-1}) \in \mathbf{R}_r^{q\mathfrak{B}_y}\}$;
(18) ${}^1\mathbf{R}_2^{2\mathfrak{A}} = \{(y, y) : y \in \mathbf{R}_0^{1\mathfrak{A}}\}$;
(19) ${}^1\mathbf{R}_0^{2\mathfrak{A}} = \{(a, y) : y \leq x \text{ and } a \in By\}$;
(20) ${}^1\mathbf{R}_0^{3\mathfrak{A}}$ is any relation such that for each $y \leq x$, $\{(a, b) : (a, b, y) \in {}^1\mathbf{R}_0^{3\mathfrak{A}}\}$ is a one–one function mapping y onto $\mathbf{R}_0^{1\mathfrak{B}_y}$;

(21) ${}^1\mathbf{R}_1^{3\mathfrak{A}} = \{(a, a, y) : y \le x \text{ and } a \in \mathbf{R}_1^{1\mathfrak{B}y}\}$;
(22) ${}^1\mathbf{R}_1^{2\mathfrak{A}} = \{(a, a) : a \in \mathbf{R}_1^{1\mathfrak{B}x}\}$;
(23) $\mathbf{R}_r^{q\mathfrak{A}} = 0$ otherwise.

Clearly $\mathfrak{A}$ is finite and $|\mathbf{R}_0^{1\mathfrak{A}}| = x$, so to finish the proof of 14.22(*i*) and hence of our lemma it remains only to check that $\mathfrak{A} \models \psi$, i.e., that (1)–(4) hold. Of these, all but (3)(a) are routinely checked. But this condition is also clear, for (7) again is easily established by induction on χ, so since $\mathfrak{B}_y \models \varphi$ for each $y \le x$, (3)(a) follows. □

Now we prove the converse of 14.23:

Theorem 14.25 (Trachtenbrot). *Every recursive function is spectrally representable.*

PROOF. Let $A = \{f : f \text{ is spectrally representable}\}$. We shall again apply Julia Robinson's theorem 3.48 in order to show that every recursive function is in A.

I. $+$ *is in* A. We can clearly write down an $\mathscr{L}_{\text{un}}$-sentence φ whose models are exactly those $\mathscr{L}_{\text{un}}$-structures $\mathfrak{A}$ for which $\mathbf{R}_0^{1\mathfrak{A}} \cap \mathbf{R}_1^{1\mathfrak{A}} = 0$ and $\mathbf{R}_2^{1\mathfrak{A}} = \mathbf{R}_0^{1\mathfrak{A}} \cup \mathbf{R}_1^{1\mathfrak{A}}$. Thus φ spectrally represents $+$.

II. $\mathfrak{s}$ *is in* A. This time take a sentence φ such that $\mathfrak{A}$ is a model of φ iff $\mathbf{R}_1^{1\mathfrak{A}}$ is a superset of $\mathbf{R}_0^{1\mathfrak{A}}$ with exactly one more element than $\mathbf{R}_0^{1\mathfrak{A}}$.

III. U_i^n *is in* A. Let φ express that $\mathbf{R}_n^{1\mathfrak{A}} = \mathbf{R}_i^{1\mathfrak{A}}$.

IV. $\mathrm{Exc} \in A$. To motivate the construction of the desired sentence φ, suppose that $\mathrm{Exc}\, x = z$. Then we can write $x = y^2 + z$ for a certain y, where $x < (y + 1)^2$. In our sentence φ, $\mathbf{R}_0^1$ is to be thought of as a set with x elements, $\mathbf{R}_1^1$ one with z elements, and $\mathbf{R}_2^1$ one with y elements. We can express that $x = y^2 + z$ by using functions (see below). Also, $\mathbf{R}_3^1$ is to have $y + 1$ elements. We can express that $x < (y + 1)^2$ by means of a one–one function that is not onto.

Getting down to details, let φ be a sentence such that $\mathfrak{A}$ is a model of φ iff the following conditions hold:

(1) $\mathbf{R}_0^{2\mathfrak{A}} : \mathbf{R}_1^{1\mathfrak{A}} \rightarrowtail \mathbf{R}_0^{1\mathfrak{A}}$;
(2) $\mathbf{R}_0^{3\mathfrak{A}} : \mathbf{R}_2^{1\mathfrak{A}} \times \mathbf{R}_2^{1\mathfrak{A}} \rightarrowtail \mathbf{R}_0^{1\mathfrak{A}}$; $\operatorname{Rng} \mathbf{R}_0^{2\mathfrak{A}} \cap \operatorname{Rng} \mathbf{R}_0^{3\mathfrak{A}} = 0$ and $\operatorname{Rng} \mathbf{R}_0^{2\mathfrak{A}} \cup \operatorname{Rng} \mathbf{R}_0^{3\mathfrak{A}} = \mathbf{R}_0^{1\mathfrak{A}}$;
(3) $\mathbf{R}_2^{1\mathfrak{A}} \subseteq \mathbf{R}_3^{1\mathfrak{A}}$, and $\mathbf{R}_3^{1\mathfrak{A}}$ has exactly one more element than $\mathbf{R}_2^{1\mathfrak{A}}$;
(4) $\mathbf{R}_1^{3\mathfrak{A}} : \mathbf{R}_0^{1\mathfrak{A}} \rightarrowtail \mathbf{R}_3^{1\mathfrak{A}} \times \mathbf{R}_3^{1\mathfrak{A}}$; but $\mathbf{R}_1^{3\mathfrak{A}}$ does not map onto $\mathbf{R}_3^{1\mathfrak{A}} \times \mathbf{R}_3^{1\mathfrak{A}}$.

To check the conditions of 14.22, let $x \in \omega$. Say $x = y^2 + \mathrm{Exc}\, x < (y + 1)^2$. Choose a finite $\mathscr{L}_{\text{un}}$-structure $\mathfrak{A}$ such that $|\mathbf{R}_0^{1\mathfrak{A}}| = x$, $|\mathbf{R}_1^{1\mathfrak{A}}| = \mathrm{Exc}\, x$, $|\mathbf{R}_2^{1\mathfrak{A}}| = y$, $\mathbf{R}_2^{1\mathfrak{A}} \subseteq \mathbf{R}_3^{1\mathfrak{A}}$, and $\mathbf{R}_3^{1\mathfrak{A}}$ has exactly one more element than $\mathbf{R}_2^{1\mathfrak{A}}$. Clearly then we can choose $\mathbf{R}_0^{2\mathfrak{A}}$, $\mathbf{R}_0^{3\mathfrak{A}}$, $\mathbf{R}_1^{3\mathfrak{A}}$ so that (1)–(4) hold. Thus 14.22(*i*) holds. Similarly, 14.22(*ii*) holds.

V. *A is closed under composition*. For, suppose $f, g_0, \ldots, g_{m-1} \in A$, where f is m-ary and each g_i is n-ary; let $h = K_n^m(f; g_0, \ldots, g_{m-1})$. Say $g_0, \ldots,$

g_{m-1}, f are spectrally represented by $\psi_0, \ldots, \psi_m$, respectively. The idea of the construction of a sentence φ spectrally representing h is this. We modify the sentences ψ_i so that pairwise they talk about entirely different symbols. Then we can introduce relationships between them in order to represent h.

Formally, the first step takes place via $m + 1$ different syntactical $\mathscr{L}_{\text{un}}$-structures in $\mathscr{L}_{\text{un}}$. We can clearly divide the symbols of $\mathscr{L}_{\text{un}}$ up into $m + 2$ pairwise disjoint parts such that each part contains infinitely many relation symbols of each rank and such that $\mathbf{R}^1_0, \ldots, \mathbf{R}^1_n$ all are in the last part. The j-ary relation symbols of the ith part $(0 \leq i \leq m + 1)$ will be denoted by ${}^i\mathbf{R}^j_0, {}^i\mathbf{R}^j_1, \ldots$, and we assume that ${}^{m+1}\mathbf{R}^1_k = \mathbf{R}^1_k$ for each $k \leq n$. For each $i \leq m$ let $\mathfrak{A}_i = ({}^{m+1}\mathbf{R}^1_{n+i+1}v_0, t_i, S_i, \Gamma_i)$ be the syntactical $\mathscr{L}_{\text{un}}$-structure in $\mathscr{L}_{\text{un}}$ such that:

(5) $\quad t_i = 0$ (since $\mathscr{L}_{\text{un}}$ has no operation symbols);

(6) $\quad S_i\mathbf{R}^j_k = {}^i\mathbf{R}^j_k v_0 \cdots v_{j-1}$;

(7) $\quad \Gamma_i = \{\exists v_0 {}^{m+1}\mathbf{R}^1_{n+i+1}v_0\}$.

By means of these syntactical structures, our sentences $\psi_0, \ldots, \psi_m$ receive translations $\psi_0^{\mathfrak{A}0}, \ldots, \psi_m^{\mathfrak{A}m}$. The relationships between these translations that we need to express are: (1) our standard argument places $\mathbf{R}^1_0, \ldots, \mathbf{R}^1_{n-1}$ correspond to the argument places in the translations $\psi_0^{\mathfrak{A}0} \cdots \psi_{m-1}^{\mathfrak{A}(m-1)}$; (2) the results obtained in $\psi_0^{\mathfrak{A}0}, \ldots, \psi_{m-1}^{\mathfrak{A}(m-1)}$ correspond to the argument places in $\psi_m^{\mathfrak{A}m}$; (3) the result obtained in $\psi_m^{\mathfrak{A}m}$ corresponds to the standard result place $\mathbf{R}^1_n$. Thus we want a sentence φ (which clearly exists) such that for any $\mathscr{L}_{\text{un}}$-structure $\mathfrak{B}$, $\mathfrak{B} \models \varphi$ iff the following conditions hold:

(8) $\quad \mathfrak{B} \models \psi_i^{\mathfrak{A}i} \wedge \exists v_0 {}^{m+1}\mathbf{R}^1_{n+i+1}v_0 \qquad$ for each $i \leq m$;

(9) $\quad \mathfrak{B} \models \bigwedge_{i<m, j \leq n} ({}^i\mathbf{R}^1_j v_0 \to {}^{m+1}\mathbf{R}^1_{n+i+1}v_0) \wedge \bigwedge_{j \leq m} ({}^m\mathbf{R}^1_j v_0 \to {}^{m+1}\mathbf{R}^1_{n+m+1}v_0)$

(10) $\quad {}^{m+1}\mathbf{R}^{2\mathfrak{B}}_{ni+k} : \mathbf{R}^{1\mathfrak{B}}_k \rightarrowtail\!\!\!\!\twoheadrightarrow {}^i\mathbf{R}^{1\mathfrak{B}}_k \qquad$ for each $i < m$ and $k < n$;

(11) $\quad {}^{m+1}\mathbf{R}^{2\mathfrak{B}}_{nm+i} : {}^i\mathbf{R}^{1\mathfrak{B}}_n \rightarrowtail\!\!\!\!\twoheadrightarrow {}^m\mathbf{R}^{1\mathfrak{B}}_i \qquad$ for each $i < m$;

(12) $\quad {}^{m+1}\mathbf{R}^{2\mathfrak{B}}_{nm+m} : {}^m\mathbf{R}^{1\mathfrak{B}}_m \rightarrowtail\!\!\!\!\twoheadrightarrow \mathbf{R}^{1\mathfrak{B}}_n$.

We check 14.22(*ii*) and leave 14.22(*i*) to the reader. Assume that $x_0, \ldots, x_{n-1} \in \omega$ and that $\mathfrak{B}$ is a model of φ such that $|\mathbf{R}^{1\mathfrak{B}}_k| = x_k$ for each $i < n$. Then by (10),

(13) $\quad |{}^i\mathbf{R}^{1\mathfrak{B}}_k| = x_k \qquad$ for each $i < m$ and $k < n$.

Now $\mathfrak{B}$ is a model of Γ_i by (8), so we can form the structure $\mathfrak{B}^{\Gamma i, \mathfrak{A}i}$ for each $i \leq m$. Note from 11.43 that

(14) $\quad \mathbf{R}^{l\mathfrak{B}(\Gamma i, \mathfrak{A}i)}_k = {}^i\mathbf{R}^{l\mathfrak{B}}_k \cap {}^l({}^{m+1}\mathbf{R}^{1\mathfrak{B}}_{n+i+1}) \qquad$ for each $i \leq m$.

Hence by (9) and (13) we have $|\mathbf{R}^{1\mathfrak{B}(\Gamma i, \mathfrak{A}i)}_k| = x_k$ for each $i < m$ and $k < n$. Furthermore, from 11.45 and (8) we know that $\mathfrak{B}^{\Gamma i, \mathfrak{A}i} \models \psi_i$ for each $i < m$. Hence by 14.19(*ii*) for ψ_i and g_i we obtain

(15) $\quad |\mathbf{R}^{1\mathfrak{B}(\Gamma i, \mathfrak{A}i)}_n| = g_i(x_0, \ldots, x_{n-1}) \qquad$ for each $i < m$.

Using (15), (14), (9), (13) we see that

(16) $$|\mathbf{R}_i^{1\mathfrak{B}(\Gamma m,\mathfrak{A}m)}| = g_i(x_0, \ldots, x_{n-1}) \qquad \text{for each } i < m.$$

Since $\mathfrak{B}^{\Gamma m,\mathfrak{A}m} \models \psi_m$, it follows that

$$|\mathbf{R}_m^{1\mathfrak{B}(\Gamma m,\mathfrak{A}m)}| = h(x_0, \ldots, x_{n-1}).$$

Hence by (14), (9), (12) we obtain $|\mathbf{R}_n^1| = h(x_0, \ldots, x_{n-1})$, as desired.

VI. *A is closed under inversion, applied to functions with range* ω. For, let $f \in A$, where f is a unary function with range ω. By Lemma 14.24, we know that f is spectrally represented by a sentence ψ satisfying the conditions of 14.24. We translate ψ so as to make room for some auxiliary symbols. Namely, we divide the symbols of $\mathscr{L}_{\text{un}}$ into equal parts denoted by ${}^0\mathbf{R}_r^q$ and ${}^1\mathbf{R}_r^q$, where ${}^1\mathbf{R}_0^1 = \mathbf{R}_0^1$ and ${}^1\mathbf{R}_1^1 = \mathbf{R}_1^1$. Let $\mathfrak{A} = ({}^1\mathbf{R}_2^1 v_0, t, S, \Gamma)$ be the syntactical $\mathscr{L}_{\text{un}}$-structure in $\mathscr{L}_{\text{un}}$ such that:

(17) $$t = 0;$$
(18) $$S\mathbf{R}_r^q = {}^0\mathbf{R}_r^q v_0 \ldots v_{q-1};$$
(19) $$\Gamma = \{\exists v_0 {}^1\mathbf{R}_2^1 v_0\}.$$

Now there clearly is a sentence χ of $\mathscr{L}_{\text{un}}$ (which we shall show spectrally represents $f^{(-1)}$) such that for any $\mathscr{L}_{\text{un}}$-structure $\mathfrak{B}$, $\mathfrak{B} \models \chi$ iff the following conditions hold:

(20) $\mathfrak{B} \models \psi^{\mathfrak{A}}$;

(21) ${}^1\mathbf{R}_0^{2\mathfrak{B}}$ is a one–one function mapping ${}^0\mathbf{R}_0^{1\mathfrak{B}}$ onto ${}^1\mathbf{R}_1^{1\mathfrak{B}}$, and ${}^1\mathbf{R}_1^{2\mathfrak{B}}$ is a one–one function mapping ${}^0\mathbf{R}_1^{1\mathfrak{B}}$ onto ${}^1\mathbf{R}_0^{1\mathfrak{B}}$; ${}^1\mathbf{R}_2^{1\mathfrak{B}} \neq 0$;

(22) if $b \in B$ and b is not the greatest element of ${}^0\mathbf{R}_2^{1\mathfrak{B}}$ under ${}^0\mathbf{R}_0^{2\mathfrak{B}}$, then either $\{(a, c) : (a, c, b) \in {}^1\mathbf{R}_0^{3\mathfrak{B}}\}$ maps ${}^1\mathbf{R}_0^{1\mathfrak{B}}$ one–one into $\{a : (a, b) \in {}^0\mathbf{R}_1^{2\mathfrak{B}}\}$ but is not an onto map, or it maps $\{a : (a, b) \in {}^0\mathbf{R}_1^{2\mathfrak{B}}\}$ one–one into ${}^1\mathbf{R}_0^{1\mathfrak{B}}$ but is not an onto map.

Now we check that the conditions of 14.22 work for $f^{(-1)}$ and χ. Suppose $x \in \omega$. To check 14.22(*i*), we first apply 14.22(*i*) for f and ψ to obtain a finite model $\mathfrak{C}$ of ψ such that $|\mathbf{R}_0^{1\mathfrak{C}}| = f^{(-1)}x$. From $\mathfrak{C}$ we construct a finite model $\mathfrak{B}$ of χ as follows. We set $B = C$, while for relation symbols we let $\mathfrak{B}$ be such that

(23) ${}^0\mathbf{R}_r^{q\mathfrak{B}} = \mathbf{R}_r^{q\mathfrak{C}}$; ${}^1\mathbf{R}_2^{1\mathfrak{B}} = B$;

(24) ${}^1\mathbf{R}_0^{1\mathfrak{B}} = \mathbf{R}_1^{1\mathfrak{C}}$ and ${}^1\mathbf{R}_1^{1\mathfrak{B}} = \mathbf{R}_0^{1\mathfrak{C}}$;

(25) ${}^1\mathbf{R}_0^{2\mathfrak{B}}$, ${}^1\mathbf{R}_1^{2\mathfrak{B}}$, and ${}^1\mathbf{R}_0^{3\mathfrak{B}}$ are arbitrary as long as (21) and (22) hold.

Clearly (25) is possible, in view of 14.24(*iii*). Clearly from the definition of $\mathfrak{B}$ we have $\mathfrak{B}$ a model of Γ and $\mathfrak{B}^{\Gamma\mathfrak{A}} = \mathfrak{C}$. Since $\mathfrak{C}$ is a model of ψ, it follows from 11.45 that $\mathfrak{B} \models \psi^{\mathfrak{A}}$, i.e., condition (20) holds. Thus by (24) we see that $\mathfrak{B}$ is a model of χ, as desired, since $|\mathbf{R}_0^{1\mathfrak{B}}| = |{}^1\mathbf{R}_0^{1\mathfrak{B}}| = |\mathbf{R}_1^{1\mathfrak{C}}| = x$.

To check 14.22(*ii*), let $\mathfrak{B}$ be any model of χ such that $|\mathbf{R}_0^{1\mathfrak{B}}| = x$. Let

$\mathfrak{C} = \mathfrak{B}^{\Gamma\mathfrak{A}}$. Then by 11.45 and (20), $\mathfrak{C}$ is a model of ψ. Note that $|\mathbf{R}_1^{1\mathfrak{C}}| = |{}^0\mathbf{R}_1^{1\mathfrak{B}}| = |{}^1\mathbf{R}_0^{1\mathfrak{B}}| = |\mathbf{R}_0^{1\mathfrak{B}}| = x$. Now we claim

(26) $$\mathbf{R}_0^{1\mathfrak{C}} \text{ is finite.}$$

For, suppose it is infinite. Since $\mathfrak{C}$ is a model of ψ, it follows by 14.24(*ii*) that there is a $y \in \mathbf{R}_2^{1\mathfrak{C}}$ such that $|\{a : (a, y) \in \mathbf{R}_0^{2\mathfrak{C}}\}| = f^{(-1)}x$. Then by 14.24(*iii*) we infer that $|\{a : (a, y) \in \mathbf{R}_1^{2\mathfrak{C}}\}| = x$. But $\mathbf{R}_1^{2\mathfrak{C}} = {}^0\mathbf{R}_1^{2\mathfrak{B}}$ and $|{}^1\mathbf{R}_0^{1\mathfrak{B}}| = |\mathbf{R}_0^{1\mathfrak{B}}| = x$, so this stands in contradiction to (22), since y cannot be the greatest element of ${}^0\mathbf{R}_2^{1\mathfrak{B}}$ under ${}^0\mathbf{R}_0^{2\mathfrak{B}}$. Thus (26) holds.

Now by (26), since ψ spectrally represents f, $f|\mathbf{R}_0^{1\mathfrak{C}}| = x$. Note that $|\mathbf{R}_1^{1\mathfrak{B}}| = |{}^1\mathbf{R}_1^{1\mathfrak{B}}| = |{}^0\mathbf{R}_0^{1\mathfrak{B}}| = |\mathbf{R}_0^{1\mathfrak{C}}|$. Thus to finish the proof of 14.25 it remains only to show that $fy \neq x$ whenever $y < |\mathbf{R}_1^{1\mathfrak{B}}|$. But if $y < |\mathbf{R}_1^{1\mathfrak{B}}|$, then $y < |\mathbf{R}_0^{1\mathfrak{C}}|$, and from 14.24(*ii*), since $\mathfrak{C}$ is a model of ψ, there is a $b \in C$ different from the greatest element of $\mathbf{R}_2^{1\mathfrak{C}}$ under $\mathbf{R}_0^{2\mathfrak{C}}$ such that $|\{a : (a, \tau) \in \mathbf{R}_0^{2\mathfrak{C}}\}| = y$. Then by 14.24(*iii*) and (22) we have

$$\begin{aligned} fy &= f|\{a : (a, b) \in \mathbf{R}_0^{2\mathfrak{C}}\}| \\ &= |\{a : (a, b) \in \mathbf{R}_1^{2\mathfrak{C}}\}| \\ &= |\{a : (a, b) \in {}^0\mathbf{R}_1^{2\mathfrak{B}}\}| \\ &\neq |{}^1\mathbf{R}_0^{1\mathfrak{B}}| = |\mathbf{R}_0^{1\mathfrak{B}}| = x. \end{aligned}$$ □

Corollary 14.26. *A function is recursive iff it is spectrally representable.*

BIBLIOGRAPHY

1. Ershov, Yu., Lavrov, I., Taimanov, A., Taitslin, M., Elementary theories. *Russian Mathematical Surveys*, **20** (1965), 35–105.
2. Shoenfield, J., *Mathematical Logic*. Reading: Addison-Wesley (1967).

EXERCISES

14.27. Let Γ be the theory of $\mathscr{L}_{\text{nos}}$ with the single axiom $\forall v_0 \forall v_1 (v_0 \equiv v_1)$. Show that every function is syntactically definable in Γ, but only countably many relations are syntactically definable in Γ. Thus the assumption $\neg(\mathbf{0} \equiv \mathbf{1}) \in \Gamma$ in 14.2 is necessary; and the assumption in 14.5 that $\neg(\mathbf{m} \equiv \mathbf{n}) \in \Gamma$ whenever $m \neq n$ is necessary.

14.28. Describe a theory in $\mathscr{L}_{\text{nos}}$ which is consistent but not ω-consistent.

14.29. Show that R is not finitely axiomatizable. *Hint*: it is enough to show that for any finite set Γ of the axioms in 14.9 for R, there is a model of Γ which is not a model of R.

14.30. Prove, without using 14.29, that R is a proper subset of Q.

14.31. Show that the sentence $\forall v_0 \neg \mathbf{s} v_0 \equiv v_0$ is in P but not in Q.

14.32. Let T be the theory in $\mathscr{L}_{\text{nos}}$ with the following axioms:

$\forall v_0 \forall v_1(\mathbf{s}v_0 = \mathbf{s}v_1 \to v_0 = v_1)$;
$\forall v_0 \neg (\mathbf{s}v_0 = \mathbf{0})$;
$[[\psi(\mathbf{0}) \wedge \forall v_0[\psi(v_0) \to \psi(\mathbf{s}v_0)] \to \forall v_0 \psi(v_0)]]$;
$\forall v_0 \forall v_1(v_0 + v_1 = \mathbf{0})$;
$\forall v_0 \forall v_1(v_0 \cdot v_1 = \mathbf{0})$.

Show that not every recursive function is syntactically definable in T. *Hint*: eliminate quantifiers and show that not every recursive set is syntactically definable in T.

14.33. Give an example of a consistent theory in $\mathscr{L}_{\text{nos}}$ in which a nonrecursive set is syntactically definable.

14.34. An m-ary function f is *strongly syntactically definable* in a theory Γ formulated in $\mathscr{L}_{\text{nos}}$ provided that there is a formula φ of $\mathscr{L}_{\text{nos}}$ with $\text{Fv}\,\varphi \subseteq \{v_0, \ldots, v_m\}$ such that the following two conditions hold:

(1) for all $x_0, \ldots, x_{m-1} \in \omega$, the sentence $\varphi(\mathbf{x}_0, \ldots, \mathbf{x}_{m-1}, \Delta f(x_0, \ldots, x_{m-1}))$ is in Γ;

(2) the sentence $\forall v_0 \cdots \forall v_{m-1} \exists v_{m+1} \forall v_m(\varphi \leftrightarrow v_m = v_{m+1})$ is in Γ. Show that if f is syntactically definable in Γ, then it is also strongly syntactically definable in Γ. *Hint*: Say f is syntactically defined by φ. Let ψ be the formula

$$\varphi \wedge \forall v_{m+1}[\varphi(v_0, \ldots, v_{m-1}, v_{m+1}) \to v_m = v_{m+1}],$$

and let χ be the formula

$$\psi \vee [\forall v_m \neg \psi(v_0, \ldots, v_m) \wedge v_m = \mathbf{0}].$$

Show that χ strongly syntactically defines f.

15 General Theory of Undecidability

In previous chapters we have introduced several concepts related to the notion of undecidable theories (complete theories, 11.9; theories, 11.29; decidable and undecidable theories, 13.1; syntactical and weak syntactical definability, 14.1; recursive axiomatizability, 14.4; spectral representability, 14.22). Our purpose in this chapter is to establish various relationships known to exist between these notions and related ones. These general theorems will be applied in the next chapter, in which numerous examples of undecidable theories are given. We proceed in this chapter from the simpler concepts to the more complicated ones.

Theorem 15.1 (Craig). *For any theory Γ in an elementary effectivized first-order language the following conditions are equivalent:*

(*i*) *Γ is axiomatizable by a set Δ such that $g^{+*}\Delta$ is elementary;*
(*ii*) *Γ is recursively axiomatizable;*
(*iii*) *$g^{+*}\Gamma$ is r.e.*

PROOF. Obviously (*i*) $\Rightarrow$ (*ii*), while (*ii*) $\Rightarrow$ (*iii*) by 10.29. Now assume that (*iii*) holds. Let f be an elementary function with range $g^{+*}\Gamma$. Set

$$\Delta = \{\varphi : \varphi \text{ is a sentence, and there is an } x \leq g^{+}\varphi \text{ such that } fx = g^{+}\psi \text{ for some sentence } \psi \text{ such that } \varphi = \psi \wedge \psi \wedge \cdots \wedge \psi\}.$$

Obviously $g^{+*}\Delta$ is elementary, and $\Delta \subseteq \Gamma$. Now suppose that $\psi \in \Gamma$. Choose x so that $fx = g^{+}\psi$. Clearly we can find a sentence φ of the form $\psi \wedge \psi \wedge \cdots \wedge \psi$ such that $g^{+}\varphi \geq x$. This follows from the simple observation that $g^{+}(\theta_0 \wedge \theta_1) > g^{+}\theta_0$ for any formulas θ_0, θ_1. Thus $\varphi \in \Delta$, and hence $\psi \in \Delta\text{-Thm}$, as desired. □

Theorem 15.1 has considerable theoretical importance. According to it, as soon as we know that a theory can be effectively enumerated we can conclude that a very effective (i.e., elementary) axiom system can be given for it. Theorem 15.1 is a kind of logical version of the equivalence of (*i*) and (*iii*) in 6.2 (equivalent definitions of r.e. sets). The following important theorem gives a logical version of the fact that a set is recursive iff both it and its complement are r.e.

Theorem 15.2. *For any complete theory* Γ *the following conditions are equivalent;*

(*i*) Γ *is decidable:*
(*ii*) Γ *is recursively axiomatizable.*

Proof. (*i*) $\Rightarrow$ (*ii*). This is obvious, since Γ axiomatizes Γ. (*ii*) $\Rightarrow$ (*i*). Assume (*ii*), say Δ axiomatizes Γ, where $\mathscr{g}^{+*}\Delta$ is recursive. A decision procedure for Γ can be described as follows. Assume that Γ is consistent. (It is obvious, and trivial, that an inconsistent theory is both decidable and recursively axiomatizable.) Given any sentence φ, start listing all sentences derivable from Δ (recall from 10.29 that $\mathscr{g}^{+*}(\Delta-\text{Thm})$ is r.e.). Since Γ is complete, either φ or $\neg\varphi$ will eventually appear on the list. If φ appears, we of course give the output $\varphi \in \Gamma$. If $\neg\varphi$ appears, by the consistency of Γ we give the output $\varphi \notin \Gamma$.

More formally, write $\mathscr{g}^{+*}(\Delta-\text{Thm}) = \text{Rng} f$, where f is recursive. Then

$$x \in \mathscr{g}^{+*}\Gamma \text{ iff } x \in \mathscr{g}^{+*}\text{Sent and } f\mu y(fy = x \text{ or } fy = \neg' x) = x. \qquad \square$$

Because of their importance, we list here two corollaries of 15.2 which are simple logical transformations of it:

Corollary 15.3. *If* Γ *is recursively axiomatizable and undecidable, then* Γ *is incomplete.*

Corollary 15.4. *If* Γ *is complete and undecidable, then* Γ *is not recursively axiomatizable.*

In Corollary 15.3 we encounter a justification for our emphasis on undecidability in this book. Even if our main concern were proving various theories incomplete, we see by 15.3 that more general results are obtained by proving undecidability (since we are mainly concerned with recursively axiomatizable theories). In 16.1 we shall see that N is an example of a theory which is complete but undecidable; thus by 15.4, N is not recursively axiomatizable.

Another important connection between our basic notions is given in

Theorem 15.5. *If* Γ *is consistent and decidable, then* $\Gamma \subseteq \Delta$ *for some consistent, decidable complete theory* Δ.

Proof. Intuitively we proceed as follows, merely effectivizing a proof of Lindenbaum's Theorem 11.13. Effectively list out all sentences. Now extend Γ by an effective, recursive procedure: at the mth step, add the mth sequence if it is consistent to do so, otherwise add nothing. In this way we extend Γ to a complete theory Δ with $\mathscr{g}^{+*}\Delta$ r.e.; by 15.2, Δ is decidable.

More formally, we proceed as follows. Let h be a recursive function with range $\mathscr{g}^{+*}\mathrm{Sent}$ and with $h0 \in \mathscr{g}^{+*}\Gamma$. Now we define a function f by recursion. Let

$$\begin{aligned} f0 &= h0 \\ f(m+1) &= hm \qquad \text{if } \Gamma \nvDash \bigwedge_{i<m} \mathscr{g}^{+-1}fi \rightarrow \neg \mathscr{g}^{+-1}hm, \\ f(m+1) &= fm \qquad \text{otherwise.} \end{aligned}$$

Let $\Delta = \mathscr{g}^{+-1*}\mathrm{Rng} f$. Now Δ is consistent. For, by induction it is clear (see the proof of the next theorem) that $\Gamma \cup \{\mathscr{g}^{+-1}fi : i < m\}$ is consistent for each $m \in \omega$; so it follows, of course, that $\Gamma \cup \Delta$ is consistent. *A fortiori*, Δ is consistent. Furthermore, $\Gamma \subseteq \Delta$. For, assume that $\varphi \in \Gamma$, say $hm = \varphi$. Clearly then, since we have already established that $\Gamma \cup \{\mathscr{g}^{+-1}fi : i < m\}$ is consistent, it follows that $f(m+1) = hm$, so $\varphi \in \Delta$. Note that Δ is a theory, since if $\Delta \vDash \varphi$, with, say, $hm = \varphi$, then obviously $f(m+1) = hm$. Next, Δ is complete, for let φ be a sentence with $\varphi \notin \Delta$, say $\varphi = hm$. Then $f(m+1) \neq hm$, so $\Gamma \vDash \bigwedge_{i<m} \mathscr{g}^{+-1}fi \rightarrow \neg\varphi$, hence $\Delta \vDash \neg\varphi$, hence $\neg\varphi \in \Delta$. Finally, since Δ is complete and $\mathscr{g}^{+*}\Delta = \mathrm{Rng} f$ is r.e., by 15.1 Δ is recursively axiomatizable and hence by 15.2 it is decidable. □

As a generalization of 15.5 we have the following result. It shows that the decidability of a theory Γ is equivalent to the existence of complete decidable extensions of Γ with certain restrictions. This result is practically useful in proving decidability of theories. The proof is just an extension of the proof of 15.5.

Theorem 15.6 (Ershov). *Let Γ be consistent and recursively axiomatizable. Then Γ is decidable iff there is a sequence $\Theta_0, \Theta_1, \ldots$ of complete consistent extensions of Γ and a binary recursive function f such that the following conditions hold:*

(*i*) $\Gamma = \bigcap_{i<\omega} \Theta_i$;

(*ii*) *For every $m \in \omega$, $\{f(m, i) : i \in \omega\} = \mathscr{g}^{+*}\Delta_m$ for some set Δ_m of sentences which axiomatizes Θ_m.*

Proof. First suppose that Γ is decidable. Let

$$\Omega = \{\varphi \in \mathrm{Sent} : \Gamma \cup \{\varphi\} \text{ is consistent}\}.$$

Since ($\Gamma \cup \{\varphi\}$ consistent iff $\neg\varphi \notin \Gamma$), $\mathscr{g}^{+*}\Omega$ is recursive. Say $\mathscr{g}^{+*}\Omega =$

Rng h, h recursive. Let k be a recursive function such that Rng $k = \mathscr{g}^{+*}$Sent. We now define our binary recursive function f. For any $m, n \in \omega$,

$$\begin{aligned} f(m, 0) &= hm \\ f(m, n+1) &= kn \qquad \text{if } \bigwedge_{i \leq n} \mathscr{g}^{+-1}f(m, i) \rightarrow \neg \mathscr{g}^{+-1}kn \notin \Gamma, \\ f(m, n+1) &= f(m, n) \qquad \text{otherwise.} \end{aligned}$$

Now let $\Delta_m = \{\mathscr{g}^{+-1}f(m, i) : i \in \omega\}$ and $\Theta_m = \{\varphi \in \text{Sent} : \Delta_m \models \varphi\}$. Thus (*ii*) automatically holds. Now

(1) $\qquad \Gamma \cup \{\mathscr{g}^{+-1}f(m, i) : i < n\}$ is consistent, for each $m, n \in \omega$.

We prove (1) by induction on n, with m fixed. Since Γ is given to be consistent, the case $n = 0$ is trivial. The definition of Ω and h immediately gives the desired result for $n = 1$. Assume inductively that $n > 1$. If (1) fails for n, then

$$\bigwedge_{i<n-1} \mathscr{g}^{+-1}f(m, i) \rightarrow \neg \mathscr{g}^{+-1}f(m, n-1) \in \Gamma$$

Hence $f(m, n-1) \neq k(n-2)$ by the way f was defined, so $f(m, n-1) = f(m, n-2)$. But then (1) fails for $n - 1$, contradiction. Thus (1) holds. Next,

(2) $\qquad \Gamma \subseteq \Delta_m \qquad$ for each $m \in \omega$.

For, let $\varphi \in \Gamma$. Say $\varphi = \mathscr{g}^{+-1}kn$. Then $\bigwedge_{i \leq n} \mathscr{g}^{+-1}f(m, i) \rightarrow \neg \mathscr{g}^{+-1}kn \notin \Gamma$; for otherwise, since also $\mathscr{g}^{+-1}kn \in \Gamma$ we would have $\Gamma \cup \{\mathscr{g}^{+-1}f(m, i) : i \leq n\}$ inconsistent, contradicting (1). Hence $f(m, n+1) = kn$, and $\varphi \in \Delta_m$. Also note the following consequence of (1):

(3) $\qquad \Theta_m$ is consistent, for each $m \in \omega$.

Next,

(4) $\qquad \Theta_m$ is complete, for each $m \in \omega$.

For, let φ be any sentence; say $\varphi = \mathscr{g}^{+-1}kn$. If $\bigwedge_{i \leq n} \mathscr{g}^{+-1}f(m, i) \rightarrow \neg \mathscr{g}^{+-1}kn \notin \Gamma$, then $f(m, n+1) = kn$ and hence $\varphi \in \Theta_m$. If $\bigwedge_{i \leq n} \mathscr{g}^{+-1}f(m, i) \rightarrow \neg \mathscr{g}^{+-1}kn \in \Gamma$, then by (2) we see that $\Delta_m \vdash \neg\varphi$, so $\neg\varphi \in \Theta_m$. Thus (4) holds.

It remains only to establish (*i*). By (2), $\Gamma \subseteq \bigcap_{i<\omega} \Theta_i$. Suppose φ is any sentence not in Γ. Then $\Gamma \cup \{\neg\varphi\}$ is consistent, so $\neg\varphi \in \Omega$. Say $\mathscr{g}^{+-1}hm = \neg\varphi$. Thus $f(m, 0) = hm$ implies that $\neg\varphi \in \Delta_m \subseteq \Theta_m$, so by (3), $\varphi \notin \Theta_m$. Hence $\varphi \notin \bigcap_{i<\omega} \Theta_i$. Thus (*i*) holds.

Conversely, suppose that $\Theta_0, \Theta_1, \ldots$ are complete consistent extensions of Γ such that (*i*) and (*ii*) hold. Since Γ is recursively axiomatizable, by 15.1 it follows that $\mathscr{g}^{+*}\Gamma$ is r.e. Hence it suffices to show that $\omega \sim \mathscr{g}^{+*}\Gamma$ is r.e. First we give an informal proof for this. Begin listing members of Θ_0 (note that $g^{+*}\Theta_m$ is r.e. for all m). After listing a few of them, list a few members of Θ_1. Then list more members of Θ_0, then of Θ_1, then of Θ_2. Then list more members of Θ_0, etc. In this listing process, if $\neg\varphi$ appears, then on a separate list put φ. Since $\Gamma \subseteq \Theta_m$ and Θ_m is consistent, each such φ fails to be in Γ. Since each Θ_m is complete and (i) holds, each sentence not in Γ eventually appears on the list.

More formally, define a partial recursive function h as follows. For any $x \in \omega$,

$$hx = \mu y(x \notin g^{+*}\text{Sent, or } x \in g^{+*}\text{Sent and for all } i \leq 1(y)_0[((y)_0)_i = f((y)_1, i)] \text{ and } (y)_2 \text{ is the Gödel number of a } 0\text{-proof of } \bigwedge_{i \leq 1(y)0} g^{+-1}((y)_0)_i \rightarrow \neg g^{+-1}x). \tag{5}$$

We claim that $\text{Dmn}\, h = \omega \sim g^{+*}\Gamma$ (as desired). For, first let $x \in \text{Dmn}\, h$. If $x \notin g^{+*}\text{Sent}$, obviously $x \in \omega \sim g^{+*}\Gamma$. Assume that $x = g^+\varphi$, where φ is a sentence. Since $x \in \text{Dmn}\, h$, we get a y as indicated in (5). Let $m = (y)_1$. Then by (ii), $\neg\varphi \in \Theta_m$, so $\varphi \notin \Theta_m$ since Θ_m is consistent. Thus $\varphi \notin \Gamma$ by (i), so $x \in \omega \sim g^{+*}\Gamma$. Second, let $x \in \omega \sim g^{+*}\Gamma$. If $x \notin g^{+*}\text{Sent}$, obviously $x \in \text{Dmn}\, h$. Suppose $x = g^+\varphi$, where φ is a sentence. Thus since $x \notin g^{+*}\Gamma$ we have $\varphi \notin \Gamma$. By (i) choose m such that $\varphi \notin \Theta_m$; hence $\neg\varphi \in \Theta_m$ by the completeness of Θ_m. By (ii) and the deduction theorem there are $\psi_1, \ldots, \psi_n \in \Delta_m$ such that $\vdash \psi_1 \wedge \cdots \wedge \psi_n \rightarrow \neg\varphi$. By (ii), choose p such that $g^+\psi_1, \ldots, g^+\psi_m \in \{f(m, i) : i \leq p\}$. Clearly then we still have $\vdash \bigwedge_{i \leq p} g^{+-1}f(m, i) \rightarrow \neg\varphi$. Now we let $q = \prod_{i \leq p} p_i^{f(m,i)}$, let r be the Gödel number of a 0-proof of $\bigwedge_{i \leq p} g^{+-1}f(m, i) \rightarrow \neg\varphi$, and let $y = 2^q \cdot 3^m \cdot 5^r$. Clearly y satisfies the conditions in (5), so $x \in \text{Dmn}\, h$. □

Some Generalizations of the Concept of Undecidability and their Relationships to Each Other and to Some Logical Notions

Definition 15.7. Let Γ be a theory in a language $\mathscr{L}$. We say that Γ is *essentially undecidable* if Γ is consistent and whenever Δ is a consistent theory in $\mathscr{L}$ with $\Gamma \subseteq \Delta$, it follows that Δ is undecidable. The theory Γ is *inseparable* if $g^{+*}\Gamma$ and $g^{+*}\{\varphi : \varphi \text{ is a sentence and } \neg\varphi \in \Gamma\}$ are effectively inseparable (see Chapter 6). And we say that Γ is *finitely inseparable* provided that $g^{+*}\text{Thm}_{\mathscr{L}}$ and $g^{+*}\{\varphi : \varphi \text{ is a sentence whose negation holds in some finite model of } \Gamma\}$ are effectively inseparable.

Some remarks on the origin and intuitions underlying these notions are in order. There are some theories which are undecidable but which can be easily extended to consistent decidable theories. An example is the theory Γ of one binary relation, which is proved undecidable in the next chapter. If we adjoin to this theory the axiom $\forall v_0 \forall v_1 (v_0 = v_1)$, we obviously get a consistent decidable theory. (See Exercise 15.24.) Thus Γ is not essentially undecidable. Most theories with this property of being undecidable but not essentially undecidable can be shown to actually be finitely inseparable. To see the logical significance of the latter notion, first note:

(*) if Γ is a finitely axiomatizable theory, then $M = \mathscr{g}^{+*}\{\varphi : \varphi$ is a sentence whose negation holds in some finite model of $\Gamma\}$ is r.e.

In fact, we can assume that the language has only finitely many nonlogical constants, and we list all members of M as follows. Begin listing all finite $\mathscr{L}$-structures with universe $m \in \omega \sim 1$ (any finite $\mathscr{L}$-structure will be isomorphic to one of this form). Simultaneously start listing all sentences. For each finite $\mathscr{L}$-structure $\mathfrak{A}$, check if $\mathfrak{A}$ is a model of Γ; since Γ is finitely axiomatizable, this can be done in finitely many steps. If $\mathfrak{A}$ is determined to be a model of Γ, check each sentence already listed for truth in $\mathfrak{A}$. For those sentences of the form $\neg\varphi$ determined to hold in $\mathfrak{A}$ we put $\mathscr{g}^{+}\varphi$ on our output list. Clearly this gives an effective enumeration of M. Thus (*) holds. A rigorous proof of (*) would be rather lengthy, but is clearly possible (weak Church's thesis).

Thus we see that if Γ is finitely inseparable, then the following is true:

(†) There is an automatic procedure P such that, given any consistent finitely axiomatizable theory Δ with the property that every finite model of Γ is a model of Δ, the procedure P produces after finitely many steps a sentence φ such that $\varphi \notin \Delta$ but φ holds in all finite models of Δ.

In fact, let f be a recursive function satisfying the definition 6.21 for the effectively inseparable pair $\mathscr{g}^{+*}\mathrm{Thm}_{\mathscr{L}}$ and $M = \mathscr{g}^{+*}\{\varphi : \varphi$ is a sentence whose negation holds in some finite model of $\Gamma\}$. Let Δ be as in (†), and set $A = \mathscr{g}^{+*}\Delta \cup \{x : x$ is not the Gödel number of a sentence$\}$ and $B = \mathscr{g}^{+*}\{\varphi :$ φ is a sentence whose negation holds in some finite model of $\Delta\}$. Thus $\mathscr{g}^{+*}\mathrm{Thm}_{\mathscr{L}} \subseteq A$, $M \subseteq B$, $A \cap B = 0$, and A and B are r.e. (using (*)). Say $A = \mathrm{Dmn}\, \varphi^1_e$ and $B = \mathrm{Dmn}\, \varphi^1_r$. Then $f(e, r) = \mathscr{g}^{+}\varphi$ for some φ satisfying the conclusion of (†).

On the other hand, it is clear from Definition 15.7 that any finitely inseparable theory Γ is undecidable. The property (†) exhibits another property of Γ: no finitely axiomatizable extension of Γ is determined by its finite models.

Roughly speaking, a finitely inseparable theory is a weak theory which is undecidable. However, there are consistent theories which are intrinsically undecidable, in the sense that all their consistent extensions are undecidable. Most such essentially undecidable theories can be shown to be inseparable, a stronger property. An important property of inseparable theories is that they are effectively incompletable, in the following sense:

(#) if Γ is an inseparable theory, then there is an automatic procedure P such that if Δ is any consistent recursively axiomatizable extension of Γ then P yields a sentence φ such that $\varphi \notin \Delta$ and $\neg\varphi \notin \Delta$.

In fact, let f be a recursive function satisfying Definition 6.21 for the effectively inseparable pair $g^{+*}\Gamma$ and $M = g^{+*}\{\varphi : \varphi \text{ is a sentence and } \neg\varphi \in \Gamma\}$. Given Δ as in (#), set $A = g^{+*}\Delta \cup \{x : x \text{ is not the Gödel number of a sentence}\}$ and $B = g^{+*}\{\varphi : \varphi \text{ is a sentence and } \neg\varphi \in \Delta\}$. Then $g^{+*}\Gamma \subseteq A$, $M \subseteq B$, $A \cap B = 0$, both A and B are r.e., say $A = \text{Dmn}\,\varphi_e^1$ and $B = \text{Dmn}\,\boldsymbol{\varphi}_r^1$. Then $f(e, r) = g^+\varphi$ for some φ as desired in (#).

The following theorem gives a rigorous version of (#):

Theorem 15.8. *Let Γ be an inseparable theory. Then there is a recursive function h such that if Δ is any consistent extension of Γ in $\mathscr{L}$ and if $g^{+*}\Delta = \text{Dmn}\,\boldsymbol{\varphi}_e^1$, then $he = g^+\varphi$ for some sentence φ such that $\varphi \notin \Delta$* and $\neg\varphi \notin \Delta$.

PROOF. Let f be as above. We define a partial recursive function k' by setting

$$k'(x, e) \simeq \mu y[(e, x, y) \in \mathrm{T}_1 \text{ or } x \notin g^{+*}\text{Sent}_{\mathscr{L}}],$$

for all $x, e \in \omega$. Say $k' = \boldsymbol{\varphi}_r^2$. Note that

$$\text{Dmn}\,k' = \{(x, e) : x \in \text{Dmn}\,\boldsymbol{\varphi}_e^1 \text{ or } x \notin g^{+*}\text{Sent}_{\mathscr{L}}\}.$$

Let $ke = \mathrm{s}_1^1(r, e)$ for all $e \in \omega$. Then for any $x \in \omega$,

$$\boldsymbol{\varphi}_{ke}^1 x \simeq \boldsymbol{\varphi}_r^2(x, e),$$

and hence

(1) $$\text{Dmn}\,\boldsymbol{\varphi}_{ke}^1 = \text{Dmn}\,\boldsymbol{\varphi}_e^1 \cup (\omega \sim g^{+*}\text{Sent}_{\mathscr{L}}).$$

Next, we define a partial recursive function l' by setting

$$l'(x, e) = \boldsymbol{\varphi}'(\neg' x, e)$$

for all $x, e \in \omega$ (recall the definitions of $\neg'$ and $\boldsymbol{\varphi}'$ from 5.8 and 10.6). Say $l' = \boldsymbol{\varphi}_s^2$. Let $le = \mathrm{s}_1^1(s, e)$ for all $e \in \omega$. Then for any $x \in \omega$,

$$\boldsymbol{\varphi}_{le}^1 \simeq \boldsymbol{\varphi}_s^2(x, e)$$

and hence

(2) $$\text{Dmn}\,\boldsymbol{\varphi}_{le}^1 = \{x : \neg' x \in \text{Dmn}\,\boldsymbol{\varphi}_e^1\}.$$

For any $e \in \omega$, let $he = f(ke, le)$. Now suppose that Δ is a consistent extension of Γ and $g^{+*}\Delta = \text{Dmn}\,\boldsymbol{\varphi}_e^1$. By (1) and (2), $\text{Dmn}\,\boldsymbol{\varphi}_{ke}^1 = A$ and $\text{Dmn}\,\boldsymbol{\varphi}_{le}^1 = B$, with A and B as in the discussion preceding this theorem. Hence $he = f(ke, le) \in \omega \sim (A \cup B)$, so $he = g^+\varphi$ for some sentence φ with $\varphi \notin \Delta$ and $\neg\varphi \notin \Delta$, as desired. □

Next we list some simple properties of the notions in 15.7.

Proposition 15.9. *Every inseparable theory is essentially undecidable, and every essentially undecidable theory is undecidable. Every finitely inseparable theory is undecidable. If Γ is finitely inseparable, then $\{g^+\varphi : \varphi$ is a sentence which holds in every finite model of $\Gamma\}$ is not recursive.*

Proof. Let Γ be an inseparable theory in $\mathscr{L}$. Thus $\mathscr{g}^{+*}\Gamma$ and $A = \mathscr{g}^{+*}\{\varphi : \varphi$ is a sentence and $\neg\varphi \in \Gamma\}$ are disjoint, by 6.21, so Γ is consistent. Suppose that Δ is a consistent theory in $\mathscr{L}$ with $\Gamma \subseteq \Delta$. Let $B = \mathscr{g}^{+*}\{\varphi :$ φ is a sentence and $\neg\varphi \in \Delta\}$. Thus $\mathscr{g}^{+*}\Gamma \subseteq \mathscr{g}^{+*}\Delta$, $A \subseteq B$, and $\mathscr{g}^{+*}\Delta \cap B = 0$. Clearly B is recursive if $\mathscr{g}^{+*}\Delta$ is recursive, so the recursive inseparability of $\mathscr{g}^{+*}\Gamma$ and A implies that $\mathscr{g}^{+*}\Delta$ is not recursive, i.e., Δ is undecidable. Hence Γ is essentially undecidable. Trivially, every essentially undecidable theory is undecidable.

Now suppose that Γ is a finitely inseparable theory in $\mathscr{L}$, and let $C = \{\mathscr{g}^{+}\varphi : \varphi$ is a sentence which holds in every finite model of $\Gamma\}$. By 15.7, $\mathscr{g}^{+*}\mathrm{Thm}_{\mathscr{L}}$ and $\mathscr{g}^{+*}\mathrm{Sent}_{\mathscr{L}} \sim C$ are effectively inseparable. If Γ is decidable, then $\mathscr{g}^{+*}\Gamma$ is a recursive set containing $\mathscr{g}^{+*}\mathrm{Thm}_{\mathscr{L}}$ and disjoint from $\mathscr{g}^{+*}\mathrm{Sent}_{\mathscr{L}} \sim C$, which is impossible. Thus Γ is undecidable. If C is recursive, C is a recursive set containing $\mathscr{g}^{+*}\mathrm{Thm}_{\mathscr{L}}$ and disjoint from $\mathscr{g}^{+*}\mathrm{Sent}_{\mathscr{L}} \sim C$, which is again impossible. □

There are examples of undecidable theories which are not essentially undecidable (see the comments following 15.7); essentially undecidable theories which are not inseparable; finitely inseparable theories which are not essentially undecidable and hence not inseparable (again see the comments following 15.7); inseparable (hence essentially undecidable and undecidable) theories which are not finitely inseparable (see the comment following 16.2); and decidable theories Γ with $\mathscr{g}^{+*}\{\varphi : \varphi$ holds in all finite models of $\Gamma\}$ nonrecursive (15.25).

Now we want to consider the effect of extending theories on the above notions. Recall the definition of extensions of theories from 11.29. The following result generalizes the definition of essentially undecidable theories.

Proposition 15.10 *Let Γ be a theory in $\mathscr{L}$, Δ a consistent theory in an effective expansion $\mathscr{L}'$ of $\mathscr{L}$, and assume that $\Gamma \subseteq \Delta$. Then Γ essentially undecidable implies Δ essentially undecidable, and Γ inseparable implies Δ inseparable.*

Proof. Assume the first sentence of 15.10. Let Γ be essentially undecidable. Suppose that Θ is a consistent decidable extension of Δ in $\mathscr{L}'$. Then $\Theta \cap \mathrm{Sent}_{\mathscr{L}}$ is a consistent decidable extension of Γ in $\mathscr{L}$, contradiction. Next, suppose Γ is inseparable. Since $\mathscr{g}^{+*}\Gamma \subseteq \mathscr{g}^{+*}\Delta$ and $\mathscr{g}^{+*}\{\varphi : \varphi$ is a sentence of $\mathscr{L}$ and $\neg\varphi \in \Gamma\} \subseteq \mathscr{g}^{+*}\{\varphi : \varphi$ is a sentence of $\mathscr{L}'$ and $\neg\varphi \in \Delta\}$, while the two sets on the right of these inclusions are disjoint, it is clear that Δ is inseparable. □

There is an example of theories Γ, Δ in a language $\mathscr{L}$ such that $\Gamma \subseteq \Delta$, Δ is inseparable, but Γ is decidable (see 15.27). On the other hand, we have:

Proposition 15.11. *If Γ and Δ are theories in a language $\mathscr{L}$, and Δ is an extension of Γ, then Δ finitely inseparable implies Γ finitely inseparable.*

Definition 15.12. Let Γ and Δ be theories in languages $\mathscr{L}$ and $\mathscr{L}'$, Δ an extension of Γ (hence $\mathscr{L}'$ an expansion of $\mathscr{L}$). We say that Δ is a *finite extension* of Γ provided that there is a finite set Θ of sentences of $\mathscr{L}'$ such that $\Gamma \cup \Theta$ axiomatizes Δ.

Proposition 15.13. *If* Γ *and* Δ *are theories in* $\mathscr{L}$ *and* Δ *is a finite extension of* Γ, *then* Δ *undecidable implies* Γ *undecidable.*

PROOF. Let Θ be a finite set of sentences such that $\Gamma \cup \Theta$ axiomatizes Δ. Then for any sentence φ, we have $\varphi \in \Delta$ iff $\bigwedge \Theta \rightarrow \varphi \in \Gamma$, so Γ is undecidable. □

The following result enables one to prove that some rather weak theories are undecidable.

Proposition 15.14. *Let* $\mathscr{L}'$ *be an effective expansion of a language* $\mathscr{L}$. *Suppose* Γ *is a theory in* $\mathscr{L}$, Δ *is a theory in* $\mathscr{L}'$, *and* $\Gamma \cup \Delta$ *is consistent. If* Γ *is finitely axiomatizable and essentially undecidable, then* Δ *is undecidable.*

PROOF. Let $\Theta = \{\varphi : \varphi$ is a sentence of $\mathscr{L}'$ and $\Gamma \cup \Delta \models \varphi\}$. Clearly Θ is a theory in $\mathscr{L}'$ which is an extension of Γ and a finite extension of Δ. By 15.10, Θ is essentially undecidable, so by 15.13, Δ is undecidable. □

In case we deal with definitional expansions, it is to be expected that we obtain complete equivalences between our notions for a theory and its expansion:

Proposition 15.15. *Let* $(\Gamma', \mathscr{L}')$ *be an effective definitional expansion of* $(\Gamma, \mathscr{L})$. *Then:*

(*i*) Γ *is decidable iff* Γ' *is decidable;*
(*ii*) Γ *is essentially undecidable iff* Γ' *is essentially undecidable;*
(*iii*) Γ *is inseparable iff* Γ' *is inseparable;*
(*iv*) Γ *is finitely inseparable iff* Γ' *is finitely inseparable.*

PROOF. Let the notation be as in 11.29 and 11.30. For any sentence φ of $\mathscr{L}$, by 11.31 we know that $\varphi \in \Gamma$ iff $\varphi \in \Gamma'$. Hence Γ' decidable $\Rightarrow$ Γ decidable. By 11.30 we know that for any sentence ψ of $\mathscr{L}'$, we have $\psi \in \Gamma'$ iff $\psi' \in \Gamma$; because the formation of ψ' is effective, Γ decidable $\Rightarrow$ Γ' decidable. Thus (*i*) holds.

The implication from left to right in (*ii*) is a special case of 15.10. Now assume that Γ' is essentially undecidable, and $\Gamma \subseteq \Delta$, where Δ is a consistent theory in $\mathscr{L}$. Let Δ' be the theory in $\mathscr{L}'$ with axioms $\Delta \cup \Gamma'$. Then Δ' is a definitional expansion of Δ; furthermore, since $\Gamma' \subseteq \Delta'$, it follows that Δ' is undecidable. Hence by (*i*) of our present theorem, proved above, Δ is undecidable.

The implication from left to right in (*iii*) is immediate from the definitions involved. Now suppose Γ' is inseparable. Let $B = g^{+*}\{\varphi : \varphi \in \mathrm{Sent}_{\mathscr{L}'}, \neg\varphi \in \Gamma'\}$, $D = g^{+*}\{\varphi : \varphi \in \mathrm{Sent}_{\mathscr{L}}, \neg\varphi \in \Gamma\}$. Thus we are given that $g^{+*}\Gamma'$ and B are effectively inseparable sets. Now by 11.30 and 11.31 we know that for any sentence φ of $\mathscr{L}'$, $\varphi \in \Gamma'$ iff $\varphi' \in \Gamma$, and $g^+\varphi \in B$ iff $g^+\varphi' \in D$. Hence from 6.25 it follows that $g^{+*}\Gamma$ and D are effectively inseparable. Hence Γ is inseparable.

For (*iv*), suppose Γ is finitely inseparable; thus $g^{+*}\mathrm{Thm}_{\mathscr{L}}$ and $B = \{\varphi : \varphi \in \mathrm{Sent}_{\mathscr{L}}, \neg\varphi$ holds in some finite model of $\Gamma\}$ are effectively inseparable. As in the proof of 11.31 we see that any model of Γ can be expanded to a model of Γ'. Hence $g^{+*}\mathrm{Thm}_{\mathscr{L}} \subseteq g^{+*}\mathrm{Thm}_{\mathscr{L}'}$, and $B \subseteq \{\varphi \in \mathrm{Sent}_{\mathscr{L}'} : \neg\varphi$ holds in some finite model of $\Gamma'\}$; it follows that Γ' is finitely inseparable. Finally, assume that Γ' is finitely inseparable. For each sentence φ of $\mathscr{L}'$, let Δ_ϕ be the set of all existence and uniqueness conditions 11.29(*ii*) for the new operation symbols occurring in φ. Let $\varphi^* = \varphi'$ (defined in the proof of 11.30) if $\Delta_\phi = 0$, and let it be $\bigwedge \Delta_\phi \rightarrow \varphi'$ if $\Delta_\phi \neq 0$. Furthermore, let Θ_ϕ be the set consisting of all definitions for the new symbols in φ. Note that by the proof of 11.30, $\Theta_\phi \models \varphi \leftrightarrow \varphi'$. Now it suffices to prove

(1) if $\varphi \in \mathrm{Sent}_{\mathscr{L}'}$ and $\models\varphi$, then $\models\varphi^*$;

(2) if $\varphi \in \mathrm{Sent}_{\mathscr{L}'}$ and $\neg\varphi$ holds in some finite model of Γ', then $\neg\varphi^*$ holds in some finite model of Γ.

To prove (1), assume that $\varphi \in \mathrm{Sent}_{\mathscr{L}'}$ and $\models\varphi$. To check that $\models\varphi^*$, let $\mathfrak{A}$ be any $\mathscr{L}$-structure which is a model of Δ_ϕ. Obviously $\mathfrak{A}$ can be expanded to an $\mathscr{L}'$-structure $\mathfrak{B}$ which is a model of Θ_ϕ. Since $\models\varphi$ and $\Theta_\phi \models \varphi \leftrightarrow \varphi'$, it follows that $\mathfrak{B} \models \varphi'$ and hence $\mathfrak{A} \models \varphi'$, as desired. To prove (2), assume that $\varphi \in \mathrm{Sent}_{\mathscr{L}'}$, and $\mathfrak{A} \models \neg\varphi$, where $\mathfrak{A}$ is a finite model of Γ'. Then $\mathfrak{A} \upharpoonright \mathscr{L}$ is a model of $\Delta_\phi \cup \Gamma$ and of $\neg\varphi'$, since $\Gamma' \models \neg\varphi \leftrightarrow \neg\varphi'$, so $\mathfrak{A} \upharpoonright \mathscr{L} \models \neg\varphi^*$. □

Next we consider the general notion of interpretation of one theory into another (11.43–11.46).

Proposition 15.16. *Let $\mathscr{L}$ and $\mathscr{L}'$ be two languages, Γ and Δ consistent theories in $\mathscr{L}$ and $\mathscr{L}'$ respectively, and suppose that $\mathfrak{A}$ is an effective interpretation of Γ in Δ. Then:*

(*i*) *Γ essentially undecidable $\Rightarrow$ Δ essentially undecidable.*
(*ii*) *Γ inseparable $\Rightarrow$ Δ inseparable.*
(*iii*) *Suppose also that for every model $\mathfrak{C}$ of Γ there is a model $\mathfrak{B}$ of Δ such that $\mathfrak{C} = \mathfrak{B}^{\Delta\mathfrak{A}}$. Then Γ undecidable $\Rightarrow$ Δ undecidable.*
(*iv*) *Suppose that for each finite model $\mathfrak{C}$ of Γ there is a finite model $\mathfrak{B}$ of Δ such that $\mathfrak{C} = \mathfrak{B}^{\Delta\mathfrak{A}}$. Then Δ finitely inseparable $\Rightarrow$ Δ finitely inseparable.*

PROOF.

(*i*) Let Δ' be any consistent theory in $\mathscr{L}'$ which extends Δ. Set $\Gamma' = \{\varphi \in \mathrm{Sent}_{\mathscr{L}} : \Delta' \models \varphi^{\mathfrak{A}}\}$. Thus $\Gamma \subseteq \Gamma'$ since $\mathfrak{A}$ is an interpretation of Γ in Δ. Now Γ' is a theory in $\mathscr{L}$, for if $\Gamma' \models \varphi$, then $\Theta \models \varphi$ for some finite subset Θ

of Γ', hence $\models \bigwedge \Theta \rightarrow \varphi$, $\models \bigwedge \{\psi^{\mathfrak{A}} : \psi \in \Theta\} \rightarrow \varphi^{\mathfrak{A}}$ by 11.46, so $\Delta' \models \varphi^{\mathfrak{A}}$ and $\varphi \in \Gamma'$. Γ' is consistent because Δ' is. Therefore, Γ' is undecidable (since we assume that Γ is essentially undecidable). The formation of $\varphi^{\mathfrak{A}}$ is recursive by 11.44, so Δ' is undecidable, as desired.

(*ii*) The implication here is essentially a trivial consequence of 6.25. Let $A = \mathcal{g}^{+*}\Gamma$ and $B = \mathcal{g}^{+*}\{\varphi : \varphi \in \mathrm{Sent}_{\mathscr{L}}, \neg\varphi \in \Gamma\}$. Thus A and B are effectively inseparable. Correspondingly, let $C = \mathcal{g}^{+*}\Delta$ and $D = \mathcal{g}^{+*}\{\varphi : \varphi \in \mathrm{Sent}_{\mathscr{L}'}, \neg\varphi \in \Delta\}$. Note that $C \cap D = 0$ since Δ is consistent. Now we define a function $f: \omega \rightarrow \omega$: for any $x \in \omega$,

$$fx = 0 \qquad \text{if } x \notin \mathcal{g}^{+*}\mathrm{Sent}_{\mathscr{L}},$$
$$fx = \mathcal{g}^{+}(\mathcal{g}^{+-1}x)^{\mathfrak{A}} \qquad \text{if } x \in \mathcal{g}^{+*}\mathrm{Sent}_{\mathscr{L}}.$$

It is easily checked that $A \subseteq f^{-1*}C$ and $B \subseteq f^{-1*}D$. Since f is recursive by 11.44, it follows that C and D are effectively inseparable, by 6.25. Thus Δ is inseparable.

(*iii*) The added hypothesis implies, as we shall see, that for any sentence φ of $\mathscr{L}$, $\varphi \in \Gamma$ iff $\varphi^{\mathfrak{A}} \in \Delta$; hence clearly Δ decidable implies Γ decidable. In fact, $\varphi \in \Gamma \Rightarrow \varphi^{\mathfrak{A}} \in \Delta$ even without the added hypothesis, by 11.46. Now assume that $\varphi^{\mathfrak{A}} \in \Delta$. Let $\mathfrak{C}$ be any model of Γ. By our added hypothesis, $\mathfrak{C} = \mathfrak{B}^{\Delta\mathfrak{A}}$ for some model $\mathfrak{B}$ of Δ. Thus $\mathfrak{B} \models \varphi^{\mathfrak{A}}$, so by 11.45, $\mathfrak{C} \models \varphi$. Hence $\Gamma \models \varphi$ and $\varphi \in \Gamma$, as desired.

(*iv*) We can prove this similarity to (*ii*); we just need to establish the following two results:

(1) for any sentence φ of $\mathscr{L}$, $\models\varphi$ implies $\models\varphi^{\mathfrak{A}}$;

(2) for any sentence φ of $\mathscr{L}$, if $\neg\varphi$ holds in some finite model of Γ, then $(\neg\varphi)^{\mathfrak{A}}$ holds in some finite model of Δ.

Now (1) is immediate from 11.46, since any syntactical $\mathscr{L}$-structure in $\mathscr{L}'$ is an interpretation of $\mathrm{Thm}_{\mathscr{L}}$ in $\mathrm{Thm}_{\mathscr{L}'}$ (obviously). Condition (2) follows from our extra hypothesis for (*iv*) along with 11.45. □

The following variant of 15.16(*iv*) will be useful in Chapter 16.

Proposition 15.17. *Let $\mathscr{L}$ and $\mathscr{L}'$ be two languages, and let Γ and Δ be consistent theories in $\mathscr{L}$ and $\mathscr{L}'$ respectively. Assume that $\mathscr{L}$ has only finitely many operation symbols (with no restriction on the number of relation symbols). Let $\mathfrak{A} = (\chi, f, R, \Gamma')$ be an effective syntactical $\mathscr{L}$-structure in $\mathscr{L}'$ such that Γ' is finite. Finally, suppose that for each finite model $\mathfrak{C}$ of Γ there is a finite model $\mathfrak{B}$ of $\Delta \cup \Gamma'$ such that $\mathfrak{C} = \mathfrak{B}^{\Gamma'\mathfrak{A}}$. Then Γ finitely inseparable $\Rightarrow$ Δ finitely inseparable.*

Proof. By virtue of 6.25 it suffices to establish the following:

(1) for any sentence φ of $\mathscr{L}$, $\models\varphi$ implies $\models \bigwedge \Gamma' \rightarrow \varphi^{\mathfrak{A}}$;

(2) for any sentence φ of $\mathscr{L}$, if $\neg\varphi$ holds in some finite model of Γ, then $\neg(\bigwedge\Gamma' \rightarrow \varphi^{\mathfrak{A}})$ holds in some finite model of Δ.

To prove (1), assume that $\models\varphi$ and that $\mathfrak{B}$ is an $\mathscr{L}'$-structure which is a model of $\bigwedge\Gamma'$. Then $\mathfrak{B}^{\Gamma'\mathfrak{A}} \models \varphi$, so by 11.45, $\mathfrak{B} \models \varphi^{\mathfrak{A}}$, as desired in (1).

To prove (2) let $\mathfrak{C}$ be a finite model of Γ such that $\mathfrak{C} \models \neg\varphi$. By hypothesis of our proposition let $\mathfrak{B}$ be a finite model of $\Delta \cup \Gamma'$ such that $\mathfrak{C} = \mathfrak{B}^{\Gamma'\mathfrak{A}}$. Thus by 11.45, $\mathfrak{B} \models \neg\varphi^{\mathfrak{A}}$, as desired in (2). □

Connecting the Notions of Implicit Definability in Chapter 14 with the Undecidability Notions

Theorem 15.18. *If Γ is a theory in $\mathscr{L}_{\text{nos}}$ in which every recursive set is weakly syntactically definable, then Γ is undecidable.*

Proof. Suppose Γ is decidable. Let

$A = \{e : e$ is the Gödel number of a formula ψ having at most v_0 free, and $\psi(\mathbf{e}) \notin \Gamma\}$.

Clearly A is recursive, by the hypothesis that Γ is decidable. Let φ weakly syntactically define A in Γ, and set $e = \mathscr{g}^+\varphi$. Then

$$\begin{aligned}\varphi(\mathbf{e}) \in \Gamma \quad &\text{iff } e \in A && \text{(since } \varphi \text{ weakly syntactically defines } A)\\ &\text{iff } \varphi(\mathbf{e}) \notin \Gamma && \text{(by definition of } A)\end{aligned}$$

This is a contradiction. □

The above proof constitutes a typical application of the diagonal method in the theory of undecidable theories. To see the idea more clearly, suppose Γ is a theory which has $\mathfrak{B} = \langle \omega, +, \cdot, \mathfrak{s}, 0\rangle$ as a model. Then if φ syntactically defines A, it follows by 14.14 that the formula φ actually elementarily defines A in $\mathfrak{B}$, in the sense of 11.7. Hence $\varphi(\mathbf{e})$ holds in $\mathfrak{B}$ iff $e \in A$, i.e., iff $\varphi(\mathbf{e}) \notin \Gamma$. Thus, in a sense, $\varphi(\mathbf{e})$ asserts of itself that it is not in Γ. (Holding in $\mathfrak{B}$ can be loosely identified with intuitive truth.)

Theorem 15.8 of course yields, along with our results of Chapter 14, that the theories R, Q, P and N are undecidable. We shall save a formal statement of this and various consequences and generalizations until the next section.

Theorem 15.19. *Let Γ be a consistent theory in $\mathscr{L}_{\text{nos}}$ with* R $\subseteq \Gamma$. *Then Γ is inseparable.*

Proof. By 6.24, let A and B be r.e. but effectively inseparable sets of integers. Since A and B are nonrecursive, they are nonempty. Hence we can write $A = \text{Rng} f$ and $B = \text{Rng}\, g$ for suitable recursive functions f and g. Note that every recursive function is syntactically definable in Γ, since this is true of R by 14.11. Say f and g are syntactically defined by formulas φ, ψ respectively in Γ. Let χ be the following formula having at most v_0 free:

$$\exists v_1\{\varphi(v_1, v_0) \wedge \forall v_2[v_2 \leq v_1 \rightarrow \neg\psi(v_2, v_0)]\}.$$

Let $hx = \mathscr{g}^+\chi(\mathbf{x})$ for each $x \in \omega$. Clearly h is a recursive function. To show that Γ is inseparable, it suffices by 6.25 to show that $A \subseteq h^{-1*}\mathscr{g}^{+*}\Gamma$ and $B \subseteq h^{-1*}\mathscr{g}^+\{\varphi \in \text{Sent} : \neg\varphi \in \Gamma\}$. That is, it suffices to prove the following two statements:

$$\text{if } x \in A, \text{ then } \Gamma \models \chi(\mathbf{x}); \tag{1}$$

$$\text{if } x \in B, \text{ then } \Gamma \models \neg\chi(\mathbf{x}). \tag{2}$$

First let $x \in A$, and choose y so that $fy = x$. Then, since φ syntactically defines f in Γ,

$$\Gamma \models \varphi(\mathbf{y}, \mathbf{x}). \tag{3}$$

Also, since $R \subseteq \Gamma$ we have

$$\Gamma \models v_2 \leq \mathbf{y} \rightarrow v_2 = \mathbf{0} \vee \cdots \vee v_2 = \mathbf{y}. \tag{4}$$

Since A and B are disjoint, $x \notin B$. Hence $\forall i \in \omega\ (gi \neq x)$, so, again since $\mathrm{R} \subseteq \Gamma$,

$$\Gamma \models \neg\Delta(gi) = \mathbf{x} \qquad \text{for all } i \in \omega. \tag{5}$$

On the other hand, by the second condition of 14.1 for ψ syntactically defining g we have

$$\Gamma \models \psi(\mathbf{i}, \mathbf{x}) \rightarrow \mathbf{x} = \Delta(gi) \qquad \text{for all } i \in \omega.$$

Putting this together with (4) and (5) we easily obtain

$$\Gamma \models v_2 \leq \mathbf{y} \rightarrow \neg\psi(v_2, \mathbf{x}).$$

Hence using (3) we easily obtain $\Gamma \models \chi(\mathbf{x})$, as desired.

Now let $x \in B$, and choose y so that $gy = x$. Since ψ syntactically defines g in Γ,

$$\Gamma \models \psi(\mathbf{y}, \mathbf{x}). \tag{6}$$

Also note that $x \notin A$, since $A \cap B = 0$. Hence $fi \neq x$ for all $i \in \omega$, and we easily infer from the second condition of φ syntactically defining f in 14.1 that

$$\Gamma \models \neg\varphi(\mathbf{i}, \mathbf{x}) \qquad \text{for all } i \in \omega. \tag{7}$$

Now let θ be the formula

$$\varphi(v_1, \mathbf{x}) \wedge \forall v_2[v_2 \leq v_1 \rightarrow \neg\psi(v_2, \mathbf{x})].$$

Clearly it suffices now to show that $\Gamma \models \neg\theta$. First note since $\mathrm{R} \subseteq \Gamma$ that

$$\Gamma \models v_1 \leq \mathbf{y} \rightarrow v_1 = \mathbf{0} \vee \cdots \vee v_1 = \mathbf{y}.$$

Hence we clearly have

$$\Gamma \models \theta \wedge v_1 \leq \mathbf{y} \rightarrow \varphi(0, \mathbf{x}) \vee \cdots \vee \varphi(\mathbf{y}, \mathbf{x}).$$

Because of (7) we then infer that

(8) $$\Gamma \models v_1 \leq \mathbf{y} \rightarrow \neg\theta.$$

On the other hand, we have

$$\Gamma \models \mathbf{y} \leq v_1 \wedge \theta \rightarrow \neg\psi(\mathbf{y}, \mathbf{x}),$$

so by (6),

(9) $$\Gamma \models \mathbf{y} \leq v_1 \rightarrow \neg\theta.$$

Now $R \subseteq \Gamma$ implies that $\Gamma \models v_1 \leq \mathbf{y} \vee \mathbf{y} \leq v_1$. Hence by (8) and (9) we have $\Gamma \models \neg\theta$, as desired. □

The notion of spectral representability will be related with our undecidability notions in the next chapter. We conclude this chapter with our first important results concerning incompleteness. The following *fixed-point*, or *self-reference* theorem is basic for the remaining results in this chapter.

Theorem 15.20. *Let Γ be a theory in $\mathscr{L}_{\text{nos}}$ in which every unary recursive function is syntactically definable, and let φ be a formula of $\mathscr{L}_{\text{nos}}$ with* $\text{Fv}\varphi \subseteq \{v_0\}$. *Then there is a sentence ψ of $\mathscr{L}_{\text{nos}}$ such that $\Gamma \models \varphi(\Delta g^+\psi) \leftrightarrow \psi$.*

Proof. For any $x \in \omega$, let

$$fx = g^+\psi(\mathbf{x}) \quad \text{if } x = g^+\psi \text{ for some formula } \psi,$$
$$fx = 0 \quad \text{otherwise.}$$

Thus f is recursive. By hypothesis, let χ be a formula of $\mathscr{L}_{\text{nos}}$ with $\text{Fv}\chi \subseteq \{v_0, v_1\}$ such that χ syntactically defines f. Let θ be the formula $\exists v_1[\varphi(v_1) \wedge \chi(v_0, v_1)]$, and let ψ be the sentence $\theta(\Delta g^+\theta)$. Thus $fg^+\theta = g^+\theta(\Delta g^+\theta) = g^+\psi$. Hence by choice of χ we have

$$\Gamma \models \chi(\Delta g^+\theta, \Delta g^+\psi).$$

Hence it obviously follows that $\Gamma \models \varphi(\Delta g^+\psi) \rightarrow \theta(\Delta g^+\theta)$, i.e., $\Gamma \models \varphi(\Delta g^+\psi) \rightarrow \psi$. On the other hand, by the second condition in 14.1 we have

$$\Gamma \models \chi(\Delta g^+\theta, v_1) \rightarrow \mathrm{v}_1 = \Delta g^+\psi,$$

and hence

$$\Gamma \models \varphi(v_1) \wedge \chi(\Delta g^+\theta, v_1) \rightarrow \varphi(\Delta g^+\psi),$$

so $\Gamma \models \psi \rightarrow \varphi(\Delta g^+\psi)$. □

As a simple application of 15.20 *let us show that Peano arithmetic, theory* P *of Chapter* 14, *is incomplete.* This result is considerably generalized in Chapter 16, so we shall not state it formally. On the other hand, it is perhaps the most interesting result so far proved in this book, and its proof is quite direct compared to our more general results. Within P it is possible to formalize essentially all of the elementary arguments found in number theory books; no exceptions are known to this author, so it is surprising that there do exist sentences of $\mathscr{L}_{\text{nos}}$ which are neither provable nor disprovable in P. From the viewpoint of the philosophy of mathematics the incompleteness of P is a very major fact, because the proof clearly works for almost any mathematical discipline. The argument that P is incomplete is based just on these previously established facts in addition to 15.20:

(1) All recursive functions and relations are syntactically definable in P.

This was proved in 14.11, 14.2 and 14.19.

(2) All sentences of P are true in $\langle \omega, +, \cdot, \mathfrak{s}, 0\rangle$.

(3) P is axiomatizable by a recursive set Δ of axioms.

The conditions (2) and (3) are obvious from the definition of P. Now we prove that P is incomplete. Let $R = \{(x, y) : x$ is the Gödel number of a Δ-proof of a formula with Gödel number $y\}$. By 10.27, R is recursive. Hence by (1) let χ be a formula with $\text{Fv}\chi \subseteq \{v_0, v_1\}$ which syntactically defines R in P. Let φ be the formula $\neg\exists v_1\chi(v_1, v_0)$. Thus $\text{Fv}\varphi \subseteq \{v_0\}$. We now apply 15.20 to obtain a sentence ψ such that $\text{P} \models \varphi(\Delta g^+\psi) \leftrightarrow \psi$. We claim that $\text{P} \not\models \psi$ and $\text{P} \not\models \neg\psi$. First suppose that $\text{P} \vdash \psi$. Let θ be a Δ-proof of ψ. Then $(g^{++}\theta, g^+\psi) \in R$, so $\text{P} \vdash \chi(\Delta g^{++}\theta, \Delta g^+\psi)$, so $\text{P} \models \exists v_1\chi(v_1, \Delta g^+\psi)$ and $\text{P} \models \neg\varphi(\Delta g^+\psi)$. This contradicts the fact that $\text{P} \models \varphi(\Delta g^+\psi) \leftrightarrow \psi$. Second, suppose that $\text{P} \vdash \neg\psi$. Then $\text{P} \models \neg\varphi(\Delta g^+\psi)$, i.e., $\text{P} \models \exists v_1\chi(v_1, \Delta g^+\psi)$. By (2), $\exists v_1\chi(v_1, \Delta g^+\psi)$ is true in $\mathfrak{A} = \langle \omega, +, \cdot, \mathfrak{s}, 0\rangle$, so $\chi(\mathbf{x}, \Delta g^+\psi)$ is true in $\mathfrak{A}$ for some $x \in \omega$. It follows that $(x, g^+\psi) \in R$, since otherwise $\text{P} \models \neg\chi(\mathbf{x}, \Delta g^+\psi)$ and $\neg\chi(\mathbf{x}, \Delta g^+\psi)$ would be true in $\mathfrak{A}$ by (2). Thus, by the definition of R, $\text{P} \vdash \psi$, contradicting the first part of this proof.

Note that ψ holds in $\langle \omega, +, \cdot, \mathfrak{s}, 0\rangle$ iff $\varphi(\Delta g^+\psi)$ holds there, i.e., as is easily seen, iff $\varphi(\Delta g^+\psi)$ is not provable in P, i.e., iff $\text{P} \not\models \psi$. (Thus ψ does hold in $\langle \omega, +, \cdot, \mathfrak{s}, 0\rangle$, and hence is intuitively true.) Thus this proof is related to the informal paradox of the man who says "I am now lying"; it is impossible that he is lying or that he is telling the truth. Of course the proof can also be considered as another instance of Cantor's diagonal method.

As another application of 15.20 we give the following important result which shows the *undefinability of the true statements of number theory.*

Theorem 15.21. (Tarski). *Let $\mathfrak{A} = \langle \omega, +, \cdot, \mathfrak{s}, 0\rangle$ and let $X = \{g^+\psi : \psi$ is a sentence of $\mathscr{L}_{\text{nos}}$ and $\mathfrak{A} \models \psi\}$. Then X is not elementarily definable in $\mathfrak{A}$.*

PROOF. Suppose X is elementarily definable by a formula φ in $\mathfrak{A}$. Thus $\text{Fv}\varphi \subseteq \{v_0\}$ and $X = {}^1\varphi^{\mathfrak{A}}$. By 15.20 choose a sentence ψ such that $\text{R} \models \neg\varphi(\Delta g^+\psi) \leftrightarrow \psi$. Since $\mathfrak{A}$ is a model of R, it follows that $\mathfrak{A} \models \neg\varphi(\Delta g^+\psi) \leftrightarrow \psi$. But then we reach a contradiction in trying to decide whether $\mathfrak{A} \models \psi$ or not: if $\mathfrak{A} \models \psi$, i.e., $g^+\psi \in X$, then $\mathfrak{A} \models \neg\varphi(\Delta g^+\psi)$ and hence $\mathfrak{A} \models \neg\varphi[g^+\psi]$, so $g^+\psi \notin X$, contradiction. Also, $\mathfrak{A} \not\models \psi$ implies $g^+\psi \notin X$, hence $\mathfrak{A} \models \neg\varphi[g^+\psi]$, $\mathfrak{A} \models \neg\varphi(\Delta g^+\psi)$ and $\mathfrak{A} \models \psi$, which is absurd, as shown above. □

Note by 14.16 and 14.17 that 15.21 really says that $\mathcal{g}^{+*}$N is not in the arithmetical hierarchy. In particular, it is not recursive, not r.e., not the complement of an r.e. set, etc.

BIBLIOGRAPHY

1. Dyson, V. H. On the decision problem for extensions of a decidable theory. *Fund. Math.*, *64* (1969), 7–70.
2. Ershov, Yu., Lavrov, I., Taimanov, A., Taitslin, M. Elementary theories. *Russian Mathematical Surveys*, *20* (1965), 35–105.
3. Hanf, W. Model-theoretic methods in the study of elementary logic. In *Model Theory*. Amsterdam: North-Holland (1965).
4. Pour-El, M. Effectively extensible theories. *J. Symb. Logic*, *33* (1968), 56–58.
5. Tarski, A., Mostowski, A., Robinson, R. M. *Undecidable Theories*. Amsterdam: North-Holland (1953).

EXERCISES

15.22. Give an example of a decidable theory with denumerably many decidable extensions. Every theory has at most countably many decidable extensions.

15.23. Let $\mathscr{L}$ be a language with unary relation symbols $\mathbf{R}_m$, $m \in \omega$, but no other nonlogical constants. Show that $\Gamma - \{\varphi \in \text{Sent}_{\mathscr{L}} : \models \varphi\}$ is decidable but has $\exp \aleph_0$ complete undecidable extensions. *Hint*: use elimination of quantifiers, as with the pure theory of equality.

15.24. (See the comments following 5.7.) Let $\mathscr{L}$ be a language with a single binary relation symbol $\mathbf{R}$, and let Γ be the theory in $\mathscr{L}$ with the axiom $\{\forall v_0 \forall v_1 (v_0 = v_1)\}$. Show that Γ is decidable.

15.25. (See the comments following 5.9.) Let $\mathscr{L}$ be a language with two unary relation symbols $\mathbf{P}$ and $\mathbf{Q}$. Let A be an r.e. set which is not recursive, say $A = \text{Rng} f$ with f recursive. Let Γ be the theory in $\mathscr{L}$ such that $\mathfrak{A} \models \Gamma$ iff the following conditions hold:

(a) $$\mathbf{P}^{\mathfrak{A}} \cap \mathbf{Q}^{\mathfrak{A}} = 0;$$
(b) $$\mathbf{P}^{\mathfrak{A}} \cup \mathbf{Q}^{\mathfrak{A}} = A;$$
(c) $$\text{if } |\mathbf{P}^{\mathfrak{A}}| = m < \omega, \text{ then } |\mathbf{Q}^{\mathfrak{A}}| = fm.$$

Show that Γ is decidable but that $\mathcal{g}^{+*}\{\varphi \in \text{Sent}_{\mathscr{L}} : \varphi$ holds in all finite models of $\mathfrak{A}\}$ is nonrecursive.

15.26. Let $\mathscr{L}'$ be an effective expansion of a language $\mathscr{L}$ obtained by adding only new individual constants, and let Γ and Γ' be theories in $\mathscr{L}$ and $\mathscr{L}'$ respectively such that $\Gamma' = \{\varphi \in \text{Sent}_{\mathscr{L}} : \Gamma \models \varphi\}$. Show:

(1) Γ undecidable $\leftrightarrow$ Γ' undecidable;
(2) Γ essentially undecidable $\leftrightarrow$ Γ' essentially undecidable;
(3) Γ inseparable $\leftrightarrow$ Γ' inseparable;
(4) Γ finitely inseparable $\leftrightarrow$ Γ' finitely inseparable.

15.27. There are theories Γ, Δ with $\Gamma \subseteq \Delta$, Δ inseparable, Γ decidable. *Hint*: take $\mathscr{L}$ and Γ as in 15.23. Let Δ have the following axioms:

$$\{\exists v_0 \mathbf{R}_m v_0 : m \in A\} \cup \{\neg\exists v_0 \mathbf{R}_m v_0 : m \in B\},$$

where A and B are effectively inseparable.

15.28′. Show in detail that the sentence ψ constructed following 15.20, such that $\mathrm{P} \nvdash \psi$ and $\mathrm{P} \nvdash \neg\psi$ is true in $(A, +, \cdot, \mathscr{s}, 0)$.

15.29. Suppose that Γ is an ω-consistent recursively axiomatizable theory in $\mathscr{L}_{\mathrm{nos}}$ in which every recursive function and relation is syntactically definable. Show that Γ is incomplete.

15.30. If Γ is finitely axiomatizable and has a finite model, then Γ is not essentially undecidable.

15.31. Any theory with only finitely many complete and decidable extensions is decidable.

15.32′. If Γ is finitely inseparable and finitely axiomatizable, then $\{\mathscr{g}^+\varphi : \varphi$ is a sentence which holds in every finite model of $\Gamma\}$ is not r.e.

Some Undecidable Theories 16

Corresponding to our list in Chapter 13 of decidable theories we begin this section with a list of undecidable theories. As we have previously indicated, most undecidable theories satisfy one of the two stronger properties of inseparability or finite inseparability, and we shall indicate these properties in the table below.

Some undecidable theories

Theory	Insepar-able?	Finitely insepar-able?	Proved by	In this book
1. $\mathrm{Th}_{\mathscr{L}}$, where (a) $\mathscr{L}$ has at least one relation symbol of rank >1, or (b) $\mathscr{L}$ has at least two unary operation symbols or (c) $\mathscr{L}$ has at least one operation symbol of rank >1.	No	Yes	Trachtenbrot, 1953	pp. 293–296
2. Theories R, Q, P, N	Yes	No	Tarski, Mostowski, Robinson, 1949	p. 280
3. Theory of $(\mathbb{Z}, +, \cdot)$	Yes	No	Tarski, Mostowski, 1949	p. 282
4. Theory of $(\mathbb{Q}, +, \cdot)$	Yes	No	J. Robinson, 1949	—
5. Theory of groups	No	Yes	Mal′cev, 1961	—

Theory	Insepar-able?	Finitely insepar-able?	Proved by	In the book
6. Theory of semi-groups	No	Yes	Mal′cev, 1961	—
7. Theory of rings	No	Yes	Mal′cev, 1961	—
8. Theory of fields	No	No	Undecidable: J. Robinson, 1949	—
9. Theory of ordered fields	No	No	Undecidable: J. Robinson, 1949	—
10. Theory of lattices	No	Yes	Taitslin, 1961	p. 297
11. Theory of distribu-tive lattices	No	Yes	Ershov, Taitslin, 1963	—
12. Theory of partial orderings	No	Yes	Taitslin, 1962	p. 297
13. Theory of two equivalence rela-tions	No	Yes	Lavrov, 1963	p. 295
14. Theory of two linear orders	No	Yes	Lavrov, 1963	—
15. Set theory ZF	Yes	No	Tarski, 1949	p. 290

In the first section of this chapter we shall be concerned with proving various theories inseparable, while in the second section we deal with finitely inseparable theories.

Immediately from 15.19 we have

Theorem 16.1. *Theories* R, Q, P, *and* N *are inseparable, and hence essentially undecidable and undecidable.*

By 15.3 we then obtain:

Theorem 16.2. *The theories* R, Q, *and* P, *as well as any of their recursively axiomatizable consistent extensions, are incomplete.*

Note that, trivially, R, Q, P and N are *not* finitely inseparable, since they do not have finite models. Less trivial examples of inseparable but not finitely inseparable theories can be found in the paper of Dyson mentioned at the end of Chapter 15.

Now we shall show how Proposition 15.16 of the last chapter can be used to take care of #3 and #15 in the above table. Intuitively speaking it is a matter of finding definitions for ω, $+$, $\cdot$, $\mathfrak{s}$ and 0 in the theory of $(\mathbb{Z}, +, \cdot)$ (for #3), and in set theory (for #15). Proceeding to the first task, we first need the following four purely number-theoretic facts, which are well known and essentially give a definition of ω in $(\mathbb{Z}, +, \cdot)$. The first is purely computational and easily checked:

Lemma 16.3. *For any integers $x_1, x_2, x_3, x_4, y_1, y_2, y_3, y_4$ we have*

$$\begin{aligned}(x_1^2 + x_2^2 + x_3^2 + x_4^2)\cdot(y_1^2 + y_2^2 + y_3^2 + y_4^2) \\ = (x_1y_1 + x_2y_2 + x_3y_3 + x_4y_4)^2 \\ + (x_1y_2 - x_2y_1 + x_3y_4 - x_4y_3)^2 \\ + (x_1y_3 - x_3y_1 + x_4y_2 - x_2y_4)^2 \\ + (x_1y_4 - x_4y_1 + x_2y_3 - x_3y_2)^2.\end{aligned}$$

Lemma 16.4. *For each prime $p > 2$ there is an m with $1 \leq m < p$ such that mp is a sum of four squares.*

PROOF. The members of $\{x^2 : 0 \leq x \leq (p-1)/2\}$ are pairwise incongruent mod p, as are the members of $\{-1 - y^2 : 0 \leq y \leq (p-1)/2\}$. There are $p + 1$ numbers in the union of these two sets, so there are x and y such that $0 \leq x, y \leq (p-1)/2$ and $x^2 \equiv -1 - y^2 \pmod p$. Say

$$x^2 + 1 + y^2 = mp.$$

Thus $1 \leq m$. Since $x, y \leq (p-1)/2$, we have

$$mp < p^2/4 + 1 + p^2/4 < p^2, \qquad \text{so } m < p,$$

as desired. □

Lemma 16.5. *For any positive prime p, p is a sum of four squares.*

PROOF. Obvious for $p = 2$ $(2 = 1 + 1 + 0 + 0)$. Suppose $p > 2$. Let m be the smallest positive integer such that mp is a sum of four squares. Thus $1 \leq m < p$ by 16.4. Say $mp = x_1^2 + x_2^2 + x_3^2 + x_4^2$. Suppose $m > 1$; we shall get a contradiction.

Now

(1) $\qquad m$ is odd.

For, suppose m is even. Then $x_1^2 + x_2^2 + x_3^2 + x_4^2$ is even, so either (1) all four of x_1, x_2, x_3, x_4 are even, (2) two are even, two are odd, or (3) all four are odd. In case (2) we may assume that x_1 and x_2 are even. Then in any of the cases (1)–(3),

$$\begin{aligned}(m/2)p = [(x_1 + x_2)/2]^2 + [(x_1 - x_2)/2]^2 \\ + [(x_3 + x_4)/2]^2 + [(x_3 - x_4)/2]^2,\end{aligned}$$

and all the entries [ ] on the right are integers. This contradicts the minimality of m. Thus (1) holds.

Now the members of $T = \{y : -(m-1)/2 \leq y \leq (m-1)/2\}$ are pairwise incongruent mod m, and $|T| = m$. Hence there exist $y_i \in T$ such that $x_i \equiv y_i \pmod m$, $i = 1, \ldots, 4$. Thus

$$y_1^2 + y_2^2 + y_3^2 + y_4^2 \equiv x_1^2 + x_2^2 + x_3^2 + x_4^2 \equiv 0 \qquad \pmod m,$$

say

$$(2) \qquad y_1^2 + y_2^2 + y_3^2 + y_4^2 = mn.$$

Then $n \neq 0$, since otherwise $y_i = 0$ for each $i = 1, \ldots, 4$ and $m|x_i$ each $i = 1, \ldots, 4$, so $m^2|x_1^2 + \cdots + x_4^2 = mp$ hence $m|p$, contradiction. Also, $mn < 4(m/2)^2 = m^2$, so $n < m$. Now $(mn)(mp)$ is a sum of four squares by Lemma 16.3 and as is easily seen, each expression ( ) on the right in 16.3 is divisible by m. Hence np is a sum of four squares. But $0 < n < m$, contradiction. □

Theorem 16.6 (LaGrange). *Let a be an integer. The following are equivalent*:

(*i*) $a \in \omega$,
(*ii*) *a is a sum of four squares.*

Using this celebrated theorem of LaGrange we can take care of the theory of $(\mathbb{Z}, +, \cdot)$:

Theorem 16.7. *Let $\mathscr{L}$ be a language appropriate for $\mathfrak{A} = (\mathbb{Z}, +, \cdot)$, and let Γ be the set of all sentences of $\mathscr{L}$ which hold in $\mathfrak{A}$. Then Γ is inseparable.*

Proof. As mentioned above, we apply 15.16 to prove this theorem; in fact, we shall define a certain effective interpretation of N into a theory which is a definitional expansion of $\mathscr{L}$. Let $\mathscr{L}'$ be a definitional expansion of $\mathscr{L}$ obtained by adjoining two operation symbols **s** (unary) and **0** (nullary). Let Δ be the definitional expansion of Γ in $\mathscr{L}'$ with the following definitions:

$$\forall v_0(\mathbf{0} = v_0 \leftrightarrow v_0 + v_0 = v_0),$$
$$\forall v \forall v_1\{\mathbf{s}v_0 = v_1 \leftrightarrow \exists v_2[\neg(v_2 + v_2 = v_2) \wedge v_2 \cdot v_2 = v_2 \wedge v_0 + v_2 = v_1]\}.$$

Now we define a syntactical $\mathscr{L}_{\mathrm{nos}}$-structure $\mathfrak{B} = (\varphi, f, R, \Delta)$ in $\mathscr{L}'$. Let φ be the formula

$$\exists v_1 \exists v_2 \exists v_3 \exists v_4(v_0 = v_1^2 + v_2^2 + v_3^2 + v_4^2)$$

(see 11.43). Set $f\mathbf{+} = \mathbf{+}$, $f\cdot = \cdot$, $f\mathbf{s} = \mathbf{s}$, $f\mathbf{0} = \mathbf{0}$, and let $R = 0$ (empty set). Clearly this *does* define a syntactical $\mathscr{L}_{\mathrm{nos}}$-structure in $\mathscr{L}'$. Let $\mathfrak{C}$ be the natural expansion of $\mathfrak{A}$ to an $\mathscr{L}'$-structure (see the proof of 11.31; the new symbols receive the denotions indicated in their definitions), and let $\mathfrak{D} = (\omega, +, \cdot, \mathfrak{s}, 0)$. Clearly $\mathfrak{C}^{\Delta\mathfrak{B}} = \mathfrak{D}$ (see 11.43), and $\Delta = \{\varphi \in \mathrm{Sent}_{\mathscr{L}} : \mathfrak{C} \models \varphi\}$, so by 11.45, $\Delta \models \varphi^{\mathfrak{B}}$ for each $\varphi \in \mathrm{N}$. Hence $\mathfrak{B}$ is an interpretation of N in Δ. By 15.16, Δ is inseparable, so by 15.15, Γ is inseparable. □

Now we want to accomplish a similar thing for set theory. Actually we shall show that the very simple set theory of the following definition is inseparable:

Definition 16.8. Let $\mathscr{L}_{\text{set}}$ be a fixed first-order language with a single non-logical constant, a binary relation symbol $\in$. Let S be the theory in $\mathscr{L}_{\text{set}}$ with the following axioms:

$$\exists v_0 \forall v_1 \neg (v_1 \in v_0), \tag{1}$$
$$\forall v_0 \forall v_1 [\forall v_2 (v_2 \in v_0 \leftrightarrow v_2 \in v_1) \rightarrow v_0 = v_1], \tag{2}$$
$$\forall v_0 \forall v_1 \exists v_2 \forall v_3 (v_3 \in v_2 \leftrightarrow v_3 \in v_0 \vee v_3 = v_1). \tag{3}$$

We shall show that the theory Q is interpretable in a definitional extension of S. To this end we must come up with definitions for ω, $+$, $\cdot$, 0 and $\mathscr{s}$. We do this by a series of definitional expansions of S. In each case it will be evident that we *do* have a definitional expansion. The notations introduced here will not be used beyond our treatment of set theory.

Definition 16.9. $\mathscr{L}_1$ is a definitional expansion of $\mathscr{L}_{\text{set}}$ with the following new symbols and axioms:

(*i*) **0**, an individual constant. New axiom:

$$\forall v_0 \neg (v_0 \in \mathbf{0}).$$

(*ii*) **Op**, a binary operation symbol. New axiom:

$$\forall v_0 \forall v_1 \forall v_2 [v_2 \in \mathbf{Op}(v_0, v_1) \leftrightarrow v_2 \in v_0 \vee v_2 = v_1].$$

(*iii*) $\{\ \}$, a unary operation symbol. New axiom:

$$\forall v_0 [\{v_0\} = \mathbf{Op}(\mathbf{0}, v_0)].$$

(*iv*) $\{\ ,\ \}$, a binary operation symbol. New axiom:

$$\forall v_0 \forall v_1 [\{v_0, v_1\} = \mathbf{Op}(\{v_0\}, v_1)].$$

(*v*) **U**, a unary operation symbol. New axiom:

$$\forall v_0 [\mathbf{U} v_0 = \mathbf{Op}(v_0, v_0)].$$

(*vi*) $\subseteq$, a binary relation symbol. New axiom:

$$\forall v_0 \forall v_1 [v_0 \subseteq v_1 \leftrightarrow \forall v_2 (v_2 \in v_0 \rightarrow v_2 \in v_1)].$$

(*vii*) **Trans**, a unary relation symbol. New axiom:

$$\forall v_0 [\mathbf{Trans}\ v_0 \leftrightarrow \forall v_1 (v_1 \in v_0 \rightarrow v_1 \subseteq v_0)].$$

(*viii*) **J**, a unary relation symbol. New axiom:

$$\forall v_0 \{\mathbf{J} v_0 \leftrightarrow \forall v_1 [v_1 \subseteq v_0 \rightarrow \forall v_2 \exists v_3 \forall v_4 (v_4 \in v_3 \leftrightarrow v_4 \in v_1 \wedge v_4 \in v_2)]\}.$$

(*ix*) $\cap$, a *ternary* operation symbol. New axiom:

$$\begin{aligned} &\forall v_0 \forall v_1 \forall v_2 \forall v_3 \{v_3 = v_0 \cap_{v_2} v_1 \\ &\quad \leftrightarrow [\mathbf{J} v_2 \wedge v_0 \subseteq v_2 \wedge \forall v_4 (v_4 \in v_3 \leftrightarrow v_4 \in v_0 \wedge v_4 \in v_1)] \\ &\qquad \vee [(\neg \mathbf{J} v_2 \vee \neg v_0 \subseteq v_2) \wedge v_3 = \mathbf{0}]\}. \end{aligned}$$

There are more definitions to come, but first a few remarks on those already made. Intuitively $\mathbf{0}$ is the empty set; $\mathbf{Op}(x, y) = x \cup \{y\}$; $\mathbf{U}x = x \cup \{x\}$. **Trans** x means that the transitive law for $\in$ holds with x on the right side: $z \in y \in x$ implies $z \in x$. The statement $\mathbf{J}x$ means that the intersection $y \cap z$ can be formed whenever $y \subseteq x$. Finally $y \cap_x z$ *is* the intersection of y and z if $\mathbf{J}x$ and $y \subseteq x$, otherwise it is 0. Note that already several complications arise because we cannot in general form the union and intersection of sets. Elementary facts such as those already mentioned will be used without proof. We formulate explicitly only the properties which are harder to prove. It is convenient in the proofs to argue informally within the given languages, and to mix logical and informal notation.

Lemma 16.10. $\mathscr{L}_1 \models \mathbf{J}v_0 \wedge v_1 \subseteq v_0 \rightarrow \mathbf{J}v_1$.

Lemma 16.11. $\mathscr{L}_1 \models \mathbf{J}v_0 \rightarrow \mathbf{JOp}(v_0, v_1)$.

Proof. Assume that $\mathbf{J}x$, y is arbitrary, $z = \mathbf{Op}(x, y)$, $w \subseteq z$, and v is arbitrary; we want to show that $w \cap_z v$ has its usual meaning, i.e., that $\exists u \forall s(s \in u \leftrightarrow s \in w$ and $s \in v)$. Since $\mathbf{J}x$, both $x \cap_x w$ and $(x \cap_x w) \cap_x v$ have their usual meanings. If $y \notin w$ or $y \notin v$, then $(x \cap_x w) \cap_x v$ is the desired set u. If $y \in w$ and $y \in v$, then $\mathbf{Op}((x \cap_x w) \cap_x v, y)$ can be taken for u. □

Definition 16.12. $\mathscr{L}_2$ is a definitional expansion of $\mathscr{L}_1$ with the following new symbols and axioms:

(*i*) **C**, a unary relation symbol. New axiom:

$$\forall v_0[\mathbf{C}v_0 \leftrightarrow \forall v_1 \exists v_2 \forall v_3(v_3 \in v_2 \leftrightarrow v_3 \in v_0 \wedge \neg(v_3 \in v_1))].$$

(*ii*) $\sim$, a binary operation symbol. New axiom:

$$\forall v_0 \forall v_1 \forall v_2\{v_2 = v_0 \sim v_1 \leftrightarrow [\mathbf{C}v_0 \wedge \forall v_3(v_3 \in v_2 \leftrightarrow v_3 \in v_0 \wedge \neg(v_3 \in v_1))] \vee [\neg \mathbf{C}v_0 \wedge v_2 = \mathbf{0}]\}.$$

(*iii*) **B**, a unary relation symbol. New axiom:

$$\forall v_0(\mathbf{B}v_0 \leftrightarrow \mathbf{J}v_0 \wedge \mathbf{C}v_0).$$

Of course $\mathbf{C}x$ means that the operation $x \sim y$ can always be performed in its usual sense. If $\mathbf{C}x$, we let $x \sim y$ have its usual sense, while $x \sim y = 0$ if $\mathbf{C}x$ fails to hold. The statement $\mathbf{B}x$ means that x admits both of the Boolean operations $\cap$ and $\sim$.

Lemma 16.13. $\mathscr{L}_2 \models \mathbf{B}v_0 \wedge v_1 \subseteq v_0 \rightarrow \mathbf{B}v_1$.

Proof. Assume that $\mathbf{B}x$ and $y \subseteq x$. Thus $\mathbf{J}x$ and $\mathbf{C}x$. Hence by 16.10, $\mathbf{J}y$. To check that $\mathbf{C}y$, let z be given; we want to show that $y \sim z$ has its usual sense. But clearly $\forall w(w \in y \cap_x (x \sim z)$ iff $w \in y$ and $w \notin z)$, since $\mathbf{J}x$, $\mathbf{C}x$, and $y \subseteq x$. □

Lemma 16.14. $\mathscr{L}_2 \models \mathbf{C}v_0 \to \mathbf{COp}(v_0, v_1)$.

PROOF. Assume that $\mathbf{C}x$, y is arbitrary, $z = \mathbf{Op}(x, y)$, and w is arbitrary; we want to show that $z \sim w$ has its usual sense. If $y \in w$, then $\forall u(u \in x \sim w$ iff $u \in z$ and $u \notin w)$, since $\mathbf{C}x$. If $y \notin w$, then $\forall u(u \in \mathbf{Op}(x \sim w, y)$ iff $u \in z$ and $u \notin w)$. □

Corollary 16.15. $\mathscr{L}_2 \models \mathbf{B}v_0 \to \mathbf{BOp}(v_0, v_1)$.

Definition 16.16. $\mathscr{L}_3$ is a definitional expansion of $\mathscr{L}_2$ with the following new symbols and axioms:

(*i*) $\mathbf{W}$, a unary relation symbol. New axiom:

$$\forall v_0(\mathbf{W}v_0 \leftrightarrow \forall v_1[v_1 \in v_0 \to \neg\exists v_2(v_1 \in v_2 \land v_2 \in v_1)] \land \forall v_1\{v_1 \subseteq v_0 \land \neg(v_1 = \mathbf{0}) \to \exists v_2[v_2 \in v_1 \land \forall v_3(v_3 \in v_1 \to v_2 \in v_3 \lor v_2 = v_3)]\} \land \forall v_1\{v_1 \subseteq v_0 \land \neg(v_1 = \mathbf{0}) \to \exists v_2[v_2 \in v_1 \land \forall v_3(v_3 \in v_1 \to v_3 \in v_2 \lor v_3 = v_2)]\}).$$

(*ii*) $\mathbf{\Omega}$, a unary relation symbol. New axiom:

$$\forall v_0[\mathbf{\Omega}v_0 \leftrightarrow \mathbf{B}v_0 \land \mathbf{Trans}\, v_0 \land \forall v_1(v_1 \in v_0 \to \mathbf{Trans}\, v_1) \land \mathbf{W}v_0].$$

The statement $\mathbf{W}x$ encodes several properties of x: each element y of x is regular, in that $y \in z \in y$ is ruled out; each nonempty subset of x has a least element under the relation $\in$, and each nonempty subset of x has a greatest element under $\in$. The statement $\mathbf{\Omega}x$ is a conventional set-theoretical definition for x being a natural number, with a few redundancies because of our very weak axioms. It is routine to check the following two lemmas.

Lemma 16.17. $\mathscr{L}_3 \models \mathbf{\Omega 0}$.

Lemma 16.18. $\mathscr{L}_3 \models \mathbf{W}v_0 \land v_1 \subseteq v_0 \to \mathbf{W}v_1$.

Lemma 16.19. $\mathscr{L}_3 \models \mathbf{J}v_0 \land \mathbf{W}v_0 \to \mathbf{WU}v_0$.

PROOF. Assume that $\mathbf{J}x$ and $\mathbf{W}x$. To check $\mathbf{WU}x$, we consider separately the three conjuncts in the definition of $\mathbf{W}$. If $y \in \mathbf{U}x$, then $y \in x$ or $y = x$. In the first case, $\neg\exists z(y \in z \in y)$ since $\mathbf{W}x$. In the second case, simply note that $x \in z \in x$ would still contradict $\mathbf{W}x$. Thus the first conjunct for $\mathbf{WU}x$ holds. For both of the other conjuncts we assume that $y \subseteq \mathbf{U}x$. If $x \notin y$, then $y \subseteq x$ and the desired result follows in both cases because $\mathbf{W}x$.

Now assume that $x \in y$. To check the second conjunct, note that it is obvious if $y = \{x\}$, while if $y \neq \{x\}$, then $x \cap_x y$ has its usual meaning (since $\mathbf{J}x$), and $0 \neq x \cap_x y \subseteq x$, so since $\mathbf{W}x$, we may choose $z \in x \cap_x y$ so that $\forall w \in x \cap_x y\ (z \in w \lor z = w)$. Thus $z \in y$, and if $w \in y$ then $w \in x$ or $w = x$, hence $w \in x \cap_x y$ or $w = x$, and in either case $z \in w \lor z = w$.

Thus the second conjunct holds. For the third conjunct, since $y \subseteq \mathbf{U}x$ we have $z \in x$ or $z = x$ for any element z of y. □

Lemma 16.20. $\mathscr{L}_3 \models \mathbf{\Omega} v_0 \leftrightarrow \mathbf{\Omega U} v_0$.

PROOF. First assume $\mathbf{\Omega}x$; thus $\mathbf{B}x$, $\mathbf{Trans}\, x$, $\forall y \in x(\mathbf{Trans}\, y)$, and $\mathbf{W}x$. Hence $\mathbf{BU}x$ and $\mathbf{WU}x$ by 16.15 and 16.19. Suppose $z \in y \in \mathbf{U}x$. Then $z \in y \in x$ or $z \in y = x$, so $z \in x$ since $\mathbf{Trans}\, x$. Thus $\mathbf{Trans\, U}x$. If $y \in \mathbf{U}x$, then $y \in x$ or $y = x$, so $\mathbf{Trans}\, y$ is clear. Thus $\mathbf{\Omega U}x$.

Now assume $\mathbf{\Omega U}x$, which means that $\mathbf{BU}x$, $\mathbf{Trans\, U}x$, $\forall y \in \mathbf{U}x(\mathbf{Trans}\, y)$, and $\mathbf{WU}x$. The condition $\forall y \in \mathbf{U}x(\mathbf{Trans}\, y)$ means that $\mathbf{Trans}\, x$ and $\forall y \in x(\mathbf{Trans}\, y)$. And we have $\mathbf{B}x$ and $\mathbf{W}x$ by 16.13 and 16.18. □

Lemma 16.21. $\mathscr{L}_3 \models \mathbf{Trans}\, v_0 \wedge \neg(v_0 \in v_0) \wedge \mathbf{U}v_0 = \mathbf{U}v_1 \rightarrow v_0 = v_1$.

PROOF. Assume that $\mathbf{Trans}\, x$, $x \notin x$, and $\mathbf{U}x = \mathbf{U}y$. Take any $z \in x$. Then $z \in \mathbf{U}x = \mathbf{U}y$, so $z \in y$ or $z = y$; we shall show that $z = y$ is impossible. Assume $z = y$; thus $y \in x$. Since $x \in \mathbf{U}x$, we have $x \in \mathbf{U}y$ and hence $x \in y$ or $x = y$. Since $\mathbf{Trans}\, x$, it follows that $x \in x$, contradiction. Thus $z = y$ *is* impossible, so by the arbitrariness of z, $x \subseteq y$. The converse, $y \subseteq x$, is similar, so $x = y$ by 16.8(2). □

Lemma 16.22. $\mathscr{L}_3 \models \mathbf{\Omega} v_0 \wedge v_1 \in v_0 \rightarrow \mathbf{\Omega} v_1$.

PROOF. Assume $\mathbf{\Omega}x$ and $y \in x$. Since $\mathbf{Trans}\, x$, it follows that $y \subseteq x$. Thus $\mathbf{W}y$ by 16.18, and $\mathbf{B}y$ by 16.13. Clearly $\mathbf{Trans}\, y$. If $z \in y$, then $z \in x$ since $\mathbf{Trans}\, x$, and hence $\mathbf{Trans}\, z$. Thus $\mathbf{\Omega}y$. □

Lemma 16.23. $\mathscr{L}_3 \models \mathbf{Trans}\, v_0 \wedge \mathbf{J}v_0 \wedge \mathbf{W}v_0 \wedge \neg(v_0 = \mathbf{0}) \rightarrow \exists v_1(v_0 = \mathbf{U}v_1)$.

PROOF. Assume $\mathbf{Trans}\, x$, $\mathbf{J}x$, $\mathbf{W}x$ and $x \neq 0$. Since $\mathbf{W}x$, choose $y \in x$ so that $\forall w \in x(w \in y$ or $w = y)$. We shall show that $x = \mathbf{U}y$; the inclusion $x \subseteq \mathbf{U}y$ has just been mentioned. If $w \in y$, then $w \in x$ since $\mathbf{Trans}\, x$. Also, $y \in x$, so $\mathbf{U}y \subseteq x$. □

Definition 16.24. $\mathscr{L}_4$ is a definitional expansion of $\mathscr{L}_3$ with $\mathbf{s}$ as a new unary operation symbol and the following new axiom:

$$\forall v_0 \forall v_1[v_1 = \mathbf{s}v_0 \leftrightarrow (\mathbf{\Omega} v_0 \wedge v_1 = \mathbf{U}v_0) \vee (\neg\mathbf{\Omega} v_0 \wedge v_1 = v_0)].$$

For technical reasons we shall interpret $\mathbf{s}$ of $\mathscr{L}_{\text{nos}}$ by $\mathbf{s}$ in $\mathscr{L}_4$ rather than by the more natural symbol $\mathbf{U}$. This is really important only when we get to the axioms involving $+$ and $\cdot$

Lemma 16.25. $\mathscr{L}_4 \models \mathbf{s}v_0 = \mathbf{s}v_1 \rightarrow v_0 = v_1$.

PROOF. Assume that $\mathbf{s}x = \mathbf{s}y$. If $\mathbf{\Omega}x$ or $\mathbf{\Omega}y$, then $\mathbf{\Omega}x$ *and* $\mathbf{\Omega}y$ by 16.24 and 16.20, and hence $x = y$ by 16.21. If neither $\mathbf{\Omega}x$ nor $\mathbf{\Omega}y$, then $x = y$ by 16.24. □

Lemma 16.26. $\mathscr{L}_4 \models \neg 0 = \mathbf{s}v_0$.

PROOF. If $\mathbf{\Omega}x$, then $\mathbf{s}x = \mathbf{U}x \neq 0$ by 16.24, since $x \in \mathbf{U}x$. If (*not* $\mathbf{\Omega}x$), then $\mathbf{s}x = x$, and $x \neq 0$ since $\mathbf{\Omega}0$ by 16.17. □

Lemma 16.27. $\mathscr{L}_4 \models \neg v_0 = \mathbf{0} \rightarrow \exists v_1(v_0 = \mathbf{s}v_1)$.

PROOF. Assume that $x \neq 0$. If $\mathbf{\Omega}x$, the desired conclusion is clear by 16.23 and 16.20. If (not $\mathbf{\Omega}x$), then $x = \mathbf{s}x$. □

Definition 16.28. $\mathscr{L}_5$ is a definitional expansion of $\mathscr{L}_4$ with the following new symbols and axioms:

(*i*) (,), a binary operation symbol. New axiom:

$$\forall v_0 \forall v_1[(v_0, v_1) = \{\{v_0\}, \{v_0, v_1\}\}].$$

(*ii*) **Rln**, a unary relation symbol. New axiom:

$$\forall v_0(\mathbf{Rln}\, v_0 \leftrightarrow \forall v_1\{v_1 \in v_0 \rightarrow \exists v_2 \exists v_3[v_1 = (v_2, v_3)]\}).$$

(*iii*) **Fcn**, a unary relation symbol. New axiom:

$$\forall v_0\{\mathbf{Fcn} v_0 \leftrightarrow \mathbf{Rln} v_0 \wedge \forall v_1 \forall v_2 \forall v_3[(v_1, v_2) \in v_0 \wedge (v_1, v_3) \in v_0 \rightarrow v_2 = v_3]\}.$$

(*iv*) **Dmn**, a binary relation symbol. New axiom:

$$\forall v_0 \forall v_1(\mathbf{Dmn}(v_0, v_1) \leftrightarrow \forall v_2\{v_2 \in v_1 \leftrightarrow \exists v_3[(v_2, v_3) \in v_0]\}).$$

(*v*) **D**, a unary relation symbol. New axiom:

$$\forall v_0\{\mathbf{D} v_0 \leftrightarrow \forall v_1[v_1 \subseteq v_0 \rightarrow \exists v_2 \mathbf{Dmn}(v_1, v_2)]\}.$$

The following lemma has its usual proof:

Lemma 16.29. $\mathscr{L}_5 \models (v_0, v_1) = (v_2, v_3) \rightarrow v_0 = v_2 \wedge v_1 = v_3$.

Lemma 16.30. $\mathscr{L}_5 \models \mathbf{B}v_0 \wedge \mathbf{D}v_0 \leftrightarrow \mathbf{BOp}(v_0, v_1) \wedge \mathbf{DOp}(v_0, v_1)$.

PROOF. Assume that $\mathbf{B}x$ and $\mathbf{D}x$, while y is arbitrary. Thus $\mathbf{BOp}(x, y)$ by 16.15. To check $\mathbf{DOp}(x, y)$, let $z \subseteq \mathbf{Op}(x, y)$. Choose w so that $\mathbf{Dmn}(x \cap_x z, w)$; this is possible since $\mathbf{D}x$; recall that $x \cap_x z$ has its usual meaning since $\mathbf{B}x$. If $y \notin z$; then $x \cap_x z = z$ and we are through. If $y \in z$ but y is not an ordered pair, then clearly $\mathbf{Dmn}(z, w)$. Finally, if $y \in z$ and $y = (u, v)$, then $\mathbf{Dmn}(z, \mathbf{Op}(w, u))$. The converse is clear. □

Definition 16.31. $\mathscr{L}_6$ is a definitional expansion of $\mathscr{L}_5$ with a new ternary relation symbol $\boldsymbol{\sigma}$ and the following new axiom:

$$\begin{aligned} \forall v_0 \forall v_1 \forall v_2(\boldsymbol{\sigma}(v_0, v_1, v_2) &\leftrightarrow \mathbf{\Omega} v_0 \wedge \mathbf{\Omega} v_1 \wedge \mathbf{\Omega} v_2 \\ &\wedge \exists v_3\{\mathbf{Fcn} v_3 \wedge \mathbf{Dmn}(v_3, \mathbf{U}v_1) \wedge (\mathbf{0}, v_0) \in v_3 \\ &\wedge \forall v_4 \forall v_5[(v_4, v_5) \in v_3 \wedge v_4 \in v_1 \rightarrow (\mathbf{U}v_4, \mathbf{U}v_5) \in v_3] \\ &\wedge (v_1, v_2) \in v_3 \wedge \mathbf{B}v_3 \wedge \mathbf{D}v_3\}). \end{aligned}$$

Lemma 16.32. $\mathscr{L}_6 \models \mathbf{\Omega} v_0 \rightarrow \boldsymbol{\sigma}(v_0, \mathbf{0}, v_0)$.

PROOF. Assume that $\mathbf{\Omega} x$. Then $\{(0, x)\}$ is easily seen to satisfy the necessary conditions on v_3 in 16.31. $\square$

Lemma 16.33. $\mathscr{L}_6 \models \boldsymbol{\sigma}(v_0, v_1, v_2) \wedge \boldsymbol{\sigma}(v_0, v_1, v_3) \rightarrow v_2 \equiv v_3$.

PROOF. Assume that $\boldsymbol{\sigma}(x, y, z)$ and $\boldsymbol{\sigma}(x, y, w)$. Let f and g be the functions mentioned in $\boldsymbol{\sigma}(x, y, z)$ and $\boldsymbol{\sigma}(x, y, w)$ respectively. Now we aim to prove

(1) $$\mathbf{Dmn}(f \cap_f g, \mathbf{U}y).$$

Note that $f \cap_f g$ has its usual meaning, since $\mathbf{J}f$ follows from our choice of f. To prove (1) choose t so that $\mathbf{Dmn}(f \cap_f g, t)$; this is possible because $\mathbf{D}f$ and $f \cap_f g \subseteq f$. Since $\mathbf{Dmn}(f, \mathbf{U}y)$ and $f \cap_f g \subseteq f$, it is clear that $t \subseteq \mathbf{U}y$. Note that $\mathbf{U}y \sim t$ has its usual meaning, since $\mathbf{\Omega U}y$ by 16.20. It now suffices to show that $\mathbf{U}y \sim t = 0$. Assume that $\mathbf{U}y \sim t \neq 0$. We may choose $u \in \mathbf{U}y \sim t$ so that $u \in v$ or $u = v$ for any $v \in \mathbf{U}y \sim t$. From 16.22 we infer that $\mathbf{\Omega} u$. Clearly $0 \in t$, so $u \neq 0$. Hence by 16.23 there is an r with $u = \mathbf{U}r$. Note that $\mathbf{\Omega} r$ by 16.20. Now we claim

(2) $$r \in y.$$

For, $r \in u \in \mathbf{U}y$, so (2) is clear since $\mathbf{\Omega} y$. Now r cannot be in $\mathbf{U}y \sim t$, since otherwise $u \in r$ or $u = r$, which along with $r \in u$ contradicts $\mathbf{\Omega} u$. Hence from (2) we infer that $r \in t$. By our choice of t we may then pick q so that $(r, q) \in f \cap_f g$. Since $r \in y$ by (2), it follows from the definition of $\boldsymbol{\sigma}$ that also $(\mathbf{U}r, \mathbf{U}q) \in f \cap_f q$. Thus $u = \mathbf{U}r \in t$, contradiction. Thus $\mathbf{U}y \sim t = 0$ after all, and (1) is established.

Since $\mathbf{Dmn}(f, \mathbf{U}y)$ and $\mathbf{Dmn}(g, \mathbf{U}y)$ also, it follows easily from (1) that $f = f \cap_f g = g$. Hence $z = fy = gy = w$, as desired. $\square$

Lemma 16.34. $\mathscr{L}_6 \models \boldsymbol{\sigma}(v_0, v_1, v_2) \rightarrow \boldsymbol{\sigma}(v_0, \mathbf{U}v_1, \mathbf{U}v_2)$.

PROOF. Assume $\boldsymbol{\sigma}(x, y, z)$, and let f be the function mentioned in $\boldsymbol{\sigma}(x, y, z)$. Then $g = \mathbf{Op}(f, (\mathbf{U}y, \mathbf{U}z))$ shows that $\boldsymbol{\sigma}(x, \mathbf{U}y, \mathbf{U}z)$, making use of 16.20 and 16.30. $\square$

Lemma 16.35. $\mathscr{L}_6 \models \boldsymbol{\sigma}(v_0, \mathbf{U}v_1, v_2) \rightarrow \exists v_3[v_2 \equiv \mathbf{U}v_3 \wedge \boldsymbol{\sigma}(v_0, v_1, v_3)]$.

PROOF. Assume that $\boldsymbol{\sigma}(x, \mathbf{U}y, z)$, and let f be the function mentioned in $\boldsymbol{\sigma}(x, \mathbf{U}y, z)$. Since $y \in \mathbf{UU}y$ and $\mathbf{Dmn}(f, \mathbf{UU}y)$, we can choose w so that $(y, w) \in f$. Then $(\mathbf{U}y, \mathbf{U}w) \in f$ also. But also $(\mathbf{U}y, z) \in f$, so from $\mathbf{Fcn}\, f$ we infer that $z = \mathbf{U}w$. Let $g = f \sim \{(\mathbf{U}y, z)\}$; here $\sim$ has its usual meaning since $\mathbf{B}f$. Thus $f = \mathbf{Op}(g, (\mathbf{U}y, z))$, so $\mathbf{B}g$ and $\mathbf{D}g$ by 16.30.

It is easily seen, then, that g establishes $\boldsymbol{\sigma}(x, y, w)$. $\square$

Definition 16.36. $\mathscr{L}_7$ is a definitional expansion of $\mathscr{L}_6$ with the following new symbols and axioms:

(*i*) **1**, an individual constant. New axiom:

$$\mathbf{1} = \mathbf{U0}.$$

(*ii*) **+**, a binary operation symbol. New axiom:

$$\forall v_0 \forall v_1 \forall v_2 \{v_0 + v_1 = v_2 \leftrightarrow \sigma(v_0, v_1, v_2) \\ \vee [\neg \exists v_2 \sigma(v_0, v_1, v_2) \wedge \Omega v_0 \wedge v_2 = \{\mathbf{1}\}] \\ \vee [\neg \exists v_2 \sigma(v_0, v_1, v_2) \wedge \neg \Omega v_0 \wedge v_2 = v_0]\}.$$

Lemma 16.37. $\mathscr{L}_7 \models \neg \Omega\{\mathbf{1}\}$.

PROOF. We have $0 \in 1 \in \{1\}$ but $0 \notin \{1\}$, so **Trans**$\{1\}$ is false. Hence $\Omega\{\mathbf{1}\}$ does not hold. □

Now we can take care of two further axioms of Q.

Lemma 16.38. $\mathscr{L}_7 \models v_0 + \mathbf{0} = v_0$.

PROOF. Let x be given. If Ωx, then $\sigma(x, 0, x)$ by 16.32, so $x + 0 = x$. If Ωx fails, then there is no z with $\sigma(x, 0, z)$, and hence $x + 0 = x$ is clear from 16.36(*ii*). □

Lemma 16.39. $\mathscr{L}_7 \models v_0 + \mathbf{s}v_1 = \mathbf{s}(v_0 + v_1)$.

PROOF. Let x and y be given. If not Ωx, then $x + \mathbf{s}y = x$ and $\mathbf{s}(x + y) = x$. So assume that Ωx. If not Ωy, then $\mathbf{s}y = y$ and for no z do we have $\sigma(x, y, z)$, so $x + \mathbf{s}y = \{1\}$, $x + y = \{1\}$, $\mathbf{s}(x + y) = \{1\}$ (using 16.37), as desired. So assume that Ωy. If not $\exists z \sigma(x, y, z)$, then by 16.35 not $\exists z \sigma(x, \mathbf{U}y, z)$, so $x + \mathbf{s}y = \{1\} = \mathbf{s}(x + y)$. Finally, assume that $\sigma(x, y, z)$. Thus $\sigma(x, \mathbf{U}y, \mathbf{U}z)$ by 16.34 and hence $x + \mathbf{s}y = x + \mathbf{U}y = \mathbf{U}z = \mathbf{U}(x + y) = \mathbf{s}(x + y)$. □

Definition 16.40. $\mathscr{L}_8$ is a definitional expansion of $\mathscr{L}_7$ with a new ternary relation symbol π and the following new axiom:

$$\forall v_0 \forall v_1 \forall v_2 [\pi(v_0, v_1, v_2) \leftrightarrow \Omega v_0 \wedge \Omega v_1 \wedge \exists v_3 \{\mathbf{Fcn} v_3 \wedge \mathbf{B} v_3 \\ \wedge \mathbf{D} v_3 \wedge \mathbf{Dmn}(v_3, \mathbf{U} v_1) \wedge (\mathbf{0}, \mathbf{0}) \in v_3 \wedge \forall v_4 \forall v_5 [(v_4, v_5) \in v_3 \\ \wedge v_4 \in v_1 \rightarrow (\mathbf{U} v_4, v_5 + v_0) \in v_3] \wedge (v_1, v_2) \in v_3\}].$$

The following four lemmas are easily established, imitating the proofs for similar facts about σ.

Lemma 16.41. $\mathscr{L}_8 \models \Omega v_0 \rightarrow \pi(v_0, \mathbf{0}, \mathbf{0})$.

Lemma 16.42. $\mathscr{L}_8 \models \pi(v_0, v_1, v_2) \wedge \pi(v_0, v_1, v_3) \rightarrow v_2 = v_3$.

Lemma 16.43. $\mathscr{L}_8 \models \pi(v_0, v_1, v_2) \rightarrow \pi(v_0, \mathbf{U} v_1, v_2 + v_0)$.

Lemma 16.44. $\mathscr{L}_8 \models \pi(v_0, \mathbf{U} v_1, v_2) \rightarrow \exists v_3 [v_2 = v_3 + v_0 \wedge \pi(v_0, v_1, v_3)]$.

Definition 16.45. $\mathscr{L}_9$ is a definitional expansion of $\mathscr{L}_8$ with a new binary operation symbol $\cdot$ and the following new axiom:

$$\begin{aligned}\forall v_0\forall v_1\forall v_2\{v_0\cdot v_1 = v_2 \leftrightarrow \pi(v_0, v_1, v_2)\\ \vee [\neg\exists v_3\pi(v_0, v_1, v_3) \wedge \Omega v_0 \wedge v_2 = \{1\}]\\ \vee [\neg\exists v_3\pi(v_0, v_1, v_3) \wedge \neg\Omega v_0 \wedge \neg v_1 = 0 \wedge v_2 = \{1\}]\\ \vee [\neg\exists v_3\pi(v_0, v_1, v_3) \wedge \neg\Omega v_0 \wedge v_1 = 0 \wedge v_2 = 0]\}.\end{aligned}$$

From 16.41 we obtain

Lemma 16.46. $\mathscr{L}_9 \models v_0\cdot 0 = 0$.

Now we can take care of the last axiom of Q.

Lemma 16.47. $\mathscr{L}_9 \models v_0\cdot \mathbf{s}v_1 = v_0\cdot v_1 + v_0$.

Proof. Let x and y be given. We consider four cases.

Case 1. $\neg\Omega x$. Clearly $\mathbf{s}y \neq 0$, so $x\cdot\mathbf{s}y = \{1\}$. Also, $x\cdot y = 0$ or $x\cdot y = \{1\}$.
Subcase 1. $x\cdot y = 0$. Then by 16.36, $x\cdot y + x = \{1\}$, as desired.
Subcase 2. $x\cdot y = \{1\}$. Then by 16.36, $x\cdot y + x = x\cdot y = x\cdot\mathbf{s}y$.

Case 2. Ωx, but $\neg\Omega y$. Then $x\cdot\mathbf{s}y = x\cdot y = \{1\} = x\cdot y + x$.

Case 3. Ωx, Ωy, but $\neg\exists z\pi(x, y, z)$. Then $x\cdot\mathbf{s}y = x\cdot\mathbf{U}y = \{1\} = x\cdot y = x\cdot y + x$, using 16.44 to infer $\neg\exists z\pi(x, \mathbf{U}y, z)$.

Case 4. Ωx, Ωy, and $\pi(x, y, z)$. Then $x\cdot\mathbf{s}y = x\cdot\mathbf{U}y = x\cdot y + x$ since $\pi(x, \mathbf{U}y, z + x)$ by 16.43. □

From our lemmas we obtain our main undecidability result on set theory by the same general argument as that used to establish 16.7:

Theorem 16.48. *The theory* S *is inseparable.*

Now we turn to the consideration of finitely inseparable theories. We do not yet have any examples at all of such theories, so our first task is to produce such an example. Then we use the method of interpretations to show that various interesting mathematical theories are finitely inseparable. Our first example will be considerably generalized later. Recall our universal language from 14.22.

Lemma 16.49. $\{\varphi : \varphi \in \mathrm{Sent}_{\mathscr{L}(\mathrm{un})}, \models\varphi\}$ *is a finitely inseparable theory.*

Proof. By Theorem 6.24, let A and B be r.e. effectively inseparable sets. By virtue of Theorem 6.25 it now suffices for the proof of our lemma to effectively associate with each $t \in \omega$ a sentence χ_t of $\mathscr{L}_{\mathrm{un}}$ such that the following two conditions hold:

(1) if $t \in A$, then $\models\chi_t$;

(2) if $t \in B$, then $\neg\chi_t$ holds in some finite $\mathscr{L}_{\mathrm{un}}$-structure.

Let f and g be recursive functions with ranges A and B respectively. By 14.24 and 14.25, let φ and ψ be sentences which spectrally represent f and g respectively and satisfy the additional conditions of 14.24. Divide the symbols of $\mathscr{L}_{\text{un}}$ into three isomorphic parts, denoted by ${}^0\mathbf{R}_r^s$, ${}^1\mathbf{R}_r^s$ and ${}^2\mathbf{R}_r^s$ respectively. We define two translations of $\mathscr{L}_{\text{un}}$ into $\mathscr{L}_{\text{un}}$. Let $\mathfrak{A} = ({}^2\mathbf{R}_0^1 v_0, 0, M, \Gamma)$, where $M\mathbf{R}_r^s = {}^0\mathbf{R}_r^s v_0 \cdots v_{s-1}$ for all s, r and $\Gamma = \{\exists v_0 {}^2\mathbf{R}_0^1 v_0\}$, and let $\mathfrak{B} = ({}^2\mathbf{R}_1^1 v_0, 0, N, \Delta)$, where $N\mathbf{R}_r^s = {}^1\mathbf{R}_r^s v_0 \cdots v_{s-1}$ for all s, r and $\Delta = \{\exists v_0 {}^2\mathbf{R}_1^1 v_0\}$. Then for any $t \in \omega$ we can define a sentence χ_t such that for any $\mathscr{L}_{\text{un}}$-structure $\mathfrak{C}$, $\mathfrak{C} \models \chi_t$ iff the following condition holds:

(3) if $\mathfrak{C} \models \varphi^{\mathfrak{A}}$, $\mathfrak{C} \models \psi^{\mathfrak{B}}$, ${}^2\mathbf{R}_0^{1\mathfrak{C}} \neq 0 \neq {}^2\mathbf{R}_1^{1\mathfrak{C}}$, ${}^1\mathbf{R}_1^{1\mathfrak{C}} \subseteq {}^2\mathbf{R}_1^{1\mathfrak{C}}$, ${}^0\mathbf{R}_0^{1\mathfrak{C}} \subseteq {}^2\mathbf{R}_0^{1\mathfrak{C}}$, $\mathfrak{C} \models \forall v_0 \forall v_1 ({}^0\mathbf{R}_1^2 v_0 v_1 \rightarrow {}^2\mathbf{R}_1^1 v_0 \wedge {}^2\mathbf{R}_0^1 v_1)$, $|{}^1\mathbf{R}_1^{1\mathfrak{C}}| = t$, and ${}^2\mathbf{R}_0^{2\mathfrak{C}}$ is a one–one function mapping ${}^1\mathbf{R}_0^{1\mathfrak{C}}$ onto ${}^0\mathbf{R}_0^{1\mathfrak{C}}$ then there is an $x \in {}^0\mathbf{R}_2^{1\mathfrak{C}}$ such that $|\{y : (y, x) \in {}^0\mathbf{R}_1^{2\mathfrak{C}}\}| = t$.

Now we check (1) and (2). For (1), Let $\mathfrak{C}$ be any $\mathscr{L}_{\text{un}}$-structure, let $t \in A$, and assume the hypothesis of (3). Then by 11.45, $\mathfrak{C}^{\Gamma\mathfrak{A}} \models \varphi$ and $\mathfrak{C}^{\Delta\mathfrak{B}} \models \psi$. Thus $|\mathbf{R}_1^{1\mathfrak{C}(\Delta,\mathfrak{B})}| = t$, so, since ψ spectrally represents g but $t \notin \operatorname{Rng} g$, it follows that $\mathbf{R}_0^{1\mathfrak{C}(\Delta,\mathfrak{B})}$ is infinite. Thus ${}^1\mathbf{R}_0^{1\mathfrak{C}}$ is infinite, so ${}^0\mathbf{R}_0^{1\mathfrak{C}} = \mathbf{R}_0^{1\mathfrak{C}(\Gamma,\mathfrak{A})}$ is infinite. Choose m so that $fm = t$, and then by 14.24(*ii*) choose $x \in \mathbf{R}_2^{1\mathfrak{C}(\Gamma,\mathfrak{A})}$ so that $|\{y : (y, x) \in \mathbf{R}_0^{2\mathfrak{C}(\Gamma,\mathfrak{A})}\}| = m$. By 14.24(*iii*), $|\{y : (y, x) \in \mathbf{R}_1^{2\mathfrak{C}(\Gamma,\mathfrak{A})}\}| = t$. Thus the conclusion of (3) holds. Hence we have shown $\models \chi_t$, as desired in (1).

Now let $t \in B$. Say $gm = t$. Let $\mathfrak{C}$ be a finite model of φ such that $|\mathbf{R}_0^{1\mathfrak{C}}| = m$, and let $\mathfrak{D}$ be a finite model of ψ such that $|\mathbf{R}_0^{1\mathfrak{D}}| = m$; this is possible by 14.22(*i*). Clearly there is a finite $\mathscr{L}_{\text{un}}$-structure $\mathfrak{E}$ such that $\mathfrak{E}^{\Gamma\mathfrak{A}} = \mathfrak{C}$, $\mathfrak{E}^{\Delta\mathfrak{B}} = \mathfrak{D}$, and ${}^1\mathbf{R}_1^{1\mathfrak{E}} \subseteq {}^2\mathbf{R}_1^{1\mathfrak{E}}$, ${}^0\mathbf{R}_0^{1\mathfrak{E}} \subseteq {}^2\mathbf{R}_0^{1\mathfrak{E}}$, $\mathfrak{E} \models \forall v_0 \forall v_1 ({}^0\mathbf{R}_1^2 v_0 v_1 \rightarrow {}^2\mathbf{R}_0^1 v_0 \wedge {}^2\mathbf{R}_0^1 v_1)$, and ${}^2\mathbf{R}_0^{2\mathfrak{E}} : {}^1\mathbf{R}_0^{1\mathfrak{E}} \rightarrowtail\!\!\!\!\rightarrow {}^0\mathbf{R}_0^{1\mathfrak{E}}$. Note that $|{}^1\mathbf{R}_1^{1\mathfrak{E}}| = |\mathbf{R}_1^{1\mathfrak{D}}| = t$ by 14.22(*ii*). Thus the hypothesis of (3) holds. The conclusion of (3) fails, however, by 14.24, since $t \notin \operatorname{Rng} f$. □

Note that the formulas χ_t in the above proof all have the same relation symbols. Hence from (1) and (2) in that proof we obtain at once:

Lemma 16.50. *There is a reduct $\mathscr{L}$ of $\mathscr{L}_{\text{un}}$, having only finitely many relation symbols, such that $\{\varphi : \varphi \in \text{Sent}_{\mathscr{L}}, \models\varphi\}$ is finitely inseparable.*

This is our basic lemma, from which we shall derive the finite inseparability of several other theories. The method we shall use for this purpose is expressed in 15.17. In many cases the application of 15.17 can be indicated by simple diagrams, from which a more rigorous proof can be written out in a routine way. This is illustrated in many examples below.

Theorem 16.51. *The theory of one binary relation is finitely inseparable.*

PROOF. The theorem is to be interpreted as saying that the theory Γ' in the language $\mathscr{L}'$ is finitely inseparable, where $\mathscr{L}'$ is a language with just one nonlogical constant, namely a binary relation symbol $\mathbf{S}$, and $\Gamma' = \{\varphi : \varphi \in \mathrm{Sent}_{\mathscr{L}'}, \models\varphi\}$.

To prove this, let $\mathscr{L}$ be the language mentioned in 16.50. Let the relation symbols of $\mathscr{L}$ be $\mathbf{R}_{n0}^{m0}, \ldots, \mathbf{R}_{ns}^{ms}$ where $s \in \omega$. Set $\Gamma = \{\varphi \in \mathrm{Sent}_{\mathscr{L}} : \models\varphi\}$. To define our syntactical $\mathscr{L}$-structure we need some notation: we define formulas $\sigma_k(v_0, v_1)$ in $\mathscr{L}'$ for each $k \in \omega \sim 1$, expressing that there is a path with k edges between the "points" v_0 and v_1:

$$\begin{aligned} \sigma_1 &= \mathbf{S}v_0v_1 \\ \sigma_{k+1} &= \exists v_2[\mathbf{S}v_0v_2 \wedge \sigma_k(v_2, v_1)]. \end{aligned}$$

Now let $\mathfrak{A} = (\chi, 0, R, \Delta)$, where χ is the formula

(1) $$\neg\exists v_1\mathbf{S}v_0v_1$$

while for each $i \leq s$, $R\mathbf{R}_{ni}^{mi}$ is the formula

(2) $$\exists v_{mi}[\sigma_{i+2}(v_{mi}, v_{mi}) \wedge \bigwedge_{j<mi} \sigma_{j+1}(v_{mi}, v_j)]$$

and $\Delta = \{\exists v_0 \neg \exists v_1\mathbf{S}v_0v_1\}$. Clearly then $\mathfrak{A}$ is a syntactical $\mathscr{L}$-structure in $\mathscr{L}'$. By Theorem 15.17 it remains to show that for each finite model $\mathfrak{C}$ of Γ there is a finite model $\mathfrak{B}$ of $\Gamma' \cup \Delta$ such that $\mathfrak{C} = \mathfrak{B}^{\Delta\mathfrak{A}}$. Note that $\mathfrak{C}$ is just an arbitrary $\mathscr{L}$-structure, and we do not have to worry about finding $\mathfrak{B}$ to be a model of Γ', this being true for any $\mathscr{L}'$-structure. An example to illustrate our construction of $\mathfrak{B}$ is given in Figure 16.51.1. This is just a small part of $\mathfrak{B}$ and is intended to illustrate what we do in case $i = 4$, $mi = 3$, and $(x_0, x_1, x_2) \in \mathbf{R}_{ni}^{mi\mathfrak{C}}$: all points except x_0, x_1, x_2 are new, we describe a circuit with six edges to identify $i = 4$, and leading off from a vertex we have paths with 1, 2, and 3 edges out to x_1, x_2, x_3 respectively to identify the ordered triple (x_0, x_1, x_2) as a member of $\mathbf{R}_{ni}^{mi\mathfrak{C}}$. From this diagram it should be evident how we construct $\mathfrak{B}$; we do this in detail here, but not in similar proofs below. Let $D = \{(i, x) : i \leq s \text{ and } x \in \mathbf{R}_{ni}^{mi\mathfrak{C}}\}$. Since $\mathfrak{C}$ is finite, so is D. With each $(i, x) \in D$ we want to associate $i + 2 + m_i(m_i - 1)/2$ new points

$$\begin{aligned} &a_{ix}^0, \ldots, a_{ix}^{i+1}, \\ &b_{ixj}^0, \ldots, b_{ixj}^{j-1} \qquad \text{for each } j < mi, j \neq 0, \end{aligned}$$

entirely new points for different members of D. To do this we can, for example, let

$$\begin{aligned} a_{ix}^k &= (i, x, k, 0, 0, C) \qquad \text{for each } k < i + 2, \\ b_{ixj}^k &= (i, x, k, j, 1, C) \qquad \text{for each } j < mi, j \neq 0, \text{ and } k < j. \end{aligned}$$

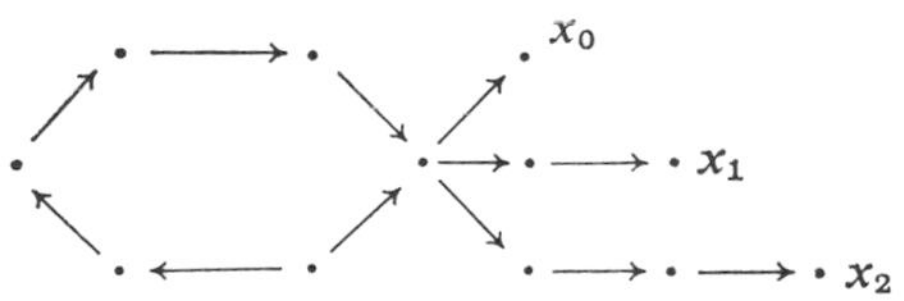

Figure 16.51.1

(The term C (=universe of $\mathfrak{C}$) is added to make sure all these elements are different from the members of C, a fact which follows from the definition of ordered sextuples in terms of ordered pairs, via the regularity axiom of set theory.) Let B be C together with all the new elements a^k_{ix}, b^k_{ixj}. Now let $\mathbf{S}^{\mathfrak{B}}$ express each of the following diagrams, for each $(i, x) \in D$:

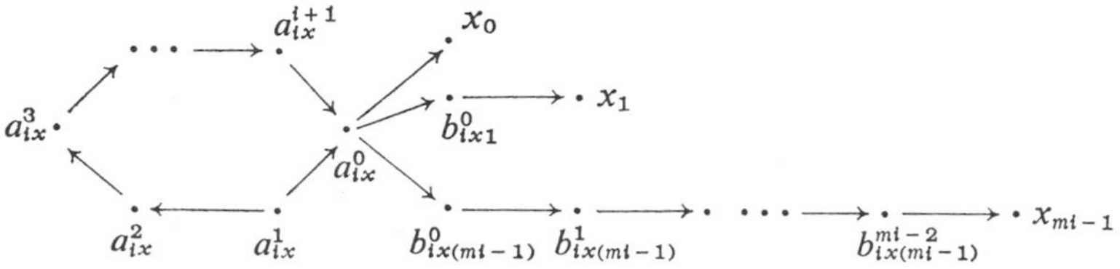

That is, let

$$\begin{aligned}\mathbf{S}^{\mathfrak{B}} = &\{(a^j_{ix}, a^{j+1}_{ix}) : (i, x) \in D, j \leq i\}\\ &\cup \{(a^{i+1}_{ix}, a^0_{ix}) : (i, x) \in D\} \cup \{(a^0_{ix}, x_0) : (i, x) \in D\}\\ &\cup \{(a^0_{ix}, b^0_{ixj}) : (i, x) \in D, 1 \leq j < mi\}\\ &\cup \{(b^k_{ixj}, b^{k+1}_{ixj}) : (i, x) \in D, 1 \leq j < mi, k < j - 1\}\\ &\cup \{(b^{j-1}_{ixj}, x_j) : (i, x) \in D, 1 \leq j < mi\}.\end{aligned}$$

This defines $\mathfrak{B}$. Obviously $\mathfrak{B} \models \chi[b]$ iff $b \in C$, so $B^{\Delta\mathfrak{A}} = C$. Now let $i \leq s$; we show that $\mathbf{R}^{mi,\mathfrak{B}(\Delta,\mathfrak{A})}_{ni} = \mathbf{R}^{mi\mathfrak{C}}_{ni}$. First let $x \in \mathbf{R}^{mi\mathfrak{C}}_{ni}$. Then $(i, x) \in D$, and the above definition of $\mathbf{S}^{\mathfrak{B}}$ yields immediately by (2) that $\mathfrak{B} \models R\mathbf{R}^{mi}_{ni}[x]$; and obviously $x \in {}^{mi}C$. Now suppose that $x \in {}^{mi}R\mathbf{R}^{mi\mathfrak{B}}_{ni} \cap {}^{mi}C$. By construction of $\mathfrak{B}$, clearly there is a $(j, y) \in D$ such that $x \subseteq y$ and $i \leq j$. The circuit in the construction assures that $i = j$ and hence $x = y$. Thus by definition of D we obtain $x \in \mathbf{R}^{mi\mathfrak{C}}_{ni}$. □

From 16.51 we can easily see that a large variety of languages have undecidable theories (a generalized form of Church's theorem; see 16.54, 16.58 for further forms).

Theorem 16.52. *If $\mathscr{L}'$ is a language with at least one relation symbol which is at least binary, then $\{\varphi \in \text{Sent}_{\mathscr{L}} : \models\varphi\}$ is finitely inseparable.*

Proof. We interpret the theory of one binary relation into $\mathscr{L}'$, i.e., we apply 15.17 as in the proof of 16.51. The following list gives the required definitions. The hypotheses of 15.17 are then clearly satisfied.

$\mathscr{L}$: language with a single binary relation symbol $\mathbf{R}$.
$\mathscr{L}'$: given language; $\mathbf{S}$ is a relation symbol of rank ≥ 2.
$\mathfrak{A} = (\chi, 0, R, \Gamma')$: here χ is $v_0 \equiv v_0$, with $R_{\mathbf{R}} = \mathbf{S}v_0v_1v_1\cdots v_1$,
$\Gamma' = \{\exists v_0(v_0 \equiv v_0)\}$. □

Theorem 16.53. *The theory of a binary operation is finitely inseparable.*

Proof. Again we interpret the theory of a binary relation. The following list and construction outline the procedure (cf. the proofs of 16.51 and 16.52).

> $\mathscr{L}$: language with a single binary relation symbol **R**.
> $\mathscr{L}'$: language with a single binary operation symbol **O**.
> $\mathfrak{C} = (\chi, 0, R, \Gamma')$: here χ is $\neg \mathbf{O}v_0v_0 \equiv v_0$;
> $R_{\mathbf{R}} = \exists v_2 \exists v_3(\mathbf{O}v_2v_2 \equiv v_2 \wedge \mathbf{O}v_2v_3 = v_2 \wedge \mathbf{O}v_3v_2 \equiv v_2 \wedge \mathbf{O}v_3v_3 \equiv v_3 \wedge \mathbf{O}v_0v_1 \equiv v_3)$;
> $\Gamma' = \exists v_0 \neg \mathbf{O}v_0v_0 \equiv v_0$.

Given any finite $\mathscr{L}$-structure $\mathfrak{A} = (A, R)$, set $B = A \cup \{(A, 0), (A, 1)\}$ and define a binary operation f on B by setting

$$
\begin{aligned}
f(a, b) &= (A, 0) && \text{if } (a, b) \notin R, \\
f(a, b) &= (A, 1) && \text{if } (a, b) \in R, \text{ for any } a, b \in A, \\
f((A, 0), (A, 0)) &= (A, 0), \\
f((A, 0), (A, 1)) &= (A, 0), \\
f((A, 1), (A, 0)) &= (A, 0), \\
f((A, 1), (A, 1)) &= (A, 1), \\
f(a, b) &= (A, 0) && \text{otherwise, for other pairs } (a, b) \in B \times B.
\end{aligned}
$$

We make B into an $\mathscr{L}'$-structure by setting $\mathbf{O}^{\mathfrak{B}} = f$. Clearly $\mathfrak{B}^{\Gamma\mathfrak{C}} = \mathfrak{A}$. □

Another large group of first order languages is taken care of in the following corollary of 16.53:

Theorem 16.54. *If $\mathscr{L}'$ is a language with at least one operation symbol which is at least binary, then $\{\varphi : \varphi \in \mathrm{Sent}_{\mathscr{L}}, \models\varphi\}$ is finitely inseparable.*

The proof is similar to the proof of 16.52. To take care of the remaining class of undecidable languages we need some auxiliary results which are interesting in themselves.

Theorem 16.55. *The theory of a symmetric binary relation is finitely inseparable.*

Proof. The following list and diagram outline the proof:

> $\mathscr{L}$: language with a single binary relation symbol **R**;
> $\Gamma = \{\varphi \in \mathrm{Sent}_{\mathscr{L}} : \models\varphi\}$.
> $\Gamma' = \{\varphi \in \mathrm{Sent}_{\mathscr{L}} : \Theta \models \varphi\}$, where $\Theta = \{\forall v_0 \forall v_1(\mathbf{R}v_0v_1 \rightarrow \mathbf{R}v_1v_0)\}$.
> $\mathfrak{A} = (\chi, 0, R, \Delta)$, where χ is the formula $\exists v_5 \forall v_2(\mathbf{R}v_2v_5 \leftrightarrow v_2 \equiv v_0)$,
> $R_{\mathbf{R}} = \exists v_2 \exists v_3 \exists v_4[\mathbf{R}v_0v_2 \wedge \mathbf{R}v_2v_3 \wedge \mathbf{R}v_3v_0 \wedge \mathbf{R}v_2v_4 \wedge \mathbf{R}v_4v_1]$,
> $\Delta = \{\exists v_0 \mathbf{R}v_0\}$.

Diagram (for representing in Δ a single pair (x, y) in an arbitrary binary relation):

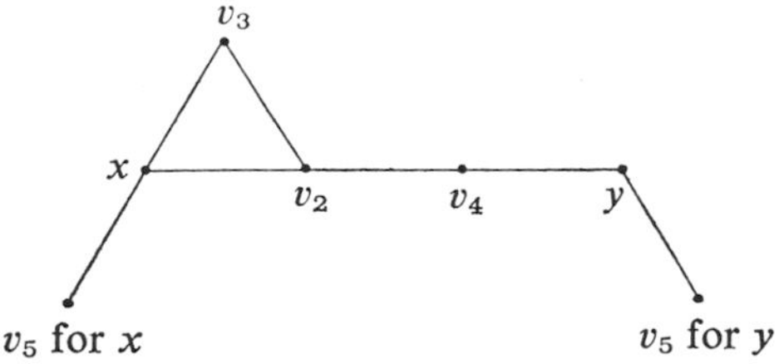

Theorem 16.56. *The theory of two equivalence relations over the universe is finitely inseparable.*

PROOF. We interpret a symmetric binary relation into this theory:

$\mathscr{L}$: language with a single binary relation symbol $\mathbf{R}$;
$\Gamma = \{\varphi \in \mathrm{Sent}_{\mathscr{L}} : \Theta \models \varphi\}$, where $\Theta = \{\forall v_0 \forall v_1(\mathbf{R}v_0v_1 \rightarrow \mathbf{R}v_1v_0)\}$.
$\mathscr{L}'$: language with two binary relation symbols $\mathbf{L}$ and $\mathbf{M}$;
$\Gamma' = \{\varphi \in \mathrm{Sent}_{\mathscr{L}'} : \Omega \models \varphi\}$, where Ω consists of the natural axioms for two equivalence relations:

$$\forall v_1 \mathbf{L}v_0v_0;$$
$$\forall v_0 \forall v_1(\mathbf{L}v_0v_1 \rightarrow \mathbf{L}v_1v_0);$$
$$\forall v_0 \forall v_1 \forall v_2(\mathbf{L}v_0v_1 \wedge \mathbf{L}v_1v_2 \rightarrow \mathbf{L}v_0v_2);$$
$$\forall v_0 \mathbf{M}v_0v_0;$$
$$\forall v_0 \forall v_1(\mathbf{M}v_0v_1 \rightarrow \mathbf{M}v_1v_0);$$
$$\forall v_0 \forall v_1 \forall v_2(\mathbf{M}v_0v_1 \wedge \mathbf{M}v_1v_2 \rightarrow \mathbf{M}v_0v_2).$$

$\mathfrak{C} = (\chi, 0, R, \Delta)$, where $\chi = \forall v_1(\mathbf{M}v_1v_0 \leftrightarrow v_1 = v_0)$,
$R_{\mathbf{R}} = \exists v_2 \exists v_3(\mathbf{L}v_0v_2 \wedge \mathbf{M}v_2v_3 \wedge \mathbf{L}v_3v_1)$, $\Delta = \{\exists v_0 \chi\}$.

Diagram (for representing in Δ a single pair (x, y) in an arbitrary symmetric binary relation, with curves for M-classes and squares for L-classes):

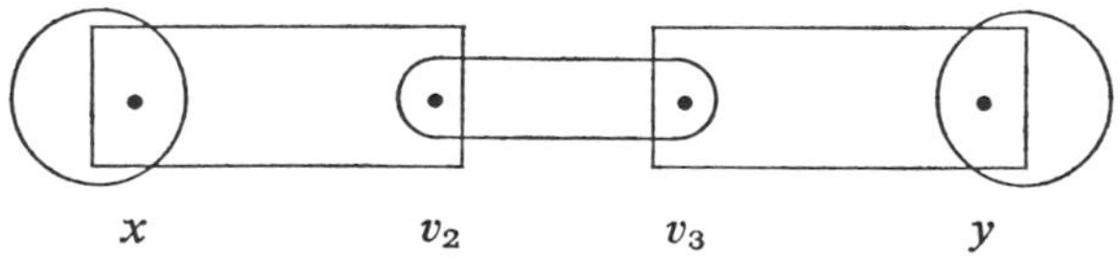

Theorem 16.57. *The theory of two total functions is finitely inseparable.*

PROOF. We interpret the theory of 16.56 into this theory:

$(\mathscr{L}', \Gamma')$: as in the proof of 16.56.
$\mathscr{L}''$: language with two unary operation symbols $\mathbf{O}$ and $\mathbf{P}$;
$\Gamma'' = \{\varphi \in \mathrm{Sent}_{\mathscr{L}''} : \models \varphi\}$.
$\mathfrak{C} = (\chi, 0, R, \Delta)$, where $\chi = v_0 = v_0$, $R_{\mathbf{L}} = \mathbf{O}v_0 = \mathbf{O}v_1$,
$R_{\mathbf{M}} = \mathbf{P}v_0 = \mathbf{P}v_1$, $\Delta = \{\exists v_0 \chi\}$.

Given a finite model $\mathfrak{A}$ of Γ', we construct a finite model $\mathfrak{B}$ of Γ'' as follows. Let $B = A$. To define $\mathbf{O}^{\mathfrak{B}}$ and $\mathbf{P}^{\mathfrak{B}}$, first let f be a choice function for non-empty subsets of A. Thus $fx \in x$ whenever $0 \neq x \subseteq A$. Now define $\mathbf{O}^{\mathfrak{B}}x = f[x]_{\mathbf{L}\mathfrak{A}}$ and $\mathbf{P}^{\mathfrak{B}}x = f[x]_{\mathbf{M}\mathfrak{B}}$ for any $x \in A$ (where $\mathfrak{A} = (A, \mathbf{L}^{\mathfrak{A}}, \mathbf{M}^{\mathfrak{A}})$); clearly $\mathfrak{B}$ is the desired interpretation of $\mathfrak{A}$. □

Theorem 16.58. *If $\mathscr{L}$ is a language with at least two unary operation symbols, then $\mathscr{L}$ is finitely inseparable.*

In Theorems 16.52, 16.54 and 16.58 we have shown that if $\mathscr{L}$ satisfies any one of the following three conditions, then $\{\varphi \in \mathrm{Sent}_{\mathscr{L}} : \models\varphi\}$ is finitely inseparable and hence undecidable:

(1) $\mathscr{L}$ has at least one relation symbol of rank at least 2;
(2) $\mathscr{L}$ has at least one operation symbol of rank at least 2;
(3) $\mathscr{L}$ has at least two unary operation symbols.

Now it is known that if $\mathscr{L}$ fails to satisfy (1), (2) or (3), which means that $\mathscr{L}$ satisfies the following condition (4), then $\{\varphi \in \mathrm{Sent}_{\mathscr{L}} : \models\varphi\}$ is decidable.

(4) $\mathscr{L}$ has only relation symbols of rank 1, no operation symbols of rank >1, and at most one unary operation symbol. (See Exercise 15.23 and Rabin [4].)

BIBLIOGRAPHY

1. Collins, G. E. and Halpern, J. D. On the interpretability of arithmetic in set theory. *Notre Dame J. Formal Logic*, *11* (1970), 477–483.
2. Ershov, Yu., Lavrov, I., Taimanov, A., Taitslin, M. Elementary theories. *Russian Mathematical Surveys*, *20* (1965), 35–105.
3. Rabin, M. A simple method for undecidability proofs and some applications. In *Logic, Methodology, Philosophy of Science*. Amsterdam: North-Holland (1965), 58–68.
4. Rabin, M. Decidability of second-order theories and automata on infinite trees. *Trans. Amer. Math. Soc.*, 141 (1969), 1–35.

EXERCISES

16.59. Show that **P** has $\exp \aleph_0$ complete extensions.

16.60. Find a finitely axiomatizable inseparable theory with $(\mathbb{Z}, +, \cdot)$ as a model.

16.61. Let $\mathscr{L}$ be a language with two nonlogical constants $\in$ (binary relation symbol) and S (unary relation symbol). Let Γ be the theory in $\mathscr{L}$ with the following axioms:

$\forall v_0[\mathrm{S}v_0 \leftrightarrow \exists v_1(v_0 \in v_1)]$,
$\exists v_0[\mathrm{S}v_0 \wedge \forall v_1 \neg (v_1 \in v_0)]$,
$\forall v_0 \forall v_1[\forall v_2(v_2 \in v_0 \leftrightarrow v_2 \in v_1) \rightarrow v_0 = v_1]$,
$\forall v_0 \forall v_1\{\mathrm{S}v_0 \wedge \mathrm{S}v_1 \rightarrow \exists v_2[\mathrm{S}v_2 \wedge \forall v_3(v_3 \in v_2 \leftrightarrow v_3 \in v_0 \vee v_3 = v_1)]\}$.

Show that Γ is inseparable. *Hint*: use 15.16(*ii*).

16.62. The theory of an infinite binary relation is undecidable.

16.63. The theory of an equivalence relation on the universe and a total function is finitely inseparable.

16.64. The theory of two supplementary equivalence relations is finitely inseparable. (Two equivalence relations E and F are *supplementary* if for any x, y in the universe of the structure there is a z such that $xEzFy$).

16.65. The theory of a partial ordering is finitely inseparable.

16.66. The theory of lattices is finitely inseparable.

17 Unprovability of Consistency

We shall prove in this chapter that in a strong theory Γ, some statements which naturally assert the consistency of Γ cannot be proved within Γ. This famous result of Gödel shows that our ordinary first-order languages have a severe limitation as far as any project for a thorough-going check on the consistency of mathematics is concerned. Historically, the theorem caused a major change of emphasis in foundational research away from a pre-occupation with consistency proofs.

It is convenient to formulate the results only for languages and theories in which a fair amount of number theory is directly expressible (without interpretations or definitions). The results carry over to other languages and theories, as long as they are sufficiently strong, in an obvious way.

Definition 17.1. Let $\mathscr{L}$ be an expansion of $\mathscr{L}_{\text{nos}}$. A theory Γ in $\mathscr{L}$ is a *strong theory* provided that for every m-ary elementary function f there is an m-ary operation symbol $\mathbf{O}$ of $\mathscr{L}$ such that for any $x_0, \ldots, x_{m-1}, y \in \omega$ the following conditions hold:

(*i*) if $f(x_0, \ldots, x_{m-1}) = y$, then $\Gamma \models \mathbf{O}(\mathbf{x}_0, \ldots, \mathbf{x}_{m-1}) \equiv \mathbf{y}$;
(*ii*) if $f(x_0, \ldots, x_{m-1}) \neq y$, then $\Gamma \models \neg\mathbf{O}(\mathbf{x}_0, \ldots, \mathbf{x}_{m-1}) \equiv \mathbf{y}$.

Thus we can find strong theories among definitional expansions of the theory R of $\mathscr{L}_{\text{nos}}$, or the set theory S (cf. 14.11 and 16.48). We shall now formulate a general theorem giving conditions on a formula $Pr(v_0)$ so that unprovability of consistency follows. Intuitively we think of $Pr(v_0)$ as saying that v_0 is provable. After the statement and proof of the general theorem we shall indicate how the conditions on $Pr(v_0)$ can be satisfied. For simplicity we write $Pr\sigma$ in place of $Pr(\sigma)$.

Theorem 17.2 (Löb). *Let Γ be a strong theory in a language $\mathscr{L}$. Assume that Pr is a formula of $\mathscr{L}$ such that* $\mathrm{Fv}\, Pr \subseteq \{v_0\}$ *and the following conditions hold for all sentences ψ, χ of $\mathscr{L}$:*

(*i*) *if* $\Gamma \models \psi$, *then* $\Gamma \models Pr\mathbf{\Delta}g^+\psi$;
(*ii*) $\Gamma \models Pr\mathbf{\Delta}g^+\psi \rightarrow Pr\mathbf{\Delta}g^+Pr\mathbf{\Delta}g^+\psi$;
(*iii*) $\Gamma \models Pr\mathbf{\Delta}g^+(\chi \rightarrow \psi) \rightarrow (Pr\mathbf{\Delta}g^+\chi \rightarrow Pr\mathbf{\Delta}g^+\psi)$.

Furthermore, let φ be a sentence of $\mathscr{L}$ such that $\Gamma \models Pr\mathbf{\Delta}g^+\varphi \rightarrow \varphi$. *Then* $\Gamma \models \varphi$.

PROOF. There clearly is an elementary function f such that for any formula ψ and any $m \in \omega$, $f(g^+\psi, m) = g^+\psi(\mathbf{m})$. Let $\mathbf{O}$ be a binary operation symbol of $\mathscr{L}$ which represents f, in the sense of 17.1. Let $a = g^+(Pr\mathbf{O}(v_0, v_0) \rightarrow \varphi)$, and let ψ be the sentence $Pr\mathbf{O}(\mathbf{a}, \mathbf{a}) \rightarrow \varphi$. Thus if χ is the formula $Pr\mathbf{O}(v_0, v_0) \rightarrow \varphi$, then $f(g^+\chi, g^+\chi) = g^+\psi$. Hence

(1)	$\Gamma \models \mathbf{O}(\mathbf{a}, \mathbf{a}) = \mathbf{\Delta}g^+\psi$;	
(2)	$\Gamma \models \psi \rightarrow (Pr\mathbf{\Delta}g^+\psi \rightarrow \varphi)$	by (1), definition of ψ
(3)	$\Gamma \models Pr\mathbf{\Delta}g^+(\psi \rightarrow (Pr\mathbf{\Delta}g^+\psi \rightarrow \varphi))$	by (2), (*i*)
(4)	$\Gamma \models Pr\mathbf{\Delta}g^+\psi \rightarrow Pr\mathbf{\Delta}g^+(Pr\mathbf{\Delta}g^+\psi \rightarrow \varphi)$	by (3), (*iii*)
(5)	$\Gamma \models Pr\mathbf{\Delta}g^+(Pr\mathbf{\Delta}g^+\psi \rightarrow \varphi) \rightarrow (Pr\mathbf{\Delta}g^+Pr\mathbf{\Delta}g^+\psi \rightarrow Pr\mathbf{\Delta}g^+\varphi)$	by (*iii*)
(6)	$\Gamma \models Pr\mathbf{\Delta}g^+\psi \rightarrow Pr\mathbf{\Delta}g^+Pr\mathbf{\Delta}g^+\psi$	by (*ii*)
(7)	$\Gamma \models Pr\mathbf{\Delta}g^+\varphi \rightarrow \varphi$	by hypothesis
(8)	$\Gamma \models Pr\mathbf{\Delta}g^+\psi \rightarrow \varphi$	by (4), (5), (6), (7), and a tautology
(9)	$\Gamma \models Pr\mathbf{O}(\mathbf{a}, \mathbf{a}) \rightarrow \varphi$	by (8), (1)
(10)	$\Gamma \models \psi$	by (9), definition of ψ
(11)	$\Gamma \models Pr\mathbf{\Delta}g^+\psi$	by (10), (*i*)
(12)	$\Gamma \models \varphi$	by (8), (11) □

Corollary 17.3. *Assume the hypothesis of* 17.2, *up to "Furthermore." Let φ be the sentence* $\neg\forall v_0(v_0 = v_0)$. *If Γ is consistent, then* $\Gamma \not\models \neg Pr\mathbf{\Delta}g^+\varphi$.

PROOF. If $\Gamma \models \neg Pr\mathbf{\Delta}g^+\varphi$, then of course $\Gamma \models Pr\mathbf{\Delta}g^+\varphi \rightarrow \varphi$, so by 17.2, $\Gamma \models \varphi$, contradicting the consistency of Γ by (*i*). □

The formula $\neg Pr\mathbf{\Delta}g^+\varphi$ in 17.3 of course intuitively expresses that Γ is consistent, if Prv_0 expresses that v_0 is provable in Γ. The content of 17.3 can be expressed in an intuitive form as follows. If Γ is a consistent theory in $\mathscr{L}$ and we have a consistency proof for Γ, then there is no formula Pr which represents our consistency proof in Γ, in the sense of the hypotheses of 17.2 and 17.3.

What has become of attempts to prove important theories consistent in a convincing way? Finitary consistency proofs for the theory P have been given. Although finitary, such proofs cannot be internalized in P. It can be seen, in fact, that the proofs involve induction exceeding induction over natural numbers, and in fact going up the first ε-number. No finitary consistency proofs for ZF (full set theory) have been given, and indeed it is hard to imagine any proof which could be called finitary which could not be formulated in ZF, in the sense of 17.2.

Now we turn to the proof of 17.2(*i*)–(*iii*) for certain natural formulas *Pr*. For this purpose we shall make some further assumptions on the theories we deal with. These additional assumptions, as is easily seen, do not really restrict the generality of the final result. The assumptions amount to an extension of our assumptions in 17.1 so as to formalize within a theory the full syntax of first-order logic.

Definition 17.4. We describe an expansion $\mathscr{L}_{el}$ of $\mathscr{L}_{nos}$ and, simultaneously, a theory $\mathscr{P}$ in $\mathscr{L}_{el}$. The following are the first axioms of $\mathscr{P}$:

(*i*) $\forall v_0(\neg \mathbf{s}v_0 = \mathbf{0})$;

(*ii*) $\forall v_0 \forall v_1(\mathbf{s}v_0 = \mathbf{s}v_1 \rightarrow v_0 = v_1)$.

(*iii*) $\forall v_0(v_0 + \mathbf{0} = v_0)$;

(*iv*) $\forall v_0 \forall v_1[v_0 + \mathbf{s}v_1 = \mathbf{s}(v_0 + v_1)]$;

(*v*) $\forall v_0(v_0 \cdot \mathbf{0} = \mathbf{0})$;

(*vi*) $\forall v_0 \forall v_1(v_0 \cdot \mathbf{s}v_1 = v_0 \cdot v_1 + v_0)$.

We introduce a binary operation symbol $|\!-\!|$ and a new axiom

(*vii*) $\forall v_0 \forall v_1[(v_0 \leq v_1 \rightarrow v_0 + |v_0 - v_1| = v_1) \wedge$
$(v_1 \leq v_0 \rightarrow v_1 + |v_0 - v_1| = v_0)]$,

We introduce a binary operation symbol $[/]$ and a new axiom

(*viii*) $\forall v_0([v_0/\mathbf{0}] = \mathbf{0}) \wedge \forall v_0 \forall v_1(\neg v_1 = \mathbf{0}$
$\rightarrow [v_0/v_1] \cdot v_1 \leq v_0 \wedge v_0 + \mathbf{s0} \leq ([v_0/v_1] + \mathbf{s0}) \cdot v_1)$.

Next we introduce new n-ary operation symbols $\mathbf{U}_i^n$ and axioms

(*ix*) $\forall v_0 \cdots \forall v_{n-1}(\mathbf{U}_i^n v_0 \cdots v_{n-1} = v_i)$,

where $i < n$. Having introduced an m-ary operation symbol $\mathbf{O}$ and m n-ary operation symbols $\mathbf{P}_0, \ldots, \mathbf{P}_{m-1}$, we introduce an n-ary operation symbol $\mathbf{C}_n^m(\mathbf{O}; \mathbf{P}_0, \ldots, \mathbf{P}_{n-1})$ and an axiom

(*x*) $\forall v_0 \cdots \forall v_{n-1}[\mathbf{C}_n^m(\mathbf{O}; \mathbf{P}_0, \ldots, \mathbf{P}_{n-1})(v_0, \ldots, v_{n-1})$
$= \mathbf{O}(\mathbf{P}_0(v_0, \ldots, v_{n-1}), \ldots, \mathbf{P}_{n-1}(v_0, \ldots, v_{n-1}))]$.

Having introduced an m-ary operation symbol $\mathbf{O}$, we introduce an m-ary operation symbol $\mathbf{\Sigma}(\mathbf{O})$ and axioms

(*xi*) $\forall v_0 \cdots \forall v_{m-2}[\mathbf{\Sigma}(\mathbf{O})(v_0, \ldots, v_{m-2}, \mathbf{0}) = \mathbf{0}]$;

(*xii*) $\forall v_0 \cdots \forall v_{m-1}[\mathbf{\Sigma}(\mathbf{O})(v_0, \ldots, v_{m-2}, \mathbf{s}v_{m-1}) =$
$\mathbf{\Sigma}(\mathbf{O})(v_0, \ldots, v_{m-1}) + \mathbf{O}(v_0, \ldots, v_{m-1})]$.

Having introduced an m-ary operation symbol $\mathbf{O}$, we introduce an m-ary operation symbol $\mathbf{\Pi}(\mathbf{O})$ and axioms

(*xiii*) $\forall v_0 \cdots \forall v_{m-2}[\mathbf{\Pi}(\mathbf{O})(v_0, \ldots, v_{m-2}, \mathbf{0}) = \mathbf{s0}]$;

(*xiv*) $\forall v_0 \cdots \forall v_{n-1}[\mathbf{\Pi}(\mathbf{O})(v_0, \ldots, v_{m-2}, \mathbf{s}v_{m-1}) =$
$\mathbf{\Pi}(\mathbf{O})(v_0, \ldots, v_{m-1}) \cdot \mathbf{O}(v_0, \ldots, v_{m-1})]$.

For any m-ary operation symbol $\mathbf{O}$ already introduced we introduce an m-ary relation symbol $\mathbf{R_O}$ and the axiom

(xv) $\forall v_0 \cdots \forall v_{m-1}(\mathbf{R_O} v_0 \cdots v_{m-1} \leftrightarrow \mathbf{O} v_0 \cdots v_{m-1} = \Delta 1)$.

This describes all of the symbols of the language. We also introduce the induction axiom schema:

(xvi) $\varphi(\mathbf{0}) \wedge \forall v_0[\varphi \rightarrow \varphi(\mathbf{s}v_0)] \rightarrow \forall v_0 \varphi$,

where φ is any formula of $\mathscr{L}_{\mathrm{el}}$.

Finally, we associate with each nonlogical constant $\mathbf{F}$ of $\mathscr{L}_{\mathrm{el}}$ a number, a number-theoretic operation or relation $^{\#}\mathbf{F}$:

$$\begin{aligned}
&{}^{\#}\mathbf{0} = 0,\\
&{}^{\#}\mathbf{s} = \mathscr{s},\\
&{}^{\#}\mathbf{+} = +,\\
&{}^{\#}\bullet = \cdot,\\
&{}^{\#}(|\!-\!|) = |-|,\\
&{}^{\#}[/] = [/],\\
&{}^{\#}\mathbf{U}_i^n = U_i^n,\\
&{}^{\#}[\mathbf{C}_m^n(\mathbf{O}; \mathbf{P}_0, \ldots, \mathbf{P}_{m-1})] = K_m^n({}^{\#}\mathbf{O}; {}^{\#}\mathbf{P}_0, \ldots, {}^{\#}\mathbf{P}_{n-1}),\\
&{}^{\#}[\mathbf{\Sigma}(\mathbf{O})] = \Sigma^{\#}\mathbf{O},\\
&{}^{\#}[\mathbf{\Pi}(\mathbf{O})] = \Pi^{\#}\mathbf{O},\\
&{}^{\#}(\mathbf{RO}) = \{(x_0, \ldots, x_{m-1}) : {}^{\#}\mathbf{O}x_0 \cdots x_{m-1} = 1\}.
\end{aligned}$$

Proposition 17.5. *$(\mathscr{P}, \mathscr{L}_{\mathrm{el}})$ is a definitional expansion of $(P, \mathscr{L}_{\mathrm{nos}})$.*

PROOF. It is clear that $(\mathscr{P}, \mathscr{L}_{\mathrm{el}})$ can be obtained as the union, in an obvious sense, of a sequence $(\mathrm{P}, \mathscr{L}_{\mathrm{nos}}) = (\mathrm{S}_0, \mathscr{L}_0), (\mathrm{S}_1, \mathscr{L}_1), \ldots$ of successive expansions, where at each step we adjoin one new symbol, its appropriate axioms from (vii)–(xv), and all new instances of (xvi). Thus it is enough to show that each $(\mathrm{S}_{i+1}, \mathscr{L}_{i+1})$ is a definitional expansion of $(\mathrm{S}_i, \mathscr{L}_i)$. Note also that the new instances of (xvi) do not constitute a problem, since once we show that the new symbol has a definition in terms of old ones, the new instances of (xvi) clearly become provable in S_{i+1}.

If the new symbol in $\mathscr{L}_{i+1}$ is covered by one of the axioms (vii)–(x), or (xv), then $(\mathrm{S}_{i+1}, \mathscr{L}_{i+1})$ is clearly a definitional expansion of $(\mathrm{S}_i, \mathscr{L}_i)$. The arguments for $\mathbf{\Sigma}$ and $\mathbf{\Pi}$ are similar, so we shall only give the argument for $\mathbf{\Sigma}$. Thus we assume that $\mathbf{\Sigma}(\mathbf{O})$ is the new symbol of $\mathscr{L}_{i+1}$, where $\mathbf{O}$ is a symbol of $\mathscr{L}_i$, say of rank m. We can eliminate this symbol because we can formalize a natural argument using Gödel's β-function (see 3.40–3.46). In fact, let $\varphi_0, \varphi_1, \varphi_2, \varphi_3, \varphi_4$ be the following formulas (intuitively representing the functions $[\sqrt{}]$, exc, L, rm, and β respectively):

(1) $\varphi_0 : v_1 \cdot v_1 \leq v_0 \wedge \mathbf{s}v_0 \leq \mathbf{s}v_1 \cdot \mathbf{s}v_1$;

(2) $\varphi_1 : \exists v_2[\varphi_0(v_0, v_2) \wedge v_1 = |v_0 - v_2 \cdot v_2|]$;

(3) $\varphi_2 : \exists v_2[\varphi_0(v_0, v_2) \wedge \varphi_1(v_2, v_1)]$;

(4) $\varphi_3 : (v_1 = \mathbf{0} \wedge v_2 = \mathbf{0}) \vee [\neg v_1 = \mathbf{0} \wedge \exists v_3 (v_0 = v_1 \cdot v_3 + v_2 \wedge \mathbf{s}v_2 \leq v_1)]$;

(5) $\varphi_4 : \exists v_3 \exists v_4[\varphi_1(v_0, v_3) \wedge \varphi_2(v_0, v_4) \wedge \varphi_3(v_3, \mathbf{s}(\mathbf{s}v_1 \cdot v_4), v_2]$.

With the aid of these formulas we can give the definition of $\boldsymbol{\Sigma}(\mathbf{O})$:

$$\begin{aligned}(6)\quad & \forall v_0 \cdots \forall v_m\{\boldsymbol{\Sigma}(\mathbf{O})(v_0, \ldots, v_{m-1}) = v_m \leftrightarrow \exists v_{m+1}[\varphi_4(v_{m+1}, \mathbf{0}, \mathbf{0}) \wedge \\ & \forall v_{m+2}\{\mathbf{s}v_{m+2} \leq v_{m-1} \rightarrow \exists v_{m+3}[\varphi_4(v_{m+1}, v_{m+2}, v_{m+3}) \wedge \\ & \varphi_4(v_{m+1}, \mathbf{s}v_{m+2}, v_{m+3} + \mathbf{O}(v_0, \ldots, v_{m-2}, v_{m+2}))]\} \\ & \varphi_4(v_{m+1}, v_{m-1}, v_m)]\}.\end{aligned}$$

To show that this is, in fact, a definition of $\boldsymbol{\Sigma}(\mathbf{O})$ amounts to proving in S_{i+1} the existence and uniqueness conditions, as well as showing from the axioms S_i that (*xi*) and (*xii*) together are equivalent to (6). Two unformalized statements each equivalent to (*xi*) and (*xii*) are as follows:

$$\begin{aligned}(7)\quad & \textstyle\sum(f)(x_0, \ldots, x_{m-1}, 0) = 0, \text{ while } \sum(f)(x_0, \ldots, x_{m-1}, y + 1) \\ & \textstyle = \sum(f)(x_0, \ldots, x_{m-1}, y) + f(x_0, \ldots, x_{m-1}, y);\end{aligned}$$

$$\begin{aligned}(8)\quad & \textstyle\sum(f)(x_0, \ldots, x_{m-1}) = y \text{ iff there is a } z \text{ such that } \beta(z, 0) = 0 \\ & \text{while for all } w \text{ with } \mathfrak{s}w \leq x_{m-1}, \beta(z, \mathfrak{s}w) = \beta(z, w) + \\ & f(x_0, \ldots, x_{m-1}), \text{ and } \beta(z, x_{m-1}) = y.\end{aligned}$$

The equivalence of (7) and (8) is proved as follows. First assume (7). To prove direction $\Rightarrow$ of (8), assume that $\sum(f)(x_0, \ldots, x_{m-1}) = y$. By Theorem 3.46 choose z so that $\beta(z, i) = \sum(f)(x_0, \ldots, x_{m-2}, i)$ for each $i \leq x_{m-1}$. Then the right side of (8) holds. Conversely, assume the right side of (8). Then by induction on i it is clear that $\beta(z, i) = \sum(f)(x_0, \ldots, x_{m-2}, i)$ for each $i \leq x_{m-1}$. In particular, $\sum(f)(x_0, \ldots, x_{m-1}) = y$. Second, assuming (8), condition (7) is clear.

The existence and uniqueness conditions for the condition following "iff" in (8) is clear by 3.46 (for existence) and by using induction (for uniqueness).

Now to establish the existence and uniqueness conditions in S_i for the formula following "iff" in (6), and to prove the equivalence of (*xi*) and (*xii*) with (6) in S_{i+1}, it is a matter of formalizing in $\mathscr{P}$ the arguments in the preceding two paragraphs. In particular, the applications of 3.46 must be formalized. We leave it to the reader to look over this proof to see that it can be formalized. The formal versions of 3.44 and 3.46 are as follows. For each $r > 1$ the following sentence is an instance of 3.44:

$$\forall v_0 \cdots \forall v_{2r-1}\Big[\bigwedge_{i<r} (\mathbf{ss0} \leq v_i) \wedge \bigwedge_{i<j<r} \varphi_5(v_i, v_j) \rightarrow \exists v_{2r} \bigwedge_{i<r} \varphi_6(v_{2r}, v_{r+i}, v_i)\Big],$$

where $\varphi_5(v_0, v_1)$ naturally expresses that v_0 and v_1 are relatively prime, and $\varphi_6(v_0, v_1, v_2)$ naturally expresses $v_0 \equiv v_1 (\text{mod } v_2)$.

Similarly, for each $n \in \omega$ the following sentence is an instance of 3.46:

$$\forall v_0 \cdots \forall v_{n-1} \exists v_n \bigwedge_{i=n} \varphi_4(v_n, \mathbf{i}, v_i).$$

This completes the proof of 17.5.

Lemma 17.6. *For each operation symbol* $\mathbf{O}$ *of* $\mathscr{L}_{el}$, *either* $^{\#}\mathbf{O} \in \omega$ *if* $\mathbf{O}$ *is* 0-*ary, or* $^{\#}\mathbf{O}$ *is an m-ary operation on* ω *if* $\mathbf{O}$ *is m-ary. In either case, for any* $x_0, \ldots, x_{m-1}, y \in \omega$ *we have:* $^{\#}\mathbf{O}x_0 \cdots x_{m-1} = y$ *implies*

$$\mathscr{P} \models \mathbf{O}(\Delta x_0, \ldots, \Delta x_{m-1}) \equiv \Delta y.$$

The proof of this lemma is a straight-forward induction on $\mathbf{O}$, following Definition 17.4. As a corollary we have:

Lemma 17.7. *If* $\mathbf{O}$ *is an m-ary operation symbol of* $\mathscr{L}_{el}$, *then* $^{\#}\mathbf{R_O}$ *is an m-ary relation on* ω *and for any* $x_0, \ldots, x_{m-1} \in \omega$,

$$(x_0, \ldots, x_{m-1}) \in {}^{\#}\mathbf{R_O} \text{ implies } \mathscr{P} \models \mathbf{R_O}(\Delta x_0, \ldots, \Delta x_{m-1}),$$
$$(x_0, \ldots, x_{m-1}) \notin {}^{\#}\mathbf{R_O} \text{ implies } \mathscr{P} \models \neg\mathbf{R_O}(\Delta x_0, \ldots, \Delta x_{m-1}).$$

In the rest of this section we work in a fixed but arbitrary expansion $\mathscr{L}'$ of $\mathscr{L}_{el}$, and with any theory Γ in $\mathscr{L}'$ which extends $\mathscr{P}$, such that Γ has a set Δ of axioms with $g^{+*}\Delta$ elementary. Our goal, of course, is to define a formula *Pr* satisfying 17.2(*i*)–(*iii*) for $(\Gamma, \mathscr{L}')$. And we want *Pr* to intuitively express provability in $(\Gamma, \mathscr{L}')$. To this end, we want to look back at our work in Chapters 2 and 10. That work was done, naturally, in our usual metalanguage for mathematics. We now want to see that most of it can be done *within* the context of $(\Gamma, \mathscr{L}')$. Our language $\mathscr{L}_{el}$ was designed to facilitate this task. We shall not carry through the task in full detail, but we shall sketch the high points. Much of the work in those Chapters was the definition of elementary functions and relations, or the proof that certain functions or relations were elementary. To systematize the notation, for each elementary operation F and relation R we denote by $^{\circ}F$ and $^{\circ}R$ the operation and relation symbols of $\mathscr{L}_{el}$ which *naturally* correspond to F and R. By "naturally correspond" we mean that the original definitions in Chapters 2 and 10 are to be formalized. For example:

(1) $$^{\circ}0 = \mathbf{0}$$
(2) $$^{\circ}\sigma = \mathbf{s}$$
(3) $$^{\circ}C_0^1 = \mathbf{C}_1^2(\mathbf{|-|}\,;\ \mathbf{U}_0^1, \mathbf{U}_0^1);$$
(4) $$^{\circ}! = \mathbf{\Pi}(\mathbf{s}).$$

One useful property of this convention is

(5) $$^{\#\circ}F = F \text{ and } ^{\#\circ}R = R.$$

It is completely routine to see what the symbols $^{\circ}F$ and $^{\circ}R$ are, as symbols of $\mathscr{L}_{el}$, for all of the functions and relations F and R explicitly shown to be elementary in Chapters 2 and 10: the proof of their elementariness is used to work out the definition of $^{\circ}F$ and $^{\circ}R$. As another example, Proposition 2.15, concerning bounded universal quantification, goes over into the following formalization:

(6) if $\mathbf{R}$ is an m-ary relation symbol of $\mathscr{L}_{el}$, then there is an m-ary relation symbol $\mathbf{T}$ of $\mathscr{L}_{el}$ such that

$$\mathscr{P} \models \forall v_0 \cdots \forall v_{m-1}\{\mathbf{T}(v_0, \ldots, v_{m-1}) \leftrightarrow \forall v_m[v_m < v_{m-1} \rightarrow \mathbf{R}(v_0, \ldots, v_{m-2}, v_m)]\}.$$

When we consider the formalization of logic, starting in Chapter 10, several remarks must be made. First, we shall of course assume that $\mathscr{L}'$ is an elementary effectivized language. As mentioned in Chapter 10, the various functions and relations shown there to be recursive can then be shown to be elementary. Hence they have corresponding formalized versions in $(\mathscr{L}', \Gamma)$. For example:

(7) ${}^\circ\neg'$ is a unary operation symbol of $\mathscr{L}_{el}$ such that $\Gamma \models {}^\circ\neg' v_0 = {}^\circ\mathrm{Cat}\,(\mathbf{\Delta}(2^{gL0+1}), v_0)$; for any expression φ of $\mathscr{L}'$, $\Gamma \models {}^\circ\neg'\mathbf{\Delta}(g^+\varphi) = \mathbf{\Delta}(g^+\neg\varphi)$;

(8) ${}^\circ\!=\!'$ is a binary operation symbol of $\mathscr{L}_{el}$ such that $\Gamma \models (v_0 \,{}^\circ\!=\!'\, v_1) = {}^\circ\mathrm{Cat}\,({}^\circ\mathrm{Cat}\,(\mathbf{\Delta}2^{gL4+1}, v_0), v_1)$; for any expressions σ and τ of $\mathscr{L}'$,

$$\Gamma \models [(\mathbf{\Delta}g^+\sigma)\,{}^\circ\!=\!'\,(\mathbf{\Delta}g^+\tau)] = \mathbf{\Delta}g^+(\sigma = \tau).$$

Since we are assuming that $g^{+*}\Delta$ is elementary, we have in $\mathscr{L}_{el}$ a unary relation symbol ${}^\circ(g^{+*}\Delta)$ corresponding to some construction of $g^{+*}\Delta$. We let Prf be the collection of all ordered pairs (m, n) such that n is the Gödel number of a Δ-proof φ and m is the Gödel number of the last member of φ. Thus ${}^\circ Prf$ is a binary relation symbol of $\mathscr{L}_{el}$ built up in a rather complicated way, using the symbol ${}^\circ(g^{+*}\Delta)$. Our desired formula Pr is the formula $\exists v_1 {}^\circ Prf(v_0, v_1)$.

Property 17.2(i) is easy to establish:

Lemma 17.8. *If* $\Gamma \models \psi$, *then* $\Gamma \models Pr\mathbf{\Delta}g^+\psi$.

Proof. Assume that $\Gamma \models \psi$. Thus there is a Δ-proof φ with last member ψ. Hence $(g^+\psi, g^{++}\varphi) \in Prf$, and $Prf = {}^{\#\circ}Prf$, so by 17.7

$$\mathscr{P} \models {}^\circ Prf(\mathbf{\Delta}g^+\psi, \mathbf{\Delta}g^{++}\varphi).$$

Since $\mathscr{P} \subseteq \Gamma$, it follows that $\Gamma \models Pr\mathbf{\Delta}g^+\psi$. □

Condition 17.2(iii) is also rather easy:

Lemma 17.9. $\Gamma \models Pr\mathbf{\Delta}g^+(\chi \rightarrow \psi) \rightarrow (Pr\mathbf{\Delta}g^+\chi \rightarrow Pr\mathbf{\Delta}g^+\psi)$.

PROOF. From the choice of $°Prf$, which is supposed to mimic the definition of Prf (see Proposition 10.27), we have

$$\begin{aligned}(9)\quad &\Gamma \models Prf(v_0, v_1) \leftrightarrow \forall v_2\{v_2 \leq {}°lv_1 \rightarrow [°(\mathscr{g}^{+*}\,\mathrm{Axm})(°(v_1)_{v2})\\ &\vee\, °(\mathscr{g}^{+*}\Delta)(°(v_1)_{v2}) \vee \exists v_3 \exists v_4[v_3 < v_2 \wedge v_4 < v_2 \wedge °(v_1)_{v3}\\ &= °(v_1)_{v4}\, °{\rightarrow}'\, °(v_1)_{v2}] \vee \exists v_3 \exists v_4[v_3 < v_2 \wedge v_4 \leq v_1 \wedge °(\mathrm{Rng}(\mathscr{g} \circ v))(v_4)\\ &\wedge\, °(v_1)_{v2} = °\forall'(°\exp(\Delta 2, \mathbf{s}v_4), °(v_1)_{v3})]\} \wedge °(v_1)°lv_1 = v_0.\end{aligned}$$

From (9) it is easy to establish the following:

$$\begin{aligned}(10)\quad &\Gamma \models Prf(\Delta\mathscr{g}^+\chi\; °{\rightarrow}'\; \Delta\mathscr{g}^+\psi, v_0) \wedge Prf(\Delta\mathscr{g}^+\chi, v_1)\\ &\qquad \rightarrow Prf(\Delta\mathscr{g}^+\psi, °\mathrm{Cat}\,(v_0, °\mathrm{Cat}\,(v_1, \Delta\,\mathrm{Exp}\,(2, \Delta\mathscr{g}^+\psi)))).\end{aligned}$$

Since $\Gamma \models (\Delta\mathscr{g}^+\chi\; °{\rightarrow}'\; \Delta\mathscr{g}^+\psi) = \Delta\mathscr{g}^+(\chi \rightarrow \psi)$, easy logic yields the desired result. □

The proof of the remaining condition, 17.2(*ii*), is based on the following lemma, for which we need a new notion. We can find in $\mathscr{L}_{\mathrm{el}}$ a unary relation symbol Nm (for "numeral"), with the following sentence provable in $\mathscr{P}$:

$$(11)\qquad Nm\mathbf{0} = \Delta 2^{\mathscr{g}\mathbf{0}} \wedge \forall v_0[Nm\mathbf{s}v_0 = °\mathrm{Cat}\,(\Delta 2^{\mathscr{g}\mathbf{s}}, Nmv_0)].$$

Thus we easily obtain:

$$(12)\qquad \mathscr{P} \models Nm\Delta m = \Delta\mathscr{g}^+\Delta m \qquad \text{for each } m \in \omega.$$

Lemma 17.10. *Let* **F** *be an n-ary operation symbol of* $\mathscr{L}_{\mathrm{el}}$. *Then*

$$\Gamma \models Pr(°\mathrm{Con}_{n+1}\,(\Delta 2^{\mathscr{g}\mathbf{F}}, Nmv_0, \ldots, Nmv_{n-1})\; °{=}'\; Nm\mathbf{F}(v_0, \ldots, v_{n-1})).$$

PROOF. We proceed by induction on the complexity of **F** as described in definition 17.4. The proof is fully illustrated by considering only the following three cases.

Case 1. $\mathbf{F} = \mathbf{0}$. Thus we must prove

$$(13)\qquad \Gamma \models Pr(°\mathrm{Con}_1\,(\Delta 2^{\mathscr{g}\mathbf{0}})\; °{=}'\; Nm\mathbf{0}).$$

Now $\Gamma \models °\mathrm{Con}_1(\Delta 2^{\mathscr{g}\mathbf{0}}) = \Delta 2^{\mathscr{g}\mathbf{0}}$ and $\Gamma \models Nm\,\mathbf{0} = \Delta 2^{\mathscr{g}\mathbf{0}}$; and by (8),

$$\Gamma \models (\Delta 2^{\mathscr{g}\mathbf{0}}\; °{=}'\; \Delta 2^{\mathscr{g}\mathbf{0}}) = \Delta\mathscr{g}^+(\mathbf{0} = \mathbf{0}).$$

Furthermore, $\Gamma \models \mathbf{0} = \mathbf{0}$, so by 17.8 $\Gamma \models Pr\Delta\mathscr{g}^+(\mathbf{0} = \mathbf{0})$. Hence (13) follows.

Case 2. $\mathbf{F} = \mathbf{s}$. Then we must prove

$$(14)\qquad \Gamma \models Pr(°\mathrm{Con}_2\,(\Delta 2^{\mathscr{g}\mathbf{s}}, Nmv_0)\; °{=}'\; Nm\mathbf{s}v_0).$$

Now by (11), $\Gamma \models Nm\mathbf{s}v_0 = °\mathrm{Cat}\,(\Delta 2^{\mathscr{g}\mathbf{s}}, Nmv_0)$. By induction within $\mathscr{P}$ we easily see that

$$(15)\qquad \Gamma \models \forall v_0[°(\mathscr{g}^{+*}\,\mathrm{Trm})(Nmv_0)].$$

Now in carrying through Chapter 10 *within* $\mathscr{P}$, the formalization of 10.34 is

$$(16)\qquad \Gamma \models °(\mathscr{g}^{+*}\,\mathrm{Trm})(v_0) \rightarrow Pr(v_0\; °{=}'\; v_0).$$

Hence (14) follows.

Case 3. $\mathbf{F} = \mathbf{+}$. In this case we need to show

(17) $$\Gamma \models Pr({}^{\circ}\mathrm{Con}_3\,(\Delta 2^{g+}, Nmv_0, Nmv_1)\ {}^{\circ}\!\!=' Nm(v_0 \mathbf{+} v_1)).$$

Here we proceed by induction on v_1 within Γ. We have

(18) $$\Gamma \models v_0 \mathbf{+} \mathbf{0} = v_0.$$

Now consider the following little argument: the sentence $\forall v_0(v_0 \mathbf{+} \mathbf{0} = v_0)$ is an axiom of $\mathscr{P}$, and hence for any term σ, $\mathscr{P} \models \sigma \mathbf{+} 0 = \sigma$ by universal specification. The formalization of this argument gives

(19) $$\Gamma \models {}^{\circ}(g^{+*}\,\mathrm{Trm})(v_0) \rightarrow Pr({}^{\circ}\mathrm{Con}_3\,(\Delta 2^{g+}, v_0, \Delta 2^{g0})\ {}^{\circ}\!\!=' v_0).$$

Now (15), (18), (19) give

(20) $$\Gamma \models Pr({}^{\circ}\mathrm{Con}_3\,(\Delta 2^{g+}, Nmv_0, Nm\mathbf{0})\ {}^{\circ}\!\!=' Nm(v_0 \mathbf{+} \mathbf{0}),$$

which is the step $v_1 = 0$. Next,

(21) $$\Gamma \models v_0 \mathbf{+} \mathbf{s}v_1 = \mathbf{s}(v_0 \mathbf{+} v_1),$$

(22) $$\Gamma \models Nm\mathbf{s}v_0 = {}^{\circ}\mathrm{Cat}\,(\Delta 2^{g\mathbf{s}}, Nmv_0).$$

Now for any terms σ, τ, $\Gamma \models \sigma \mathbf{+} \mathbf{s}\tau = \mathbf{s}(\sigma \mathbf{+} \tau)$. Formalizing,

$$\Gamma \models {}^{\circ}(g^{+*}\,\mathrm{Trm})(v_0) \wedge {}^{\circ}(g^{+*}\,\mathrm{Trm})v_1 \rightarrow Pr({}^{\circ}\mathrm{Con}_3\,(\Delta 2^{g+}v_0, {}^{\circ}\mathrm{Cat}\,(\Delta 2^{g\mathbf{s}}, v_1))\ {}^{\circ}\!\!=' {}^{\circ}\mathrm{Con}_4\,(\Delta 2^{g\mathbf{s}}, \Delta 2^{g+}, v_0, v_1)).$$

Since by (15), $\Gamma \models {}^{\circ}(g^{+*}\,\mathrm{Trm})(Nmv_0) \wedge {}^{\circ}(g^{+*}\,\mathrm{Trm})(Nmv_1)$, it follows using (22) that

(23) $$\Gamma \models Pr({}^{\circ}\mathrm{Con}_3\,(\Delta 2^{g+}, Nmv_0, Nm\mathbf{s}v_1)\ {}^{\circ}\!\!=' {}^{\circ}\mathrm{Con}_4\,(\Delta 2^{g\mathbf{s}}, \Delta 2^{g+}, Nmv_0, Nmv_1).$$

Also, $\Gamma \models {}^{\circ}\mathrm{Con}_4\,(\Delta 2^{g\mathbf{s}}, \Delta 2^{g+}, Nmv_0, Nmv_1)\ {}^{\circ}\!\!=' {}^{\circ}\mathrm{Cat}\,(\Delta 2^{g\mathbf{s}}, {}^{\circ}\mathrm{Con}_3\,(\Delta 2^{g+}, Nmv_0, Nmv_1))$, so we easily obtain from (23) and (21) that

$$\Gamma \models Pr({}^{\circ}\mathrm{Con}_3\,(\Delta 2^{g+}, Nmv_0, Nmv_1)\ {}^{\circ}\!\!=' Nm(v_0 \mathbf{+} v_1)) \rightarrow Pr({}^{\circ}\mathrm{Con}_3\,(\Delta 2^{g+}, Nmv_0, Nm\mathbf{s}v_1)\ {}^{\circ}\!\!=' Nm(v_0 \mathbf{+} \mathbf{s}v_1)).$$

This is the induction step, so (17) follows by induction. □

From Lemma 17.10 our final condition 17.2(*ii*) easily follows:

Lemma 17.11. $\Gamma \models Pr\Delta g^{+}\psi \rightarrow Pr\Delta g^{+}Pr\Delta g^{+}\psi$.

PROOF. Let ${}^{\circ}Prf = \mathbf{R_F}$, where $\mathbf{F}$ is a binary operation symbol of $\mathscr{L}_{\mathrm{el}}$. By axiom (*xv*) of $\mathscr{P}$,

(24) $$\Gamma \models {}^{\circ}Prf(\Delta g^{+}\psi, v_1) \leftrightarrow \mathbf{F}(\Delta g^{+}\psi, v_1) = \Delta 1,$$

while by Lemma 17.10,

(25) $$\Gamma \models Pr({}^{\circ}\mathrm{Con}_3\,(\Delta 2^{g\mathbf{F}}, Nm\Delta g^{+}\psi, Nmv_1)\ {}^{\circ}\!\!=' Nm\mathbf{F}(\Delta g^{+}\psi, v_1)).$$

Now from axiom (*xv*) of $\mathscr{P}$, for any terms σ, τ of $\mathscr{L}'$ we have $\Gamma \models {}^{\circ}Prf(\sigma, \tau) \leftrightarrow \mathbf{F}(\sigma, \tau) = \Delta 1$. The formalization of this fact is

$$(26)\quad \Gamma \models {}^{\circ}(g^{+*}\,\mathrm{Trm})(v_0) \wedge {}^{\circ}(g^{+*}\,\mathrm{Trm})(v_1) \rightarrow Pr({}^{\circ}\mathrm{Con}_3\,(\Delta 2^{g{}^{\circ}Prf}, v_0, v_1)\ {}^{\circ}\!\leftrightarrow'\ {}^{\circ}\mathrm{Con}_3\,(\Delta 2^{g\mathbf{F}}, v_0, v_1)\ {}^{\circ}\!=' Nm\Delta 1).$$

Now $\Gamma \models (g^{+*}\,\mathrm{Trm})(Nm\Delta g^{+}\psi) \wedge {}^{\circ}(g^{+*}\,\mathrm{Trm})(Nmv_1)$, so (26) yields

$$(27)\quad \Gamma \models Pr({}^{\circ}\mathrm{Con}_3\,(\Delta 2^{g{}^{\circ}Prf}, Nm\Delta g^{+}\psi, Nmv_1)\ {}^{\circ}\!\leftrightarrow'\ {}^{\circ}\mathrm{Con}_3\,(\Delta 2^{g\mathbf{F}}, Nm\Delta g^{+}\psi, Nmv_1)\ {}^{\circ}\!=' Nm\Delta 1).$$

But (24) and (25) imply

$$\Gamma \models {}^{\circ}Prf(\Delta g^{+}\psi, v_1) \rightarrow Pr({}^{\circ}\mathrm{Con}_3\,(\Delta 2^{g\mathbf{F}}, Nm\Delta g^{+}\psi, Nmv_1)\ {}^{\circ}\!=' Nm\Delta 1),$$

so by (27) we have

$$(28)\quad \Gamma \models {}^{\circ}Prf(\Delta g^{+}\psi, v_1) \rightarrow Pr({}^{\circ}\mathrm{Con}_3\,(\Delta 2^{g{}^{\circ}Prf}, Nm\Delta g^{+}\psi, Nmv_1)).$$

Clearly

$$\Gamma \models Pr({}^{\circ}\mathrm{Con}_3\,(\Delta 2^{g{}^{\circ}Prf}, Nm\Delta g^{+}\psi, Nmv_1)) \rightarrow Pr({}^{\circ}\exists'(\Delta g v1, {}^{\circ}\mathrm{Con}_3\,(\Delta 2^{g{}^{\circ}Prf}, Nm\Delta g^{+}\psi, \Delta 2^{g v1}))),$$

and

$$\Gamma \models {}^{\circ}\exists'(\Delta g v1, {}^{\circ}\mathrm{Con}_3\,(\Delta 2^{g{}^{\circ}Prf}, Nm\Delta g^{+}\psi, \Delta 2^{g v1}))\ {}^{\circ}\!=' \Delta g^{+}Pr\Delta g^{+}\psi,$$

so by (28) we have

$$\Gamma \models {}^{\circ}Prf(\Delta g^{+}\psi, v_1) \rightarrow Pr\Delta g^{+}Prg^{+}\psi.$$

Now an easy logical transformation gives the desired result. □

BIBLIOGRAPHY

1. Feferman, S. Arithmetization of metamathematics in a general setting. *Fund. Math.*, *49* (1960), 35–92.

2. Löb, M. H. Solution of a problem of Leon Henkin, *J. Symb. Logic*, *20* (1955), 115–118.

EXERCISES

17.12. Apply 15.20 in $\mathscr{P}$ to get a sentence ψ such that $\mathscr{P} \models Pr(\Delta g^{+}\psi) \leftrightarrow \psi$. Show that ψ holds in the natural model. [Note that ψ "says" of itself that it is provable, and sure enough it is. The corresponding nonlogical statement would be "I am telling the truth," a nonparadoxical statement which could be true or false.]

17.13. Let φ be any sentence of $\mathscr{L}_{\mathrm{el}}$. Apply 15.20 to $\mathscr{P}$ to get a sentence ψ such that

$$\mathscr{P} \models [Pr(\Delta g^{+}\psi) \rightarrow \varphi] \leftrightarrow \psi.$$

Show that $\mathscr{P} \models \varphi$ *iff* $\mathscr{P} \models \psi$. [The informal paradox behind this construction is as follows. Given *any* sentence φ, let ψ be the sentence

If this sentence is true, then φ is true.

Clearly ψ is true. Hence φ is true. Thus every sentence is true.]

17.14. Apply 15.20 to obtain a sentence ψ such that

$$\mathscr{P} \models Pr({}^\circ\neg'\Delta_{\mathscr{g}}{}^+\psi) \leftrightarrow \psi.$$

Show that neither $\mathscr{P} \models \psi$ nor $\mathscr{P} \models \neg\psi$. This is another way of interpreting the Liar paradox different from that given following 15.20.

17.15. Show that there is a sentence ψ such that

$$\mathscr{P} \models Pr(\Delta_{\mathscr{g}}{}^+ Pr({}^\circ\neg'\Delta_{\mathscr{g}}{}^+\psi)) \leftrightarrow \psi.$$

Show that neither $\mathscr{P} \models \psi$ nor $\mathscr{P} \models \neg\psi$. [The intuitive paradox is as follows. Let φ be the sentence

It is true that it is true that this sentence is false.

Since φ is just a slight reformulation of the liar, a paradox arises as usual.]

17.16. Construct a sentence ψ such that neither $\mathscr{P} \models \psi$ nor $\mathscr{P} \models \neg\psi$ by working from the following paradox. Let φ be the sentence

It is true that it is false that this sentence is true.

Again a paradox arises.

PART IV

Model Theory

We now turn to the detailed study of the relationships between syntactical and semantic properties of structures. We consider various ways of constructing structures: the usual algebraic ones of products, homomorphic images and substructures, and some new ones especially designed for their logical applications. In much of this part we shall be concerned with questions about the relationship between syntactical properties of sentences and algebraic properties of structures; for example, we show in Chapter 24 that a class **K** is the class of all models of a set of equations iff **K** is closed under the algebraic operations of formation of products, homomorphic images, and subalgebras. Using model-theoretic methods, we can prove various concrete theories to be complete and decidable. Finally, the model theory of formulas with free variables is treated in Chapter 27. For this entire part there now is in print a comprehensive treatise, to which we refer the reader for further information:

Chang, C. C., and Keisler, H. J. *Model Theory*. Amsterdam: North-Holland (1973).

Construction of Models 18

In this chapter we shall give two methods of constructing models which form the main tools for *all* of our later work. The first method, the *consistency family* method, is an abstraction from the proof of the completeness theorem given in Chapter 11, and is due to Henkin, Smullyan, Makkai, and Keisler. The second method, of *ultraproducts*, is an algebraic kind of construction of a model from old models, and is mainly due to Łoś.

The consistency family method will be of use mainly for countable languages, for which the main definitions and results are simpler. We begin with a preliminary definition.

Definition 18.1. Let $\mathscr{L}$ be a first-order language. With any formula φ of $\mathscr{L}$ we associate a new formula $\varphi^{\rightarrow}$ as follows. If φ is atomic, let $\varphi^{\rightarrow} = \neg\varphi$. Further, let

$$\begin{aligned}(\neg\varphi)^{\rightarrow} &= \varphi\\ (\varphi \vee \psi)^{\rightarrow} &= \neg\varphi \wedge \neg\psi\\ (\varphi \wedge \psi)^{\rightarrow} &= \neg\varphi \vee \neg\psi\\ (\forall\alpha\varphi)^{\rightarrow} &= \exists\alpha\neg\varphi\end{aligned}$$

We may say that $\varphi^{\rightarrow}$ is obtained from $\neg\varphi$ by shoving the negation sign one step inside, if possible, and eliminating an initial double negation. The purpose of this procedure is to enable us to do a special kind of induction on formulas (see the proof of 18.9). Clearly we have:

Proposition 18.2. $\models \neg\varphi \leftrightarrow \varphi^{\rightarrow}$.

Definition 18.3. Let $\mathscr{L}$ be a first-order language. An expansion $\mathscr{L}'$ of $\mathscr{L}$ is *rich by C* if $\mathscr{L}'$ is obtained from $\mathscr{L}$ by adjoining a set C of individual constants with $|C| = |\mathrm{Fmla}_{\mathscr{L}}|$. We shall denote $\mathscr{L}'$-structures frequently by $(\mathfrak{A}, a_c)_{c \in C}$, where $\mathfrak{A}$ is an $\mathscr{L}$-structure. A *primitive term of* $\mathscr{L}'$ is a term of the form $\mathbf{O}\mathbf{c}_0 \cdots \mathbf{c}_{m-1}$ where $\mathbf{O}$ is an operation symbol of rank m and $\mathbf{c} \in {}^mC$. In particular ($m = 0$), any individual constant is a primitive term.

In this definition we have made a little more definite part of our procedure in Chapter 11. Now we come to the main concept:

Definition 18.4. Let $\mathscr{L}'$ be a rich expansion of $\mathscr{L}$ by C. Let S be a family of sets of sentences of $\mathscr{L}'$. Then S is a *consistency family* (*for* $\mathscr{L}$, $\mathscr{L}'$) iff for each $\Gamma \in S$ all of the following hold, for all sentences φ, ψ of $\mathscr{L}'$:

(C0) If $\Delta \subseteq \Gamma$, then $\Delta \in S$.
(C1) $\varphi \notin \Gamma$ or $\neg\varphi \notin \Gamma$.
(C2) If $\neg\varphi \in \Gamma$, then $\Gamma \cup \{\varphi^{\rightarrow}\} \in S$.
(C3) If $\varphi \wedge \psi \in \Gamma$, then $\Gamma \cup \{\varphi\} \in S$ and $\Gamma \cup \{\psi\} \in S$.
(C4) If $\varphi \vee \psi \in \Gamma$, then $\Gamma \cup \{\varphi\} \in S$ or $\Gamma \cup \{\psi\} \in S$.
(C5) If $\forall \alpha\varphi \in \Gamma$, then for all $\mathbf{c} \in C$, $\Gamma \cup \{\mathrm{Subf}^{\alpha}_{\mathbf{c}}\varphi\} \in S$.
(C6) If $\exists \alpha\varphi \in \Gamma$, then for some $\mathbf{c} \in C$, $\Gamma \cup \{\mathrm{Subf}^{\alpha}_{\mathbf{c}}\varphi\} \in S$.
(C7) If $\mathbf{c}, \mathbf{d} \in C$, and $(\mathbf{c} = \mathbf{d}) \in \Gamma$, then $\Gamma \cup \{\mathbf{d} = \mathbf{c}\} \in S$.
(C8) If $\mathbf{c} \in C$, τ is a primitive term, $\mathbf{c} = \tau \in \Gamma$ and $\mathrm{Subf}^{\alpha}_{\tau}\varphi \in \Gamma$, then $\Gamma \cup \{\mathrm{Subf}^{\alpha}_{\mathbf{c}}\varphi\} \in S$.
(C9) For any primitive term τ there is a $\mathbf{c} \in C$ such that $\Gamma \cup \{\mathbf{c} = \tau\} \in S$.
(C10) If α is a limit ordinal $< |\mathrm{Fmla}_{\mathscr{L}}|$, $\Gamma \in {}^{\alpha}S$, $\Gamma_\beta \subseteq \Gamma_\gamma$ whenever $\beta < \gamma < \alpha$, and $|\{\mathbf{c} \in C\colon \mathbf{c} \text{ occurs in } \varphi \text{ for some } \varphi \in \bigcup_{\beta<\alpha} \Gamma_\beta\}| < |C|$, then $\bigcup_{\beta<\alpha} \Gamma_\beta \in S$.
(C11) $|\{\mathbf{c} : \mathbf{c} \text{ occurs in some } \varphi \in \Gamma\}| < |C|$.

We may think of the members of a consistency family as being consistent sets of sentences in $\mathscr{L}'$. They are so related in the family that they can be extended using the properties (C1)–(C11) just as we extended our consistent set in in Chapter 11 to a complete rich consistent set. As we shall see, analyzing that construction in this way enables us to get "inside" the construction and make modifications which will assure special desirable properties of the resulting model; see the later proofs of 22.1, 27.4, and 28.6, for example. Note that if $|\mathrm{Fmla}_{\mathscr{L}}| = \aleph_0$, then condition (C10) drops away. Under the assumption $|\mathrm{Fmla}_{\mathscr{L}}| = \aleph_0$, assumption (C11) can also be dropped and the main theorem 18.9 remains valid; and by this assumption (C0) can also be dropped (see 18.7).

Now we shall give a couple of examples of consistency families. More will be found in the exercises and in later chapters.

Proposition 18.5. *Let $\mathscr{L}'$ be a rich expansion of $\mathscr{L}$ by C. Let S be the set of all formally consistent sets Γ of sentences of $\mathscr{L}'$ such that $|\{\mathbf{c} \in C\colon \mathbf{c}$ occurs in some $\varphi \in \Gamma\}| < |C|$. Then S is a consistency family.*

PROOF. We need to check conditions (C0)–(C11) for an arbitrary $\Gamma \in S$. (C0) and (C1) are clear since Γ is consistent. (C2) follows from 18.2. Since $\varphi \wedge \psi \in \Gamma$ implies that $\Gamma \models \varphi$ and $\Gamma \models \psi$, (C3) is clear. For (C4), assume that $\Gamma \cup \{\varphi\} \notin S$ and $\Gamma \cup \{\psi\} \notin S$. Then $\Gamma \models \neg\varphi$ and $\Gamma \models \neg\psi$, so $\Gamma \models \neg(\varphi \vee \psi)$ and hence $\varphi \vee \psi \notin \Gamma$ since Γ is consistent. (C5) is clear since $\Gamma \models \mathrm{Subf}^{\alpha}_{\mathbf{c}}\varphi$. For (C6), assume that $\Gamma \cup \{\mathrm{Subf}^{\alpha}_{\mathbf{c}}\varphi\} \notin S$ for all $\mathbf{c} \in C$. Choose $\mathbf{c} \in C$ so that $\mathbf{c}$ does not occur in φ or in any sentence of Γ. Then $\Gamma \models \neg\mathrm{Subf}^{\alpha}_{\mathbf{c}}\varphi$. In a proof of $\neg\mathrm{Subf}^{\alpha}_{\mathbf{c}}\varphi$ from Γ, replace $\mathbf{c}$ be a new variable β. Thus $\Gamma \models \neg\mathrm{Subf}^{\alpha}_{\beta}\varphi$, hence $\Gamma \models \forall\beta \neg \mathrm{Subf}^{\alpha}_{\beta}\varphi$, hence $\Gamma \models \forall\alpha \neg \varphi$. Thus $\exists\alpha\varphi \notin \Gamma$. (C7) and (C8) are clear. For (C9), choose $\mathbf{c} \in C$ not occurring in τ or in any sentence of Γ. If $\Gamma \cup \{\mathbf{c} = \tau\} \notin S$, then $\Gamma \models \neg(\mathbf{c} = \tau)$. In a proof of $\neg(\mathbf{c} = \tau)$ from Γ, replace $\mathbf{c}$ by a new variable α. Thus $\Gamma \models \neg(\alpha \equiv \tau)$, hence $\Gamma \models \forall\alpha \neg (\alpha \equiv \tau)$, hence $\Gamma \models \neg(\tau \equiv \tau)$, hence Γ is inconsistent, contradiction. (C10) is clear, as is (C11). □

Proposition 18.6. *Let $\mathscr{L}'$ be a rich expansion of $\mathscr{L}$ by C, where $|\mathrm{Fmla}_{\mathscr{L}}| = \aleph_0$. Let S be the set of all sets Γ of sentences such that Γ has a model $(\mathfrak{A}, a_{\mathbf{c}})_{\mathbf{c}\in C}$ with $A = \{a_{\mathbf{c}} : \mathbf{c} \in C\}$ and* (C11) *holds. Then S is a consistency family.*

This proposition is clear. Another lemma which will be used in applying our main theorem is as follows.

Lemma 18.7. *If S satisfies conditions* (C1)–(C9), *then $S' = \{\Gamma : \Gamma \subseteq \Delta$ for some $\Delta \in S\}$ satisfies* (C0)–(C9).

PROOF. Assume that $\Gamma \in S'$, say $\Gamma \subseteq \Delta \in S$. Both (C0) and (C1) are clear for Γ. To check (C2), suppose $\neg\varphi \in \Gamma$. Thus $\neg\varphi \in \Delta$, so by (C2) for S, $\Delta \cup \{\varphi^{\rightarrow}\} \in S$. Now $\Gamma \cup \{\varphi^{\rightarrow}\} \subseteq \Delta \cup \{\varphi^{\rightarrow}\}$, so $\Gamma \cup \{\varphi^{\rightarrow}\} \in S'$. (C3)–(C9) are established similarly. □

The concept that really enables us to "get inside" the construction is as follows:

Definition 18.8. Let $\mathscr{L}'$ be a rich expansion of $\mathscr{L}$ by C, and let S be a consistency family. A function $f\colon S \to S$ is *admissible* over S provided that for any $\Gamma \in S$ the following two conditions hold:

(*i*) $\Gamma \subseteq f\Gamma$;

(*ii*) $|\{\mathbf{c} \in C\colon \mathbf{c}$ occurs in some $\varphi \in f\Gamma\}| = |\{\mathbf{c} \in C\colon \mathbf{c}$ occurs in some $\varphi \in \Gamma\}| + m$ for some $m \in \omega$.

Theorem 18.9. (Model existence theorem). *Let $\mathscr{L}'$ be a rich expansion of $\mathscr{L}$ by C, let S be a consistency family, and let $\langle f_\alpha : \alpha < |\mathrm{Fmla}_{\mathscr{L}}|\rangle$ be a family of admissible functions over S. Then for any $\Gamma \in S$ there is a model*

$(\mathfrak{A}, a_c)_{c\in C}$ *of* Γ *satisfying the following two conditions:*

(*i*) $A = \{a_c : \mathbf{c} \in C\}$ *and so* $|A| \leq |\mathrm{Fmla}_{\mathscr{L}}|$;
(*ii*) *for each* $\alpha < |\mathrm{Fmla}_{\mathscr{L}}|$ *there is a* $\Delta \in S$ *such that*

$$\Gamma \subseteq f_\alpha \Delta \subseteq \{\varphi : (\mathfrak{A}, a_c)_{c\in C} \models \varphi\}.$$

PROOF. Set $\mathfrak{m} = |\mathrm{Fmla}_{\mathscr{L}}|$ for brevity. Let $\varphi: \mathfrak{m} \rightarrowtail\!\!\!\!\twoheadrightarrow \mathrm{Sent}_{\mathscr{L}'}$, let τ map $\mathfrak{m}$ one-one onto the set of primitive terms of $\mathscr{L}'$, and let a well-ordering of C be fixed. Let $\Gamma \in S$. We now define a sequence $\langle \Theta_\alpha : \alpha \leq \mathfrak{m}\rangle$. Let $\Theta_0 = \Gamma$. Suppose $\Theta_\alpha \in S$ has been defined, where $\alpha < \mathfrak{m}$, and the following condition holds:

(0) $|\{\mathbf{c} : \mathbf{c} \text{ occurs in some } \varphi \in \Theta_\alpha\}| \leq |\{\mathbf{c} : \mathbf{c} \text{ occurs in some } \varphi \in \Gamma\}| + |\alpha| + \aleph_0$.

We now define $\Theta_{\alpha+1}$. Let

$\Theta'_\alpha = \Theta_\alpha$ if $\Theta_\alpha \cup \{\varphi_\alpha\} \notin S$,
$\Theta'_\alpha = \Theta_\alpha \cup \{\varphi_\alpha\}$ otherwise.

$\Theta''_\alpha = \Theta'_\alpha \cup \{\varphi_s\}$ if $\Theta_\alpha \cup \{\varphi_\alpha\} \in S$, $\varphi_\alpha = \varphi_s \vee \varphi_t$, and $\Theta'_\alpha \cup \{\varphi_s\} \in S$,
$\Theta''_\alpha = \Theta'_\alpha \cup \{\varphi_t\}$ if $\Theta_\alpha \cup \{\varphi_\alpha\} \in S$, $\varphi_\alpha = \varphi_s \vee \varphi_t$, and $\Theta'_\alpha \cup \{\varphi_s\} \notin S$,
$\Theta''_\alpha = \Theta'_\alpha$ otherwise.

$\Theta'''_\alpha = \Theta''_\alpha \cup \{\mathrm{Subf}^\alpha_{\mathbf{c}}\psi\}$ if $\Theta_\alpha \cup \{\varphi_\alpha\} \in S$, $\varphi_\alpha = \exists\alpha\psi$, and $\mathbf{c}$ is the $\leq$-least member of C such that $\Theta_\alpha \cup \{\varphi_\alpha, \mathrm{Sub\,f}^\alpha_{\mathbf{c}}\psi\} \in S$,
$\Theta'''_\alpha = \Theta''_\alpha$ otherwise.

$\Theta^{\mathrm{iv}}_\alpha = \Theta'''_\alpha \cup \{\mathbf{c} \equiv \tau_\alpha\}$ where $\mathbf{c}$ is $\leq$-least such that $\Theta'''_\alpha \cup \{\mathbf{c} = \tau_\alpha\} \in S$.
$\Theta_{\alpha+1} = f_\alpha \Theta^{\mathrm{iv}}_\alpha$.

Clearly $\Theta_{\alpha+1} \in S$ again, and (0) holds for $\alpha + 1$. Now suppose that λ is a limit ordinal $\leq \mathfrak{m}$. We set $\Theta_\lambda = \bigcup_{\alpha<\lambda} \Theta_\alpha$. For $\lambda < \mathfrak{m}$ we have $\Theta_\lambda \in S$ by (C10), using (0), and clearly (0) still holds for λ. This completes the definition of $\langle \Theta_\alpha : \alpha \leq \mathfrak{m}\rangle$. Clearly $\Theta_\alpha \in S$ for each $\alpha < \mathfrak{m}$. Furthermore, all of the following hold, for all sentences φ, ψ of $\mathscr{L}'$:

(1) $\varphi \notin \Theta_{\mathfrak{m}}$ or $\neg\varphi \notin \Theta_{\mathfrak{m}}$.
(2) If $\neg\varphi \in \Theta_{\mathfrak{m}}$, then $\varphi^{\rightarrow} \in \Theta_{\mathfrak{m}}$.
(3) If $\varphi \wedge \psi \in \Theta_{\mathfrak{m}}$, then $\varphi, \psi \in \Theta_{\mathfrak{m}}$.
(4) If $\varphi \vee \psi \in \Theta_{\mathfrak{m}}$, then $\varphi \in \Theta_{\mathfrak{m}}$ or $\psi \in \Theta_{\mathfrak{m}}$.
(5) If $\forall\alpha\varphi \in \Theta_{\mathfrak{m}}$, then for all $\mathbf{c} \in C$, $\mathrm{Subf}^\alpha_{\mathbf{c}}\varphi \in \Theta_{\mathfrak{m}}$.
(6) If $\exists\alpha\varphi \in \Theta_{\mathfrak{m}}$, then for some $\mathbf{c} \in C$, $\mathrm{Subf}^\alpha_{\mathbf{c}}\varphi \in \Theta_{\mathfrak{m}}$.
(7) If $\mathbf{c}, \mathbf{d} \in C$ and $\mathbf{c} \equiv \mathbf{d} \in \Theta_{\mathfrak{m}}$, then $\mathbf{d} \equiv \mathbf{c} \in \Theta_{\mathfrak{m}}$.
(8) If $\mathbf{c} \in C$, τ is a primitive term, $\mathbf{c} \equiv \tau \in \Theta_{\mathfrak{m}}$, and $\mathrm{Subf}^\alpha_{\mathbf{c}}\varphi \in \Theta_{\mathfrak{m}}$, then $\mathrm{Subf}^\alpha_{\mathbf{c}}\varphi \in \Theta_{\mathfrak{m}}$.
(9) For any primitive term τ there is a $\mathbf{c} \in C$ such that $\mathbf{c} \equiv \tau \in \Theta_{\mathfrak{m}}$.
(10) For each $\alpha < \mathfrak{m}$ there is a $\Delta \in S$ such that $\Gamma \subseteq \mathrm{f}_\alpha\Delta \subseteq \Theta_{\mathfrak{m}}$.

All of these conditions are easily established; we check (1), (5), (6), and (10) as examples. Suppose φ, $\neg\varphi \in \Theta_{\mathfrak{m}}$. Then there is an $\alpha < \mathfrak{m}$ such that φ, $\neg\varphi \in \Theta_\alpha$, contradicting $\Theta_\alpha \in S$. Thus (1) holds. Assume $\forall\alpha\varphi \in \Theta_{\mathfrak{m}}$ and $\mathbf{c} \in C$. Let

$\mathrm{Subf}_{\mathbf{c}}^{\alpha}\varphi = \varphi_\beta$. Choose $\gamma \geq \beta$ with $\forall\alpha\varphi \in \Theta_\gamma$. Now by (C5), $\Theta_\gamma \cup \{\mathrm{Subf}_{\mathbf{c}}^{\alpha}\varphi\} \in S$, so by (C0), $\Theta_\beta \cup \{\mathrm{Subf}_{\mathbf{c}}^{\alpha}\varphi\} \in S$. Hence $\mathrm{Subf}_{\mathbf{c}}^{\alpha}\varphi \in \Theta_{\beta+1} \subseteq \Theta_{\mathfrak{m}}$, as desired for (5). Next, assume that $\exists\alpha\varphi \in \Theta_{\mathfrak{m}}$. Say $\exists\alpha\varphi = \varphi_\beta$. Choose $\gamma \geq \beta$ so that $\exists\alpha\varphi \in \Theta_\gamma$. Thus $\Theta_\beta \cup \{\varphi_\beta\} \subseteq \Theta_\gamma$, so by (C0), $\Theta_\beta \cup \{\varphi_\beta\} \in S$. Hence by construction there is a $\mathbf{c} \in C$ such that $\mathrm{Subf}_{\mathbf{c}}^{\alpha}\varphi \in \Theta_{\beta+1} \subseteq \Theta_{\mathfrak{m}}$. Hence (6) holds. Finally, suppose $\alpha < \mathfrak{m}$. Then $\Gamma \subseteq \Theta_\alpha \subseteq \Theta_\alpha^{\mathrm{iv}} \subseteq \mathrm{f}_\alpha\Theta_\alpha^{\mathrm{iv}} = \Theta_{\alpha+1} \subseteq \Theta_{\mathfrak{m}}$. Thus (10) holds.

Now let $\equiv$ be the binary relation on C such that for any $\mathbf{c}, \mathbf{d} \in C$, $\mathbf{c} \equiv \mathbf{d}$ iff $\mathbf{c} = \mathbf{d} \in \Theta_{\mathfrak{m}}$. Then

(11) $\equiv$ is an equivalence relation on C.

In fact, $\equiv$ is symmetric by (7). If $\mathbf{c} \equiv \mathbf{d}$ and $\mathbf{d} \equiv \mathbf{e}$, let φ be the formula $v_0 = \mathbf{e}$. Then $\mathbf{c} = \mathbf{d}$, $\mathrm{Subf}_{\mathbf{d}}^{v0}\varphi \in \Theta_{\mathfrak{m}}$, so by (8), $\mathrm{Subf}_{\mathbf{c}}^{v0}\varphi \in \Theta_{\mathfrak{m}}$, i.e., $\mathbf{c} = \mathbf{e} \in \Theta_{\mathfrak{m}}$. Thus $\equiv$ is transitive. By (9), given any $\mathbf{d} \in C$ there is a $\mathbf{c} \in C$ with $\mathbf{c} \equiv \mathbf{d}$. Thus $\equiv$ is an equivalence relation on C.

(12) If φ is a formula of $\mathscr{L}'$ with $\mathrm{Fv}\varphi \subseteq \{v_0, \ldots, v_{m-1}\}$, if $\mathbf{c} \in {}^mC$, $\mathbf{d} \in {}^mC$ with $\mathbf{c}_i \equiv \mathbf{d}_i$ for all $i < m$, and if $\varphi(\mathbf{c}_0, \ldots, \mathbf{c}_{m-1}) \in \Theta_{\mathfrak{m}}$, then $\varphi(\mathbf{d}_0, \ldots, \mathbf{d}_{j-1}, \mathbf{c}_j, \ldots, \mathbf{c}_{m-1}) \in \Theta_{\mathfrak{m}}$ for each $j \leq m$.

We prove (12) by induction on j. The case $j = 0$ is trivial. Assume (12) for j, with $j + 1 \leq m$, and suppose that $\varphi(\mathbf{c}_0, \ldots, \mathbf{c}_{m-1}) \in \Theta_{\mathfrak{m}}$. By the induction hypothesis, $\varphi(\mathbf{d}_0, \ldots, \mathbf{d}_{j-1}, \mathbf{c}_j, \ldots, \mathbf{c}_{m-1}) \in \Theta_{\mathfrak{m}}$. Let ψ be the formula $\varphi(\mathbf{d}_0, \ldots, \mathbf{d}_{j-1}, v_0, \mathbf{c}_{j+1}, \ldots, \mathbf{c}_{m-1})$. Thus $\mathbf{d}_j = \mathbf{c}_j \in \Theta_{\mathfrak{m}}$ (by (7)) and $\mathrm{Subf}_{\mathbf{c}j}^{v0}\psi \in \Theta_{\mathfrak{m}}$. By (8), $\varphi(\mathbf{d}_0, \ldots, \mathbf{d}_j, \mathbf{c}_{j+1}, \ldots, \mathbf{c}_{m-1}) \in \Theta_{\mathfrak{m}}$, as desired. By (12) with $j = m$,

(13) if φ is a formula of $\mathscr{L}'$ with $\mathrm{Fv}\varphi \subseteq \{v_0, \ldots, v_{m-1}\}$, if $\mathbf{c} \in {}^mC$, $\mathbf{d} \in {}^mC$ with $\mathbf{c}_i \equiv \mathbf{d}_i$ for all $i < m$, and if $\varphi(\mathbf{c}_0, \ldots, \mathbf{c}_{m-1}) \in \Theta_{\mathfrak{m}}$, then $\varphi(\mathbf{d}_0, \ldots, \mathbf{d}_{m-1}) \in \Theta_{\mathfrak{m}}$.

Now let $A = C/{\equiv}$. We define an $\mathscr{L}$-structure $\mathfrak{A}$ with universe A. In what follows, $[\mathbf{c}]$ denotes the equivalence class of $\mathbf{c}$ under $\equiv$.

(14) If $\mathbf{O}$ is an m-ary operation symbol of $\mathscr{L}$, then $\mathbf{O}^{\mathfrak{A}} = \{(a, b) : a \in {}^mA, b \in A$, and there exist $\mathbf{c} \in {}^mC$ and $\mathbf{d} \in b$ with $\mathbf{c}_i \in a_i$ for all $i < m$ and $\mathbf{d} = \mathbf{O}\mathbf{c}_0 \cdots \mathbf{c}_{m-1} \in \Theta_{\mathfrak{m}}\}$;

(15) if $\mathbf{R}$ is an m-ary relation symbol of $\mathscr{L}$, then $\mathbf{R}^{\mathfrak{A}} = \{a \in {}^mA$: there exists $\mathbf{c} \in {}^mC$ with $\mathbf{c}_i \in a_i$ for all $i < m$ such that $\mathbf{R}\mathbf{c}_0 \cdots \mathbf{c}_{m-1} \in \Theta_{\mathfrak{m}}\}$.

Then

(16) if $\mathbf{O}$ is an m-ary operation symbol of $\mathscr{L}$, then $\mathbf{O}^{\mathfrak{A}}$ is an m-ary operation on A.

For, let $a \in {}^mA$; say $\mathbf{c} \in {}^mC$ with $\mathbf{c}_i \in a_i$ for all $i < m$. Then by (9) there is a $\mathbf{d} \in C$ with $\mathbf{d} = \mathbf{O}\mathbf{c}_0 \cdots \mathbf{c}_{m-1} \in \Theta_{\mathfrak{m}}$. Thus $(a, [\mathbf{d}]) \in \mathbf{O}^{\mathfrak{A}}$. Next, suppose $(s, t) \in \mathbf{O}^{\mathfrak{A}}$ and $(s, u) \in \mathbf{O}^{\mathfrak{A}}$. Say $\mathbf{w} \in {}^mC$, $\mathbf{x} \in t$, $\mathbf{w}_i \in s_i$ for all $i < m$, and $\mathbf{x} = \mathbf{O}\mathbf{w}_0 \cdots \mathbf{w}_{m-1} \in \Theta_{\mathfrak{m}}$; also say $\mathbf{y} \in {}^mC$, $\mathbf{z} \in u$, $\mathbf{y}_i \in s_i$ for all $i < m$, and $\mathbf{z} = \mathbf{O}\mathbf{y}_0 \cdots \mathbf{y}_{m-1} \in \Theta_{\mathfrak{m}}$. Thus $\mathbf{w}_i \equiv \mathbf{y}_i$ for each $i < m$, so by (13), $\mathbf{x} = \mathbf{O}\mathbf{y}_0 \cdots \mathbf{y}_{m-1} \in \Theta_{\mathfrak{m}}$. Let

ψ be $\mathbf{x} = v_0$. Then $\mathbf{z} = \mathbf{O}\mathbf{y}_0 \cdots \mathbf{y}_{m-1} \in \Theta_m$ and $\mathrm{Subf}^{v0}_{\mathbf{O}\mathbf{y}0\cdots\mathbf{y}(m-1)}\psi \in \Theta_m$ .Thus by (8) we infer that $\mathbf{x} = \mathbf{z} \in \Theta_m$. Thus $\mathbf{x} \equiv \mathbf{z}$, so $t = u$, as desired. Thus (16) holds. Next,

$$(17) \qquad \mathbf{O}^{\mathfrak{A}}([\mathbf{c}_0], \ldots, [\mathbf{c}_{m-1}]) = [\mathbf{d}] \text{ iff } (\mathbf{d} = \mathbf{O}\mathbf{c}_0 \cdots \mathbf{c}_{m-1}) \in \Theta_m.$$

In fact, $\Leftarrow$ is obvious by the definition of $\mathbf{O}^{\mathfrak{A}}$. For $\Rightarrow$, there exist $\mathbf{e} \in {}^mC$, $\mathbf{s} \in [\mathbf{d}]$ with $\mathbf{e}_i \equiv \mathbf{c}_i$ for all $i < m$ and $\mathbf{s} = \mathbf{O}\mathbf{e}_0 \cdots \mathbf{e}_{m-1} \in \Theta_m$. Thus $\mathbf{d} \equiv \mathbf{s}$, and by (13) $\mathbf{s} = \mathbf{O}\mathbf{c}_0 \cdots \mathbf{c}_{m-1} \in \Theta_m$. Let ψ be $v_0 = \mathbf{O}\mathbf{c}_0 \cdots \mathbf{c}_{m-1}$. Then $\mathbf{d} = \mathbf{s}$, $\mathrm{Subf}^{v0}_{\mathbf{s}}\psi \in \Theta_m$, so using (8) we infer that $\mathbf{d} = \mathbf{O}\mathbf{c}_0 \cdots \mathbf{c}_{m-1} \in \Theta_m$, as desired.

Using (13),

$$(18) \qquad \langle[\mathbf{c}_0], \ldots, [\mathbf{c}_{m-1}]\rangle \in \mathbf{R}^{\mathfrak{A}} \text{ iff } \mathbf{R}\mathbf{c}_0 \cdots \mathbf{c}_{m-1} \in \Theta_m.$$

Now let $\mathfrak{A}' = (\mathfrak{A}.\ [\mathbf{c}])_{\mathbf{c}\in C}$.

(19) If φ is an atomic sentence and $\varphi \in \Theta_m$, then φ holds in $\mathfrak{A}'$.

We prove (19) by induction on the number p of operation symbols $\notin C$ which occur in φ. If $p = 0$, then (19) follows from (18) and the definition of $\equiv$. Now assume (19) true for p. Let φ be an atomic sentence with $p + 1$ operation symbols $\notin C$ occurring in φ. Then some primitive term $\mathbf{O}\mathbf{c}_0 \cdots \mathbf{c}_{m-1}$ occurs in φ where it is not the case that $m = 0$ and $\mathbf{O} \in C$; let ψ be obtained from φ by replacing an occurrence of $\mathbf{O}\mathbf{c}_0 \cdots \mathbf{c}_{m-1}$ by v_0. Say $\mathbf{O}^{\mathfrak{A}}([\mathbf{c}_0], \ldots. [\mathbf{c}_{m-1}]) = [\mathbf{d}]$. Then by (17), $\mathbf{d} = \mathbf{O}\mathbf{c}_0 \cdots \mathbf{c}_{m-1} \in \Theta_m$. Also, $\mathrm{Subf}^{v0}_{\mathbf{O}(\mathbf{c}0\cdots\mathbf{c}(m-1))}\psi \in \Theta_m$ since $\mathrm{Subf}^{v0}_{\mathbf{O}(\mathbf{c}0\cdots\mathbf{c}(m-1))}\psi = \varphi$. Hence using (8), $\mathrm{Subf}^{v0}_{\mathbf{d}}\psi \in \Theta_m$. The induction hypothesis applies to $\mathrm{Subf}^{v0}_{\mathbf{d}}\psi$, which hence holds in $\mathfrak{A}'$. Since $\mathbf{O}^{\mathfrak{A}}([\mathbf{c}_0], \ldots [\mathbf{c}_{m-1}]) = [\mathbf{d}]$, it follows that φ holds in $\mathfrak{A}'$. This completes the inductive proof of (19).

(20) If φ is an atomic sentence and $\neg\varphi \in \Theta_m$, then $\neg\varphi$ holds in $\mathfrak{A}'$.

This statement is proved almost verbatim like (19). Now we come to our final statement:

(21) for any sentence φ of $\mathscr{L}'$, if $\varphi \in \Theta_m$ then φ holds in $\mathfrak{A}'$, and if $\neg\varphi \in \Theta_m$ then $\neg\varphi$ holds in $\mathfrak{A}'$.

We prove (21) by induction on the number of logical connectives—$\neg$, $\vee$, $\wedge$, $\forall$—occurring in φ. If φ is atomic, (21) is given by (19) and (20). Now suppose φ satisfies (21). To show that $\neg\varphi$ satisfies (21) it obviously suffices to show that if $\neg\neg\varphi \in \Theta_m$ then φ holds in $\mathfrak{A}'$. By (2), $(\neg\varphi)^{\rightarrow} \in \Theta_m$, i.e., $\varphi \in \Theta_m$, so φ holds in $\mathfrak{A}'$ by the induction assumption. Now suppose φ and ψ satisfy (21). First suppose $\varphi \wedge \psi \in \Theta_m$. By (3), $\varphi, \psi \in \Theta_m$ and hence by the induction assumption φ and ψ hold in $\mathfrak{A}'$, hence $\varphi \wedge \psi$ holds in $\mathfrak{A}'$. Second suppose $\neg(\varphi \wedge \psi) \in \Theta_m$. Then by (2), $(\varphi \wedge \psi)^{\rightarrow} \in \Theta_m$, i.e., $\neg\varphi \vee \neg\psi \in \Theta_m$. Hence by (3), $\neg\varphi \in \Theta_m$ or $\neg\psi \in \Theta_m$, so by the induction assumption $\neg\varphi$ holds in $\mathfrak{A}'$ or $\neg\psi$ holds in $\mathfrak{A}'$, so $\neg(\varphi \wedge \psi)$ holds in $\mathfrak{A}'$. Similarly, $\varphi \vee \psi$ satisfies (21) if φ and ψ do. Finally, suppose $\forall\alpha\varphi \in \Theta_m$. Then by (5), $\mathrm{Subf}^{\alpha}_{\mathbf{c}}\varphi \in \Theta_m$ for all $\mathbf{c} \in C$, so by the induction assumption $\mathrm{Subf}^{\alpha}_{\mathbf{c}}\varphi$ holds in $\mathfrak{A}'$ for all

$\mathbf{c} \in C$, hence $\forall\alpha\varphi$ holds in $\mathfrak{A}'$. On the other hand, suppose $\neg\forall\alpha\varphi \in \Theta_{\mathfrak{m}}$. By (2), $(\forall\alpha\varphi)^{\rightarrow} \in \Theta_{\mathfrak{m}}$, i.e., $\exists\alpha \neg \varphi \in \Theta_{\mathfrak{m}}$. By (6) choose $\mathbf{c} \in C$ so that $\mathrm{Subf}^{\alpha}_{\mathbf{c}} \neg \varphi \in \Theta_{\mathfrak{m}}$. By the induction assumption, $\mathrm{Subf}^{\alpha}_{\mathbf{c}} \neg \varphi$ holds in $\mathfrak{A}'$, so $\neg\forall\alpha\varphi$ holds in $\mathfrak{A}'$. This completes the proof of (21).

Since $\Gamma \subseteq \Theta_{\mathfrak{m}}$, it follows from (21) that $\mathfrak{A}'$ is a model of Γ. Clearly $\mathfrak{A}'$ has the form $(\mathfrak{A}, [\mathbf{c}])_{\mathbf{c}\in C}$ with $A = \{[\mathbf{c}] : \mathbf{c} \in C\}$. Finally, if $\alpha < \mathfrak{m}$ then by (10) and (21) there is a $\Delta \in S$ such that $\Gamma \subseteq \mathrm{f}_{\alpha}\Delta \subseteq \Theta_{\mathfrak{m}} \subseteq \{\varphi : \mathfrak{A}' \models \varphi\}$. □

Taking all of the admissible functions in 18.9 to be the identity we obtain

Corollary 18.10. *Let $\mathscr{L}'$ be a rich expansion of $\mathscr{L}$ by C, and let S be a consistency family. Then each member of S has a model of power* $\leq |\mathrm{Fmla}_{\mathscr{L}}|$.

Note that if $|\mathrm{Fmla}_{\mathscr{L}}| = \aleph_0$, then condition (C11) is not needed in the proof of 18.9 (in its proof, (0) can then be omitted). Hence we obtain

Corollary 18.11. *Let $\mathscr{L}'$ be a rich expansion of $\mathscr{L}$ by C, assume that* $|\mathrm{Fmla}_{\mathscr{L}}| = \aleph_0$, *and let S satisfy* (C0)–(C9). *Let $\langle f_{\alpha} : \alpha < \omega\rangle$ be a family of admissible functions over S. Then each $\Gamma \in S$ has a model $(\mathfrak{A}, a_{\mathbf{c}})_{\mathbf{c}\in C}$ with $A = \{a_{\mathbf{c}} : \mathbf{c} \in C\}$ and hence $|A| \leq \aleph_0$, such that for each $\alpha < \omega$ there is some $\Delta \in S$ with $\Gamma \subseteq f_{\alpha}\Delta \subseteq \{\varphi : (\mathfrak{A}, a_{\mathbf{c}})_{\mathbf{c}\in C} \models \varphi\}$.*

Finally, using 18.7 we obtain the simplest form of the model existence theorem:

Corollary 18.12. *Let $\mathscr{L}'$ be a rich expansion of $\mathscr{L}$ by C, assume that* $|\mathrm{Fmla}_{\mathscr{L}}| = \aleph_0$, *and let S satisfy* (C1)–(C9). *Then each $\Gamma \in S$ has a model $(\mathfrak{A}, a_{\mathbf{c}})_{\mathbf{c}\in C}$ with $A = \{a_{\mathbf{c}} : \mathbf{c} \in C\}$ and hence $|A| \leq \aleph_0$.*

Applying 18.9 to the consistency family given in 18.5, we obtain the completeness theorem 11.19. Theorem 18.9 will be used for many of our later results, as we have mentioned.

The construction method given in the model existence theorem is of a somewhat syntactical nature, and yields a model internally, so to speak, with no models given to start with. The method of constructing models to which we now turn, the *ultraproduct method*, does not involve syntactical notions at all, but proceeds from "partial" models directly to a "full" model. The ultraproduct of structures is an operation of a general algebraic nature, and it also plays a significant role in the general theory of algebras. In this chapter we shall only give the basic definitions and results concerning ultraproducts. More detailed connections with logic will be indicated as we proceed in this part. As the name suggests, ultraproducts are formed from *products* of structures by use of *ultrafilters* over certain sets. We first give the definition of products. A more detailed investigation of products and their relationships to logic will come later.

Definition 18.13. Let $\langle \mathfrak{A}_i : i \in I\rangle$ be a system of $\mathscr{L}$-structures, where $\mathscr{L}$ is any first-order language. By the *product* $\mathsf{P}_{i\in I} \mathfrak{A}_i$ of the system $\langle \mathfrak{A}_i : i \in I\rangle$ we mean the $\mathscr{L}$-structure $\mathscr{L}$ with universe $B = \mathsf{P}_{i\in I} A_i = \{f : f$ is a function, $\mathrm{Dmn}\, f = I$, and for each $i \in I$, $f_i \in A_i\}$, and with relations and operations as follows:

(*i*) if $\mathbf{O}$ is an m-ary operation symbol and $x_0, \ldots, x_{m-1} \in B$, then for any $i \in I$,

$$[\mathbf{O}^{\mathfrak{B}}(x_0, \ldots, x_{m-1})]_i = \mathbf{O}^{\mathfrak{A}_i}(x_{0i}, \ldots, x_{m-1,i});$$

(*ii*) if $\mathbf{R}$ is an m-ary relation symbol, then

$$\mathbf{R}^{\mathfrak{B}} = \{(x_0, \ldots, x_{m-1}) \in {}^mB : \forall i \in I\, (x_{0i}, \ldots, x_{m-1,i}) \in \mathbf{R}^{\mathfrak{A}_i}\}.$$

If $\mathfrak{A}_i = \mathfrak{C}$ for all $i \in I$ we write ${}^I\mathfrak{C}$ instead of $\mathsf{P}_{i\in I} \mathfrak{A}_i$; ${}^I\mathfrak{C}$ is the Ith *direct power* of $\mathfrak{C}$. In case I has exactly two elements i, j, we write $\mathfrak{A}_i \times \mathfrak{A}_j$ instead of $\mathsf{P}_{i\in I} \mathfrak{A}_i$. Thus the product of two structures $\mathfrak{C}$ and $\mathfrak{D}$ has been defined and is denoted by $\mathfrak{C} \times \mathfrak{D}$; we think of it as $\mathsf{P}_{i\in 2} \mathfrak{A}_i$, where $\mathfrak{A}_0 = \mathfrak{C}$ and $\mathfrak{A}_1 = \mathfrak{D}$. One final bit of notation for the general product $\mathsf{P}_{i\in I} \mathfrak{A}_i$: for each $i \in I$ we denote by $\mathrm{pr}_i^{\mathfrak{A}}$, or simply pr_i when $\mathfrak{A}$ is understood from the context, the *projection function* whose domain is $\mathsf{P}_{i\in I} A_i$ such that for any $x \in \mathsf{P}_{i\in I} A_i$,

$$\mathrm{pr}_i\, x = x_i.$$

The following useful proposition is easily established by induction on σ:

Proposition 18.14. *Let* $\mathfrak{B} = \mathsf{P}_{i\in I} \mathfrak{A}_i$. *For any term* σ, *any* $x \in {}^\omega\mathsf{P}_{i\in I} A_i$, *and any* $i \in I$, $(\sigma^{\mathfrak{A}}x)_i = \sigma^{\mathfrak{A}_i}(\mathrm{pr}_i \circ x)$.

Now we give the basic definitions and a few basic facts concerning ultrafilters. Filters and ultrafilters are used in various parts of mathematics, especially in general topology. Their usefulness in logic is mainly connected with the ultraproduct construction and with certain large cardinals that arise in the study of infinitary languages and the metamathematics of set theory.

Definition 18.15. Let I be any set and F a collection of subsets of I.

(*i*) F has the *finite intersection property* if $a_0 \cap \cdots \cap a_{m-1} \neq 0$ whenever $m \in \omega$ and $a \in {}^mF$.

(*ii*) F is a *filter over* I provided $F \neq 0$ and:

(*a*) $b \supseteq a \in F$ implies $b \in F$;

(*b*) $a, b \in F$ implies $a \cap b \in F$.

(*iii*) F is an *ultrafilter over* I provided F is a filter over I, $0 \notin F$, and for any $J \subseteq I$, either $J \in I$ or $I \sim J \in F$.

The following proposition is obvious.

Proposition 18.16. *If F is a filter and $0 \notin F$, then F has the finite intersection property. For any filter F over I we have $I \in F$. For any filter F on I, $0 \in F$ iff F is the set of all subsets of I.*

Proposition 18.17. *Let F be a filter over I with $0 \notin F$. Then the following conditions are equivalent:*

(i) F is an ultrafilter.
(ii) for all $a, b \subseteq I$, if $a \cup b \in F$, then $a \in F$ or $b \in F$.
(iii) for any filter G on I, if $F \subset G$, then $0 \in G$.

PROOF. (*i*) $\Rightarrow$ (*ii*). Assume that $a \cup b \in F$ while $a \notin F$. Then by (*i*), $I \sim a \in F$. Now $(I \sim a) \cap (a \cup b) \subseteq b$, so $b \in F$. (*ii*) $\Rightarrow$ (*iii*). Say $a \in G \sim F$. Then $a \cup (I \sim a) = I \in F$, so by (*ii*) $I \sim a \in F \subseteq G$. Hence $0 = a \cap (I \sim a) \in G$. (*iii*) $\Rightarrow$ (*i*). Assume that $a \subseteq I$ and $a \notin F$. Let $G = \{b \subseteq I$: there is an $x \in F$ with $x \cap a \subseteq b\}$. Clearly $F \subseteq G$, $a \in G$, and G is a filter. In particular $F \subset G$, so $0 \in G$ by (*iii*). Hence $x \cap a = 0$ for some $x \in F$; hence $x \subseteq I \sim a$, so $I \sim a \in F$. □

The following is the basic existence principle for ultrafilters.

Proposition 18.18. *If F is a collection of subsets of $I \neq 0$ with the finite intersection property, then there is an ultrafilter G such that $F \subseteq G$.*

PROOF. Let $\mathscr{A}$ be the collection of all filters G such that $F \subseteq G$ and $0 \notin G$. Then $\mathscr{A}$ is nonempty; for, let $H = \{x \subseteq I$: there exist $m \in \omega$ and $y \in {}^mF$ with $y_0 \cap \cdots \cap y_{m-1} \subseteq x\}$. Clearly $H \in \mathscr{A}$. If $0 \neq \mathscr{B} \subseteq \mathscr{A}$ is simply ordered by inclusion, clearly $\bigcup \mathscr{B} \in \mathscr{A}$. By Zorn's lemma, let G be a maximal element of $\mathscr{A}$. By 18.17, G is an ultrafilter. □

Definition 18.19. Let $\mathfrak{A} = \langle \mathfrak{A}_i : i \in I \rangle$ be a system of $\mathscr{L}$-structures, and let F be an ultrafilter on I. We define

$$\bar{F}^{\mathfrak{A}} = \{(x, y) \in {}^2\mathsf{P}_{i \in I} A_i : \{i : x_i = y_i\} \in F\}.$$

We write $\bar{F}$ if $\mathfrak{A}$ is understood. Obviously $\bar{F}$ depends only on the system $\langle A_i : i \in I \rangle$ and not at all on the language $\mathscr{L}$.

Proposition 18.20. *Under the assumptions of* 18.19, *$\bar{F}$ is an equivalence relation on $\mathsf{P}_{i \in I} A_i$. Let $\mathfrak{B} = \mathsf{P}_{i \in I} \mathfrak{A}_i$. Then*

(i) if $\mathbf{O}$ is an m-ary operation symbol, and if $x_t \bar{F} y_t$ for all $t < m$, then $\mathbf{O}^{\mathfrak{B}} x \bar{F} \mathbf{O} y^{\mathfrak{B}}$;
(ii) if $\mathbf{R}$ is an m-ary relation symbol, and if $x_t \bar{F} y_t$ for all $t < m$, then $\{i \in I : \langle x_{0i}, \ldots, x_{m-1,i} \rangle \in \mathbf{R}^{\mathfrak{A}i}\} \in F$ iff $\{i \in I : \langle y_{0i}, \ldots, y_{m-1,i} \rangle \in \mathbf{R}^{\mathfrak{A}i}\} \in F$.

PROOF. The proof is routine, and we just check (i) and the transitivity of $\bar{F}$ as examples. Assume that $x\bar{F}y\bar{F}z$. Thus $\{i \in I : x_i = y_i\} \in F$ and $\{i \in I : y_i = z_i\} \in F$. But

$$\{i \in I : x_i = y_i\} \cap \{i \in I : y_i = z_i\} \subseteq \{i \in I : x_i = z_i\},$$

so also $\{i \in I : x_i = z_i\} \in F$, and so $x\bar{F}z$. To check (i), note that

$$\{i \in I : \forall t < m(x_{ti} = y_{ti})\} = \bigcap_{t<m} \{i \in I : x_{ti} = y_{ti}\} \in F,$$

and hence

$$\{i \in I : \forall t < m(x_{ti} = y_{ti})\} \subseteq \{i \in I : (\mathbf{O}^{\mathfrak{B}}x)_i = (\mathbf{O}^{\mathfrak{B}}y)_i\} \in F,$$

so $\mathbf{O}^{\mathfrak{B}}x\bar{F}\mathbf{O}^{\mathfrak{B}}y$. □

We may think of the members of an ultrafilter F as "big" subsets of I. Thus the passage from a member $x \in \mathrm{P}_{i \in I} A_i$ to its equivalence class under $\bar{F}$ amounts to *identifying* all functions which are equal to x on a "big" subset of I. As usual, $[x]_{\bar{F}}$, or simply $[x]$ if $\bar{F}$ is understood, denotes the equivalence class of x under $\bar{F}$. Proposition 18.20 justifies the definition of ultraproducts:

Definition 18.21. Let $\mathfrak{A} = \langle \mathfrak{A}_i : i \in I \rangle$ be a system of $\mathscr{L}$-structures, set $\mathfrak{B} = \mathrm{P}_{i \in I} \mathfrak{A}_i$, and let F be an ultrafilter on I. The *ultraproduct of* $\mathfrak{A}$ *over* $\bar{F}$, denoted by $\mathfrak{A}/\bar{F}$ or $\mathrm{P}_{i \in I} \mathfrak{A}_i/\bar{F}$, is the structure $\mathfrak{C}$ with universe $C = \mathrm{P}_{i \in I} A_i/\bar{F}$ (the collection of all equivalence classes under $\bar{F}$) and with operations and relations given as follows:

(i) if $\mathbf{O}$ is an m-ary operation symbol and $x \in {}^mB$, then $\mathbf{O}^{\mathfrak{C}}([x_0], \ldots, [x_{m-1}]) = [\mathbf{O}^{\mathfrak{B}}(x_0, \ldots, x_{m-1})]$;

(ii) if $\mathbf{R}$ is an m-ary relation symbol, then we let $\mathbf{R}^{\mathfrak{C}}$ consist of all m-tuples of the form $\langle [x_0], \ldots, [x_{m-1}] \rangle$ such that $\{i : \langle x_{0i}, \ldots, x_{m-1,i} \rangle \in \mathbf{R}^{\mathfrak{A}i}\} \in F$.

Of particular importance later will be the case when all factors $\mathfrak{A}_i$ of an ultraproduct are equal to some structure $\mathfrak{C}$. Then the ultraproduct $\mathrm{P}_{i \in I}\mathfrak{A}_i/\bar{F}$ is denoted of course by ${}^I\mathfrak{C}/\bar{F}$, and is called an *ultrapower* of $\mathfrak{C}$. Sometimes we omit the bar on F.

We shall give only a few basic properties of ultraproducts here. Our first very simple property uses the notion of an isomorphism, which we shall now briefly discuss.

Definition 18.22. Let $\mathfrak{A}$ and $\mathfrak{B}$ be $\mathscr{L}$-structures. An *isomorphism from* $\mathfrak{A}$ *into* $\mathfrak{B}$, or an *embedding of* $\mathfrak{A}$ *into* $\mathfrak{B}$ is a one-one function f mapping A into B such that

$$f\mathbf{O}^{\mathfrak{A}}(a_0, \ldots, a_{m-1}) = \mathbf{O}^{\mathfrak{B}}(fa_0, \ldots, fa_{m-1})$$

whenever $\mathbf{O}$ is an m-ary operation symbol and $a_0, \ldots, a_{m-1} \in A$, while

$$\langle a_0, \ldots, a_{m-1} \rangle \in \mathbf{R}^{\mathfrak{A}} \text{ iff } \langle fa_0, \ldots, fa_{m-1} \rangle \in \mathbf{R}^{\mathfrak{B}}$$

whenever $\mathbf{R}$ is an m-ary relation symbol and $a_0, \ldots, a_{m-1} \in A$. We say that f is an isomorphism from $\mathfrak{A}$ onto $\mathfrak{B}$ if the function f maps onto B. Finally, we write $\mathfrak{A} \cong \mathfrak{B}$ if there *is* an isomorphism from $\mathfrak{B}$ onto $\mathfrak{B}$.

The following proposition, easily established by induction on σ and φ, is the basic fact about isomorphism as far as first-order languages are concerned. Actually a similar theorem holds for any of the languages which have been considered by logicians. The result says roughly that isomorphic structures are indistinguishable in first-order logic. For this reason, most of our definitions and results extend automatically to isomorphic structures.

Proposition 18.23. *If f is an isomorphism from $\mathfrak{A}$ onto $\mathfrak{B}$, $x \in {}^{\omega}A$, σ is a term, and φ is a formula, then:*

$$f\sigma^{\mathfrak{A}}x = \sigma^{\mathfrak{B}}(f \circ x);$$
$$\mathfrak{A} \models \varphi[x] \quad \textit{iff}\ \mathfrak{B} \models \varphi[f \circ x].$$

Thus for any sentence φ, φ holds in $\mathfrak{A}$ iff φ holds in $\mathfrak{B}$.

Definition 18.24. An ultrafilter over I is *principal* (or *fixed*) if there is an $i \in I$ such that $F = \{a \subseteq I : i \in a\}$.

It is trivial to check that for any set I and any $i \in I$, $\{a \subseteq I : i \in a\}$ *is* an ultrafilter. Nonprincipal ultrafilters are sometimes called *free*. As we shall shortly see, ultraproducts with principal ultrafilters are not interesting.

Proposition 18.25. *If I is a finite set, then every ultrafilter over I is principal.*

Proof. Let F be an ultrafilter over I. Then $\bigcup_{i \in I}\{i\} = I \in F$, so by 18.13(*ii*), since I is finite, there is an $i \in I$ such that $\{i\} \in F$. Then, in fact, $F = \{a \subseteq I : i \in a\}$. For, if $i \in a \subseteq I$, then $a \supseteq \{i\} \in F$, so $a \in F$. Conversely, if $a \in F$, then $a \cap \{i\} \in F$ and so, since $0 \notin F$, we have $i \in a$. □

Proposition 18.26. *If I is an infinite set, then there is a non-principal ultrafilter over I.*

Proof. Let $F = \{I \sim \{i\} : i \in I\}$. Since I is infinite, it is clear that F has the finite intersection property. Let G be an ultrafilter containing F, by 18.18. Clearly G is nonprincipal. □

Proposition 18.27. *If $\langle \mathfrak{A}_i : i \in I \rangle$ is a system of $\mathscr{L}$-structures and F is a principal ultrafilter over I, say $F = \{a \subseteq I : j \in a\}$, then $\mathfrak{A}_j \cong \mathsf{P}_{i \in I}\, \mathfrak{A}_i/\bar{F}$.*

Proof. Fix $x \in \mathsf{P}_{i \in I} A_i$. We define $fa = [x_a^j]$ for each $a \in A_j$. Thus f maps A_j into $\mathsf{P}_{i \in I} A_i$. Now f is one-one; for if $a, b \in A_j$ and $a \neq b$, then $(x_a^j) = a \neq$

$b = (x_b^j)_j$ and hence $j \notin \{i : (x_a^j)_i = (x_b^j)_i\}$, so not $(x_a^j \bar{F} x_b^j)$ and $fa = [x_a^j] \neq [x_b^j] = fb$. Also, f maps onto $\mathsf{P}_{i \in I} A_i / \bar{F}$; for, if $y \in \mathsf{P}_{i \in I} A_i$, then $(x_{yj}^j)_j = y_j$ and hence $\{j\} \subseteq \{i : (x_{yj}^j)_i = y_i\} \in F$ and so $fy_j = [x_{yj}^j] = [y]$. Next, f preserves operations, for if $\mathbf{O}$ is an m-ary operation symbol, $a_0, \ldots, a_{m-1} \in A_j$, and $b = \mathbf{O}^{\mathfrak{A}j}(a_0, \ldots, a_{m-1})$, then with $\mathfrak{B} = \mathsf{P}_{i \in I} \mathfrak{A}_i$,

$$(\mathbf{O}^{\mathfrak{B}}(x_{a0}^j, \ldots, x_{a(m-1)}^j))_j = b = (x_b^j)_j,$$

and hence $[\mathbf{O}^{\mathfrak{B}}(x_{a0}^j, \ldots, x_{a(m-1)}^j)] = [x_b^j]$. Thus with $\mathfrak{C} = \mathsf{P}_{i \in I} \mathfrak{A}_i / \bar{F}$,

$$\begin{aligned} f\mathbf{O}^{\mathfrak{A}_j}(a_0, \ldots, a_{m-1}) &= fb = [x_b^j] = [\mathbf{O}^{\mathfrak{B}}(x_{a0}^j, \ldots, x_{a(m-1)}^j)] \\ &= \mathbf{O}^{\mathfrak{C}}([x_{a0}^j], \ldots, [x_{a(m-1)}^j]) \\ &= \mathbf{O}^{\mathfrak{C}}(fa_0, \ldots, fa_{m-1}). \end{aligned}$$

Finally, f preserves relations. For, if $\mathbf{R}$ is an m-ary relation symbol, $a_0, \ldots, a_{m-1} \in A_j$, and $\mathfrak{C}$ is as above, then

$$\begin{aligned} \langle a_0, \ldots, a_{m-1} \rangle \in \mathbf{R}^{\mathfrak{A}_j} \quad & \text{iff } \{i : \langle x_{a0}^j i, \ldots, x_{a(m-1)}^j i \rangle \in \mathbf{R}^{\mathfrak{A}_i}\} \in F \\ & \text{iff } \langle [x_{a0}^j], \ldots, [x_{a(m-1)}^j] \rangle \in \mathbf{R}^{\mathfrak{C}} \qquad \text{(by 18.20(}ii\text{))} \\ & \text{iff } \langle fa_0, \ldots, fa_{m-1} \rangle \in \mathbf{R}^{\mathfrak{C}}. \end{aligned}$$

□

By 18.27, ultraproducts are interesting only with respect to nonprincipal ultrafilters, and hence by 18.25, only over infinite sets. In fact, an ultraproduct can be considered as a kind of *limit* of its factors (see 18.29 below). The following useful proposition follows immediately from the definitions of the notions involved.

Proposition 18.28. *Let $\mathscr{L}'$ be an expansion of $\mathscr{L}$, and let $\langle \mathfrak{A}_i : i \in I \rangle$ be a system of $\mathscr{L}'$-structures. Then $(\mathsf{P}_{i \in I} \mathfrak{A}_I / \bar{F}) \upharpoonright \mathscr{L} = \mathsf{P}_{i \in I} (\mathfrak{A}_i \upharpoonright \mathscr{L}) / \bar{F}$.*

Now we shall prove a result showing the logical nature of ultraproducts. In almost all of our applications of ultraproducts this is the theorem which is actually applied.

Theorem 18.29. (Basic theorem on ultraproducts). *Let $\langle \mathfrak{A}_i : i \in I \rangle$ be a system of $\mathscr{L}$-structures, and F an ultrafilter over I. Let $[\ ]$ be the function assigning to each $x \in \mathsf{P}_{i \in I} A_i$ the equivalence class $[x]$ of x relative to $\bar{F}$. Then for any formula φ of $\mathscr{L}$ and any $x \in {}^{\omega}\mathsf{P}_{i \in I} A_i$ the following conditions are equivalent:*

(*i*) $\mathsf{P}_{i \in I} \mathfrak{A}_i / \bar{F} \models \varphi[[\ ] \circ x]$;
(*ii*) $\{i \in I : \mathfrak{A}_i \models \varphi[\mathrm{pr}_i \circ x]\} \in F$.

Proof. Set $\mathfrak{B} = \mathsf{P}_{i \in I} \mathfrak{A}_i$ and $\mathfrak{C} = \mathsf{P}_{i \in I} \mathfrak{A}_i / \bar{F}$. By induction on σ one easily shows

(1) for any term σ and any $x \in {}^{\omega}\mathsf{P}_{i \in I} A_i$, $\sigma^{\mathfrak{C}}([\ ] \circ x) = [\sigma^{\mathfrak{B}} x]$.

We now prove the theorem itself by induction on φ. First suppose that φ is $\sigma \equiv \tau$. Then

$$\begin{aligned}\mathfrak{C} \models \varphi[[\ ] \circ x] \quad & \text{iff } \sigma^{\mathfrak{C}}([\ ] \circ x) = \tau^{\mathfrak{C}}([\ ] \circ x) \\ & \text{iff } [\sigma^{\mathfrak{B}}x] = [\tau^{\mathfrak{B}}x] \qquad \text{by (1)} \\ & \text{iff } \{i : (\sigma^{\mathfrak{B}}x)_i = (\tau^{\mathfrak{B}}x)_i\} \in F \\ & \text{iff } \{i : \sigma^{\mathfrak{A}_i}(\mathrm{pr}_i \circ x) = \tau^{\mathfrak{A}_i}(\mathrm{pr}_i \circ x)\} \in F \qquad \text{by 18.14} \\ & \text{iff } \{i : \mathfrak{A}_i \models \varphi[\mathrm{pr}_i \circ x]\} \in F.\end{aligned}$$

The other atomic case is entirely analogous. We now check the result for $\varphi = \psi \vee \chi$; the proof is similar for $\neg$ and $\wedge$.

$$\begin{aligned}\mathfrak{C} \models \varphi[[\ ] \circ x] \quad & \text{iff } \mathfrak{C} \models \psi[[\ ] \circ x] \text{ or } \mathfrak{C} \models \chi[[\ ] \circ x] \\ & \text{iff } \{i \in I : \mathfrak{A}_i \models \psi[\mathrm{pr}_i \circ x]\} \in F \\ & \quad \text{or } \{i \in I : \mathfrak{A}_i \models \chi[\mathrm{pr}_i \circ x]\} \in F \qquad \text{by induction hypothesis} \\ & \text{iff } \{i \in I : \mathfrak{A}_i \models \psi[\mathrm{pr}_i \circ x]\} \cup \{i \in I : \mathfrak{A}_i \models \chi[\mathrm{pr}_i \circ x]\} \in F \\ & \qquad \text{by 18.17} \\ & \text{iff } \{i \in I : \mathfrak{A}_i \models \varphi[\mathrm{pr}_i \circ x]\} \in F.\end{aligned}$$

Finally, suppose φ is $\forall v_j\psi$. Let $M = \{i \in I : \mathfrak{A}_i \models \varphi[\mathrm{pr}_i \circ x]\}$. First suppose that $M \in F$; then we want to show that $\mathfrak{C} \models \varphi[[\ ] \circ x]$. So, let $a \in \mathrm{P}_{i \in I} A_i$; we want to show that $\mathfrak{C} \models \psi[([\ ] \circ x)^j_{[a]}]$. Now for each $i \in M$, $\mathfrak{A}_i \models \psi[(\mathrm{pr}_i \circ x)^j_{ai}]$. Note that $(\mathrm{pr}_i \circ x)^j_{ai} = \mathrm{pr}_i \circ x^j_a$, so

$$\{i \in I : \mathfrak{A}_i \models \varphi[\mathrm{pr}_i \circ x^j_a]\} \supseteq M \in F,$$

and hence $\{i \in I : \mathfrak{A}_i \models \psi[\mathrm{pr}_i \circ x]^j_a\} \in F$. By the induction hypothesis, $\mathfrak{C} \models \psi[[\ ] \circ x^j_a]$. Clearly $[\ ] \circ x^j_a = ([\ ] \circ x)^j_{[a]}$, so $\mathfrak{C} \models \psi[([\ ] \circ x)^j_{[a]}]$.

Second, suppose $M \notin F$. Then $I \sim M \in F$, i.e., $I \sim M = \{i \in I : \mathfrak{A}_i \models \exists v_j \neg \psi[\mathrm{pr}_i \circ x]\} \in F$. For each $i \in I \sim M$ choose $a_i \in A_i$ so that $\mathfrak{A}_i \models \neg\psi[(\mathrm{pr}_i \circ x)^j_{ai}]$, and for $i \in M$ choose $a_i \in A_i$ arbitrarily. Again $(\mathrm{pr}_i \circ x)^j_{ai} = \mathrm{pr}_i \circ x^j_a$, so

$$\{i \in I : \mathfrak{A}_i \models \neg\psi[\mathrm{pr}_i \circ x^j_a]\} \supseteq I \sim M \in F,$$

and hence $\{i \in I : \mathfrak{A}_i \models \neg\psi[\mathrm{pr}_i \circ x^j_a]\} \in F$. By the induction hypothesis, $\mathfrak{C} \models \neg\psi[[\ ] \circ x^j_a]$. Again we have $[\ ] \circ x^j_a = ([\ ] \circ x)^j_{[a]}$. Hence $\mathfrak{C} \not\models \varphi[[\ ] \circ x]$. □

Corollary 18.30. *If φ is a sentence, then φ holds in $\mathrm{P}_{i \in I} \mathfrak{A}_i/\bar{F}$ iff $\{i : \varphi$ holds in $\mathfrak{A}_i\} \in F$.*

Corollary 18.31. *If a sentence φ holds in $\mathfrak{A}_i$ for each $i \in I$, then φ holds in $\mathrm{P}_{i \in I} \mathfrak{A}_i/\bar{F}$.*

These two corollaries are used more frequently than the fundamental theorem itself. As an illustration of the use of ultraproducts we now give another proof of the compactness theorem. In this proof the final model is obtained rather clearly as a limit of the models of finite subsets, so this proof appears much more constructive than the proof using the completeness theorem.

Assume that each finite subset Δ of Γ has a model $\mathfrak{A}_\Delta$. Let $I = \{\Delta : \Delta$ is a finite subset of $\Gamma\}$. For each $\Delta \in I$ let $G_\Delta = \{\Theta : \Delta \subseteq \Theta \in I\}$. If $\Delta_0, \ldots, \Delta_{m-1} \in I$, then $G_{\Delta 0} \cap \cdots \cap G_{\Delta(m-1)} = G_{\Delta 0 \cup \cdots \cup \Delta(m-1)} \neq 0$. Thus $\{G_\Delta : \Delta \in I\}$ has the finite intersection property, so by Proposition 18.18 let F be an ultrafilter over I such that $G_\Delta \in F$ for each $\Delta \in I$. Then $\mathsf{P}_{\Delta \in I} \mathfrak{A}_\Delta/\bar{F}$ is a model of Γ. In fact, let $\varphi \in \Gamma$. Then $G_{\{\varphi\}} \subseteq \{\Delta \in I : \varphi$ holds in $\mathfrak{A}_\Delta\}$, and $G_{\{\varphi\}} \in F$, so $\{\Delta \in I : \varphi$ holds in $\mathfrak{A}_\Delta\} \in F$. Hence by Corollary 18.30, φ holds in $\mathsf{P}_{\Delta \in I} \mathfrak{A}_\Delta/\bar{F}$, as desired.

In the remainder of this section we shall discuss briefly the cardinality of ultraproducts. Our main purpose is to show that given any infinite structure, we can find arbitrarily large ultrapowers of it.

Proposition 18.32. *If $n \in \omega$ and $\{i \in I : |A_i| = n\} \in F$, then $|\mathsf{P}_{i \in I} A_i/\bar{F}| = n$.*

Proof. Let φ be a sentence (involving equality only) which holds in a structure $\mathfrak{B}$ iff $|B| = n$. Thus $\{i \in I : \varphi$ holds in $\mathfrak{A}_i\} \in F$, so by 18.30, φ holds in $\mathsf{P}_{i \in I} \mathfrak{A}_i/\bar{F}$. Thus $|\mathsf{P}_{i \in I} A_i/\bar{F}| = n$. □

The converse of 18.32 is established similarly:

Proposition 18.33. *If $|\mathsf{P}_{i \in I} A_i/\bar{F}| < \aleph_0$, then $\{i \in I : |A_i| = n\} \in F$ for some $n \in \omega$.*

Actually it is rarely the case that $|\mathsf{P}_{i \in I} A_i/\bar{F}| = \aleph_0$, but we shall not go into that. Now we define a special class of ultrafilters with respect to which ultrapowers of infinite structures are "big."

Definition 18.34. An ultrafilter F over I is *regular* if there is an $E \subseteq F$ such that $|E| = |I|$ and $\bigcap G = 0$ whenever $G \subseteq E$ and G is infinite.

To show that regular ultrafilters exist on infinite sets we need the following well-known set-theoretical fact:

Proposition 18.35. *If I is infinite, then $|\{J : J \subseteq I, J$ finite$\}| = |I|$.*

Proof. For each $m \in \omega$ and $f \in {}^mI$ let $Ff = \text{Rng} f$. Thus F maps $\bigcup_{m \in \omega} {}^mI$ onto $\{J \subseteq I : J$ *finite*$\}$, so

$$|\{J \subseteq I : J \textit{ finite}\}| \leq \left|\bigcup_{m \in \omega} {}^mI\right| \leq \sum_{m \in \omega} |{}^mI|$$
$$= \sum_{m \in \omega} |I| = |I| \cdot \aleph_0 = |I|. \qquad \square$$

Proposition 18.36. *If I is an infinite set, then there is a regular ultrafilter over I.*

PROOF. Let $J = \{X \subseteq I: X \text{ is finite}\}$. It clearly suffices to prove the proposition for J in place of I, since $|I| = |J|$ by 18.35. For each $X \in J$ let $G_X = \{Y \in J: X \subseteq Y\}$. If $X_0, \ldots, X_{m-1} \in J$, then

$$G_{X0} \cap \cdots \cap G_{X(m-1)} = G_{X0 \cup \cdots \cup X(m-1)} \neq 0,$$

so $E = \{G_X : X \in J\}$ has the finite intersection property and hence is included in an ultrafilter F over J. Note that $G_X \neq G_Y$ if $X, Y \in J$ and $X \neq Y$, so $|E| = |J|$. If $H \subseteq E$ and H is infinite, clearly $\bigcap H = 0$. □

The above proof is of course similar to a part of our proof of the compactness theorem using ultraproducts; the ultrafilter constructed there is regular.

Theorem 18.37. *If A and I are infinite sets and F is a regular ultrafilter over I, then $|{}^IA/\bar{F}| \geq 2^{|I|}$.*

PROOF. By Definition 18.34 choose $E \subseteq F$ so that $|E| = |I|$ and $\bigcap G = 0$ whenever $G \subseteq E$ and G is infinite. For each $i \in I$ let $H_i = \{a \in E : i \in a\}$. Thus by our choice of E, H_i cannot be infinite; hence we can choose a one-one function f_i mapping $\mathbf{S}H_i$ $(= \{z : z \subseteq H_i\})$ into A. Now for each $X \subseteq E$ we define a function $g_X: I \to A$ by setting

$$(g_X)_i = f_i(X \cap H_i)$$

for all $i \in I$. Since $|E| = |I|$ and hence $|\mathbf{S}E| = 2^{|I|}$, it suffices now to show that if $X, Y \subseteq E$ and $X \neq Y$ then not $(g_X\bar{F}g_Y)$. Say $a \in X \sim Y$. Then

(1) $$a \subseteq \{i \in I : (g_X)_i \neq (g_Y)_i\}.$$

In fact, if $i \in a$, then $a \in H_i$ and hence $a \in (X \cap H_i) \sim (Y \cap H_i)$ and so $(g_X)_i \neq (g_Y)_i$. Hence (1) holds. Since $a \in F$, it follows that not $(g_X\bar{F}g_y)$. □

Note with regard to 18.37 that $|{}^IA| = 2^{|I|}$ if $\aleph_0 \leq |A| \leq |I|$. The following corollary of 18.37 gives an important fact about ultraproducts.

Corollary 18.38. *If A is infinite and $\mathfrak{m} \geq \aleph_0$, then there is a set I and an ultrafilter F over I such that $|{}^IA/\bar{F}| \geq \mathfrak{m}$.*

EXERCISES

18.39. Let $\mathscr{L}'$ be a rich expansion of $\mathscr{L}$ by C. Let S consist of all $\Gamma \subseteq \text{Sent}_{\mathscr{L}}$ such that every finite subset of Γ has a model and $|\bigcup_{\varphi\in\Gamma} \{\mathbf{c} \in C : \mathbf{c} \text{ occurs in } \varphi\}| < |C|$. Show that S is a consistency family. Hence derive the compactness theorem directly from 18.9.

18.40. Let F be an ultrafilter over I, let $J \in F$, and set $G = \{X \in F: X \subseteq J\}$. Show that $G = \{X \cap J : X \in F\}$ and that G is an ultrafilter over J.

18.41. Let F, J, G be as in 18.40. Show that

$$\mathsf{P}_{i\in I}\,\mathfrak{A}_i/\bar{F} \simeq \mathsf{P}_{j\in J}\,\mathfrak{A}_j/\bar{G}.$$

18.42. Let $\langle J_k : k \in K \rangle$ be a partition of I into nonempty sets. For each $k \in K$ let F_k be an ultrafilter over J_k, and let G be an ultrafilter over K. Define

$$H = \{T \subseteq I : \{k \in K : T \cap J_k \in F_k\} \in G\}.$$

Show that H is an ultrafilter over I.

18.43. Under the assumptions of 18.42 show that

$$\mathsf{P}_{k \in K} (\mathsf{P}_{j \in Jk} \mathfrak{A}_j / \bar{F}_k)/\bar{G} \simeq \mathsf{P}_{i \in I} \mathfrak{A}_i / \bar{H}.$$

18.44. Let P be the set of all positive prime numbers, and let F be a nonprincipal ultrafilter over P. For each $p \in P$, let $\mathfrak{A}_p$ be the group $\mathbb{Z}_p$ of residues of integers mod p. Show that $\mathsf{P}_{\in P} \mathfrak{A}_p/\bar{F}$ is a group all elements of which have infinite order.

18.45. Let P and F be as in 18.44. For each $p \in P$ let $\mathfrak{B}_p$ be the prime field of characteristic p. Show that $\mathsf{P}_{p \in P} \mathfrak{B}_p/\bar{F}$ is a field of characteristic 0.

18.46. Let $\mathfrak{A}$ be the ring of integers and F a nonprincipal ultrafilter over ω. Show that ${}^{\omega}\mathfrak{A}/\bar{F}$ is an integral domain which is not a unique factorization domain.

18.47. Let $\mathfrak{A}$ be a field and I a non-empty set. For any ultrafilter F over I let $M_F = \{x \in {}^I A : x\bar{F}0\}$, where $0 \in {}^I A$ is the zero function, $0i = 0$ for all $i \in I$. Show that M is a one-one function from the set of all ultrafilters over I onto the set of all maximal ideals in the ring ${}^I\mathfrak{A}$.

Elementary Equivalence 19

We now introduce and study the basic logical relation between structures —elementary equivalence, as well as some related notions.

Definition 19.1. Two $\mathscr{L}$-structures $\mathfrak{A}$ and $\mathfrak{B}$ are *elementary equivalent*, in symbols $\mathfrak{A} \equiv_{ee} \mathfrak{B}$, if $(\mathfrak{A} \models \varphi \Leftrightarrow \mathfrak{B} \models \varphi)$ for every sentence φ.

Thus two elementarily equivalent structures are indistinguishable by first-order means. Note from 18.19 that isomorphic structures are automatically elementarily equivalent. We shall see that the converse is far from true. The following useful theorem is easy to establish:

Proposition 19.2. *For any theory* Γ *the following conditions are equivalent:*

(*i*) Γ *is complete;*
(*ii*) *any two models of* Γ *are elementarily equivalent.*

Proposition 19.3. *If* $\mathfrak{A}_i \equiv_{ee} \mathfrak{B}_i$ *for each* $i \in I$, *and* F *is an ultrafilter over* I, *then* $\mathsf{P}_{i \in I} \mathfrak{A}_i/\bar{F} \equiv_{ee} \mathsf{P}_{i \in I} \mathfrak{B}_i/\bar{F}$.

Proof. Let φ be a sentence which holds in $\mathsf{P}_{i \in I} \mathfrak{A}_i/\bar{F}$. Then by the basic theorem on ultraproducts, $\{i \in I : \mathfrak{A}_i \models \varphi\} \in F$. But $\{i \in I : \mathfrak{A}_i \models \varphi\} = \{i \in I : \mathfrak{B}_i \models \varphi\}$ by hypothesis, so by the basic theorem on ultraproducts, $\mathsf{P}_{i \in I} \mathfrak{B}_i/\bar{F} \models \varphi$. Taking $\neg\varphi$ for φ, we see that the converse holds also. □

In Chapter 26 we shall make a deeper study of elementary equivalence; in particular, we provide there mathematical equivalents of this logical notion. We shall be concerned for most of this section with a stronger form of elementary equivalence which can hold between two structures when one is a substructure of the other. So we now turn to a brief discussion of the general algebraic notion of a substructure.

Definition 19.4. Let $\mathfrak{A}$ and $\mathfrak{B}$ be two $\mathscr{L}$-structures. We say that $\mathfrak{A}$ is a *substructure of* $\mathfrak{B}$, and $\mathfrak{B}$ is an *extension of* $\mathfrak{A}$, $\mathfrak{A} \subseteq \mathfrak{B}$ or $\mathfrak{B} \supseteq \mathfrak{A}$, provided that the following conditions hold:

(*i*) $A \subseteq B$;
(*ii*) for each operation symbol $\mathbf{O}$ (say m-ary), $\mathbf{O}^{\mathfrak{A}} = \mathbf{O}^{\mathfrak{B}} \restriction {}^mA$;
(*iii*) for each relation symbol $\mathbf{R}$ (say m-ary), $\mathbf{R}^{\mathfrak{A}} = \mathbf{R}^{\mathfrak{B}} \cap {}^mA$.

Note that if $\mathfrak{B}$ is an $\mathscr{L}$-structure, $0 \neq A \subseteq B$, and A is closed under all of the operations of $\mathfrak{B}$, then there is a unique $\mathscr{L}$-structure $\mathfrak{A}$ with universe A such that $\mathfrak{A} \subseteq \mathfrak{B}$. In case $\mathscr{L}$ has no operation symbols, each nonempty subset of B is the universe of some substructure of $\mathfrak{B}$. This is no longer true in general when there are operation symbols. For example, $\mathfrak{B} = (\omega, \mathsf{s}, 0)$ has no proper subalgebras, and in particular no finite substructures.

The following simple proposition will be useful later.

Proposition 19.5. *Suppose* $\mathfrak{A} \subseteq \mathfrak{B}$, $x \in {}^{\omega}A$, *and* σ *is a term. Then* $\sigma^{\mathfrak{A}}x = \sigma^{\mathfrak{B}}x$.

Proposition 19.6. *If* $\mathfrak{A}$ *and* $\mathfrak{B}$ *are* $\mathscr{L}$*-structures, a function* f *is an embedding of* $\mathfrak{A}$ *into* $\mathfrak{B}$ *iff* f *is an isomorphism of* $\mathfrak{A}$ *onto a substructure of* $\mathfrak{B}$.

The following is a basic result on embeddings which is usually implicit in basic courses in algebra.

Proposition 19.7. *If* f *is an embedding of* $\mathfrak{A}$ *into* $\mathfrak{B}$*, then there is an* $\mathscr{L}$*-structure* $\mathfrak{C}$ *and an isomorphism* g *of* $\mathfrak{C}$ *onto* $\mathfrak{B}$ *such that* $\mathfrak{A} \subseteq \mathfrak{C}$ *and* $f \subseteq g$*. This is indicated by the following diagram, where* $\mathfrak{D}$ *is the image of* f*:*

$$\begin{array}{ccc} \mathfrak{C} & \overset{g}{\rightarrowtail\!\!\!\!\twoheadrightarrow} & \mathfrak{B} \\ \cup\!\!| & & \cup\!\!| \\ \mathfrak{A} & \underset{f}{\rightarrowtail} & \mathfrak{D} \end{array}$$

PROOF. Let $C = A \cup \{(A, x) : x \in B \sim D\}$. Note that $A \cap \{(A, x) : x \in B \sim D\} = 0$; for if $(A, x) \in A$, then $A \in \{A\} \in (A, x) \in A$, contradicting the regularity axiom of set theory. Define $g: C \to B$ by: $ga = fa$ for all $a \in A$, and $g(A, x) = x$ for all $x \in B \sim D$. Clearly g is a one-one function mapping C onto B. We define an $\mathscr{L}$-structure $\mathfrak{C}$ with universe C so that g is automatically on isomorphism from $\mathfrak{C}$ onto $\mathfrak{B}$: for $\mathbf{O}$ an m-ary operation symbol and for $c_0, \ldots, c_{m-1} \in C$,

$$\mathbf{O}^{\mathfrak{C}}(c_0, \ldots, c_{m-1}) = g^{-1}\mathbf{O}^{\mathfrak{B}}(gc_0, \ldots, gc_{m-1})$$

while for $\mathbf{R}$ an m-ary relation symbol,

$$\mathbf{R}^{\mathfrak{C}} = \{(c_0, \ldots, c_{m-1}) : (gc_0, \ldots, gc_{m-1}) \in \mathbf{R}^{\mathfrak{B}}\}.$$

Since $f \subseteq g$, it remains only to check that $\mathfrak{A} \subseteq \mathfrak{C}$. For an m-ary operation symbol $\mathbf{O}$ of $\mathscr{L}$ and for any $a_0, \ldots, a_{n-1} \in A$,

$$\begin{aligned}\mathbf{O}^{\mathfrak{C}}(a_0, \ldots, a_{m-1}) &= g^{-1}\mathbf{O}^{\mathfrak{B}}(ga_0, \ldots, ga_{m-1}) = g^{-1}\mathbf{O}^{\mathfrak{B}}(fa_0, \ldots, fa_{m-1}) \\ &= g^{-1}\mathbf{O}^{\mathfrak{D}}(fa_0, \ldots, fa_{m-1}) = g^{-1}f\mathbf{O}^{\mathfrak{A}}(a_0, \ldots, a_{m-1}) \\ &= \mathbf{O}^{\mathfrak{A}}(a_0, \ldots, a_{m-1}).\end{aligned}$$

The case of relation symbols is similar. □

We now introduce the technique of diagrams, due to Henkin and A. Robinson, which gives a method for dealing with substructures in a logical context. There are many variations on this important technique, and we will meet with some of them later.

Definition 19.8. (Diagrams). Let X be any set. An *X-expansion* of $\mathscr{L}$ is an expansion $\mathscr{L}'$ of $\mathscr{L}$ obtained from $\mathscr{L}$ by adding new distinct individual constants $\mathbf{c}_x$ for all $x \in X$. We sometimes denote $\mathscr{L}'$ by $(\mathscr{L}, \mathbf{c}_x)_{x \in X}$, and $\mathscr{L}'$ structures are denoted by $(\mathfrak{A}, l_x)_{x \in X}$, where $\mathfrak{A}$ is an $\mathscr{L}$-structure and l_x is a member of A for each $x \in X$; l_x is the denotation in the structure of $\mathbf{c}_x$.

Let $\mathfrak{A}$ be an $\mathscr{L}$-structure, and let $\mathscr{L}'$ be an A-expansion of $\mathscr{L}$. The $\mathscr{L}'$-*diagram* of $\mathfrak{A}$ is the set of all sentences of $\mathscr{L}'$ of the following forms:

$\neg \mathbf{c}_a \equiv \mathbf{c}_b$ for $a, b \in A$ and $a \neq b$;
$\mathbf{O}\mathbf{c}_{a0} \cdots \mathbf{c}_{a(m-1)} \equiv \mathbf{c}_b$ if $\mathbf{O}$ is an operation symbol of $\mathscr{L}$ of rank m, $a \in {}^mA$, and $\mathbf{O}^{\mathfrak{A}}a_0 \cdots a_{m-1} = b$;
$\mathbf{R}\mathbf{c}_{a0} \cdots \mathbf{c}_{a(m-1)}$ if $\mathbf{R}$ is an m-ary relation symbol of $\mathscr{L}$, and $\langle a_0, \ldots, a_{m-1}\rangle \in \mathbf{R}^{\mathfrak{A}}$;
$\neg\mathbf{R}\mathbf{c}_{a0} \cdots \mathbf{c}_{a(m-1)}$ if $\mathbf{R}$ is an m-ary relation symbol of $\mathscr{L}$, and $\langle a_0, \ldots, a_{m-1}\rangle \notin \mathbf{R}^{\mathfrak{A}}$.

Diagrams are essentially a logical expression of the notion of substructure, or embedding. The basic properties of diagrams are given in the next two theorems, which essentially show that the models of the diagram of $\mathfrak{A}$ are exactly the structures in which $\mathfrak{A}$ can be embedded.

Proposition 19.9. *Let $\mathfrak{A}$ and $\mathfrak{B}$ be $\mathscr{L}$-structures with f an embedding of $\mathfrak{A}$ into $\mathfrak{B}$. Let $\mathscr{L}'$ be an A-expansion of $\mathscr{L}$. Then $(\mathfrak{B}, fa)_{a \in A}$ is a model of the $\mathscr{L}'$-diagram of $\mathfrak{A}$.*

PROOF. The proof is essentially trivial, and we only illustrate it by verifying that a member $\varphi = \neg\mathbf{R}\mathbf{c}_{a0} \cdots \mathbf{c}_{a(m-1)}$ of the diagram holds in $(\mathfrak{B}, fa)_{a \in A}$, where $\mathbf{R}$ is an m-ary relation symbol of $\mathscr{L}$ and $\langle a_0, \ldots, a_{m-1}\rangle \notin \mathbf{R}^{\mathfrak{A}}$.

$$\begin{aligned}(\mathfrak{B}, fa)_{a \in A} \models \varphi \quad &\text{iff not } (\mathfrak{B}, fa)_{a \in A} \models \mathbf{R}\mathbf{c}_{a0} \cdots \mathbf{c}_{a(m-1)} \\ &\text{iff } \langle fa_0, \ldots, fa_{m-1}\rangle \notin \mathbf{R}^{\mathfrak{B}} \\ &\text{iff } \langle a_0, \ldots, a_{m-1}\rangle \notin \mathbf{R}^{\mathfrak{A}},\end{aligned}$$

so $(\mathfrak{B}, fa)_{a \in A} \models \varphi$. □

Note that 19.9 applies in particular if $\mathfrak{A} \subseteq \mathfrak{B}$, where f is just the identity on A. Then $(\mathfrak{B}, a)_{a\in A}$ is a model of the diagram of $\mathfrak{A}$. More particularly, $(\mathfrak{A}, a)_{a\in A}$ is a model of the diagram of $\mathfrak{A}$.

Proposition 19.10. *Let $\mathfrak{A}$ be an $\mathscr{L}$-structure, $\mathscr{L}'$ an A-expansion of $\mathscr{L}$, and $(\mathfrak{B}, l_a)_{a\in A}$ a model of the $\mathscr{L}'$-diagram of $\mathfrak{A}$. Then l is an embedding of $\mathfrak{A}$ into $\mathfrak{B}$.*

Proof. Again the proof is almost trivial, and we will just check that l is one-one and that it preserves operations. If $a, b \in A$ and $a \neq b$, then $\neg \mathbf{c}_a = \mathbf{c}_b$ is in the diagram of $\mathfrak{A}$, so $(\mathfrak{B}, l_a)_{a\in A} \models \neg \mathbf{c}_a \equiv \mathbf{c}_b$ and hence $la \neq lb$. Now let $\mathbf{O}$ be an m-ary operation symbol, suppose that $a_0, \ldots, a_{m-1} \in A$, and set $\mathbf{O}^{\mathfrak{A}}(a_0, \ldots, a_{m-1}) = b$. Then $\mathbf{O}(\mathbf{c}_{a0}, \ldots, \mathbf{c}_{a(m-1)}) \equiv \mathbf{c}_b$ is in the diagram of $\mathfrak{A}$, and hence

$$(\mathfrak{B}, la)_{a\in A} \models \mathbf{O}(\mathbf{c}_{a0}, \ldots, \mathbf{c}_{a(m-1)}) = \mathbf{c}_b.$$

It follows that $\mathbf{O}^{\mathfrak{B}}(la_0, \ldots, la_{m-1}) = lb$. □

The method of diagrams enables us to prove a mathematically useful embedding theorem. To formulate it we need a definition.

Definition 19.11. Let $\mathfrak{A}$ be an $\mathscr{L}$-structure and $0 \neq X \subseteq A$. The *subuniverse of $\mathfrak{A}$ generated by X* is the closure of X under all of the operations $\mathbf{O}^{\mathfrak{A}}$ of $\mathfrak{A}$; clearly it is the universe of a uniquely determined substructure of $\mathfrak{A}$. A substructure $\mathfrak{B}$ of $\mathfrak{A}$ is *finitely generated* if B is generated by a finite non-empty subset of A.

Note that if the language contains no operation symbols then any non-empty subset X of A coincides with the subuniverse of $\mathfrak{A}$ generated by X, and every finitely generated substructure of $\mathfrak{A}$ is finite. The following simple cardinality result will be useful later.

Proposition 19.12. *Let $\mathfrak{A}$ be an $\mathscr{L}$-structure, $0 \neq X \subseteq A$, and let B be the subuniverse of $\mathfrak{A}$ generated by X. Then $|B| \leq |X| + |\mathrm{Fmla}_{\mathscr{L}}|$.*

Proof. Let $C_0 = X$, and for each $m \in \omega$ let

$$C_{m+1} = C_m \cup \{\mathbf{O}^{\mathfrak{A}}x : \mathbf{O} \text{ is an } n\text{-ary operation symbol of } \mathscr{L} \text{ and } x \in {}^nC_m\}.$$

Clearly $B = \bigcup_{m\in\omega} C_m$, $|C_m| \leq |X| + |\mathrm{Fmla}_{\mathscr{L}}|$ for each $m \in \omega$, and hence $|B| \leq |X| + |\mathrm{Fmla}_{\mathscr{L}}|$ (recall that $\aleph_0 \leq |\mathrm{Fmla}_{\mathscr{L}}|$). □

Theorem 19.13 (Henkin's embedding theorem). *Let $\mathbf{K}$ be the class of all models of a set Γ of sentences and let $\mathfrak{A}$ be an $\mathscr{L}$-structure. Suppose that every finitely generated substructure of $\mathfrak{A}$ can be embedded in a member of $\mathbf{K}$. Then $\mathfrak{A}$ can be embedded in a member of $\mathbf{K}$.*

Proof. Let $\mathscr{L}'$ be an A-expansion of $\mathscr{L}$, and let Δ be the $\mathscr{L}'$-diagram of $\mathfrak{A}$. Then every finite subset Θ of $\Gamma \cup \Delta$ has a model. In fact, let $\mathbf{c}_{b0}, \ldots, \mathbf{c}_{b(m-1)}$

be all of the new individual constants that occur in sentences of Θ. Let $\mathfrak{B}$ be the substructure of $\mathfrak{A}$ generated by $\{b_0, \ldots, b_{m-1}\}$, or let $\mathfrak{B}$ be any finitely generated substructure of $\mathfrak{A}$ if $m = 0$. Set $\mathfrak{C} = (\mathfrak{B}, l_a)_{a\in A}$, where $la = a$ for a of the form b_i, la any element of $\mathbf{B}$ otherwise. Since $(\mathfrak{A}, a)_{a\in A}$ is a model of Δ, it follows that $\mathfrak{C}$ is a model of $\Delta \cap \Theta$. Now by the hypothesis of the theorem let f be an embedding of $\mathfrak{B}$ into a member $\mathfrak{C}$ of $\mathbf{K}$. It is easily verified that $(\mathfrak{C}, fl_a)_{a\in A}$ is a model of Θ.

Thus by the compactness theorem, $\Gamma \cup \Delta$ has a model $(D, k_a)_{a\in A} = \mathfrak{D}$. From Proposition 19.10 we see that k is an embedding of $\mathfrak{A}$ into $\mathfrak{D}$. □

Theorem 19.13 can also be established using ultraproducts; see Exercise 19.42. As an example of an application of Henkin's embedding theorem we show that any simple ordering can be embedded in a discrete ordering. Recall that a *discrete ordering* is an ordering $(A, \leq)$ such that every element of A has an immediate successor if it is not the last element of A, and an immediate predecessor if it is not the first element of A. The hypothesis of 19.13 is trivially satisfied, since for the logic implicit here there are no operation symbols, and so finite generation just means finite, and every finite simple ordering is discrete. Thus 19.13 immediately gives the desired result. Of course this result can be proved directly, rather easily. See Exercises 19.43 and 19.44 for further applications.

Now we define the stronger form of elementary equivalence alluded to at the beginning of this chapter.

Definition 19.14. Let $\mathfrak{A}$ and $\mathfrak{B}$ be $\mathscr{L}$-structures. We say that $\mathfrak{A}$ is an *elementary substructure* of $\mathfrak{B}$ and that $\mathfrak{B}$ is an *elementary extension* of $\mathfrak{A}$, in symbols $\mathfrak{A} \preccurlyeq \mathfrak{B}$ or $\mathfrak{B} \succcurlyeq \mathfrak{A}$, if the following two conditions hold:

(*i*) $A \subseteq B$;
(*ii*) for any formula φ and any $x \in {}^{\omega}A$, $\mathfrak{A} \models \varphi[x]$ iff $\mathfrak{B} \models \varphi[x]$.

It is obvious that $\mathfrak{A} \preccurlyeq \mathfrak{B}$ implies $\mathfrak{A} \equiv_{\mathrm{ee}} \mathfrak{B}$ (take φ to be a sentence in (*ii*)). A useful general observation is that in checking (*ii*) one only needs to prove one direction of the indicated equivalence, since if it holds in one direction, passage from φ to $\neg\varphi$ gives the other direction. We have formulated 19.14 without insisting that $\mathfrak{A} \subseteq \mathfrak{B}$, but this *is* a consequence of the definition:

Corollary 19.15. *If* $\mathfrak{A} \preccurlyeq \mathfrak{B}$, *then* $\mathfrak{A} \subseteq \mathfrak{B}$.

PROOF. Let $\mathbf{O}$ be an m-ary operation symbol, and let $a_0, \ldots, a_m \in A$. Then

$$\begin{aligned}\mathbf{O}^{\mathfrak{A}}(a_0, \ldots, a_{m-1}) = a_m \quad & \text{iff } \mathfrak{A} \models \mathbf{O}v_0 \cdots v_{m-1} \mathbf{=} v_m[a_0, \ldots, a_m] \\ & \text{iff } \mathfrak{B} \models \mathbf{O}v_0 \cdots v_{m-1} \mathbf{=} v_m[a_0, \ldots, a_m] \\ & \text{iff } \mathbf{O}^{\mathfrak{B}}(a_0, \ldots, a_{m-1}) = a_m.\end{aligned}$$

Relations are treated analogously. □

The following proposition gives a useful criterion for $\mathfrak{A} \preccurlyeq \mathfrak{B}$, in which one only has to talk about satisfaction in the larger structure.

Proposition 19.16. *Let $\mathfrak{A}$ and $\mathfrak{B}$ be $\mathscr{L}$-structures, and assume that $\mathfrak{A} \subseteq \mathfrak{B}$. Then the following two conditions are equivalent:*

(*i*) $\mathfrak{A} \preccurlyeq \mathfrak{B}$;

(*ii*) *for every formula φ, every $k \in \omega$, and every $x \in {}^{\omega}A$, if $\mathfrak{B} \models \exists v_k\varphi[x]$, then there is an $a \in A$ such that $\mathfrak{B} \models \varphi[x_a^k]$.*

Proof. (*i*) $\Rightarrow$ (*ii*). By the hypothesis of (*ii*) and the meaning of (*i*), $\mathfrak{A} \models \exists v_k\varphi[x]$. Hence by the definition of satisfaction there is an $a \in A$ such that $\mathfrak{A} \models \varphi[x_a^k]$. Hence by (*i*) again, $\mathfrak{B} \models \varphi[x_a^k]$.

(*ii*) $\Rightarrow$ (*i*). Assume (*ii*). By induction on φ we show

(1) for every formula φ and every $x \in {}^{\omega}A$, $\mathfrak{A} \models \varphi[x]$ iff $\mathfrak{B} \models \varphi[x]$.

If φ is an atomic formula $\mathbf{R}\sigma_0 \cdots \sigma_{m-1}$, then

$$\begin{aligned}\mathfrak{A} \models \varphi[x] \quad & \text{iff } \langle \sigma_0^{\mathfrak{A}}x, \ldots, \sigma_{m-1}^{\mathfrak{A}}x\rangle \in \mathbf{R}^{\mathfrak{A}} \\ & \text{iff } \langle \sigma_0^{\mathfrak{B}}x, \ldots, \sigma_{m-1}^{\mathfrak{B}}x\rangle \in \mathbf{R}^{\mathfrak{B}} \text{ since } \mathfrak{A} \subseteq \mathfrak{B}, \text{ using 19.5} \\ & \text{iff } \mathfrak{B} \models \varphi[x],\end{aligned}$$

and atomic equality formulas are similar. The induction steps for $\neg$, $\vee$, $\wedge$ are obvious. Now suppose (1) is true for φ, and let $k \in \omega$, $x \in {}^{\omega}A$. First suppose that not ($\mathfrak{A} \models \forall v_k\varphi[x]$). Thus there must be an $a \in A$ so that not($\mathfrak{A} \models \varphi[x_a^k]$). Then by (1) for φ, not ($\mathfrak{B} \models \varphi[x_a^k]$), so not ($\mathfrak{B} \models \forall v_k\varphi[x]$). Second suppose that not ($\mathfrak{B} \models \forall v_k\varphi[x]$). Thus $\mathfrak{B} \models \exists v_k \neg \varphi[x]$, so by (*ii*) choose $a \in A$ so that $\mathfrak{B} \models \neg\varphi[x_a^k]$. Thus by (1) for φ, not ($\mathfrak{A} \models \varphi[x_a^k]$, so not ($\mathfrak{A} \models \forall v_k\varphi[x]$). □

Now we come to another of the fundamental results of model theory. We already know from our general form of the completeness theorem 11.19 that a consistent set of sentences in a language $\mathscr{L}$ has a model of cardinality $\leq |\text{Fmla}_{\mathscr{L}}|$. The following result is a kind of generalization of this cardinality condition.

Theorem 19.17 (Downward Löwenheim–Skolem theorem). *Let $\mathfrak{B}$ be an $\mathscr{L}$-structure, let $\mathfrak{m}$ be a cardinal such that $|\text{Fmla}_{\mathscr{L}}| \leq \mathfrak{m} \leq |B|$, and let C be a subset of B such that $|C| \leq \mathfrak{m}$.* Then *there is an elementary substructure $\mathfrak{A}$ of $\mathfrak{B}$ such that $C \subseteq A$* and $|A| = \mathfrak{m}$.

Proof. Let $\mathscr{C}$ be a choice function for nonempty subsets of B. We now define a sequence $D_0, \ldots, D_m, \ldots, m \in \omega$, by recursion. Let D_0 be any subset of B such that $C \subseteq D_0$ and $|D_0| = \mathfrak{m}$. Fix $d \in D_0$.

Now suppose D_m has been defined. Let J_m be the set of all quadruples (F, φ, k, x) satisfying the following conditions:

(1) F is a finite subset of D_m;
(2) φ is a formula of $\mathscr{L}$;
(3) $k < \omega$;
(4) $x \in {}^\omega F$, $x_j = d$ whenever v_j does not occur in φ, and $\mathfrak{B} \models \exists v_k[x]$.

Note that for (F, φ, k) satisfying (1)–(3) there are only finitely many x satisfying (4). Hence

(5) $$|J_m| \leq |\{F : F \text{ is a finite subset of } D_m\}| \cdot |\text{Fmla}_{\mathscr{L}}| \cdot \aleph_0.$$

Now let $D_{m+1} = D_m \cup \{\mathscr{C}\{a : \mathfrak{B} \models \varphi[x_a^k]\} : (F, \varphi, k, x) \in J_m\}$. This completes the definition of the sequence $D_0, D_1, \ldots$. Let $A = \bigcup_{m\in\omega} D_m$. Thus $C \subseteq A$ since $C \subseteq D_0$. By induction an m, using (5), we see that $|D| = \mathfrak{m}$ for each $m \in \omega$, and hence $|A| = \mathfrak{m}$. Next,

(6) A is closed under each operation $\mathbf{O}^{\mathfrak{B}}$.

For, suppose that $\mathbf{O}$ has rank m, and let $b_0, \ldots, b_{m-1} \in A$. Then there is an $n \in \omega$ with $b_0, \ldots, b_{m-1} \in D_n$. Let

$$\begin{aligned} F &= \{b_0, \ldots, b_{m-1}, d\} \\ \varphi &= \mathbf{O}v_0 \cdots v_{m-1} \equiv v_m, \\ k &= m, \\ x &\in {}^\omega F \text{ with } x_0 = b_0, \ldots, x_{m-1} = b_{m-1}, x_j = d \text{ for } j \geq m. \end{aligned}$$

Since $\models \exists v_m \mathbf{O}v_0 \cdots v_{m-1} \equiv v_m$, it follows that $\mathfrak{B} \models \exists v_m\varphi[x]$. Hence $(F, \varphi, k, x) \in J_n$. Now $\{a \in B : \mathfrak{B} \models \varphi[x_0, \ldots, x_{m-1}, a]\}$ has exactly one element, namely $\mathbf{O}^{\mathfrak{B}}x_0 \cdots x_{m-1}$. It thus follows that $\mathbf{O}^{\mathfrak{B}}b_0 \cdots b_{m-1} \in D_{n+1} \subseteq A$. Hence (6) holds.

By (6), A is the universe of a uniquely determined substructure $\mathfrak{A}$ of $\mathfrak{B}$. It remains only to show that $\mathfrak{A} \preccurlyeq \mathfrak{B}$, and to do this we shall apply 19.16. Let φ be a formula, and let $k \in \omega$, $x \in {}^\omega A$, and assume that $\mathfrak{B} \models \exists v_k\varphi[x]$. Define $F = \{d\} \cup \{x_j : v_j \text{ occurs in } \varphi\}$, and choose $y \in {}^\omega F$ with $y_j = x_j$ for v_j occurring in φ, with $y_j = d$ otherwise. Choose m so that $F \subseteq D_m$. Thus $(F, \varphi, k, y) \in J_m$ so, letting $b = \mathscr{C}\{a : \mathfrak{B} \models \varphi[y_a^k]\}$, we see that $\mathfrak{B} \models \varphi[y_b^k]$ and $b \in D_{m+1} \subseteq A$. Thus $\mathfrak{B} \models \varphi[x_b^k]$. □

It can be shown by easy examples (see Exercise 19.48) that 19.17 cannot be strengthened by dropping the assumption that $|\text{Fmla}_{\mathscr{L}}| \leq \mathfrak{m}$. The proof given above works with minor modifications for many other languages besides first-order languages; see Part V.

Theorem 19.17 has the following two obvious corollaries. They also are obvious on the basis of the completeness theorem 11.19, but it should be noted that the proof above of 19.17 was direct and much simpler than the proof of 11.19.

Corollary 19.18. *If a set Γ of sentences has a model, then Γ has a model of power $\leq |\text{Fmla}_{\mathscr{L}}|$.*

Corollary 19.19. *Let $\mathscr{L}$ be any first-order language (with no restriction on $|\mathrm{Fmla}_{\mathscr{L}}|$). For a sentence φ of $\mathscr{L}$, φ is universally valid iff φ holds in every countable $\mathscr{L}$-structure.*

Applying 19.18 to a theory Γ of sets, for example to $\Gamma = \mathrm{ZF}$, we obtain the so-called *Skolem paradox*: if ZF has a model (which clearly must be infinite), then it has a denumerable model. This is true even though in ZF one can prove the existence of nondenumerable sets.

Theorem 19.17 can be proved quickly using Skolem functions (see 11.33–11.38). But of course this new proof is not as elementary as the one above, since it is based on the somewhat complicated apparatus of Skolem expansions. The two proofs should be compared to one another; each helps explain the other. The new proof is based upon the following proposition.

Proposition 19.20. *Let $\mathscr{L}'$ be a Skolem expansion of $\mathscr{L}$, and let $\mathfrak{B}$ be an $\mathscr{L}'$ structure which is a model of the Skolem set of $\mathscr{L}'$ over $\mathscr{L}$. Then for any $\mathscr{L}'$-structure $\mathfrak{A}$, the following conditions are equivalent:*

(*i*) $\mathfrak{A} \subseteq \mathfrak{B}$;
(*ii*) $\mathfrak{A} \preccurlyeq \mathfrak{B}$.

Proof. The implication (*ii*) $\Rightarrow$ (*i*) is trivial. Now assume (*i*). We shall apply 19.16 in order to prove (*ii*). To this end, assume that $x \in {}^{\omega}A$ and $\mathfrak{B} \models \exists v_i\varphi[x]$. Then, since $\mathfrak{B}$ is a model of the Skolem set of $\mathscr{L}'$ over $\mathscr{L}$ we have $\mathfrak{B} \models \varphi(v_0, \ldots, v_{i-1}, \sigma)[x]$, where σ is the term

$$S^j_{\exists vi\varphi}\alpha_0 \cdots \alpha_{m-1},$$

with $m = |\mathrm{Fv}\, \exists v_i\varphi|$, $\mathrm{Fv}\, \exists v_i\varphi = \{\alpha_0, \ldots, \alpha_{m-1}\}$, $v^{-1}\alpha_0 < \cdots < v^{-1}\alpha_{m-1}$, and j is minimal such that $\exists v_i\varphi$ is a formula of $\mathscr{L}_j$ (see Definition 11.33). Since $\mathfrak{A} \subseteq \mathfrak{B}$, we have

$$(S^j_{\exists vi\varphi}\alpha_0 \cdots \alpha_{m-1})^{\mathfrak{B}}x = a \in A.$$

Thus $\mathfrak{B} \models \varphi[x^i_a]$. □

Now we can give the new proof of 19.17. Assume the hypothesis of 19.17. Let $\mathscr{L}'$ be a Skolem expansion of $\mathscr{L}$, and let $\mathfrak{B}'$ be an $\mathscr{L}'$-expansion of $\mathfrak{B}$ which is a model of the Skolem set of $\mathscr{L}'$ over $\mathscr{L}$ (using 11.34 and 11.36). Let A' be the subuniverse of $\mathfrak{B}'$ generated by some subset D of B such that $C \subseteq D$ and $|D| = \mathfrak{m}$ (recall 19.11). Thus A' is the universe of a substructure $\mathfrak{A}'$ of $\mathfrak{B}'$, and by 19.12, $|A'| = \mathfrak{m}$. By 19.20, $\mathfrak{A}' \preccurlyeq \mathfrak{B}'$, so $\mathfrak{A}' \restriction \mathscr{L} \preccurlyeq \mathfrak{B}$, as desired.

Now we want to prove an "upward" version of 19.17. To this end we need the notion of elementary embedding, defined analogously to ordinary embeddings:

Definition 19.21. If $\mathfrak{A}$ and $\mathfrak{B}$ are $\mathscr{L}$-structures, an *elementary embedding* of $\mathfrak{A}$ into $\mathfrak{B}$ is an isomorphism of $\mathfrak{A}$ onto an elementary substructure of $\mathfrak{B}$.

The following proposition is proved analogously to 19.7.

Proposition 19.22. *If f is an elementary embedding of $\mathfrak{A}$ into $\mathfrak{B}$, then there is an $\mathscr{L}$-structure $\mathfrak{C}$ and an isomorphism g of $\mathfrak{C}$ onto $\mathfrak{B}$ such that $\mathfrak{A} \preccurlyeq \mathfrak{C}$ and $f \subseteq g$. Pictorially (where $\mathfrak{D}$ is the image of f):*

$$\begin{array}{ccc} \mathfrak{C} & \overset{g}{\rightarrowtail\!\!\twoheadrightarrow} & \mathfrak{B} \\ \curlyvee\!\!| & & \curlyvee\!\!| \\ \mathfrak{A} & \underset{f}{\rightarrowtail} & \mathfrak{D} \end{array}$$

Another very important property of ultraproducts is given in the following proposition:

Proposition 19.23. *Let $\mathfrak{A}$ be an $\mathscr{L}$-structure, I a non-empty set, and F an ultrafilter over I. For each $a \in A$ set $fa = [\langle a : i \in I \rangle]_{\bar{F}}$. Then f is an elementary embedding of $\mathfrak{A}$ into $I_{\mathfrak{A}/\bar{F}}$.*

Proof. First, f is one-one. For, suppose $fa = fb$. Thus $\langle a : i \in I\rangle \bar{F} \langle b : i \in I\rangle$, i.e., $\{i \in I : a = b\} \in F$. But $\{i : a = b\}$ is either empty or all of I depending upon whether $a \neq b$ or $a = b$ respectively. Since $0 \notin F$, it follows that $a = b$.

Since f is one-one, we can use the procedure of the proof of 19.7 to make the image C of f into an $\mathscr{L}$-structure, and f into an isomorphism of $\mathfrak{A}$ onto $\mathfrak{B}$. Thus for any $c \in {}^mC$,

$$\mathbf{O}^{\mathfrak{C}} c_0 \cdots c_{m-1} = f\mathbf{O}^{\mathfrak{A}} f^{-1}c_0 \cdots f^{-1}c_{m-1}, \text{ and } \mathbf{R}^{\mathfrak{C}} = \{c \in {}^mC : f^{-1} \circ c \in \mathbf{R}^{\mathfrak{A}}\},$$

for m-ary operation and relation symbols $\mathbf{O}$ and $\mathbf{R}$. Now we prove $\mathfrak{C} \preccurlyeq {}^I\mathfrak{A}/\bar{F}$ by checking 19.14(*ii*). Let φ be a formula and $x \in {}^\omega C$. Choose $y \in {}^\omega A$ with $x = f \circ y$. Set $z_m = \langle y_m : i \in I\rangle$ for each $m \in \omega$. Then $x = [\ ] \circ z$, so

$$\begin{array}{lll} {}^I\mathfrak{A}/\bar{F} \models \varphi[x] & \text{iff } \{i : \mathfrak{A} \models \varphi[\mathrm{pr}_i \circ z]\} \in F & \text{by 18.29} \\ & \text{iff } \{i : \mathfrak{A} \models \varphi[y]\} \in F & \\ & \text{iff } \mathfrak{A} \models \varphi[y] & \\ & \text{iff } \mathfrak{C} \models \varphi[x] & \text{by 18.23} \end{array}$$ □

With the help of 19.23 we can now prove the upward analog of 19.17, which, too, is a major result in model theory.

Theorem 19.24 (Upward Löwenheim-Skolem theorem). *Let $\mathfrak{m}$ be an infinite cardinal $\geq |\mathrm{Fmla}_{\mathscr{L}}|$, and let $\mathfrak{A}$ be an infinite $\mathscr{L}$-structure with $|A| \leq \mathfrak{m}$. Then $\mathfrak{A}$ has an elementary extension $\mathfrak{B}$ such that $|B| = \mathfrak{m}$.*

Proof. By 18.36 find 18.37 let I be a set and F an ultrafilter over I such that $|{}^IA/\bar{F}| \geq \mathfrak{m}$. By 19.23 we obtain an elementary embedding of $\mathfrak{A}$ into ${}^I\mathfrak{A}/\bar{F}$, and by 19.22 we obtain an elementary extension $\mathfrak{C}$ of $\mathfrak{A}$ isomorphic to ${}^I\mathfrak{A}/\bar{F}$. By the downward Löwenheim–Skolem theorem there is an elementary substructure $\mathfrak{B}$ of $\mathfrak{C}$ with $A \subseteq B$ and $|B| = \mathfrak{m}$. Clearly $\mathfrak{A} \preccurlyeq \mathfrak{B}$. □

Regarding the possibility of improving 19.24 by dropping the assumption that $\mathfrak{m} \geq |\mathrm{Fmla}_{\mathscr{L}}|$, the situation is somewhat complex. The theorem in general can no longer be proved then, but some strengthenings of 19.24 *are* known; see Exercise 19.49 for example.

The following two corollaries follow immediately from 19.24.

Corollary 19.25. *If a set* Γ *of sentences has an infinite model, then* Γ *has a model of each cardinality* $\geq |\mathrm{Fmla}_{\mathscr{L}}|$.

From the compactness theorem it is easy to see that if Γ has models of arbitrarily large finite cardinalities, then Γ has an infinite model, and hence by 19.25 has models of each cardinality $\geq |\mathrm{Fmla}_{\mathscr{L}}|$.

Corollary 19.26. *For any sentence* φ *(in any language* $\mathscr{L}$*),* φ *either has no infinite models or else has a model of each infinite cardinality.*

The situation is unknown concerning the cardinalities of finite models of a sentence φ. For each sentence φ, let $\mathrm{Sp}\,\varphi = \{m : m \in \omega \sim 1$, and φ has a model of power $m\}$; we call $\mathrm{Sp}\,\varphi$ the *spectrum* of φ. It is known that each set $\mathrm{Sp}\,\varphi$ is elementary, but the collection of all sets $\mathrm{Sp}\,\varphi$ is not the collection of all elementary sets. It is not known whether this collection is closed under complementation.

For later purposes we want to put the upward Löwenheim–Skolem theorem in a more general context. In fact, a weakened form of this theorem holds for many languages besides first-order languages, and is even applicable in nonlinguistic contexts. The following definition gives a very general setting for this phenomenon. It is based on the set-class distinction in set theory.

Definition 19.27. A *Hanf system* is a ternary relation R with the following properties:

(*i*) $\mathrm{pr}_0^* R$ is a set (not a proper class). Recall that $\mathrm{pr}_0^* R = \{x : \exists y, z\ (x, y, z) \in R\}$.
(*ii*) $\mathrm{pr}_2^* R$ is a class of cardinals.
(*iii*) For any Γ, $\mathfrak{A}$, there is at most one $\mathfrak{m}$ such that $(\Gamma, \mathfrak{A}, \mathfrak{m}) \in R$.

If $\mathscr{L}$ is a first-order language, then the *Hanf system* $\mathrm{H}_{\mathscr{L}}$ *of* $\mathscr{L}$ is the relation

$$\{(\Gamma, \mathfrak{A}, \mathfrak{m}) : \Gamma \text{ is a set of sentences of } \mathscr{L}, \mathfrak{A} \text{ is a model of } \Gamma, \text{ and } \mathfrak{m} = |A|\}.$$

Just as for first-order languages, one can introduce in a natural way a Hanf system for any of the more general languages considered in Part V. Here are two examples of nonlinguistic Hanf systems:

Example 1. $R = \{(i, \aleph_i, \aleph_i) : i \in \omega\} \cup \{(\omega, \mathfrak{m}, \mathfrak{m}) : \mathfrak{m} \geq \aleph_\omega\}$.

Example 2. $R = \{(0, \mathfrak{A}, |A|) : \mathfrak{A}$ is an Archimedean-ordered field$\} \cup \{(1, \mathfrak{A}, |A|) : \mathfrak{A}$ is a non-Archimedean-ordered field$\}$.

The basic theorem on Hanf systems is as follows.

Theorem 19.28. *Let R be a Hanf system. Then there is a cardinal $\mathfrak{m}$ such that for all $\Gamma \in \mathrm{pr}_0^* R$, condition (i) implies condition (ii):*

(i) there exist $\mathfrak{A}$, $\mathfrak{n}$ with $(\Gamma, \mathfrak{A}, \mathfrak{n}) \in R$ and $\mathfrak{n} \geq \mathfrak{m}$;

(ii) for every cardinal $\mathfrak{p}$ there is an $\mathfrak{n} \geq \mathfrak{p}$ and an $\mathfrak{A}$ such that $(\Gamma, \mathfrak{A}, \mathfrak{n}) \in R$ and $\mathfrak{n} \geq \mathfrak{m}$.

PROOF. For each $\Gamma \in \mathrm{pr}_0^* R$, let

$$f_\Gamma = 0 \quad \text{if } \forall\mathfrak{p}\exists\mathfrak{A}\exists\mathfrak{n}[(\Gamma, \mathfrak{A}, \mathfrak{n}) \in R \text{ and } \mathfrak{n} \geq \mathfrak{p}].$$
$$f_\Gamma = \text{least } \mathfrak{p} \text{ such that } \forall\mathfrak{A}\forall\mathfrak{n}[(\Gamma, \mathfrak{A}, \mathfrak{n}) \in R \Rightarrow \mathfrak{n} < \mathfrak{p}) \text{ otherwise.}$$

Let $\mathfrak{m} = (\bigcup \{f_\Gamma : \Gamma \in \mathrm{pr}_0^* R\})^+$. Then $\mathfrak{m}$ satisfies the condition of the theorem. In fact, suppose $\Gamma \in \mathrm{pr}_0^* R$ and (i) holds, say $(\Gamma, \mathfrak{A}, \mathfrak{n}) \in R$ with $\mathfrak{n} \geq \mathfrak{m}$. If the second condition in the definition of f_Γ holds, we would have $\mathfrak{n} < f_\Gamma < \mathfrak{m} \leq \mathfrak{n}$, contradiction. Thus the first condition holds. □

Theorem 19.28 justifies the following definition.

Definition 19.29. If R is a Hanf system, the *Hanf number* of R is the least cardinal $\mathfrak{m}$ satisfying the condition of 19.28.

Corollary 19.30. *For any first-order language $\mathscr{L}$, the Hanf number of* $\mathrm{H}_{\mathscr{L}}$ is $\aleph_0$.

This corollary is just a restatement of the upward Löwenheim–Skolem theorem in a general framework. The Hanf numbers of the systems in Examples 1 and 2 above are $\aleph_\omega$ and $(\exp \aleph_0)^+$ respectively (for Example 2, we use the fact that any Archimedean-ordered field is an ordered subfield of the reals). For each cardinal $\mathfrak{m}$ there is a Hanf system with Hanf number $\mathfrak{m}$ (see Exercise 19.50).

We proved the upward Löwenheim–Skolem theorem above by using ultraproducts. Now we want to reprove it using the compactness theorem, or more specifically, by extending the notion of diagram.

Definition 19.31. Let $\mathfrak{A}$ be an $\mathscr{L}$-structure, and $\mathscr{L}'$ an A-expansion of $\mathscr{L}$. The *elementary $\mathscr{L}'$-diagram* of $\mathfrak{A}$ is the set of *all* sentences of $\mathscr{L}'$ which hold in $(\mathfrak{A}, a)_{a \in A}$. If $\mathscr{L}''$ is an expansion of $\mathscr{L}'$, the elementary $\mathscr{L}'$-diagram of $\mathfrak{A}$ is also called the elementary $\mathscr{L}''$-diagram of $\mathfrak{A}$.

We then obtain analogs of our two theorems on diagrams; models of the elementary diagram of $\mathfrak{A}$ essentially are just those structures in which $\mathfrak{A}$ can be elementarily embedded.

Proposition 19.32. *Let $\mathfrak{A}$ be an $\mathscr{L}$-structure, $\mathfrak{B}$ an elementary extension of $\mathfrak{A}$, and $\mathscr{L}'$ an A-expansion of $\mathscr{L}$. Then $(\mathfrak{B}, a)_{a \in A}$ is a model of the elementary $\mathscr{L}'$-diagram of $\mathfrak{A}$.*

PROOF. Let the new constants of $\mathscr{L}'$ be $\mathbf{c}_a$ for $a \in A$. The following is easily proved from the definition of satisfaction:

(1) for any $x \in {}^{\omega}A$, any $m \in \omega$, and any formula φ of $\mathscr{L}$ such that $\mathrm{Fv}\,\varphi \subseteq \{v_0, \ldots, v_{m1}\}$, $\mathfrak{A} \models \varphi[x]$ iff $(\mathfrak{A}, a)_{a\in A} \models \varphi(\mathbf{c}_{x0}, \ldots, \mathbf{c}_{x(m-1)})$.

An analogous statement holds with $\mathfrak{A}$ replaced by $\mathfrak{B}$ (but no other changes). Now any sentence of $\mathscr{L}'$ can be written in the form $\varphi(\mathbf{c}_{x0}, \ldots, \mathbf{c}_{x(m-1)})$, where $m \in \omega$, φ is a formula of $\mathscr{L}$, $x \in {}^{m}A$, and $\mathrm{Fv}\,\varphi \subseteq \{v_0, \ldots, v_{m-1}\}$. If $\varphi(\mathbf{c}_{x0}, \ldots, \mathbf{c}_{x(m-1)})$ is a member of the elementary $\mathscr{L}'$-diagram of $\mathfrak{A}$, then $\mathfrak{A} \models \varphi[x]$ by (1), hence $\mathfrak{B} \models \varphi[x]$ since $\mathfrak{A} \preccurlyeq \mathfrak{B}$, and finally $\mathfrak{B} \models \varphi(\mathbf{c}_{x0}, \ldots, \mathbf{c}_{x(m-1)})$ by the $\mathfrak{B}$-analog of (1). □

Proposition 19.33. *Let $\mathfrak{A}$ be an $\mathscr{L}$-structure, $\mathscr{L}'$ an A-expansion of $\mathscr{L}$, and $(\mathfrak{B}, la)_{a\in A}$ a model of the elementary $\mathscr{L}'$-diagram Γ of $\mathfrak{A}$. Then l is an elementary embedding of $\mathfrak{A}$ into $\mathfrak{B}$.*

PROOF. Again, let the new constants of $\mathscr{L}'$ be $\mathbf{c}_a$ for $a \in A$. By Proposition 19.10, l is an embedding of $\mathfrak{A}$ into $\mathfrak{B}$. Now for any $\mathscr{L}$-formula φ, say with $\mathrm{Fv}\,\varphi \subseteq \{v_0, \ldots, v_{m-1}\}$, and for any $x \in {}^{\omega}A$,

$$\begin{aligned}\mathfrak{A} \models \varphi[x] &\Rightarrow (\mathfrak{A}, a)_{a\in A} \models \varphi(\mathbf{c}_{x0}, \ldots, \mathbf{c}_{x(m-1)})\\ &\Rightarrow \varphi(\mathbf{c}_{x0}, \ldots, \mathbf{c}_{x(m-1)}) \in \Gamma\\ &\Rightarrow (\mathfrak{B}, la)_{a\in A} \models \varphi(\mathbf{c}_{x0}, \ldots, \mathbf{c}_{x(m-1)})\\ &\Rightarrow \mathfrak{B} \models \varphi[l \circ x].\end{aligned}$$ □

Now we give the new proof of the upward Löwenheim–Skolem theorem 19.24. Assume the hypothesis of 19.24: $\mathfrak{m}$ is an infinite cardinal $\geq |\mathrm{Fmla}_{\mathscr{L}}|$, and $\mathfrak{A}$ is an infinite $\mathscr{L}$-structure with $|A| \leq \mathfrak{m}$. Let $\mathscr{L}'$ be an A-expansion of $\mathscr{L}$, and expand $\mathscr{L}'$ further to $\mathscr{L}''$ by adding new individual constants $\mathbf{k}_\xi$ for $\xi < \mathfrak{m}$. Let Γ be the elementary $\mathscr{L}'$-diagram of $\mathfrak{A}$ together with all sentences

(*) $$\neg(\mathbf{k}_\xi = \mathbf{k}_\eta) \qquad \text{for } \xi < \eta < \mathfrak{m}.$$

Every finite subset of Γ has as a model a suitable expansion of $\mathfrak{A}$. By the compactness theorem, Γ has a model $(\mathfrak{B}', la, \xi)_{a\in A, \xi<\mathfrak{m}}$. By the proof of the compactness theorem, or by applying the downward Löwenheim–Skolem theorem, we may assume that $|B'| \leq \mathfrak{m}$. Then since the sentences (*) are in Γ, $|B'| = \mathfrak{m}$. By 19.33, l is an elementary embedding of $\mathfrak{A}$ into $\mathfrak{B}$. An application of 19.22 completes the proof.

The last topic we take up in this chapter is the notion of the union of a set of structures.

Definition 19.34. Let $\mathbf{K}$ be a set of $\mathscr{L}$-structures directed by $\subseteq$, i.e., such that if $\mathfrak{A}, \mathfrak{B} \in \mathbf{K}$ then there is a $\mathfrak{C} \in \mathbf{K}$ with $\mathfrak{A} \subseteq \mathfrak{C}$ and $\mathfrak{B} \subseteq \mathfrak{C}$. Then $\bigcup \mathbf{K}$, the *union* of the members of $\mathbf{K}$, is the following $\mathscr{L}$-structure. Its universe is $\bigcup_{\mathfrak{A}\in\mathbf{K}} A$. If $\mathbf{O}$ is an m-ary operation symbol and $a \in {}^{m}\bigcup_{\mathfrak{A}\in\mathbf{K}} A$, then

$\mathbf{O}^{\cup\mathbf{K}}a_0 \cdots a_{m-1} = \mathbf{O}^{\mathfrak{A}}a_0 \cdots a_{m-1}$, where $\mathfrak{A}$ is any member of $\mathbf{K}$ such that $a_0, \ldots, a_{m-1} \in A$. If $\mathbf{R}$ is an m-ary relation symbol, then $\mathbf{R}^{\cup\mathbf{K}} = \bigcup_{\mathfrak{A}\in\mathbf{K}} \mathbf{R}^{\mathfrak{A}}$.

This definition, as well as the following basic proposition, are easily justified.

Proposition 19.35. *If* $\mathbf{K}$ *is a set of* $\mathscr{L}$*-structures directed by* $\subseteq$, *then* $\mathfrak{A} \subseteq \bigcup \mathbf{K}$ *for each* $\mathfrak{A} \in \mathbf{K}$.

The basic logical result concerning unions is the following analog of 19.35 for $\preccurlyeq$:

Theorem 19.36 (Tarski). *If* $\mathbf{K}$ *is a set of* $\mathscr{L}$*-structures directed by* $\preccurlyeq$, *then* $\mathfrak{A} \preccurlyeq \bigcup \mathbf{K}$ *for each* $\mathfrak{A} \in \mathbf{K}$.

Proof. We proceed by induction on formulas to show that for every formula φ, every $\mathfrak{A} \in \mathbf{K}$, and every $x \in {}^{\omega}A$, $\mathfrak{A} \models \varphi[x]$ iff $\bigcup \mathbf{K} \models \varphi[x]$. The case φ atomic is trivial, since $\mathfrak{A} \subseteq \bigcup \mathbf{K}$, and the induction steps for $\neg$, $\vee$, $\wedge$ are obvious. Now suppose $\mathfrak{A} \in \mathbf{K}$, $x \in {}^{\omega}A$, and $\mathfrak{A} \models \forall v_i\varphi[x]$. To show that $\bigcup \mathbf{K} \models \forall v_i\varphi[x]$, let $b \in \bigcup \mathbf{K}$ be arbitrary, say $b \in \mathfrak{B} \in \mathbf{K}$. Since $\mathbf{K}$ is directed by $\preccurlyeq$, choose $\mathfrak{C} \in \mathbf{K}$ with $\mathfrak{A} \preccurlyeq \mathfrak{C}$, $\mathfrak{B} \preccurlyeq \mathfrak{C}$. Thus $\mathfrak{A} \preccurlyeq \mathfrak{C}$ yields $\mathfrak{C} \models \forall v_i\varphi[x]$, so $\mathfrak{C} \models \varphi[x_b^i]$. By the induction assumption, $\bigcup \mathbf{K} \models \varphi[x_b^i]$, as desired. Conversely, assume that $\bigcup \mathbf{K} \models \forall v_i\varphi[x]$. For any $a \in A$, $\bigcup \mathbf{K} \models \varphi[x_a^i]$, and hence by the induction assumption, $\mathfrak{A} \models \varphi[x_a^i]$. □

EXERCISES

19.37. Let $\mathscr{L}$ have finitely many non-logical constants. Suppose that $\mathfrak{A}$ and $\mathfrak{B}$ are $\mathscr{L}$-structures, $\mathfrak{A} \equiv_{\mathrm{ee}} \mathfrak{B}$, and $\mathfrak{A}$ is finite. Show that $\mathfrak{A} \cong \mathfrak{B}$.

19.38.* Prove the result of 19.37 with $\mathscr{L}$ arbitrary.

19.39. For any language $\mathscr{L}$ there are $\mathscr{L}$-structures $\mathfrak{A}$, $\mathfrak{B}$ with $\mathfrak{A} \equiv_{\mathrm{ee}} \mathfrak{B}$ but not $(\mathfrak{A} \cong \mathfrak{B})$.

19.40.* Give an example of systems $\langle \mathfrak{A}_i : i \in I\rangle$ and $\langle \mathfrak{B}_i : i \in I\rangle$ of $\mathscr{L}$-structures and an ultrafilter F over I such that $\mathsf{P}_{i\in I}\,\mathfrak{A}_i/\bar{F} \cong \mathsf{P}_{i\in I}\,\mathfrak{B}_i/\bar{F}$ but $\forall i \in I$ not $(\mathfrak{A}_i \equiv_{\mathrm{ee}} \mathfrak{B}_i)$.

19.41. If $\mathfrak{A}$ is finite and $\mathscr{L}$ has finitely many non-logical constants, then there is a sentence φ of $\mathscr{L}$ such that

$$\{\mathfrak{B} : \mathfrak{B} \models \varphi\} = \{\mathfrak{B} : \mathfrak{A} \text{ is not embeddable in } \mathfrak{B}\}.$$

19.42. Let $\mathfrak{A}$ be an infinite $\mathscr{L}$-structure. Let $I = \{F : 0 \neq F \subseteq A \text{ and } |F| < \aleph_0\}$. For each $F \in I$ let $\mathfrak{B}_F$ be the substructure of $\mathfrak{A}$ generated by F. For each $F \in I$ let $M_F = \{G \in I : F \subseteq G\}$. Show that there is an ultrafilter $\mathscr{F}$ over I such that $M_F \in \mathscr{F}$ for each $F \in I$. Show that for any such $\mathscr{F}$, $\mathfrak{A}$ can be embedded in $\mathsf{P}_{F\in I}\,\mathfrak{B}_F/\bar{\mathscr{F}}$. From this give a new proof of Henkin's embedding theorem 19.13.

19.43. Using Henkin's embedding theorem, prove that any partial ordering can be embedded in a simple ordering.

19.44. Using Henkin's embedding theorem, prove that any Abelian group can be embedded in a divisible Abelian group.

19.45. Prove that if $\mathfrak{A} \preccurlyeq \mathfrak{C}$, $\mathfrak{B} \preccurlyeq \mathfrak{C}$ and $A \subseteq B$, then $\mathfrak{A} \preccurlyeq \mathfrak{B}$.

19.46. Give an example of structures $\mathfrak{A}$, $\mathfrak{B}$, $\mathfrak{C}$ with $\mathfrak{A} \preccurlyeq \mathfrak{B}$, $\mathfrak{A} \preccurlyeq \mathfrak{C}$, $\mathfrak{B} \subseteq \mathfrak{C}$, but not $\mathfrak{B} \preccurlyeq \mathfrak{C}$.

19.47. Give an example of structures $\mathfrak{A}$, $\mathfrak{B}$ with $\mathfrak{A} \subseteq \mathfrak{B}$, $\mathfrak{A} \cong \mathfrak{B}$, but not $\mathfrak{A} \preccurlyeq \mathfrak{B}$.

19.48. For any language $\mathscr{L}$, give an example of a set Γ of sentences of $\mathscr{L}$ such that Γ has a model $\mathfrak{A}$ of power $|\mathrm{Fmla}_{\mathscr{L}}|$ but Γ has no model of power $< |\mathrm{Fmla}_{\mathscr{L}}|$.

19.49. Using ultraproducts and unions, and assuming the generalized continuum hypothesis, show that in any language $\mathscr{L}$, any infinite $\mathscr{L}$-structure $\mathfrak{A}$ has an elementary extension of each cardinality $> |A|$.

19.50. For each cardinal $\mathfrak{m}$ (finite or infinite) there is a Hanf system with Hanf number $\mathfrak{m}$.

Nonstandard Mathematics 20*

We shall describe in this chapter the beginnings of nonstandard number theory and nonstandard analysis. Essentially we deal here with logically very simple applications of model theory. No big logical theorems will be applied, just the simplest notions of satisfaction, elementary extensions, and ultraproducts. Nonstandard mathematics was extensively developed by A. Robinson, who used a system of type theory in most of his work in this area. We shall restrict ourselves to a much simpler framework.

Let $\mathbb{R}$ be the set of real number and F any nonprincipal ultrafilter over ω. Then we can form the ultrapower ${}^{\omega}\mathbb{R}/F$ independent of any logical considerations. We know from 19.23 that the map u such that $ur = [\langle r : i \in \omega\rangle]$ for all $r \in \mathbb{R}$ is a one-one map of $\mathbb{R}$ into ${}^{\omega}\mathbb{R}/F$. Hence there is a set ${}^*\mathbb{R} \supseteq \mathbb{R}$ and a bijection $t : {}^{\omega}\mathbb{R}/F \rightarrowtail\!\!\!\!\twoheadrightarrow {}^*\mathbb{R}$ such that $t \circ u$ is the identity on $\mathbb{R}$. *Throughout this chapter we shall work with fixed F, u, ${}^*\mathbb{R}$, t satisfying these conditions.* Given any relation S on $\mathbb{R}$, say m-ary, we can take a suitable first-order language for the structure $\langle\mathbb{R}, S\rangle$ and again form the ultrapower ${}^{\omega}\langle\mathbb{R}, S\rangle/F$. There is then an m-ary relation ${}^*S$ on ${}^*\mathbb{R}$ such that t is an isomorphism from ${}^{\omega}\langle\mathbb{R}, S\rangle/F$ onto $\langle{}^*\mathbb{R}, {}^*S\rangle$. Thus u is then an elementary embedding of $\langle\mathbb{R}, S\rangle$ into ${}^{\omega}\langle\mathbb{R}, S\rangle/F$, and $\langle\mathbb{R}, S\rangle \preccurlyeq \langle{}^*\mathbb{R}, {}^*S\rangle$. In a completely analogous way operations on $\mathbb{R}$ can be extended to operations on ${}^*\mathbb{R}$. Combining any number of such extensions we obtain structures on ${}^*\mathbb{R}$ which are elementary extensions of the original structures on $\mathbb{R}$, and we can then carry over first order facts from $\mathbb{R}$ to ${}^*\mathbb{R}$ or vice-versa. In the literature the presuperscript * is frequently omitted in various contexts. For complete clarity we shall not follow this custom here. Members of ${}^*\mathbb{R}$ are called *nonstandard real numbers.*

We begin with a discussion of nonstandard number theory. Members of ${}^*\omega$ will be called *nonstandard natural numbers*. Note that if we extend the ordinary ordering $\leq$ on $\mathbb{R}$ to ${}^*\!\leq$ on ${}^*\mathbb{R}$, then ${}^*\mathbb{R}$ is linearly ordered by

$^*\leq$, since there is a sentence φ true in any structure $\langle A, S\rangle$ iff S linearly orders A. An element r of $^*\mathbb{R}$ is *infinite* if $^*|r|\ {}^*\!> m$ for each $m \in \omega$; otherwise it is *finite*.

Proposition 20.1. *There is an infinite nonstandard natural number.*

PROOF. Define $x \in {}^\omega\omega$ by $xi = i$ for all $i \in \omega$. For each $m \in \omega$, $\{i : m < xi\} = \omega \sim m \in F$. Hence $[\langle m : i \in \omega\rangle] < [x]$ in ${}^\omega\langle\mathbb{R}, <\rangle/F$, so $m\ {}^*\!< t[x]$ in $^*\langle\mathbb{R}, <\rangle$. Since m is arbitrary, this shows that $t[x]$ is infinite. Also, $\{i : xi \in \omega\} = \omega \in F$, so $t[x] \in {}^*\omega$. □

For the next result, and only for it, we need a special kind of construction. If S densely orders A (in the $\leq$ sense) with no first or last element, let

$$\begin{aligned} A' &= \{(i, 0, 0) : i \in \omega\} \cup \{(m, a, 1) : m \in \mathbb{Z}, a \in A\}, \\ S' &= \{((i, 0, 0), (j, 0, 0)) : i \leq j \text{ and } i, j \in \omega\} \\ &\quad \cup \{((i, 0, 0), (m, a, 1)) : i \in \omega, m \in \mathbb{Z}, a \in A\} \\ &\quad \cup \{((m, a, 1), (n, a, 1)) : m \leq n, m, n \in \mathbb{Z}, a \in A\} \\ &\quad \cup \{((m, a, 1), (n, b, 1)) : m, n \in \mathbb{Z}, a, b \in A, aSb, a \neq b\}. \end{aligned}$$

It is easily seen that S' linearly orders A'. Intuitively the ordering consists of a copy of ω on the left, and on the right a copy of the original ordering $\langle A, S\rangle$ with each element of A replaced by a copy of $\mathbb{Z}$ in its natural ordering. In the general theory of order types discussed in set theory, the ordering of $\langle A', S'\rangle$ has order type

$$\omega + (\omega^* + \omega)\cdot\theta,$$

where θ is the order type of $\langle A, S\rangle$.

Proposition 20.2. *The structure $\langle {}^*\omega, {}^*\!\leq\rangle$ is order-isomorphic to $\langle A', S'\rangle$ for some structure $\langle A, S\rangle$ densely ordered with no first or last element.*

PROOF. The following two facts about $\langle {}^*\omega, {}^*\!\leq\rangle$ are basic to the proof. They are true since they are true in $\langle\omega, \leq\rangle$.

(1) Every element of $^*\omega$ has an immediate successor.
(2) Every element of $^*\omega$ except 0 has an immediate predecessor.

Now let I be the set of all infinite elements of $^*\omega$. For $m, n \in I$ we define $m \equiv n$ iff ($m\ {}^*\!\leq n$ and $\{p : m\ {}^*\!< p\ {}^*\!< n\}$ is finite) or ($n\ {}^*\!\leq m$ and $\{p : n\ {}^*\!< p\ {}^*\!< m\}$ is finite). Clearly $\equiv$ is an equivalence relation on I. Let $A = I/{\equiv}$. Let $S = \{(a, b) : a, b \in A$ and there exist $m \in a$ and $n \in b$ with $m\ {}^*\!\leq n\}$. Now

(3) if aSb and $a \neq b$, then $m\ {}^*\!< n$ for all $m \in a$ and $n \in b$.

For, assume aSb and $a \neq b$. Choose $p \in a$ and $q \in b$ so that $p\ {}^*\!< q$. Since $a \neq b$, it follows that $\{s : p\ {}^*\!< s\ {}^*\!< q\}$ is infinite. Hence for all $m \in a$ and $n \in b$, not only $m\ {}^*\!< n$ but also $\{s : m\ {}^*\!< s\ {}^*\!< n\}$ is infinite.

(4) For each $m \in I$, $^*\!\leq$ orders $[m]_\equiv$ similarly to $\mathbb{Z}$.

To prove (4), for each $n \in [m]_{\equiv}$ let

$$fn = |\{p : m \mathbin{*<} p \mathbin{*<} n\}| \qquad \text{if } m \mathbin{*<} n,$$
$$fn = -|\{p : n \mathbin{*<} p \mathbin{*<} m\}| \qquad \text{if } n \mathbin{*<} m.$$

By induction using (1) and (2) one easily shows:

(5) for each $n \in \omega$ there is a unique $p \in [m]_{\equiv}$ such that $fp = n$.
(6) for each $n \in \omega$ there is a unique $p \in [m]_{\equiv}$ such that $fp = -n$.

Thus (4) follows. By (4), choose for each $a \in A$ an order isomorphism t_a of a under $*\leq$ onto $\mathbb{Z}$. Now we are ready to define the isomorphism demanded in the proposition. Take any $m \in {}^*\omega$. If m is finite, it is in ω since for each $n \in \omega$ the sentence

$$\forall v_0[v_0 \leq \mathbf{c}_n \rightarrow \bigwedge_{i \leq n} v_0 = \mathbf{c}_i]$$

holds in $\langle \omega, \leq, 0, 1, \ldots \rangle$ and hence in $\langle {}^*\omega, {}^*\leq, 0, 1, \ldots \rangle$. In this case we let $gm = (m, 0, 0)$. If m is infinite, we let $gm = (t_{[m]}m, [m], 1)$. Clearly g is an order isomorphism of $\langle {}^*\omega, {}^*\leq \rangle$ onto $\langle A', S' \rangle$. It remains to show that $\langle A, S \rangle$ is densely ordered with no first or last element. To show that $\langle A, S \rangle$ has no first element, take any $a \in A$. Say $m \in a$. Choose $n \mathbin{*<} m$ such that $n \mathbin{*+} n = m$ or $n \mathbin{*+} n \mathbin{*+} 1 = m$. Thus n is infinite since m is. For each $i \in \omega \sim 1$ we have $0 \mathbin{*<} i \mathbin{*<} n$ and hence $n \mathbin{*<} n \mathbin{*+} i \mathbin{*<} n \mathbin{*+} n$. [For the last results we have used certain obvious first order properties carrying over from ω to ${}^*\omega$.] Hence $\langle n \mathbin{*+} i : i \in \omega \sim 1 \rangle$ is a one-one function from $\omega \sim 1$ onto $\{p : n \mathbin{*<} p \mathbin{*<} m\}$, so the latter is infinite. Thus $[n]S[m]$ and $[n] \neq [m]$. Hence A has no first element. It is similarly seen that A has no last element. Finally, suppose that $a, b \in A$, aSb, and $a \neq b$. Choose $m \in a$ and $n \in b$. Thus $m \mathbin{*<} n$ and $\{p : m \mathbin{*<} p \mathbin{*<} n\}$ is infinite [see above, proof of (3)]. Suppose that $m \mathbin{*+} n$ is even; the case $m \mathbin{*+} n$ odd is treated similarly. Choose p so that $p \mathbin{*+} p = m \mathbin{*+} n$. If $p \mathbin{*\leq} m$, then

$$p \mathbin{*+} p \mathbin{*\leq} m \mathbin{*+} m \mathbin{*<} m \mathbin{*+} n = p \mathbin{*+} p,$$

contradiction. Thus $m \mathbin{*<} p$, and similarly $p \mathbin{*<} n$. Suppose $m \equiv p$. Then $p = m \mathbin{*+} i$ for some $i \in \omega$, so $m \mathbin{*+} n = p \mathbin{*+} p = m \mathbin{*+} m \mathbin{*+} 2i$ and hence $n = m \mathbin{*+} 2i$. But then $n \equiv m$, contradiction. Thus not $(m \equiv p)$, and similarly not $(p \equiv n)$. Hence $[p]$ is strictly inbetween $[m]$ and $[n]$, in the S-sense. □

An element p of ${}^*\omega$ is a *prime* if $1 \mathbin{*<} p$ and no element of ${}^*\omega$ strictly between 1 and p divides p. Thus the finite primes are exactly the usual primes. The following result illustrates the way in which number-theoretic statements can be given a nonstandard formulation.

Proposition 20.3. *The following conditions are equivalent:*

(*i*) *there are infinitely many pairs of* (*ordinary*) *primes* p, q *such that* $q = p + 2$;
(*ii*) *there is at least one pair* (p, q) *of infinite primes* p, q *such that* $q = p \mathbin{*+} 2$.

At this time the statement 20.3(i) is an open problem, usually called the *twin-prime conjecture*. Examples of twin primes are (5, 7), (29, 31), (137, 139).

PROOF. (i) $\Rightarrow$ (ii). Assume (i). Let $f \in {}^{\omega}\omega$ be such that for each i, f_i is a prime such that $f_i + 2$ is a prime, and $f_i + 2 < f_{i+1}$. Clearly $t[f]$ is an infinite prime such that $t[f] *+ 2$ is also a prime. (ii) $\Rightarrow$ (i). Say $t[g]$ is an infinite prime such that t$[g] *+ 2$ is also a prime. We show that for any $m \in \omega$ there is a prime $p > m$ such that $p + 2$ is also a prime. Since $[g]$ is infinite, $I_0 = \{i : m < gi\} \in$ F. Since $[g]$ is a prime, $I_1 = \{i : g_i \text{ is a prime}\} \in F$. Since $[g] *+ 2$ is a prime, $I_2 = \{i : g_i + 2 \text{ is a prime}\} \in F$. Thus $I_0 \cap I_1 \cap I_2 \in F$. For any $i \in I_0 \cap I_1 \cap I_2$, g_i is a prime $>m$ such that $g_i + 2$ is a prime. □

We now discuss briefly the properties of primes in $*\omega$. Clearly there are infinite primes. If a nonstandard natural number can be written as a (finite) product of primes, it can be so written in a unique way, up to order. But there are nonstandard numbers which cannot be written as a finite product of primes. An interesting example is $[\langle 2^n : n \in \omega \rangle]$, which is divisible only by the prime 2, but is not a (finite) power of 2. There is no smallest infinite prime, since if $[f]$ is infinite we may find an infinite prime $[g] < [f]$ as follows. Set

$$gi = \text{greatest prime} < f_i \qquad \text{if } f_i > 2,$$
$$gi = 0 \qquad \text{if } f_i \leq 2.$$

Now $\{i : f_i > 2\} \in F$ since $[f]$ is infinite, and

$$\{i : fi > 2\} \subseteq \{i : g_i \text{ is a prime} < f_i\},$$

so $[g]$ is a prime $<[f]$. For any prime p, $\{i : fi > p\} \in F$, and $\{i : fi > p\} \subseteq \{i : p \leq gi\}$, so $[g]$ is infinite. There are numbers divisible by infinitely many primes; an example is $[\langle n! : n \in \omega \rangle]$.

Now we turn to nonstandard analysis. A nonstandard real number r is *infinitesimal* if $*|r| *< t$ for every standard positive real number t. The reciprocal of an infinite real number is obviously infinitesimal; thus by 20.1 infinitesimals exist.

Proposition 20.4. *For any finite nonstandard real number r there is a unique standard real number s such that r *− s is infinitesimal.*

PROOF. We may assume that $r * \geq 0$. Since r is finite, there is a natural number m such that $r *\leq m$. Let s be the inf of $\{t : t \text{ is a real number and } r *\leq t\}$. Thus s is a standard real number. If $r *- s$ is not infinitesimal, choose a standard $\varepsilon > 0$ with $\varepsilon *\leq *|r *- s|$. Then $s - \varepsilon/2 *< r$, so $*|r *- s| *< \varepsilon/2$, contradiction. Thus $r *- s$ *is* infinitesimal. Suppose that t is a standard real $<s$. Let $\varepsilon = (s - t)/2$. Then $t + \varepsilon < s$, so $t + \varepsilon *< r$ and $*|r *- t| *> \varepsilon$, so $r *- t$ is not infinitesimal. Similarly, if $t > s$, t standard, then $r *- t$ is not infinitesimal. So s is unique. □

For any finite nonstandard real number r, the unique standard real s such that $r \mathbin{*-} s$ is infinitesimal is called the *standard part* of r and is denoted by st r. Two nonstandard reals a and b are said to be *infinitely close*, in symbols $a \simeq b$, if $a \mathbin{*-} b$ is infinitesimal. Clearly the difference and product of finite nonstandard reals are again finite. Hence the finite nonstandard reals form a subring ${}^{f}\mathbb{R}$ of ${}^*\mathbb{R}$. Every infinitesimal is finite. A sum or difference of infinitesimals is infinitesimal, and the product of an infinitesimal and a finite nonstandard real is again infinitesimal. The relation $\simeq$ is an equivalence relation on ${}^*\mathbb{R}$, and each equivalence class contains at most one standard real.

An infinite sequence $x = \langle x_n : n \in \omega \rangle$ of real numbers can be given two nonstandard interpretations. First, it is an element of ${}^{\omega}R$, and hence $t[x]$ is an element of ${}^*\mathbb{R}$. But also x can be considered as the set of all ordered pairs (n, xn) with $n \in \omega$, and hence as a binary relation on $\mathbb{R}$. As such it receives an interpretation ${}^*x$ as a binary relation on ${}^*\mathbb{R}$. Clearly ${}^*x$ maps ${}^*\omega$ into ${}^*\mathbb{R}$.

Proposition 20.5. *For any* $x \in {}^{\omega}\mathbb{R}$ *and* $s \in \mathbb{R}$ *the following conditions are equivalent*:

(*i*) $\lim_{n\to\infty} x_n = s$;
(*ii*) ${}^*x_n \simeq s$ *for every infinite natural number n.*

Proof. (*i*) $\Rightarrow$ (*ii*). Let $\varepsilon > 0$, ε standard. Then by (*i*) there is an $m \in \omega$ such that $|x_n - s| \leq \varepsilon$ for all $n \geq m$. Thus the following formula φ is satisfied in $\mathfrak{A} = \langle \mathbb{R}, \leq, |-|, \omega, x \rangle$ by m, s and ε assigned to v_0, v_1 and v_2, where **S** and **T** are interpreted by ω and x respectively.

$$\forall v_3 \forall v_4 [v_0 \leq v_3 \wedge \mathbf{S}v_3 \wedge \mathbf{T}(v_3, v_4) \rightarrow |v_4 - v_1| \leq v_2].$$

Since $\mathfrak{A} \preccurlyeq \langle {}^*\mathbb{R}, {}^*\leq, {}^*|-|, {}^*\omega, {}^*x \rangle = {}^*\mathfrak{A}$, it is still satisfied in ${}^*\mathfrak{A}$ by the same elements. Thus ${}^*|{}^*x_n \mathbin{*-} s| \mathbin{*\leq} \varepsilon$ for any infinite natural number n. Since ε is arbitrary, (*ii*) holds.

(*ii*) $\Rightarrow$ (*i*). Assume (*ii*). Thus for any standard $\varepsilon > 0$ the formula

$$\exists v_0(\mathbf{S}v_0 \wedge \varphi)$$

is satisfied in ${}^*\mathfrak{A}$ by s and ε assigned to v_1 and v_2, so it is satisfied in $\mathfrak{A}$ by the same assignment. Thus there is an $m \in \omega$ such that $|x_n - s| \leq \varepsilon$ whenever $n \geq m$. Since $\in$ is arbitrary, (*i*) holds. □

Corollary 20.6. *Assume that* $\lim_{n\to\infty} x_n = s$ and $\lim_{n\to\infty} y_n = t$. *Then*

(*i*) $\lim_{n\to\infty} (x_n + y_n) = s + t$;
(*ii*) $\lim_{n\to\infty} (x_n \cdot y_n) = s \cdot t$.

Proof. Let n be an infinite natural number. Then by 20.5, ${}^*x_n \mathbin{*-} s$ and ${}^*y_n \mathbin{*-} t$ are infinitesimal. Hence ${}^*(x + y)_n \mathbin{*-} (s + t) = {}^*x_n \mathbin{*+} {}^*y_n \mathbin{*-} s \mathbin{*-} t$ is also infinitesimal. So (*i*) holds by 20.5. Also,

$$\begin{aligned} {}^*(x \cdot y)_n \mathbin{*-} s \cdot t &= {}^*x_n \mathbin{*\cdot} {}^*y_n \mathbin{*-} {}^*x_n \mathbin{*\cdot} t \mathbin{*+} {}^*x_n \mathbin{*\cdot} t \mathbin{*-} s \mathbin{*\cdot} t \\ &= {}^*x_n({}^*y_n \mathbin{*-} t) \mathbin{*+} ({}^*x_n \mathbin{*-} s) \mathbin{*\cdot} t \end{aligned}$$

is infinitesimal since finite·infinitesimal + infinitesimal·finite = infinitesimal. Hence (*ii*) is true by 20.5. □

Corollary 20.6 is our first example of a standard theorem proved by nonstandard means. Note that no ε, δ-methods were involved in this proof, although of course they were essential to connect standard and nonstandard notions, in 20.5. Another example of such a proof is as follows.

Proposition 20.7 (Bolzano–Weierstrauss). *A bounded infinite sequence has at least one limit point.*

PROOF. Let $x \in {}^{\omega}\mathbb{R}$ be bounded. Thus there is an $M \in \mathbb{R}$ such that $|x_n| \leq M$ for all $n \in \omega$. Hence $*|t[x]|\ {*\leq}\ M$ also, since $\omega = \{n : |x_n| \leq M\} \in F$. Let $y = \operatorname{st} t[x]$; we claim that y is a limit point of x. Given $\varepsilon > 0$ and any positive integer m, we must find $n \geq m$ such that $|x_n - y| \leq \varepsilon$. Now $*|t[x]\ {*-}\ y|\ {*\leq}\ \varepsilon$ since $t[x]\ {*-}\ y$ is infinitesimal, so $\{n : |x_n - y| \leq \varepsilon\} \in F$. In particular, there are infinitely many n so that $|x_n - y| \leq \varepsilon$, as desired. □

The notion of a bounded sequence can be given a nonstandard formulation as follows.

Proposition 20.8. *Let $x \in {}^{\omega}\mathbb{R}$. Then the following conditions are equivalent*:

(*i*) *x is a bounded sequence*;
(*ii*) *$*x_n$ is finite for every infinite natural number n.*

PROOF. (*i*) ⇒ (*ii*). Choose a positive real number M such that $|x_n| \leq M$ for all n. Since $*\mathbb{R}$ is an elementary extension of $\mathbb{R}$ for any structures over $\mathbb{R}$, $*|{*x_n}|\ {*\leq}\ M$ for all infinite natural numbers n also. Thus (*ii*) holds.

(*ii*) ⇒ (*i*). Suppose (*i*) fails. Thus for every positive real number M there is an $m \in \omega$ such that $|x_m| \geq M$. The nonstandard version of this holds in $*\mathbb{R}$; taking a positive infinite M we obtain an $m \in {*\omega}$ such that $*|{*x_m}|\ {*\geq}\ M$. Thus $*x_m$ is infinite. □

Next we shall deal with functions mapping $\mathbb{R}$ into $\mathbb{R}$. We start with a nonstandard formulation of continuity:

Proposition 20.9. *For any f: $\mathbb{R} \to \mathbb{R}$ the following conditions are equivalent:*

(*i*) *f is continuous at a;*
(*ii*) *for all $x \simeq a$ we have $*fx \simeq fa$.*

PROOF. First assume (*i*). Thus, in $\mathbb{R}$,

(1) for every $\varepsilon > 0$ there is a $\delta > 0$ such that for all x, if $|x - a| < \delta$ then $|fx - fa| < \varepsilon$.

Now assume that $x \simeq a$. Let ε be any positive standard real number. By (1), still in $\mathbb{R}$, choose a standard $\delta > 0$ such that, in $\mathbb{R}$,

(2) for all y, if $|y - a| < \delta$ then $|fy - fa| < \varepsilon$.

Since $x *\!- a$ is infinitesimal, of course $*|x *\!- a| *\!< \delta$. Hence by (2) in $*\mathbb{R}$, $*|*fx *\!- fa| *\!< \varepsilon$. Since ε is arbitrary, $*fx \simeq fa$.

Now assume (*ii*), and let ε be any positive standard real. Let ζ be a positive infinitesimal. Then for any $x \in *\mathbb{R}$, $*|x *\!- a| *\!< \zeta$ implies that $*|x *\!- a|$ itself is infinitesimal, or 0, so $x \simeq a$ and hence by (*ii*) $*fx \simeq fa$. Thus, in $*\mathbb{R}$

(3) $\exists\zeta * > 0\, \forall x \in *\mathbb{R}(*|x *\!- a| *\!< \zeta \Rightarrow *|*fx *\!- fa| *\!< \varepsilon)$

Since the same statement holds in $\mathbb{R}$, f is continuous at a. □

We shall give a mixed, standard and nonstandard, proof for the intermediate value theorem for continuous functions:

Proposition 20.10. *Let f be a real-valued continuous function defined on a closed interval $[a, b]$, such that $fa < 0$ and $fb > 0$. Then there is a $c \in [a, b]$ such that $fc = 0$.*

PROOF. Let $c = \inf\{x \in [a, b] : fx \geq 0\}$. Thus

(1) $\forall x(x < c \Rightarrow fx < 0)$,
(2) $\forall x[c < x \Rightarrow \exists y(c \leq y < x \text{ and } fy \geq 0)]$.

Let i be a positive infinitesimal. By (1), $*f(c *\!- i) *\!< 0$. By (2), choose y so that $c *\!\leq y *\!< c *\!+ i$ and $*fy *\!\geq 0$. Now $c *\!- i \simeq c \simeq y$, so by 20.9 $*f(c *\!- i) \simeq fc \simeq *fy$. Since $*f(c *\!- i) < 0 \leq *fy$, it follows that $fc \simeq 0$; hence $fc = 0$. □

Lemma 20.11. *Let f be a real-valued continuous function defined on a closed interval $[a, b]$. Then f is bounded. Hence $*fc$ is finite for each non-standard $c \in [a, b]$.*

PROOF. Suppose f is not bounded. Then it is easy to define a sequence $x_0, x_1, \ldots$ of members of $[a, b]$ such that $|fx_i| \geq i$ for all $i \in \omega$. Let n be an infinite natural number. Since the sentence $\forall i(|fx_i| \geq i)$ holds in $\mathbb{R}$, it holds in $*\mathbb{R}$. It follows that $*f*x_n$ is infinite. By 20.8, $*x_n$ is finite. Let $y = \text{st}\, *x_n$. Then $y \simeq *x_n$, so by 20.9 $fy \simeq *f*x_n$, which is impossible since $*f*x_n$ is infinite. Thus f is bounded, i.e., $\forall c \in [a, b]\, |fc| \leq M$ for a suitable M. Since this holds in $*\mathbb{R}$, it follows that $*|*fc| *\!\leq M$ for all $c \in *[a, b]$, standard or not. □

Proposition 20.12. *Let f be a real-valued continuous function defined on a closed interval $[a, b]$. Then f attains a maximum value on this interval.*

PROOF. For any two integers $i, j \in \omega$ with $j \neq 0$, let $x_{ij} = a + (i/j)(b - a)$. Let m be an infinite natural number. Now

(1) for any standard $c \in [a, b]$ there is an $i *\!\leq m$ such that $c \simeq *x_{im}$.

For, the following statement holds in $\mathbb{R}$:

$$\forall j \in \omega \sim 1 \forall c \in [a, b] \exists i \leq j(x_{ij} \leq c \leq x_{i+1,j}).$$

This holds in $^*\mathbb{R}$ also, so there is an $i \mathbin{^*\!\le} m$ such that $^*x_{im} \mathbin{^*\!\le} c \mathbin{^*\!\le} {}^*(i \mathbin{^*\!+} 1, m)$. Now clearly $c \mathbin{^*\!-} {}^*x_{im} \mathbin{^*\!\le} 1^*/m$, and $1^*/m$ is infinitesimal, so $c \simeq {}^*x_{im}$.

Next, note that the following statement holds in R:

$$\forall j \in \omega \sim 1 \exists i \le j \forall k \le j (fx_{kj} \le fx_{ij}).$$

The statement holds in $^*\mathbb{R}$ also, so we can choose $i \mathbin{^*\!\le} m$ such that $\forall k \mathbin{^*\!\le} m$ $({}^*f{}^*x_{km} \mathbin{^*\!\le} {}^*f{}^*x_{im})$. Let $c = \operatorname{st} {}^*x_i$; note that $^*x_i \in [a, b]$ since $\forall j \in \omega \sim 1$ $\forall i \le j\, x_{ij} \in [a, b]$. Hence $c \in [a, b]$. Since $c \simeq {}^*x_i$, by 20.9 we have $fc \simeq {}^*f{}^*x_i$. Now for any $y \in [a, b]$, choose by (1) $k \mathbin{^*\!\le} m$ such that $y \simeq {}^*x_{km}$. Then by 20.9, $fy \simeq {}^*f{}^*x_{km}$, while by the above $^*f{}^*x_{km} \mathbin{^*\!\le} {}^*f{}^*x_{im} \simeq fc$. Hence $fy \le fc$. □

The above results indicate something of the flavor of nonstandard mathematics. The subject has now been rather extensively developed; one can do nonstandard algebra, complex analysis, functional analysis, topology, and even nonstandard logic.

BIBLIOGRAPHY

1. Robinson, A. *Non-standard Analysis*. Amsterdam: North-Holland (1966).

EXERCISES

20.13. Let $\mathfrak{A}$ be any model of the set of all sentences holding in $\langle \omega, \le \rangle$. Show that the order type of $\mathfrak{A}$ has the form $\omega + (\omega^* + \omega) \cdot \tau$ for some type τ. Also prove the converse.

20.14. If a nonstandard natural number is divisible by all finite primes, then it is divisible by some infinite prime.

20.15. $^*\mathbb{Z}$ is an integral domain which fails to satisfy the ascending chain condition or the descending chain condition. It is not a principal ideal domain.

20.16.* $^*\mathbb{Q}$ is isomorphic to the quotient field of $^*\mathbb{Z}$. Every element of $^*\mathbb{Q}$ is transcendental over $\mathbb{Q}$. $^*\mathbb{Q}$ is not a pure transcendental extension of $\mathbb{Q}$.

20.17. Let S be the ring of finite elements of $^*\mathbb{Q}$, and let I be the ideal of infinitesimal elements of S. Show that $S/I \cong \mathbb{R}$.

20.18. Let $x \in {}^\omega\mathbb{R}$ be bounded. Then $\operatorname{st} {}^*x_n$ is a limit point of x for every infinite natural number n.

20.19. Using 20.9, show that the sum and product of continuous functions are continuous.

20.20. A sequence $x \in {}^\omega\mathbb{R}$ converges iff $^*x_m \simeq {}^*x_n$ for all infinite m and n.

Complete Theories 21

Our main purpose in this chapter is to give concrete examples of complete theories. As a byproduct we obtain several new examples of decidable theories. We shall discuss some general model-theoretical methods for proving theories complete, mainly m-elementary extensions, the Łoś-Vaught test, and model completeness. In our examples of complete theories we shall need to assume some familiarity with several standard algebraic notions and results. No single example is crucial for succeeding chapters. The notion of a complete theory is closely connected historically and philosophically with the notion of categoricity. As we shall see, the latter notion is trivial for first-order languages, but a cardinality version of it is quite interesting.

Definition 21.1. A theory Γ is *categorical* if any two models of Γ are isomorphic. Let $\mathfrak{m}$ be a nonzero cardinal. A theory Γ is $\mathfrak{m}$-*categorical* if any two models of Γ of power $\mathfrak{m}$ are isomorphic.

Since isomorphism implies elementary equivalence, 21.1 combined with 19.2 yields

Corollary 21.2. *Any categorical theory is complete.*

Now we show that categorical theories are trivial:

Theorem 21.3. *For any consistent theory* Γ, *the following conditions are equivalent:*

(*i*) Γ *is categorical;*
(*ii*) *there is a positive integer* m *such that* Γ *is* m*-categorical and every model of* Γ *has power* m.

PROOF. Obviously (*ii*) $\Rightarrow$ (*i*). Now assume (*i*). Since any two models of Γ are isomorphic, they all have the same cardinality $\mathfrak{m}$, and Γ is $\mathfrak{m}$-categorical. Suppose $\mathfrak{m} \geq \aleph_0$. Let $\mathfrak{A}$ be a model of Γ, and by the upward Löwenheim-Skolem theorem let $\mathfrak{B}$ be an elementary extension of $\mathfrak{A}$ of power $>\mathfrak{m}$. Then $\mathfrak{A} \equiv_{ee} \mathfrak{B}$, so $\mathfrak{B}$ is also a model of Γ, contradiction. Hence $\mathfrak{m} < \aleph_0$. □

Much of the importance of the notion of $\mathfrak{m}$-categoricity stems from the following result.

Theorem 21.4 (Łoś–Vaught test for completeness). *If Γ is a theory with only infinite models and Γ is categorical in some power $\mathfrak{m} \geq |\mathrm{Fmla}_{\mathfrak{B}}|$, then Γ is complete.*

PROOF. We need to show that any two models $\mathfrak{A}$, $\mathfrak{B}$ of Γ are elementarily equivalent. By the Löwenheim–Skolem theorems there are $\mathscr{L}$-structures $\mathfrak{A}'$, $\mathfrak{B}'$ of power $\mathfrak{m}$ with $\mathfrak{A} \equiv_{ee} \mathfrak{A}'$ and $\mathfrak{B} \equiv_{ee} \mathfrak{B}'$. Thus $\mathfrak{A}'$ and $\mathfrak{B}'$ are models of Γ, so $\mathfrak{A}' \cong \mathfrak{B}'$ by assumption. Hence $\mathfrak{A} \equiv_{ee} \mathfrak{B}$. □

There are many theories which satisfy the hypothesis of 21.4, which hence gives a practical method for showing many theories complete. The assumption of categoricity in power $\geq |\mathrm{Fmla}_{\mathscr{L}}|$ is necessary in 21.4; see Exercise 21.38. As we shall see later, there are many complete theories which are not categorical in any power. We now give a few applications of 21.4. More are found in Exercises 21.39 and 21.40.

A linearly ordered set A is *densely linearly ordered* if for any $x, y \in A$ with $x < y$, $\exists a(x < a < y)$. The argument proving the following set-theoretical result is frequently used in some form in model-theoretic arguments (see also 9.47).

Theorem 21.5 (Cantor). *Let A and B be two denumerable densely linearly ordered sets, each without first or last elements. Then $A \cong B$.*

PROOF. Let $<$ and $<'$ be the dense linear orderings of A and B respectively. Say $A = \{a_n : n \in \omega\}$ and $B = \{b_n : n \in \omega\}$, with a and b one–one. We define a sequence $\langle (x_n, y_n) : n \in \omega \rangle$ by induction. Let $(x_0, y_0) = (a_0, b_0)$. Suppose $(x_0, y_0), \ldots, (x_n, y_n)$ have already been defined so that $x_0, \ldots, x_n$ are all distinct and $y_0, \ldots, y_n$ are all distinct. We distinguish two cases.

Case 1. n is odd. Let $x_{n+1} = a_m$, m the least element of $\{p : a_p \notin \{x_0, \ldots, x_n\}\}$.

We define y_{n+1} by considering several subcases.

Subcase 1. $a_m < x_i$ for all $i \leq n$. Since B does not have a first element, the set $M = \{b_p : b_p <' y_i$ for all $i \leq n\}$ is nonempty; let y_{n+1} be any element of it.

Subcase 2. $x_i < a_m$ for all $i \leq n$. This case is similar.

Subcase 3. $x_i < a_m < x_j$ for certain $i, j \leq n$. There are unique s, t such that x_s is the greatest element of $\{x_i : i \leq n, x_i < a_m\}$ and x_t is the least element of $\{x_j : j \leq n, a_m < x_j\}$. If $y_s \geq' y_t$, let $y_{n+1} = b_0$. If $y_s <' y_t$, choose y_{n+1} such that $y_s < y_{n+1} < y_t$.

Case 2. n even. Interchange A and B in Case 1.

By an easy induction on n it is established that

(1) for every $n < \omega$, $\{(x_0, y_0), \ldots, (x_n, y_n)\}$ is a one–one function such that $x_i < x_j$ iff $y_i < y_j$; $\{a_0, \ldots, a_n\} \subseteq \{x_0, \ldots, x_{2n}\}$; $\{b_0, \ldots, b_n\} \subseteq \{y_0, \ldots, y_{2n+1}\}$

Let $f = \{(x_i, y_i) : i < \omega\}$. f is the desired isomorphism. □

Since we can obviously write down finitely many axioms expressing that $<$ is a dense linear ordering on A without first element or last element, the theory of such structures is recursively axiomatizable. Hence by the Łoś–Vaught test and 21.5 we have:

Corollary 21.6. *The theory of dense linear order without first or last elements is complete and decidable.*

By Theorem 9.48 we know that any two denumerable atomless Boolean algebras are isomorphic. Hence:

Theorem 21.7. *The theory of nontrivial atomless Boolean algebras is complete and decidable.*

It is well known that for any $\mathfrak{m} > \aleph_0$, any two divisible torsion-free Abelian groups of power $\mathfrak{m}$ are isomorphic. Hence:

Theorem 21.8. *The theory of nontrivial divisible torsion-free Abelian groups is complete and decidable.*

Any field can be obtained by a pure transcendental extension of its prime field followed by a pure algebraic extension. Hence for any $\mathfrak{m} > \aleph_0$, any two algebraically closed fields of power $\mathfrak{m}$ that have the same characteristic are isomorphic. Hence:

Theorem 21.9. *For any p (p a prime or p* $= 0$), *the theory of algebraically closed fields of characteristic p is complete and decidable.*

Theorem 21.9 forms a partial justification for the heuristic Lefschetz principle in algebraic geometry, according to which results over the complex field generalize to results over any algebraically closed field of characteristic 0. Note that, by 21.9, to establish a first-order statement for all algebraically closed fields of characteristic 0 it suffices to prove it for the complex field, where one has available all the tools of classical analysis.

Next, let F be any field. We can describe a first-order language $\mathscr{L}$ appropriate for discussing vector spaces over F. Namely, $\mathscr{L}$ is to have the binary operation symbol $\mathbf{+}$, the individual constant $\mathbf{0}$, and for each $a \in F$ a unary operation symbol $\mathbf{s}_a$. Then the theory of vector spaces over F is the set of all consequences of the axioms for Abelian groups $\langle A, +, 0\rangle$ together with all sentences of the following forms:

$$\forall v_0 \forall v_1[\mathbf{s}_a(v_0 \mathbf{+} v_1) \mathbf{=} \mathbf{s}_a v_0 \mathbf{+} \mathbf{s}_a v_1] \qquad \text{for each } a \in F;$$
$$\forall v_0(\mathbf{s}_{a+b} v_0 \mathbf{=} \mathbf{s}_a v_0 \mathbf{+} \mathbf{s}_b v_0) \qquad \text{for all } a, b \in F;$$
$$\forall v_0(\mathbf{s}_{ab} v_0 \mathbf{=} \mathbf{s}_a \mathbf{s}_b v_0) \qquad \text{for all } a, b \in F;$$
$$\forall v_0(\mathbf{s}_1 v_0 \mathbf{=} v_0).$$

Now any two vector spaces over F of the same dimension are isomorphic. Hence the theory of vector spaces over F is $(|F|^+ \cup \aleph_0)$-categorical, and hence:

Theorem 21.10. *For any field F, the theory of infinite vector spaces over F is complete.*

All of the above theories Γ have the following property: if Γ is $\mathfrak{m}$-categorical in some power $\mathfrak{m} > |\mathrm{Fmla}_{\mathscr{L}}|$, then Γ is $\mathfrak{m}$-categorical in every such power. One of the deepest results in model theory, due to Morley and Shelah, is that this implication holds for all theories Γ.

We shall next present a modified notion of elementary equivalence which has been used by Dana Scott to show that certain theories are decidable or complete.

Definition 21.11. $\mathfrak{A}$ is called an *m-elementary extension* of $\mathfrak{B}$, and $\mathfrak{B}$ an *m-elementary substructure* of $\mathfrak{A}$, where $m \in \omega$, in symbols $\mathfrak{A} \succcurlyeq_m \mathfrak{B}$ or $\mathfrak{B} \preccurlyeq_m \mathfrak{A}$, if $\mathfrak{B} \subseteq \mathfrak{A}$ and if for every formula φ containing at most m distinct variables, $\varphi^{\mathfrak{B}} = \varphi^{\mathfrak{A}} \cap {}^{\omega}B$.

The following sufficient condition for m-elementary extensions will play a central role in our applications of Scott's method. It is formulated in terms of automorphisms. An *automorphism* of a structure $\mathfrak{A}$ is simply an isomorphism of $\mathfrak{A}$ onto $\mathfrak{A}$. There are many connections between logical properties and properties of automorphisms. The following proposition gives a typical connection.

Proposition 21.12. *Assume that $m \in \omega$, $\mathfrak{B} \subseteq \mathfrak{A}$, and that for any $X \subseteq B$ with $|X| < m$ and any $a \in A \sim B$ there is an automorphism f of $\mathfrak{A}$ such that $f \restriction X = \mathbf{I} \restriction X$ and $fa \in B$. Then $\mathfrak{B} \preccurlyeq_m \mathfrak{A}$.*

Proof. We proceed by induction on φ to show that $\varphi^{\mathfrak{B}} = \varphi^{\mathfrak{A}} \cap {}^{\omega}B$ for any formula φ with at most m distinct variables. The only nontrivial step is the induction step to $\varphi = \forall v_i \psi$. Obviously $\varphi^{\mathfrak{A}} \cap {}^{\omega}B \subseteq \varphi^{\mathfrak{B}}$ by the induction assumption. Now suppose that $x \in {}^{\omega}B$ but $x \notin \varphi^{\mathfrak{A}}$. Say $a \in A$ and $x_a^i \notin \psi^{\mathfrak{A}}$. Let $X = \{x_j : v_j \text{ occurs in } \psi, j \neq i\}$. Obviously we may assume that v_i actually occurs in ψ, so $|X| < m$ by our assumption on φ. Hence by hypothesis let f be an automorphism of $\mathfrak{A}$ such that $f \restriction X = \mathbf{I} \restriction X$ and $fa \in B$. By the basic theorem on isomorphisms it follows that $x^i_{fa} \notin \psi^{\mathfrak{A}}$. Since $x^i_{fa} \in {}^\omega B$, the induction hypothesis yields $x^i_{fa} \notin \psi^{\mathfrak{B}}$. Thus $x \notin \varphi^{\mathfrak{B}}$, as desired. □

Corollary 21.13. *Assume that $\mathfrak{B} \subseteq \mathfrak{A}$, and that for any $X \subseteq B$ with $|X| < \omega$, and any $a \in A$, there is an automorphism f of $\mathfrak{A}$ such that $f \restriction X = \mathbf{I} \restriction X$ and $fa \in B$. Then $\mathfrak{B} \preccurlyeq \mathfrak{A}$.*

We shall apply 21.12 to show that the theory of one equivalence relation is decidable. This is of course to be contrasted with 16.56, according to which the theory of two equivalence relations is undecidable.

Definition 21.14

(*i*) Γ_{equiv} is the set of all logical consequences of the following sentences, in a language with a single binary relation symbol $\mathbf{R}$:

$$\forall v_0 \mathbf{R} v_0 v_0$$
$$\forall v_0 \forall v_1 (\mathbf{R} v_0 v_1 \rightarrow \mathbf{R} v_1 v_0)$$
$$\forall v_0 \forall v_1 \forall v_2 (\mathbf{R} v_0 v_1 \wedge \mathbf{R} v_1 v_2 \rightarrow \mathbf{R} v_0 v_2)$$

(*ii*) For $m \in \omega \sim 1$, a model $\mathfrak{A} = \langle A, \mathbf{R}\rangle$ of Γ_{equiv} is called an *m-basic model of* Γ_{equiv} provided that for every $n \leq m$ A has at most m equivalence classes with exactly n elements, and A has no equivalence classes with $m + 1$ or more elements.

Lemma 21.15

(*i*) *For any $m \in \omega \sim 1$, Γ_{equiv} has exactly $(m + 1)^m - 1$ nonisomorphic m-basic models.*

(*ii*) *For any $m \in \omega \sim 1$, any m-basic model of Γ_{equiv} has at most $[m^2(m + 1)]/2$ elements.*

Proof

(*i*) For each n with $1 \leq n \leq m$ we allow any number $< m + 1$ of equivalence classes with exactly n elements. There must be at least one equivalence class. Hence (*i*) follows.

(*ii*) The largest m-basic model has for each $n \leq m$, $n \neq 0$, exactly m equivalence classes with exactly n elements. Thus the number of elements together is

$$m \cdot \sum_{1 \leq i \leq m} i = [m^2(m + 1)]/2. \qquad \square$$

Lemma 21.16. *Let $\mathfrak{A}$ be a model of Γ_{equiv}, and let $m \in \omega \sim 1$. Then there is a model $\mathfrak{B}$ of Γ_{equiv} such that $\mathfrak{B} \preccurlyeq_m \mathfrak{A}$ and each $\mathfrak{B}$-equivalence class contains at most m elements.*

Proof. Let f be a choice function for nonempty sets of subsets of A. For each $X \in A/R$ we define

$$gX = X \qquad \text{if } |X| \leq m,$$
$$gX = f\{Y : Y \subseteq X, |Y| = m\} \qquad \text{if } |X| > m.$$

Then let $B = \bigcup_{X \in A/R} gX$, $S = {}^2B \cap R$, $\mathfrak{B} = \langle B, S \rangle$. Thus $\mathfrak{B}$ is a model of Γ_{equiv}, $\mathfrak{B} \subseteq \mathfrak{A}$, and each $\mathfrak{B}$-equivalence class contains at most m elements. We now prepare to apply 21.12. Assume that $X \subseteq B$ and $|X| < m$, and also that $a \in A \sim B$. Since $a \in A \sim B$, it follows from our construction that $|[a]_R| > m$. Choose $b \in g[a] \sim X$, and let h be the transposition (a, b). Clearly h is an automorphism of $\mathfrak{A}$ and $ha \in B$. Hence by 21.12, $\mathfrak{B} \preccurlyeq_m \mathfrak{A}$. □

Lemma 21.17. *Let $\mathfrak{A}$ be a model of Γ_{equiv}, $m \in \omega \sim 1$, and suppose that each $\mathfrak{A}$-equivalence class has at most m elements. Then there is an m-basic model $\mathfrak{B}$ of Γ_{equiv} such that $\mathfrak{B} \preccurlyeq_m \mathfrak{A}$.*

PROOF. Let $\mathfrak{A} = \langle A, R \rangle$. For $i \in \{1, \ldots, m\}$ let $C_i = \{X : X \in A/R, |X| = i\}$. Let f be a choice function for nonempty sets of subsets of $\mathbf{S}A$. For each $i \in \{1, \ldots, m\}$ let

$$\begin{aligned} gi &= C_i \qquad \text{if } |C_i| \leq m \\ gi &= f\{\mathscr{X} : \mathscr{X} \subseteq C_i, |\mathscr{X}| = m\} \qquad \textit{otherwise.} \end{aligned}$$

Then let $B = \bigcup_{1 \leq i \leq m} \bigcup_{X \in gi} X$, $S = {}^2B \cap R$, $\mathfrak{B} = \langle B, S \rangle$. Thus $\mathfrak{B}$ is an m-basic model of Γ_{equiv} and $\mathfrak{B} \subseteq \mathfrak{A}$. Again we prepare to apply 21.12. Assume that $X \subseteq B$ and $|X| < m$, and also that $a \in A \sim B$. Let $|[a]_R| = i$. Thus $1 \leq i \leq m$ by our assumption on $\mathfrak{A}$. Since $a \in A \sim B$, it follows by our construction that $|C_i| > m$. Thus $|gi| = m$, and hence, since $|X| < m$, there is a $Y \in gi$ such that $Y \cap X = 0$. Let h map Y one–one onto $[a]_R$, $h^{-1} = h$, h the identity outside $Y \cup [a]_R$. Then h is an automorphism of $\mathfrak{A}$ and $ha \in B$, as desired. □

Lemma 21.18. *If $\mathfrak{A}$ is any model of Γ_{equiv} and $m \in \omega \sim 1$, then there is an m-basic model $\mathfrak{B}$ of Γ_{equiv} such that $\mathfrak{B} \preccurlyeq_m \mathfrak{A}$.*

Theorem 21.19. *Γ_{equiv} is decidable.*

PROOF. We first claim

(1) If φ is a formula with at most m distinct variables, then $\varphi \in \Gamma_{\text{equiv}}$ iff φ holds in every m-basic model of Γ_{equiv}.

In fact, $\Rightarrow$ is trivial, while $\Leftarrow$ is an obvious consequence of 21.18.

Now the decision method runs as follows. Given a sentence φ, determine the number m of distinct variables occurring in φ. Check through all $\mathscr{L}$-structures with universe $\subseteq [m^2(m + 1)]/2$; there are only finitely many. For each such $\mathscr{L}$-structure, check whether it is a model of Γ_{equiv}. This is possible since Γ_{equiv} is finitely axiomatizable. For each such model of Γ_{equiv} check whether φ holds. If it fails, then $\varphi \notin \Gamma_{\text{equiv}}$. If it holds for all of them, then $\varphi \in \Gamma_{\text{equiv}}$ by (1) and 21.15. □

We now turn to the notion of *model-completeness*, a widely used notion due to A. Robinson. First we give some general facts about this notion and then we give two applications to proving theories complete.

Definition 21.20. A set Γ of sentences is *model-complete* if Γ is consistent and for any two models $\mathfrak{A}$ and $\mathfrak{B}$ of Γ, $\mathfrak{A} \subseteq \mathfrak{B}$ iff $\mathfrak{A} \preccurlyeq \mathfrak{B}$ (cf. 19.20).

We want to give some equivalent forms of this definition, and to this end we need to discuss further universal and existential formulas, defined in 11.37 and 11.39.

Proposition 21.21. *Let Δ be the smallest set of formulas containing all atomic formulas and their negations and closed under the operations $\vee$, $\wedge$ and $\forall v_i$ for each $i < \omega$. Then for each $\varphi \in \Delta$ there is a universal formula ψ with* $\mathrm{Fv}\,\varphi = \mathrm{Fv}\,\psi$ *and* $\models \varphi \leftrightarrow \psi$.

PROOF. Let Θ be the set of φ for which there is such a ψ. Obviously $\Delta \subseteq \Theta$ since Θ is trivially seen to satisfy the above conditions on Δ. □

The following proposition is proved similarly:

Proposition 21.22. *Let Δ be the smallest set of formulas containing all atomic formulas and their negations and closed under the operations $\vee$, $\wedge$ and $\exists v_i$ for each $i < \omega$. Then for each $\varphi \in \Delta$ there is an existential formula ψ with* $\mathrm{Fv}\,\varphi = \mathrm{Fv}\,\psi$ *and* $\models \varphi \leftrightarrow \psi$.

Some trivial but useful properties of universal and existential formulas are given in the next proposition:

Proposition 21.23. *Assume that* $\mathfrak{A} \subseteq \mathfrak{B}$, φ is a formula, and $x \in {}^{\omega}A$.

(*i*) *if φ is universal and $\mathfrak{B} \models \varphi[x]$, then $\mathfrak{A} \models \varphi[x]$;*
(*ii*) *if φ is a universal sentence and φ holds in $\mathfrak{B}$, then φ holds in $\mathfrak{A}$;*
(*iii*) *if φ is existential and $\mathfrak{A} \models \varphi[x]$, then $\mathfrak{B} \models \varphi[x]$;*
(*iv*) *if φ is an existential sentence and φ holds in $\mathfrak{A}$, then φ holds in $\mathfrak{B}$.*

The following lemma on diagrams will be useful below.

Lemma 21.24. *Let $\mathfrak{A}$ be an $\mathscr{L}$-structure, $\mathscr{L}'$ an A-expansion of $\mathscr{L}$, Δ the $\mathscr{L}'$-diagram of $\mathfrak{A}$, and $\Gamma \cup \{\varphi\} \subseteq \mathrm{Sent}_{\mathscr{L}}$. Assume that $\Gamma \cup \Delta \models \varphi$. Then there is an existential sentence ψ of $\mathscr{L}$ such that $\mathfrak{A} \models \psi$ and $\Gamma \models \psi \rightarrow \varphi$.*

PROOF. By the deduction theorem for sentences, $\Gamma \models \chi \rightarrow \varphi$ for χ a conjunction of certain members of Δ. Let χ' be obtained from χ by replacing all of the new individual constants $\mathbf{c}_{a0}, \ldots, \mathbf{c}_{a(m-1)}$ occurring in χ by new variables $v_{i0}, \ldots, v_{i(m-1)}$. Since no new constants appear in Γ, it is clear by a simple semantic argument that $\Gamma \models \chi' \rightarrow \varphi$. Hence $\Gamma \models \exists v_{i0} \cdots \exists v_{i(m-1)} \chi' \rightarrow \varphi$. Obviously $\exists v_{i0} \cdots \exists v_{i(m-1)} \chi'$ is existential and holds in $\mathfrak{A}$. □

Lemma 21.25. *Let Γ be a model-complete set in a language $\mathscr{L}$, and let $\mathscr{L}'$ be an expansion of $\mathscr{L}$ by adjoining new individual constants. Then Γ is model-complete in $\mathscr{L}'$.*

PROOF. Let $\mathfrak{A}' = (\mathfrak{A}, l_k)_{k\in K}$ and $\mathfrak{B}' = (\mathfrak{B}, s_k)_{k\in K}$ be $\mathscr{L}'$-structures which are models of Γ such that $\mathfrak{A}' \subseteq \mathfrak{B}'$, where $\mathfrak{A}$ and $\mathfrak{B}$ are the underlying $\mathscr{L}$-structures. Thus by the definition of substructure, $\mathfrak{A} \subseteq \mathfrak{B}$ and $l_k = s_k$ for each $k \in K$. Let φ by any formula of $\mathscr{L}'$ and $x \in {}^{\omega}A$. Let $\mathbf{c}_{t0}, \ldots, \mathbf{c}_{t(m-1)}$ be the new individual constants occurring in φ, let $v_{i0}, \ldots, v_{i(m-1)}$ be new individual variables, and let φ' be the formula of $\mathscr{L}$ which is the result of substituting for $\mathbf{c}_{t0}, \ldots, \mathbf{c}_{t(m-1)}$ in φ respectively $v_{i0}, \ldots, v_{i(m-1)}$. Let $y_{ij} = l_{tj}$ for each $j < m$ and $y_h = x_h$ if $h \in \omega \sim \{i_0, \ldots, i_{m-1}\}$. Then

$$\begin{aligned}\mathfrak{A}' \models \varphi[x] \quad & \text{iff } \mathfrak{A}' \models \varphi[y] \quad \text{iff } \mathfrak{A} \models \varphi'[y]\\ & \text{iff } \mathfrak{B} \models \varphi'[y] \text{ (since } \Gamma \text{ is model-complete in } \mathscr{L})\\ & \text{iff } \mathfrak{B}' \models \varphi'[y] \quad \text{iff } \mathfrak{B}' \models \varphi[x].\end{aligned}$$ □

The following kind of formula plays a special role in the theory of model-completeness.

Definition 21.26. A formula φ is *primitive* if it has the form $\exists v_{i0} \cdots \exists v_{i(m-1)}\psi$, where ψ is a conjunction of formulas of the form $\mathbf{R}v_{j0} \cdots v_{j(n-1)}$, $v_i \equiv v_j$, $\mathbf{O}v_{j0} \cdots v_{j(n-1)} \equiv v_{jn}$, or their negations.

Thus a primitive formula is a special kind of existential formula. Note that the formula given by the proof of 21.24 is primitive.

Now we are ready to give our equivalent versions of the notion of model completeness.

Theorem 21.27. *Let Γ be a consistent set of sentences in a language $\mathscr{L}$. Then the following conditions are equivalent:*

(*i*) *Γ is model-complete;*
(*ii*) *for every model $\mathfrak{A}$ of Γ and every A-expansion $\mathscr{L}'$ of $\mathscr{L}$, $\Gamma \cup$ ($\mathscr{L}'$-diagram of $\mathfrak{A}$) is complete (hence the name* model-complete);
(*iii*) *if $\mathfrak{A}$ and $\mathscr{L}$ are models of Γ, $\mathfrak{A} \subseteq \mathfrak{B}$, φ is a universal formula, $x \in {}^{\omega}A$, and $\mathfrak{A} \models \varphi[x]$, then $\mathfrak{B} \models \varphi[x]$;*
(*iv*) *for any formula φ of $\mathscr{L}$ there is a universal formula ψ of $\mathscr{L}$ with* $\mathrm{Fv}\,\varphi = \mathrm{Fv}\,\psi$ *and $\Gamma \models \varphi \leftrightarrow \psi$;*
(*v*) *if $\mathfrak{A}$ and $\mathfrak{B}$ are models of Γ, $\mathfrak{A} \subseteq \mathfrak{B}$, φ is a primitive formula, $x \in {}^{\omega}A$, and $\mathfrak{B} \models \varphi[x]$, then $\mathfrak{A} \models \varphi[x]$.*

PROOF

(*i*) ⇒ (*ii*). Assume (*i*), let $\mathfrak{A}$ be a model of Γ, let $\mathscr{L}'$ be an A-expansion of $\mathscr{L}$, and let Δ be $\Gamma \cup$ (*$\mathscr{L}'$-diagram of* $\mathfrak{A}$). We shall show that any model $\mathfrak{B}' = (\mathfrak{B}, l_a)_{a\in A}$ of Δ is elementarily equivalent to $\mathfrak{A}' = (\mathfrak{A}, a)_{a\in A}$, which by 19.2 shows that Δ is complete. By 19.10, l is an embedding of $\mathfrak{A}$ into $\mathfrak{B}$. Hence by the definition of model completeness l is an elementary embedding, so by 19.32, $\mathfrak{B}'$ is a model of the elementary $\mathscr{L}'$-diagram of $\mathfrak{A}$. In particular, $\mathfrak{A}' \equiv \mathfrak{B}'$.

(*ii*) ⇒ (*iii*). Assume (*ii*), and let $\mathfrak{A}$, $\mathfrak{B}$, φ, and x be as in (*iii*), in particular, assume that $\mathfrak{A} \models \varphi[x]$. Choose m so that $\mathrm{Fv}\,\varphi \subseteq \{v_0, \ldots, v_{m-1}\}$. Let $\mathscr{L}'$ be an A-expansion of $\mathscr{L}$, and let Δ be $\Gamma \cup$ ($\mathscr{L}'$*-diagram of* $\mathfrak{A}$). Then $(\mathfrak{A}, a)_{a\in A} \models \varphi(\mathbf{c}_{x0}, \ldots, \mathbf{c}_{x(m-1)})$, and $(\mathfrak{A}, a)_{a\in A}$ is a model of Δ, so by (*ii*), $\Delta \models \varphi(\mathbf{c}_{x0}, \ldots, \mathbf{c}_{x(m-1)})$. But by 19.9 $(\mathfrak{B}, a)_{a\in A}$ is also a model of Δ, so $(\mathfrak{B}, a)_{a\in A} \models \varphi(\mathbf{c}_{x0}, \ldots, \mathbf{c}_{x(m-1)})$. Thus $\mathfrak{B} \models \varphi[x]$.

(*iii*) $\Rightarrow$ (*iv*). Assume (*iii*). First we establish

(1) for any existential formula φ there is a universal formula ψ such that $\mathrm{Fv}\,\varphi = \mathrm{Fv}\,\psi$ and $\Gamma \models \varphi \leftrightarrow \psi$.

Indeed, let $\mathrm{Fv}\,\varphi = \{v_{i0}, \ldots, v_{i(m-1)}\}$ with $i_0 < \cdots < i_{m-1}$. Expand $\mathscr{L}$ to $\mathscr{L}'$ by adjoining distinct new individual constants $\mathbf{c}_0, \ldots, \mathbf{c}_{m-1}$, and let $\varphi' = \mathrm{Subf}^{vi0}_{\mathbf{c}0} \cdots \mathrm{Subf}^{vi(m-1)}_{\mathbf{c}(m-1)}\varphi$. Finally, let $\Delta = \{\psi : \psi$ is a universal sentence in $\mathscr{L}'$ and $\Gamma \models \varphi' \rightarrow \varphi\}$. The following statement will be used to prove (1):

(2) $$\Gamma \cup \Delta \cup \{\neg\varphi'\} \text{ is inconsistent.}$$

To establish (2), assume the contrary; then $\Gamma \cup \Delta \cup \{\neg\varphi'\}$ has a model $\mathfrak{A}' = (\mathfrak{A}, a_i)_{i<m}$. Let $\mathscr{L}''$ be an A-expansion of $\mathscr{L}'$, and let Θ be the $\mathscr{L}''$-diagram of $\mathfrak{A}'$. To establish a contradiction yielding (2) we need in turn

(3) $$\Gamma \cup \Theta \models \neg\varphi'.$$

To prove this, let $\mathfrak{B}' = (\mathfrak{B}, s_i, t_a)_{i<m, a\in A}$ be any model of $\Gamma \cup \Theta$. Thus by 19.10, t is an embedding of $\mathfrak{A}'$ into $(\mathfrak{B}, s_i)_{i<m}$. Choose any $x \in {}^{\omega}A$ such that $x_{ij} = a_j$ for each $j < m$. Since $\mathfrak{A}' \models \neg\varphi'$, it follows that $\mathfrak{A} \models \neg\varphi[x]$. Now φ is existential, so $\models \neg\varphi \leftrightarrow \psi$ for some universal formula ψ with $\mathrm{Fv}\,\varphi = \mathrm{Fv}\,\psi$. Thus $\mathfrak{A} \models \psi[x]$. Hence by (*iii*), $\mathfrak{B} \models \psi[t \circ x]$ hence $\mathfrak{B} \models \neg\varphi[t \circ x]$, hence $\mathfrak{B}' \models \neg\varphi'$. Thus we have established (3).

From (3), a contradiction easily follows. Namely, by (3) and Lemma 21.24 there is an existential sentence θ of $\mathscr{L}'$ such that $\mathfrak{A}' \models \theta$ and $\Gamma \models \theta \rightarrow \neg\varphi'$. Hence $\Gamma \models \varphi' \rightarrow \neg\theta$. But $\models \neg\,\theta \leftrightarrow \xi$ for some universal sentence ξ of $\mathscr{L}'$, so $\Gamma \models \varphi' \rightarrow \xi$. Thus ξ is a member of our set Δ. Since $\mathfrak{A}'$ is a model of Δ, it follows that $\mathfrak{A}' \models \xi$ and hence $\mathfrak{A}' \models \neg\theta$, in contradiction to the choice of θ. Thus (2) holds after all.

From (2) and the deduction theorem we obtain $\Gamma \models \psi_0 \wedge \cdots \wedge \psi_{m-1} \rightarrow \varphi'$ for certain $\psi_0, \ldots, \psi_{m-1} \in \Delta$. But there clearly is a universal sentence μ of $\mathscr{L}'$ such that $\models \psi_0 \wedge \cdots \wedge \psi_{m-1} \leftrightarrow \mu$. Also, the presence of $\psi_0, \ldots, \psi_{m-1}$ in Δ means that $\Gamma \models \varphi' \rightarrow \psi_i$ for each $i < m$. Hence $\Gamma \models \mu \leftrightarrow \varphi'$. Replacing $\mathbf{c}_0, \ldots, \mathbf{c}_{m-1}$ by $v_{i0}, \ldots, v_{i(m-1)}$ (after perhaps changing bound variables in μ), an easy semantic argument then yields (1) for our formula φ.

To prove the full result (*iv*) involves a simple induction on φ; the atomic case and the passages using $\vee$, $\wedge$, and $\forall v_i$ are trivial, while for the passage using $\neg$ one uses (1).

(*iv*) $\Rightarrow$ (*i*). Assume (*iv*). Let $\mathfrak{A}$ and $\mathfrak{B}$ be models of Γ such that $\mathfrak{A} \subseteq \mathfrak{B}$, let $x \in {}^{\omega}A$, and let φ be any formula. By (*iv*), let ψ be a universal formula such that $\mathrm{Fv}\,\varphi = \mathrm{Fv}\,\psi$ and $\Gamma \models \varphi \leftrightarrow \psi$. Then

> $\mathfrak{B} \models \varphi[x]$ implies $\mathfrak{B} \models \psi[x]$ (since $\mathfrak{B} \models \Gamma$) implies $\mathfrak{A} \models \psi[x]$ (by 21.23(*i*)) implies $\mathfrak{A} \models \varphi[x]$.

(*iii*) ⇒ (*v*). Trivial.

(*v*) ⇒ (*iii*). To prove this implication, it suffices to establish the following quite general logical fact: for any existential formula φ there is a system $\langle\psi_i : i < m\rangle$ of primitive formulas such that $\models\varphi \leftrightarrow \bigvee_{i<m}\psi_i$. The procedure for showing this is similar to our method for eliminating operation symbols in Chapter 11; see 11.26. Namely, φ has the form $\exists v_{i0}\cdots\exists v_{i(m-1)}\varphi'$, where φ' is quantifier-free. Now we may assume that φ' is in disjunctive normal form, and

(4) the atomic parts of φ' have the forms $\mathbf{R}v_{j0}\cdots v_{j(n-1)}$ or $\mathbf{O}v_{j0}\cdots v_{j(n-1)} = v_{jn}$.

This is easily seen from the following logical validities, where $\alpha_0, \ldots, \alpha_{n-1}, \beta$ are new variables:

$$\models\mathbf{R}\sigma_0\cdots\sigma_{n-1} \leftrightarrow \exists\alpha_0\cdots\exists\alpha_{n-1}(\mathbf{R}\alpha_0\cdots\alpha_{n-1} \wedge \sigma_0 = \alpha_0 \wedge\cdots\wedge \sigma_{n-1} = \alpha_{n-1});$$
$$\models\neg\mathbf{R}\sigma_0\cdots\sigma_{n-1} \leftrightarrow \exists\alpha_0\cdots\exists\alpha_{n-1}(\neg\mathbf{R}\alpha_0\cdots\alpha_{n-1} \wedge \sigma_0 = \alpha_0 \wedge\cdots\wedge \sigma_{n-1} = \alpha_{n-1});$$
$$\models\sigma = \tau \leftrightarrow \exists\beta(\sigma = \beta \wedge \tau = \beta);$$
$$\models\neg(\sigma = \tau) \leftrightarrow \exists\beta(\sigma = \beta \wedge \neg(\tau = \beta));$$
$$\models\mathbf{O}\sigma_0\cdots\sigma_{n-1} = \beta \leftrightarrow \exists\alpha_0\cdots\exists\alpha_{n-1}(\mathbf{O}\alpha_0\cdots\alpha_{n-1} = \beta \wedge \sigma_0 = \alpha_0 \wedge\cdots\wedge \sigma_{n-1} = \alpha_{n-1});$$
$$\models\neg(\mathbf{O}\sigma_0\cdots\sigma_{n-1} = \beta) \leftrightarrow \exists\alpha_0\cdots\exists\alpha_{n-1}(\neg\mathbf{O}\alpha_0\cdots\alpha_{n-1} = \beta \wedge \sigma_0 = \alpha_0 \wedge\cdots\wedge \sigma_{n-1} = \alpha_{n-1}).$$

From (4), making use of the validity $\models\exists\alpha(\varphi \vee \psi) \leftrightarrow \exists\alpha\varphi \vee \exists\alpha\psi$, our logical fact above follows. □

By 21.27(*iv*) we see that model completeness is related to elimination of quantifiers: quantifiers can "almost" be eliminated. In the exercises it is shown that a theory can be model complete without being complete, and complete without being model complete. We now turn, however, to conditions which, when added to model completeness, give completeness. These conditions are important in their own right.

Definition 21.28. $\mathfrak{A}$ is a *prime model* of Γ if $\mathfrak{A}$ is a model of Γ and $\mathfrak{A}$ can be embedded in any model of Γ.

Proposition 21.29. *If Γ is model complete and has a prime model, then Γ is complete.*

Proof. Let $\mathfrak{A}$ and $\mathfrak{B}$ be any two models of Γ; we show that $\mathfrak{A} \equiv \mathfrak{B}$. Let $\mathfrak{C}$ be a prime model of Γ. Thus there are embeddings f and g of $\mathfrak{C}$ into $\mathfrak{A}$ and $\mathfrak{B}$ respectively. Since Γ is model complete, f and g are actually elementary embeddings. Hence $\mathfrak{A} \equiv \mathfrak{C} \equiv \mathfrak{B}$. □

Definition 21.30. A theory Γ has the *joint extension property* if any two models of Γ can be embedded in a model of Γ.

The following proposition is proved just like 21.29:

Proposition 21.31. *If* Γ *is model complete and has the joint extension property, then* Γ *is complete.*

As our final theoretical result concerning model completeness, we shall use the concept to give a highly useful purely mathematical condition for completeness.

Theorem 21.32. *Let* Γ *be a theory satisfying the following two conditions:*

(*i*) *if* $\mathfrak{A}$ *and* $\mathfrak{B}$ *are any two models of* Γ, *then every finitely generated substructure of* $\mathfrak{A}$ *is embeddable in* $\mathfrak{B}$;

(*ii*) *if* $\mathfrak{A}$ *and* $\mathfrak{B}$ *are any two models of* Γ, $\mathfrak{A} \subseteq \mathfrak{B}$, *and* $\mathfrak{B}'$ *is a finitely generated substructure of* $\mathfrak{B}$, *then there is an embedding* f *of* $\mathfrak{B}'$ *into* $\mathfrak{A}$ *such that* $f \upharpoonright A \cap B' = \mathbf{I} \upharpoonright A \cap B'$.

Then Γ *is complete.*

Proof. First we show that Γ is model-complete; to prove this we shall apply 21.27(*iii*). Assume, then, that $\mathfrak{A}$ and $\mathfrak{B}$ are models of Γ, $\mathfrak{A} \subseteq \mathfrak{B}$, $x \in {}^{\omega}A$, φ is a universal formula, and not ($\mathfrak{B} \models \varphi[x]$); we want to show that not ($\mathfrak{A} \models \varphi[x]$). Say $\varphi = \forall v_{i0} \cdots \forall v_{i(m-1)}\psi$ with ψ quantifier free. Then there is a $y \in {}^{\omega}B$ with not ($\mathfrak{B} \models \psi[y]$) and $y_j = x_j$ for all $j \in \omega \sim \{i_0, \ldots, i_{m-1}\}$. Let $M = \{y_j : v_j \text{ occurs in } \varphi\}$, and let $\mathfrak{B}'$ be the substructure of $\mathfrak{B}$ generated by M. Choose $z \in {}^{\omega}B'$ with $z_j = y_j$ whenever v_j occurs in φ. By our assumption (*ii*), there is an embedding f of $\mathfrak{B}'$ into $\mathfrak{A}$ such that $f \upharpoonright A \cap B' = \mathbf{I} \upharpoonright A \cap B'$. Since ψ is quantifier-free, $\mathfrak{B}' \models \neg\psi[z]$ and so $\mathfrak{A} \models \neg\psi[f \circ z]$ and $\mathfrak{A} \models \neg\varphi[f \circ z]$. Since $(f \circ z)_j = x_j$ for v_j occurring free in φ, $\mathfrak{A} \models \neg\varphi[x]$, as desired.

Now to show that Γ is complete it suffices by 21.27(*iv*) to show that if a universal sentence holds in one model of Γ then it holds in any model of Γ. Let $\varphi = \forall v_{i0} \cdots \forall v_{i(m-1)}\psi$ be a universal sentence, with ψ quantifier-free; suppose that $\mathfrak{A}$ and $\mathfrak{B}$ are models of Γ and $\mathfrak{A} \models \neg\varphi$. Choose $x \in {}^{\omega}A$ such that $\mathfrak{A} \models \neg\psi[x]$, and let $\mathfrak{A}'$ be the substructure of $\mathfrak{A}$ generated by $\{x_j : v_j \text{ occurs in } \psi\}$. We may assume that $x \in {}^{\omega}A'$. By our condition (*i*), let f be an embedding of $\mathfrak{A}'$ into $\mathfrak{B}$. Then $\mathfrak{A}' \models \neg\psi[x]$, hence $\mathfrak{B} \models \neg\psi[f \circ x]$, hence $\mathfrak{B} \models \neg\varphi$. □

We finish this chapter with two important examples of complete theories proved via model completeness: the theory of infinite atomic Boolean algebras, and the theory of real-closed fields.

To fix the notation for the logical theory of Boolean algebras we introduce the following definitions.

Definition 21.33

(*i*) $\mathscr{L}_{\mathrm{BA}}$ is a first-order language for Boolean algebras; the only non-logical constants are $+$, $\cdot$, $-$, $\mathbf{0}$, $\mathbf{1}$, operation symbols of ranks 2, 2, 1, 0, 0 respectively.

(*ii*) Γ_{BA}, the theory of Boolean algebras, has as axioms the formal sentences corresponding to the axioms given in 9.3.

(*iii*) φ_{At} is the following formula of $\mathscr{L}_{\mathrm{BA}}$:

$$\neg(v_0 = 0) \wedge \forall v_1(v_1 \cdot v_0 = v_1 \rightarrow v_1 = \mathbf{0} \vee v_1 = v_0).$$

Thus Fv $\varphi_{\mathrm{At}} = \{v_0\}$ and for any BA $\mathfrak{A}$, ${}^1\varphi^{\mathfrak{A}}$ is the set of all atoms of $\mathfrak{A}$.

(*iv*) Γ_{At} is the extension of Γ_{BA} with the additional axiom

$$\forall v_1\{\neg(v_1 = \mathbf{0}) \rightarrow \exists v_0[\varphi_{\mathrm{At}}(v_0) \wedge v_0 \cdot v_1 = v_0]\}.$$

Thus $\mathfrak{A}$ is a model of Γ_{At} iff $\mathfrak{A}$ is an atomic BA.

(*v*) $\Gamma_{\mathrm{At}}^{\infty}$ is the extension of Γ_{At} with the following additional axioms (for each $m \in \omega \sim 1$):

$$\exists v_0 \cdots \exists v_{m-1} \bigwedge_{i<j<m} \neg(v_i = v_j).$$

Thus $\mathfrak{A}$ is a model of $\Gamma_{\mathrm{At}}^{\infty}$ iff $\mathfrak{A}$ is an infinite atomic **BA**.

(*vi*) $\mathscr{L}_{\mathrm{At}}$ is an expansion of $\mathscr{L}_{\mathrm{BA}}$ obtained by adjoining a new unary relation symbol **P**.

(*vii*) ${}_0\Gamma_{\mathrm{At}}$ (resp. ${}_0\Gamma_{\mathrm{At}}^{\infty}$) is obtained from Γ_{At} (resp. $\Gamma_{\mathrm{At}}^{\infty}$) by adjoining the following sentence as an axiom:

$$\forall v_0[\mathbf{P}v_0 \leftrightarrow \varphi_{\mathrm{At}}(v_0)].$$

It is obvious that $(\mathscr{L}_{\mathrm{At}}, {}_0\Gamma_{\mathrm{At}})$ is a definitional expansion of $(\mathscr{L}_{\mathrm{BA}}, \Gamma_{\mathrm{At}})$, and $(\mathscr{L}_{\mathrm{At}}, {}_0\Gamma_{\mathrm{At}}^{\infty})$ is a definitional expansion of $(\mathscr{L}_{\mathrm{BA}}, \Gamma_{\mathrm{At}}^{\infty})$. We shall actually show that ${}_0\Gamma_{\mathrm{At}}^{\infty}$ is complete, from which the completeness of $\Gamma_{\mathrm{At}}^{\infty}$ is clear. To prove ${}_0\Gamma_{\mathrm{At}}^{\infty}$ complete we apply 21.32, which shows that ${}_0\Gamma_{\mathrm{At}}^{\infty}$ is also model-complete (see the proof of 21.32). It is easy to see that $\Gamma_{\mathrm{At}}^{\infty}$ itself is *not* model-complete (see Exercise 21.47). The trick of adding a defined symbol to convert a theory into a model-complete theory is rather common.

Theorem 21.34. *The theory of infinite atomic Boolean algebras is complete.*

Proof. As indicated above, it suffices to use 21.32 to show that ${}_0\Gamma_{\mathrm{At}}^{\infty}$ is complete. First we verify 21.32(*i*). To this end, let $(\mathfrak{A}, P)$ and $(\mathfrak{B}, Q)$ be two models of ${}_0\Gamma_{\mathrm{At}}^{\infty}$, and let $(\mathfrak{A}', P')$ be a finitely generated substructure of $(\mathfrak{A}, P)$. By exercise 9.65 we know that A' is finite, so by 9.29 $\mathfrak{A}'$ is atomic. Let $a_0, \ldots, a_{m-1}$ be all of the atoms of $\mathfrak{A}'$; say $a_0, \ldots, a_{n-1}$ are also atoms of $\mathfrak{A}$, while $a_n, \ldots, a_{m-1}$ are not atoms of $\mathfrak{A}$. Thus $n = 0$ is possible, but $n < m$, since otherwise $a_0, \ldots, a_{m-1}$ would be all of the atoms of $\mathfrak{A}$ and hence, as is easily seen, $\mathfrak{A}$ would be finite. Let $R_0, \ldots, R_{m-1}$ be a partition of Q into nonempty sets such that $|R_0| = \cdots = |R_{n-1}| = 1$, $|R_n| = \cdots = |R_{m-2}| = 2$, and hence $|R_{m-1}| = |Q| \geq \aleph_0$. For each $j < m - 1$, let $b_j = \Sigma_{x \in Rj}\, x$, and let $b_{m-1} = -(b_0 + \cdots + b_{m-2})$. Then

(1) $$b_j \cdot b_k = 0 \qquad \text{if } j < k < m$$

(2) $$\Sigma_{i<m}\, b_i = 1.$$

Now for any $x \in A'$ we set $fx = \Sigma\{b_j : a_j \leq x\}$. It is easily seen that f embeds $\mathfrak{A}'$ into $\mathfrak{B}$. We have to also check that f preserves **P**. For any $x \in A'$,

$$\begin{aligned} x \in P' \quad & \text{iff } x \in P \quad \text{iff } x \text{ is an atom of } \mathfrak{A} \\ & \text{iff } \exists j < n\,(x = a_j) \quad \text{iff } fx \text{ is an atom of } \mathfrak{B} \\ & \text{iff } fx \in Q. \end{aligned}$$

Thus 21.32(*i*) holds.

To verify 21.32(*ii*), assume that $(\mathfrak{A}, P)$ and $(\mathfrak{B}, Q)$ are models of ${}_0\Gamma^{\infty}_{\mathrm{At}}$, that $(\mathfrak{A}, P) \subseteq (\mathfrak{B}, Q)$, and that $(\mathfrak{B}', Q')$ is a finitely generated substructure of $(\mathfrak{B}, Q)$. Thus, again, $\mathfrak{B}'$ is a finite BA. Now $A' = A \cap B'$ is the universe of a subalgebra $\mathfrak{A}'$ of $\mathfrak{A}$. Let $a_0, \ldots, a_{m-1}$ be a list of all of the atoms of $\mathfrak{A}'$. Then by 9.34 we have $\Sigma_{i<m} a_i = 1$. Hence by 9.44 we have:

$$(3) \qquad \begin{aligned} & \mathfrak{A} \cong \mathsf{P}_{i<m}\, \mathfrak{A} \restriction a_i;\ \mathfrak{A}' \cong \mathsf{P}_{i<m}\, \mathfrak{A}' \restriction a_i;\ \mathfrak{B} \cong \mathsf{P}_{i<m}\, \mathfrak{B} \restriction a_i, \text{ and} \\ & \mathfrak{B}' = \mathsf{P}_{i<m}\, \mathfrak{B}' \restriction a_i. \end{aligned}$$

Now

$$(4) \qquad \begin{aligned} & \text{for each } i < m, \text{ either } \mathfrak{A} \restriction a_i \cong \mathfrak{B} \restriction a_i \text{ or else both } \mathfrak{A} \restriction a_i \text{ and} \\ & \mathfrak{B} \restriction a_i \text{ are infinite atomic BA's.} \end{aligned}$$

For, let $i < m$; there are then two cases which present themselves.

Case 1. $\mathfrak{A} \restriction a_i$ is finite. Then $\mathfrak{A} \restriction a_i$ is atomic; let $e_0, \ldots, e_{n-1}$ be all of the atoms of $\mathfrak{A} \restriction e_i$. Thus $\mathfrak{A} \models \varphi_{\mathrm{At}}[e_j]$ for each $j < n$, so $e_j \in P \subseteq \mathrm{Q}$ and hence e_j is an atom of $\mathfrak{B}$. Also, $\Sigma_{j<n} e_j = a_i$. Hence $\mathfrak{B} \restriction a_i$ is also an atomic BA with exactly n atoms. Therefore $\mathfrak{A} \restriction a_i \cong \mathfrak{B} \restriction b_i$.

Case 2. $\mathfrak{A} \restriction a_i$ is infinite. Then $\mathfrak{B} \restriction a_i$ is also infinite; and both $\mathfrak{A} \restriction a_i$ and $\mathfrak{B} \restriction a_i$ are atomic.

Hence (4) holds. By (4), for each $i < m$ the BA $\mathfrak{B}' \restriction a_i$ can be embedded in $\mathfrak{A} \restriction a_i$. Hence by (3) $\mathfrak{B}'$ can be embedded in $\mathfrak{A}$ so that elements of $B' \cap A$ are fixed. □

Now we turn to the theory of real-closed ordered fields. We shall establish the famous result of Tarski that this theory, too, is complete. We assume only a knowledge of real-closed ordered fields such as is found in van der Waerden. The language $\mathscr{L}$ which we deal with has operation symbols $\mathbf{+}$, $\mathbf{\cdot}$, $\mathbf{-}$, $\mathbf{0}$, $\mathbf{1}$ of ranks 2, 2, 1, 0, 0 respectively, and a binary relation symbol $\mathbf{\leq}$. We begin with a purely algebraic lemma not found in van der Waerden.

Lemma 21.35. *Let $\mathfrak{A}$ be a real-closed ordered field and let $\mathfrak{A}(b)$ be a simple transcendental ordered extension of $\mathfrak{A}$; denote the order of $\mathfrak{A}(b)$ by $<$. If $\prec$ is any order on $\mathfrak{A}(b)$ which makes it into an ordered field, and if $\forall a \in A$ $(a < b \Leftrightarrow a \prec b)$, then $< \, = \, \prec$.*

Proof. It suffices to show that $\mathfrak{A}(b)$ has the same positive elements under $<$ and $\prec$, since $a < c$ iff $0 < c - a$ and similarly for $\prec$. Note that for any $a \in A$,

$$\begin{aligned} b < a \quad & \text{iff not}(a < b), \text{ since } a \neq b; \\ & \text{iff not}(a \prec b), \text{ by assumption}; \\ & \text{iff } b \prec a. \end{aligned}$$

Now consider any nonzero element c of $A(b)$. We may write

$$c = a[f(b)/g(b)],$$

where $a \in A$ and $f(x)$ and $g(x)$ are monic polynomials over $\mathfrak{A}$. Since $\mathfrak{A}$ is real-closed, $f(x)$ and $g(x)$ split over $\mathfrak{A}$ into linear and quadratic factors; say

$$f(x) = \prod_{i<m} (x - d_i) \cdot \prod_{j<n} h_j(x),$$

$$g(x) = \prod_{i<p} (x - e_i) \cdot \prod_{j<q} k_j(x),$$

where $d_i, e_i \in A$ and $h_j(x)$, $k_j(x)$ are monic irreducible quadratic polynomials over $\mathfrak{A}$. Now we claim

(1) $\qquad h_j(b) > 0 \qquad$ for each $j < n$.

For, write $h_j(x) = x^2 + ux + v$. If $s \in A$ and $s > |u| + |v| + 1$, then

$$s^2 + us + v = s(s + u) + v > s + v > 0.$$

Thus $h_j(s) > 0$. Since h_j has no roots in $\mathfrak{A}$, it follows by the Weierstrauss Nullstellensatz that $h_j(s) > 0$ for all $s \in A$. Now write $h_j(x) = (x + \frac{1}{2}u)^2 + v - \frac{1}{4}u^2$. Then

$$0 < h_j(-\tfrac{1}{2}u) = v - \tfrac{1}{4}u^2.$$

From this it follows that $h_j(b) = (b + \frac{1}{2}u)^2 + v - \frac{1}{4}u^2 \geq v - \frac{1}{4}u^2 > 0$, as desired in (1).

Similarly to (1) we have

(2) $\quad h_j(b) \succ 0$ for each $j < n$; $k_j(b) > 0$ and $k_j(b) \succ 0$ for each $j < q$.

But from (1) and (2) it follows that the positiveness of $f(b)$ in the $<$ sense is determined by the order of b among the d_i in the $<$ sense, and similarly for $\prec$ and for $g(b)$. The order in $\mathfrak{A}$ is unique, so the lemma follows. □

Recall that the ordered field of real algebraic numbers can be embedded in any real-closed field. Thus by Proposition 21.29 we only need to show model-completeness:

Theorem 21.36. *The theory of real-closed ordered fields is model complete and hence complete and decidable.*

PROOF. We shall apply 21.27(*v*). To this end, let $\mathfrak{A}$ and $\mathfrak{B}$ be real-closed ordered fields with $\mathfrak{A} \subseteq \mathfrak{B}$; assume that $x \in {}^{\omega}A$, φ is a primitive formula, and $\mathfrak{B} \models \varphi[x]$; we want to show that $\mathfrak{A} \models \varphi[x]$. Say φ is the formula $\exists v_{i0} \cdots \exists v_{i(m-1)}\psi$, with ψ quantifier free, as in 21.26. Say $\mathfrak{B} \models \psi[y]$, where $x_j = y_j$ for all $j \notin \{i_0, \ldots, i_{m-1}\}$. Let $\mathfrak{B}'$ be the subfield of $\mathfrak{B}$ generated by $A \cup \{y_{i0}, \ldots, y_{i(m-1)}\}$. Then, of course, $\mathfrak{B}'$ has finite transcendence degree over $\mathfrak{A}$. Let $b_0, \ldots, b_{n-1}$ be algebraically independent over $\mathfrak{A}$ in $\mathfrak{B}'$, with $\mathfrak{B}'$ an algebraic extension of $\mathfrak{A}(b_0, \ldots, b_{n-1})$. Now we define a sequence

$\mathfrak{C}_0, \ldots, \mathfrak{C}_n$ of subfields of $\mathfrak{B}$. Let $\mathfrak{C}_0 = \mathfrak{A}$. Having defined $\mathfrak{C}_i$, let $\mathfrak{C}_{i+1}$ be the real-closure in $\mathfrak{B}$ of $\mathfrak{C}_i(b_i)$. Thus $\mathfrak{C}_0, \ldots, \mathfrak{C}_n$ are real-closed ordered fields, $\mathfrak{C}_0 = \mathfrak{A}$, $\mathfrak{C}_0 \subseteq \cdots \subseteq \mathfrak{C}_n$, $\mathfrak{C}_{i+1}$ has transcendence degree one over $\mathfrak{C}_i$, and $\mathfrak{C}_n \models \psi[y]$. We show by (downward) induction that $\mathfrak{C}_i \models \varphi[x]$ for each $i \leq n$. This is true for $i = n$; now assume it true for $i > 0$, and we prove it for $i - 1$.

Let $\mathscr{L}'$ be a C_{i-1}-expansion of $\mathscr{L}$, and $\mathscr{L}''$ a $\{b_{i-1}\}$-expansion of $\mathscr{L}'$. Let $X = C_{i-1} \cup \{b_{i-1}\}$. Let Γ be a set of axioms for real-closed ordered fields (in $\mathfrak{L}$), let Δ be the $\mathscr{L}'$-diagram of $\mathfrak{C}_{i-1}$, and let Θ be the set of all sentences

$$\begin{array}{ll} \mathbf{c}_z < \mathbf{c}_{b(i-1)} & \text{for } z \in C_{i-1},\ z < b_{i-1}, \\ \mathbf{c}_{b(i-1)} < \mathbf{c}_z & \text{for } z \in C_{i-1},\ b_{i-1} < z. \end{array}$$

Let φ' be φ with each free variable v_i replaced by $\mathbf{c}_{xi}$. We now claim

(1) $$\Gamma \cup \Delta \cup \Theta \models \varphi'.$$

In fact, let $(\mathfrak{D}, l_z)_{z\in X}$ be any model of $\Gamma \cup \Delta \cup \Theta$. Then by 19.10, $l \restriction C_{i-1}$ is an isomorphism of $\mathfrak{A}$ onto a subfield $\mathfrak{E}$ of $\mathfrak{D}$. Clearly $l_{b(i-1)} \notin E$, and both $\mathfrak{E}$ and $\mathfrak{D}$ are real-closed, so $l_{b(i-1)}$ is transcendental over $\mathfrak{E}$. Now $l \restriction C_{i-1}$ extends to a field isomorphism t of $C_{i-1}(b_{i-1})$ onto $\mathfrak{E}(l_{b(i-1)})$, and, since $(\mathfrak{D}, l_z)_{z\in X}$ is a model of Θ it follows that t is also an order isomorphism. Therefore, t extends on up to an (order-) isomorphism between C_i and the real-closure $\mathfrak{F}$ of $\mathfrak{E}(l_{b(i-1)})$ in $\mathfrak{D}$. Thus $(\mathfrak{F}, l_z)_{z\in X} \models \varphi'$ and hence, since φ' is existential, $(\mathfrak{D}, l_z)_{z\in X} \models \varphi'$ also. Thus (1) holds.

From (1) we obtain

(2) $$\Gamma \cup \Delta \models \chi \rightarrow \varphi',$$

where χ is a conjunction of members of Θ. Say χ is the sentence

$$\bigwedge_{a\in y} (\mathbf{c}_a < \mathbf{c}_{b(i-1)}) \wedge \bigwedge_{a\in z} (\mathbf{c}_{b(i-1)}) < \mathbf{c}_a).$$

Since $\mathbf{c}_{b(i-1)}$ does not occur in formulas of $\Gamma \cup \Delta \cup \{\varphi'\}$, it follows from (2) that

(3) $$\Gamma \cup \Delta \models \exists v_0[\bigwedge_{a\in y} (\mathbf{c}_a < v_0) \wedge \bigwedge_{a\in z} (v_0 < \mathbf{c}_a)] \rightarrow \varphi'.$$

But $(\mathfrak{C}_i, z)_{z\in X} \models \psi$ for each $\psi \in \Theta$, so $(\mathfrak{C}_i, z)_{z\in X} \models \bigwedge_{a\in Y} \bigwedge_{b\in Z} (\mathbf{c}_a < \mathbf{c}_b)$. Furthermore, the ordering of $\mathfrak{C}_{i-1}$ is dense, so there is a $d \in C_{i-1}$ such that $\forall a \in Y \forall b \in Z (a < d < b)$. Hence from (3), $\Gamma \cup \Delta \models \varphi'$. It follows that $\mathfrak{C}_{i-1} \models \varphi[x]$. □

Since the reals are the most important example of a real-closed field, it follows from 21.36 that the theory of the ordered field of real numbers is decidable. This has many practical consequences, since many practical problems in mathematics can be so formulated as to apply this decision procedure or some of its consequences.

EXERCISES

21.37. Give an example of a theory which has infinite models, is categorical in some power $\geq |\mathrm{Fmla}_{\mathscr{L}}|$, but is incomplete.

21.38. Construct an incomplete theory Γ in a language with $\aleph_1$ symbols such that Γ has only infinite models, Γ has denumerable models, and Γ is $\aleph_0$-categorical.

21.39. Using the Łoś–Vaught test, show that the theory of dense linear order with first and last elements is complete and decidable.

21.40. Let $\mathscr{L}$ be the language with nonlogical constants **s** and **0**, unary and 0-ary operation symbols respectively. Let Γ be the theory with the following axioms:

$$\begin{aligned}
&\forall v_0 \neg (\mathbf{s}v_0 = \mathbf{0})\\
&\forall v_0 \forall v_1 (\mathbf{s}v_0 = \mathbf{s}v_1 \rightarrow v_0 = v_1)\\
&\forall v_0 \neg \mathbf{s}^n v_0 = v_0 \qquad \text{for each } n \in \omega \sim 1,\\
&\forall v_0 [\neg v_0 = \mathbf{0} \rightarrow \exists v_1 (\mathbf{s}v_1 = v_0)].
\end{aligned}$$

Using the Łoś–Vaught test, show that Γ is complete and decidable (cf. Chapter 13).

21.41. Show that the theory of one equivalence relation is not $\mathfrak{m}$-categorical for any $\mathfrak{m} > 1$.

21.42. The theory of algebraically closed fields is model complete but not complete.

21.43. The theory of dense linear order is decidable.

21.44. The theory of discrete linear order with first but no last element is complete and decidable. *Hint*: the axioms are:

Linear order
First element
No last element
$\forall v_0 \forall v_1 [v_0 < v_1 \rightarrow \exists v_2 (v_2$ immediately follows $v_0) \wedge \exists v_2 (v_1$ immediately follows $v_2)]$

Take a definitional expansion $(\mathscr{L}, \Gamma)$ with **0** and **s**. Show that Γ is model-complete and has a prime model.

21.45. The theory of infinite atomic BA's is not $\mathfrak{m}$-categorical for any infinite $\mathfrak{m}$.

21.46. The theory in 21.44 is not $\mathfrak{m}$-categorical for any $\mathfrak{m}$.

21.47. The theory of infinite atomic BA's is not model-complete.

21.48. If Γ is complete, then Γ has the joint extension property.

The Interpolation Theorem 22

In this chapter we shall prove several theorems which involve, loosely speaking, elimination of superfluous notions. These results are considerably deeper than the similar sounding facts exposed in Chapter 11, and the reader would be well advised to review that chapter before beginning this one. The theorems of this chapter are based on the following fundamental theorem.

Theorem 22.1 (Craig's interpolation theorem). *Let φ and ψ be sentences such that $\models\varphi \rightarrow \psi$. Then there is a sentence χ such that:*

(*i*) $\models\varphi \rightarrow \chi$ *and* $\models\chi \rightarrow \psi$;
(*ii*) *every nonlogical constant occurring in χ occurs in both φ and ψ.*

PROOF. Clearly we may assume that the underlying language $\mathscr{L}$ is countable (i.e., that $|\mathrm{Fmla}_{\mathscr{L}}| = \aleph_0$). We shall use the model existence theorem (Corollary 18.12). To this end, let $\mathscr{L}'$ be an expansion of $\mathscr{L}$ rich by C. Now for each sentence χ of $\mathscr{L}$ let

$$\Gamma_\chi = \{\theta : \theta \text{ is a sentence of } \mathscr{L}', \text{ and the nonlogical constants of } \mathscr{L} \text{ which occur in } \theta \text{ also occur in } \chi\}.$$

Let S be the collection of all $\Delta \subseteq \mathrm{Sent}_{\mathscr{L}'}$ for which there exist finite subsets $\Theta_0 \subseteq \Gamma_\varphi$ and $\Theta_1 \subseteq \Gamma_\psi$ such that $\Delta = \Theta_0 \cup \Theta_1$ and for all $\chi_0, \chi_1 \in \Gamma_\varphi \cap \Gamma_\psi$, if $\models \bigwedge \Theta_0 \rightarrow \chi_0$ and $\models \bigwedge \Theta_1 \rightarrow \chi_1$, then $\chi_0 \wedge \chi_1$ has a model. We shall now establish (C1)–(C9) of 18.4. So, assume that $\Delta \in S$, with Θ_0 and Θ_1 as above.

(C1). Suppose θ is a sentence of $\mathscr{L}'$ such that $\theta, \neg\theta \in \Delta$. If $\theta, \neg\theta \in \Theta_0$, let χ_0 be the sentence $\neg\exists v_0(v_0 = v_0)$ and let χ_1 be the sentence $\exists v_0(v_0 = v_0)$. Our assumptions above imply that $\chi_0 \wedge \chi_1$ has a model, contradiction. The assumption θ, $\neg\theta \in \Theta_1$ similarly gives a contradiction. Now assume that $\theta \in \Theta_0$ and $\neg\theta \in \Theta_1$. It follows then that θ, $\neg\theta \in \Gamma_\varphi \cap \Gamma_\psi$, and our assumption on Θ_0, Θ_1 then implies that $\theta \wedge \neg\theta$ has a model, contradiction. Similarly, $\theta \in \Theta_1$ and $\neg\theta \in \Theta_0$ gives a contradiction. Thus there is no such sentence θ that θ, $\neg\theta \in \Delta$.

(C2), (C3). These conditions are obvious.

(C4). Suppose that $\theta_0 \vee \theta_1 \in \Delta$; say, without loss of generality that $\theta_0 \vee \theta_1 \in \Theta_0$. Let $\Theta_0' = \Theta_0 \cup \{\theta_0\}$ and $\Theta_0'' = \Theta_0 \cup \{\theta_1\}$. Assume that $\Theta_0' \cup \Theta_1 \notin S$ and $\Theta_0'' \cup \Theta_1 \notin S$. Then there exist $\chi_0', \chi_1', \chi_0'', \chi_1'' \in \Gamma_\varphi \cap \Gamma_\psi$ such that $\models \bigwedge \Theta_0' \rightarrow \chi_0'$, $\models \bigwedge \Theta_1 \rightarrow \chi_1'$, $\models \bigwedge \Theta_0'' \rightarrow \chi_0''$, $\models \bigwedge \Theta_1 \rightarrow \chi_1''$, while $\chi_0' \wedge \chi_1'$ has no model and $\chi_0'' \wedge \chi_1''$ has no model. Thus $\models \bigwedge \Theta_0 \wedge \theta_0 \rightarrow \chi_0'$ and $\models \bigwedge \Theta_0 \wedge \theta_1 \rightarrow \chi_0''$, so, since $\theta_0 \vee \theta_1 \in \Theta_0$,

(1) $$\models \bigwedge \Theta_0 \rightarrow \chi_0' \vee \chi_0''.$$

We also have

$$\models \bigwedge \Theta_1 \rightarrow \chi_1' \wedge \chi_1'',$$

so by (1) and our choice of Θ_0, Θ_1 we infer that $(\chi_0' \vee \chi_0'') \wedge (\chi_1' \wedge \chi_1'')$ has a model. This contradicts the fact that neither $\chi_0' \wedge \chi_1'$ nor $\chi_0'' \wedge \chi_1''$ has a model.

(C5) This is obvious.

(C6). Assume that $\exists\alpha\theta \in \Delta$. Choose $\mathbf{c} \in C$ so that $\mathbf{c}$ does not occur in $\bigwedge\Theta_0 \wedge \bigwedge \Theta_1$. Say without loss of generality that $\exists\alpha\theta \in \Theta_0$. Let $\Theta_0' = \Theta_0 \cup \{\mathrm{Subf}^\alpha_\mathbf{c}\, \theta\}$. We claim that $\Theta_0' \cup \Theta_1 \in S$ (as desired). To prove this, assume that we have $\chi_0, \chi_1 \in \Gamma_\varphi \cap \Gamma_\psi \models \bigwedge \Theta_0 \rightarrow \{\mathrm{Subf}^\alpha_\mathbf{c}\theta \rightarrow \chi_0\}$ and $\models \bigwedge \Theta_1 \rightarrow \chi_1$. We may assume that α does not occur in $\chi_0 \wedge \chi_1$. Let $\chi_0' \wedge \chi_1'$ be obtained from χ_0, χ_1 respectively by replacing $\mathbf{c}$ by α. Then we easily obtain

$$\models \bigwedge \Theta_0 \rightarrow \exists\alpha\chi_0', \qquad \models \bigwedge \Theta_1 \rightarrow \forall\alpha\chi_1'.$$

Hence $\exists\alpha\chi_0' \wedge \forall\alpha\chi_1'$ has a model. Such a model obviously yields a model of $\chi_0 \wedge \chi_1$, as desired.

(C7). This is obvious.

(C8). Suppose that $\mathbf{c} \in C$, τ is a primitive term, and $\mathbf{c} \equiv \tau$, $\mathrm{Subf}^\alpha_\tau\theta \in \Delta$. If both $\mathbf{c} \equiv \tau$ and $\mathrm{Subf}^\alpha_\tau\theta \in \Theta_0$, or both $\in \Theta_1$, the desired conclusion is obvious. So suppose, say, that $\mathbf{c} \equiv \tau \in \Theta_0$ while $\mathrm{Subf}^\alpha_\tau\theta \in \Theta_1$. Let $\Theta_1' = \Theta_1 \cup \{\mathrm{Subf}^\alpha_\mathbf{c}\theta\}$. We claim that $\Theta_0 \cup \Theta_1' \in S$. To show this, assume that $\chi_0, \chi_1 \in \Gamma_\varphi \cap \Gamma_\psi$ and $\models \bigwedge \Theta_0 \rightarrow \chi_0$, $\models \bigwedge \Theta_1' \rightarrow \chi_1$. Thus $\models \bigwedge \Theta_1 \rightarrow (\mathrm{Subf}^\alpha_\mathbf{c}\theta \rightarrow \chi_1)$. Hence

(2) $$\models \bigwedge \Theta_0 \rightarrow \mathbf{c} \equiv \tau \wedge \chi_0,$$

(3) $$\models \bigwedge \Theta_1 \rightarrow (\mathbf{c} \equiv \tau \rightarrow \chi_1).$$

Since we may assume that α occurs free in θ, it follows that τ occurs in $\mathrm{Subf}^\alpha_\tau\theta$. Thus since $\mathbf{c} \equiv \tau \in \Theta_0$ and $\mathrm{Subf}^\alpha_\tau\theta \in \Theta_1$ it follows that $\mathbf{c} \equiv \tau \in \Gamma_\varphi \cap \Gamma_\psi$. Hence from (2) and (3) we infer from our choice of Θ_0 and Θ_1

that $(\mathbf{c} \equiv \tau \wedge \chi_0) \wedge (\mathbf{c} \equiv \tau \rightarrow \chi_1)$ has a model. It is clearly also a model of $\chi_0 \wedge \chi_1$.

(C9). Let τ be a primitive term. Let $\mathbf{c} \in C$ be such that $\mathbf{c}$ does not occur in τ or in $\bigwedge\Theta_0 \wedge \bigwedge\Theta_1$. Let $\Theta_0' = \Theta_0 \cup \{\mathbf{c} \equiv \tau\}$. Suppose $\chi_0, \chi_1 \in \Gamma_\varphi \cap \Gamma_\psi$ and $\models \bigwedge \Theta_0' \rightarrow \chi_0$, $\models \bigwedge \Theta_1 \rightarrow \chi_1$. Then $\models \bigwedge \Theta_0 \rightarrow (\mathbf{c} \equiv \tau \rightarrow \chi_0)$ and hence we easily obtain $\models \bigwedge \Theta_0 \rightarrow (\exists\alpha(\alpha \equiv \tau) \rightarrow \exists\alpha\chi_0')$ and $\models \bigwedge \Theta_1 \rightarrow \forall\alpha\chi_1'$, where α is a new variable and χ_0', χ_1' are obtained from χ_0, χ_1 by replacing $\mathbf{c}$ by α. Since $\models\exists\alpha(\alpha \equiv \tau)$, it follows that $\models \bigwedge \Theta_0 \rightarrow \exists\alpha\chi_0'$. Hence $\exists\alpha\chi_0' \wedge \forall\alpha\chi_1'$ has a model, which easily yields a model of $\chi_0 \wedge \chi_1$.

Thus we have checked (C1)–(C9) of 18.4. Now our assumption, that $\models\varphi \rightarrow \psi$, implies that $\{\varphi, \neg\psi\}$ does not have a model. Hence by 18.12, $\{\varphi, \neg\psi\} \notin S$. Thus there exist $\chi_0, \chi_1 \in \Gamma_\varphi \cap \Gamma_\psi$ such that $\models\varphi \rightarrow \chi_0$, $\models \neg\psi \rightarrow \chi_1$, while $\chi_0 \wedge \chi_1$ does not have a model. Thus $\models \neg\chi_1 \rightarrow \psi$ and $\models\chi_0 \rightarrow \neg\chi_1$, so $\models\varphi \rightarrow \chi_0$ and $\models\chi_0 \rightarrow \psi$. Let $\mathbf{c}_0, \ldots, \mathbf{c}_{m-1}$ be all of the members of C occurring in χ_0. Then clearly there exist new variables $\alpha_0, \ldots, \alpha_{m-1}$ such that

$$\models\varphi \rightarrow \forall\alpha_0 \cdots \forall\alpha_{m-1}\chi_0'$$
$$\models\forall\alpha_0 \cdots \forall\alpha_{m-1}\chi_0' \rightarrow \psi$$

where χ_0' is obtained from χ_0 by replacing $\mathbf{c}_0, \ldots, \mathbf{c}_{m-1}$ by $\alpha_0, \ldots, \alpha_{m-1}$ respectively. □

The next theorem is of an entirely different character from the interpolation theorem. According to this theorem, if one conservatively extends a given theory in two different ways, the two ways are consistent with each other.

Theorem 22.2 (A. Robinson's consistency theorem). Assume *that* $\mathscr{L}_0$, $\mathscr{L}_1$, $\mathscr{L}_2$, $\mathscr{L}_3$, Γ_0, Γ_1, Γ_2 *are given satisfying the following conditions:*

(*i*) $\mathscr{L}_0$, $\mathscr{L}_1$, $\mathscr{L}_2$, $\mathscr{L}_3$ *are first-order languages;* $\mathscr{L}_1$ *and* $\mathscr{L}_2$ *are expansions of* $\mathscr{L}_0$, *and* $\mathscr{L}_3$ *is an expansion of both* $\mathscr{L}_1$ *and* $\mathscr{L}_2$;
(*ii*) *any nonlogical symbol common to* $\mathscr{L}_1$ *and* $\mathscr{L}_2$ *is a symbol of* $\mathscr{L}_0$.
(*iii*) Γ_0, Γ_1, Γ_2 *are consistent theories in* $\mathscr{L}_0$, $\mathscr{L}_1$, $\mathscr{L}_2$ *respectively;*
(*iv*) Γ_1 *and* Γ_2 *are conservative extensions of* Γ_0.

Then $\Gamma_1 \cup \Gamma_2$ *is consistent.*

The various assumptions in the hypothesis of 22.2 are indicated in the following diagram:

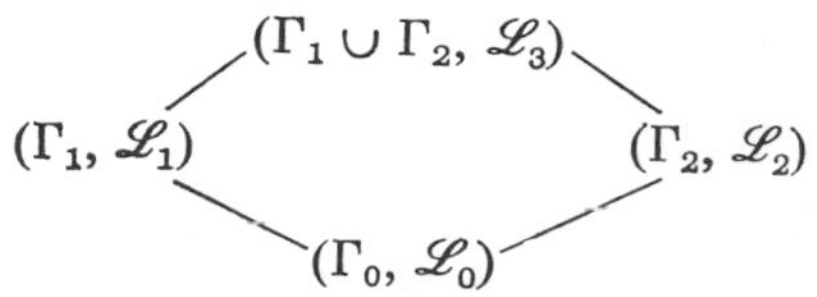

Proof. Suppose that $\Gamma_1 \cup \Gamma_2$ is not consistent. Then, by the compactness theorem, there are finite subsets Δ_1 and Δ_2 of Γ_1 and Γ_2 respectively such that $\Delta_1 \cup \Delta_2$ has no model; and we may assume that $\Delta_1 \neq 0 \neq \Delta_2$. Thus we have

$$\models \bigwedge \Delta_1 \rightarrow \bigvee \{\neg\varphi : \varphi \in \Delta_2\},$$

and hence by (*ii*) and the interpolation theorem there is a $\chi \in \text{Sent}\,\mathscr{L}_0$ such that

$$\models \bigwedge \Delta_1 \rightarrow \chi \quad \text{and} \quad \models \chi \rightarrow \bigvee \{\neg\varphi : \varphi \in \Delta_2\}.$$

Hence $\chi \in \Gamma_1$ and $\models \bigwedge \Delta_2 \rightarrow \neg\chi$, so $\neg\chi \in \Gamma_2$. But then by (*iv*), χ, $\neg\chi \in \Gamma_0$, contradicting the consistency of Γ_0 (condition (*iii*)). □

Note that if Γ_0 is complete, then condition (*iv*) in 22.2 follows from (*iii*). It is of some interest that we can easily derive Craig's theorem from Robinson's theorem. In fact, assume that $\models \varphi \rightarrow \psi$. Let $\mathscr{L}_0$, $\mathscr{L}_1$, $\mathscr{L}_2$, $\mathscr{L}_3$ be respectively the logics with nonlogical constants those occurring in (1) both φ and ψ, (2) φ, (3) ψ, (4) $\varphi \rightarrow \psi$. Let $\Gamma = \{\theta \in \text{Sent}_{\mathscr{L}_0} : \models \varphi \rightarrow \theta\}$. Now we claim

(1) $\Gamma \cup \{\neg\psi\}$ does not have a model.

For suppose, on the contrary, that $\Gamma \cup \{\neg\psi\}$ has a model $\mathfrak{A}$ (an $\mathscr{L}_2$-structure). Let $\Delta = \{\theta \in \text{Sent}_{\mathscr{L}_0} : \theta \text{ holds in } \mathfrak{A}\}$. Then Δ is a complete theory in $\mathscr{L}_0$, and $\Delta \cup \{\neg\psi\}$ has a model (namely $\mathfrak{A}$). *If* $\Delta \cup \{\varphi\}$ had a model, A. Robinson's consistency theorem would yield a model of $\Delta \cup \{\varphi, \neg\psi\}$, contradicting $\models \varphi \rightarrow \psi$. Thus $\Delta \cup \{\varphi\}$ has no model, so by the compactness theorem $\models \varphi \rightarrow \bigvee_{i<m} \neg\chi_i$ for some $m \in \omega$ and some $\chi \in {}^m\Delta$. Hence $\bigvee_{i<m} \neg\chi_i \in \Gamma$, so $\bigvee_{i<m} \neg\chi_i$ holds in $\mathfrak{A}$, which is impossible.

Thus (1) holds. By the compactness theorem, $\models \bigwedge_{i<n} \chi_i \rightarrow \psi$ for some $n \in \omega$ and some $\chi \in {}^n\Gamma$. Clearly also $\models \varphi \rightarrow \bigwedge_{i<n} \chi_i$, as desired.

Another application of Craig's theorem is in the theory of definition. This application is in proving a converse of the following easy and classical result, which is frequently applied to show that certain notions cannot be defined in terms of others.

Theorem 22.3 (Padoa's method). *Let Γ be a theory in a language $\mathscr{L}$, and let $\boldsymbol{\pi}$ be a nonlogical constant of $\mathscr{L}$. Suppose that $\mathfrak{A}$ and $\mathfrak{B}$ are two models of Γ such that $A = B$, $\boldsymbol{\sigma}^{\mathfrak{A}} = \boldsymbol{\sigma}^{\mathfrak{B}}$ for any nonlogical constant $\boldsymbol{\sigma} \neq \boldsymbol{\pi}$, while $\boldsymbol{\pi}^{\mathfrak{A}} \neq \boldsymbol{\pi}^{\mathfrak{B}}$.*

Then $\boldsymbol{\pi}$ is not definable, with respect to Γ, in terms of the other nonlogical constants of $\mathscr{L}$. That is, there is no theory Δ, in the reduct $\mathscr{L}'$ of $\mathscr{L}$ obtained by deleting $\boldsymbol{\pi}$, such that Γ is a definitional expansion of Δ.

Proof. We take the case of a relation symbol $\boldsymbol{\pi}$; operation symbols are treated similarly. Suppose there is such a theory Δ, and let φ be a possible definition of $\boldsymbol{\pi}$, with $\Gamma \models \forall v_0 \cdots \forall v_{m-1}(\boldsymbol{\pi} v_0 \cdots v_{m-1} \leftrightarrow \varphi)$. (See Definition 11.29.) Then for any $x \in {}^mA$, $x \in \boldsymbol{\pi}^{\mathfrak{A}}$ iff $\mathfrak{A} \models \varphi[x]$ iff $\mathfrak{B} \models \varphi[x]$ (since $\boldsymbol{\pi}$ does not occur in φ) iff $x \in \boldsymbol{\pi}^{\mathfrak{B}}$, so $\boldsymbol{\pi}^{\mathfrak{A}} = \boldsymbol{\pi}^{\mathfrak{B}}$, contradiction. □

Here is an application of Padoa's method. Let Γ be the theory of the structure $\langle \mathbb{Z}, + \rangle$, i.e., $\Gamma = \{\varphi : \varphi$ is a sentence and φ holds in $\langle \mathbb{Z}, + \rangle\}$. Then $<$ is not definable in Γ. To prove this, let Γ' be the theory of $\langle \mathbb{Z}, +, < \rangle$. Then $\langle \mathbb{Z}, +, > \rangle$ is also a model of Γ', since $\langle -x : x \in \mathbb{Z} \rangle$ is an isomorphism from $\langle \mathbb{Z}, +, < \rangle$ onto $\langle \mathbb{Z}, +, > \rangle$. It follows by Padoa's method that $<$ is not definable in Γ'.

Beth's theorem is the converse of Padoa's method. It is a kind of completeness theorem in the theory of definition.

Theorem 22.4 (Beth's theorem). *Let* Γ *be a theory in* $\mathscr{L}$, *and assume that* $\boldsymbol{\pi}$ *is a nonlogical constant of* $\mathscr{L}$ *such that*

(*i*) *if* $\mathfrak{A}$ *and* $\mathfrak{B}$ *are two models of* Γ *such that* $A = B$ *and* $\boldsymbol{\sigma}^{\mathfrak{A}} = \boldsymbol{\sigma}^{\mathfrak{B}}$ *for any nonlogical constant* $\boldsymbol{\sigma} \neq \boldsymbol{\pi}$, *then* $\boldsymbol{\pi}^{\mathfrak{A}} = \boldsymbol{\pi}^{\mathfrak{B}}$.

Then $\boldsymbol{\pi}$ *can be defined in terms of the other nonlogical constants, i.e.,*

(*ii*) *if* $\mathscr{L}'$ *is the reduct of* $\mathscr{L}$ *obtained by deleting* $\boldsymbol{\pi}$, *then there is a theory* Δ *in* $\mathscr{L}'$ *such that* $(\Gamma, \mathscr{L})$ *is a definitional expansion of* $(\Delta, \mathscr{L}')$.

Proof. First we treat the case in which $\boldsymbol{\pi}$ is a relation symbol, say of rank m. Expand $\mathscr{L}$ to $\mathscr{L}^*$ by adjoining a new m-ary relation symbol $\mathbf{S}$ and individual constants $\mathbf{c}_0, \ldots, \mathbf{c}_{m-1}$. Let Θ be obtained from Γ by replacing $\boldsymbol{\pi}$ by $\mathbf{S}$ in each sentence of Γ. We then claim

(1) $$\Gamma \cup \Theta \models \boldsymbol{\pi}\mathbf{c}_0 \cdots \mathbf{c}_{m-1} \rightarrow \mathbf{S}\mathbf{c}_0 \cdots \mathbf{c}_{m-1}.$$

Indeed, let $\mathfrak{B} = (\mathfrak{A}, R, S, a_i)_{i<m}$ be any model of $\Gamma \cup \Theta$, with $\mathfrak{A}$ the underlying $\mathscr{L}'$-structure, $\boldsymbol{\pi}^{\mathfrak{B}} = R$, $\mathbf{S}^{\mathfrak{B}} = S$, and $\mathbf{c}_i^{\mathfrak{B}} = a_i$ for each $i < m$. Clearly then $(\mathfrak{A}, R)$ and $(\mathfrak{A}, S)$ both are models of Γ. Hence by (*i*) $R = S$, so $\boldsymbol{\pi}\mathbf{c}_0 \cdots \mathbf{c}_{m-1} \rightarrow \mathbf{S}\mathbf{c}_0 \cdots \mathbf{c}_{m-1}$ holds in $\mathfrak{B}$. Thus (1) holds.

By (1) there are finite subsets Γ' and Θ' of Γ and Θ respectively such that

$$\Gamma' \cup \Theta' \models \boldsymbol{\pi}\mathbf{c}_0 \cdots \mathbf{c}_{m-1} \rightarrow \mathbf{S}\mathbf{c}_0 \cdots \mathbf{c}_{m-1}.$$

Let γ and θ be the conjunctions of all members of Γ' and Θ' respectively. Thus $\{\gamma, \theta\} \models \boldsymbol{\pi}\mathbf{c}_0 \cdots \mathbf{c}_{m-1} \rightarrow \mathbf{S}\mathbf{c}_0 \cdots \mathbf{c}_{m-1}$, and hence we easily obtain

$$\models \gamma \wedge \boldsymbol{\pi}\mathbf{c}_0 \cdots \mathbf{c}_{m-1} \rightarrow (\theta \rightarrow \mathbf{S}\mathbf{c}_0 \cdots \mathbf{c}_{m-1}).$$

Note that the sentence $\gamma \wedge \boldsymbol{\pi}\mathbf{c}_0 \cdots \mathbf{c}_{m-1}$ does not involve $\mathbf{S}$, and the sentence $\theta \rightarrow \mathbf{S}\mathbf{c}_0 \cdots \mathbf{c}_{m-1}$ does not involve $\boldsymbol{\pi}$. Hence by the interpolation theorem there is a sentence ξ not involving $\boldsymbol{\pi}$ or $\mathbf{S}$ that the following two conditions hold:

(2) $$\models \gamma \wedge \boldsymbol{\pi}\mathbf{c}_0 \cdots \mathbf{c}_{m-1} \rightarrow \xi;$$
(3) $$\models \xi \rightarrow (\theta \rightarrow \mathbf{S}\mathbf{c}_0 \cdots \mathbf{c}_{m-1}).$$

Next we claim

(4) $$\Gamma \models \boldsymbol{\pi}\mathbf{c}_0 \cdots \mathbf{c}_{m-1} \leftrightarrow \xi.$$

In fact, $\Gamma \models \boldsymbol{\pi}\mathbf{c}_0 \cdots \mathbf{c}_{m-1} \rightarrow \xi$ by (2), since γ is a conjunction of members of Γ. To prove the converse, suppose $\mathfrak{B} = (\mathfrak{A}, R, a_i)_{i<m}$ is any model of $\Gamma \cup \{\xi\}$, where $\mathfrak{A}$ is an $\mathscr{L}'$-structure, $\boldsymbol{\pi}^{\mathfrak{B}} = R$, and $\mathbf{c}_i^{\mathfrak{B}} = a_i$ for all $i < m$. Let $\mathfrak{C}$ be the expansion of $\mathfrak{B}$ to an $\mathscr{L}^*$-structure with $\mathbf{S}^{\mathfrak{C}} = R$ also. Clearly $\mathfrak{C}$ is a model of $\Gamma \cup \Theta \cup \{\xi\}$. Since θ is a conjunction of members of Θ, it follows that $\mathfrak{C} \models \theta$. Hence by (3), $\mathfrak{C} \models \mathbf{S}\mathbf{c}_0 \cdots \mathbf{c}_{m-1}$, and hence $\mathfrak{C} \models \boldsymbol{\pi}\mathbf{c}_0 \cdots \mathbf{c}_{m-1}$. Thus $\mathfrak{B} \models \boldsymbol{\pi}\mathbf{c}_0 \cdots \mathbf{c}_{m-1}$, as desired: (4) holds.

From (4) we infer that

$$(5) \qquad \Gamma \models \forall v_0 \cdots \forall v_{m-1}(\boldsymbol{\pi} v_0 \cdots v_{m-1} \leftrightarrow \xi'),$$

where ξ' is obtained from ξ by replacing $\mathbf{c}_0, \ldots, \mathbf{c}_{m-1}$ by $v_0, \ldots, v_{m-1}$ respectively, after changing bound variables to avoid clashes. Now let $\Gamma' = \{\theta \in \mathrm{Sent}_{\mathscr{L}'} : \Gamma \models \theta\}$, and let φ be the sentence of (5). It remains only to show that $\Gamma' \cup \{\varphi\}$ axiomatizes Γ. Obviously $\Gamma' \cup \{\varphi\} \subseteq \Gamma$. Recalling the proof of 11.30, we easily show

(6) for any formula ψ of $\mathscr{L}$ there is a formula ψ' of $\mathscr{L}'$ with the same free variables as ψ such that $\{\varphi\} \models \psi \leftrightarrow \psi'$.

Now if $\psi \in \Gamma$, choose by (6) a sentence ψ' of $\mathscr{L}'$ such that $\{\varphi\} \models \psi \leftrightarrow \psi'$. By (5), $\Gamma \models \psi'$, so $\psi' \in \Gamma'$ by the definition of Γ'. Hence $\Gamma' \cup \{\varphi\} \models \psi$, as desired. This completes the proof in case $\boldsymbol{\pi}$ is a relation symbol.

The case of an operation symbol $\boldsymbol{\pi}$ is similar: we begin by adjoining an m-ary operation symbol $\mathbf{O}$ and $m + 1$ individual constants $\mathbf{c}_0, \ldots, \mathbf{c}_m$, obtaining $\mathscr{L}^*$. Let Θ be obtained from Γ by replacing $\boldsymbol{\pi}$ by $\mathbf{O}$ in each sentence of Γ. Then we obtain formulas γ, θ, ξ analogous to the above:

$$(1) \qquad \Gamma \cup \Theta \models \boldsymbol{\pi}\mathbf{c}_0 \cdots \mathbf{c}_{m-1} \equiv \mathbf{c}_m \rightarrow \mathbf{O}\mathbf{c}_0 \cdots \mathbf{c}_{m-1} \equiv \mathbf{c}_m;$$
$$(2) \qquad \models \gamma \wedge \boldsymbol{\pi}\mathbf{c}_0 \cdots \mathbf{c}_{m-1} \equiv \mathbf{c}_m \rightarrow \xi;$$
$$(3) \qquad \models \xi \rightarrow (\theta \rightarrow \mathbf{O}\mathbf{c}_0 \cdots \mathbf{c}_{m-1} \equiv \mathbf{c}_m);$$
$$(4) \qquad \Gamma \models \boldsymbol{\pi}\mathbf{c}_0 \cdots \mathbf{c}_{m-1} \equiv \mathbf{c}_m \leftrightarrow \xi;$$
$$(5) \qquad \Gamma \models \forall v_0 \cdots \forall v_m(\boldsymbol{\pi} v_0 \cdots v_{m-1} \equiv v_m \leftrightarrow \xi).$$

It remains only to check the existence and uniqueness conditions for ξ. But both conditions are clear by (5). □

As our final application of the interpolation theorem we shall discuss the problem of independently axiomatizing theories.

Definition 22.5. A set of Γ of sentences is *independent* if for every $\varphi \in \Gamma$ we have not $(\Gamma \sim \{\varphi\} \models \varphi)$. A theory Γ is *independently axiomatizable* if there is an independent set which axiomatizes Γ.

Our goal is to show that every theory is independently axiomatizable. The case of countable languages is rather obvious, and will be useful in the general case:

Lemma 22.6. *If Γ is a theory in a countable language, then Γ is independently axiomatizable.*

PROOF. Write $\Gamma = \{\varphi_i : i \in \omega\}$. For each $i \in \omega$ let ψ_i be the sentence $\bigwedge_{j<i} \varphi_j \rightarrow \varphi_i$, where by convention ψ_0 is φ_0. Let $\Delta = \{\psi_i : \text{not}(\models \psi_i)\}$. We claim that Δ independently axiomatizes Γ. Obviously $\Gamma \models \psi_i$ for each i, and hence $\Gamma \models \chi$ for each $\chi \in \Delta$. By induction on i it is clear that $\models \bigwedge_{j \le i} \psi_i \rightarrow \varphi_i$ for each $i \in \omega$. Hence $\Delta \models \varphi_i$ for each $i \in \omega$. Hence $\Gamma = \{\psi : \Delta \models \psi\}$, i.e., Δ axiomatizes Γ. It remains to show that Δ is independent. First note:

(1) $\qquad$ if $\psi_i, \psi_j \in \Delta$ and $i \neq j$, $\qquad$ then $\models \psi_i \vee \psi_j$.

In fact, suppose $i < j$. Clearly $\models \neg \psi_j \rightarrow \bigwedge_{k<j} \varphi_k$, and hence $\models \neg \psi_j \rightarrow \varphi_i$ and $\models \neg \psi_j \rightarrow \psi_i$. Thus (1) holds.

Now assume that $\chi \in \Delta$ and $\Delta \sim \{\chi\} \models \chi$. Thus $\vdash \bigwedge \Theta \rightarrow \chi$ for some finite subset Θ of $\Delta \sim \{\chi\}$. But for each $\theta \in \Theta$ we have $\models \theta \vee \chi$ by (1), so $\models \bigwedge \Theta \vee \chi$. Hence $\models \chi$, contradiction. $\square$

Lemma 22.7. *Suppose* $\Gamma, \Delta \subseteq \text{Sent}_{\mathscr{L}}$ *and:*

(i) $|\Gamma| \le |\Delta|$, *and* $\Gamma \cap \Delta = 0$
(ii) for every $\varphi \in \Delta$, *not*$(\Gamma \cup \Delta \sim \{\varphi\} \models \varphi)$

Then there is an independent set Θ *which axiomatizes* $\{\varphi \in \text{Sent}_{\mathscr{L}} : \Gamma \cup \Delta \models \varphi\}$.

PROOF. Let $\psi : \Gamma \rightarrowtail \Delta$. Let $\Theta = \{\varphi \wedge \psi_\varphi : \varphi \in \Gamma\} \cup (\Delta \sim \text{Rng}\,\psi)$. Then obviously $\Gamma \cup \Delta \subseteq \{\varphi : \Theta \models \varphi\}$, and $\Theta \subseteq \{\varphi : \Gamma \cup \Delta \models \varphi\}$, so Θ axiomatizes $\Gamma \cup \Delta$.

Now to show that Θ is independent, let $\chi \in \Theta$, and suppose, to try to reach a contradiction, that $\Theta \sim \{\chi\} \models \chi$. First suppose that χ has the form $\varphi \wedge \psi_\varphi$ where $\varphi \in \Gamma$. If $\theta \in \Gamma$ and $\theta \neq \varphi$, then $\theta \neq \psi_\varphi$ since $\Gamma \cap \Delta = 0$, so $(\Gamma \cup \Delta) \sim \{\psi_\varphi\} \models \theta$; also $\psi_\theta \neq \psi_\varphi$, so $(\Gamma \cup \Delta) \sim \{\psi_\varphi\} \models \psi_\theta$; it follows that $(\Gamma \cup \Delta) \sim \{\psi_\varphi\} \models \theta \wedge \psi_\theta$ whenever $\theta \in \Gamma$, $\theta \neq \varphi$. If $\theta \in \Delta \sim \text{Rng}\,\psi$, then $\theta \neq \psi_\varphi$ and $(\Gamma \cup \Delta) \sim \{\psi_\varphi) \models \theta$. Thus $(\Gamma \cup \Delta) \sim \{\psi_\varphi\} \models \theta$ for each $\theta \in \Theta \sim \{\chi\}$, so, by our assumption that $\Theta \sim \{\chi\} \models \chi$, $(\Gamma \cup \Delta) \sim \{\psi_\varphi\} \models \chi$ and hence $(\Gamma \cup \Delta) \sim \{\psi_\varphi\} \models \psi_\varphi$. This contradicts (*ii*). If $\chi \in \Delta \sim \text{Rng}\,\psi$, a contradiction is similarly reached. $\square$

Lemma 22.8. *Let* $\mathscr{L}$ *be a language with infinitely many nonlogical constants. Suppose that* Γ *is a set of sentences in* $\mathscr{L}$ *satisfying the following condition:*

(i) if $\varphi \in \Gamma$, $m \in \omega \sim 1$, $\psi \in {}^m\Gamma$, *and there is a nonlogical constant occurring in* φ *but not in* $\bigwedge_{i<m} \psi_i$, *then not*$(\models \bigwedge_{i<m} \psi_i \rightarrow \varphi)$.

Then $\{\varphi : \Gamma \models \varphi\}$ *is independently axiomatizable.*

PROOF. Let $\mathscr{L}^-$ be the reduct of $\mathscr{L}$ to all nonlogical constants occurring in some sentence of Γ. Now

(1) $\quad$ if $\Delta \subseteq \text{Sent}_{\mathscr{L}^-}$ independently axiomatizes $\{\varphi \in \text{Sent}_{\mathscr{L}^-} : \Gamma \models \varphi\}$ in $\mathscr{L}^-$, then Δ independently axiomatizes $\{\varphi : \Gamma \models \varphi\}$ in $\mathscr{L}$.

For, assume the hypothesis of (1). Then $\Delta \models \varphi$ for each $\varphi \in \Gamma$, so Δ axiomatizes $\{\varphi : \Gamma \models \varphi\}$. Obviously Δ is still independent in the $\mathscr{L}$-sense.

By (1) it suffices to show that $\{\varphi \in \mathrm{Sent}_{\mathscr{L}^-} : \Gamma \models \varphi\}$ is independently axiomatizable. Thus, returning to the original statement of the theorem, we may assume

(2) every nonlogical constant of $\mathscr{L}$ occurs in some sentence of Γ.

By 22.6 we still may assume that $\mathscr{L}$ has infinitely many nonlogical constants. Let a well-ordering of Γ be fixed, so that we can speak of the *first member* of any nonempty subset of Γ. Let $\mathfrak{m} = |\mathrm{Fmla}_{\mathscr{L}}|$. Clearly $\mathfrak{m}$ is the cardinality of the set C of all nonlogical constants of $\mathscr{L}$, and it is also the cardinality of Γ. Choose $\mathbf{c}$ so that $\mathbf{c} : \mathfrak{m} \rightarrowtail\!\!\!\!\twoheadrightarrow C$. For each sentence φ of $\mathscr{L}$ let $C\varphi$ be the set of all nonlogical constants appearing in φ.

We now define a sequence $\langle \varphi_\alpha : \alpha < \mathfrak{m}\rangle$ of sentences of Γ. Assume that $\beta < \mathfrak{m}$ and that φ_α has been defined for all $\alpha < \beta$. Thus clearly $|\bigcup_{\alpha<\beta} C\varphi_\alpha| < \mathfrak{m}$. Let $\mathbf{d}$ be the first nonlogical constant in the list $\mathbf{c}_0, \mathbf{c}_1, \ldots$ not in $\bigcup_{\alpha<\beta} C\varphi_\alpha$, and let φ_β be the first member of Γ in which $\mathbf{d}$ occurs. This defines the sequence $\langle \varphi_\alpha : \alpha < \mathfrak{m}\rangle$.

Next we define a sequence $\langle \Delta_\alpha : \alpha < \mathfrak{m}\rangle$ of subsets of Γ, namely $\Delta_0 = \{\psi \in \Gamma : C\psi \subseteq C\varphi_0\}$, while for any $\beta \neq 0$ we set

$$\Delta_\beta = \{\psi \in \Gamma : C\psi \subseteq \bigcup_{\alpha<\beta} C\varphi_\alpha \text{ and } C\psi \cap C\varphi_\beta \sim \bigcup_{\alpha<\beta} C\varphi_\alpha \neq 0\}.$$

Now we note some properties of the sequences φ and Δ. By induction on β we clearly have

(3) $$\mathbf{c}_\beta \in \textstyle\bigcup_{\alpha\leq\beta} C\varphi_\alpha \qquad \text{for each } \beta < \mathfrak{m}.$$

Hence it follows that

(4) $$\bigcup_{\beta<\mathfrak{m}} C\varphi_\beta = C.$$

(5) $$\text{If } \beta < \gamma < \mathfrak{m}, \text{ then } \Delta_\beta \cap \Delta_\gamma = 0.$$

For, if $\beta < \gamma < \mathfrak{m}$ and $\psi \in \Delta_\beta$, then $C\psi \subseteq \bigcup_{\alpha\leq\beta} C\varphi_\alpha$, while for each $\chi \in \Delta_\gamma$ we have $C\chi \nsubseteq \bigcup_{\alpha<\gamma} C\varphi_\alpha \supseteq \bigcup_{\alpha<\beta} C\varphi_\alpha$, so $\psi \notin \Delta_\gamma$.

(6) $$\bigcup_{\beta<\mathfrak{m}} \Delta_\beta = \Gamma.$$

For, $\subseteq$ is trivial, so assume that $\psi \in \Gamma$. If $C\psi = 0$, then $\psi \in \Delta_0$. Assume that $C\psi \neq 0$. By (4) we may choose $\gamma < \mathfrak{m}$ minimum so that $C\psi \subseteq \bigcup_{\beta<\gamma} C\varphi_\beta$. Note that $\gamma \neq 0$ and γ is not a limit ordinal, since $0 < |C\psi| < \aleph_0$. Say $\gamma = \delta + 1$. Thus $C\psi \nsubseteq \bigcup_{\beta<\delta} C\varphi_\beta$. It follows that $\psi \in \Delta_\gamma$, as wished in (6).

For any $\beta < \mathfrak{m}$, let $T_\beta = C\varphi_\beta \sim \bigcup_{\alpha<\beta} C\varphi_\alpha$. Then the following two conditions are obvious:

(7) $$\text{if } \beta < \gamma < \mathfrak{m}, \text{ then } T_\beta \cap T_\gamma = 0;$$

(8) $$\bigcup_{\beta<\mathfrak{m}} T_\beta = C.$$

Since (7) holds, it follows that for any $\psi \in \Gamma$ the set $\{\beta : C\psi \cap T_\beta \neq 0\}$ is finite. This enables us to define, for any $\beta < \mathfrak{m}$,

$$\psi_\beta = \bigwedge \{\varphi_\alpha : \alpha < \beta \text{ and } C\varphi_\beta \cap T_\alpha \neq 0\} \to \varphi_\beta$$

where if $C\varphi_\beta \cap T_\alpha = 0$ for all $\alpha < \beta$ we understand that ψ_β is the sentence $\forall v_0(v_0 = v_0) \to \varphi_\beta$. If $\chi \in \Gamma \sim \{\varphi_\alpha : \alpha < \mathfrak{m}\}$, we define a sentence χ' as follows:

$$\chi' = \bigwedge \{\varphi_\alpha : \alpha < \mathfrak{m} \text{ and } C\chi \cap T_\alpha \neq 0\} \to \chi;$$

again, we let χ' be the sentence $\forall v_0(v_0 = v_0) \to \chi$ if $C\chi \cap T_\alpha = 0$ for all $\alpha < \mathfrak{m}$ (this can happen, by (8), only if χ has *no* nonlogical constants).

Let $\Theta = \{\psi_\beta : \beta < \mathfrak{m}\}$ and $\Omega = \{\chi' : \chi \in \Gamma \sim \{\varphi_\alpha : \alpha < \mathfrak{m}\}\}$. The rest of this proof is devoted to showing that $\Omega \cup \Theta$ axiomatizes $\{\varphi : \Gamma \models \varphi\}$ and that Ω and Θ satisfy the conditions of Lemma 22.7; hence by that lemma, our lemma follows. Obviously $\Omega \cup \Theta \subseteq \{\varphi : \Gamma \models \varphi\}$. By induction on β it is clear that $\Omega \cup \Theta \models \varphi_\beta$ for each $\beta < \mathfrak{m}$. Hence it follows easily that $\Omega \cup \Theta \models \chi$ for each $\chi \in \Gamma \sim \{\varphi_\alpha : \alpha < \mathfrak{m}\}$. Thus $\Omega \cup \Theta$ axiomatizes Γ.

It remains only to check the conditions of 22.7. It is clear that $\varphi_\beta \in \Delta_\beta$ for each $\beta < \mathfrak{m}$, and hence by (5) φ is one–one. Therefore ψ is also one–one, and hence $|\Omega| \leq |\Theta|$. Furthermore, it is obvious that $\Omega \cap \Theta = 0$. To check condition (*ii*) of 22.7, assume that $\beta < \mathfrak{m}$ and that $(\Omega \cup \Theta) \sim \{\psi_\beta\} \models \psi_\beta$; we shall reach a contradiction. By this assumption, there exist $m, n \in \omega$ and $\alpha \in {}^m(\mathfrak{m} \sim \{\beta\})$, $\chi \in {}^n(\Gamma \sim \{\varphi_\gamma : \gamma < \mathfrak{m}\})$ such that

$$\text{(9)}\quad \models \bigwedge_{i<m} (\bigwedge \{\varphi_\gamma : \gamma < \alpha_i \text{ and } C\varphi_{\alpha i} \cap T_\gamma \neq 0\} \to \varphi_{\alpha i})$$
$$\wedge \bigwedge_{j<n} (\bigwedge \{\varphi_\gamma : \gamma < \mathfrak{m} \text{ and } C\chi_j \cap T_\gamma \neq 0\} \to \chi_j)$$
$$\to (\bigwedge \{\varphi_\gamma : \gamma < \beta \text{ and } C\varphi_\beta \cap T_\gamma \neq 0\} \to \varphi_\beta)),$$

where for simplicity we ignore the case when some of these expressions should be replaced by $\forall v_0(v_0 = v_0)$. We now construct a structure which is *not* a model of the sentence in (9). Let $I = \{i < m : C\varphi_{\alpha i} \cap T_\beta = 0\}$ and $J = \{j < n : C\chi_j \cap T_\beta = 0\}$. Note that if $\gamma < \beta$ then $C\varphi_\gamma \cap T_\beta = 0$, while $C\varphi_\beta \cap T_\beta \neq 0$. Thus

$$C\varphi_\beta \nsubseteq \bigcup_{i\in I} C\varphi_{\alpha i} \cup \bigcup_{j\in J} C\chi_j \cup \bigcup \{C\varphi_\gamma : \gamma < \beta, C\varphi_\beta \cap T_\gamma \neq 0\}.$$

It follows from assumption (*i*) of our lemma that there is an $\mathscr{L}$-structure $\mathfrak{A}$ which is a model of the sentence

$$\text{(10)}\qquad \bigwedge_{i\in I} \varphi_{\alpha i} \wedge \bigwedge_{j\in J} \chi_j \wedge \bigwedge \{\varphi_\gamma : \gamma < \beta, C\varphi_\beta \cap T_\gamma \neq 0\} \wedge \neg\varphi_\beta.$$

Now if $i \in m \sim I$, then $C\varphi_{\alpha i} \cap T_\beta \neq 0$; since $\alpha_i \neq \beta$ it is clear that $\beta < \alpha_i$, and hence $\mathfrak{A}$ is a model of the sentence

$$\text{(11)}\qquad \neg \bigwedge \{\varphi_\gamma : \gamma < \alpha_i \text{ and } C\varphi_{\alpha i} \cap T_\gamma \neq 0\}$$

for each $i \in m \sim I$. Similarly, $\mathfrak{A}$ is a model of the sentence

(12) $$\neg \bigwedge \{\varphi_\gamma : \gamma < m \text{ and } C\chi_j \cap T_\gamma \neq 0\}$$

for each $j \in n \sim J$. Thus $\mathfrak{A}$ is a model of all of the sentences in (10), (11), (12). This contradicts (9). □

Theorem 22.9. (Reznikoff). *Every theory is independently axiomatizable.*

Proof. Again, we let $C\varphi$ denote the set of all nonlogical constants occurring in φ. By 22.6 we may assume that our language has infinitely many nonlogical constants. Let Γ be a theory. We define $\langle \Delta_i : i < \omega \rangle$ by recursion:

$$\Delta_n = \left\{\varphi \in \Gamma : \text{not}\left(\bigcup_{i<n} \Delta_i \models \varphi\right) \text{ and } |C\varphi| = n\right\}.$$

Let $\Delta_\omega = \bigcup_{i<\omega} \Delta_i$. Then Δ_ω axiomatizes Γ. This follows from the following obvious fact:

$$\text{if } \varphi \in \Gamma \text{ and } |C\varphi| = n, \qquad \text{then } \bigcup_{i \le n} \Delta_i \models \varphi.$$

This fact also implies that

(1) $$\text{if } \varphi \in \Delta_\omega, \chi \in \Gamma, \text{ and } |C\chi| < |C\varphi|, \qquad \text{then not}(\models \chi \rightarrow \varphi).$$

Now to show that Γ is independently axiomatizable, it suffices to show that Δ_ω satisfies the condition (i) of Lemma 22.8. Hence assume that $\varphi \in \Delta_\omega$, $m \in \omega \sim 1$, $\psi \in {}^m\Delta_\omega$, $C\varphi \nsubseteq C \bigwedge_{i<m} \psi_i$, while $\models \bigwedge_{i<m} \psi_i \rightarrow \varphi$. By the interpolation theorem let χ be a sentence such that $\models \bigwedge_{i<m} \psi_i \rightarrow \chi$, $\models \chi \rightarrow \varphi$, and $C\chi \subseteq C\varphi \cap C \bigwedge_{i<m} \psi_i$. Since $C\varphi \nsubseteq C \bigwedge_{i<m} \psi_i$, it follows that $C\chi \subset C\varphi$ and hence $|C\chi| < |C\varphi|$. This contradicts (1). So the condition (i) of 22.8 holds, and 22.8 implies that Γ is independently axiomatizable. □

EXERCISES

22.10. Prove the interpolation theorem for *formulas* φ, ψ and χ.

22.11. Let φ be the sentence

$$\exists v_0 \forall v_1 (v_1 = v_0 \vee v_1 = \mathbf{O} v_0 \vee v_1 = \mathbf{OO} v_0)$$

and let ψ be the sentence

$$\forall v_0 \neg \mathbf{R} v_0 v_0 \wedge \forall v_0 \forall v_1 \forall v_2 (\mathbf{R} v_0 v_1 \wedge \mathbf{R} v_1 v_2 \rightarrow \mathbf{R} v_0 v_2) \rightarrow \exists v_0 \forall v_1 \neg \mathbf{R} v_0 v_1.$$

Show that $\models \varphi \rightarrow \psi$, and find a sentence χ involving only equality such that $\models \varphi \rightarrow \chi$ and $\models \chi \rightarrow \psi$.

22.12. Let φ be the sentence

$$\forall v_0 \forall v_1 (v_0 = v_1) \wedge \exists v_0 \mathbf{S} v_0$$

and let ψ be the sentence

$$\forall v_0 \mathbf{S} v_0 \wedge \forall v_0 \forall v_1 (\mathbf{R} v_0 \leftrightarrow \mathbf{R} v_1).$$

Show that $\models \varphi \rightarrow \psi$, but there is no sentence χ involving neither **R** nor equality such that $\models \varphi \rightarrow \chi$ and $\models \chi \rightarrow \psi$.

22.13. Show that 22.2(*ii*) cannot be omitted.

22.14. Show that 22.2(*iv*) cannot be omitted.

22.15. Under the hypotheses of 22.2, show that $\Gamma_1 \cup \Gamma_2$ is a conservative extension of Γ_0.

22.16. Let $\mathscr{L}_0$, Γ_0, $\mathscr{L}_i$, Γ_i for $i \in I$ be given, satisfying the following conditions:

(*i*) $\mathscr{L}_0$ and each $\mathscr{L}_i$ are first-order languages, and $\mathscr{L}_i$ is an expansion of $\mathscr{L}_0$;

(*ii*) for distinct $i, j \in I$, any nonlogical symbol common to $\mathscr{L}_i$ and $\mathscr{L}_j$ is a symbol of $\mathscr{L}_0$;

(*iii*) Γ_i is a consistent theory in $\mathscr{L}_i$ (each $i \in I$), and Γ_0 is a consistent theory in $\mathscr{L}_0$;

(*iv*) for each $i \in I$, Γ_i is a conservative extension of Γ_0.

Show that $\bigcup_{i \in I} \Gamma_i$ is consistent.

22.17. Let $\mathscr{L}$ be a language with nonlogical constants **O** (unary operation symbol) and **P** (unary relation symbol). Let Γ be the theory in $\mathscr{L}$ with the following axioms:

$$\mathbf{O} \text{ is one–one}$$
$$\forall v_0(\mathbf{O}^n v_0 = v_0 \rightarrow \mathbf{P} v_0), \text{ each } n \in \omega \sim 1,$$

where $\mathbf{O}^n$ is **O** repeated n times: Show that **P** is not definable in terms of **O** in $(\mathscr{L}, \Gamma)$.

22.18. For any structure $\mathfrak{A}$, give explicitly an independent axiomatization of the diagram of $\mathfrak{A}$.

22.19. If $\mathfrak{A}$ is an infinite free Boolean algebra and I is an ideal of $\mathfrak{A}$, then I has an irredundant set of generators. That is, there is a subset X of I which generates I but is such that no proper subset generates I.

23* Generalized Products

In this chapter we want to present some interesting logical properties of products. The proofs of the results use a very general notion of product, which we shall develop only enough to obtain results concerning ordinary products (but see the exercises). The idea of the generalization is to consider two languages, one language for the factors and one somehow connected with the index set of the product. A third language, for the product, is derived from these. The basic definitions are as follows.

Definition 23.1. Let two first-order languages $\mathscr{L}_{\text{fac}}$ and $\mathscr{L}_{\text{ind}}$ be given. Suppose that $\langle \mathfrak{A}_i : i \in I \rangle$ is a system of $\mathscr{L}_{\text{fac}}$-structures. For any formula φ of $\mathscr{L}_{\text{fac}}$ and any $f \in {}^{\omega}\mathsf{P}_{i \in I} A_i$ we set

$$K_{\varphi}^{\mathfrak{A}} f = \{i \in I : \mathfrak{A}_i \models \varphi[\mathrm{pr}_i \circ f]\}.$$

An $\mathscr{L}_{\text{fac}}$, $\mathscr{L}_{\text{ind}}$-sequence is a sequence $\langle \varphi, \psi_0, \ldots, \psi_m \rangle$ such that φ is a formula of $\mathscr{L}_{\text{ind}}$ with $\mathrm{Fv}\, \varphi \subseteq \{v_0, \ldots, v_m\}$ and $\psi_0, \ldots, \psi_m$ are formulas of $\mathscr{L}_{\text{fac}}$. We let $\rho\langle \varphi, \psi_0, \ldots, \psi_m \rangle$ be the least n such that $\mathrm{Fv} \bigvee_{i \leq m} \psi_i \subseteq \{v_0, \ldots, v_n\}$. The sequence $\langle \varphi, \psi_0, \ldots, \psi_m \rangle$ is *partitioning* if the following formulas are tautologies:

$$\bigvee_{i \leq m} \psi_i;$$
$$\neg(\psi_i \wedge \psi_j) \qquad \text{for } i \neq j.$$

We suppose now that $\mathscr{L}_{\text{fac}}$ and $\mathscr{L}_{\text{ind}}$ each have only finitely many nonlogical constants, and then we fix upon an enumeration $\zeta_0, \zeta_1, \ldots$ of all $\mathscr{L}_{\text{fac}}$, $\mathscr{L}_{\text{ind}}$-sequences. In case $\mathscr{L}_{\text{fac}}$ and $\mathscr{L}_{\text{ind}}$ are effectivized, we suppose that the associated sequence of Godel numbers $\mathscr{g}^+\zeta_0, \mathscr{g}^+\zeta_1, \ldots$ is recursive, where $\mathscr{g}^+$ is defined in the usual way. Let $\mathscr{L}_{\text{prod}}$ be a first-order language whose nonlogical constants are, for each $n \in \omega$, a ρ_n-ary relation constant $\mathbf{R}_n$, where $\rho_n = \rho\zeta_n$.

Now if we are given $\mathfrak{A}$ as above and we are also given an $\mathscr{L}_{\text{ind}}$ structure $\mathfrak{B}$ with universe $\mathbf{S}I$, we define the *generalized product* $\mathfrak{C} = \mathsf{P}^{\mathfrak{B}}_{i\in I}\,\mathfrak{A}_i$, which is to be an $\mathscr{L}_{\text{prod}}$-structure. Its universe is $\mathsf{P}_{i\in I} A_i$, while for each $n \in \omega$,

$\mathbf{R}_n^{\mathfrak{C}} = \{f \in {}^{\rho n}\mathsf{P}_{i\in I} A_i$: there is a $g \in {}^{\omega}\mathsf{P}_{i\in I} A_i$ such that $f \subseteq g$ and
$\mathfrak{B} \models \varphi[K^{\mathfrak{A}}_{\psi 0}g, \ldots, K^{\mathfrak{A}}_{\psi m}g]\}$,
where $\zeta_n = \langle \varphi, \psi_0, \ldots, \psi_m\rangle$.

We will also assume below that $\mathscr{L}_{\text{ind}}$ is a language extending a language for Boolean algebras, and that $\mathfrak{B}$ is an expansion of the Boolean algebra $(\mathbf{S}I, \cup, \cap, \sim, 0, I)$.

Before proceeding to the main theorem let us see how to conceive of direct products as a special case of this notion. Let $\mathfrak{B}$ be the Boolean algebra $(\mathbf{S}(I), \cup, \cap, \sim, 0, I)$. Now $\varphi = v_0 \equiv 1$ is an $\mathscr{L}_{\text{ind}}$ formula with $\text{Fv}\,\varphi = \{v_0\}$, and, if $\mathbf{S}$ is an n-ary relation symbol of $\mathscr{L}_{\text{fac}}$, then $\psi = \mathbf{S}v_0 \cdots v_{n-1}$ is a formula of $\mathscr{L}_{\text{fac}}$. Thus (φ, ψ) is an $\mathscr{L}_{\text{fac}}, \mathscr{L}_{\text{ind}}$-sequence, say $(\varphi, \psi) = \zeta_p$. Then for any $f \in {}^{n}\mathsf{P}_{i\in I} A_i$,

$$\begin{aligned} f \in \mathbf{R}_p^{\mathfrak{C}} \quad & \text{iff } \exists g \in {}^{\omega}\mathsf{P}_{i\in I} A_i \text{ such that } f \subseteq g \text{ and } \mathfrak{B} \models \varphi[K^{\mathfrak{A}}_{\psi}g] \\ & \text{iff } \exists g \in {}^{\omega}\mathsf{P}_{i\in I} A_i \text{ such that } f \subseteq g \text{ and } K^{\mathfrak{A}}_{\psi}g = I \\ & \text{iff } \forall i \in I(f_{0i}, \ldots, f_{n-1,i}) \in \mathbf{S}^{\mathfrak{A}i}. \end{aligned}$$

Thus $\mathbf{R}_p^{\mathfrak{C}}$ is the same as the relation of $\mathsf{P}_{i\in I}\,\mathfrak{A}_i$ corresponding to $\mathbf{S}$. Similarly, for an n-ary operation symbol $\mathbf{O}$ there is a $p \in \omega$ such that $\mathbf{R}_p^{\mathfrak{C}}$ is an $(n + 1)$-ary relation of $\mathfrak{C}$ which is identical with the n-ary operation that is the operation of $\mathsf{P}_{i\in I}\,\mathfrak{A}_i$ corresponding to $\mathbf{O}$ (take φ as above and $\psi = \mathbf{O}v_0 \cdots v_{n-1} \equiv v_n$).

Lemma 23.2. *If $(\varphi, \psi_0, \ldots, \psi_m)$ is an $\mathscr{L}_{\text{fac}}, \mathscr{L}_{\text{ind}}$-sequence then one can effectively find a partitioning $\mathscr{L}_{\text{fac}}, \mathscr{L}_{\text{ind}}$-sequence $(\chi, \theta_0, \ldots, \theta_n)$ such that* $\text{Fv} \bigwedge_{i\le m} \psi_i = \text{Fv} \bigwedge_{i \le n} \theta_i$ *and for any $\mathfrak{A}$ and $\mathfrak{B}$ as above, and $f \in {}^{\omega}\mathsf{P}_{i\in I} A_i$, $\mathfrak{B} \models \varphi[K^{\mathfrak{A}}_{\psi 0}f, \ldots, K^{\mathfrak{A}}_{\psi m}f]$ iff $\mathfrak{B} \models \chi[K^{\mathfrak{A}}_{\theta 0}f, \ldots, K^{\mathfrak{A}}_{\theta n}f]$.*

PROOF. Let $n = 2^{m+1} - 1$, and let $r_0, \ldots, r_n$ be a list of all subsets of $\{0, \ldots, m\}$. For each $k \le n$ let

$$\theta_k = \bigwedge_{j\in rk} \psi_j \wedge \bigwedge_{j \in (m+1)\sim rk} \neg\psi_j.$$

and set

$$\chi = \varphi(\mathbf{\Sigma}\{v_k : k \le n, 0 \in r_k\}, \ldots, \mathbf{\Sigma}\{v_k : k \le n, m \in r_k\}).$$

Clearly $(\chi, \theta_0, \ldots, \theta_n)$ is a partitioning $\mathscr{L}_{\text{fac}}, \mathscr{L}_{\text{ind}}$-sequence and $\text{Fv} \bigwedge_{i\le m} \psi_i = \text{Fv} \bigwedge_{i\le n} \theta_i$. Now note that

$$\begin{aligned} &\mathfrak{B} \models \chi[K^{\mathfrak{A}}_{\theta 0}f, \ldots, K^{\mathfrak{A}}_{\theta n}f] \quad \text{iff} \\ &\quad \mathfrak{B} \models [\bigcup\{K^{\mathfrak{A}}_{\theta k}f : k \le n, 0 \in r_k\}, \ldots, \bigcup\{K^{\mathfrak{A}}_{\theta k}f : k \le n, m \in r_k\}]. \end{aligned}$$

But it is easily seen that if $l \le m$, then $\bigcup\{K^{\mathfrak{A}}_{\theta k}f : k \le n, l \in r_k\} = K^{\mathfrak{A}}_{\psi l}f$, so the desired equivalence follows. □

The main theorem on generalized products is as follows:

Theorem 23.3 (Feferman, Vaught). *Assume the notation above. Then for each formula χ of $\mathscr{L}_{\text{prod}}$ there is a partitioning $\mathscr{L}_{\text{fac}}$, $\mathscr{L}_{\text{ind}}$-sequence $\xi = (\varphi, \psi_0, \ldots, \psi_n)$ with the following properties (for any system $\langle \mathfrak{A}_i : i \in I \rangle$ of $\mathscr{L}_{\text{fac}}$-structures and any $\mathscr{L}_{\text{ind}}$-structure $\mathfrak{B}$):*

(i) v_i is free in χ iff it is free in some ψ_j;

(ii) ξ is effectively obtained from χ;

(iii) $\mathsf{P}^{\mathfrak{B}}_{i \in I} \mathfrak{A}_i \models \chi[f]$ iff $\mathfrak{B} \models \varphi[K^{\mathfrak{A}}_{\psi 0} f, \ldots, K^{\mathfrak{A}}_{\psi m} f]$.

Proof. We proceed by induction on χ, where by 23.2 we do not have to produce a partitioning sequence. First suppose that χ is $\mathbf{R}_n v_{k0} \cdots v_{k(\rho n-1)}$. Let $\zeta_n = (\varphi, \psi_0, \ldots, \psi_m)$. For each $j \leq m$ let $\chi_j = \psi_j(v_{k0}, \ldots, v_{k(\rho n-1)})$. Consider the $\mathscr{L}_{\text{fac}}$, $\mathscr{L}_{\text{ind}}$-sequence $(\varphi, \chi_0, \ldots, \chi_m)$. We have, with $\mathfrak{C} = \mathsf{P}^{\mathfrak{B}}_{i \in I} \mathfrak{A}_i$,

$\mathsf{P}^{\mathfrak{B}}_{i \in I} \mathfrak{A}_i \models \chi[f]$ iff $(f_{k0}, \ldots, f_{k(\rho n-1)}) \in \mathbf{R}^{\mathfrak{C}}_n$ iff there is a $g \supseteq (f_{k0}, \ldots, f_{k(\rho n-1)})$ such that $g \in {}^{\omega}\mathsf{P}_{i \in I} A_i$ and $\mathfrak{B} \models \varphi[K^{\mathfrak{A}}_{\psi 0} g, \ldots, K^{\mathfrak{A}}_{\psi m} g]$.

Now for each $j \leq m$, if g is as above, then

$$\begin{aligned} K^{\mathfrak{A}}_{\psi j} g &= \{i \in I : \mathfrak{A}_i \models \psi_j[\mathrm{pr}_i \circ g]\} \\ &= \{i \in I : \mathfrak{A}_i \models \psi_j[f_{k0i}, \ldots, f_{k(\rho n-1)i}]\} \\ &= \{i \in I : \mathfrak{A}_i \models \chi_j[\mathrm{pr}_i \circ f]\} \\ &= K^{\mathfrak{A}}_{\chi j} f. \end{aligned}$$

Thus, by the above,

$$\mathsf{P}^{\mathfrak{B}}_{i \in I} \mathfrak{A}_i \models \chi[f] \quad \text{iff } \mathfrak{B} \models \varphi[K^{\mathfrak{A}}_{\chi 0} f, \ldots, K^{\mathfrak{A}}_{\chi m} f].$$

Second, suppose that χ is $v_j \equiv v_k$, let φ be $v_0 \equiv \mathbf{1}$, and let ψ be $v_j \equiv v_k$. Thus (φ, ψ) is an $\mathscr{L}_{\text{fac}}$, $\mathscr{L}_{\text{ind}}$-sequence, and

$$\begin{aligned} \mathsf{P}^{\mathfrak{B}}_{i \in I} \mathfrak{A}_i \models \chi[f] \quad &\text{iff } f_j = f_k \quad &&\text{iff } \forall i \in I (f_{ji} = f_{ki}) \\ &\text{iff } K^{\mathfrak{A}}_{\psi} = I &&\text{iff } \mathfrak{B} \models \varphi[K^{\mathfrak{A}}_{\psi} f]. \end{aligned}$$

Now we proceed to the induction steps. Suppose that $(\varphi, \psi_0, \ldots, \psi_m)$ is correlated with χ as in the statement of the theorem. Then $(\neg\varphi, \psi_0, \ldots, \psi_m)$ is an $\mathscr{L}_{\text{fac}}$, $\mathscr{L}_{\text{ind}}$-sequence, and

$$\begin{aligned} \mathsf{P}^{\mathfrak{B}}_{i \in I} \mathfrak{A}_i \models \neg\chi[f] \quad &\text{iff not}(\mathsf{P}^{\mathfrak{B}}_{i \in I} \mathfrak{A}_i \models \chi[f]) \\ &\text{iff not}(\mathfrak{B} \models \varphi[K^{\mathfrak{A}}_{\psi 0} f, \ldots, K^{\mathfrak{A}}_{\psi m} f]) \\ &\text{iff } \mathfrak{B} \models \neg\varphi[K^{\mathfrak{A}}_{\psi 0} f, \ldots, K^{\mathfrak{A}}_{\psi 0} f]. \end{aligned}$$

Next, suppose that $(\varphi, \psi_0, \ldots, \psi_m)$ is correlated with χ_0 and $(\theta, \tau_0, \ldots, \tau_n)$ with χ_1, as in the statement of the theorem. Then $(\varphi \vee \theta(v_{m+1}, \ldots, v_{m+n+1}), \psi_0, \ldots, \psi_m, \tau_0, \ldots, \tau_n)$ is an $\mathscr{L}_{\text{fac}}$, $\mathscr{L}_{\text{ind}}$-sequence, and it is straightforward to check that it works for $\chi_0 \vee \chi_1$. Conjunction is treated similarly.

Finally, suppose that a formula $\forall v_j \chi$ is given, and that a sequcnce

$\zeta = (\varphi, \psi_0, \ldots, \psi_m)$ is correlated with χ (in particular, ζ is a partitioning sequence). For each $k \leq m$ let μ_k be the formula $\exists v_j \psi_k$, and let θ be the formula,

$$\forall v_{m+1} \cdots \forall v_{2m+1}[v_{m+1} + \cdots + v_{2m+1} = 1 \wedge \bigwedge_{k<l\leq m} (v_{m+k+1} \cdot v_{m+l+1} = 0) \wedge \bigwedge_{k \leq m} (v_{m+k+1} \cdot v_k = v_{m+k+1}) \rightarrow \varphi(v_{m+1}, \ldots, v_{2m+1})].$$

Thus $(\theta, \mu_0, \ldots, \mu_m)$ is an $\mathscr{L}_{\text{fac}}$, $\mathscr{L}_{\text{ind}}$-sequence. Now suppose that $\mathsf{P}^{\mathfrak{B}}_{i\in I} \mathfrak{A}_i \models \forall v_j \chi[f]$. In order to check that $\mathfrak{B} \models \theta[K^{\mathfrak{A}}_{\mu 0} f, \ldots, K^{\mathfrak{A}}_{\mu(n-1)} f]$, assume that $J_0, \ldots, J_m \in B$ forms a partition of I, and that $J_k \subseteq K^{\mathfrak{A}}_{\mu k} f$ for each $k \leq m$. We now define $g \in \mathsf{P}_{i\in I} A_i$. Given $i \in I$, there is a unique $k \leq m$ such that $i \in J_k$, and hence $i \in K^{\mathfrak{A}}_{\mu k} f$, which means that $\mathfrak{A}_i \models \mu_k[\mathrm{pr}_i \circ f]$. Since $\mu_k = \exists v_j \psi_k$, we may hence choose $g_i \in A_i$ such that $\mathfrak{A}_i \models \psi_k[(\mathrm{pr}_i \circ f)^j_{gi}]$. This defines g. Thus for any $i \in I$, with k as above, $\mathfrak{A}_i \models \psi_k[\mathrm{pr}_i \circ f^j_g]$, so $i \in K^{\mathfrak{A}}_{\psi k} f^j_g$. That is, $J_k \subseteq K^{\mathfrak{A}}_{\psi k} f^j_g$. Now since ζ is a partitioning sequence, it follows that $K^{\mathfrak{A}}_{\psi m} f^j_g, \ldots, K^{\mathfrak{A}}_{\psi m} f^j_g$ is a partition of I. Hence $J_k = K^{\mathfrak{A}}_{\psi k} f^j_g$ for all $k \leq m$. Also, $\mathsf{P}^{\mathfrak{B}}_{i\in I} \mathfrak{A}_i \models \chi[f^j_g]$, so by the induction hypothesis, $\mathfrak{B} \models \varphi[J_0, \ldots, J_m]$, as desired.

Conversely, suppose that $\mathfrak{B} \models \theta[K^{\mathfrak{A}}_{\mu 0} f, \ldots, K^{\mathfrak{A}}_{\mu n} f]$, and let $g \in \mathsf{P}_{i\in I} A_i$. Then $K^{\mathfrak{A}}_{\psi k} f^j_g \subseteq K^{\mathfrak{A}}_{\mu k} f$ for all $k \leq m$, and $K^{\mathfrak{A}}_{\psi m} f^j_g$ is a partition of I, so the definition of θ yields $\mathfrak{B} \models \varphi[K^{\mathfrak{A}}_{\psi 0} f^j_g, \ldots, K^{\mathfrak{A}}_{\psi m} f^j_g]$. By the induction hypothesis, $\mathsf{P}^{\mathfrak{B}}_{i\in I} \mathfrak{A}_i \models \chi[f^j_g]$. □

Now we shall give some interesting consequences of 23.3.

Theorem 23.4. *If $\mathfrak{A}_i \equiv_{\text{ee}} \mathfrak{B}_i$ for each $i \in I$, then $\mathsf{P}_{i\in I} \mathfrak{A}_i \equiv_{\text{ee}} \mathsf{P}_{i\in I} \mathfrak{B}_i$.*

Proof. Let $\mathfrak{C}$ be the Boolean algebra of all subsets of I. As we saw following Definition 23.1, any sentence χ of our given language can be considered as a sentence of $\mathscr{L}_{\text{prod}}$. Thus by 23.3, $\mathsf{P}_{i\in I} \mathfrak{A}_i \models \chi$ iff $\mathfrak{C} \models \varphi[K^{\mathfrak{A}}_{\psi 0}, \ldots, K^{\mathfrak{A}}_{\psi(m-1)}]$ iff $\mathfrak{C} \models \varphi[K^{\mathfrak{B}}_{\psi 0}, \ldots, K^{\mathfrak{B}}_{\psi(m-1)}]$ iff $\mathsf{P}_{i\in I} \mathfrak{B}_i \models \chi$, where $(\varphi, \psi_0, \ldots, \psi_m)$ is correlated with χ by 23.3, and $K^{\mathfrak{A}}_{\psi k} = K^{\mathfrak{B}}_{\psi k}$ since $\mathfrak{A}_i \equiv_{\text{ee}} \mathfrak{B}_i$ for all $i \in I$. □

Theorem 23.5. *If $\mathfrak{A} \equiv_{\text{ee}} \mathfrak{B}$ and I and J are index sets such that either $|I| = |J| < \aleph_0$ or else $|I|, |J| \geq \aleph_0$, then ${}^I\mathfrak{A} \equiv_{\text{ee}} {}^I\mathfrak{B}$.*

Proof. By Theorem 21.34, the hypothesis implies that the Boolean algebras $\mathbf{S}I$ and $\mathbf{S}J$ are elementarily equivalent. Now if ψ is a sentence of our language, then $K_\psi^{\langle \mathfrak{A}: i\in I\rangle}$ is I or 0, and $K^{\langle \mathfrak{B}: j\in J\rangle}$ is J or 0. Furthermore, the first is I iff the second is J, since $\mathfrak{A} \equiv_{\text{ee}} \mathfrak{B}$. Hence for any sentence χ, if $(\varphi, \psi_0, \ldots, \psi_m)$ is correlated by 23.3 we have

$$\begin{aligned} {}^I\mathfrak{A} \models \chi \quad & \text{iff } \mathbf{S}I \models \varphi[K^{\langle \mathfrak{A}: i\in I\rangle}_{\psi 0}, \ldots, K^{\langle \mathfrak{A}: i\in I\rangle}_{\psi m}] \\ & \text{iff } \mathbf{S}J \models \varphi[K^{\langle \mathfrak{B}: j\in J\rangle}_{\psi 0}, \ldots, K^{\langle \mathfrak{A}: i\in I\rangle}_{\psi m}] \\ & \text{iff } {}^J\mathfrak{B} \models \chi. \end{aligned}$$ □

Theorem 23.6 (Vaught). *If $\mathsf{P}_{i\in F} \mathfrak{A}_i \models \chi$ for each finite $F \subseteq I$, then $\mathsf{P}_{i\in I} \mathfrak{A}_i \models \chi$.*

Proof. We may assume that I is infinite. Assume the hypothesis of 23.6. Let $(\varphi, \psi_0, \ldots, \psi_m)$ be correlated with χ by 23.3. Set $L = \{k \leq m : K^{\mathfrak{A}}_{\psi k}$ is finite$\}$. For each $s \in \omega$ let θ_s be the following sentence of the theory of Boolean algebras.

$$\forall v_0 \cdots \forall v_m[(\langle v_0, \ldots, v_m\rangle \text{ forms a partition of the universe}) \wedge \bigwedge_{k\in L} (\text{there are exactly } |K^{\mathfrak{A}}_{\psi k}| \text{ atoms } \leq v_k) \wedge \bigwedge_{k\in(m+1)\sim L} (\text{there are at least } s \text{ atoms } \leq v_k) \rightarrow \varphi].$$

We claim

(1) there is an $s \in \omega$ such that θ_s holds in every finite **BA**.

Otherwise, for each $s \in \omega$ let $\mathfrak{B}_s$ be a finite BA in which θ_s fails; say $b^s_0, \ldots, b^s_m \in B_s$ satisfying the hypothesis of the implication above but not φ in $\mathfrak{B}_s$. Choose a finite $L_s \subseteq I$ so that $K^{\mathfrak{A}}_{\psi k} \subseteq L_s$ for each $k \in L$, while $L_s \cap K^{\mathfrak{A}}_{\psi k}$ has at least as many elements as there are atoms of $\mathfrak{B}_s \leq b^s_k$, for $k \in (m + 1) \sim L$. Note that $K^{\mathfrak{A}\upharpoonright Ls}_{\psi k} = L_s \cap K^{\mathfrak{A}}_{\psi k}$ for each $k \leq m$. Let $\mathfrak{C}_s = \mathbf{S}L_s$. Clearly there is an isomorphism f_s of $\mathfrak{B}_s$ into $\mathfrak{C}_s$ such that $f_s b^s_k = K^{\mathfrak{A}\upharpoonright Ls}_{\psi k}$ for each $k \leq m$. Thus $\mathfrak{B}_s \models \neg\varphi[b^s_0, \ldots, b^s_m]$, while $\mathfrak{C}_s \models \varphi[f_s b^s_0, \ldots, f_s b^s_m]$, by the hypothesis of the theorem. Let F be a nonprincipal ultrafilter over ω. Then $\mathsf{P}_{s\in\omega} \mathfrak{B}_s/F$ and $\mathsf{P}_{s\in\omega} \mathfrak{C}_s/F$ are infinite atomic BA's. Let $b'_k = [\langle b^s_k : s \in \omega\rangle]$. Then there is an isomorphism g of $\mathsf{P}_{s\in\omega} \mathfrak{B}_s/F$ into $\mathsf{P}_{s\in\omega} \mathfrak{C}_s/F$ such that $g[x] = [\langle f_s x_s : s \in \omega\rangle]$ for all $x \in \mathsf{P}_{s\in\omega} B_s$. Now let $\mathfrak{B}'$ and $\mathfrak{C}'$ be the expansions of $\mathsf{P}_{s\in\omega} \mathfrak{B}_s/F$ and $\mathsf{P}_{s\in\omega} \mathfrak{C}_s/F$ respectively to models of ${}_0\Gamma^{\infty}_{\text{At}}$. By the proof of 21.34, ${}_0\Gamma^{\infty}_{\text{At}}$ is model complete. Hence by 21.27(*iv*) there is a universal formula ψ in ${}_0\Gamma^{\infty}_{\text{At}}$ such that ${}_0\Gamma^{\infty}_{\text{At}} \models \varphi \leftrightarrow \psi$ and $\text{Fv}\,\psi \subseteq \{v_0, \ldots, v_m\}$. Since $\mathfrak{C}' \models \varphi[gb'_0, \ldots, gb'_m]$, it follows that $\mathfrak{C}' \models \psi[gb_0, \ldots, gb_m]$, hence $\mathfrak{B}' \models \psi[b'_0, \ldots, b'_m]$ and $\mathfrak{B} \models \varphi[b_0, \ldots, b_m]$. But also $\mathfrak{B} \models \neg\varphi[b'_0, \ldots, b'_m]$, contradiction. Thus (1) holds.

From (1) and the completeness of the theory of infinite atomic BA's it follows that θ_s holds in all infinite BA's, for some s. Choosing such an s, we see that $\mathbf{S}I \models \varphi[K^{\mathfrak{A}}_{\psi 0}, \ldots, K^{\mathfrak{A}}_{\psi m}]$. Hence by 23.3, $\mathsf{P}_{i\in I} \mathfrak{A}_i \models \chi$. □

Corollary 23.7. *If* Γ *is a set of sentences and* Mod $\Gamma = \{\mathfrak{A} : \mathfrak{A}$ *is a model of* $\Gamma\}$ *is closed under products with two, or zero, factors, then* Mod Γ *is closed under arbitrary products.*

Finally, we give some applications to decidable theories.

Theorem 23.8. *If* $\mathfrak{A}$ *has a decidable theory, then so does* ${}^I\mathfrak{A}$.

Proof. From 21.34, $\mathbf{S}I$ has a decidable theory. The decision procedure for $\{\chi : {}^I\mathfrak{A} \models \chi\}$ goes as follows. Given χ, determine $(\varphi, \psi_0, \ldots, \psi_m)$ by 23.3. Note that $K^{\mathfrak{A}}_{\psi k} = I$ or 0 for each $k \leq m$. Hence we can determine if $\mathbf{S}I \models \varphi[K^{\mathfrak{A}}_{\psi 0}, \ldots, K^{\mathfrak{A}}_{\psi m}]$ by first deciding whether or not $\mathfrak{A} \models \psi_k$ for each $k \leq m$, then using the decision procedure for $\mathbf{S}I$. By 23.3 this gives the decision procedure for ${}^I\mathfrak{A} \models \chi$. □

Theorem 23.9. *If $\{\chi : \mathbf{K} \models \chi\}$ is decidable, then so is $\{\chi : \mathbf{PK} \models \chi\}$, where* **PK** *is the class of all products of members of* **K**.

PROOF. The decision procedure for $\{\chi : \mathbf{PK} \models \chi\}$ is as follows. Given χ, let $(\varphi, \psi_0, \ldots, \psi_m)$ be determined as in 23.3. Then let $T = \{k \leq m : \mathbf{K} \models \neg\psi_k\}$, which can be effectively determined, by assumption. Let θ be the following sentence of the theory of Boolean algebras:

$$\forall v_0 \cdots \forall v_m[v_0 + \cdots + v_m = 1 \wedge \bigwedge_{i<j<m} (v_i \cdot v_j = 0) \wedge \bigwedge_{k\in T} v_k = 0 \rightarrow \varphi].$$

We claim

(1) $\qquad \mathbf{PK} \models \chi \qquad$ iff θ holds in all atomic BA's.

By 21.34 and 15.6, this completes the decision procedure.

To prove (1), first suppose that $\mathbf{PK} \models \chi$. By 21.34 it suffices to show that θ holds in any BA of the form $\mathbf{S}I$. Assume that $J_0, \ldots, J_m \subseteq I$, they form a partition of I, and $J_k = 0$ whenever $k \in T$; we must show that $\mathbf{S}I \models \varphi[J_0, \ldots, J_m]$. Now for $k \in (m + 1) \sim T$ there is a structure $\mathfrak{A}_k \in \mathbf{K}$ such that $\mathfrak{A}_k \models \psi_k$. Define $\mathfrak{B}: I \rightarrow \mathbf{K}$ as follows. Given $i \in I$, there is a unique $k \in (m + 1) \sim T$ such that $i \in J_k$, and we set $\mathfrak{B}_i = \mathfrak{A}_k$. Thus $K^{\mathfrak{B}}_{\psi k} \supseteq J_k$ for each $k \in m + 1$, and hence, since $(\varphi, \psi_0, \ldots, \psi_m)$ is partitioning, $K^{\mathfrak{B}}_{\psi k} = J_k$ for all $k \in m + 1$. But $\mathbf{PK} \models \chi$, so by 23.3, $\mathbf{S}I \models \varphi[K^{\mathfrak{B}}_{\psi 0}, \ldots, K^{\mathfrak{B}}_{\psi m}]$, as desired.

Conversely, suppose that θ holds in all atomic BA's. To show that $\mathbf{PK} \models \chi$, let $\mathfrak{A}: I \rightarrow \mathbf{K}$ be given. Then $K^{\mathfrak{A}}_{\psi 0}, \ldots, K^{\mathfrak{A}}_{\psi m}$ forms a partition of I, and $K^{\mathfrak{A}}_{\psi k} = 0$ if $k \in T$. Hence, since θ holds in $\mathbf{S}I$, we see that $\mathbf{S}I \models \varphi[K^{\mathfrak{A}}_{\psi 0}, \ldots, K^{\mathfrak{A}}_{\psi m}]$. Hence, by 23.3, $\mathsf{P}_{i\in I} \mathfrak{A}_i \models \chi$. □

We conclude this chapter with two simple applications of the above results.

Corollary 23.10. *The theory of Boolean algebras which have a maximal nonzero atomless element, but infinitely many atoms, is complete and decidable.*

PROOF. Let $\mathfrak{A}$ and $\mathfrak{B}$ be two such Boolean algebras. Then we can write $\mathfrak{A} \cong \mathfrak{C}_0 \times \mathfrak{C}_1$ and $\mathfrak{B} \cong \mathfrak{D}_0 \times \mathfrak{D}_1$, where $\mathfrak{C}_0$ and $\mathfrak{D}_0$ are atomless, while $\mathfrak{C}_1$ and $\mathfrak{D}_1$ are infinite atomic. By 21.7 and 21.34 we have $\mathfrak{C}_i \equiv_{\text{ee}} \mathfrak{D}_i$ for $i < 2$. Hence $\mathfrak{A} \equiv_{\text{ee}} \mathfrak{B}$ by 23.4. □

The following is immediate from 23.6:

Corollary 23.11. *For any structure $\mathfrak{A}$, a sentence φ holds in all powers ${}^I\mathfrak{A}$ of $\mathfrak{A}$ iff φ holds in all finite powers ${}^i\mathfrak{A}$ of $\mathfrak{A}$, $i \in \omega$.*

EXERCISES

23.12. Let $\mathscr{L}_{\text{fin}}$ be an expansion of the language for Boolean algebras obtained by adjoining a unary relation symbol **Fin**. Given any formula φ of $\mathscr{L}_{\text{fin}}$

and any $m \in \omega$ with $\mathrm{Fv}\,\varphi \subseteq \{v_0, \ldots, v_{m-1}\}$ one can effectively find $M \in \omega$, functions $p^k \in {}^{Sm}\omega$ for each $k < m$ and partitions $(U_1^k, U_2^k, U_3^k, U_4^k)$ of Sm for each $k < M$, such that for any $\mathscr{L}$-structure of the form $\mathfrak{B} = (SI, \cup, \cap, \sim, 0, I, F_I)$, where F_I is the collection of all finite subsets of I, and given $x \in {}^{\omega}B$ the following conditions are equivalent:

(*i*) $\mathfrak{B} \models \varphi[x]$;

(*ii*) there exists $k < M$ such that for each $r \subseteq m$, the set $\bigcap_{i \in r} x_i \cap \bigcap_{i \in m \sim r}(I \sim x_i)$ has:

(a) exactly $p^k r$ elements, if $r \in U_1^k$;

(b) at least $p^k r$ elements, if $r \in U_2^k$;

(c) at least $p^k r$ elements, but only finitely many elements, if $r \in U_3^k$;

(d) infinitely many elements, if $r \in U_4^k$.

Hint: proceed by induction on φ, beginning with atomic formulas of the form $v_i \equiv v_j$, $v_i + v_j \equiv v_k$, etc.

23.13. The theory of a structure of the form $\mathfrak{B}$ in 23.12 is decidable.

23.14. Let a language $\mathscr{L}$ have a certain individual constant **c** in it. Given a system $\langle(\mathfrak{A}_i, a_i) : i \in I\rangle$ of $\mathscr{L}$-structures such that for each $i \in I$, $\{a_i\}$ is the universe of a substructure of $\mathfrak{A}_i$, we form the *weak direct product* $\mathsf{P}^{\mathrm{w}}_{i \in I}\mathfrak{A}_i$ as follows. Its universe is $\{f \in \mathsf{P}_{i \in I} A_i : \{i : f_i \neq a_i\}$ is finite$\}$. This set is clearly the universe of a substructure $\mathsf{P}^{\mathrm{w}}_{i \in I}\mathfrak{A}_i$ of $\mathsf{P}_{i \in I}\mathfrak{A}_i$.

We now indicate how to encompass such products under the general notion of product of 23.1. We let $\mathscr{L}_{\mathrm{fac}} = \mathscr{L}$ and $\mathscr{L}_{\mathrm{ind}} = \mathscr{L}_{\mathrm{fin}}$ in 23.1. Let $\zeta_m = (\mathbf{Fin}\, v_0, \neg v_0 \equiv \mathbf{c})$. We define a translation from certain formulas χ of $\mathscr{L}$ to formulas χ^* of $\mathscr{L}_{\mathrm{prod}}$. Set $(v_i \equiv v_j)^* = v_i \equiv v_j$. For an n-ary relation symbol **S** of $\mathscr{L}$, set $\zeta_p = (v_0 \equiv \mathbf{1}, \mathbf{S}v_0 \cdots v_{n-1})$ and let

$$(\mathbf{S}v_{i0} \cdots v_{i(n-1)})^* = \mathbf{R}_p v_{i0} \cdots v_{i(n-1)}.$$

If **O** is an n-ary operation symbol of $\mathscr{L}$, set $\zeta_q = (v_0 \equiv \mathbf{1}, \mathbf{O}v_0 \cdots v_{n-1} \equiv v_n)$ and let

$$(\mathbf{O}v_{i0} \cdots v_{i(n-1)} \equiv v_{in})^* = \mathbf{R}_q v_{i0} \cdots v_{in}.$$

Further, let $(\neg\varphi)^* = \neg\varphi^*$, $(\varphi \vee \psi)^* = \varphi^* \vee \psi^*$, $(\varphi \wedge \psi)^* = \varphi^* \wedge \psi^*$, and

$$(\forall v_i \varphi)^* = \forall v_i(\mathbf{R}_m v_i \rightarrow \varphi^*).$$

Now let $\langle(\mathfrak{A}_i, a_i) : i \in I\rangle$ be as above, let $\mathfrak{B}$ be as in 23.12, and show that for any formula χ of $\mathscr{L}$ and any $x \in {}^{\omega}\mathsf{P}^{\mathrm{w}}_{i \in I} A_i$,

$$\mathsf{P}^{\mathfrak{B}}_{i \in I}\mathfrak{A}_i \models \chi^*[x] \quad \text{iff } \mathsf{P}^{\mathrm{w}}_{i \in I}\mathfrak{A}_i \models \chi[x].$$

23.15. If $\mathfrak{A}$ has a decidable theory, then so does ${}^{I}\mathfrak{A}^{\mathrm{w}}$ (i.e., $\mathsf{P}^{\mathrm{w}}_{i \in I}\mathfrak{B}_i$ where $\mathfrak{B}_i = \mathfrak{A}$ for each $i \in I$).

23.16 (Skolem). The structure $(\omega \sim 1, \cdot, 1)$ has a decidable theory. *Hint:* use 23.15, and recall that $(\omega, +, 0)$ has a decidable theory.

23.17. The structure $(\omega, \cdot)$ has a decidable theory. *Hint:* if φ is any formula built up from atomic formulas $v_i \equiv v_j$ and $v_i \cdot v_j \equiv v_k$, and $y : \{i : v_i$ occurs in $\varphi\} \rightarrow 2$, we can effectively associate a formula φ_y as follows:

$(v_i \equiv v_j)_y = v_i \equiv v_j$ if $y_i, y_j \neq 0$,
$(v_i \equiv v_j)_y = \forall v_0(v_0 \equiv v_0)$ if $y_i = y_j = 0$,
$(v_i \equiv v_j)_y = \neg\forall v_0(v_0 \equiv v_0)$ otherwise,
$(v_i \cdot v_j \equiv v_k)_y = v_i \cdot v_j \equiv v_k$ if $y_i, y_j, y_k \neq 0$,
$(v_i \cdot v_j \equiv v_k)_y = \forall v_0(v_0 \equiv v_0)$ if $y_i = y_k = 0$ or $y_j = y_k = 0$,
$(v_i \cdot v_j \equiv v_k)_y = \neg\forall v_0(v_0 \equiv v_0)$ otherwise,
$(\neg\varphi)_y = \neg\varphi_y$, $(\varphi \vee \psi)_y = \varphi_y \vee \psi_y$, $(\varphi \wedge \psi)_y = \varphi_y \wedge \psi_y$,
$(\forall v_i\varphi)_y = \varphi_{y \sim \{(i,yi)\}}$ if v_i does not occur in φ,
$(\forall v_i\varphi)_y = \forall v_i\varphi_z \wedge \varphi_w$ if v_i occurs in φ, where $z = y^i_1$, $w = y^i_0$.

Note that if $y_i = 0$, then v_i does not occur free in φ_y. Now show that for any φ, y, x, z, if φ and y are as above, $x \in {}^\omega\omega$, $y \subseteq \mathrm{sg} \circ x$, $z \in {}^\omega(\omega \sim 1)$, $zi = xi$ if $xi \neq 0$, and $zi = 1$ if $xi = 0$, then $(\omega, \cdot) \models \varphi[x]$ iff $(\omega \sim 1, \cdot) \models \varphi_y[z]$. Then use 23.16.

23.18. The theory of any free Abelian group is decidable.

23.19. Theorem 23.6 does not carry over for weak direct products. *Hint:* consider Boolean algebras.

24 Equational Logic

In this chapter we shall prove several theorems of a truly model-theoretical character, relating structures and languages, but restricting ourselves to very simple first-order sentences, namely equations. The main result is a purely algebraic characterization of those classes of structures definable by equations. The results of this chapter show in a simplified situation the kind of model-theoretic connections with which we deal in more complicated settings in the two following chapters. Also, a great deal of algebraic machinery needed later is introduced. The material of this chapter really lies on the borderline between logic and general algebra.

Definition 24.1. A language $\mathscr{L}$ is *algebraic* if it has no relation symbols. In any language, an atomic equality formula will also be called an *equation*. An (*equational*) *identity* is the universal closure of an equation. A *variety* is a class of $\mathscr{L}$-structures which is the class of all models of a set of identities, in an algebraic language. For any class $\mathbf{K}$ of $\mathscr{L}$-structures, let

$$\mathbf{SK} = \{\mathfrak{A} : \mathfrak{A} \rightarrowtail \mathfrak{B} \in \mathbf{K} \text{ for some } \mathfrak{B}\};$$
$$\mathbf{PK} = \{\mathfrak{A} : \mathfrak{A} \cong \mathsf{P}_{i \in I} \mathfrak{B}_i \text{ for some } \mathfrak{B} \in {}^I\mathbf{K}\}.$$

Note that $\mathbf{PK}$ is always nonempty, even if $\mathbf{K}$ is empty, since one can choose $I = 0$ in 24.1. Thus $\mathbf{K} = \mathbf{PK}$ implies $\mathbf{K} \neq 0$.

Lemma 24.2. *If* $\mathbf{K}$ *is a variety, then* $\mathbf{K} = \mathbf{SK} = \mathbf{PK}$.

Proof. Say $\mathbf{K}$ is the class of all models of the set Γ of identities. Obviously $\mathbf{K} \subseteq \mathbf{SK}$ and $\mathbf{K} \subseteq \mathbf{PK}$. Now assume that $f\colon \mathfrak{A} \rightarrowtail \mathfrak{B}$, $\mathfrak{B} \in \mathbf{K}$, and $[[\sigma = \tau]] \in \Gamma$. for any $x \in {}^{\omega}A$ we have, using 19.5,

$$f\sigma^{\mathfrak{A}}x = \sigma^{\mathfrak{B}}(f \circ x) = \tau^{\mathfrak{B}}(f \circ x) = f\tau^{\mathfrak{A}}x.$$

Thus $[[\sigma = \tau]]$ holds in $\mathfrak{A}$, and it is an arbitrary member of Γ, so $\mathfrak{A} \in \mathbf{K}$. Next, assume that $f: \mathfrak{A} \rightarrowtail\!\!\!\!\twoheadrightarrow \mathsf{P}_{i \in I} \mathfrak{B}_i = \mathfrak{C}$ with $\mathfrak{B} \in {}^I\mathbf{K}$, and $[[\sigma = \tau]] \in \Gamma$. For any $x \in {}^\omega A$ and any $i \in I$ we have using 18.14

$$\begin{aligned}[\sigma^{\mathfrak{C}}(f \circ x)]_i &= \sigma^{\mathfrak{B}i}(\mathrm{pr}_i \circ f \circ x) = \tau^{\mathfrak{B}i}(\mathrm{pr}_i \circ f \circ x) \\ &= [\tau^{\mathfrak{C}}(f \circ x)]_i.\end{aligned}$$

Since i is arbitrary, $\sigma^{\mathfrak{C}}(f \circ x) = \tau^{\mathfrak{C}}(f \circ x)$. Then by 18.22 we have $\sigma^{\mathfrak{A}}x = \tau^{\mathfrak{A}}x$. Thus $[[\sigma = \tau]]$ holds in $\mathfrak{A}$. Hence $\mathfrak{A} \in \mathbf{K}$. □

We need to consider another algebraic operation under which varieties are closed. It is important for languages in general, not only for algebraic languages, so we do not restrict the language.

Definition 24.3. Let $\mathfrak{A}$ and $\mathfrak{B}$ be $\mathscr{L}$-structures. A *homomorphism from* $\mathfrak{A}$ *into* $\mathscr{L}$ is a function f mapping A into B such that:

(*i*) if $\mathbf{O}$ is an operation symbol, say of rank m, and if $a \in {}^mA$, then

$$f\mathbf{O}^{\mathfrak{A}}(a_0, \ldots, a_{m-1}) = \mathbf{O}^{\mathfrak{B}}(fa_0, \ldots, fa_{m-1});$$

(*ii*) if $\mathbf{R}$ is an operation symbol, say of rank m, and if $a \in {}^mA$, then

$$\langle a_0, \ldots, a_{m-1}\rangle \in \mathbf{R}^{\mathfrak{A}} \Rightarrow \langle fa_0, \ldots, fa_{m-1}\rangle \in \mathbf{R}^{\mathfrak{B}}.$$

We say that f is a homomorphism of $\mathfrak{A}$ *onto* $\mathfrak{B}$ if it maps onto B. We use the respective notations $f: \mathfrak{A} \rightarrow \mathfrak{B}$ and $f: \mathfrak{A} \twoheadrightarrow \mathfrak{B}$ for these two cases.

Note that in (*ii*) above we do *not* insist that $\langle fa_0, \ldots, fa_{m-1}\rangle \in \mathbf{R}^{\mathfrak{B}} \Rightarrow \langle a_0, \ldots, a_{m-1}\rangle \in \mathbf{R}^{\mathfrak{A}}$. Hence a one–one homomorphism is not necessarily an isomorphism. We state without proof two very easy consequences of the definition of homomorphism; the first proposition is a generalization of a part of 18.23.

Proposition 24.4. *Let f be a homomorphism from $\mathfrak{A}$ into $\mathfrak{B}$; let σ be a term, and let $x \in {}^\omega A$. Then $f\sigma^{\mathfrak{A}}x = \sigma^{\mathfrak{B}}(f \circ x)$.*

Proposition 24.5. *Let $\langle \mathfrak{A}_i : i \in I\rangle$ be a system of $\mathscr{L}$-structures. Then* pr_i *is a homomorphism from $\mathsf{P}_{i \in I} \mathfrak{A}_i$ onto $\mathfrak{A}_i$.*

Now we introduce a useful notion related to the notion of homomorphism; in fact, it can be considered as a kind of *internal* version of homomorphisms.

Definition 24.6. Let $\mathscr{L}$ be an algebraic language, and let $\mathfrak{A}$ be an $\mathscr{L}$-structure. A *congruence relation on* $\mathfrak{A}$ is an equivalence relation R on A such that for any operation symbol $\mathbf{O}$ (say m-ary) and any $a, b \in {}^mA$, the condition $\forall i < m(a_iRb_i)$ implies $\mathbf{O}^{\mathfrak{A}}(a_0, \ldots, a_{m-1})R\,\mathbf{O}^{\mathfrak{A}}(b_0, \ldots, b_{m-1})$.

Given such a relation R, we can define a uniquely determined $\mathscr{L}$-structure $\mathfrak{A}/R$ whose universe is the set A/R of R-equivalence classes and whose operations are given by

$$\mathbf{O}^{\mathfrak{A}/R}([a_0], \ldots, [a_{m-1}]) = [\mathbf{O}^{\mathfrak{A}}(a_0, \ldots, a_{m-1})];$$

this structure is denoted by $\mathfrak{A}/R$.

If f is any function, then the *kernel* of f, $\ker f$, is $\{(x, y) : x, y \in \text{Dmn} f$ and $fx = fy\}$.

The basic facts relating homomorphisms and congruence relations are as follows:

Proposition 24.7. *Let $\mathscr{L}$ be an algebraic language, and let $\mathfrak{A}$ and $\mathscr{L}$ be $\mathscr{L}$-structures.*

(*i*) *If R is a congruence relation on $\mathfrak{A}$, then $[\ ]_R$ is a homomorphism from $\mathfrak{A}$ onto $\mathfrak{A}/R$.*

(*ii*) *If f is a homomorphism from $\mathfrak{A}$ into $\mathfrak{B}$, then* $\ker f$ *is a congruence relation on $\mathfrak{A}$.*

(*iii*) *If f is a homomorphism from $\mathfrak{A}$ onto $\mathfrak{B}$, then $\mathfrak{B} \cong \mathfrak{A}/R$; in fact, there is an isomorphism g from $\mathfrak{B}$ onto $\mathfrak{A}/R$ such that $g \circ f = [\ ]_{\ker f}$:*

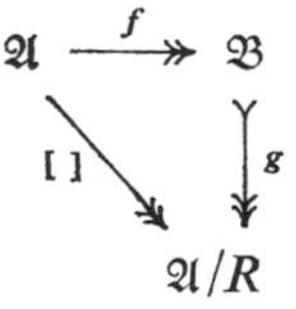

PROOF. Conditions (*i*) and (*ii*) are completely straightforward. Now assume the hypothesis of (*iii*). Let $g = \{(fa, [a]) : a \in A\}$. To see that g is a function, assume that $fa = fc$ with $a, c \in A$; we must show that $[a] = [c]$—but this is obvious. Thus $g: B \twoheadrightarrow A/R$. Similarly, g is one–one. If $\mathbf{O}$ is m-ary and $a \in {}^mA$, then

$$\begin{aligned} g\,\mathbf{O}^{\mathfrak{B}}(fa_0, \ldots, fa_{m-1}) &= gf\,\mathbf{O}^{\mathfrak{A}}(a_0, \ldots, a_{m-1}) \\ &= [\mathbf{O}^{\mathfrak{A}}(a_0, \ldots, a_{m-1})] \\ &= \mathbf{O}^{\mathfrak{A}/R}([a_0], \ldots, [a_{m-1}]) \\ &= \mathbf{O}^{\mathfrak{A}/R}(gfa_0, \ldots, gfa_{m-1}). \end{aligned}$$

Hence $g: \mathfrak{B} \rightarrowtail\!\!\!\twoheadrightarrow \mathfrak{A}/R$. Obviously $g \circ f = [\ ]$. □

Definition 24.8. Let $\mathbf{K}$ be a class of $\mathscr{L}$-structures, where $\mathscr{L}$ is not necessarily algebraic. We set

$$\begin{aligned} \mathbf{HK} &= \{\mathfrak{A} : \mathfrak{A} \text{ is a homomorphic image of some } \mathfrak{B} \in \mathbf{K}\}; \\ \mathbf{UpK} &= \{\mathfrak{A} : \mathfrak{A} \text{ is isomorphic to an ultraproduct of members of } \mathbf{K}\}. \end{aligned}$$

Note that *all* of the classes **SK**, **PK**, **HK** and **UpK** are closed under taking isomorphic images.

Proposition 24.9. *If* $\mathbf{K}$ *is a variety, then* $\mathbf{K} = \mathbf{HK}$.

PROOF. Obvious from 24.4. □

Thus a variety is closed under all three operators **H**, **S**, and **P**. The main theorem of this chapter is that the converse holds. Before beginning the proof of this, we want to develop the basics of general algebra and equational logic a little further. The following proposition is easy to prove.

Proposition 24.10

(*i*) $\mathbf{SHK} \subseteq \mathbf{HSK}$.
(*ii*) $\mathbf{PSK} \subseteq \mathbf{SPK}$.
(*iii*) $\mathbf{PHK} \subseteq \mathbf{HPK}$.

Easy examples can be given to show that none of the inclusions in 24.10 can be replaced by equalities, in general; see Exercises 24.25–24.27. These inequalities lead to the following useful equivalences.

Proposition 24.11. *The following conditions are equivalent:*

(*i*) $\mathbf{K} = \mathbf{SK} = \mathbf{HK} = \mathbf{PK}$.
(*ii*) $\mathbf{K} = \mathbf{HSPK}$.
(*iii*) $\mathbf{K} = \mathbf{HSPL}$ *for some* $\mathbf{L}$.

PROOF. Obviously (*i*) ⇒ (*ii*) ⇒ (*iii*). Now assume (*iii*). Clearly $\mathbf{HK} = \mathbf{K}$ and $\mathbf{K} \subseteq \mathbf{SK}$, $\mathbf{K} \subseteq \mathbf{PK}$. Next,

$$\begin{aligned}\mathbf{SK} = \mathbf{SHSPL} &\subseteq \mathbf{HSSPL} && \text{by 24.10}(i)\\ &= \mathbf{HSPL} = \mathbf{K}.\end{aligned}$$

Finally,

$$\begin{aligned}\mathbf{PK} &= \mathbf{PHSPL} \subseteq \mathbf{HPSPL} \subseteq \mathbf{HSPPL} && \text{by 24.10}(ii), (iii)\\ &= \mathbf{HSPL} = \mathbf{K}.\end{aligned}$$ □

To prove our main theorem, we need the notion of an absolutely free algebra. As will be seen, this notion is very analogous to the construction in 11.12 of an "internal" model for a consistent set of sentences. It is really just a part of that construction.

Definition 24.12. Let $\mathscr{L}$ be an algebraic language. We construct an $\mathscr{L}$-structure $\mathfrak{Fr}_{\mathscr{L}}$ which will be called the *absolutely free* $\mathscr{L}$*-algebra*. Its universe is $\mathrm{Trm}_{\mathscr{L}}$, and for any operation symbol $\mathbf{O}$ of $\mathscr{L}$,

$$\mathbf{O}^{\mathfrak{Fr}_{\mathscr{L}}}(\sigma_0, \ldots, \sigma_{m-1}) = \mathbf{O}\sigma_0 \cdots \sigma_{m-1}.$$

Now the definition of satisfaction yields the following basic fact about $\mathfrak{Fr}_{\mathscr{L}}$:

Proposition 24.13. *If* $\mathscr{L}$ *is algebraic,* $\mathfrak{A}$ *is an* $\mathscr{L}$*-structure, and* $x \in {}^{\omega}A$*, then* $\langle \sigma^{\mathfrak{A}}x : \sigma \in \mathrm{Trm}_{\mathscr{L}} \rangle$ *is a homomorphism of* $\mathfrak{Fr}_{\mathscr{L}}$ *into* $\mathfrak{A}$.

A very useful congruence relation on $\mathfrak{Fr}_{\mathscr{L}}$ is introduced in the following definition.

Definition 24.14. If $\mathscr{L}$ is an algebraic language and Γ is a set of equations of $\mathscr{L}$, we let

$$\equiv_\Gamma = \{(\sigma, \tau) : \sigma, \tau \in \mathrm{Trm}_{\mathscr{L}} \text{ and } \Gamma \models \sigma = \tau\}.$$

Proposition 24.15. *Under the assumptions of* 24.14, $\equiv_\Gamma$ *is a congruence relation on* $\mathfrak{Fr}_{\mathscr{L}}$.

The following simple proposition is proved by induction on σ:

Proposition 24.16. *If g is a homomorphism from $\mathfrak{Fr}_{\mathscr{L}}$ into $\mathfrak{A}$, and if $x_i = gv_i$ for every $i < \omega$, then $\sigma^{\mathfrak{A}}x = g\sigma$ for every term σ.*

Recall that, by the completeness theorem, the model-theoretic condition $\Gamma \models \varphi$ is equivalent to the proof-theoretic condition $\Gamma \vdash \varphi$. If we apply this fact when φ is an equation and Γ is a set of equations, it seems a little unsatisfactory, since the proof-theoretic condition involves the logical axioms and hence the whole apparatus of first order logic. We now describe a proof-theoretic condition in which only equations appear—no quantifiers and not even any sentential connectives.

Definition 24.17. Let Γ be a set of equations in an algebraic language. Then Γ-eqthm is the intersection of all sets Δ of equations such that the following conditions hold:

(*i*) $\Gamma \subseteq \Delta$;
(*ii*) $v_0 \mathbf{=} v_0 \in \Delta$;
(*iii*) if $\varphi \in \Delta$, $i < \omega$, σ is a term, and ψ is obtained from φ by replacing v_i throughout φ by σ, then $\psi \in \Delta$;
(*iv*) if $\sigma \mathbf{=} \tau \in \Delta$ and $\rho \mathbf{=} \tau \in \Delta$, then $\sigma \mathbf{=} \rho \in \Delta$;
(*v*) if $\sigma \mathbf{=} \tau \in \Delta$, $\mathbf{O}$ is an operation symbol, say of rank m, $i < m$, and $\alpha_0, \ldots, \alpha_{m-2}$ are variables, then the following equation is in Δ:
$$\mathbf{O}(\alpha_0, \ldots, \alpha_{i-1}, \sigma, \alpha_i, \ldots, \alpha_{m-2}) \mathbf{=} \mathbf{O}(\alpha_0, \ldots, \alpha_{i-1}, \tau, \alpha_i, \ldots, \alpha_{m-2}).$$

We write $\Gamma \vdash_{\mathrm{eq}} \varphi$ instead of $\varphi \in \Gamma$-eqthm.

We shall prove an analog of the completeness theorem for this notion. First a technical lemma:

Lemma 24.18. *Let Γ be a set of equations in an algebraic language. Then for any terms σ, τ, ρ,*

(*i*) $\Gamma \vdash_{eq} \sigma \equiv \sigma$;
(*ii*) *if* $\Gamma \vdash_{eq} \sigma \equiv \tau$, *then* $\Gamma \vdash_{eq} \tau \equiv \sigma$;
(*iii*) *if* $\Gamma \vdash_{eq} \sigma \equiv \tau$ *and* $\Gamma \vdash_{eq} \tau \equiv \rho$, *then* $\Gamma \vdash_{eq} \sigma \equiv \rho$;
(*iv*) *if* $\mathbf{O}$ *is an operation symbol, say of rank* m, *and if* $\sigma_0, \ldots, \sigma_{m-1}, \tau_0, \ldots, \tau_{m-1}$ *are terms such that* $\Gamma \vdash_{eq} \sigma_i \equiv \tau_i$ *for each* $i < m$, *then* $\Gamma \vdash_{eq} \mathbf{O}\sigma_0 \cdots \sigma_{m-1} \equiv \mathbf{O}\tau_0 \cdots \tau_{m-1}$.

PROOF. Condition (*i*) is obvious from 24.17(*ii*) and 24.17(*iii*). For (*ii*): assume that $\Gamma \vdash_{eq} \sigma \equiv \tau$. We also have $\Gamma \vdash_{eq} \tau \equiv \tau$ by (*i*), so $\Gamma \vdash_{eq} \tau \equiv \sigma$ by 24.17(*iv*). Condition (*iii*) clearly follows from (*ii*) and 24.17(*iv*). To prove (*iv*), let $\alpha_0, \ldots, \alpha_{m-1}$ be distinct variables not occurring in any of $\sigma_0, \ldots, \sigma_{m-1}, \tau_0, \ldots, \tau_{m-1}$. Then

$$\Gamma \vdash_{eq} \mathbf{O}(\alpha_0, \ldots, \alpha_{i-1}, \sigma_i, \alpha_{i+1}, \ldots, \alpha_{m-1}) \equiv \mathbf{O}(\alpha_0, \ldots, \alpha_{i-1}, \tau_i, \alpha_{i+1}, \ldots, \alpha_{m-1})$$

for each $i < m$. Applications of 24.17(*iii*) then give, for each $i < m$,

$$\Gamma \vdash_{eq} \mathbf{O}(\tau_0, \ldots, \tau_{i-1}, \sigma_i, \sigma_{i+1}, \ldots, \sigma_{m-1}) \equiv \mathbf{O}(\tau_0, \ldots, \tau_{i-1}, \tau_i, \sigma_{i+1}, \ldots, \sigma_{m-1}).$$

Now several applications of (*iv*) give the desired result. □

Theorem 24.19 (Completeness theorem for equational logic). *Let* $\Gamma \cup \{\varphi\}$ *be a set of equations in an algebraic language. Then* $\Gamma \models \varphi$ *iff* $\Gamma \vdash_{eq} \varphi$.

PROOF. It is easily checked that $\Gamma \vdash_{eq} \varphi \Rightarrow \Gamma \models \varphi$. Now suppose that not($\Gamma \vdash_{eq} \varphi$); we shall construct a model of $\Gamma \cup \{\neg[[\varphi]]\}$. In fact, the desired model is simply $\mathfrak{A} = \mathfrak{Fr}/\equiv_\Gamma$. To show that $\mathfrak{A}$ is a model of Γ, let $\sigma \equiv \tau$ be an arbitrary element of Γ, and let $x \in {}^\omega\mathrm{Trm}_{\mathscr{L}}$; we want to show that $\sigma^{\mathfrak{A}}([\ ] \circ x) = \tau^{\mathfrak{A}}([\ ] \circ x)$. To do this, for any $\rho \in \mathrm{Trm}_{\mathscr{L}}$ let $S\rho$ be the result of simultaneously replacing v_i by x_i in ρ, for each $i < \omega$. Then

(1) $$\Gamma \vdash_{eq} S\sigma \equiv S\tau.$$

To prove (1), let $v_{i0}, \ldots, v_{i(m-1)}$ be all of the variables occurring in the equation $\sigma \equiv \tau$. Let $\beta_0, \ldots, \beta_{m-1}$ be new variables, not appearing in $\sigma \equiv \tau$, different from $v_{i0}, \ldots, v_{i(m-1)}$ and not appearing in $x_{i0}, \ldots, x_{i(m-1)}$. Let $\sigma' \equiv \tau'$ be obtained from $\sigma \equiv \tau$ by replacing in succession v_{i0} throughout by β_0, v_{i1} throughout by $\beta_1, \ldots, v_{i(m-1)}$ throughout by β_{m-1}. By 24.17(*iii*) we have $\Gamma \vdash_{eq} \sigma' \equiv \tau'$. Since $\beta_0, \ldots, \beta_{m-1}$ do not occur in $\sigma \equiv \tau$ and they are different from $v_{i0}, \ldots, v_{i(m-1)}$, the equation $\sigma' \equiv \tau'$ is also obtainable from $\sigma \equiv \tau$ by *simultaneously* replacing $v_{i0}, \ldots, v_{i(m-1)}$ by $\beta_0, \ldots, \beta_{m-1}$ respectively. Now by replacing β_0 throughout by x_{i0}, then β_1 throughout by $x_{i1}, \ldots, \beta_{m-1}$ throughout by $x_{i(m-1)}$ we obtain $S\sigma \equiv S\tau$ and (1) follows. Note that $\beta_0, \ldots, \beta_{m-1}$ are introduced to reduce simultaneous substitution to a sequence of simple substitutions.

Next, note that

(2) for any term ρ we have $\rho^{\mathfrak{A}}([\ ] \circ x) = [S\rho]$.

Condition (2) is easily proven by induction on ρ.

From (1) and (2) we obtain

$$\sigma^{\mathfrak{A}}([\ ]\circ x) = [S\sigma] = [S\tau] = \tau^{\mathfrak{A}}([\ ]\circ x).$$

Thus $\sigma \equiv \tau$ holds in $\mathfrak{A}$, so $\mathfrak{A}$ is a model of Γ. However, $\mathfrak{A}$ is not a model of φ. For, say φ is the equation $\sigma_0 \equiv \tau_0$. Let $y_i = [v_i]$ for each $i \in \omega$. Then by 24.16 we obtain $\sigma_0^{\mathfrak{A}}y = [\sigma_0]$ and $\tau_0^{\mathfrak{A}}y = [\tau_0]$. Since not($\Gamma \vdash_{\text{eq}} \sigma_0 \equiv \tau_0$), we have $[\sigma_0] \neq [\tau_0]$. Thus $\sigma_0 \equiv \tau_0$ fails to hold in $\mathfrak{A}$. □

Now we return to our discussion of free algebras, and give a lemma needed for our main characterization theorem.

Lemma 24.20. *Let* **K** *be a class of* $\mathscr{L}$*-structures, where* $\mathscr{L}$ *is an algebraic language. Let* $\Gamma = \{\sigma \equiv \tau$: *the equation* $\sigma \equiv \tau$ *holds in each member of* **K**$\}$ *Then* $\mathfrak{Fr}_{\mathscr{L}}/{\equiv_\Gamma} \in \mathbf{SPK}$.

Proof. Let $I = {}^2\text{Trm} \sim \{(\sigma, \tau) : \sigma \equiv \tau \in \Gamma\}$. Then for each $(\sigma, \tau) \in I$ there is a structure $\mathfrak{A}_{\sigma\tau} \in \mathbf{K}$ such that $\sigma \equiv \tau$ fails to hold in $\mathfrak{A}_{\sigma\tau}$; hence, further, there is an $x_{\sigma\tau} \in {}^{\omega}A_{\sigma\tau}$ such that $\sigma^{\mathfrak{A}_{\sigma\tau}}x_{\sigma\tau} \neq \tau^{\mathfrak{A}_{\sigma\tau}}x_{\sigma\tau}$. Let y be the unique element of ${}^{\omega}\mathsf{P}_{i\in I}\,\mathfrak{A}_i$ such that for each $m \in \omega$ and $i \in I$, $(y_m)_i = (x_i)_m$. Let $\mathfrak{B} = \mathsf{P}_{i\in I}\,\mathfrak{A}_i$. Then

(1) $$\equiv_\Gamma \subseteq \{\sigma \equiv \tau : \sigma^{\mathfrak{B}}y = \tau^{\mathfrak{B}}y\}.$$

For, suppose $\sigma \equiv_\Gamma \tau$. Thus $\Gamma \models \sigma \equiv \tau$, so it follows that $\sigma \equiv \tau$ holds in every member of **K**. From 18.14 we infer that $\sigma \equiv \tau$ holds in $\mathfrak{B}$, as desired.

By (1) there is a mapping f of $Fr_{\mathscr{L}}/{\equiv_\Gamma}$ into B such that $f[\sigma] = \sigma^{\mathfrak{B}}y$ for each term σ. It is easily checked that f is a homomorphism of $\mathfrak{Fr}_{\mathscr{L}}/{\equiv_\Gamma}$ into $\mathfrak{B}$. It only remains to check that f is one–one. Suppose $[\sigma] \neq [\tau]$. Thus not ($\Gamma \models \sigma \equiv \tau$), so in particular $\sigma \equiv \tau \notin \Gamma$. Thus $(\sigma, \tau) \in I$. Hence

$$\begin{aligned}(\sigma^{\mathfrak{B}}y)_{\sigma\tau} &= \text{pr}_{\sigma\tau}(\sigma^{\mathfrak{B}}y) = \sigma^{\mathfrak{A}_{\sigma\tau}}(\text{pr}_{\sigma\tau}\circ y)\\ &= \sigma^{\mathfrak{A}_{\sigma\tau}}x_{\sigma\tau} \neq \tau^{\mathfrak{A}_{\sigma\tau}}x_{\sigma\tau} = (\tau^{\mathfrak{B}}y)_{\sigma\tau}.\end{aligned}$$

Hence $\sigma^{\mathfrak{B}}y \neq \tau^{\mathfrak{B}}y$, so $f[\sigma] \neq f[\tau]$. □

We now find it convenient to prove Exercise 19.42.

Lemma 24.21. *Suppose* $\mathbf{K} = \mathbf{SK} = \mathbf{UpK}$, *in a not necessarily algebraic language* $\mathscr{L}$. *If* $\mathfrak{A}$ *is an* $\mathscr{L}$*-structure every finitely generated substructure of which is in* **K**, *then* $\mathfrak{A}$ *is in* **K**.

Proof. Let $I = \{F : 0 \neq F \subseteq A,\ F$ finite]. For each $F \in I$, let $\mathfrak{A}_F$ be the substructure of $\mathfrak{A}$ generated by F. Thus $\mathfrak{A}_F \in \mathbf{K}$ by assumption. For each $F \in I$ let $M_F = \{G \in I : F \subseteq G\}$. If $F, G \in I$, then $M_F \cap M_G = M_{F\cup G}$. It follows that there is an ultrafilter $\mathfrak{F}$ over I such that $M_F \in \mathfrak{F}$ for all $F \in I$. Thus $\mathsf{P}_{F\in I}\,\mathfrak{A}_F/\mathfrak{F} \in \mathbf{K}$, so it suffices to define an embedding f of $\mathfrak{A}$ into $\mathsf{P}_{F\in I}\,\mathfrak{A}_F/\mathfrak{F}$.

By the axiom of choice let $x \in \mathsf{P}_{F \in I} A_F$. For each $a \in A$ define $g_a \in \mathsf{P}_{F \in I} A_F$ by

$$\begin{aligned} g_a F &= a \qquad \text{iff } a \in A_F, \\ g_a F &= x_F \qquad \text{otherwise,} \end{aligned}$$

and let $fa = [ga]_{\mathfrak{F}}$. The desired properties of f are easily checked. □

Theorem 24.22 (Birkhoff). **K** *is a variety iff* **K** = **HSPK.**

Proof. The trivial direction ⇒ is given by 24.2 and 24.9. Now assume that **K** = **HPSK**. By 24.11, **K** is closed under **H**, **S**, and **P**; and so it is also obviously closed under **Up**. Let $\Gamma = \{\sigma \equiv \tau : \sigma \equiv \tau$ holds in every member of **K**$\}$. We claim that **K** is exactly the class of all models of Γ. Obviously each member of **K** is a model of Γ. Now let $\mathfrak{A}$ be a model of Γ. By 24.21 it now suffices to take any finitely generated substructure $\mathfrak{B}$ of $\mathfrak{A}$, say generated by $F \neq 0$, F finite, and show that $\mathfrak{B} \in$ **K**.

Choose $x \in {}^{\omega}B$ so that $\operatorname{Rng} x = F$. Then

(1) for every $b \in B$ there is a term σ such that $\sigma^{\mathfrak{B}}x = b$.

Indeed, let $C = \{\sigma^{\mathfrak{A}}x : \sigma \in \operatorname{Trm}_{\mathscr{L}}\}$. It is easily seen that $F \subseteq C$ and C is a subuniverse of $\mathfrak{A}$, so $B \subseteq C$, as desired in (1). Now $\mathfrak{A}$ is a model of Γ and $\mathfrak{B} \subseteq \mathfrak{A}$, so $\mathfrak{B}$ is a model of Γ. Hence there is a function f mapping $Fr_{\mathscr{L}}/{\equiv_\Gamma}$ into B such that $f[\sigma] = \sigma^{\mathfrak{B}}x$ for all $\sigma \in \operatorname{Trm}_{\mathscr{L}}$. Clearly f is a homomorphism from $\mathfrak{Fr}_{\mathscr{L}}/{\equiv_\Gamma}$ into $\mathfrak{B}$, and it is onto by (1). Hence $\mathfrak{B} \in$ **K** by 24.20. □

Theorem 24.22 is a typical *characterization* theorem (it was historically the first such). A logical concept (variety) is characterized in terms of non-logical notions (closure under **H**, **S** and **P**). It may of course also be viewed as a characterization of closure under **H**, **S**, and **P** in terms of the logical notion of a variety. Such equivalences between notions in widely separated domains are typical of a large part of model theory and illustrate a healthy trend toward unification in mathematics.

A closely related result is the following *preservation* theorem.

Theorem 24.23. *Let φ be a sentence, in an algebraic language, which is preserved under* **H**, **S**, *and* **P**. *That is, assume about φ the following three things:*

(*i*) *if $\mathfrak{A} \models \varphi$ and $\mathfrak{A} \twoheadrightarrow \mathfrak{B}$, then $\mathfrak{B} \models \varphi$;*
(*ii*) *if $\mathfrak{A} \models \varphi$ and $\mathfrak{B} \subseteq \mathfrak{A}$, then $\mathfrak{B} \models \varphi$;*
(*iii*) *if $\mathfrak{A}_i \models \varphi$ for each $i \in I$, then $\mathsf{P}_{i \in I} \mathfrak{A}_i \models \varphi$.*

Then there is a conjunction ψ of equational identities such that $\models \varphi \leftrightarrow \psi$.

Proof. Let **K** be the class of all models of φ. The conditions (*i*)–(*iii*) imply that **K** = **HSPK**. Hence by 24.22 there is a set Γ of identities such that **K** is the class of all models of Γ. Thus $\Gamma \models \varphi$, so $\models \psi \rightarrow \varphi$ for some finite conjunction ψ of members of Γ. But obviously $\models \varphi \rightarrow \psi$ also, so $\models \varphi \leftrightarrow \psi$. □

Theorem 24.23 is a typical preservation theorem. Such a theorem has the following form: if a sentence φ is preserved under certain algebraic operations or relations, then φ is equivalent to a sentence ψ of a specified syntactical form. In such cases it is usually obvious that such a sentence ψ is preserved under the given operations. Thus, for example, any identity is obviously preserved under **H**, **S**, and **P**.

BIBLIOGRAPHY

1. Gratzer, G. *Universal Algebra*. Princeton: van Nostrand (1968).
2. Jonsson, B. *Topics in Universal Algebra*. Berlin: Springer (1972).
3. Tarski, A. Equational logic and equational theories of algebras. In *Contributions to Mathematical Logic*. Amsterdam: North Holland (1968).

EXERCISES

24.24.* If we consider sequences of operators **H**, **S**, and **P**, there are exactly 18 distinct sequences (with the empty sequence allowed).

24.25. Give an example of a class **K** such that **SHK** $\neq$ **HSK**.

24.26. Give an example of a class **K** such that **PSK** $\neq$ **SPK**.

24.27. Give an example of a class **K** such that **PHK** $\neq$ **HPK**.

24.28. For any equation φ, $\models\varphi$ iff φ has the form $\sigma \mathbf{=} \sigma$ for some term σ.

24.29. In an algebraic language, a class **K** is a variety iff $\mathbf{K} = \mathbf{HSP}(\mathfrak{Fr}/\equiv_\Gamma)$, where $\Gamma = \{(\sigma, \tau) : \sigma \mathbf{=} \tau$ holds in each member of $\mathbf{K}\}$.

24.30. Birkhoff's theorem cannot be improved by eliminating **H**, **S**, or **P**.

24.31. An algebra $\mathfrak{A}$ is a *subdirect product* of the system $\langle \mathfrak{B}_i : i \in I\rangle$ provided that $\mathfrak{A} \subseteq \mathsf{P}_{i\in I}\mathfrak{B}_i$ and $\mathrm{pr}_i^* A = B_i$ for all $i \in I$. Show that **K** is a variety iff **K** is closed under **H** and under subdirect products. *Hint*: if $\mathfrak{A} \subseteq \mathfrak{B} \in \mathbf{K}$, consider $\{b \in {}^{\omega}B : \exists a \in A\{i \in \omega : b_i \neq a\}$ is finite$\}$.

24.32. A class **K** is a variety iff **K** is closed under **H**, **S**, **Up**, and finite products.

24.33. Let $\Gamma \cup \{\varphi\}$ be a set of sentences in an algebraic language. Assume that φ is preserved under **HSP** relative to Γ, i.e., assume that if **K** is the set of all models of Γ and **L** is the set of all models of $\Gamma \cup \{\varphi\}$, then $\mathfrak{A} \in \mathbf{K} \cap \mathbf{HSPL}$ implies $\mathfrak{A} \models \varphi$. Show that there is a conjunction ψ of equational identities such that $\Gamma \models \varphi \leftrightarrow \psi$.

Preservation and Characterization Theorems 25

In this chapter we shall prove several more theorems of the general type of 24.22 and 24.23. We shall consider universal sentences, quasivarieties, universal-existential sentences, and positive sentences. We have not tried to give uniform proofs for these results, but we have instead chosen the proofs that seem most natural to us.

Universal Sentences

Definition 25.1. For any set Γ of sentences, Mod Γ is the class of all models of Γ. A class $\mathbf{K}$ is a *universal class* if $\mathbf{K} = \text{Mod}\ \Gamma$ for some set Γ of universal sentences. $\mathbf{K}$ is an *elementary* class if $\mathbf{K} = \text{Mod}\ \Gamma$ for some set Γ of sentences. We also say that Γ *characterizes* Mod Γ.

We shall give two characterizations of universal classes. The first is due to Łoś, and is valid in any language; its proof is similar to that of 24.22 in many respects. The second is due to Tarski. In its simplest form, which is all we shall give here, it applies only to languages with no operation symbols and only finitely many relation symbols. Elementary classes will be discussed in the next chapter.

Theorem 25.2 (Łoś). *A class* $\mathbf{K}$ *is universal iff* $\mathbf{SK} = \mathbf{K}$ *and* $\mathbf{UpK} = \mathbf{K}$.

Proof. The direction $\Rightarrow$ is trivial. Now assume that $\mathbf{SK} = \mathbf{K}$ and $\mathbf{UpK} = \mathbf{K}$. Let $\Gamma = \{\varphi : \varphi$ is a universal sentence which holds in each $\mathfrak{A} \in \mathbf{K}\}$. Clearly $\mathbf{K} \subseteq \text{Mod}\ \Gamma$. Thus the main part of our proof is to take any $\mathfrak{A} \in \text{Mod}\ \Gamma$ and show that $\mathfrak{A} \in \mathbf{K}$. To do this it is enough, by our assumptions, to show that $\mathfrak{A} \in \mathbf{SUpK}$.

Let $<$ be a well-ordering of A. Let $\mathscr{L}'$ be an A-expansion of $\mathscr{L}$, and let Δ

be the $\mathscr{L}'$-diagram of $\mathfrak{A}$. We shall form an ultraproduct over the index set $I = \{\Theta : \Theta$ is a finite subset of $\Delta\}$. Given any $\Theta \in I$, let $\mathbf{c}a(\Theta, 0), \ldots, \mathbf{c}a(\Theta, m_\Theta - 1)$ be all of the new constants which occur in the sentences of Θ, where $a(\Theta, 0) < \cdots < a(\Theta, m_\Theta - 1)$. Let ψ_Θ be obtained from $\bigwedge \Theta$ by replacing all occurrences of $\mathbf{c}a(\Theta, 0), \ldots, \mathbf{c}a(\Theta, m_\Theta - 1)$ by $v0, \ldots, v(m_\Theta - 1)$ respectively, and let φ_Θ be the universal sentence $\forall v_0 \cdots \forall v(m_\Theta - 1) \neg \psi_\Theta$. Thus $\mathfrak{A} \models \neg\varphi_\Theta$ so, since $\mathfrak{A}$ is a model of Γ, $\varphi_\Theta \notin \Gamma$. Hence we may pick a structure $\mathfrak{B}_\Theta \in \mathbf{K}$ such that $\mathfrak{B}_\Theta \models \neg\varphi_\Theta$. Choose $x_\Theta \in {}^\omega B_\Theta$ such that $\mathfrak{B}_\Theta \models \psi_\Theta[x_\Theta]$. For any $\Theta \in I$ let

$$M_\Theta = \{\Omega \in I : \Theta \subseteq \Omega\}.$$

Thus $M_\Theta \cap M_\Omega = M_{\Theta \cup \Omega}$ for any $\Theta, \Omega \in I$, so the family $\{M_\Theta : \Theta \in I\}$ has the finite intersection property. Let F be an ultrafilter over I such that $M_\Theta \in F$ for each $\Theta \in I$. Thus $\mathsf{P}_{\Theta \in I} \mathfrak{B}_\Theta/F \in \mathbf{UpK}$. It remains to define an embedding f of $\mathfrak{A}$ into this ultraproduct.

Define $g: A \to \mathsf{P}_{\Theta \in I} B_\Theta$ by setting, for any $b \in A$,

$$\begin{aligned} (gb)_\Theta &= x_\Theta i \qquad \text{if } b = a(\Theta, i), \\ (gb)_\Theta &= x_\Theta 0 \qquad \text{if } b \neq a(\Theta, i) \text{ for all } i < m_\Theta. \end{aligned}$$

Let $fb = [gb]$ for all $b \in A$. Now we check that $f: \mathfrak{A} \rightarrowtail \mathsf{P}_{\Theta \in I} \mathfrak{B}_\Theta/F$. First, f is one-one. For, suppose that b and c are distinct elements of A. Then $\neg(\mathbf{c}_b \equiv \mathbf{c}_c) \in \Delta$. Let $\Theta = \{\neg(\mathbf{c}_b \equiv \mathbf{c}_c)\}$. If Ω is any member of M_Θ, we may choose i, j so that $a(\Omega, i) = b$ and $a(\Omega, j) = c$. Now $\neg(\mathbf{c}_b \equiv \mathbf{c}_c)$ is a conjunct of $\bigwedge \Omega$, so $\neg(v_i \equiv v_j)$ (or $\neg(v_j \equiv v_i)$) is a conjunct of ψ_Ω. It follows that

$$(gb)_\Omega = x_\Omega i \neq x_\Omega j = (gc)_\Omega.$$

Hence $\{\Omega : (gb)_\Omega \neq (gc)_\Omega\} \supseteq M_\Theta \in F$, so $fb \neq fc$.

We check that f preserves operations; a similar argument shows that f preserves relations. Let $\mathbf{O}$ be an operation symbol of rank n. Set $\mathfrak{C} = \mathsf{P}_{\Theta \in I} \mathfrak{B}_\Theta/F$, and let $b_0, \ldots, b_{n-1} \in A$. Let $c = \mathbf{O}^{\mathfrak{A}} b_0 \cdots b_{n-1}$, and set

$$\Theta = \{\mathbf{O}\mathbf{c}_{b0} \cdots \mathbf{c}_{b(n-1)} \equiv \mathbf{c}_c\}.$$

Suppose that Ω is any member of M_Θ. Then we may choose $i_0, \ldots, i_n < m_\Omega$ such that $a(\Omega, i_j) = b_j$ for each $j < n$ and $a(\Omega, i_n) = c$. Thus the formula

$$\mathbf{O}(vi_0, \ldots, vi_{n-1}) \equiv vi_n$$

is a conjunct of ψ_Ω, and hence

$$\mathbf{O}^{\mathfrak{B}\Omega}((gb_0)_\Omega, \ldots, (gb_{n-1})_\Omega) = \mathbf{O}^{\mathfrak{B}\Omega}(x_\Omega i_0, \ldots, x_\Omega i_{n-1}) = x_\Omega i_n = (gc)_\Omega.$$

Since this is true for each $\Omega \in M_\Theta$, and $M_\Theta \in F$, it follows easily that

$$\mathbf{O}^{\mathfrak{C}}(fb_0, \ldots, fb_{n-1}) = fc. \qquad \square$$

Now we want to draw some conclusions from 25.2. For this purpose the following trivial proposition will be needed; it is analogous to 24.10.

Proposition 25.3. $\mathbf{UpSK} \subseteq \mathbf{SUpK}$.

Recall from 11.14 the notion of a reduct of a structure, for which we introduced the notation $\mathfrak{A} \upharpoonright \mathscr{L}$. We now need to extend this notation to classes of structures:

Definition 25.4. Let $\mathscr{L}'$ be an expansion of $\mathscr{L}$, and let $\mathbf{K}$ be a class of $\mathscr{L}'$-structures. Then we let $\mathbf{K} \upharpoonright \mathscr{L}$ denote the class of all $\mathscr{L}$-reducts of members of $\mathbf{K}$.

The following corollary of 25.2 is very important.

Corollary 25.5. *Let* $\mathbf{K}$ *be a class of* $\mathscr{L}'$*-structures and let* $\mathscr{L}$ *be a reduct of* $\mathscr{L}'$. *If* $\mathbf{K}$ *can be characterized by first-order sentences, then* $\mathbf{S}(\mathbf{K} \upharpoonright \mathscr{L})$ *is a universal class.*

PROOF. By 25.3 and 18.29, $\mathbf{S}(\mathbf{K} \upharpoonright \mathscr{L})$ is closed under $\mathbf{S}$ and $\mathbf{Up}$. Hence 25.5 is immediate from 25.2. □

Of course, Corollary 25.5 is valid for $\mathscr{L} = \mathscr{L}'$ also. Some applications of 25.5 are as follows:

1. The class of all semigroups embeddable in groups is universal. Specific universal sentences for this purpose have been found by Malcev; his sentences are infinite in number and cannot be replaced by a finite set of sentences.
2. The class of all rings which can be ordered is universal.
3. The class of all rings embeddable in fields is universal.
4. The class of all structures $\langle A, R\rangle$ embeddable in linear orderings $\langle B, <\rangle$ is universal.

A somewhat stronger form of 25.5 can be given which depends on the following important model-theoretic notion.

Definition 25.6. Let $\mathbf{U}$ be a unary relation symbol in a language $\mathscr{L}$. The $\mathscr{L}$-*closure conditions* for $\mathbf{U}$ are all sentences of the following forms:

$$\exists v_0 \mathbf{U} v_0$$
$$\forall v_0 \cdots \forall v_{m-1}\left(\bigwedge_{j<m} \mathbf{U} v_i \rightarrow \mathbf{UO} v_0 \cdots v_{m-1}\right)$$

for each (m-ary) operation symbol $\mathbf{O}$.

Let $\mathscr{L}^{-\mathbf{U}}$ be the reduct of $\mathscr{L}$ obtained by deleting $\mathbf{U}$. Now with each $\mathscr{L}$-structure $\mathfrak{A}$ which is a model of the $\mathscr{L}$-closure conditions for $\mathbf{U}$ we associate the uniquely determined substructure $\mathfrak{A} \upharpoonright \mathbf{U}$ of $\mathfrak{A} \upharpoonright \mathscr{L}^{-\mathbf{U}}$ with universe $\mathbf{U}^{\mathfrak{A}}$. And for each class $\mathbf{K}$ of such structures $\mathfrak{A}$ we let $\mathbf{K} \upharpoonright \mathbf{U} = \{\mathfrak{A} \upharpoonright \mathbf{U} : \mathfrak{A} \in \mathbf{K}\}$. Given any expansion $\mathscr{L}'$ of $\mathscr{L}$ and any $\mathscr{L}'$-structure $\mathfrak{B}$ which is a model of the $\mathscr{L}$-closure conditions for $\mathbf{U}$, the $\mathscr{L}^{-\mathbf{U}}$ structure $(\mathfrak{B} \upharpoonright \mathscr{L}) \upharpoonright \mathbf{U}$ is called a *relativized reduct of* $\mathfrak{B}$.

Proposition 25.7. *With notation as in 25.6, suppose that for each $i \in I$, an $\mathscr{L}$-structure $\mathfrak{C}_i$ is a model of the $\mathscr{L}$-closure conditions for* **U**. *Then*

$$\mathsf{P}_{i\in I} (\mathfrak{C}_i \restriction \mathbf{U})/F \cong (\mathsf{P}_{i\in I} \mathfrak{C}_i/F) \restriction \mathbf{U}.$$

PROOF For any $x \in \mathsf{P}_{i\in I} (\mathfrak{C}_i \restriction \mathbf{U})$, let $fx = [x]$, the equivalence class of x under F in the structure $\mathsf{P}_{i\in I} \mathfrak{C}_i/F$. It is easy to see that f induces the desired isomorphism. The induced map is, in fact, only in a subtle way different from the identity map. □

Corollary 25.8. *Assume that $\mathscr{L}$,* **U**, *and $\mathscr{L}'$ are as in 25.6, and that Γ is a set of sentences of $\mathscr{L}'$ including the set of $\mathscr{L}$-closure conditions for* **U**. *Then* $\mathbf{S}((\mathrm{Mod}\ \Gamma) \restriction \mathscr{L}) \restriction \mathbf{U})$ *is a universal class.*

PROOF By 25.2 and 25.3 it suffices to show that $((\mathrm{Mod}\ \Gamma) \restriction \mathscr{L}) \restriction \mathbf{U}$ is closed under ultraproducts. Suppose that $\mathfrak{A}_i \in ((\mathrm{Mod}\ \Gamma) \restriction \mathscr{L}) \restriction \mathbf{U}$ for each $i \in I$, say $\mathfrak{A}_i = (\mathfrak{B}_i \restriction \mathscr{L}) \restriction \mathbf{U}$ where $\mathfrak{B}_i \in \mathrm{Mod}\ \Gamma$. Then

$$\begin{aligned} \mathsf{P}_{i\in I} \mathfrak{A}_i/F &\cong \left(\mathsf{P}_{i\in I} (\mathfrak{B}_i \restriction \mathscr{L})/F\right) \restriction \mathbf{U} && \text{by 25.7} \\ &= \left(\left(\mathsf{P}_{i\in I} \mathfrak{B}_i/F\right) \restriction \mathscr{L}\right) \restriction \mathbf{U} && \text{by 18.28} \end{aligned}$$

and $\mathsf{P}_{i\in I} \mathfrak{B}_i/F \in \mathrm{Mod}\ \Gamma$, so $\mathsf{P}_{i\in I} \mathfrak{A}_i/F \in ((\mathrm{Mod}\ \Gamma) \restriction \mathscr{L}) \restriction \mathbf{U}$. □

We want to give one nontrivial application of 25.8. Let $\mathscr{L}$ be a language with two nonlogical constants, a binary operation symbol $\cdot$ and a unary operation symbol $^{\cup}$. We say that an $\mathscr{L}$-structure $\mathfrak{A} = (A, \cdot, ^{\cup})$ is a *relation system* provided that A is a collection of binary relations on some set X, and for any $R, S \in A$,

$$R \cdot S = \{(a, b) : \exists c(aRcSb)\},$$
$$R^{\cup} = \{(a, b) : (b, a) \in R\}.$$

Let **K** be the set of all $\mathscr{L}$-structures isomorphic to a relation system. Then **K** is a universal class. In fact, expand $\mathscr{L}$ first by adding a unary relation symbol **U**, forming $\mathscr{L}'$. Then expand $\mathscr{L}'$ further to $\mathscr{L}''$ by adjoining **V** (another unary relation symbol) and **F** (a ternary relation symbol). Let Γ be a collection of sentences in $\mathscr{L}''$ expressing the following:

(1) $\mathscr{L}'$-closure conditions for **U**;
(2) **U** and **V** are disjoint;
(3) **V** is nonempty;
(4) **F** establishes an isomorphism from $(\mathbf{U}, \cdot, ^{\cup})$ onto a relation system of relations on **V**.

For example, four sentences can express (4):

$$\forall v_0 \forall v_1 \forall v_2 (\mathbf{F} v_0 v_1 v_2 \to \mathbf{U} v_0 \wedge \mathbf{V} v_1 \wedge \mathbf{V} v_2),$$
$$\forall v_0 \forall v_1 \{\mathbf{U} v_0 \wedge \mathbf{U} v_1 \wedge \neg v_0 = v_1 \to \exists v_2 \exists v_3$$
$$[(\mathbf{F} v_0 v_2 v_3 \wedge \neg \mathbf{F} v_1 v_2 v_3) \vee (\mathbf{F} v_1 v_2 v_3 \wedge \neg \mathbf{F} v_0 v_2 v_3)]\},$$
$$\forall v_0 \forall v_1 \forall v_2 \forall v_3 [\mathbf{F}(v_0 \cdot v_1, v_2, v_3) \leftrightarrow \exists v_4 (\mathbf{F} v_0 v_2 v_4 \wedge \mathbf{F} v_1 v_4 v_3)],$$
$$\forall v_0 \forall v_2 \forall v_3 (\mathbf{F} v_0 v_2 v_3 \leftrightarrow \mathbf{F} v_0^{\cup} v_3 v_2).$$

Clearly $\mathbf{S}(((\text{Mod }\Gamma) \restriction \mathscr{L}') \restriction \mathbf{U}) = \mathbf{K}$. Hence by 25.8, **K** is a universal class. This class **K** has been studied in connection with algebraic logic and in the theory of semigroups. A "nice" set of axoms for **K** has not been given. It is known that **K** is axiomatizable by a set whose corresponding set of Gödel numbers is elementary, but it is unknown whether it is finitely axiomatizable.

Another corollary of 25.2 is the following characterization of universal classes.

Corollary 25.9. *A class* **K** *is universal iff* $\mathbf{SK} = \mathbf{K}$ *and* $\mathbf{K} = \text{Mod }\Gamma$ *for some set* Γ *of sentences.*

And a corollary is the following preservation theorem which is analogous to 24.23.

Corollary 25.10. *Let* $\Gamma \cup \{\varphi\}$ *be a collection of sentences. Then the following conditions are equivalent:*

(*i*) φ *is preserved under substructures relative to* Γ, *i.e., for all* $\mathfrak{A}, \mathfrak{B} \in \text{Mod }\Gamma$, *if* $\mathfrak{A} \subseteq \mathfrak{B}$ *and* $\mathfrak{B} \models \varphi$, *then* $\mathfrak{A} \models \varphi$;

(*ii*) *there is a universal sentence* ψ *such that* $\Gamma \models \varphi \leftrightarrow \psi$.

PROOF. It is trivial that (*ii*) $\Rightarrow$ (*i*). Now assume that (*i*) holds. Now by 25.3, $\mathbf{S}\text{Mod}(\Gamma \cup \{\varphi\})$ is closed under **Up**, so by 25.2 we may write $\mathbf{S}\text{Mod}(\Gamma \cup \{\varphi\}) = \text{Mod }\Delta$ for some set Δ of universal sentences. Now by (*i*) $\text{Mod}\Gamma \cap \mathbf{S}\text{Mod}(\Gamma \cup \{\varphi\}) \subseteq \text{Mod }\{\varphi\}$, so $\Gamma \cup \Delta \models \varphi$. Hence $\Gamma \models \psi \rightarrow \varphi$ for some finite conjunction ψ of members of Δ. But $\text{Mod }(\Gamma \cup \{\varphi\}) \subseteq \text{Mod }\Delta$, so $\Gamma \cup \{\varphi\} \models \psi$, hence $\Gamma \models \varphi \leftrightarrow \psi$. Clearly $\models \psi \leftrightarrow \chi$ for some universal sentence χ. □

Now we shall give Tarski's characterization of universal classes, which is simpler than the ultraproduct characterization and easier to prove.

Theorem 25.11 (Tarski). *Let* $\mathscr{L}$ *be a language with no operation symbols, and with only finitely many relation symbols. Let* **K** *be a class of* $\mathscr{L}$ *structures. Then the following conditions are equivalent:*

(*i*) **K** *is a universal class;*

(*ii*) $\mathbf{SK} = \mathbf{K}$, *and for every subset* $\mathscr{A}$ *of* **K** *directed by* $\subseteq$, $\bigcup \mathscr{A} \in \mathbf{K}$;

(*iii*) $\mathbf{SK} = \mathbf{K}$, *and for every* $\mathscr{L}$*-structure* $\mathfrak{A}$, *if every finite substructure of* $\mathfrak{A}$ *is in* **K**, *then* $\mathfrak{A}$ *is in* **K**.

PROOF

(*i*) $\Rightarrow$ (*ii*). Assume (*i*) and the hypothesis of (*ii*); say $\mathbf{K} = \text{Mod }\Gamma$, with Γ a set of universal sentences. Let $\varphi = \forall v_{i0} \cdots \forall v_{i(m-1)}\psi$ be any member of Γ, where ψ is quantifier-free. To show that $\bigcup \mathscr{A} \models \varphi$, let $x \in {}^{\omega}\bigcup \mathscr{A}$ be arbitrary. Then, since $\mathscr{A}$ is directed, choose $\mathfrak{A} \in \mathscr{A}$ such that $x_{ij} \in A$ for each $j < m$. Thus $\mathfrak{A} \models \psi[y]$ for any $y \in {}^{\omega}A$ with $x_{ij} = y_{ij}$ for each $j < m$, since $\mathfrak{A} \in \text{Mod }\Gamma$. Hence $\bigcup \mathscr{A} \models \psi[y]$ and so $\bigcup \mathscr{A} \models \varphi$. Since φ is arbitrary. $\bigcup \mathscr{A} \in \text{Mod }\Gamma = \mathbf{K}$.

(*ii*) $\Rightarrow$ (*iii*). This is obvious, since any structure is the union of its finite substructures.

Finally, assume (*iii*). Let $\Gamma = \{\varphi : \varphi$ is a universal sentence and $\mathfrak{A} \models \varphi$ for each $\mathfrak{A} \in \mathbf{K}\}$. Obviously $\mathbf{K} \subseteq \operatorname{Mod} \Gamma$. To prove the converse, let $\mathfrak{A} \in \operatorname{Mod} \Gamma$. Let $\mathfrak{B}$ be any finite substructure of $\mathfrak{A}$. Let $\mathscr{L}'$ be a B-expansion of $\mathscr{L}$, and let Δ be the $\mathscr{L}'$-diagram of $\mathfrak{B}$. Thus Δ is finite, by our special assumptions on the language $\mathscr{L}$. Let φ be a conjunction of all members of Δ, let φ' be obtained from φ by replacing all constants $\mathbf{c}_b$ for $b \in B$ by variables, and let ψ be obtained from φ' by prefixing existential quantifiers in front of φ' for all these variables. Then ψ holds in $\mathfrak{A}$. Hence $\neg\psi$ does not hold in $\mathfrak{A}$ and, since $\models \neg\psi \leftrightarrow \chi$ for some universal sentence χ and $\mathfrak{A} \in \operatorname{Mod} \Gamma$, $\neg\psi$ fails in some $\mathfrak{C} \in \mathbf{K}$, i.e., $\mathfrak{C} \models \psi$. Hence $\mathfrak{B}$ can be isomorphically embedded in $\mathfrak{C}$. Hence $\mathfrak{B} \in \mathbf{K}$ since $\mathbf{SK} = \mathbf{K}$ by (*iii*). Since $\mathfrak{B}$ is arbitrary, $\mathfrak{A} \in \mathbf{K}$ by (*iii*).

Quasivarieties

As the name suggests, quasivarieties are closely related to varieties. Instead of equations, we deal with simple implications, as defined in the following definition.

Definition 25.12. A *Horn formula* is a formula φ of the form

$$\mathbf{Q}_0 \cdots \mathbf{Q}_{m-1} \bigwedge_{i<n} \psi_i,$$

where each $\mathbf{Q}_j$ is a quantifier $\forall v_k$ or $\exists v_k$, and each ψ_i has one of the following three forms:

(*i*) χ, χ atomic;

(*ii*) $\neg\chi_0 \vee \cdots \vee \neg\chi_{pi}$, each χ_s atomic;

(*iii*) $\chi_0 \wedge \cdots \wedge \chi_{pi} \rightarrow \chi_q$; each χ_s atomic.

We say that φ is a *universal Horn formula* (or an *implicational identity*) if each $\mathbf{Q}_i$ is universal. A *quasivariety* is a class characterized by universal Horn sentences.

We obtain a logically equivalent definition of Horn formulas if in 25.12 we say that each formula ψ_i is a disjunction of atomic formulas and their negations, having at most one unnegated disjunct. In a language with no relation symbols, every variety is a quasivariety, but there are examples in which the converse fails. The theory of quasivarieties is not as well developed as the theory of varieties. We give here, again, characterization and preservation theorems.

Theorem 25.13. *The following conditions are equivalent:*

(*i*) $\mathbf{K}$ *is a quasivariety;*

(*ii*) $\mathbf{K}$ *is a universal class, and* $\mathbf{PK} = \mathbf{K}$*;*

(*iii*) $\mathbf{SK} = \mathbf{K}$, $\mathbf{UpK} = \mathbf{K}$, *and* $\mathbf{PK} = \mathbf{K}$*;*

(*iv*) $\mathbf{SK} = \mathbf{K}$, $\mathbf{UpK} = \mathbf{K}$, *and* $\mathfrak{A} \times \mathfrak{B} \in \mathbf{K}$ *whenever* $\mathfrak{A} \in \mathbf{K}$ *and* $\mathfrak{B} \in \mathbf{K}$.

PROOF

(*i*) $\Rightarrow$ (*ii*). Let $\mathbf{K} = \operatorname{Mod} \Gamma$, Γ a set of universal Horn sentences. Let $\mathfrak{A} \in {}^I\mathbf{K}$; we wish to show that $\mathsf{P}_{i \in I} \mathfrak{A}_i = \mathfrak{B} \in \mathbf{K}$. So, let φ be any member of Γ; say φ

has the form 25.12(*iii*). Let $x \in {}^{\omega}B$, and assume that $\mathfrak{B} \models (\chi_0 \wedge \cdots \wedge \chi_{pi})[x]$. Now since each χ_j is atomic, it follows from 18.13 and 18.14 that $\mathfrak{A}_k \models (\chi_0 \wedge \cdots \wedge \chi_{pi})[\mathrm{pr}_k \circ x]$ for each $k \in I$. Since $\mathfrak{A}_k \in \mathbf{K} = \mathrm{Mod}\,\Gamma$, it follows that $\mathfrak{A}_k \models \chi_q[\mathrm{pr}_k \circ x]$ for each $k \in I$. Hence by 18.13 and 18.14 again, $\mathfrak{B} \models \chi_q[x]$. Thus $\mathfrak{B} \models \varphi$. Since φ is arbitrary, $\mathfrak{B} \in \mathbf{K}$.

(*ii*) ⇒ (*iii*) and (*iii*) ⇒ (*iv*) are obvious.

(*iv*) ⇒ (*i*). Assume (*iv*). By Łoś's theorem 25.2, $\mathbf{K}$ is a universal class, say $\mathbf{K} = \mathrm{Mod}\,\Gamma$, where Γ is a set of universal sentences. We may assume that each member φ of Γ has the form

$$\mathbf{Q}_{\varphi}[\chi_{\varphi} \to \theta(0, \varphi) \vee \cdots \vee \theta(m_{\varphi} - 1, \varphi)],$$

where $\mathbf{Q}_{\varphi}$ is a string of universal quantifiers, χ_{φ} is a conjunction of atomic formulas, and each $\theta(i, j)$ is atomic (perhaps χ_{φ} does not occur, or $m_{\varphi} = 0$; then the notation should be changed, but the argument to follow is still o.k.). For each $\varphi \in \Gamma$ and each $j < m_{\varphi}$ let φ_j be the universal Horn sentence $\mathbf{Q}_{\varphi}(\chi_{\varphi} \to \theta(j, \varphi))$. Clearly

(1) $\models \varphi_j \to \varphi$ whenever $\varphi \in \Gamma$ and $j < m_{\varphi}$.

(2) For each $\varphi \in \Gamma$ with $m_{\varphi} \neq 0$ there is a $j < m_{\varphi}$ such $\Gamma \models \varphi_j$.

For, assume the contrary: say $\varphi \in \Gamma$, $m_{\varphi} \neq 0$, and for all $j < m_{\varphi}$ not $(\Gamma \models \varphi_j)$. Then for each $j < m_{\varphi}$ let $\mathfrak{A}_j$ be a model of Γ in which not $(\mathfrak{A}_j \models \varphi_j)$. Say $x^j \in {}^{\omega}A_j$ and not $\mathfrak{A}_j \models [\chi_{\varphi} \to \theta(j, \varphi)][x^j]$. Let $y: \omega \to \mathsf{P}_{j<m\varphi}\,\mathfrak{A}_j$ be defined by setting $(ym)_j = x^j m$ for all $m \in \omega$ and $j < m_{\varphi}$. Then by 18.13 and 18.14, not

$$\left(\mathsf{P}_{j<m\varphi}\,\mathfrak{A}_j \models [\chi_{\varphi} \to \theta(0, \varphi) \vee \cdots \vee \theta(m_{\varphi} - 1, \varphi)][y]\right).$$

Since $\mathsf{P}_{j<m\varphi}\,\mathfrak{A}_j \in \mathbf{K}$ by (*iv*), this is a contradiction. Thus (2) holds.

For each $\varphi \in \Gamma$ with $m_{\varphi} \neq 0$, let $f\varphi$ be the least natural number $< m_{\varphi}$ such that $\Gamma \models \varphi_{f\varphi}$. Clearly for each $\varphi \in \Gamma$ with $m\varphi = 0$ there is a set Θ_{φ} of identities with $\mathrm{Mod}\,\varphi = \mathrm{Mod}\,\Theta_{\varphi}$. Let $\Delta = \bigcup\{\Theta_{\varphi} : \varphi \in \Gamma, m_{\varphi} = 0\} \cup \{\varphi_{f\varphi} : \varphi \in \Gamma, m\varphi \neq 0\}$. Clearly Δ is a set of universal Horn sentences, and by (1) and (2), $\mathbf{K} = \mathrm{Mod}\,\Gamma$. □

Our preservation theorem for quasivarieties is as follows.

Theorem 25.14. *Let $\Gamma \cup \{\varphi\}$ be a set of sentences such that* $\mathbf{P}\,\mathrm{Mod}\,\Gamma = \mathrm{Mod}\,\Gamma$. *Then the following two conditions are equivalent:*

(*i*) *φ is preserved under substructures and finite products relative to Γ, i.e., for all $\mathfrak{A}, \mathfrak{B} \in \mathrm{Mod}\,\Gamma$, if $\mathfrak{A} \subseteq \mathfrak{B}$ and $\mathfrak{B} \models \varphi$ then $\mathfrak{A} \models \varphi$, and if $\mathfrak{A} \models \varphi$ and $\mathfrak{B} \models \varphi$ then $\mathfrak{A} \times \mathfrak{B} \models \varphi$;*

(*ii*) *there is a conjunction ψ of universal Horn sentences such that $\Gamma \models \varphi \leftrightarrow \psi$.*

Proof. It is trivial that (*ii*) ⇒ (*i*). To prove (*i*) ⇒ (*ii*), assume (*i*). By 25.10 let χ be a universal sentence such that $\Gamma \models \varphi \leftrightarrow \chi$. The method of proof of 25.13 then gives the desired sentence ψ. □

Universal–Existential Sentences

Definition 25.15. A formula is *universal-existential* if it has the form

$$\forall\alpha_0\cdots\forall\alpha_{m-1}\exists\beta_0\cdots\exists\beta_{n-1}\psi,$$

where ψ is quantifier-free.

Obviously every universal sentence is also universal-existential. A class characterized by universal-existential sentences is not in general closed under substructures. For an example, take the class of linearly ordered structures with no last element. There is a natural algebraic operation which characterizes universal-existential sentences, however. Our characterization theorem in this case has the following form.

Theorem 25.16 (Chang; Łoś, Suszko). *Let Γ be a set of sentences. Then the following conditions are equivalent:*

(*i*) Mod Γ *can be characterized by a set of universal-existential sentences;*
(*ii*) Mod Γ *is closed under unions of directed subsets;*
(*iii*) Mod Γ *is closed under unions of chains of type* ω.

PROOF

(*i*) $\Rightarrow$ (*ii*). Suppose Mod Γ = Mod Δ, where Δ is a set of universal-existential sentences. Suppose $\mathscr{A}$ is a subset of Mod Γ directed by $\subseteq$. Let φ be any member of Δ; say φ is the sentence

(1) $$\forall v_{i0}\cdots\forall v_{i(m-1)}\exists v_{im}\cdots\exists v_{i(n-1)}\psi.$$

To show that φ holds in $\bigcup\mathscr{A}$, take any $x\in{}^{\omega}\bigcup\mathscr{A}$. Then by directedness there is an $\mathfrak{A}\in\mathscr{A}$ such that $x_{ij}\in A$ for each $j<m$. Since $\mathfrak{A}\models\varphi$, choose $y\in{}^{\omega}A$ with $y\restriction\omega\sim\{i_j:m\leq j<n\}=x\restriction\omega\sim\{i_j:m\leq j<n\}$ such that $\mathfrak{A}\models\psi[y]$. But $\mathfrak{A}\subseteq\bigcup\mathscr{A}$, so $\bigcup\mathscr{A}\models\psi[y]$ also. Hence $\bigcup\mathscr{A}\models\exists v_{im}\cdots\exists v_{i(n-1)}\psi[x]$. Since x is arbitrary, $\bigcup\mathscr{A}\models\varphi$. Hence, φ being arbitrary, $\bigcup\mathscr{A}\in$ Mod Δ = Mod Γ.

(*ii*) $\Rightarrow$ (*iii*) is obvious. Now assume (*iii*). Let $\Delta=\{\varphi:\varphi$ is a universal existential sentence and $\Gamma\models\varphi\}$. Obviously Mod $\Gamma\subseteq$ Mod Δ. Now take any $\mathfrak{A}\in$ Mod Δ; it suffices to show that $\mathfrak{A}\in$ Mod Γ. To this end we shall construct a chain $\mathfrak{A}=\mathfrak{B}_0\subseteq\mathfrak{B}_1\subseteq\cdots$ of structures whose union will be in Mod Γ by (*iii*). The following notation will be useful. We write $\mathfrak{A}\subseteq_{\pi\sigma}\mathfrak{B}$ provided that $\mathfrak{A}\subseteq\mathfrak{B}$ and for every universal-existential formula φ and every $x\in{}^{\omega}A$, if $\mathfrak{B}\models\varphi[x]$ then $\mathfrak{A}\models\varphi[x]$. To get our chain started we need the following statement:

(2) there is a model $\mathfrak{B}_1$ of Γ such that $\mathfrak{A}\subseteq_{\pi\sigma}\mathfrak{B}_1$.

To prove (2), let $\mathscr{L}'$ be an A-expansion of $\mathscr{L}$. Let Δ' be the set of all universal sentences in $\mathscr{L}'$ which hold in $(\mathfrak{A},a)_{a\in A}$. Then

(3) $\Delta'\cup\Gamma$ is consistent.

For, otherwise there is a conjunction φ of members of Δ' such that $\Gamma \models \neg\varphi$. Now $\models \neg\varphi \leftrightarrow \psi$ for some existential sentence ψ of $\mathscr{L}'$. We can replace the constants $\mathbf{c}_a$, for $a \in A$, which appear in ψ by variables and obtain a universal-existential sentence χ of $\mathscr{L}$ such that $\models \chi \rightarrow \psi$ and $\Gamma \models \chi$. Thus $\chi \in \Delta$, so $\mathfrak{A} \models \chi$. On the other hand, $\neg\psi$ holds in $(\mathfrak{A}, a)_{a \in A}$, so $\neg\chi$ holds in $\mathfrak{A}$, contradiction. Thus (3) holds.

Let $(\mathfrak{B}, l_a)_{a \in A}$ be a model of $\Delta' \cup \Gamma$. Noticing that the $\mathscr{L}'$-diagram of $\mathfrak{A}$ is a subset of Δ', we see that $l\colon \mathfrak{A} \rightarrowtail \mathfrak{B}$. Now suppose that φ is a universal-existential formula of $\mathscr{L}$, $x \in {}^{\omega}A$, and $\mathfrak{A} \models \neg\varphi[x]$. Say φ has the form (1). Choose $y \in {}^{\omega}A$ such that $y \upharpoonright \omega \sim \{i_j : j < m\} = x \upharpoonright \omega \sim \{i_j : j < m\}$ and $\mathfrak{A} \models \forall v_{im} \cdots \forall v_{i(n-1)} \neg \psi[y]$. Let $\mathrm{Fv}\, \forall v_{im} \cdots \forall v_{i(n-1)} \neg \psi \subseteq \{v_0, \ldots, v_{p-1}\}$. Then $(\mathfrak{A}, a)_{a \in A} \models \forall v_{im} \cdots \forall v_{i(n-1)} \neg \psi(\mathbf{c}_{y0}, \ldots, \mathbf{c}_{y(p-1)})$, so the *sentence* $\forall v_{im} \cdots \forall v_{i(n-1)} \neg \psi(\mathbf{c}_{y0}, \ldots, \mathbf{c}_{y(p-1)})$ of $\mathscr{L}'$ is in Δ' and so holds in $(\mathfrak{B}, l_a)_{a \in A}$. Hence $\mathfrak{B} \models \forall v_{im} \cdots \forall v_{i(n-1)} \neg \psi[l \circ y]$ and so $\mathfrak{B} \models \neg\varphi[l \circ x]$. Now a simple replacement yields (2).

Let $\mathfrak{B}_0 = \mathfrak{A}$, and let $\mathfrak{B}_1$ be as in (2). We now suppose that $\mathfrak{B}_{2n}$ and $\mathfrak{B}_{2n+1}$ have been defined so that $\mathfrak{B}_{2n} \subseteq_{\pi\sigma} \mathfrak{B}_{2n+1}$. Now we define $\mathfrak{B}_{2n+2}$ and $\mathfrak{B}_{2n+3}$. We shall find $\mathfrak{B}_{2n+2}$ so that

(4) $$\mathfrak{B}_{2n} \preccurlyeq \mathfrak{B}_{2n+2} \quad \text{and } \mathfrak{B}_{2n+1} \subseteq \mathfrak{B}_{2n+2}.$$

To do this, let $\mathscr{L}'$ be a B_{2n+1}-expansion of $\mathscr{L}$, let Δ' be the elementary $\mathscr{L}'$-diagram of $\mathfrak{B}_{2n}$, and let Γ' be the $\mathscr{L}'$-diagram of $\mathfrak{B}_{2n+1}$. Then

(5) $\Delta' \cup \Gamma'$ is consistent.

For, otherwise there is a finite conjunction φ of members of Γ' such that $\Delta' \models \neg\varphi$. Replace constants $\mathbf{c}_b$ for $b \in B_{2n+1} \sim B_{2n}$ occurring in φ by variables, and generalize; we obtain a universal sentence ψ of $\mathscr{L}'$, involving only constants from $\mathfrak{B}_{2n}$, such that $\Delta' \models \psi$ and $\models \psi \rightarrow \neg\varphi$. There is then a universal formula χ of $\mathscr{L}$ such that $\psi = \chi(\mathbf{c}_{b0}, \ldots, \mathbf{c}_{b(m-1)})$ for some m. Hence $\mathfrak{B}_{2n} \models \chi[b_0, \ldots, b_{m-1}]$. Since $\mathfrak{B}_{2n} \subseteq_{\pi\sigma} \mathfrak{B}_{2n+1}$, it is clear then that $\mathfrak{B}_{2n+1} \models \chi[b_0, \ldots, b_{m-1}]$. But $(\mathfrak{B}_{2n+1}, b)_{b \in B(2n+1)} \models \varphi$, and hence $(\mathfrak{B}_{2n+1}, b)_{b \in B(2n)} \models \neg\psi$ and $\mathfrak{B}_{2n+1} \models \neg\chi[b_0, \ldots, b_{m-1}]$, contradiction. Thus (5) holds after all.

By (5), let $(\mathfrak{C}, l_b)_{b \in B(2n+1)}$ be a model of $\Delta' \cup \Gamma'$. Thus l is an embedding of $\mathfrak{B}_{2n+1}$ into $\mathfrak{C}$ and $l \upharpoonright B_{2n}$ is an elementary embedding of $\mathfrak{B}_{2n}$ into $\mathfrak{C}$. Let $\mathfrak{B}_{2n+2}$ be an $\mathscr{L}$-structure and s a map such that $\mathfrak{B}_{2n+1} \subseteq \mathfrak{B}_{2n+2}$, $s\colon \mathfrak{C} \rightarrowtail\!\!\!\!\twoheadrightarrow \mathfrak{B}_{2n+2}$, and $s \circ l = \mathbf{I} \upharpoonright B_{2n+1}$. Clearly then also $\mathfrak{B}_{2n} \preccurlyeq \mathfrak{B}_{2n+2}$ as desired in (4).

Next, we shall find $\mathfrak{B}_{2n+3}$ so that

(6) $$\mathfrak{B}_{2n+1} \equiv_{ee} \mathfrak{B}_{2n+3} \quad \text{and } \mathfrak{B}_{2n+2} \subseteq_{\pi\sigma} \mathscr{L}_{2n+3}.$$

To this end, let $\mathscr{L}'$ be a B_{2n+2}-expansion of $\mathscr{L}$. Let Δ' be the set of all sentences holding in $\mathfrak{B}_{2n+1}$, and let Γ' be the set of universal sentences of $\mathscr{L}'$ holding in $(\mathfrak{B}_{2n+2}, b)_{b \in B(2n+2)}$. Then

(7) $\Delta' \cup \Gamma'$ is consistent.

For, otherwise there is a finite conjunction φ of members of Γ' such that $\Delta' \models \neg\varphi$. Replacing the constants $\mathbf{c}_b$, $b \in B_{2n+2}$, which occur in φ by variables and generalizing, we obtain a sentence ψ equivalent to a universal-existential sentence such that $\models \psi \rightarrow \neg\varphi$ and $\Delta' \models \psi$. Hence $\mathfrak{B}_{2n+1} \models \psi$. But $(\mathfrak{B}_{2n+2}, b)_{b\in B(2n+2)} \models \varphi$, and hence $\mathfrak{B}_{2n+2} \models \neg\psi$. By (4) it follows that $\mathfrak{B}_{2n} \models \neg\psi$, and hence the assumption $\mathfrak{B}_{2n} \subseteq_{\pi\sigma} \mathfrak{B}_{2n+1}$ yields $\mathfrak{B}_{2n+1} \models \neg\psi$, contradiction. Thus (7) holds. Then (6) easily follows, using the argument that (3) implies (2).

Thus we have constructed our chain $\mathfrak{A} = \mathfrak{B}_0 \subseteq \mathfrak{B}_1 \subseteq \mathfrak{B}_2 \subseteq \cdots$ so that $\mathfrak{B}_1$ is a model of Γ, $\mathfrak{B}_1 \equiv_{ee} \mathfrak{B}_3 \equiv_{ee} \mathfrak{B}_5 \equiv_{ee} \cdots$, and $\mathfrak{B}_0 \preccurlyeq \mathfrak{B}_2 \preccurlyeq \mathfrak{B}_4 \preccurlyeq \cdots$. Hence $\mathfrak{B}_{2n+1}$ is a model of Γ for each $n \in \omega$, so that by (*iii*), $\bigcup_{n\in\omega} \mathfrak{B}_{2n+1}$ is a model of Γ. And by 19.36, we have $\mathfrak{A} \preccurlyeq \bigcup_{n\in\omega} \mathfrak{B}_{2n}$. But obviously $\bigcup_{n\in\omega} \mathfrak{B}_{2n+1} = \bigcup_{n\in\omega} \mathfrak{B}_n = \bigcup_{n\in\omega} \mathfrak{B}_{2n}$, so $\mathfrak{A}$ has been elementarily embedded in a model of Γ. Thus $\mathfrak{A}$ itself is a model of Γ. □

Corollary 25.17. *Let* $\Gamma \cup \{\varphi\}$ *be a set of sentences such that* Mod Γ *is closed under directed unions. Then the following conditions are equivalent:*

(*i*) φ *is preserved under directed union relative to* Γ, *i.e.*, Mod $(\Gamma \cup \{\varphi\})$ *is closed under directed unions;*

(*ii*) *there is a universal-existential sentence* ψ *such that* $\Gamma \models \varphi \leftrightarrow \psi$.

The corollary is proved just like 25.10.

Positive Sentences

Definition 25.18. A formula φ is *positive* if it is built up from atomic formulas using only $\wedge$, $\vee$, $\forall$, and $\exists$ (no negation). We write $\mathfrak{A}$ pos $\mathfrak{B}$ provided that every positive sentence holding in $\mathfrak{A}$ also holds in $\mathfrak{B}$.

The main theorem concerning this notion is proved very similarly to 25.16 above. First we need three lemmas.

Lemma 25.19. *If* $\mathfrak{A}$ pos $\mathfrak{B}$, *then there is an elementary extension* $\mathfrak{B}' \succcurlyeq \mathfrak{B}$ *and a homomorphism* $f\colon \mathfrak{A} \to \mathfrak{B}'$ *such that* $(\mathfrak{A}, a)_{a\in A}$ pos $(\mathfrak{B}', fa)_{a\in A}$ (in a suitable A-expansion of our language).

Proof. Let $\mathscr{L}$ be our initial language, $\mathscr{L}'$ an A-expansion of $\mathscr{L}$, and $\mathscr{L}''$ a B-expansion of $\mathscr{L}$; say we have new constants $\mathbf{c}_a (a \in A)$ in $\mathscr{L}'$ and new constants $\mathbf{d}_b (b \in B)$ in $\mathscr{L}''$. Let Γ be the set of all positive sentences of $\mathscr{L}'$ holding in $(\mathfrak{A}, a)_{a\in A}$, and Δ the elementary $\mathscr{L}''$-diagram of $(\mathfrak{B}, b)_{b\in B}$. Our assumption $\mathfrak{A}$ pos $\mathfrak{B}$ implies, by the argument usual in this chapter, that $\Gamma \cup \Delta$ has a model $\mathfrak{C}' = (\mathfrak{C}, la, sb)_{a\in A, b\in B}$. Thus by a basic fact concerning elementary diagrams, there is an isomorphism t of $\mathfrak{C}$ onto an elementary extension $\mathfrak{B}' \succcurlyeq \mathfrak{B}$ such that $t \circ s = \mathbf{I} \restriction B$. Clearly $(\mathfrak{A}, a)_{a\in A}$ pos $(\mathfrak{B}', tla)_{a\in A}$ (because of Γ holding in $\mathfrak{C}'$) and hence, as is easily seen, $t \circ l\colon \mathfrak{A} \to \mathfrak{B}'$ is a homomorphism. □

Lemma 25.20. *If* $\mathfrak{A}$ pos $\mathfrak{B}$, *then there is an elementary extension* $\mathfrak{A}' \succcurlyeq \mathfrak{A}$ *and a one-one mapping* $g\colon B \to A'$ *such that* $(\mathfrak{A}', gb)_{b\in B}$ pos $(\mathfrak{B}, b)_{b\in B}$.

PROOF Let $\mathscr{L}'$ and $\mathscr{L}''$ be as in the proof of 25.19. Let Γ be the elementary $\mathscr{L}'$-diagram of $\mathfrak{A}$, and let Δ be the set $\{\neg\varphi : \varphi$ is a positive sentence of $\mathscr{L}''$ and $(\mathfrak{B}, b)_{b\in B} \models \neg\varphi\}$. The rest of the proof proceeds as with the proof of 25.19. □

Similarly we have:

Lemma 25.21. *Assume that* $\mathfrak{B}$ *is a model of* $\{\varphi : \varphi$ *is a positive sentence and* $\Gamma \models \varphi\}$. *Then* Γ *has a model* $\mathfrak{A}$ *such that* $\mathfrak{A}$ pos $\mathfrak{B}$.

PROOF. Consider $\Gamma \cup \{\neg\varphi : \varphi$ is a positive sentence and $\Gamma \models \neg\varphi\}$. □

Theorem 25.22 (Lyndon). *For any set* Γ *of sentences the following conditions are equivalent:*

(*i*) Mod Γ *can be characterized by a set of positive sentences;*

(*ii*) Mod Γ *is closed under homomorphisms, i.e., if* $\mathfrak{A} \in$ Mod Γ *and* $\mathfrak{A} \twoheadrightarrow \mathfrak{B}$, *then* $\mathfrak{B} \in$ Mod Γ.

PROOF. The trivial direction is $(i) \Rightarrow (ii)$: assume (i). To show (ii) it suffices to prove

(1) if φ is any positive formula, $f: \mathfrak{A} \twoheadrightarrow \mathfrak{B}$, $x \in {}^{\omega}A$ and $\mathfrak{A} \models \varphi[x]$, then $\mathfrak{B} \models \varphi[f \circ x]$.

This statement is proved by an obvious induction on φ, using 24.4; we take one atomic case and the passage from φ to $\forall v_i\varphi$ as examples. Suppose first that φ is $\mathbf{R}\sigma_0 \cdots \sigma_{m-1}$. Since $\mathfrak{A} \models \varphi[x]$, we have $(\sigma_0^{\mathfrak{A}}x, \ldots, \sigma_{m-1}^{\mathfrak{A}}x) \in \mathbf{R}^{\mathfrak{A}}$. Hence $(f\sigma_0^{\mathfrak{A}}x, \ldots, f\sigma_{m-1}^{\mathfrak{A}}x) \in \mathbf{R}^{\mathfrak{B}}$, so by 24.4, $(\sigma_0^{\mathfrak{B}}(f \circ x), \ldots, \sigma_{m-1}^{\mathfrak{A}}(f \circ x)) \in \mathbf{R}^{\mathfrak{B}}$. Thus $\mathfrak{B} \models \varphi[f \circ x]$. Now assume (1) for φ, and suppose that $\mathfrak{A} \models \forall v_i\varphi[x]$. Let b be any element of B. Say $b = fa$. Then $\mathfrak{A} \models \varphi[x_a^i]$, so by the induction hypothesis $\mathfrak{B} \models \varphi[f \circ x_a^i]$, i.e., $\mathfrak{B} \models \varphi[(f \circ x)_b^i]$, as desired.

$(ii) \Rightarrow (i)$. As usual, let $\Delta = \{\varphi : \varphi$ is a positive sentence holding in each member of Mod $\Gamma\}$. Thus Mod $\Gamma \subseteq$ Mod Δ, so it suffices to take any model $\mathfrak{B}$ of Δ and show that $\mathfrak{B}$ is a model of Γ. We shall do this by constructing structures, $\mathfrak{A}_\omega$ and $\mathfrak{B}_\omega$ such that $\mathfrak{B} \preccurlyeq \mathfrak{B}_\omega$, $\mathfrak{A}_\omega$ is a model of Γ, and $\mathfrak{A}_\omega \twoheadrightarrow \mathfrak{B}_\omega$, which obviously shows by (ii) that $\mathfrak{B}$ is a model of Γ.

To construct $\mathfrak{A}_\omega$ and $\mathfrak{B}_\omega$ we construct certain sequences $\langle\mathfrak{A}_n : n \in \omega\rangle$, $\langle\mathfrak{B}_n : n \in \omega\rangle$, $\langle f_n : n \in \omega\rangle$, and $\langle g_n : n \in \omega \sim 1\rangle$. Let $\mathfrak{B}_0 = \mathfrak{B}$, and choose $\mathfrak{A}_0 \in$ Mod Γ so that $\mathfrak{A}_0$ pos $\mathfrak{B}_0$, by 25.21. By 25.19 choose $\mathfrak{B}_1$ and f_0 so that $\mathfrak{B}_0 \preccurlyeq \mathfrak{B}_1$, f_0 is a homomorphism of $\mathfrak{A}_0$ into $\mathfrak{B}_1$, and $(\mathfrak{A}_0, a)_{a\in A0}$ pos $(\mathfrak{B}_1, f_0a)_{a\in A0}$. Then by 25.20 choose $\mathfrak{A}_1$ and g_1 so that $(\mathfrak{A}_0, a)_{a\in A0} \preccurlyeq (\mathfrak{A}_1, a)_{a\in A0}$, g_1 is a one-one map of B_1 into A_1, and $(\mathfrak{A}_1, a, g_1b)_{a\in A0, b\in B1}$ pos $(\mathfrak{B}_1, f_0a, b)_{a\in A0, b\in B1}$. Now suppose that $n \geq 1$, $\mathfrak{A}_i$, $\mathfrak{B}_i$, and g_i have been defined for all $i \leq n$, and f_i has been defined for all $i < n$, so that the following condition holds:

(2) $$(\mathfrak{A}_n, a, g_nb)_{a\in A(n-1), b\in Bn} \text{ pos } (\mathfrak{B}_n, f_{n-1}a, b)_{a\in A(n-1), b\in Bn}.$$

We now define $\mathfrak{B}_{n+1}$, f_n, $\mathfrak{A}_{n+1}$, and g_{n+1}. Choose $\mathfrak{B}_{n+1}$ and f_n by 25.19 so that $(\mathfrak{B}_n, f_{n-1}a, b)_{a\in A(n-1), b\in Bn} \preccurlyeq (\mathfrak{B}_{n+1}, f_{n-1}a, b)_{a\in A(n-1), b\in Bn}$, f_n is a homomorphism of $(\mathfrak{A}_n, a, g_n b)_{a\in A(n-1), b\in Bn}$ into $(\mathfrak{B}_{n+1}, f_{n-1}a, b)_{a\in A(n-1), b\in Bn}$, and $(\mathfrak{A}_n, a, g_n b, a')_{a\in A(n-1), b\in Bn, a'\in An}$ pos $(\mathfrak{B}_{n+1}, f_{n-1}a, b, f_n a')_{a\in A(n-1), b\in Bn, a'\in An}$. Then we use 25.20 to find $\mathfrak{A}_{n+1}$ and g_{n+1} so that $(\mathfrak{A}_n, a)_{a\in An} \preccurlyeq (\mathfrak{A}_{n+1}, a)_{a\in An}$, g_{n+1} is a one-one mapping of B_{n+1} into A_{n+1}, and $(\mathfrak{A}_{n+1}, a, g_{n+1}b)_{a\in An, b\in B(n+1)}$ pos $(\mathfrak{B}_{n+1}, f_n a, b)_{a\in An, b\in B(n+1)}$.

This completes our inductive definition. Thus we have the following situation:

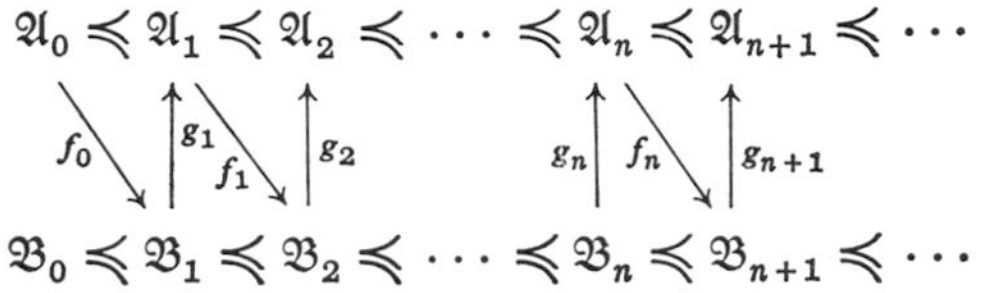

Since f_n is a homomorphism of $(\mathfrak{A}_n, a, g_n b)_{a\in A(n-1), b\in Bn}$ into

$$(\mathfrak{B}_{n+1}, f_{n-1}a, b)_{a\in A(n-1), b\in Bn},$$

it follows that $f_{n-1} \subseteq f_n$ and $g_n^{-1} \subseteq f_n$. Let $\mathfrak{A}_\omega = \bigcup_{n\in\omega} \mathfrak{A}_n$ and $\mathfrak{B}_\omega = \bigcup_{n\in\omega} \mathfrak{B}_n$. By 19.36 we have $\mathfrak{A} = \mathfrak{A}_0 \preccurlyeq \mathfrak{A}_\omega$ and $\mathfrak{B} = \mathfrak{B}_0 \preccurlyeq \mathfrak{B}_\omega$. Since $\mathfrak{A}$ is a model of Γ, so is $\mathfrak{A}_\omega$. Let $f_\omega = \bigcup_{n\in\omega} f_n$. Then clearly f_ω is a homomorphism from $\mathfrak{A}_\omega$ into $\mathfrak{B}_\omega$. But $g_n^{-1} \subseteq f_n$ for all $n \in \omega$, so f_ω maps onto B_ω. □

Corollary 25.23. *Let Γ be a set of sentences such that* Mod Γ *is closed under* **H**, *and let φ be a sentence. Then the following conditions are equivalent:*

(*i*) *φ is preserved under homomorphism relative to Γ;*

(*ii*) *there is a positive sentence ψ such that $\Gamma \models \varphi \leftrightarrow \psi$.*

EXERCISES

25.24. For any class **K**, the class **SUpK** is the smallest universal class containing **K**.

25.25. A class **K** can be characterized by a single universal sentence iff **SK** = **K**, **UpK** = **K**, and **Up**($\sim$**K**) = $\sim$**K**.

25.26. Let $\mathscr{L}$ be a language with one nonlogical constant, a binary operation symbol. Let **K** be the class of all structures isomorphic to structures $(B, \circ)$, where B is a set of one-one functions (with not necessarily equal domains) closed under composition $\circ$ of functions. Show that **K** is universal.

25.27. Let $\mathscr{L}'$ have nonlogical constants $<$ (binary relation symbol) and s (unary operation symbol), and let $\mathscr{L}$ be the reduct of $\mathscr{L}'$ to s. Let Γ be the following set of sentences in $\mathscr{L}'$:

sentences saying $<$ is a linear ordering

$\forall v_0(v_0 < \mathrm{s}v_0)$.

Thus by 25.5 **S**(Mod $\Gamma \upharpoonright \mathscr{L}$) is a universal class. Find explicit universal axioms for it, and show that it is not finitely axiomatizable.

25.28. Let $\mathscr{L}'$ be an expansion of $\mathscr{L}$, Γ a set of sentences of $\mathscr{L}'$, and let $\mathfrak{A}$ be an $\mathscr{L}$-structure. Then the following conditions are equivalent:

(1) $\mathfrak{A}$ can be embedded in $\mathfrak{B} \upharpoonright \mathscr{L}$ for some model $\mathfrak{B}$ of Γ.
(2) for every universal sentence φ of $\mathscr{L}$, the condition $\Gamma \models \varphi$ implies that $\mathfrak{A} \models \varphi$.

25.29. Let **K** be the class of all Abelian groups in which every element is of finite order. Show that 25.11(*ii*) and 25.11(*iii*) hold but 25.11(*i*) fails.

25.30. Give an example of an elementary class **K** such that **PK** is not elementary.

25.31. Let **K** be the class of all commutative rings satisfying $\forall v_0(v_0 \cdot v_0 \equiv v_0 \rightarrow v_0 \equiv \mathbf{O})$. Show that **K** is a quasivariety but not a variety.

25.32. Give an example of a universal class which is not a quasivariety.

25.33. Give an example of a system $\langle \mathfrak{A}_i : i \in \omega \rangle$ of structures such that $\mathfrak{A}_i \subseteq \mathfrak{A}_{i+1}$ for all $i \in \omega$, $\mathfrak{A}_i \equiv \mathfrak{A}_{i+1}$ for all i, but not $(\mathfrak{A}_0 \equiv \bigcup_{i \in \omega} \mathfrak{A}_i)$.

25.34. If **K** is an elementary class closed under directed unions and under products, then **K** can be characterized by universal-existential Horn sentences. (Such sentences are defined in a natural way along the lines of 25.12 and 25.15.)

25.35. An occurrence of a symbol in φ is *positive* if it is within the scope of an even number of negation symbols; otherwise it is *negative*. Now let $\mathscr{L}$ be a language with no operation symbols. Assume that φ and ψ are sentences in which $\equiv$ does not occur, and that $\models \varphi \rightarrow \psi$, not $(\models \neg \varphi)$, and not $(\models \psi)$. Then there is a sentence χ in which $\equiv$ does not occur such that $\models \varphi \rightarrow \chi$ and $\models \chi \rightarrow \psi$, and such that any relation symbol occurring positively (negatively) in χ also occurs positively (negatively) in both φ and ψ.

Hint: Use the model existence theorem, as follows. For each sentence χ of $\mathscr{L}$ let

$$\Gamma_\chi = \{\theta : \theta \text{ is a sentence of } \mathscr{L}', \text{ and each relation symbol which occurs positively (negatively) in } \theta \text{ also occurs positively (negatively) in } \chi\}.$$

Also, let $\Gamma' = \{\theta : \theta$ is a sentence of $\mathscr{L}$, and $\equiv$ does not occur in $\theta\}$. Let S be the collection of all $\Delta \subseteq \mathrm{Sent}_{\mathscr{L}'}$ such that there exist finite subsets $\Theta_0 \subseteq \Gamma_\varphi$ and $\Theta_1 \subseteq \Gamma_{\neg\psi}$ with $\Delta = \Theta_0 \cup \Theta_1$ such that:
(1) if $\chi \in \Delta$ and $\equiv$ occurs in χ, then χ is $\mathbf{c} \equiv \mathbf{c}$ for some $\mathbf{c} \in C$;
(2) both Θ_0 and Θ_1 have models;
(3) for all $\chi_0 \in \Gamma_\varphi \cap \Gamma_\psi \cap \Gamma'$ and $\chi_1 \in \Gamma_{\neg\varphi} \cap \Gamma_{\neg\psi} \cap \Gamma'$, if $\models \bigwedge \Theta_0 \rightarrow \chi_0$ and $\models \bigwedge \Theta_1 \rightarrow \chi_1$, then $\chi_0 \wedge \chi_1$ has a model.

25.36. Let Γ be a consistent theory in $\mathscr{L}$, and let Δ be a set of sentences of $\mathscr{L}$ such that for each finite subset Θ of Δ there is a $\varphi \in \Delta$ such that $\models \bigvee \Theta \leftrightarrow \varphi$. Then the following conditions are equivalent:
(1) there is a $\Theta \subseteq \Delta$ which axiomatizes Γ;
(2) for any model $\mathfrak{A}$ of Γ and any $\mathscr{L}$-structure $\mathfrak{B}$, if every $\varphi \in \Delta$ which holds in $\mathfrak{A}$ holds in $\mathfrak{B}$, then $\mathfrak{B}$ is a model of Γ.

This lemma can be used to shorten the proofs of 25.16 and 25.22.

26 Elementary Classes and Elementary Equivalence

We now investigate the basic notions of first-order logic. We shall give mathematical characterizations of elementary classes due to Fraissé, Ehrenfeucht, and Keisler and Shelah. The fundamental results of first-order logic are distinguished from more general languages by a theorem of Lindstrom.

Definition 26.1. Let **K** be a class of $\mathscr{L}$-structures.

(*i*) We set $\Theta\rho\mathbf{K} = \{\varphi \in \mathrm{Sent}_{\mathscr{L}}: \varphi$ holds in each member of $\mathbf{K}\}$; we call $\Theta\rho$ the *theory* of **K**. We set $\Theta\rho\mathfrak{A} = \Theta\rho\{\mathfrak{A}\}$.

(*ii*) **K** is *elementarily closed* iff for all $\mathscr{L}$-structures $\mathfrak{A}$, $\mathfrak{B}$, if $\mathfrak{A} \equiv_{\mathrm{ee}} \mathfrak{B} \in \mathbf{K}$ then $\mathfrak{A} \in \mathbf{K}$.

(*iii*) **K** is a *projective* (or *pseudo-elementary*) *class* iff there is an expansion $\mathscr{L}'$ of $\mathscr{L}$ and an elementary class **L** in $\mathscr{L}'$ such that $\mathbf{K} = \{\mathfrak{A} \upharpoonright \mathscr{L}: \mathfrak{A} \in \mathbf{L}\}$.

(*iv*) **K** is *compact* iff for all $\Gamma \subseteq \mathrm{Sent}_{\mathscr{L}}$, if every finite subset of Γ has a model in **K**, then Γ has a model in **K**.

Relationships between these notions are given in the following theorem, which has an important corollary for the rest of our work in this chapter.

Theorem 26.2. *Let* **K** *be a class of* $\mathscr{L}$*-structures. Each of the following conditions implies the succeeding one. If* **K** *is elementarily closed, then they are all equivalent.*

(*i*) **K** *is an elementary class;*
(*ii*) **K** *is a projective class;*
(*iii*) $\mathbf{UpK} = \mathbf{K}$*;*
(*iv*) **K** *is compact.*

PROOF. Obviously $(i) \Rightarrow (ii) \Rightarrow (iii) \Rightarrow (iv)$. Now suppose that $\mathbf{K}$ is elementarily closed and compact. We show that $\mathbf{K} = \text{Mod}\,\Theta\rho\mathbf{K}$, so that (i) holds. Obviously $\mathbf{K} \subseteq \text{Mod}\,\Theta\rho\mathbf{K}$, so we need to take any $\mathfrak{A} \in \text{Mod}\,\Theta\rho\mathbf{K}$ and show that $\mathfrak{A} \in \mathbf{K}$. Now

(1) $$\text{Mod}\,\Theta\rho\{\mathfrak{A}\} \cap \mathbf{K} \neq 0.$$

For, otherwise $\mathbf{K}$ compact implies that there is a finite $\Delta \subseteq \theta\rho\{\mathfrak{A}\}$ such that Δ does not have a model in $\mathbf{K}$. Thus $\neg \bigwedge \Delta \in \mathbf{K}$, so $\neg \bigwedge \Delta \in \Theta\rho\mathbf{K}$ and $\mathfrak{A} \models \neg \bigwedge \Delta$, which is impossible.

So (1) holds. Hence $\mathfrak{A} \equiv_{\text{ee}} \mathfrak{B} \in \mathbf{K}$ for some $\mathfrak{B}$, so $\mathfrak{A} \in \mathbf{K}$ since $\mathbf{K}$ is elementarily closed. □

There is a projective class which is not elementary. In fact, let $\mathscr{L}$ be any language, and let $\mathbf{K}$ be the class of all $\mathscr{L}$-structures with $> |\text{Fmla}_{\mathscr{L}}|$ elements. Then by the downward Löwenheim–Skolem theorem, $\mathbf{K}$ is not elementary. To see that $\mathbf{K}$ is projective, let $\mathscr{L}'$ be an expansion of $\mathscr{L}$ obtained by adjoining new nonlogical constants $\mathbf{c}_\alpha$ for all $\alpha < |\text{Fmla}_{\mathscr{L}}|^+$, and let Γ be the set of all sentences of the form $\neg(\mathbf{c}_\alpha = \mathbf{c}_\beta)$, where $\alpha < \beta$. Obviously $\mathbf{K} = (\text{Mod}\,\Gamma) \upharpoonright \mathscr{L}$, as desired. Examples of classes $\mathbf{K}$ with $\mathbf{UpK} = \mathbf{K}$ but $\mathbf{K}$ not projective appear to be known only under the assumption that no measurable cardinals exist. Finally, a compact $\mathbf{K}$ with $\mathbf{UpK} \neq \mathbf{K}$ is found by letting $\mathbf{K} = \{\mathfrak{A} : \mathfrak{A}$ is an $\mathscr{L}$-structure and $|A| = |\text{Fmla}_{\mathscr{L}}|$.

Corollary 26.3. *For any class* $\mathbf{K}$ *the following conditions are equivalent:*

(*i*) $\mathbf{K}$ *is an elementary class;*
(*ii*) $\mathbf{UpK} = \mathbf{K}$, *and* $\mathbf{K}$ *is elementarily closed.*

This corollary gives rise to the possibility of mathematically characterizing elementary classes by means of mathematically characterizing elementary equivalence. We shall do this in two different ways, due respectively to Ehrenfeucht and Fraissé and to Keisler and Shelah.

The characterization of Ehrenfeucht and Fraissé can be given in many forms. We shall give four of these forms here and prove their equivalence. Since $\mathfrak{A} \equiv_{\text{ee}} \mathfrak{B}$ iff $\mathfrak{A} \upharpoonright \mathscr{L}' \equiv_{\text{ee}} \mathfrak{B} \upharpoonright \mathscr{L}'$ for every reduct $\mathscr{L}'$ of $\mathscr{L}$ to finitely many symbols, we may assume that our language has only finitely many symbols. It is also clear that $\mathfrak{A} \equiv_{\text{ee}} \mathfrak{B}$ iff $\mathfrak{A}' \equiv_{\text{ee}} \mathfrak{B}'$, where $\mathfrak{A}'$ and $\mathfrak{B}'$ are the relational versions of $\mathfrak{A}$ and $\mathfrak{B}$ respectively (see Theorem 11.28). So, we may also assume that our language has no operation symbols. These assumptions will be useful for our main theorem, but they are not needed for some of the lemmas.

Definition 26.4. Let $\mathscr{L}$ be any first-order language, and let $\mathfrak{A}$ be an $\mathscr{L}$-structure. If $\mathscr{L}$ has individual constants, we let $\mathfrak{A}^-$ be the substructure of $\mathfrak{A}$ generated by $\{\mathbf{c}^{\mathfrak{A}} : \mathbf{c}$ is an individual constant of $\mathscr{L}\}$.

Now we define some equivalence relations $\equiv_m$ between structures in various languages; however, $\mathfrak{A} \equiv_m \mathfrak{B}$ shall always imply, implicitly, that

$\mathfrak{A}$ and $\mathfrak{B}$ are $\mathscr{L}$-structures for the same $\mathscr{L}$. These relations are defined by induction on m.

$\mathfrak{A} \equiv_0 \mathfrak{B}$ iff $\mathfrak{A}$ and $\mathfrak{B}$ are $\mathscr{L}$-structures and either $\mathscr{L}$ has no individual constants or else it does and $\mathfrak{A}^- \cong \mathfrak{B}^-$;

$\mathfrak{A} \equiv_{m+1} \mathfrak{B}$ iff for every $a \in A$ there is a $b \in B$ such that $(\mathfrak{A}, a) \equiv_m (\mathfrak{B}, b)$ and similarly with the roles of $\mathfrak{A}$ and $\mathfrak{B}$ interchanged.

Lemma 26.5. *Let φ be a sentence in prenex normal form with m initial quantifiers, and suppose that* $\mathfrak{A} \equiv_m \mathfrak{B}$. Then $\mathfrak{A} \models \varphi$ *iff* $\mathfrak{B} \models \varphi$.

Proof. We proceed by induction on m. For $m = 0$, φ is a sentence with no variables, and hence $\mathscr{L}$ has individual constants. Thus the assumption $\mathfrak{A} \equiv_0 \mathfrak{B}$ means that $\mathfrak{A}^- \cong \mathfrak{B}^-$. Hence, obviously, $\mathfrak{A} \models \varphi$ iff $\mathfrak{B} \models \varphi$.

Now assume the result true for m, for all logics $\mathscr{L}$ and all pairs of $\mathscr{L}$-structures. Suppose φ is a prenex sentence with $m + 1$ initial quantifiers and $\mathfrak{A} \equiv_{m+1} \mathfrak{B}$. We take only the case where the first quantifier of φ is existential, assume that $\mathfrak{A} \models \varphi$, and prove that $\mathfrak{B} \models \varphi$. Say φ is $\exists v_i \psi$. Choose $a \in A$ so that $\mathfrak{A} \models \psi[a, a, \ldots]$. Thus $(\mathfrak{A}, a) \models \mathrm{Subf}^{vi}_{\mathbf{ca}}\psi$ in a suitable expansion of the language of $\mathfrak{A}$. Since $\mathfrak{A} \equiv_{m+1} \mathfrak{B}$, choose $b \in B$ so that $(\mathfrak{A}, a) \equiv_m (\mathfrak{B}, b)$. Then by the induction hypothesis, $(\mathfrak{B}, b) \models \mathrm{Subf}^{vi}_{\mathbf{ca}}\psi$, i.e., $\mathfrak{B} \models \psi[b, b, \ldots]$, i.e., $\mathfrak{B} \models \varphi$. □

According to this lemma, if $\mathfrak{A} \equiv_m \mathfrak{B}$ for all m (a purely mathematical condition), then $\mathfrak{A} \equiv_{\mathrm{ee}} \mathfrak{B}$. This is true for any language whatsoever. The converse fails in general (see Exercise 26.47), but it does hold for languages with no operation symbols and only finitely many relation symbols. We shall prove this via a slight reformulation of the definition of $\equiv_m$.

Definition 26.6

(*i*) Let $\mathscr{L}$ be a language with no operation symbols and only finitely many relation symbols. We define some sets Δ^i_n of formulas of $\mathscr{L}$, where $0 \leq i \leq n > 0$, by induction on i. Let Δ^0_n be the set of all atomic formulas of $\mathscr{L}$ with free variables among $v_0, \ldots, v_{n-1}$. Our assumptions on $\mathscr{L}$ imply that Δ^0_n is finite. Suppose $i < n$ and Δ^i_n has been defined so that it is a finite set of formulas with free variables among $v_0, \ldots, v_{n-1-i}$. Let Δ^{i+1}_n consist of all formulas

$$\exists v_{n-1-i}\Big(\bigwedge_{\varphi \in \Theta} \varphi \wedge \bigwedge_{\varphi \in \Delta(i,n) \sim \Theta} \neg \varphi\Big)$$

for Θ a subset of Δ^i_n.

(*ii*) Let $\alpha \in \omega \cup \{\omega\}$, and let $\mathfrak{A}$ and $\mathfrak{B}$ be two $\mathscr{L}$-structures ($\mathscr{L}$ any first-order language). An *α-sequence for* $\mathfrak{A}$, $\mathfrak{B}$ is a sequence $I = \langle I_m : m \in \alpha \rangle$ such that the following conditions hold:

(1) $I_m \subseteq {}^mA \times {}^mB$ for each $m \in \alpha$;

(2) $0 I_0 0$;

(3) If $m + 1 \in \alpha$, $xI_m y$, and $a \in A$, then there is a $b \in B$ such that $\langle x_0, \ldots, x_{m-1}, a\rangle I_{m+1}\langle y_0, \ldots, y_{m-1}, b\rangle$, and similarly with the roles of A and B exchanged;
(4) If $\alpha \in \omega \sim 1$ and $xI_{\alpha-1}y$, then $\mathfrak{A} \models \varphi[x]$ iff $\mathfrak{B} \models \varphi[y]$ for any atomic formula φ with variables among $\{v_i : i \in \alpha - 1\}$, and if $\alpha = \omega$ then $\mathfrak{A} \models \varphi[x]$ iff $\mathfrak{B} \models \varphi[y]$ for any $x \in {}^mA$ and $y \in {}^mA$ with $m \in \omega$, $xI_m y$, and any atomic formula φ with $\mathrm{Fv}\,\varphi \subseteq \{v_0, \ldots, v_{m-1}\}$.

Note that there is, trivially, always a 1-sequence for any two structures $\mathfrak{A}$, $\mathfrak{B}$ in a language with no operation symbols, since then 26.6(*ii*)(4) holds vacuously, there being no atomic formulas without variables.

Lemma 26.7. *Let $\mathscr{L}$ be a language with no operation symbols and only finitely many relation symbols. Assume that $\mathfrak{A}$ and $\mathfrak{B}$ are $\mathscr{L}$-structures, $n \in \omega \sim 1$, and that for any $\varphi \in \Delta_n^n$, $\mathfrak{A} \models \varphi$ iff $\mathfrak{B} \models \varphi$. Then there is an $(n+1)$-sequence for $\mathfrak{A}$, $\mathfrak{B}$.*

PROOF. For each $i \le n$ let

$$I_i = \{(x, y) : x \in {}^iA,\ y \in {}^iB, \text{ and } \mathfrak{A} \models \varphi[x] \text{ iff } \mathfrak{B} \models \varphi[y] \text{ for all } \varphi \in \Delta_n^{n-i}\}.$$

Obviously 26.6(*ii*)(1),(2),(4) hold. To check (3), suppose that $m + 1 < n + 1$, $xI_m y$, and $a \in A$. Let

$$\Theta = \{\varphi \in \Delta_n^{n-m-1} : \mathfrak{A} \models \varphi[x_0, \ldots, x_{m-1}, a]\}.$$

Then the formula $\exists v_m(\bigwedge_{\varphi\in\Theta}\varphi \wedge \bigwedge_{\varphi\in\Delta(n-m-1,n)\sim\Theta} \neg\varphi)$ is in Δ_n^{n-m} and it holds in $\mathfrak{A}$, so it also holds in $\mathfrak{B}$. This gives us an element $b \in B$ such that $\langle x_0, \ldots, x_{m-1}, a\rangle I_{m+1}\langle y_0, \ldots, y_{m-1}, b\rangle$, as desired. Of course the argument is similar if $\mathfrak{A}$ and $\mathfrak{B}$ are interchanged. Hence 26.6(*ii*)(3) holds. □

The following lemma completes the proof of equivalence of $\equiv_{ee}$ with our mathematical notions in 26.4 and 26.6.

Lemma 26.8. *Let $n \in \omega$. If there is an $(n+1)$-sequence for $\mathfrak{A}$ and $\mathfrak{B}$, then $\mathfrak{A} \equiv_n \mathfrak{B}$.*

PROOF. Let $\langle I_m : m \le n\rangle$ be an $(n+1)$-sequence for $\mathfrak{A}$, $\mathfrak{B}$. We now prove the following statement by induction on i; the case $i = n$ gives the desired result.

(1) $$\text{if } xI_{n-i}y, \text{ then } (\mathfrak{A}, x_t)_{t<n-i} \equiv_i (\mathfrak{B}, y_t)_{t<n-i}.$$

The case $i = 0$ is obvious. Now assume (1) for i, and suppose that $xI_{n-i-1}y$. To prove that $(\mathfrak{A}, x_t)_{t<n-i-1} \equiv_{i+1} (\mathfrak{B}, y_t)_{t<n-i-1}$, by symmetry take $a \in A$. From the definition of I, 26.6(*ii*), choose $b \in B$ such that $\langle x_0, \ldots, x_{n-i-2}, a\rangle I_{n-i}\langle y_0, \ldots, y_{n-i-2}, b\rangle$. By our induction assumption, $(\mathfrak{A}, x_t, a)_{t<n-i-1} \equiv_i (\mathfrak{B}, y_t, b)_{t<n-i-1}$. □

Now we introduce another slight variation of the notion $\equiv_m$.

Definition 26.9. Let $\mathfrak{A}$ and $\mathfrak{B}$ be $\mathscr{L}$-structures. A *partial isomorphism of* $\mathfrak{A}$ *into* $\mathfrak{B}$ is a one-one function f mapping a subset of A into B such that if $m \leq |\text{Dmn}\, f|$, φ is an atomic formula, $\text{Fv}\,\varphi \subseteq \{v_0, \ldots, v_{m-1}\}$, and $x \in {}^m\text{Dmn}\, f$, then $\mathfrak{A} \models \varphi[x]$ iff $\mathfrak{B} \models \varphi[f \circ x]$. For any $\alpha \in \omega \cup \{\omega\}$, an α-system of partial isomorphisms of $\mathfrak{A}$ into $\mathfrak{B}$ is a system $\langle I_m : m \in \alpha \rangle$ satisfying the following conditions:

(*i*) each I_m is a nonempty set of partial isomorphism of $\mathfrak{A}$ into $\mathfrak{B}$;
(*ii*) if $m + 1 \in \alpha$, then $I_{m+1} \subseteq I_m$;
(*iii*) if $m + 1 \in \alpha$, $f \in I_{m+1}$, and $a \in A$ (resp. $b \in B$), then there is a $g \in I_m$ such that $f \subseteq g$ and $a \in \text{Dmn}\, g$ (resp. $b \in \text{Rng}\, g$).

In view of the above lemmas, the following lemma shows that elementary equivalence implies this new notion.

Lemma 26.10. *Let* $\mathfrak{A}$ *and* $\mathfrak{B}$ *be* $\mathscr{L}$*-structures and let* $n \in \omega$. *If there is an* $(n + 1)$*-sequence for* $\mathfrak{A}$ *and* $\mathfrak{B}$, *then there is an* $(n + 1)$*-system of partial isomorphisms of* $\mathfrak{A}$ *into* $\mathfrak{B}$.

PROOF. Let $\langle J_m : m \in n + 1 \rangle$ be an $(n + 1)$-sequence for $\mathfrak{A}$, $\mathfrak{B}$. For each $m \in n + 1$ we set

$$I_m = \{\{(x_i, y_i) : i < k\} : k \leq n - m \text{ and } xJ_k y\}.$$

Thus 26.9(*ii*) is obvious. By 26.6(*ii*)(2),(3) each J_k is nonempty, so each I_m is nonempty. Also, from 26.6(*ii*)(2),(3),(4) we infer that for each $k \leq n$, if $xJ_k y$ then $\mathfrak{A} \models \varphi[x]$ iff $\mathfrak{B} \models \varphi[y]$, for each atomic formula φ with variables among $v_0, \ldots, v_{k-1}$. It follows that each I_m is a set of partial isomorphisms of $\mathfrak{A}$ into $\mathfrak{B}$. Finally, 26.6(*ii*)(3) immediately yields (*iii*). □

Our final equivalent for elementary equivalence is not quite such a simple reformulation of the relations $\equiv_m$. It is formulated in game terminology, and we shall first give an intuitive account of the game.

Let $\mathfrak{A}$ and $\mathfrak{B}$ be $\mathscr{L}$-structures, and let two players I and II be given. Player I begins the game by picking a positive integer n, an $\varepsilon \in 2$, and if $\varepsilon = 0$ an element a_0 of A, while if $\varepsilon = 1$ an element b_0 of B. The game continues with II and I moving in turn. At the ith move of *I*, he chooses an $\varepsilon \in 2$, and if $\varepsilon = 0$ an element a_i of A while if $\varepsilon = 1$ he chooses an element b_i of B. Then II chooses an element b_i of B if $\varepsilon = 0$, or an element a_i of A if $\varepsilon = 1$. The game ends after $2n$ moves, at which point two sequences $a \in {}^nA$ and $b \in {}^nB$ have been constructed. The rule of the game is that II wins provided that $\{(a_i, b_i) : i < n\}$ is a partial isomorphism of $\mathfrak{A}$ into $\mathfrak{B}$. As we shall show, elementary equivalence is equivalent to the existence of a winning strategy for II. It is clear intuitively what we mean by "winning strategy": there must exist a completely deterministic method for II to make a move, given what has happened so far in the game, so that at the end II always wins. The precise definition runs as follows:

Definition 26.11. Let $\mathfrak{A}$ and $\mathfrak{B}$ be $\mathscr{L}$-structures. We say that II *has a winning strategy for the m-elementary game over* $\mathfrak{A}$, $\mathfrak{B}$ provided that there is a function F with the following two properties:

(*i*) If $k < m$, $x \in {}^kA$, and $y \in {}^kB$, then $F(x, y, 0, a) \in B$ for each $a \in A$ and $F(x, y, 1, b) \in A$ for each $b \in B$.

(*ii*) For every $z \in {}^m[(\{0\} \times A) \cup (\{1\} \times B)]$, define $x \in {}^mA$ and $y \in {}^mB$ by induction as follows. Suppose $k < m$ and $x \restriction k, y \restriction k$ have been defined. If $(zk)_0 = 0$, let $x_k = (zk)_1$ and $y_k = F(x \restriction k, y \restriction k, 0, x_k)$. If $(zk)_0 = 1$, let $y_k = (zk)_1$ and $x_k = F(x \restriction k, y \restriction k, 1, y_k)$. Then for the so defined sequences x and y it is the case that $\{(x_i, y_i)\} : i < m\}$ is a partial isomorphism of $\mathfrak{A}$ into $\mathfrak{B}$.

Lemma 26.12. *Let* $\mathfrak{A}$ *and* $\mathfrak{B}$ *be* $\mathscr{L}$*-structures, and let m be a positive integer. If there is an* $(m + 1)$*-system of partial isomorphisms of* $\mathfrak{A}$ *into* $\mathfrak{B}$*, then player* II *has a winning strategy for the m-elementary game over* $\mathfrak{A}$, $\mathfrak{B}$.

PROOF. Assume the hypothesis of 26.12, and let $\langle I_k : k \leq m \rangle$ be an $(m + 1)$-system of partial isomorphisms of $\mathfrak{A}$ into $\mathfrak{B}$. We define a function satisfying 26.11(*i*) as follows. Let a well-ordering of A, B, and all sets I_k for $k < m$ be given. Let $k < m$, $x \in {}^kA$, and $y \in {}^kB$. Set $f = \{(x_i, y_i) : i < k\}$. Let $a \in A$ and $b \in B$. If there is no $g \in I_{m-k}$ such that $f \subseteq g$, we let $F(x, y, 0, a)$ be the first element of B and $F(x, y, 1, b)$ be the first element of A. Now suppose there is such a g, and let g be the first such. By 26.9(*iii*) let h be the first member of I_{m-k-1} such that $g \subseteq h$ and $a \in \text{Dmn}\, h$, and let l be the first member of I_{m-k-1} such that $g \subseteq l$ and $b \in \text{Rng}\, l$. Then we set $F(x, y, 0, a) = ha$ and $F(x, y, 1, b) = l^{-1}b$. Thus F is constructed so that 26.11(*i*) holds.

To check 26.11(*ii*), let $z \in {}^m[(\{0\} \times A) \cup (\{1\} \times B)]$ be given, and construct x and y as in 26.11(*ii*). It is straightforward to check by induction on k that whenever $k \leq m$ there is an $g \in I_{m-k}$ such that $\{(x_i, y_i) : i < k\} \subseteq g$. Applying this for $k = m$, we obtain the conclusion of 26.11(*ii*). □

Our final lemma completes the circle of implications between the above notions.

Lemma 26.13. *Let* $\mathfrak{A}$ *and* $\mathfrak{B}$ *be* $\mathscr{L}$*-structures, and let m be a positive integer. If player* II *has a winning strategy for the m-elementary game over* $\mathfrak{A}$, $\mathfrak{B}$, *then* $\mathfrak{A} \models \varphi$ *iff* $\mathfrak{B} \models \varphi$ *whenever* φ *is a prenex sentence with m initial quantifiers.*

PROOF. Let F be as in 26.11, and for each $z \in {}^m[(\{0\} \times A) \cup (\{1\} \times B)]$ let x_z and y_z be constructed as in 26.11(*ii*). Let φ be a sentence $\mathbf{Q}_0 v_0 \cdots \mathbf{Q}_{m-1} v_{m-1} \psi$ where ψ is quantifier free and each $\mathbf{Q}_i$ is $\forall$ or $\exists$. We now prove the following statement by *downward* induction on k from m to 0.

(1) for all $k \leq m$ and all $z \in {}^m[(\{0\} \times A) \cup (\{1\} \times B)]$, $\mathfrak{A} \models \mathbf{Q}_k v_k \cdots \mathbf{Q}_{m-1} v_{m-1} \psi[x_z 0, \ldots, x_z(k-1)]$ iff $\mathfrak{B} \models \mathbf{Q}_k v_k \cdots \mathbf{Q}_{m-1} v_{m-1} \psi[y_z 0, \ldots, y_z(k-1)]$.

The case $k = m$ is true because of the conclusion of 26.11(*ii*). Now suppose that (1) is true for $k + 1$; we prove it for k. Suppose that $z \in {}^m[(\{0\} \times A) \cup (\{1\} \times B)]$. We take only the case $\mathbf{Q}_k = \forall$, and argue from satisfaction in $\mathfrak{A}$ to satisfaction in $\mathfrak{B}$. Assume, then, that $\mathfrak{A} \models \mathbf{Q}_k v_k \cdots \mathbf{Q}_{m-1} v_{m-1} \psi[x_z 0, \ldots, x_z(k-1)]$. Let b be any element of B. Let w be like z except that $w_k = (1, b)$. Clearly (by 26.11(*ii*)) $x_z \restriction k = x_w \restriction k$, so $\mathfrak{A} \models \mathbf{Q}_k v_k \cdots \mathbf{Q}_{m-1} v_{m-1} \psi[x_w 0, \ldots, x_w(k-1)]$. Since $\mathbf{Q}_k = \forall$, we have $\mathfrak{A} \models \mathbf{Q}_{k+1} v_{k+1} \cdots \mathbf{Q}_{m-1} v_{m-1} \psi[x_w 0, \ldots, x_w k]$. By the induction hypothesis, $\mathfrak{B} \models \mathbf{Q}_{k+1} v_{k+1} \cdots \mathbf{Q}_{m-1} v_{m-1} \psi[y_w 0, \ldots, y_w k]$. But $y_w k = b$ and $y_w \restriction k = y_z \restriction k$, so $\mathfrak{B} \models \mathbf{Q}_{k+1} v_{k+1} \cdots \mathbf{Q}_{m-1} v_{m-1} \psi[y_z 0, \ldots, y_z(k-1), b]$. Since b is arbitrary, it follows that $\mathfrak{B} \models \mathbf{Q}_k v_k \cdots \mathbf{Q}_{m-1} v_{m-1} \psi[y_z 0, \ldots, y_z(k-1)]$. Hence (1) holds. But condition (1) for $k = 0$ is exactly what we want to prove. □

Now we can give the main theorem characterizing elementary equivalence by the *back-and-forth* method:

Theorem 26.14 (Fraissé, Ehrenfeucht). *Let $\mathscr{L}$ be a language with no operation symbols and only finitely many relation symbols. Let $\mathfrak{A}$ and $\mathfrak{B}$ be two $\mathscr{L}$-structures. Then the following conditions are equivalent:*

(*i*) $\mathfrak{A} \equiv_{ee} \mathfrak{B}$;

(*ii*) $\mathfrak{A} \equiv_m \mathfrak{B}$ *for each* $m \in \omega$;

(*iii*) *for each* $m \in \omega$, *there is an m-sequence for* $\mathfrak{A}$, $\mathfrak{B}$;

(*iv*) *for each* $m \in \omega$, *there is an m-system of partial isomorphisms of* $\mathfrak{A}$ *into* $\mathfrak{B}$;

(*v*) *for each* $m \in \omega$, *player* II *has a winning strategy for the m-elementary game over* $\mathfrak{A}$, $\mathfrak{B}$.

As mentioned following 26.3, this theorem immediately gives rise to mathematical characterizations of the logical notion of an elementary class, using 26.3. We shall not state these characterizations explicitly.

We now want to give an application of 26.14. We shall give a simple characterization for the elementary equivalence of two structures $(\alpha, <)$, $(\beta, <)$, where α and β are nonzero ordinals. For the next few pages $\mathscr{L}$ will denote a fixed first-order language with only one nonlogical constant, a binary relation symbol $\mathbf{<}$, and we shall implicitly work within $\mathscr{L}$. We assume a very modest acquaintance with the arithmetic of ordinals. In what follows we deal with the ordinal operations of addition, multiplication, and exponentiation.

The *division algorithm* for ordinals plays an important role in what follows. According to it, if α and β are any two ordinals with $\beta \neq 0$, then there exist unique ordinals γ and δ such that $\alpha = \beta \cdot \gamma + \delta$ and $\delta < \beta$; we call δ the *remainder* upon dividing α by β. Given ordinals α, β and a nonzero ordinal γ, we say that α and β are *congruent modulo* γ, in symbols $\alpha \equiv \beta \pmod{\gamma}$, if α and β have the same remainder upon division by γ. Our mathematical characterization of elementary equivalence of $(\alpha, <)$ and $(\beta, <)$ is that $\alpha = \beta < \omega^\omega$ or else $\alpha, \beta \geq \omega^\omega$ and $\alpha \equiv \beta \pmod{\omega^\omega}$. We proceed to prove one implication:

Lemma 26.15. *If* $(\alpha, <) \equiv_{ee} (\beta, <)$, *then* $\alpha \equiv \beta \pmod{\omega^\omega}$, *and if* $\alpha < \omega^\omega$ *or* $\beta < \omega^\omega$ *then* $\alpha = \beta$.

PROOF. The idea of the proof is to show that each ordinal $<\omega^\omega$ is elementarily definable; the theorem then easily follows. To prove this, we need to express some simple concepts concerning ordinals. Let "$v_0 - v_1$ is a successor" be the formula

$$v_1 < v_0 \wedge \exists v_2[v_2 < v_0 \wedge \neg\exists v_3(v_2 < v_3 \wedge v_3 < v_0)].$$

Clearly we have

(1) $(\alpha, <) \models$ "$v_0 - v_1$ is a successor" $[\beta, \gamma]$ iff $\gamma < \beta < \alpha$ and $\beta - \gamma$ is a successor ordinal.

Next, let "$v_0 - v_1$ is divisible by ω" be the formula

$$v_1 < v_0 \wedge \neg(\text{“}v_0 - v_1 \text{ is a successor”}).$$

Thus

(2) $(\alpha, <) \models$ "$v_0 - v_1$ is divisible by ω" $[\beta, \gamma]$ iff $\gamma < \beta < \alpha$ and $\beta - \gamma$ is divisible by ω.

For any positive integer m, having defined the formula "$v_0 - v_1$ is divisible by ω^m", let "$v_0 - v_1$ is divisible by ω^{m+1}" be the formula

(3) $\exists v_2(v_2 < v_0 \wedge$ "$v_2 - v_1$ is divisible by ω^m"$) \wedge \forall v_2[v_2 < v_0 \rightarrow \exists v_3(v_2 < v_3 \wedge v_3 < v_0 \wedge$ "$v_3 - v_1$ is divisible by ω^m"$)]$.

We now prove:

(4) $(\alpha, <) \models$ "$v_0 - v_1$ is divisible by ω^m" $[\beta, \gamma]$ iff $\gamma < \beta < \alpha$ and $\beta - \gamma$ is divisible by ω^m.

We show this by induction on m. The case $m = 1$ is given by (2). Now suppose that (4) holds for m; we verify it for $m + 1$. First suppose that $\gamma < \beta < \alpha$ and $\beta - \gamma$ is divisible by ω^{m+1}; say $\beta - \gamma = \omega^{m+1} \cdot \delta$. Thus $\delta \neq 0$. Hence $\gamma + \omega^m < \beta$ and $(\gamma + \omega^m) - \gamma = \omega^m$ is divisible by ω^m, and the induction hypothesis (4) yields the satisfaction by (β, γ) in $(\alpha, <)$ of the first conjunct of (3) for $m + 1$. Now let $\varepsilon < \beta$ be arbitrary. If $\varepsilon \leq \gamma$, then $\varepsilon < \gamma + \omega^m < \beta$ and $(\gamma + \omega^m) - \gamma = \omega^m$ is divisible by ω^m, giving satisfaction of the second conjunct of (3). If $\gamma < \varepsilon$, write $\varepsilon - \gamma = \omega^{m+1} \cdot \zeta + \eta$ with $\eta < \omega^{m+1}$, and $\eta = \omega^m \cdot n + \theta$ where $n \in \omega$ and $\theta < \omega^m$. Then $\varepsilon < \gamma + \omega^{m+1} \cdot \zeta + \omega^m \cdot (n + 1) \leq \beta$ and $(\gamma + \omega^{m+1} \cdot \zeta + \omega^m \cdot (n + 1)) - \gamma = \omega^{m+1} \cdot \zeta + \omega^m \cdot (n + 1)$ is divisible by ω^m, so again the second conjunct of (3) is satisfied.

For the converse of (4), assume that $(\alpha, <) \models$ "$v_0 - v_1$ is divisible by ω^{m+1}" $[\beta, \gamma]$. Clearly $\gamma < \beta$. Write $\beta - \gamma = \omega^{m+1} \cdot \delta + \varepsilon$ with $\varepsilon < \omega^{m+1}$, and write $\varepsilon = \omega^m \cdot n + \zeta$ with $n \in \omega$ and $\zeta < \omega^m$. Suppose $\zeta \neq 0$. Then $\gamma + \omega^{m+1} \cdot \delta + \omega^m \cdot n < \beta$, so by (3) and the induction hypothesis there is an η with $\gamma + \omega^{m+1} \cdot \delta + \omega^m \cdot n < \eta < \beta$ and $\eta - \gamma$ divisible by ω^m. Say $\eta - \gamma = \omega^m \cdot \theta$. Since

$$\gamma + \omega^m \cdot (\omega \cdot \delta + n) = \gamma + \omega^{m+1} \cdot \delta + \omega^m \cdot n < \eta = \gamma + \omega^m \cdot \theta,$$

it follows that $\omega \cdot \delta + n < \theta$. But

$$\begin{aligned}\beta &= \gamma + \omega^{m+1} \cdot \delta + \varepsilon = \gamma + \omega^{m+1} \cdot \delta + \omega^m \cdot n + \zeta \\ &< \gamma + \omega^{m+1} \cdot \delta + \omega^m \cdot n + \omega^m = \gamma + \omega^m \cdot (\omega \cdot \delta + n + 1) \\ &\leq \gamma + \omega^m \cdot \theta = \eta < \beta,\end{aligned}$$

a contradiction. Thus $\zeta = 0$. Suppose $n \neq 0$. Then $\gamma + \omega^{m+1} \cdot \delta + \omega^m \cdot (n - 1) < \beta$, so by (3) and the induction hypothesis there is an ι with $\gamma + \omega^{m+1} \cdot \delta + \omega^m \cdot (n - 1) < \iota < \beta$ and $\iota - \gamma$ divisible by ω^m. Say $\iota - \gamma = \omega^m \cdot \kappa$. Then

$$\begin{aligned}\gamma + \omega^m \cdot (\omega \cdot \delta + n - 1) &= \gamma + \omega^{m+1} \cdot \delta + \omega^m \cdot (n - 1) < \iota \\ &= \gamma + \omega^m \cdot \kappa,\end{aligned}$$

so $\omega \cdot \delta + n - 1 < \kappa$. Hence

$$\begin{aligned}\beta &= \gamma + \omega^{m+1} \cdot \delta + \omega^m \cdot n = \gamma + \omega^m \cdot (\omega \cdot \delta + n) \leq \gamma + \omega^m \cdot \kappa \\ &= \iota < \beta,\end{aligned}$$

again a contradiction. So $n = 0$ also, and we have $\beta - \gamma = \omega^{m+1} \cdot \delta$. This completes the proof of (4).

Now for each $m \in \omega$ let "$v_0 - v_1 = m$" be the formula saying that $v_1 < v_0$ and there are exactly m v_2 such that $v_1 < v_2 < v_0$. For $m, n \in \omega \sim 1$, let "$v_0 - v_1 = \omega^m \cdot n$" be the formula

(5) "$v_0 - v_1$ is divisible by ω^m," and there are exactly $n - 1$ v_2 such that $v_1 < v_2 < v_0$ and "$v_2 - v_1$ is divisible by ω^m".

Let $k \in \omega \sim 1$, $m \in {}^k(\omega \sim 1)$, $n \in {}^k(\omega \sim 1)$, and $m_0 > \cdots > m_{k-1}$. We define "$v_0 - v_1 = \omega^{m0} \cdot n_0 + \cdots + \omega^{m(k-1)} \cdot n_{k-1}$" by induction on k. The case $k = 1$ is taken care of by (5). Assuming the formula defined for k, we let "$v_0 - v_1 = \omega^{m0} \cdot n_0 + \cdots + \omega^{mk} \cdot n_k$" be the formula

$$\exists v_2(v_1 < v_2 \wedge v_2 < v_0 \wedge \text{"}v_2 - v_1 = \omega^{m0} \cdot n_0 + \cdots + \omega^{m(k-1)} \cdot n_{k-1}\text{"} \\ \wedge \text{"}v_0 - v_2 = \omega^{mk} \cdot n_k\text{"}).$$

Clearly we have

(6) if $m \in \omega$, then $(\alpha, <)$ "$v_0 - v_1 = m$" $[\beta, \gamma]$ iff $\gamma < \beta < \alpha$ and $\beta - \gamma = m$;

(7) if $k \in \omega \sim 1$, $m \in {}^k(m \sim 1)$, $n \in {}^k(m \sim 1)$, and $m_0 > \cdots > m_{k-1}$, then $(\alpha, <) \models$ "$v_0 - v_1 = \omega^{m0} \cdot n_0 + \cdots + \omega^{m(k-1)} \cdot n_{k-1}$" $[\beta, \gamma]$ iff $\gamma < \beta < \alpha$ and $\beta - \gamma = \omega^{m0} \cdot n_0 + \cdots + \omega^{m(k-1)} \cdot n_{k-1}$.

Recall that for any $\beta < \omega^\omega$, either $\beta < \omega$ or β can be written in the form $\beta = \omega^{m0} \cdot n_0 + \cdots + \omega^{m(k-1)} \cdot n_{k-1}$ with $m_0 > \cdots > m_{k-1}$. Thus we have now shown that any ordinal $< \omega^\omega$ is elementarily definable.

To prove the lemma, assume that $(\alpha, <) \equiv_{ee} (\beta, <)$. Write $\alpha = \omega^\omega \cdot \gamma + \delta$ and $\beta = \omega^\omega \cdot \varepsilon + \zeta$ with $\delta, \zeta < \omega^\omega$. Suppose $\delta \neq \zeta$, say $\delta < \zeta$. Choose $m \in \omega \sim 1$ so that $\zeta < \omega^m$. Now clearly

$$(\alpha, <) \models \exists v_1[\forall v_2 \neg (v_2 \leq v_1) \wedge \text{"}v_0 - v_1 \text{ is divisible by } \omega^m\text{"}] \wedge \neg \exists v_1(\text{"}v_1 - v_0 = \delta\text{"})[\omega^\omega \cdot \gamma].$$

Hence the following sentence holds in $(\alpha, <)$:

$$\exists v_0\{\exists v_1[\forall v_2 \neg (v_2 < v_1) \wedge \text{“}v_0 - v_1 \text{ is divisible by } \omega^m\text{”}] \wedge \neg\exists v_1(\text{“}v_1 - v_0 = \delta\text{”})\}.$$

Hence it also holds in $(\beta, <)$, so there is an $\eta < \beta$ such that η is divisible by ω^m, but there is no θ such that $\theta - \eta = \delta$. This latter statement implies that $\eta + \delta \geq \beta$. Say $\eta = \omega^m \cdot \iota$. Then

$$\omega^m \cdot \iota = \eta < \beta = \omega^\omega \cdot \varepsilon + \zeta < \omega^\omega \cdot \varepsilon + \omega^m$$
$$= \omega^m \cdot \omega^\omega \cdot \varepsilon + \omega^m = \omega^m \cdot (\omega^\omega \cdot \varepsilon + 1),$$

so $\iota \leq \omega^\omega \cdot \varepsilon$. But then

$$\beta \leq \eta + \delta = \omega^m \cdot \iota + \delta < \omega^m \cdot \omega^\omega \cdot \varepsilon + \zeta = \omega^\omega \cdot \varepsilon + \zeta = \beta,$$

contradiction. Thus $\delta = \zeta$ after all. Thus $\alpha \equiv \beta \pmod{\omega^\omega}$.

Now suppose, say, that $\alpha < \omega^\omega$ while $\beta \geq \omega^\omega$. Then

$$(\beta, <) \models \exists v_1(\forall v_2 \neg (v_2 \leq v_1) \wedge \text{“}v_0 - v_1 = \alpha\text{”})[\alpha].$$

Hence

$$(\alpha, <) \models \exists v_0 \exists v_1(\forall v_2 \neg (v_2 \leq v_1) \wedge \text{“}v_0 - v_1 = \alpha\text{”}),$$

which is impossible. □

To obtain the converse of 26.15 we need to define another equivalence relation:

Definition 26.16. Let $\langle \alpha_0, \ldots, \alpha_{n-1} \rangle$ and $\langle \beta_0, \ldots, \beta_{n-1} \rangle$ be sequences of ordinals, where $n \in \omega \sim 1$, and let $m \in \omega$. We write $\langle \alpha_0, \ldots, \alpha_{n-1} \rangle \sim_m \langle \beta_0, \ldots, \beta_{n-1} \rangle$ provided the following two conditions hold for all $i, j < m$:

(*i*) $\alpha_i < \alpha_j$ iff $\beta_i < \beta_j$;
(*ii*) if $\alpha_i < \alpha_j$, then either $\alpha_j - \alpha_i = \beta_j - \beta_i$, or else both are $\geq \omega^m$ and $\alpha_j - \alpha_i \equiv \beta_j - \beta_i \pmod{\omega^m}$.

Note that $\langle \alpha \rangle \sim_m \langle \beta \rangle$ for any ordinals α, β and any $m \in \omega$. If $m = 0$, then condition (*ii*) does not say anything. Also note that if $m > n$ then $\alpha \equiv \beta \pmod{\omega^m}$ implies that $\alpha \equiv \beta \pmod{\omega^n}$. The converse of 26.15 follows from the following lemma:

Lemma 26.17. *Let* $\mathfrak{A} = \langle \alpha_0, <, \alpha_2, \ldots, \alpha_{n-1} \rangle$ *and* $\mathfrak{B} = \langle \beta_0, <, \beta_2, \ldots, \beta_{n-1} \rangle$, *where* $\alpha_2, \ldots, \alpha_{n-1} \in \alpha_0 \neq 0$ *and* $\beta_2, \ldots, \beta_{n-1} \in \beta_0 \neq 0$. *Assume that* $\langle \alpha_0, \ldots, \alpha_{n-1} \rangle \sim_m \langle \beta_0, \ldots, \beta_{n-1} \rangle$, *where* $\alpha_1 = \beta_1 = 0$. *Then* $\mathfrak{A} \equiv_m \mathfrak{B}$.

Proof. We proceed by induction on m. The case $m = 0$ is clear. Now assume the lemma for m, and assume the hypothesis of the lemma for $m + 1$. To verify that $\mathfrak{A} \equiv_{m+1} \mathfrak{B}$ it suffices, by symmetry, to take any $\alpha_n \in \alpha_0$ and find $\beta_n \in \beta_0$ so that $(\mathfrak{A}, \alpha_n) \equiv_m (\mathfrak{B}, \beta_n)$. If $\alpha_n = \alpha_i$ for some $i = 1, \ldots, n - 1$, let

$\beta_n = \beta_i$; from 26.16 it is then clear that $(\mathfrak{A}, \alpha_n) \equiv_m (\mathfrak{B}, \beta_n)$. So suppose that $\alpha_n \neq \alpha_i$ for all $i < n$. Choose α_j maximum so that $\alpha_j < \alpha_n$, and choose α_k minimum so that $\alpha_n < \alpha_k$; α_k always exist (perhaps $j = 1$ or $k = 0$). We shall now find β_n so that

(1) $\beta_n - \beta_j = \alpha_n - \alpha_j$ or both are $\geq \omega^m$ and $\beta_n - \beta_j \equiv \alpha_n - \alpha_j$ (mod ω^m), and $\beta_k - \beta_n = \alpha_k - \alpha_n$ or both are $\geq \omega^m$ and $\beta_k - \beta_n \equiv \alpha_k - \alpha_n$ (mod ω^m).

Because $\langle \alpha_0, \ldots, \alpha_{n-1} \rangle \sim_{m+1} \langle \beta_0, \ldots, \beta_{n-1} \rangle$, we have two possible cases:

Case 1. $\alpha_k - \alpha_j = \beta_k - \beta_j$. Let $\beta_n = \beta_j + (\alpha_n - \alpha_j)$. Then (1) is clear.

Case 2. $\alpha_k - \alpha_j$ and $\beta_k - \beta_j$ are both $\geq \omega^{m+1}$, and $\alpha_k - \alpha_j \equiv \beta_k - \beta_j$ (mod ω^{m+1}). We consider three subcases.

Subcase 1. $\alpha_n - \alpha_j < \omega^m$. Let $\beta_n = \beta_j + (\alpha_n - \alpha_j)$. Obviously then $\beta_n - \beta_j = \alpha_n - \alpha_j$. Since $\alpha_k - \alpha_j$ and $\beta_k - \beta_j$ are both $\geq \omega^{m+1}$, and ω^{m+1} absorbs all smaller ordinals, we have

$$\beta_k - \beta_j = (\beta_n - \beta_j) + (\beta_k - \beta_n)$$

and hence $\beta_k - \beta_n \geq \omega^{m+1}$, so

$$\begin{aligned} \beta_k - \beta_j &= (\beta_n - \beta_j) + \omega^{m+1} + ((\beta_k - \beta_n) - \omega^{m+1}) \\ &= \omega^{m+1} + ((\beta_k - \beta_n) - \omega^{m+1}) \\ &= \beta_k - \beta_n, \end{aligned}$$

and similarly $\alpha_k - \alpha_j = \alpha_k - \alpha_n \geq \omega^{m+1}$. Hence also $\alpha_k - \alpha_n \equiv \beta_k - \beta_n$ (mod ω^m).

Subcase 2. $\alpha_k - \alpha_n < \omega^m$. (Note that since $(\alpha_n - \alpha_j) + (\alpha_k - \alpha_n) = \alpha_k - \alpha_j \geq \omega^{m+1}$, this subcase is mutually exclusive of subcase 1.) By the assumption of this case we may write $\alpha_k = \alpha_j + \omega^{m+1} \cdot \gamma + \delta$ and $\beta_k = \beta_j + \omega^{m+1} \cdot \varepsilon + \delta$ with $\delta < \omega^{m+1}$ and $\gamma, \varepsilon \neq 0$. Now

$$\alpha_j + \omega^{m+1} \cdot \gamma \leq \alpha_n. \tag{2}$$

For, otherwise $\alpha_n < \alpha_j + \omega^{m+1} \cdot \gamma$; write $\alpha_n = \alpha_j + \zeta$. Then $\zeta < \omega^{m+1} \cdot \gamma$, and if we write $\zeta = \omega^{m+1} \cdot \eta + \theta$ with $\theta < \omega^{m+1}$ we have $\eta < \gamma$. But then

$$\begin{aligned} \alpha_k - \alpha_j &= (\alpha_n - \alpha_j) + (\alpha_k - \alpha_n) = \zeta + (\alpha_k - \alpha_n) \\ &= \omega^{m+1} \cdot \eta + \theta + (\alpha_k - \alpha_n), \end{aligned}$$

and $\theta + (\alpha_k - \alpha_n) < \omega^{m+1}$. But γ and δ are uniquely determined by α_k and α_j, so $\eta = \gamma$, a contradiction. Hence (2) holds.

Write $\alpha_n = \alpha_j + \omega^{m+1} \cdot \gamma + \iota$ with $\iota < \omega^{m+1}$. Then

$$\begin{aligned} \omega^{m+1} \cdot \gamma + \delta = \alpha_k - \alpha_j &= (\alpha_n - \alpha_j) + (\alpha_k - \alpha_n) \\ &= \omega^{m+1} \cdot \gamma + \iota + (\alpha_k - \alpha_n), \end{aligned}$$

so $\delta = \iota + (\alpha_k - \alpha_n)$. Let $\beta_n = \beta_j + \omega^{m+1} \cdot \varepsilon + \iota$. Clearly $\beta_k - \beta_n = \alpha_k - \alpha_n$, while $\beta_n - \beta_j, \alpha_n - \alpha_j \geq \omega^m$ and

$$\begin{aligned} \beta_n - \beta_j &\equiv \iota && (\text{mod } \omega^m) \\ &\equiv \alpha_n - \alpha_j && (\text{mod } \omega^m) \end{aligned}$$

Subcase 3. $\alpha_n - \alpha_j, \alpha_k - \alpha_n \geq \omega^m$. Write $\alpha_n - \alpha_j = \omega^m \cdot \kappa + \lambda$ with $\lambda < \omega^m$. Let $\beta_n = \beta_j + \omega^m\kappa + \lambda$. Thus $\alpha_n - \alpha_j, \beta_n - \beta_j \geq \omega^m$ and both are congruent to λ mod ω^m. Now since $\alpha_k - \alpha_n \geq \omega^m$ and hence $\alpha_k - \alpha_n$ absorbs λ, we have

$$\begin{aligned}\alpha_k - \alpha_n &\equiv \omega^m \cdot \kappa + \lambda + (\alpha_k - \alpha_n) && (\text{mod } \omega^m)\\ &= (\alpha_n - \alpha_j) + (\alpha_k - \alpha_n)\\ &= \alpha_k - \alpha_j \equiv \beta_k - \beta_j && (\text{mod } \omega^m)\\ &= (\beta_n - \beta_j) + (\beta_k - \beta_n)\\ &= \omega^m + \lambda + (\beta_k - \beta_n) \equiv \beta_k - \beta_n && (\text{mod } \omega^m)\end{aligned}$$

This completes the proof of (1).

Now suppose that $\alpha_n < \alpha_l \neq \alpha_k$. Then $\alpha_l - \alpha_n = (\alpha_k - \alpha_n) + (\alpha_l - \alpha_k)$, and either $\alpha_l - \alpha_k = \beta_l - \beta_k$ or else both are $\geq \omega^{m+1}$; and $\alpha_l - \alpha_k \equiv \beta_l - \beta_k \pmod{\omega^{m+1}}$. Hence clearly either $\alpha_l - \alpha_n = \beta_l - \beta_n$ or else both are $\geq \omega^m$ and $\alpha_l - \alpha_n \equiv \beta_l - \beta_n \pmod{\omega^m}$. We argue similarly if $\alpha_l < \alpha_n$. Hence we have now shown that $\langle \alpha_0, \ldots, \alpha_n \rangle \sim_m \langle \beta_0, \ldots, \beta_n \rangle$. Hence by the induction hypothesis $\langle \alpha_0, <, \alpha_2, \ldots, \alpha_n \rangle \equiv_m \langle \beta_0, <, \beta_2, \ldots, \beta_n \rangle$. Thus $\mathfrak{A} \equiv_{m+1} \mathfrak{B}$. □

Combining Lemmas 26.15 and 26.17 with the Ehrenfeucht–Fraissé theorem we obtain

Theorem 26.18. *For any nonzero ordinals* α, β, *we have* $(\alpha, <) \equiv_{ee} (\beta, <)$ *iff* $\alpha = \beta < \omega^\omega$ *or both* $\alpha, \beta \geq \omega^\omega$ *and* $\alpha \equiv \beta \pmod{\omega^\omega}$.

Now we turn to the second main topic of this section, Lindstrom's characterization of first-order logic. Roughly speaking, only first-order logic has the compactness theorem and the downward Löwenheim–Skolem theorem. To make this precise we must, of course, have some general notion of logic. We introduce this notion in the following definition.

Definition 26.19

(*i*) Let $\mathscr{L}$ and $\mathscr{L}'$ be two first-order languages. A function f is an *isomorphism of* $\mathscr{L}$ *into* $\mathscr{L}'$ provided that f maps the nonlogical constants of $\mathscr{L}$ one-one into the nonlogical constants of $\mathscr{L}'$, taking relation symbols to relation symbols and operation symbols to operation symbols, and f preserves the ranks of all symbols. We call f an isomorphism *onto* if f maps onto all nonlogical constants of $\mathscr{L}'$.

(*ii*) Given $\mathscr{L}, \mathscr{L}', f$ as in (*i*), with f onto, with each $\mathscr{L}$-structure $\mathfrak{A}$ we associate an $\mathscr{L}'$-structure $\mathfrak{A}_f$ by setting

$$A_f = A;$$

$\mathbf{\Pi}^{\mathfrak{A}f} = (f^{-1}\mathbf{\Pi})^{\mathfrak{A}}$ for each nonlogical constant $\mathbf{\Pi}$ of $\mathscr{L}'$.

(*iii*) For any first-order language $\mathscr{L}$, we let $\mathbf{S}_{\mathscr{L}}$ be the class of all $\mathscr{L}$-structures.

(*iv*) A *weak general logic* is a class L of quadruples $(\mathscr{L}, \Gamma, \mathfrak{A}, \varphi)$ such that $\mathscr{L}$ is a first-order language, with no operation symbols and only finitely many relation symbols, Γ is a class, $\mathfrak{A} \in \mathbf{S}_{\mathscr{L}}$, and $\varphi \in \Gamma$, such that Γ is uniquely determined by $\mathscr{L}$ alone, i.e., if $(\mathscr{L}, \Gamma, \mathfrak{A}, \varphi)$ and also $(\mathscr{L}, \Gamma', \mathfrak{A}'', \varphi')$ are in L, then $\Gamma = \Gamma'$. We sometimes write $\mathfrak{A} \models_{L,\mathscr{L}} \varphi$ when $(\mathscr{L}, \Gamma, \mathfrak{A}, \varphi) \in L$. An L, $\mathscr{L}$-sentence is an object φ such that $\varphi \in \Gamma$, where $(\mathscr{L}, \Gamma, \mathfrak{B}, \psi) \in L$ for some $\mathfrak{B}$, ψ. For any L, $\mathscr{L}$-sentence φ we let

$$\mathrm{Mod}_{\mathscr{L},L}\, \varphi = \{\mathfrak{A} : \mathfrak{A} \models_{L,\mathscr{L}} \varphi\}.$$

A class $\mathbf{K}$ is an L-elementary class provided that $\mathbf{K} = \mathrm{Mod}_{\mathscr{L},L}\, \varphi$ for some $\mathscr{L}$ and φ; we also call it an L, $\mathscr{L}$-elementary class.

(v) A *general logic* is a weak general logic L which satisfies the following conditions:

(1) Every L-elementary class is closed under isomorphism.
(2) If $\mathscr{L}, \mathscr{L}', f$ are as in (i), f onto, and $\mathbf{K}$ is an L, $\mathscr{L}$-elementary class, then $\{\mathfrak{A}_f : \mathfrak{A} \in \mathbf{K}\}$ is an L, $\mathscr{L}'$-elementary class.
(3) If $\mathbf{K}$ is an L, $\mathscr{L}$-elementary class and $\mathscr{L}'$ is an expansion of $\mathscr{L}$, then $\{\mathfrak{A} \in \mathbf{S}_{\mathscr{L}'} : \mathfrak{A} \restriction \mathscr{L} \in \mathbf{K}\}$ is an L, $\mathscr{L}'$-elementary class.
(4) If $\mathbf{K}$ is an L, $\mathscr{L}$-elementary class, then $\mathbf{S}_{\mathscr{L}} \sim \mathbf{K}$ is an L, $\mathscr{L}$-elementary class.
(5) If $\mathbf{K}$ and $\mathbf{M}$ are L, $\mathscr{L}$-elementary classes, then so is $\mathbf{K} \cap \mathbf{M}$.

(vi) Given a first-order language $\mathscr{L}$, and a general logic L, a class $\mathbf{K}$ of $\mathscr{L}$-structures is an *L-projective class*, or an *L, $\mathscr{L}$-projective class* provided that there is an expansion $\mathscr{L}'$ of $\mathscr{L}$ such that $\mathbf{K} = \{\mathfrak{A} \restriction \mathscr{L} : \mathfrak{A} \in \mathbf{M}\}$ for some L, $\mathscr{L}'$-elementary class $\mathbf{M}$.

(vii) If L_0 and L_1 are general logics, we write $L_0 \subseteq L_1$ (respectively $L_0 \subseteq_{\mathrm{inf}} L_1$) provided that every L_0-elementary class is also an L_1-elementary class (respectively, every L_0-elementary class has the same infinite members as some L_1-elementary class). Also, we write $L_0 \equiv L_1$ provided that $L_0 \subseteq L_1 \subseteq L_0$.

($viii$) L_{fo} is the class of all quadruples $(\mathscr{L}, \mathrm{Sent}_{\mathscr{L}}, \mathfrak{A}, \varphi)$ such that $\mathscr{L}$ is a first-order language with no operation symbols and only finitely many relation symbols, $\mathfrak{A} \in \mathbf{S}_{\mathscr{L}}$, $\varphi \in \mathrm{Sent}_{\mathscr{L}}$, and $\mathfrak{A} \models \varphi$. Clearly L_{fo} is a general logic.

This definition is rather long, but the notions introduced in it are all very intuitive. The properties of a general logic given in 26.19(v) are very minimal, and apply to all of the known extensions of first-order logic in which the notion of model is still first order; thus the various infinitary languages and the **Q**-languages of Chapters 30 and 31 fall under this heading. Note that the notion of elementary class corresponds to the notion of a finitely axiomatizable elementary class in the usual sense: a class is L_{fo}-elementary in the sense of 26.19(iv) iff it is a finitely axiomatizable elementary class in the usual sense. Since we *are* dealing with finitely axiomatizable elementary classes, and since operation symbols can be replaced by relation symbols as indicated in Chapter 11, it does not seem to be a loss of generality to restrict ourselves as indicated, to first-order languages with no operation symbols and only finitely many relation symbols.

We shall prove two theorems about general logics here. One is that a general logic in which the compactness theorem and downward Löwenheim–Skolem theorem hold is equivalent to L_{fo} in the sense of 26.4(*viii*). The other is that an effectivized general logic, as defined below, in which the downward Löwenheim–Skolem theorem holds is effectively equivalent to L_{fo}, in the sense indicated below. The proofs of these results depend on a main lemma, 26.22 below. First we need two small lemmas. The first one gives another useful fact about the Ehrenfeucht–Fraissé construction:

Lemma 26.20. *If $\mathfrak{A}$ and $\mathfrak{B}$ are denumerable $\mathscr{L}$-structures and there is an ω-sequence for $\mathfrak{A}$, $\mathfrak{B}$, then $\mathfrak{A} \cong \mathfrak{B}$.*

PROOF. Let $A = \{a_i : i \in \omega\}$ and $B = \{b_i : i \in \omega\}$. Let $\langle I_m : m \in \omega\rangle$ be an ω-sequence for $\mathfrak{A}$, $\mathfrak{B}$. Now we define sequences $x \in {}^{\omega}A$, $y \in {}^{\omega}B$ by recursion. Suppose that x_i and y_i have been defined for all $i < 2m$ in such a way that $\langle x_0, \ldots, x_{2m-1}\rangle I_{2m} \langle y_0, \ldots, y_{2m-1}\rangle$ (note that this holds if $m = 0$). Choose i minimum such that $a_i \notin \{x_0, \ldots, x_{2m-1}\}$, and let $x_{2m} = a_i$. Then choose j minimum such that $\langle x_0, \ldots, x_{2m}\rangle I_{2m+1} \langle y_0, \ldots, y_{2m-1}, b_j\rangle$ and set $y_{2m} = b_j$. To define y_{2m+1} and x_{2m+1} we interchange the roles of A and B. This completes the definition of x and y. Thus

(1) $$\langle x_0, \ldots, x_{m-1}\rangle I_m \langle y_0, \ldots, y_{m-1}\rangle \text{ for all } m \in \omega.$$

Furthermore, by induction on i,

(2) $$a_i \in \{x_0, \ldots, x_{2i}\} \text{ for all } i \in \omega, \text{ and } b_i \in \{y_0, \ldots, y_{2i+1}\}$$

for all $i \in \omega$. Hence by 26.6(*ii*)(4), $\{(x_i, y_i) : i \in \omega\}$ is the desired isomorphism. □

Lemma 26.21. *Let $\mathscr{L}$ be a first-order language with no operation symbols and only finitely many relation symbols, and let* **K** *be a class of $\mathscr{L}$-structures. Suppose that $\mathfrak{m}$ is any cardinal, and that there is no finitely axiomatizable elementary class which has the same members of power $\mathfrak{m}$ as* **K**. *Then for every $n \in \omega$ there are $\mathscr{L}$-structures $\mathfrak{A} \in \mathbf{K}$ and $\mathfrak{B} \in \mathbf{S}_{\mathscr{L}} \sim \mathbf{K}$ of power $\mathfrak{m}$ and an $(n + 1)$-sequence for $\mathfrak{A}$, $\mathfrak{B}$.*

PROOF. Clearly we may assume that $n \neq 0$. We shall apply 26.7. For each $\mathfrak{A} \in \mathbf{K}$ let $\varphi_{\mathfrak{A}}$ be the sentence

$$\bigwedge \{\varphi \in \Delta_n^n : \mathfrak{A} \models \varphi\} \wedge \bigwedge \{\neg\varphi : \varphi \in \Delta_n^n, \mathfrak{A} \models \neg\varphi\},$$

and let χ be the sentence

$$\bigvee \{\varphi_{\mathfrak{A}} : \mathfrak{A} \in \mathbf{K}, \text{ and } \mathfrak{A} \text{ has power } \mathfrak{m}\}.$$

Note that **K** has members of power $\mathfrak{m}$, since otherwise **K** and Mod $\{\varphi \wedge \neg\varphi\}$ would have the same members of power $\mathfrak{m}$, for any $\varphi \in \text{Sent}_{\mathscr{L}}$, contradicting our hypothesis. Since each member of $\mathbf{K}$ of power $\mathfrak{m}$ is in Mod $\{\chi\}$, but by hypothesis the converse fails, choose $\mathfrak{B} \in \text{Mod}\{\chi\} \sim \mathbf{K}$. From the construction of χ, $\mathfrak{B} \models \varphi_{\mathfrak{A}}$ for some $\mathfrak{A} \in \mathbf{K}$. Clearly $\mathfrak{A} \models \psi$ iff $\mathfrak{B} \models \psi$ for all $\psi \in \Delta_n^n$, so our lemma follows by 26.7. □

Now we come to the main lemma:

Lemma 26.22. *Let L be a general logic such that $L_{\text{fo}} \subseteq L$ but $L \nsubseteq_{\text{inf}} L_{\text{fo}}$. Assume also that L satisfies the downward Löwenheim–Skolem theorem, in the sense that any L-elementary class with an infinite member has a denumerable member.*

Furthermore, let $\mathscr{L}$ be a first-order language with exactly one nonlogical constant, a unary relation symbol $\mathbf{P}$. Then there is a class $\mathbf{K}$ of $\mathscr{L}$-structures satisfying the following conditions:

(*i*) *every member of $\mathbf{K}$ is infinite;*
(*ii*) *for any $\mathfrak{A} = (A, \mathbf{P}^{\mathfrak{A}}) \in \mathbf{K}$, the set $\mathbf{P}^{\mathfrak{A}}$ is finite and nonempty;*
(*iii*) *for every $n \in \omega \sim 1$ there is a countable $\mathfrak{A} \in \mathbf{K}$ such that $|\mathbf{P}^{\mathfrak{A}}| = n$;*
(*iv*) *$\mathbf{K}$ is L-projective.*

Proof. By assumption there is an L-elementary class $\mathbf{M}$, say in a language $\mathscr{L}'$, such that there is no finitely axiomatizable elementary class in the usual sense with the same infinite members as $\mathbf{M}$. Then

(1) there is no finitely axiomatizable elementary class $\mathbf{N}$ with the same denumerable members as $\mathbf{M}$.

For let $\mathbf{N}$ be any elementary class. Since $L_{\text{fo}} \subseteq L$, we also have that $\mathbf{N}$ is an L-elementary class. Now let $\mathbf{Q} = (\mathbf{N} \sim \mathbf{M}) \cup (\mathbf{M} \sim \mathbf{N})$. Then by 26.19(*v*)(4), (5) we see that $\mathbf{Q}$ is an L, $\mathscr{L}'$-elementary class. Now by our choice of $\mathbf{M}$, there is an infinite $\mathfrak{A} \in \mathbf{Q}$. Hence by the downward Löwenheim–Skolem theorem for L, $\mathbf{Q}$ has a denumerable member $\mathfrak{B}$. Thus $\mathfrak{B} \in (\mathbf{M} \sim \mathbf{N}) \cup (\mathbf{N} \sim \mathbf{M})$, as desired in (1).

Let the nonlogical constants of $\mathscr{L}'$ be $\mathbf{R}_0, \ldots, \mathbf{R}_{m-1}$, of ranks $n_0, \ldots, n_{m-1}$ respectively. Let $\mathscr{L}''$ be an expansion of $\mathscr{L}'$ in which we have additional relation symbols $\mathbf{S}_0, \ldots, \mathbf{S}_{m-1}, \mathbf{T}_0, \mathbf{T}_1, \mathbf{T}_2, \mathbf{T}_3, \mathbf{T}_4$ of ranks $n_0, \ldots, n_{m-1}$, 1, 2, 2, 3, 3 respectively. Let $\mathscr{L}'''$ be the reduct of $\mathscr{L}''$ to $\mathbf{T}_0$. It suffices now to verify the conclusion of the lemma for $\mathscr{L}'''$ in place of $\mathscr{L}$, since $\mathscr{L}''$ is isomorphic to an expansion of $\mathscr{L}$ via a mapping that takes $\mathbf{T}_0$ to $\mathbf{P}$ [see 26.19(*i*),(*ii*),(*v*)(2)].

Clearly there is a sentence φ of $\mathscr{L}''$ such that for any $\mathscr{L}''$-structure $\mathfrak{A}$, $\mathfrak{A} \models \varphi$ iff the following conditions hold:

(2) $\mathbf{T}_0^{\mathfrak{A}} \neq 0$;
(3) $\mathbf{T}_1^{\mathfrak{A}}$ is a one-one function mapping A onto a proper subset of A;
(4) $\mathbf{T}_2^{\mathfrak{A}}$ is a linear ordering of $\mathbf{T}_0^{\mathfrak{A}}$ such that $\mathbf{T}_0^{\mathfrak{A}}$ has a $\mathbf{T}_2^{\mathfrak{A}}$-first element, while every member of $\mathbf{T}_0^{\mathfrak{A}}$ which has a $\mathbf{T}_2^{\mathfrak{A}}$-successor has an immediate $\mathbf{T}_2^{\mathfrak{A}}$-successor;

(5) for every $a \in A$, the relation $f_a = \{(x, y) : (a, x, y) \in \mathbf{T}_3\}$ is a function mapping $\mathbf{T}_0^{\mathfrak{A}}$ into A;
(6) if x is the $\mathbf{T}_2^{\mathfrak{A}}$-first element of $\mathbf{T}_0^{\mathfrak{A}}$, then there are a, b such that $(x, a, b) \in \mathbf{T}_4^{\mathfrak{A}}$;
(7) if $(x, a, b) \in \mathbf{T}_4^{\mathfrak{A}}$, $x \in \mathbf{T}_0^{\mathfrak{A}}$, y is the immediate $\mathbf{T}_2^{\mathfrak{A}}$-successor of x, and $z \in A$, then there exist c, d, u such that $(y, c, d) \in \mathbf{T}_4^{\mathfrak{A}}$, $f_c x = z$, $f_d x = u$, while for every $v \in \mathbf{T}_0^{\mathfrak{A}}$, if $v \neq x$ then $f_c v = f_a v$ and $f_d v = f_b v$;
(8) like (7), but with a and b interchanged in its conclusion;
(9) if $(x, a, b) \in \mathbf{T}_2^{\mathfrak{A}}$, $i < m$, and $y = \langle y_0, \ldots, y_{ni-1}\rangle$ is a sequence of $\mathbf{T}_i^{\mathfrak{A}}$-predecessors of x, then $f_a \circ y \in \mathbf{R}_i^{\mathfrak{A}}$ iff $f_b \circ y \in \mathbf{S}_i^{\mathfrak{A}}$;
(10) if $(x, a, b) \in \mathbf{T}_4^{\mathfrak{A}}$ and y and z are $\mathbf{T}_2^{\mathfrak{A}}$-predecessors of x, then $f_a y = f_a z$ iff $f_b y = f_b z$.

Next, let $\mathscr{L}^{\text{iv}}$ be the reduct of $\mathscr{L}''$ to the symbols $\mathbf{S}_0, \ldots, \mathbf{S}_{m-1}$. Clearly $\mathscr{L}'$ is isomorphic to $\mathscr{L}^{\text{iv}}$, via the isomorphism g taking $\mathbf{R}_i$ to $\mathbf{S}_i$ for each $i < m$. Let $\mathbf{Q}_0$ be the class of $\mathscr{L}''$-structures such that $\mathfrak{A} \restriction \mathscr{L}' \in \mathbf{M}$; thus by 26.19($v$)(3), $\mathbf{Q}_0$ is an L-elementary class. Let $\mathbf{Q}_1 = \{\mathfrak{A}_g : \mathfrak{A} \in \mathbf{S}_{\mathscr{L}'} \sim \mathbf{M}\}$; by 26.19($v$)(4),(2), $\mathbf{Q}_1$ is an L-elementary class. Hence so are $\mathbf{Q}_2 = \{\mathfrak{A} \in \mathbf{S}_{\mathscr{L}''} : \mathfrak{A} \restriction \mathscr{L}' \in \mathbf{Q}_1\}$ and $\mathbf{Q}_3 = \mathbf{Q}_0 \cap \mathbf{Q}_2$. By the assumption $L_{\text{fo}} \subseteq L$, Mod $\{\varphi\}$ is an L-elementary class. Let $\mathbf{Q}_4 = \mathbf{Q}_3 \cap \text{Mod}\,\{\varphi\}$. Thus $\mathbf{Q}_4$ is also an L-elementary class. It is $\mathbf{Q}_4$ that we work with below, $\mathbf{Q}_0$ through $\mathbf{Q}_3$ being mentioned just to construct it; and the properties of $\mathbf{Q}_4$ that we need below are just the following: $\mathfrak{A} \in \mathbf{Q}_4$ iff the following hold:

(11) $\mathfrak{A} \models \varphi$;
(12) $\mathfrak{A} \restriction \mathscr{L}' \in \mathbf{M}$;
(13) $(\mathfrak{A} \restriction \mathscr{L}^{\text{iv}})_{g^{-1}} \in \mathbf{S}_{\mathscr{L}'} \sim \mathbf{M}$.

Now we shall verify that $\mathbf{K} = \mathbf{Q}_4 \restriction \mathscr{L}'''$ satisfies the conditions (i)–(iv). Indeed, (i) is immediate from (11) and (3), and (iv) is obvious. Next we check (iii). To this end, let n be any positive integer. By (1) and Lemma 26.21 there are $\mathscr{L}'$-structures $\mathfrak{B} \in \mathbf{M}$ and $\mathfrak{C} \in \mathbf{S}_{\mathscr{L}'} \sim \mathbf{M}$, both denumerable, and an n-sequence $\langle I_i : i < n\rangle$ for $\mathfrak{B}$, $\mathfrak{C}$. We now construct an $\mathscr{L}''$-structure $\mathfrak{A}$ whose $\mathscr{L}'''$-reduct will satisfy (iii). We may assume that $B = C = \omega$ (using 26.19(v)(1)). Let $A = \omega$, $\mathbf{R}_i^{\mathfrak{A}} = \mathbf{R}_i^{\mathfrak{B}}$, $\mathbf{S}_i^{\mathfrak{A}} = \mathbf{R}_i^{\mathfrak{C}}$ for all $i < m$. Set $\mathbf{T}_0^{\mathfrak{A}} = n$. Let $\mathbf{T}_1^{\mathfrak{A}}$ be any one-one function mapping ω onto a proper subset of ω. Let $\mathbf{T}_2^{\mathfrak{A}}$ be the natural ordering of $\mathbf{T}_0^{\mathfrak{A}}$. In order to define $\mathbf{T}_3^{\mathfrak{A}}$ and $\mathbf{T}_4^{\mathfrak{A}}$, let g be a one-one function mapping ω onto ω^n. Then we set

$$\mathbf{T}_3^{\mathfrak{A}} = \{(m, x, y) : m, y \in \omega \text{ and } x \in n \text{ and } g_m x = y\};$$
$$\mathbf{T}_4^{\mathfrak{A}} = \{(m, x, y) : m \in n \text{ and } x, y \in \omega \text{ and } \langle g_x 0, \ldots, g_x(m-1)\rangle I_m \langle g_y 0, \ldots, g_y(m-1)\rangle\}.$$

In the definition of $\mathbf{T}_4^{\mathfrak{A}}$ it is to be understood that $(0, x, y) \in \mathbf{T}_4^{\mathfrak{A}}$ for all $x, y \in \omega$. It is now a straightforward matter to check that $\mathfrak{A} \in \mathbf{Q}_4$ (which is all that is needed to check (iii), since $|\mathbf{T}_0^{\mathfrak{A}}| = n$ and $|A| = \aleph_0$). In fact, $\mathfrak{A} \restriction \mathscr{L}' = \mathfrak{B}$, and $(\mathfrak{A} \restriction \mathscr{L}^{\text{iv}})_{g^{-1}} = \mathfrak{C}$, so (12) and (13) hold. To check (11) we must look at the conditions (2)–(10). Of these conditions, (2), (3), (4), (5), and (6) are obvious. Now assume the hypotheses of (7). Thus

$$\langle g_a0, \ldots, g_a(x-1)\rangle I_x \langle g_b0, \ldots, g_b(x-1)\rangle,$$

so there is a $u \in \omega$ such that $\langle g_a0, \ldots, g_a(x-1), z\rangle I_{x+1}\langle g_b0, \ldots, g_b(x-1, u\rangle$. Choose $c, d \in \omega$ so that $g_c = (g_a)^x_z$ and $g_d = (g_b)^x_u$. Then the conclusion of (7) is clear. Condition (8) is checked similarly. Next, assume the hypothesis of (9). Thus $\langle g_a0, \ldots, g_a(x-1)\rangle T_x \langle g_b0, \ldots, g_b(x-1)\rangle$. By 26.6(*ii*)(3),(4), it follows that for any atomic formula φ' with variables among $v_0, \ldots, v_{x-1}$ we have $\mathfrak{A} \models \varphi'[g_a0, \ldots, g_a(x-1)]$ iff $\mathfrak{B} \models \varphi'[g_b0, \ldots, g_b(x-1)]$. Hence, noting that $f_t = g_t$ for all $t \in \omega$,

$$\begin{aligned} f_a \circ y \in \mathbf{R}_i^{\mathfrak{A}} \quad & \text{iff } \langle g_a y_0, \ldots, g_a y_{ni-1}\rangle \in \mathbf{R}_i^{\mathfrak{A}} \\ & \text{iff } \mathfrak{A} \models \mathbf{R}_i v_{y0} \cdots v_{y(ni-1)}[g_a0, \ldots, g_a(x-1)] \\ & \text{iff } \mathfrak{B} \models \mathbf{R}_i v_{y0} \cdots v_{y(ni-1)}[g_b0, \ldots, g_b(x-1)] \\ & \text{iff } \langle g_b y_0, \ldots, g_b y_{ni-1}\rangle \in \mathbf{R}_i^{\mathfrak{B}} \\ & \text{iff } f_b \circ y \in \mathbf{R}_i^{\mathfrak{B}}. \end{aligned}$$

Thus (9) holds. Finally, (10) is proved just like (9). We have now checked condition (*iii*) of our lemma.

It remains only to verify condition (*ii*). Suppose, to the contrary, that we have a member $\mathfrak{B}$ of $\mathbf{Q}_4$ such that $\mathbf{T}_0^{\mathfrak{B}}$ is infinite (note, by (11) and (2) that $\mathbf{T}_0^{\mathfrak{B}} \neq 0$ always). Let $\mathscr{L}^{\mathrm{v}}$ be an expansion of $\mathscr{L}''$ by adjoining a new binary relation symbol $\mathbf{T}_5$. Let $\mathbf{Q}_5$ be the elementary class of $\mathscr{L}^{\mathrm{v}}$-structures $\mathfrak{A}$ such that $\mathbf{T}_5^{\mathfrak{A}}$ is a one-one function mapping $\mathbf{T}_0^{\mathfrak{A}}$ onto a proper subset of $\mathbf{T}_0^{\mathfrak{A}}$. Since $L_{\mathrm{fo}} \subseteq L$, $\mathbf{Q}_5$ is also an L-elementary class. Let $\mathbf{Q}_6 = \{\mathfrak{A} \in \mathbf{Q}_5 : \mathfrak{A} \restriction \mathscr{L}'' \in \mathbf{Q}_4\}$. Clearly [by 26.19(*v*)(3),(5)] $\mathbf{Q}_6$ is an L-elementary class. For any $\mathfrak{A} \in \mathbf{Q}_6$ we have properties (11)–(13) as well as

(14) $\mathbf{T}_5^{\mathfrak{A}}$ is a one-one function mapping $\mathbf{T}_0^{\mathfrak{A}}$ onto a proper subset of $\mathbf{T}_0^{\mathfrak{A}}$.

Now clearly $\mathfrak{B}$ can be expanded to give us a member of $\mathbf{Q}_6$, so, in particular, $\mathbf{Q}_6$ has an infinite member. By the downward Löwenheim–Skolem theorem for L, let $\mathfrak{A}$ be a denumerable member of $\mathbf{Q}_6$. Let $\mathfrak{C}$ and $\mathfrak{D}$ be the two $\mathscr{L}'$-structures $\mathfrak{C} = (A, \mathbf{R}_i^{\mathfrak{A}})_{i<m}$ and $\mathfrak{D} = (A, \mathbf{S}_i^{\mathfrak{A}})_{i<m}$. By (12) and (13) we have

(15) $\mathfrak{C} \in \mathbf{M}$ and $\mathfrak{D} \in \mathbf{S}_{\mathscr{L}'} \sim \mathbf{M}$.

But we shall now apply Lemma 26.20 to show that $\mathfrak{C} \cong \mathfrak{D}$. In view of 26.19(*v*), (1), this is the contradiction we have been seeking.

To apply 26.20, first, in view of (11) we may choose f as in (5). Now $\mathbf{T}_0^{\mathfrak{A}}$ is infinite by (14), so by (4) we can choose for each $n \in \omega$ a member a_n of $\mathbf{T}_0^{\mathfrak{A}}$ which has exactly n $\mathbf{T}_2$-predecessors. Now for each $n \in \omega$ we set

(16) $I_n = \{(x, y) : x \in {}^nA, y \in {}^nA$, and there exist $c, d \in A$ such that $(a_n, c, d) \in \mathbf{T}_4^{\mathfrak{A}}$ and $\forall i < n\ (f_c a_i = x_i$ and $f_d a_i = y_i)\}$.

Thus to apply 26.20 and hence finish the proof it remains only to check that $\langle I_n : n \in \omega\rangle$ is an ω-sequence for $\mathfrak{C}, \mathfrak{D}$. Conditions 26.6(*ii*)(1),(2) are obvious.

Now assume that xI_py and $b \in C$ (the case $b \in D$ is treated similarly). Thus by (16) there exist $c, d \in A$ such that $(a_p, c, d) \in \mathbf{T}_4^{\mathfrak{A}}$ and $\forall i < p\ (f_c a_i = x_i$ and $f_d a_i = y_i)$. It follows from (7) that there are $c', d' \in A$ such that $(a_{p+1}, c', d') \in \mathbf{T}_4^{\mathfrak{A}}, f_{c'} a_p = b$, while for any $v \in \mathbf{R}_0^{\mathfrak{A}}$ different from a_p we have $f_{c'} v = f_c v$ and $f_{d'} v = f_d v$. Hence $\langle x_0, \ldots, x_{p-1}, b\rangle I_{p+1} \langle y_0, \ldots, y_{p-1}, f_{d'} a_p\rangle$, as desired in 26.6(*ii*)(3). Finally, assume that $p \in \omega$, $x, y \in {}^pA$, xI_py, and $\mathrm{Fv}\varphi' \subseteq \{v_0, \ldots, v_{p-1}\}$, where φ' is an atomic formula. Choose c, d as in (16). If φ' is $v_i \equiv v_j$, then

$$\begin{aligned} \mathfrak{C} \models v_i \equiv v_j[x] \quad & \text{iff } x_i = x_j \quad \text{iff } f_c a_i = f_c a_j \\ & \text{iff } f_d a_i = f_d a_j \text{ (by (10))} \\ & \text{iff } \mathscr{D} \models v_i \equiv v_j[y] \text{ similarly.} \end{aligned}$$

If φ' is $\mathbf{R}_i v_{j0} \cdots v_{j(ni-1)}$ with $i < m$, then

$$\begin{aligned} \mathfrak{C} \models \varphi'[x] \quad & \text{iff } \langle x_{j0}, \ldots, x_{j(ni-1)}\rangle \in \mathbf{R}_i^{\mathfrak{C}} \\ & \text{iff } \langle x_{j0}, \ldots, x_{j(ni-1)}\rangle \in \mathbf{R}_i^{\mathfrak{A}} \\ & \text{iff } \langle f_c a_{j0}, \ldots, f_c a_{j(ni-1)}\rangle \in \mathbf{R}_i^{\mathfrak{A}} \text{ by (16)} \\ & \text{iff } \langle f_d a_{j0}, \ldots, f_d a_{j(ni-1)}\rangle \in \mathbf{R}_i^{\mathfrak{A}} \text{ by (9)} \\ & \text{iff } \mathscr{D} \models \varphi'[y] \text{ similarly.} \end{aligned}$$

□

The following form of this lemma will be needed.

Lemma 26.23. *Let L be a general logic such that $L_{\mathrm{fo}} \subseteq L$ but $L \nsubseteq L_{\mathrm{fo}}$. Assume also that L satisfies the downward Löwenheim–Skolem theorem, in the sense of 26.22. Let $\mathscr{L}$ be as in 26.22. Then there is a class $\mathbf{K}$ of $\mathscr{L}$-structures satisfying the conditions (ii)–(iv) of 26.22.*

Proof. By 26.22 we may assume that $L \subseteq_{\inf} L_{\mathrm{fo}}$. Let $\mathbf{K}$ be an L-elementary class which is not a finitely axiomatizable elementary class in the usual sense. Then there is a finitely axiomatizable elementary class $\mathbf{L}$ with the same infinite members as $\mathbf{K}$. Then

(1) for every $m \in \omega$ there is an $n > m$ such that $\mathbf{K}$ and $\mathbf{L}$ do not have the same members of power n.

For, otherwise choose $m \in \omega$ such that for all $n > m$, $\mathbf{K}$ and $\mathbf{L}$ have the same members of power n. Say $\mathbf{L} = \mathrm{Mod}\,\{\varphi\}$. For each $n \le m$, let $\mathbf{M}_n$ be a set of members of $\mathbf{K}$ containing exactly one member from each isomorphism type of members of $\mathbf{K}$ of power n. For each $\mathfrak{A}$ of power $\le n$ let $\psi_{\mathfrak{A}}$ be a sentence whose models are exactly all structures isomorphic to $\mathfrak{A}$. Then the sentence χ:

$$(\exists \text{ at least } m + 1 \text{ things, and } \varphi) \wedge \bigwedge_{n \le m} \bigvee_{\mathfrak{A} \in \mathbf{M}n} \psi_{\mathfrak{A}}$$

characterizes $\mathbf{K}$, i.e., $\mathbf{K} = \mathrm{Mod}\,\{\chi\}$, which is impossible. Thus (1) holds. Let $\mathscr{L}$ be the language of $\mathbf{K}$, let $\mathscr{L}'$ be an expansion of $\mathscr{L}$ by adjoining a new unary relation symbol $\mathbf{P}$, and let $\mathscr{L}''$ be the reduct of $\mathscr{L}'$ to $\mathbf{P}$. It suffices to verify our desired conclusions for $\mathscr{L}''$ in place of $\mathscr{L}$ (by 26.19(v)(2)). Let φ be the formula $\exists v_0(\mathbf{P}v_0)$. Then $\mathrm{Mod}\,\{\varphi\}$ is an $L, \mathscr{L}'$-elementary class, since $L_{\mathrm{fo}} \subseteq L$. Set

(2) $$\mathbf{M} = \mathrm{Mod}\,\{\varphi\} \cap ((\mathbf{K} \sim \mathbf{L}) \cup (\mathbf{L} \sim \mathbf{K})).$$

Then $\mathbf{M} \restriction \mathscr{L}''$ satisfies (*ii*) and (*iv*) of 26.22, obviously. Given $n \in \omega \sim 1$, choose by (1) a structure $\mathfrak{A} \in (\mathbf{K} \sim \mathbf{L}) \cup (\mathbf{L} \sim \mathbf{K})$ with $|A| \geq n$, and let $\mathfrak{B}$ be any expansion of $\mathfrak{A}$ with $|\mathbf{P}^{\mathfrak{A}}| = n$. From (2) it is clear that $\mathfrak{B} \in \mathbf{M}$. Thus 26.22(*iii*) holds. □

Now we can prove the first main theorem of Lindstrom.

Theorem 26.24 (Lindstrom). *Let L be a general logic such that $L_{\text{fo}} \subseteq L$. Assume that L satisfies the downward Löwenheim–Skolem theorem and the countable compactness theorem; that is, assume the following conditions:*

(*i*) *any L-elementary class with an infinite member has a denumerable member;*

(*ii*) *if $\mathscr{K}$ is a countable collection of L-elementary classes and $\bigcap \mathscr{F} \neq 0$ for every finite $\mathscr{F} \subseteq \mathscr{K}$, then $\bigcap \mathscr{K} \neq 0$.*

Then $L_{\text{fo}} \equiv L$.

Proof. Assume the hypothesis, but suppose that $L \nsubseteq L_{\text{fo}}$. By Lemma 26.23, with $\mathscr{L}$ as in 26.22, there is an expansion $\mathscr{L}'$ of $\mathscr{L}$ and an L-elementary class $\mathbf{K}_0$ of $\mathscr{L}'$-structures such that $\mathbf{K}_0 \restriction \mathscr{L}$ satisfies (*ii*)–(*iv*) of 26.22. For each $n \in \omega \sim 1$ let

$$\mathbf{K}_n = \text{Mod}\left\{\exists v_0 \cdots \exists v_{n-1} \bigwedge_{i<j<n} [\neg(v_i = v_j) \wedge \mathbf{P}v_i \wedge \mathbf{P}v_j]\right\}.$$

Then for each $m \in \omega$, $\bigcap_{n \leq m} \mathbf{K}_n \neq 0$ by 26.22(*iii*), but $\bigcap_{n<\omega} \mathbf{K}_n = 0$ by 26.22(*ii*), contradiction. □

The notation used in this theorem as well as its proof is a little bit illegal. Since L-elementary classes are proper classes if they are nonempty [by 26.19(*v*)(1)], we cannot really speak of a collection of L-elementary classes. But each L-elementary class is determined by a single object φ, so we could make these formulations legal by talking about the φ's instead. Thus 26.24(*ii*) could be reformulated as follows:

Let $\mathscr{L}$ be a first-order language, let $(\mathscr{L}, \Gamma, \mathfrak{B}, \psi) \in L$, and let T be a countable subset of Γ. Suppose that for each finite subset U of T there is an $\mathfrak{A}$ with $\mathfrak{A} \models_{L,\mathscr{L}} \varphi$ for each $\varphi \in U$. Then there is an $\mathfrak{A}$ such that $\mathfrak{A} \models_{L,\mathscr{L}} \varphi$ for each $\varphi \in T$.

We think that the formulation actually given in 26.24 is clearer from an intuitive point of view.

The second theorem of Lindstrom depends upon an effectivization of our notion of general logic:

Definition 26.25

(*i*) A *weak effectivized general logic* is a collection L of quintuples $(\mathscr{L}, \Gamma, \mathfrak{A}, \varphi, h)$ such that $(\mathscr{L}, \Gamma, \mathfrak{A}, \varphi)$ satisfies the conditions of 26.19(*iv*), h depends only on $\mathscr{L}$, and h is a one-one function mapping Γ into ω.

(*ii*) An *effectivized general logic* is a weak effectivized general logic L such that the following conditions hold:

(1) There is a unary recursive function Comp_L such that if $(\mathscr{L}, \Gamma, \mathfrak{A}, \varphi, h) \in L$ and $\mathscr{L}$ is a reduct of $\mathscr{L}_{\mathrm{un}}$ (see 14.22), then for any $\varphi \in \Gamma$, $\mathrm{Comp}_L\, h\varphi = h\psi$ for some $\psi \in \Gamma$ such that $\mathrm{Mod}_{\mathscr{L},L}\, \psi = \mathbf{S}_{\mathscr{L}} \sim \mathrm{Mod}\mathscr{L}_{,L}\, \varphi$;

(2) There is a binary recursive function Int_L such that if $\mathscr{L}$, Γ, h are as in (1) and $\varphi, \psi \in \Gamma$, then $\mathrm{Int}_L\,(h\varphi, h\psi) = h\chi$ for some $\chi \in \Gamma$ such that $\mathrm{Mod}_{\mathscr{L},L}\, \chi = \mathrm{Mod}_{\mathscr{L},L}\, \varphi \cap \mathrm{Mod}_{\mathscr{L}L}\, \psi$;

(3) The set $\{h\varphi : (\mathscr{L}, \Gamma, \mathfrak{A}, \psi, h) \in L, \varphi \in \Gamma, \mathscr{L}$ is a reduct of $\mathscr{L}_{\mathrm{un}}\}$ is recursive.

(*iii*) $L_{\mathrm{fo}}^{\mathrm{eff}}$ is the set of all quintuples $(\mathscr{L}, \mathrm{Sent}_{\mathscr{L}}, \mathfrak{A}, \varphi, h)$ such that $(\mathscr{L}, \mathrm{Sent}_{\mathscr{L}}, \mathfrak{A}, \varphi)$ is as in 26.19(*viii*) and for some g, $(\mathscr{L}, g)$ is an effectivized first-order language and $h = g^+ \restriction \mathrm{Sent}_{\mathscr{L}}$; if $\mathscr{L}$ is a reduct of $\mathscr{L}_{\mathrm{un}}$, then we insist that g is the restriction of the Gödel-numbering function of $\mathscr{L}_{\mathrm{un}}$. Clearly then $L_{\mathrm{fo}}^{\mathrm{eff}}$ is an effectivized general logic.

(*iv*) If L and L' are effectivized general logics, we write $L \subseteq_{\mathrm{eff}} L'$ provided that $L \subseteq L'$, and there is a unary recursive function $\mathrm{Trans}_{LL'}$ such that if $(\mathscr{L}, \Gamma, \mathfrak{A}, \varphi, h) \in L$, $(\mathscr{L}, \Gamma', \mathfrak{A}', \varphi', h') \in L'$ and $\mathscr{L}$ is a reduct of $\mathscr{L}_{\mathrm{un}}$, then for any $\psi \in \Gamma$, $\mathrm{Trans}_{LL'}\, h\psi = h'\chi$ for some $\chi \in \Gamma'$ such that $\mathrm{Mod}_{\mathscr{L},L}\, \psi = \mathrm{Mod}_{\mathscr{L},L'}\, \chi$. We write $L \equiv_{\mathrm{eff}} L'$ if $L \subseteq_{\mathrm{eff}} L' \subseteq_{\mathrm{eff}} L$.

(*v*) An effectivized general logic L is *recursively enumerable* provided that the following set is r.e.:

$\{h\psi : (\mathscr{L}, \Gamma, \mathfrak{A}, \varphi, h) \in L, \mathscr{L}$ a reduct of $\mathscr{L}_{\mathrm{un}}, \psi \in \Gamma$, and $\mathfrak{A} \models_{L,\mathscr{L}} \psi$ for all $\mathfrak{A} \in S_{\mathscr{L}}\}$.

Note that the functions Comp_L, Int_L, and $\mathrm{Trans}_{LL'}$ above do not depend on the particular language $\mathscr{L}$. Also note for later reference that under the assumptions of (*ii*) a kind of disjunction can be formed. Namely, there is a binary recursive function Un_L such that if $\varphi, \psi \in \Gamma$, then $\mathrm{Un}_L(h\varphi, h\psi) = h\chi$ for some $\chi \in L$ such that

$$\mathrm{Mod}_{\mathscr{L},L}\, \chi = \mathrm{Mod}_{\mathscr{L},L}\, \varphi \cup \mathrm{Mod}_{\mathscr{L},L}\, \psi.$$

Namely, for any $i, j \in \omega$ let

$$\mathrm{Un}_L\,(i, j) = \mathrm{Comp}_L\,(\mathrm{Int}_L, (\mathrm{Comp}_L\, i, \mathrm{Comp}_L\, j)).$$

The last clause of 26.25 says, roughly speaking, that a logic is recursively enumerable if its set of universally valid sentences is (uniformly) recursively enumerable.

The main theorem on effectivized general logics can now be proved:

Theorem 26.26 (Lindstrom). *If L is a recursively enumerable effectivized general logic, $L_{\mathrm{fo}}^{\mathrm{eff}} \subseteq_{\mathrm{eff}} L$, and L satisfies the downward Löwenheim–Skolem theorem, then $L_{\mathrm{fo}}^{\mathrm{eff}} \equiv_{\mathrm{eff}} L$.*

Proof. First we show that $L \subseteq L_{\mathrm{fo}}$. Assume not. Let $\mathscr{L}$ be a language with only one nonlogical constant, a unary relation symbol $\mathbf{P}$. Then by 26.23

there is an expansion $\mathscr{L}'$ of $\mathscr{L}$, still with only finitely many relation symbols and no operation symbols, and there is an L, $\mathscr{L}'$-elementary class $\mathrm{Mod}_{L,\mathscr{L}'}\,\psi$, such that the following conditions hold:

(1) For any $\mathfrak{A} \in \mathrm{Mod}_{\mathscr{L}',L}\,\psi$, $\mathbf{P}^{\mathfrak{A}}$ is finite and nonempty.
(2) For every $n \in \omega \sim 1$ there is a countable $\mathfrak{A} \in \mathrm{Mod}_{\mathscr{L}',L}\,\psi$ such that $|\mathbf{P}^{\mathfrak{A}}| = n$.

By 26.19(v)(2) we may assume that $\mathscr{L}'$ is a reduct of $\mathscr{L}_{\mathrm{un}}$. Let $\mathscr{L}''$ be an expansion of $\mathscr{L}'$ by adjoining a binary relation symbol $\mathbf{R}$, with $\mathscr{L}''$ still a reduct of $\mathscr{L}_{\mathrm{un}}$. Say the symbols of $\mathscr{L}''$ are $\mathbf{S}_0, \ldots, \mathbf{S}_{m-1}, \mathbf{R}, \mathbf{P}$, of ranks $n_0, \ldots, n_{m-1}, 2, 1$ respectively. By 26.19(iv)(3) let χ be an L, $\mathscr{L}''$-sentence such that

(3) $\mathrm{Mod}_{\mathscr{L}'',L}\,\chi = \{\mathfrak{A} \in \mathbf{S}_{\mathscr{L}''} : \mathfrak{A} \upharpoonright \mathscr{L}' \in \mathrm{Mod}_{\mathscr{L}',L}\,\psi\}$.

Let $\mathscr{L}'''$ be the reduct of $\mathscr{L}''$ to $\mathbf{R}$. We recall the following fact about $\mathscr{L}'''$ which is a consequence of 16.52 and 15.32.

(4) $g^{+*}\{\theta \in \mathrm{Sent}_{\mathscr{L}'''} : \theta$ holds in all finite $\mathscr{L}'''$-structures$\}$ is not r.e.

For each formula θ of $\mathscr{L}'''$ let $s\theta$ be obtained from φ by relativizing all quantifiers to $\mathbf{P}$ (replacing $\forall v_i$ by $\forall v_i(\mathbf{P}v_i \rightarrow\ )$. Clearly the formation of $s\theta$ is an effective procedure, so there is a recursive function s' such that $s'g^+\theta = g^+s\theta$ for every formula θ of $\mathscr{L}'''$. For any $\mathscr{L}''$-structure $\mathfrak{A}$ with $\mathbf{P}^{\mathfrak{A}} \neq 0$ let $\mathfrak{A}^{\mathbf{P}}$ be the $\mathscr{L}'''$-structure $(\mathbf{P}^{\mathfrak{A}}, \mathbf{R}^{\mathfrak{A}} \cap {}^2\,\mathbf{P}^{\mathfrak{A}})$. By induction on θ one easily shows that for any $\mathscr{L}''$-structure $\mathfrak{A}$ with $\mathbf{P}^{\mathfrak{A}} \neq 0$, any $x \in {}^{\omega}\mathbf{P}^{\mathfrak{A}}$, and any formula θ of $\mathscr{L}'''$,

$$\mathfrak{A} \models s\theta[x] \qquad \text{iff } \mathfrak{A}^{\mathbf{P}} \models \theta[x].$$

Hence for any sentence θ of $\mathscr{L}'''$ and any $\mathscr{L}''$-structure $\mathfrak{A}$ with $\mathbf{P}^{\mathfrak{A}} \neq 0$ we have

(5) $\mathfrak{A} \models s\theta$ iff $\mathfrak{A}^{\mathbf{P}} \models \theta$.

Now for any $x \in \omega$ let

$$tx = \mathrm{Un}_L\,(\mathrm{Comp}_L\, g^+\chi, \mathrm{Trans}_{L(\mathrm{fo,eff}),L}\, s'x).$$

We claim

(6) for any sentence θ of $\mathscr{L}'''$, θ holds in all finite $\mathscr{L}'''$-structures iff $tg^+\theta = h\psi$ for some $\psi \in \Gamma$ and $\forall \mathfrak{A} \in \mathbf{S}_{\mathscr{L}'''}(\mathfrak{A} \models_{L,\mathscr{L}'''} \psi)$, where $(\mathscr{L}''', \Gamma, \mathfrak{B}, \mu, h) \in L$ for some $\mathfrak{B}$, μ.

By 26.25(iv), this will contradict (4) (cf. 6.10). To establish (6), note from the construction of t that for any sentence θ of $\mathscr{L}'''$, $tg^+\theta$ has the form $h\sigma$, where

(7) $\mathrm{Mod}_{\mathscr{L}'',L}\,\sigma = (\mathbf{S}_{\mathscr{L}''} \sim \mathrm{Mod}_{\mathscr{L}'',L}\,\chi) \cup \mathrm{Mod}_{\mathscr{L}''}\,\{s\theta\}$.

Assume that $\mathfrak{A} \models_{\mathscr{L}'',L} \sigma$ for all $\mathscr{L}''$-structures $\mathfrak{A}$. If $\mathfrak{B}$ is any finite $\mathscr{L}'''$-structure, we can, by (2), find an $\mathscr{L}''$-structure $\mathfrak{A}$ such that $\mathfrak{A} \upharpoonright \mathscr{L}' \models_{\mathscr{L}',L} \psi$ and $\mathfrak{A}^{\mathbf{P}} = \mathfrak{B}$. Thus by (3) $\mathfrak{A} \in \mathrm{Mod}_{\mathscr{L}'',L}\,\chi$, so the assumption $\mathfrak{A} \models_{\mathscr{L}'',L} \sigma$ yields via (7) that $\mathfrak{A} \models s\theta$. Hence by (5) $\mathfrak{B} \models \theta$. Similarly reasoning gives the converse of (6). Hence (6) holds, and a contradiction has been reached. Hence $L \subseteq L_{\mathrm{fo}}$.

Now we show that $L \subseteq_{\text{eff}} L_{\text{fo}}^{\text{eff}}$. Let f be a recursive function with range $\mathcal{g}^{+*}\,\text{Sent}_{\mathscr{L}(\text{un})}$. Now for any $x, y \in \omega$ we set

$$k(x, y) = \text{Un}_L(\text{Int}_L\,(x, \text{Trans}_{L(\text{fo,eff}),L}\,fy), \text{Int}_L\,(\text{Comp}_L\,x, \\ \text{Comp}_L\,\text{Trans}_{L(\text{fo,eff}),L}\,fy)).$$

Thus k is a recursive function, and if $(\mathscr{L}, \Gamma, \mathfrak{A}, \varphi, h) \in L$, $\psi \in \Gamma$, $\mathscr{L}$ is a reduct of $\mathscr{L}_{\text{un}}$, $n \in \omega$, and $fn = \mathcal{g}^{+}\chi$, then $k(h\psi, n) = h\theta$ for some $\theta \in \Gamma$ such that

$$\text{Mod}_{\mathscr{L},L}\,\theta = (\text{Mod}_{\mathscr{L},L}\,\psi \cap \text{Mod}_{\mathscr{L},L(\text{fo,eff})}\,\chi) \cup [(\mathbf{S}_{\mathscr{L}} \sim \text{Mod}_{\mathscr{L},L}\,\psi) \\ \cap (\mathbf{S}_{\mathscr{L}} \sim \text{Mod}_{\mathscr{L},L(\text{fo,eff})}\,\chi)].$$

Thus

(8) $\mathfrak{A} \models_{\mathscr{L},L} \theta$ for all $\mathfrak{A} \in \mathbf{S}_{\mathscr{L}}$ iff $\text{Mod}_{\mathscr{L},L}\,\psi = \text{Mod}_{\mathscr{L},L(\text{fo,eff})}\,\chi$.

Now by the assumption of the theorem and 26.25(*iv*), let l be a recursive function whose range is the set indicated there. Then our desired function t showing that $L \subseteq_{\text{eff}} L_{\text{fo}}^{\text{eff}}$ is defined as follows:

$tx = 0 \qquad$ if x is not in the set of 26.25(*i*)(3),
$tx = f(\mu n(l(n)_0 = k(x, (n)_1)))_1 \qquad$ if x is in the set of 26.25(*i*)(3).

In fact, t is defined by cases over a recursive set by 26.25(*i*)(3). The μ operator is always applicable, by (8) and the fact that $L \subseteq L_{\text{fo}}$. So, t is recursive, and the preceding facts also show that it works as desired. □

The last main topic of this chapter is the theorem of Keisler and Shelah characterizing elementary equivalence and hence elementary classes. The theorem, 26.42 below, is that $\mathfrak{A} \equiv_{\text{ee}} \mathfrak{B}$ iff some ultrapower of $\mathfrak{A}$ is isomorphic to some ultrapower of $\mathfrak{B}$. There is no restriction on the language here as there is in the Ehrenfeucht–Fraisse characterization. The theorem was first proved assuming GCH by Keisler in 1961; his proof was relatively short, being a fairly natural back-and-forth argument. Shelah in 1972 showed that the GCH is not needed, using an entirely different and much more intricate argument. The argument of Shelah, given below, is based upon the idea of introducing set-theoretical combinatorial conditions in the construction of ultrafilters, an idea first introduced in a simpler setting by Kunen.

Recall the notion of a filter from 18.15. The following proposition is obvious and justifies the definition given after it.

Proposition 26.27. *If $\mathscr{F}$ is a nonempty collection of filters over I, then $\bigcap \mathscr{F}$ is a filter over I.*

Proposition 26.28. If K is a collection of subsets of I, we set

$$\mathscr{F}\mathcal{g}K = \bigcap \{F : K \subseteq F, F \text{ is a filter over } I\};$$

$\mathscr{F}\mathcal{g}K$ is called the *filter generated by K*.

The following three propositions are easy to establish, but are quite useful later on. We shall use them without specific reference to them.

Proposition 26.29. *If K is a nonempty collection of subsets of I, then for any $X \subseteq I$ the following conditions are equivalent:*

(*i*) $X \in \mathscr{F}g K$;
(*ii*) $\exists m \in \omega \sim 1 \, \exists Y \in {}^m K (Y_0 \cap \cdots \cap Y_{m-1} \subseteq X)$.

Proposition 26.30. *Let F be a filter over I and let $Z \subseteq I$. Then for any $X \subseteq I$ the following conditions are equivalent:*

(*i*) $X \in \mathscr{F}g(F \cup \{Z\})$;
(*ii*) $\exists Y \in F(Y \cap Z \subseteq X)$.

Proposition 26.31. *If K is a collection of subsets of I, then the following conditions are equivalent:*

(*i*) *K has the finite intersection property;*
(*ii*) $0 \notin \mathscr{F}g K$.

It is convenient to have a special notation for the following operation on cardinals:

Definition 26.32. For any infinite cardinal $\mathfrak{m}$ let $\mathfrak{m}^\partial$ be the least cardinal $\mathfrak{n}$ such that $\mathfrak{m}^\mathfrak{n} > \mathfrak{m}$.

Let us note a few facts about this operation; they will be used below without proof. Since $\mathfrak{m}^\mathfrak{n} = \mathfrak{m}$ for $\mathfrak{n}$ finite but $\mathfrak{m}^\mathfrak{m} > \mathfrak{m}$, we clearly have $\aleph_0 \leq \mathfrak{m}^\partial \leq \mathfrak{m}$. Obviously $\aleph_0^\partial = \aleph_0$. If GCH is assumed and $\mathfrak{m}$ is regular, then $\mathfrak{m}^\partial = \mathfrak{m}$. We have $\aleph_\omega^\partial = \aleph_0$, since

$$\aleph_\omega = \Sigma_{m \in \omega} \aleph_m < \Pi_{m \in \omega} \aleph_{m+1} \leq \aleph_\omega^{\aleph_0}.$$

Again assuming GCH, $\mathfrak{m}^\partial = \operatorname{cf} \mathfrak{m}$ for all infinite $\mathfrak{m}$, where $\operatorname{cf} \mathfrak{m}$ is the smallest cardinality of a family of ordinals $< \mathfrak{m}$ with union $\mathfrak{m}$. It is consistent with ZFC to assume that $\aleph_1^\partial = \aleph_0$, or that $\aleph_1^\partial = \aleph_1$. The main fact used below is that $\aleph_0 \leq \mathfrak{m}^\partial \leq \mathfrak{m}$.

The following definition is the main one for all that follows. It expresses the combination of some set-theoretical conditions with the existence of a filter.

Definition 26.33. Let $\mathfrak{m}$ and $\mathfrak{n}$ be cardinal numbers, with $\mathfrak{n}$ infinite. A triple (F, G, D) is $\mathfrak{m}$-*consistent over* $\mathfrak{n}$ provided that the following conditions hold:

(*i*) $F \subseteq {}^\mathfrak{n}(\mathfrak{n}^\partial)$ (i.e., F is a set of functions mapping $\mathfrak{n}$ into $\mathfrak{n}^\partial$);
(*ii*) $G \subseteq \bigcup_{\rho < \mathfrak{n}^\partial} {}^\mathfrak{n}\rho$ (i.e. G is a set of functions f with $\operatorname{Dmn} f = \mathfrak{n}$ and $\operatorname{Rng} f \subseteq \rho$ for some $\rho < \mathfrak{n}^\partial$);

(*iii*) D is a filter over $\mathfrak{n}$, and D is generated by a set of cardinality $\leq \mathfrak{m}$;
(*iv*) for all $\rho < \mathfrak{n}^{\partial}$, for all one-one $f \in {}^{\rho}F$, for all $\alpha \in {}^{\rho}(\mathfrak{n}^{\partial})$, for all $n \in \omega$, for all one-one $h \in {}^{n}F$ with $\operatorname{Rng} h \cap \operatorname{Rng} f = 0$, and for all $k \in {}^{n}G$, the following set is *not* in D:
$\mathfrak{n} \sim \{\beta < \mathfrak{n} : \forall \gamma < \rho(f_\gamma \beta = \alpha_\gamma) \text{ and } \forall m < n(h_m \beta = k_m \beta)\}$.

To understand this definition a little bit, think of the members of D as being "large," $|F|$ large, and G as a small set. The set in braces in 26.33(*iv*) is then to be considered large. Thus intuitively speaking the members of F take on many common values and agree with finitely many members of G at a large number of places. Note that D is a proper filter because of condition (*iv*) (with $\rho = f = \alpha = n = h = k = 0$). The notion will be used as follows. Given $\mathfrak{A} \equiv_{ee} \mathfrak{B}$, the cardinal $\mathfrak{n}$ will be chosen large relative to $|A|$, $|B|$, and $|\mathrm{Fmla}_{\mathscr{L}}|$. Then we shall construct a long sequence of triples $\mathfrak{m}$-consistent over $\mathfrak{n}$, in which we are only really interested in the third component, the filter over $\mathfrak{n}$. The first triple has $|F|$ very large, $G = 0$, $\mathfrak{m} = 2$, and $D = \{\mathfrak{n}\}$. As the construction proceeds, D is gradually extended to an ultrafilter over $\mathfrak{n}$ with some desirable properties; F is reduced in size, G increases, and $\mathfrak{m}$ increases. The final ultrafilter D over $\mathfrak{n}$ will be such that ${}^{\mathfrak{n}}\mathfrak{A}/D \cong {}^{\mathfrak{n}}\mathfrak{B}/D$. The sets F, G, $\mathfrak{m}$ serve only an auxiliary purpose, to obtain our desired properties of D. What these "desired properties" are will perhaps become clear as we proceed. The main parts of the construction are expressed in several lemmas. The first few lemmas are purely set-theoretical. We begin with the lemma that enables the above construction to be started.

Lemma 26.34. *Let* $\mathfrak{n} \geq \aleph_0$. *Then there is an* $F \subseteq {}^{\mathfrak{n}}(\mathfrak{n}^{\partial})$ *with* $|F| = 2^{\mathfrak{n}}$ *such that* $(F, 0, \{\mathfrak{n}\})$ *is 2-consistent over* $\mathfrak{n}$.

Proof. Since G is empty, we really only have to find a "big" F whose members take on many common values; in our case the condition 26.33(*iv*) simply reads

(1) for all $\rho < \mathfrak{n}^{\partial}$ and for all one-one $g \in {}^{\rho}F$ and all $\alpha \in {}^{\rho}(\mathfrak{n}^{\partial})$ there is a $\beta < \mathfrak{n}$ such that $\forall \gamma < \rho(g_\gamma \beta = \alpha_\gamma)$.

The other conditions are trivial. Let H be the set of all pairs (A, h) such that:

(2) $A \subseteq \mathfrak{n}$, and $|A| < \mathfrak{n}^{\partial}$,
(3) h is a function, $\operatorname{Dmn} h \subseteq \mathbf{S}A$, $|\operatorname{Dmn} h| < \mathfrak{n}^{\partial}$, and $\operatorname{Rng} h \subseteq \mathfrak{n}^{\partial}$.

Now the number of sets A satisfying (2) is

$$\sum_{\mathfrak{p} < \mathfrak{n}^{\partial}} \mathfrak{n}^{\mathfrak{p}} \leq \sum_{\mathfrak{p} < \mathfrak{n}^{\partial}} \mathfrak{n} \qquad \text{(by definition of } \mathfrak{n}^{\partial}\text{)}$$
$$\leq \mathfrak{n} \cdot \mathfrak{n}^{\partial} = \mathfrak{n} \qquad \text{(since } \mathfrak{n}^{\partial} \leq \mathfrak{n}\text{)}.$$

For each such A, the number of $X \subseteq \mathbf{S}A$ with $|X| < \mathfrak{n}^{\partial}$ is at most

$$\sum_{\mathfrak{p} < \mathfrak{n}^{\partial}} (2^{|A|})^{\mathfrak{p}} \leq \sum_{\mathfrak{p} < \mathfrak{n}^{\partial}} (\mathfrak{n}^{|A|})^{\mathfrak{p}} \leq \sum_{\mathfrak{p} < \mathfrak{n}^{\partial}} \mathfrak{n}$$
$$\leq \mathfrak{n} \cdot \mathfrak{n}^{\partial} = \mathfrak{n} \qquad \text{(since } |A|, \mathfrak{p} < \mathfrak{n}^{\partial}\text{)}.$$

And for each such A and X, the number of functions h mapping X into $\mathfrak{n}^{\partial}$ is at most $(\mathfrak{n}^{\partial})^{|X|} \leq \mathfrak{n}^{|X|} \leq \mathfrak{n}$. Thus $|H| \leq \mathfrak{n}$. But clearly $(\{\alpha\}, \{(\{\alpha\}, 0)\}) \in H$ for each $\alpha < \mathfrak{n}$, so actually $|H| = \mathfrak{n}$. Hence we may write

$$H = \{(A_\alpha, h_\alpha) : \alpha < \mathfrak{n}\}.$$

Now for each $B \subseteq \mathfrak{n}$ define $f_B : \mathfrak{n} \to \mathfrak{n}^{\partial}$ as follows. For any $\alpha < \mathfrak{n}$,

$$f_B\alpha = h_\alpha(A_\alpha \cap B) \qquad \text{if } A_\alpha \cap B \in \operatorname{Dmn} h_\alpha,$$

$$f_B\alpha = 0 \qquad \text{otherwise.}$$

Let $F = \{f_B : B \subseteq \mathfrak{n}\}$. We claim now that $(F, 0, \{\mathfrak{n}\})$ is 2-consistent over $\mathfrak{n}$ and $|F| = 2^{\mathfrak{n}}$, as desired in the lemma. To prove this, first note:

(4) if $B, C \subseteq \mathfrak{n}$ and $B \neq C$, then $f_B \neq f_C$.

For, say $B \nsubseteq C$, say $\alpha \in B \sim C$. Then $(\{\alpha\}, \{(\{\alpha\}, 1)\}) \in H$, by (2) and (3), so there is a $\beta < \mathfrak{n}$ such that $A_\beta = \{\alpha\}$ and $h_\beta = \{(\{\alpha\}, 1)\}$. Thus $A_\beta \cap B = \{\alpha\} \in \operatorname{Dmn} h_\beta$, so $f_B\beta = h_\beta(A_\beta \cap B) = 1$. Clearly, though, $f_C\beta = 0$. Hence $f_B \neq f_C$, and (4) holds.

From (4) we know that $|F| = 2^{\mathfrak{n}}$. To check (1), which is all that remains, assume its hypothesis. Say $g_\beta = f_{B\beta}$ for each $\beta < \rho$. Since g is one-one, the function B in the subscript of f here is also one-one. Hence whenever $\beta < \gamma < \rho$ we can choose $\delta_{\beta\gamma} \in B_\beta \Delta B_\gamma$ (the symmetric difference of B_β and B_γ). Let $C = \{\delta_{\beta\gamma} : \beta < \gamma < \rho\}$. Thus $|C| < \mathfrak{n}^{\partial}$, and $B_\beta \cap C \neq B_\gamma \cap C$ whenever $\beta < \gamma < \rho$. Let l be the function with domain $\{B_\beta \cap C : \beta < \rho\}$ such that $l(B_\beta \cap C) = \alpha\beta$ for each $\beta < \rho$. Thus by (2) and (3), $(C, l) \in H$; say $(C, l) = (A_\beta, h_\beta)$. Then for every $\gamma < \rho$ we have

$$g_\gamma\beta = f_{B\gamma}\beta = h_\beta(A_\beta \cap B_\gamma) = l(C \cap B_\gamma) = \alpha\gamma. \qquad \square$$

The following obvious lemma will be used without specific citation:

Lemma 26.35. *If (F, G, D) is $\mathfrak{m}$-consistent over $\mathfrak{n}$ and $F' \subseteq F$, $G' \subseteq G$, and $\mathfrak{m} \leq \mathfrak{m}'$, then (F', G', D) is $\mathfrak{m}'$-consistent over $\mathfrak{n}$.*

The next lemma will be used at limit stages in our construction:

Lemma 26.36. *Let δ be a limit ordinal $<\mathfrak{m}^+$ and suppose that $(F_\beta, G_\beta, D_\beta)$ is $\mathfrak{m}_\beta$-consistent over $\mathfrak{n}$ for each $\beta < \delta$. Also assume that if $\beta < \gamma < \delta$ then $F_\beta \supseteq F_\gamma$, $G_\beta \subseteq G_\gamma$, and $D_\beta \subseteq D_\gamma$. Set $F_\delta = \bigcap_{\beta<\delta} F_\beta$, $G_\delta = \bigcup_{\beta<\delta} G_\beta$, $D_\delta = \bigcup_{\beta<\delta} D_\beta$, and $\mathfrak{m}_\delta = (\bigcup_{\beta<\delta} \mathfrak{m}_\beta)\cdot|\delta|$.*

Then $(f_\delta, G_\delta, D_\delta)$ is $\mathfrak{m}_\delta$-consistent over $\mathfrak{n}$.

PROOF. The conditions (*i*)–(*iii*) of 26.33 are obvious. Now assume the hypothesis of 26.33(*iv*), but assume that the conclusion fails:

(1) $\mathfrak{n} \sim \{\beta < \mathfrak{n} : \forall\gamma < \rho(f_\gamma\beta = \alpha_\gamma) \text{ and } \forall m < n(h_m\beta = k_m\beta)\} \in D_\delta$.

Say that this set is in D_ε, where $\varepsilon < \delta$. Clearly $k \in {}^nG_\xi$ for some $\xi < \delta$. Let $\eta = \varepsilon \cup \xi$, which is the maximum of ε and ξ. Then $f \in {}^\rho F_\eta$, $h \in {}^n F_\eta$, $k \in {}^n G_\eta$ and the set of (1) is in D_η, contradicting the fact that (f_η, G_η, D_η) is $\mathfrak{m}_\eta$-consistent over $\mathfrak{n}$. □

The following lemma enables us to add to the set G in certain circumstances.

Lemma 26.37. *Let $\mathfrak{n} \geq \aleph_0$, and assume that $\mathfrak{n}^\partial \leq \mathfrak{m} \geq \aleph_0$. Suppose that $(F, 0, D)$ is $\mathfrak{m}$-consistent over $\mathfrak{n}$ and $|F| > \mathfrak{m}$. Suppose that G is a finite subset of $\bigcup_{\rho < \mathfrak{n}^\partial} {}^{\mathfrak{n}}\rho$.*

Then there is an $F' \subseteq F$ with $|F'| \leq \mathfrak{m}$ such that $(F \sim F', G, D)$ is $\mathfrak{m}$-consistent over $\mathfrak{n}$.

PROOF. Let $G = \{g_0, \ldots, g_n\}$, where g is one-one. Suppose there is no F' of the kind indicated. Then we shall define $\rho \in {}^{\mathfrak{m}^+}(\mathfrak{n}^\partial)$, $f \in \mathsf{P}_{\beta < \mathfrak{m}^+} {}^{\rho\beta} F$, $\chi \in \mathsf{P}_{\beta < \mathfrak{m}^+} {}^{\rho\beta}(\mathfrak{n}^\partial)$, $q \in {}^{\mathfrak{m}^+}\omega$, $h \in \mathsf{P}_{\alpha < \mathfrak{m}^+} {}^{q\beta} F$, and $k \in \mathsf{P}_{\beta < \mathfrak{m}^+} {}^{q\beta} G$ by induction. Suppose that $\beta < \mathfrak{m}^+$ and $\rho\gamma$, $f\gamma$, $\chi\gamma$, $q\gamma$, $h\gamma$, and $k\gamma$ have been defined for all $\gamma < \beta$. Let $F' = \{f_\gamma\delta : \gamma < \beta, \delta < \rho\gamma\} \cup \{h_{\gamma m} : \gamma < \beta, m < q\gamma\}$. Clearly

$$|F'| \leq \sum_{\gamma < \beta} |\rho\gamma| + \sum_{\gamma < \beta} \aleph_0 \leq \mathfrak{m} \cdot \mathfrak{n}^\partial + \mathfrak{m} \cdot \aleph_0 \leq \mathfrak{m}.$$

Hence by our supposition, $(F \sim F', G, D)$ is not $\mathfrak{m}$-consistent over $\mathfrak{n}$. This means that we may choose a $\rho_\beta < \mathfrak{n}^\partial$, a one-one $f_\beta \in {}^{\rho\beta}(F \sim F')$, a $\chi_\beta \in {}^{\rho\beta}(\mathfrak{n}^\partial)$, a $q_\beta \in \omega$, a one-one $h_\beta \in {}^{q\beta}(F \sim F')$ with $\operatorname{Rng} h_\beta \cap \operatorname{Rng} f_\beta = 0$, and a $k_\beta \in {}^{q\beta}G$ such that $\mathfrak{n} \sim A_\beta \in D$, where

$$A_\beta = \{\delta < \mathfrak{n} : \forall\gamma < \rho_\beta(f_{\beta\gamma}\delta = \chi_{\beta\gamma}) \text{ and } \forall m < q_\beta(h_{\beta m}\delta = k_{\beta m}\delta)\}.$$

This completes the definition of ρ, f, χ, q, h, and k.

Since $k \in \mathsf{P}_{\beta < \mathfrak{m}^+} {}^{q\beta}G$, we can choose for each $\beta < \mathfrak{m}^+$ and $m < q_\beta$ an integer $i_{\beta m} \leq n$ such that $k_{\beta m} = g_{i(\beta,m)}$. Since $(F, 0, D)$ is $\mathfrak{m}$-consistent over $\mathfrak{n}$, by 26.33(*iii*) write $D = \mathscr{F}g\{J_\beta : \beta < \mathfrak{m}\}$, where $\{J_\beta : \beta < \mathfrak{m}\}$ is closed under intersections. Since $\mathfrak{n} \sim A_\beta \in D$ for each $\beta < \mathfrak{m}^+$, choose $\alpha \in {}^{\mathfrak{m}^+}\mathfrak{m}$ such that $J_{\alpha\beta} \subseteq \mathfrak{n} \sim A_\beta$ for every $\beta < \mathfrak{m}^+$. Now

$$\mathfrak{m}^+ = \bigcup_{\gamma < \mathfrak{m}} \bigcup_{\mu < \mathfrak{n}^\partial} \bigcup_{s \in \omega} \{\beta \in \mathfrak{m}^+ : \alpha_\beta = \gamma, \rho_\beta = \mu, q_\beta = s\},$$

and all of $\mathfrak{m}$, $\mathfrak{n}^\partial$, ω are $\leq \mathfrak{m}$, so there exist $\gamma_0 < \mathfrak{m}$, $\mu_0 < \mathfrak{n}^\partial$ and $s_0 \in \omega$ so that $|B| = \mathfrak{m}^+$, where

$$B = \{\beta \in \mathfrak{m}^+ : \alpha_\beta = \gamma_0, \rho_\beta = \mu_0, q_\beta = s_0\}.$$

Let ψ be a one-one function mapping $\mathfrak{m}^+$ onto B. Now since G is finite and $G \subseteq \bigcup_{\tau < \mathfrak{n}^\partial} {}^{\mathfrak{n}}\tau$, there is a $\tau_0 < \mathfrak{n}^\partial$ such that $g_m \in {}^{\mathfrak{n}}\tau_0$ for all $m \leq n$. Let $\{\langle \theta_0^\beta, \ldots, \theta_n^\beta \rangle : \beta < \tau_1\}$ be an enumeration of all $(n + 1)$-termed sequences of ordinals $\leq \tau_0$ (thus $\tau_1 \leq \tau_0 + \aleph_0$, and $\tau_1 = \tau_0$ if τ_0 is infinite; note that $\tau_1 < \mathfrak{n}^{\partial}$). Now let

$$C = \{\delta < \mathfrak{n} : \forall \beta < \tau_1 \, \forall \gamma < \mu_0(f_{\psi\beta,\gamma}\delta = \chi_{\psi\beta,\gamma}) \text{ and}$$
$$\forall \beta < \tau_1 \, \forall m < s_0(h_{\psi\beta,m}\delta = \theta^{\beta}_{i(\psi\beta,m)})\}.$$

Note that if $\beta, \beta' < \tau_1$, $\gamma, \gamma' < \mu_0$, and $(\beta, \gamma) \neq (\beta', \gamma')$, then $f_{\psi\beta,\gamma} \neq f_{\psi\beta',\gamma'}$. Further, if $\beta, \beta' < \tau_1$, $\gamma < \mu_0$, and $m < s_0$, then $f_{\psi\beta,\gamma} \neq h_{\beta'm}$. And if β, $\beta' < \tau_1$, $m, m' < s_0$, and $(\beta, m) \neq (\beta', m')$, then $h_{\psi\beta,m} \neq h_{\psi\beta',m'}$. Since $|\tau_1| \cdot |\mu_0| + |\tau_1| \cdot |s_0| < \mathfrak{n}^{\partial}$, and since $(F, 0, D)$ is $\mathfrak{m}$-consistent over $\mathfrak{n}$, it follows that $\mathfrak{n} \sim C \notin D$. In particular $J_{\gamma 0} \nsubseteq \mathfrak{n} \sim C$, so we may choose $\delta \in J_{\gamma 0} \cap C$. Choose $\beta < \tau_1$ such that $\langle g_0\delta, \ldots, g_n\delta\rangle = \langle \theta^{\beta}_0, \ldots, \theta^{\beta}_n\rangle$. Then since $\rho_{\psi\beta} = \mu_0$ and $\delta \in C$ we have for any $m < s_0$

$$h_{\psi\beta,m}\delta = \theta^{\beta}_{i(\psi\beta,m)} = g_{i(\psi\beta,m)}\delta = k_{\psi\beta,m}\delta.$$

Since $q_{\psi\beta} = s_0$, the fact that $\delta \in C$ yields further that $\delta \in A_{\psi\beta}$. But $\alpha_{\psi\beta} = \gamma_0$, so $J_{\gamma 0} = J_{\alpha\psi\beta} \subseteq \mathfrak{n} \sim A_{\psi\beta}$; but $\delta \in J_{\gamma 0}$, contradiction. □

We now extend this lemma slightly:

Lemma 26.38. *Let $\mathfrak{n} \geq \aleph_0$. Assume that $G \subseteq \bigcup_{\rho < \mathfrak{n}^{\partial}} {}^{\mathfrak{n}}\rho$ and that $\mathfrak{n}^{\partial} + |G| \leq \mathfrak{m} \geq \aleph_0$. Suppose that $(F, 0, D)$ is $\mathfrak{m}$-consistent over $\mathfrak{n}$, with $|F| > \mathfrak{m}$.*

Then there is an $F' \subseteq F$ with $|F'| \leq \mathfrak{m}$ such that $(F \sim F', G, D)$ is $\mathfrak{m}$-consistent over $\mathfrak{n}$.

Proof. Let $I = \{H : H \subseteq G, H \text{ finite}\}$. For each $H \in I$ choose $F_H \subseteq F$ by the preceding lemma so that $|F_H| \leq \mathfrak{m}$ and $(F \sim F_H, H, D)$ is $\mathfrak{m}$-consistent over $\mathfrak{n}$. Then clearly $\bigcup_{H \in I} F_H \subseteq F$, $|\bigcup_{H \in I} F_H| \leq \mathfrak{m}$, and $(F \sim \bigcup_{H \in I} F_H, G, D)$ is $\mathfrak{m}$-consistent over $\mathfrak{n}$. □

The next, very important, lemma enables us to slowly extend D to an ultrafilter.

Lemma 26.39. *Suppose that (F, G, D) is $\mathfrak{m}$-consistent over $\mathfrak{n}$ and $\Gamma \subseteq \mathfrak{n}$. Then there is an $F' \subseteq F$ with $|F - F'| < \mathfrak{n}^{\partial}$ such that either $(F', G, \mathscr{F}\mathscr{g}(D \cup \{\Gamma\}))$ or $(F', G, \mathscr{F}\mathscr{g}(D \cup \{\mathfrak{n} \sim \Gamma\}))$ is $\mathfrak{m}$-consistent over $\mathfrak{n}$.*

Proof. Let $D_1 = \mathscr{F}\mathscr{g}(D \cup \{\Gamma\})$ and $D_2 = \mathscr{F}\mathscr{g}(D \cup \{\mathfrak{n} \sim \Gamma\})$. Assume that (F, G, D_1) is not $\mathfrak{m}$-consistent over $\mathfrak{n}$. This means that there is a $\rho < \mathfrak{n}^{\partial}$, a one-one $f \in {}^{\rho}F$, an $\alpha \in {}^{\rho}(\mathfrak{n}^{\partial})$, an $n \in \omega$, a one-one $h \in {}^{n}F$ with $\text{Rng}\, h \cap \text{Rng} f = 0$, and a $k \in {}^{n}G$ such that $\mathfrak{n} \sim B \in D_1$, where

$$B = \{\beta < \mathfrak{n} : \forall \gamma < \rho(f_{\gamma}\beta = \alpha_{\gamma}) \text{ and } \forall m < n(h_m\beta = k_m\beta)\}.$$

Choose $\Delta \in D$ so that $\Gamma \cap \Delta \subseteq \mathfrak{n} \sim B$ (using 26.30). Now set $F'' = \text{Rng} f \cup \text{Rng}\, h$. Clearly $|F''| < \mathfrak{n}^{\partial}$, so it suffices now to show that $(F \sim F'', G, D_2)$ is $\mathfrak{m}$-consistent over $\mathfrak{n}$. Suppose not: then there is a $\rho' < \mathfrak{n}^{\partial}$, a one-one

$f' \in {}^{\rho}(F \sim F'')$, an $\alpha' \in {}^{\rho}(\mathfrak{n}^{\partial})$, an $n' \in \omega$, a one-one $h' \in {}^{n'}(F \sim F'')$ with $\operatorname{Rng} h' \cap \operatorname{Rng} f' = 0$, and a $k' \in {}^{n'}G$ such that $\mathfrak{n} \sim C \in D_2$, where

$$C = \{\beta < \mathfrak{n} : \forall\gamma < \rho'(f'_\gamma\beta = \alpha'_\gamma) \text{ and } \forall m < n(h'_m\beta = k'_m\beta)\}.$$

Choose $\Theta \in D$ so that $(\mathfrak{n} \sim \Gamma) \cap \Theta \subseteq \mathfrak{n} \sim C$. Thus $\Delta \cap \Theta \subseteq \mathfrak{n} \sim (B \cap C)$, so $\mathfrak{n} \sim (B \cap C) \in D$. But this clearly contradicts the fact that (F, G, D) $\mathfrak{m}$-consistent over $\mathfrak{n}$. □

An easy argument gives a generalization of the lemma just proved:

Lemma 26.40. *Suppose that (F, G, D) is $\mathfrak{m}$-consistent over $\mathfrak{n}$ and that $\Gamma_\alpha \subseteq \mathfrak{n}$ for every $\alpha < \mathfrak{m}$. Assume that $\mathfrak{n}^{\partial} \leq \mathfrak{m}$. Then there exist $F' \subseteq F$ with $|F \sim F'| \leq \mathfrak{m}$, and $D' \supseteq D$ such that (F', G, D') is $\mathfrak{m}$-consistent over $\mathfrak{n}$ and for every $\alpha < \mathfrak{m}$, either $\Gamma_\alpha \in D'$ or $\mathfrak{n} \sim \Gamma_\alpha \in D'$.*

Proof. We define $\langle H_\alpha : \alpha < \mathfrak{m}\rangle$ and $\langle E_\alpha : \alpha < \mathfrak{m}\rangle$ by transfinite recursion. Suppose H_β and E_β have been defined for all $\beta < \alpha$, where $\alpha < \mathfrak{m}$, in such a way that $H_\beta \subseteq F$, $|H_\beta| \leq \mathfrak{m}$, $(F \sim H_\beta, G, E_\beta)$ is $\mathfrak{m}$-consistent over $\mathfrak{n}$, and $H_\beta \subseteq H_\gamma$ and $E_\beta \subseteq E_\gamma$ whenever $\beta < \gamma < \alpha$. Then by Lemma 26.36, $(F \sim \bigcup_{\beta<\alpha} H_\beta, G, \bigcup_{\beta<\alpha} E_\beta)$ is $\mathfrak{m}$-consistent over $\mathfrak{n}$. By 26.39 there is an $F' \subseteq F \sim \bigcup_{\beta<\alpha} H_\beta$ with $|(F \sim \bigcup_{\beta<\alpha} H_\beta) \sim F'| < \mathfrak{n}^{\partial}$ such that (F', G, E_α) is $\mathfrak{m}$-consistent over $\mathfrak{n}$, where either $E_\alpha = \mathscr{F}\mathscr{g}(\bigcup_{\beta<\alpha} E_\beta \cup \{\Gamma_\alpha\})$ or $E_\alpha = \mathscr{F}\mathscr{g}(\bigcup_{\beta<\alpha} E_\beta \cup \{\mathfrak{n} \sim \Gamma_\alpha\})$. Let $H_\alpha = \bigcup_{\beta<\alpha} H_\beta \cup ((F \sim \bigcup_{\beta<\alpha} H_\beta) \sim F')$. Note that $F \sim H_\alpha = F'$, so the induction assumptions are met and the definition of H and E is complete. Since $\mathfrak{n}^{\partial} \leq \mathfrak{m}$, an argument by transfinite induction shows that $|H_\alpha| \leq \mathfrak{m}$ for each $\alpha < \mathfrak{m}$. Applying 26.36 again, we see that $(F \sim \bigcup_{\alpha<\mathfrak{m}} H_\alpha, G, \bigcup_{\alpha<\mathfrak{m}} E_\alpha)$ is $\mathfrak{m}$-consistent over $\mathfrak{n}$. Clearly $|\bigcup_{\alpha<\mathfrak{m}} H_\alpha| \leq \mathfrak{m}$ and the conditions of the lemma are met. □

The next lemma, whose formulation is somewhat involved, essentially ensures the desirable properties of the final ultrafilter which yield the isomorphism between the ultrapowers.

Lemma 26.41. *Assume that $\mathfrak{n} \geq \aleph_0$ and $\mathfrak{n}^{\partial} \leq \mathfrak{m}$. Suppose that $(F, 0, D)$ is $\mathfrak{m}$-consistent over $\mathfrak{n}$, and $|F| > \mathfrak{m}$. Let $\mathfrak{A}$ be an $\mathscr{L}$-structure, $|A| < \mathfrak{n}^{\partial}$. Suppose $\mu < \mathfrak{m}^+$, $a: \mu \times \omega \times \mathfrak{n} \to A$, and $\varphi \in {}^{\mu}\mathrm{Fmla}_{\mathscr{L}}$. Suppose that for every finite $M \subseteq \mu$ there is a $\nu < \mu$ such that for all $x \in A$ and all $\delta < \mathfrak{n}$ the following condition holds:*

(*i*) *$\mathfrak{A} \models \varphi_\nu[\langle a_{\nu j\delta} : j < \omega\rangle^0_x]$ iff for all $\sigma \in M$ $\mathfrak{A} \models \varphi_\sigma[a_{\nu j\delta} : j < \omega\rangle^0_x]$.*

Furthermore, assume that for every $\nu < \mu$ we have

(*ii*) *$\{\delta < \mathfrak{n} : \mathfrak{A} \models (\exists v_0\varphi_\nu)[\langle a_{\nu j\delta} : j < \omega\rangle]\} \in D$.*

Under all of these assumptions it follows that there are $b \in {}^{\mathfrak{n}}A$, $F' \subseteq F$ and $D' \supseteq D$ such that $|F \sim F'| \leq \mathfrak{m}$, the triple $(F', 0, D')$ is $\mathfrak{m}$-consistent over $\mathfrak{n}$, and for every $\nu < \mu$ we have

(*iii*) *$\{\delta < \mathfrak{n} : \mathfrak{A} \models \varphi_\nu[\langle a_{\nu j\delta} : j < \omega\rangle^0_{b\delta}]\} \in D'$.*

Note that the essence of this lemma is that in the final ultraproduct certain small sets of formulas can be satisfied simultaneously by the same elements, a general phenomenon which will be investigated more thoroughly in the next two sections. The condition (*i*) of 26.41 is rather innocuous, saying roughly that the system of formulas φ is closed under conjunction.

PROOF OF 26.41. Let $A = \{c_\gamma : \gamma < \rho\}$, where $\rho < \mathfrak{n}^{\mathfrak{d}}$. For each $\nu < \mu$ let

$$B_\nu = \{\delta < \mathfrak{n} : \mathfrak{A} \models (\exists v_0 \varphi_\nu)[\langle a_{\nu j \delta} : j < \omega\rangle]\}.$$

Thus $B_\nu \in D$ by 26.41(*ii*). Now for every $\nu < \mu$ choose $g_\nu : \mathfrak{n} \to \rho$ such that for every $\delta \in B_\nu$ we have $\mathfrak{A} \models \varphi_\nu[\langle a_{\nu j \delta} : j < \omega\rangle^0_{cg\nu\delta}]$. Let $G = \{g_\nu : \nu < \mu\}$. Since $(F, 0, D)$ is $\mathfrak{m}$-consistent over $\mathfrak{n}$ and $|F| > \mathfrak{m}$, by Lemma 26.38 choose a nonempty $F_1 \subseteq F$ such that $|F_1| \leq \mathfrak{m}$ and $(F \sim F_1, G, D)$ is $\mathfrak{m}$-consistent over $\mathfrak{n}$. Fix $f \in F \sim F_1$. Let

$$F' = (F \sim F_1) \sim \{f\},$$

and for each $\delta < \mathfrak{n}$ let

$$\begin{aligned} b_\delta &= c_{f\delta} && \text{if } f\delta < \rho \\ b_\delta &= c_0 && \text{otherwise.} \end{aligned}$$

For each $\nu < \mu$ let

$$C_\nu = \{\delta < \mathfrak{n} : \mathfrak{A} \models \varphi_\nu[\langle a_{\nu j \delta} : j < \omega\rangle^0_{b\delta}]\},$$

and let $D' = \mathscr{F}\!g(D \cup \{C_\nu : \nu < \mu\})$. We claim, now, that $(F', 0, D')$ is $\mathfrak{m}$-consistent over $\mathfrak{n}$, and hence the lemma holds. Since $\mu < \mathfrak{m}^+$, clearly D' is generated by $\leq \mathfrak{m}$ elements. Suppose that $(F', 0, D')$ is not $\mathfrak{m}$-consistent over $\mathfrak{n}$. Then there is a $\xi < \mathfrak{n}^{\mathfrak{d}}$, a one-one $h \in {}^\xi F'$, and an $\alpha \in {}^\xi(\mathfrak{n}^{\mathfrak{d}})$ such that $\mathfrak{n} \sim E \in D'$, where

(1) $$E = \{\delta < \mathfrak{n} : \forall \gamma < \xi(h_\gamma \delta = \alpha_\gamma)\}.$$

Now by (*i*) it is clear that if M is any finite subset of μ then there is a $\nu < \mu$ such that $\bigcap_{\sigma \in M} C_\sigma = C_\nu$. Hence from our assumption $\mathfrak{n} \sim E \in D'$ and by the definition of D', there exist $X \in D$ and $\nu < \mu$ such that $X \cap C_\nu \subseteq \mathfrak{n} \sim E$. Hence

(2) $$X \cap B_\nu \subseteq \mathfrak{n} \sim \{\delta < \mathfrak{n} : \forall \gamma < \xi(h_\gamma \delta = \alpha_\gamma), \text{ and } f\delta = g_\nu \delta\}.$$

In fact, let $\delta \in X \cap B_\nu$. Thus $\mathfrak{A} \models \varphi_\nu[\langle a_{\nu j \delta} : j < \omega\rangle^0_{cg\gamma\delta}]$. Assume that δ is in the set in braces in (2). In particular, $\delta \in E$ and $f\delta = g_\nu \delta < \rho$, so $b_\delta = c_{f\delta} = c_{g\nu\delta}$. So $\mathfrak{A} \models \varphi_\nu[\langle a_{\nu j \delta} : j < \omega\rangle^0_{b\delta}]$; but this contradicts the fact that $X \cap C_\nu \subseteq \mathfrak{n} \sim E$. Hence (2) does hold. But $B_\nu \in D$, so the right-hand side of (2) is in D. This contradicts the fact that $(F \sim F_1, G, D)$ is $\mathfrak{m}$-consistent over $\mathfrak{n}$. □

Now we can give the promised ultrapower characterization of elementary equivalence:

Theorem 26.42 (Keisler, Shelah). *The following conditions are equivalent:*

(*i*) $\mathfrak{A} \equiv_{ee} \mathfrak{B}$;
(*ii*) *there exist I and D such that* ${}^I\mathfrak{A}/D \cong {}^I\mathfrak{B}/D$.

PROOF. Obviously (*ii*) ⇒ (*i*), so we begin immediately with the hard direction (*i*) ⇒ (*ii*). Assume that $\mathfrak{A} \equiv_{ee} \mathfrak{B}$. Let

$$\mathfrak{n} = 2^{|A| + |B| + |\mathrm{Fmla}_{\mathscr{L}}|}.$$

The set I of the theorem will be this cardinal number $\mathfrak{n}$. Note that $\mathfrak{n}^{|A|} = \mathfrak{n}^{|B|} = \mathfrak{n}^{|\mathrm{Fmla}_{\mathscr{L}}|} = \mathfrak{n}$, and hence $|A|, |B|, |\mathrm{Fmla}_{\mathscr{L}}| < \mathfrak{n}^{\partial}$. Let ${}^{\mathfrak{n}}A = \{a_\alpha : \alpha < 2^{\mathfrak{n}}\}$ and ${}^{\mathfrak{n}}B = \{b_\alpha : \alpha < 2^{\mathfrak{n}}\}$. Now the set

$$\{(0, a_\alpha) : \alpha < 2^{\mathfrak{n}}\} \cup \{(1, b_\alpha) : \alpha < 2^{\mathfrak{n}}\} \cup \{(2, \Delta) : \Delta \subseteq \mathfrak{n}\}$$

has cardinality $2^{\mathfrak{n}}$; let R be a one-one function from $2^{\mathfrak{n}}$ onto this set, with $R_0 = (0, a_0)$.

Now we shall define by recursion on $\gamma \leq 2^{\mathfrak{n}}$ a set of functions F_γ, a filter D_γ over $\mathfrak{n}$, and a relation H_γ such that the following conditions hold:

(1) $(F_\gamma, 0, D_\gamma)$ is $(\mathfrak{n} + |\gamma|)$-consistent over $\mathfrak{n}$, $|F_0| = 2^{\mathfrak{n}}$, $D_0 = \{\mathfrak{n}\}$, $|F_0 \sim F_\gamma| \leq \mathfrak{n} + |\gamma|$, and for any $\beta < \gamma$, $F_\beta \supseteq F_\gamma$, and $D_\beta \subseteq D_\gamma$;
(2) $H_\gamma \subseteq {}^{\mathfrak{n}}A \times {}^{\mathfrak{n}}B$, $|H_\gamma| \leq |\gamma|$, and for any $\beta < \gamma$, $H_\beta \subseteq H_\gamma$;
(3) if $c_i H_\gamma d_i$ for all $i < \omega$ and φ is any formula, then $\{\delta < \mathfrak{n} : \mathfrak{A} \models \varphi[\langle c_i\delta : i < \omega\rangle]\} \in D_\gamma$ iff $\{\delta < \mathfrak{n} : \mathfrak{B} \models \varphi[\langle d_i\delta : i < \omega\rangle]\} \in D_\gamma$;
(4) if $c \in {}^{\omega}\mathrm{Dmn}\, H_\gamma$ and φ is any formula, then either $\{\delta < \mathfrak{n} : \mathfrak{A} \models \varphi[\langle c_i\delta : i < \omega\rangle]\} \in D_\gamma$ or $\{\delta < \mathfrak{n} : \mathfrak{A} \models \neg\varphi[\langle c_i\delta : i < \omega\rangle]\} \in D_\gamma$;
(5) if $R_\gamma = (0, a_\alpha)$, then $a_\alpha \in \mathrm{Dmn}\, H_{\gamma+1}$;
(6) if $R_\gamma = (1, b_\alpha)$, then $b_\alpha \in \mathrm{Rng}\, H_{\gamma+1}$;
(7) if $R_\gamma = (2, \Delta)$, then $\Delta \in D_{\gamma+1}$ or $\mathfrak{n} \sim \Delta \in D_{\gamma+1}$.

The importance of these various conditions for our construction should be fairly clear. Condition (3) assures that in the final ultrapowers H_γ-related elements will have the same first-order properties, while (5) and (6) assure that the limit of the relations H_γ will induce a function with domain ${}^{\mathfrak{n}}A/D$ and range ${}^{\mathfrak{n}}B/D$. Condition (7) allows a slow extension to an ultrafilter. Condition (4) has a technical character, while (1) embodies the combinatorial conditions that make the construction possible. Note that the construction again has a back-and-forth character.

Suppose that F_η, D_η, and H_η have been defined for all $\eta < \gamma$ so that (1)–(7) hold whenever the relevant indices are $\leq \eta$, where $\gamma \leq 2^{\mathfrak{n}}$. We now define F_γ, D_γ, and H_γ by distinguishing several cases.

Case 1. $\gamma = 0$. By Lemma 26.34, choose $F_0 \subseteq {}^{\mathfrak{n}}(\mathfrak{n}^{\partial})$ with $|F_0| = 2^{\mathfrak{n}}$ such that $(F_0, 0, \{\mathfrak{n}\})$ is 2-consistent over $\mathfrak{n}$. Let $H_0 = 0$, $D_0 = \{\mathfrak{n}\}$. Clearly (1), (2), and (4) hold. Condition (3) holds since $\mathfrak{A} \equiv_{ee} \mathfrak{B}$. Conditions (5)–(7) are not relevant to this case.

Case 2. γ is a limit ordinal. Let

$$F_\gamma = \bigcap_{\eta<\gamma} F_\eta, \qquad D_\gamma = \bigcup_{\eta<\gamma} D_\eta, \qquad H_\gamma = \bigcup_{\eta<\gamma} H_\eta.$$

By Lemma 26.36, $(F_\gamma, 0, D_\gamma)$ is $(\mathfrak{n} + |\gamma|)$-consistent over $\mathfrak{n}$. Also, we have

$$|F_0 \sim F_\gamma| = \left|\bigcup_{\eta<\gamma} (F_0 \sim F_\eta)\right| \leq \sum_{\eta<\gamma} |F_0 - F_\eta|$$
$$\leq \sum_{\eta<\gamma} (\mathfrak{n} + |\eta|) \leq (\mathfrak{n} + |\gamma|)\cdot|\gamma| = \mathfrak{n} + |\gamma|.$$

Thus (1) holds for γ. Clearly (2) holds. In (3) and (4) we may assume that Rng c and Rng d are finite, and then these conditions are obvious. Conditions (5)–(7) are not relevant to this case.

Case 3. $\gamma = \beta + 1$, $R_\beta = (0, a_\alpha)$. The construction here is in two steps: first to insure that (4) holds, and second, to take care of (3). For every formula φ and every $c \in {}^\omega(\text{Dmn } H_\beta \cup \{a_\alpha\})$ for which there is a $d \in \text{Dmn } H_\beta \cup \{a_\alpha\}$ with $c_i = d$ for all but finitely many $i < \omega$ let

$$\Gamma_{\varphi c} = \{\delta < \mathfrak{n} : \mathfrak{A} \models \varphi[\langle c_i\delta : i < \omega\rangle]\}.$$

Clearly the number of such sets $\Gamma_{\varphi c}$ is

$$\leq |\text{Fmla}_{\mathscr{L}}|\cdot\left(\sum_{m\in\omega} |\gamma|^m\right)\cdot|\gamma| \qquad \text{(using (2) for } \beta)$$
$$= |\text{Fmla}_{\mathscr{L}}|\cdot|\gamma|\cdot\aleph_0 \leq \mathfrak{n} + |\beta|.$$

Hence we may apply 26.40 to $(F_\beta, 0, D_\beta)$ and the set of all $\Gamma_{\varphi c}$. We thus obtain $(F', 0, D')$ such that $F' \subseteq F_\beta$, $D_\beta \subseteq D'$, $|F_\beta \sim F'| \leq \mathfrak{n}\,|\beta|$, $(F', 0, D')$ is $(\mathfrak{n} + |\beta|)$-consistent over $\mathfrak{n}$, and for all φ, c as above, $\Gamma_{\varphi c} \in D'$ or $\mathfrak{n} \sim \Gamma_{\varphi c} \in D'$. To do the second step of the construction in this case, it is convenient to consider two subcases.

Subcase 1. $\gamma = 1$. Thus $\beta = 0$ and $\alpha = 0$. Let Θ consist of all φ such that $\text{Fv}\varphi \subseteq \{v_0\}$ and $\{\delta < \mathfrak{n} : \mathfrak{A} \models \varphi[a_0\delta]\} \in D'$.

Say $\Theta = \{\varphi_\nu : \nu < \mathfrak{n}\}$ (obviously $0 \neq |\Theta| \leq \mathfrak{n}$). Pick $b : \mathfrak{n} \times \omega \times \mathfrak{n} \to B$ arbitrarily. We shall now verify the hypotheses of 26.41 with $\mathfrak{n}$, $\mathfrak{m}$, F, D, $\mathfrak{A}$, μ, a, φ replaced by $\mathfrak{n}$, $\mathfrak{n} + |\beta|$, F', D', $\mathfrak{B}$, $\mathfrak{n}$, b, φ respectively. Now $\mathfrak{A} \models \exists v_0\varphi$ for each $\varphi \in \Theta$, and $\mathfrak{A} \equiv_{\text{ee}} \mathfrak{B}$, so $\mathfrak{B} \models \exists v_0\varphi$ for each $\varphi \in \Theta$. If $M \subseteq \mathfrak{n}$ is finite, then clearly $\bigwedge_{\sigma\in M} \varphi_\sigma \in \Theta$, say $\bigwedge_{\sigma\in M} \varphi_\sigma = \varphi_\nu$; clearly then 26.41($i$) holds for $\mathfrak{B}$, b. Thus the hypotheses of 26.41 are met. Hence we can choose $b \in {}^{\mathfrak{n}}B$, $F_1 \subseteq F'$, and $D_1 \supseteq D'$ such that $|F' \sim F_1| \leq \mathfrak{n}$, the triple $(F_1, 0, D_1)$ is $\mathfrak{n}$-consistent over $\mathfrak{n}$, and for every $\nu < \mathfrak{n}$, $\{\delta < \mathfrak{n} : \mathfrak{B} \models \varphi_\nu[b\delta]\} \in D_1$. Set $H_1 = \{(a_0, b)\}$. Then all of the conditions (1)–(7) except (3) and (4) are clear; and (4) is immediate from our construction of D'. Now suppose that $c_i H_1 d_i$ for all $i < \omega$, and φ is any formula. Thus $c_i = a_0$ and $d_i = b$ for all $i < \omega$. Let $\psi = \varphi(v_0, \ldots, v_0)$, with enough v_0's to take care of all free variables of φ. Then for any structure $\mathfrak{C}$ and any $c \in C$, $\mathfrak{C} \models \varphi[c, c, \ldots]$ iff $\mathfrak{C} \models \psi[c]$. If $\{\delta < \mathfrak{n} : \mathfrak{A} \models \varphi[\langle c_i\delta : i < \omega\rangle]\} \in D_1$, then $\{\delta < \mathfrak{n} : \mathfrak{A} \models \psi[c_0\delta]\} \in D_1$ and hence $\{\delta < \mathfrak{n} : \mathfrak{A} \models \psi[c_0\delta]\} \in D'$ by our choice of D', since $D_1 \supseteq D'$. So $\psi \in \Theta$, and we infer that $\{\delta < \mathfrak{n} : \mathfrak{B} \models \psi[d_0\delta]\} \in D_1$. Hence $\{\delta < \mathfrak{n} : \mathfrak{B} \models \varphi[\langle d_i\delta : i < \omega\rangle]\} \in D_1$, as desired.

Subcase 2. $\gamma > 1$. From (5) and the fact that $R_0 = (0, a_0)$ we know that

$H_\beta \neq 0$. Let Θ consist of all pairs (φ, c) such that φ is a formula, $c \in {}^\omega\text{Dmn } H_\beta$, for some $d \in \text{Dmn } H_\beta$ we have $|\{i : c_i \neq d\}| < \aleph_0$, and

(8) $\{\delta < \mathfrak{n} : \mathfrak{A} \models \varphi[\langle c_i\delta : i < \omega\rangle^0_{a\alpha\delta}]\} \in D'$.

Clearly $0 \neq |\Theta| \leq \mathfrak{n} + |\beta|$. Say $\Theta = \{(\varphi_\nu, c_\nu) : \nu < \mathfrak{n} + |\beta|\}$. Now for any $\nu < \mathfrak{n} + |\beta|$ we have $\{\delta < \mathfrak{n} : \mathfrak{A} \models \exists v_0\varphi_\nu[\langle c_{\nu i}\delta : i < \omega\rangle]\} \in D' \supseteq D_\beta$ so, by (4) for β, $\{\delta < \mathfrak{n} : \mathfrak{A} \models \exists v_0\varphi_\nu[\langle c_{\nu i}\delta : i < \omega\rangle]\} \in D_\beta$. For each $x \in \text{Dmn } H_\beta$ choose d_x so that $xH_\beta d_x$. Then by (3) for β we obtain for each $\nu < \mathfrak{n} + |\beta|$

$$\{\delta < \mathfrak{n} : \mathfrak{B} \models \exists v_0\varphi_\nu[\langle d_{c\nu i}\delta : i < \omega\rangle]\} \in D_\beta.$$

We now check the hypotheses of 26.41 with $\mathfrak{n}$, $\mathfrak{m}$, F, D, $\mathfrak{A}$, μ, a, φ replaced by $\mathfrak{n}$, $\mathfrak{n} + |\beta|$, F', D', $\mathfrak{B}$, $\mathfrak{n} + |\beta|$, $\langle d_{c\nu i}\delta : \nu < \mathfrak{n} + |\beta|, i \in \omega, \delta < \mathfrak{n}\rangle$, φ respectively. In fact, it remains only to check (i). So suppose $M \subseteq \mathfrak{n} + |\beta|$ is finite. For each $\nu \in M$ let $J_\nu = \{j \in \omega : v_j$ occurs free in $\varphi_\nu\} \sim \{0\}$. Choose $\langle K_\nu : \nu \in M\rangle$ so that each $K_\nu \subseteq \omega \sim 1$, the K_ν's are pairwise disjoint, $|J_\nu| = |K_\nu|$, and for all $j \in \bigcup_{\nu\in M} K_\nu$, the variable v_j does not occur in any of the φ_ν for $\nu \in M$. Say $u_\nu : J_\nu \rightarrowtail\!\!\!\twoheadrightarrow K_\nu$. Let φ'_ν be obtained from φ_ν by replacing each free occurrence of v_j by $v_{u\nu j}$ for each $j \in J_\nu$. Let $\psi = \bigwedge_{\nu\in M} \varphi'_\nu$. Define $e \in {}^\omega\text{Dmn } H_\beta$ as follows, where $z \in \text{Dmn } H_\beta$ is fixed: for each $j \in \omega$,

$$e_j = c_\nu u_\nu^{-1} j \qquad \text{if } j \in K_\nu \text{ and } \nu \in M,$$
$$e_j = z \qquad \text{otherwise.}$$

Now the value assigned to $u\nu j$ under $\langle e_i\delta : i < \omega\rangle^0_x$ is $c_{\nu j}\delta$, for any $j \in J_\nu$, so for any $x \in A$, $\delta < \mathfrak{n}$, and $\nu \in M$ we have

$$\mathfrak{A} \models \varphi_\nu[\langle c_{\nu i}\delta : i < \omega\rangle^0_x] \quad \text{iff} \quad \mathfrak{A} \models \varphi'_\nu[\langle e_i\delta : i < \omega\rangle^0_x].$$

Hence it is clear that $\mathfrak{A} \models \psi[\langle e_i\delta : i < \omega\rangle^0_x]$ iff for all $\nu \in M$, $\mathfrak{A} \models \varphi_\nu[\langle c_{\nu i}\delta : i < \omega\rangle^0_x]$. Now by (8) for any $\nu \in M$ we have $\{\delta < \mathfrak{n} : \mathfrak{A} \models \varphi_\nu[\langle c_{\nu i}\delta : i < \omega\rangle^0_{a\alpha\delta}]\} \in D'$, so $\{\delta < \mathfrak{n} : \mathfrak{A} \models \psi[\langle e_i\delta : i < \omega\rangle^0_{a\alpha\delta}]\} \in D'$. Hence $(\psi, e) \in \Theta$, say $(\psi, e) = (\varphi_\sigma, c_\sigma)$, where $\sigma < \mathfrak{n} + |\beta|$. Now for each $\nu \in M$ and $\delta < \mathfrak{m}$, the value assigned to $u\nu j$ by $\langle d_{c\sigma i}\delta : i < \omega\rangle$ is $d_{c\sigma u\nu j}\delta = d_{eu\nu j}\delta = d_{c\nu j}\delta$, which is the same value the sequence $\langle d_{c\nu i}\delta : i < \omega\rangle$ assigns to j, where $j \in K_\nu$. Hence for any $y \in B$ we have

$$\mathfrak{B} \models \varphi_\sigma[\langle d_{c\sigma i}\delta : i < \omega\rangle^0_y] \quad \text{iff } \forall\nu \in M$$
$$\mathfrak{B} \models \varphi'_\nu[\langle d_{c\sigma i}\delta : i < \omega\rangle^0_y] \quad \text{iff } \forall\nu \in M\ \mathfrak{B} \models \varphi_\nu[\langle d_{c\nu i}\delta : i < \omega\rangle^0_y],$$

which is as desired in 26.41(i). We now apply 26.41 to obtain $b \in {}^\mathfrak{n}B$, $F_\gamma \subseteq F'$ and $D_\gamma \supseteq D'$ so that $|F' \sim F_\gamma| \leq \mathfrak{n} + |\gamma|$, $(F_\gamma, 0, D_\gamma)$ is $(\mathfrak{n} + |\gamma|)$-consistent over $\mathfrak{n}$, and for every $\nu < \mathfrak{n} + |\beta|$ we have

(9) $\{\delta < \mathfrak{n} : \mathfrak{B} \models \varphi_\nu[\langle d_{c\nu i}\delta : i < \omega\rangle^0_{b\delta}]\} \in D_\gamma$.

Let $H_\gamma = H_\beta \cup \{(a_\alpha, b)\}$. Clearly now (1), (2), and (5) hold, while (6) and (7) are not relevant to this case. To check (4), suppose $x \in {}^\omega\text{Dmn } H_\gamma$ and θ is any formula. We may assume that there is a $y \in \text{Dmn } H_\gamma$ such that $x_i = y$ for all but finitely many $i \in \omega$. Thus by our choice of D', we have either $\Gamma_{\theta x} \in D' \subseteq D_\gamma$ or $\mathfrak{n} \sim \Gamma_{\theta x} \in D' \subseteq D_\gamma$, as desired.

To check (3), suppose that $x_iH_\gamma y_i$ for each $i < \omega$, that θ is any formula, and that

10 $\{\delta < \mathfrak{n} : \mathfrak{A} \models \theta[\langle x_i\delta : i < \omega\rangle]\} \in D_\gamma$.

We may assume that $a_\alpha \notin \mathrm{Dmn}\, H_\beta$, and hence that $y_i = b$ whenever $x_i = a_\alpha$. To apply the result of our construction, we need to reformulate things so that only the variable v_0 has the new value $a_\alpha\delta$ assigned to it. To this end, let θ' be obtained from θ by replacing each free occurrence of v_0 by a new variable v_s, and changing all bound occurrences of v_0 to some still newer variable. Let $x' = x^s_{x0}$ and $y' = y^s_{y0}$. Thus still $x'_iH_\gamma y'_i$ for each $i < \omega$, and by (10),

(11) $\{\delta < \mathfrak{n} : \mathfrak{A} \models \theta'[\langle x'_i\delta : i < \omega\rangle]\} \in D_\gamma$.

Let $X = \{i : x'_i = a_\alpha,\ v_i \text{ occurs in } \theta'\}$, and let ψ be obtained from θ' by replacing v_i by v_0 for each $i \in X$. Choose w so that $a_0H_\beta w$, and define for each $i \in \omega$

$$\begin{aligned} z_i &= x'_i && \text{if } x'_i \neq a_\alpha \text{ and } v_i \text{ occurs in } \theta', \\ z_i &= a_0 && \text{otherwise}, \\ u_i &= y'_i && \text{if } x'_i \neq a_\alpha \text{ and } v_i \text{ occurs in } \theta', \\ u_i &= w && \text{otherwise}. \end{aligned}$$

Thus $z_iH_\beta u_i$ for all $i < \omega$. Moreover, $z_i = a_0 \in \mathrm{Dmn}\, H_\beta$ for all but finitely many $i \in \omega$. Furthermore, for any $\delta < \mathfrak{n}$ we have $\mathfrak{A} \models \theta'[\langle x'_i\delta : i < \omega\rangle]$ iff $\mathfrak{A} \models \psi[\langle z_i\delta : i < \omega\rangle^0_{a\alpha\delta}]$. Hence from (11) we obtain $E \in D_\gamma$, where $E = \{\delta \in \mathfrak{n} : \mathfrak{A} \models \psi[\langle z_i\delta : i < \omega\rangle^0_{a\alpha\delta}]\}$. By our choice of D' it is then clear that $E \in D'$. Thus $(\psi, z) \in \Theta$, say $(\psi, z) = (\varphi_\nu, c_\nu)$. Thus (9) holds for this ν. Now for all $i < \omega$ we have $z_iH_\beta u_i$ and $z_iH_\beta d_{c\nu i}$, and $\{\delta < \mathfrak{n} : \mathfrak{A} \models (v_0 \equiv v_1)[z_i\delta, z_i\delta]\} = \mathfrak{n} \in D_\beta$, so by (3) for β, $\{\delta < \mathfrak{n} : \mathfrak{B} \models (v_0 \equiv v_1)[u_i\delta, d_{c\nu i}\delta]\} \in D_\beta$. Thus for any $i \in \omega$,

$$\{\delta < \mathfrak{n} : u_i\delta = d_{c\nu i}\delta\} \in D_\gamma.$$

Now if we intersect all of these sets where v_i occurs in ψ with (9) we easily obtain

$$\{\delta < \mathfrak{n} : \mathfrak{B} \models \psi[\langle u_i\delta : i < \omega\rangle^0_{b\delta}]\} \in D_\gamma.$$

It follows easily that

$$\{\delta < \mathfrak{n} : \mathfrak{B} \models \theta[\langle y_i\delta : i < \omega\rangle]\} \in D_\gamma,$$

as desired.

Case 4. $\gamma = \beta + 1$, $R_\beta = (1, b_\alpha)$. Like Case 3, with A and B interchanged. Subcase 1 is unnecessary.

Case 5. $\gamma = \beta + 1$, $R_\beta(2, \Delta)$ where $\Delta \subseteq \mathfrak{n}$. By Lemma 26.39 choose $F_\gamma \subseteq F_\beta$ and $D_\gamma \supseteq D_\beta$ so that $|F_\beta \sim F_\gamma| < \mathfrak{n}^\partial$, $D_\gamma = \mathscr{F}\!g(D_\beta \cup \{\Delta\})$ or $D_\gamma = \mathscr{F}\!g(D_\beta \cup \{\mathfrak{n} \sim \Delta\})$, and $(F_\gamma, 0, D_\gamma)$ is $(\mathfrak{n} + |\gamma|)$-consistent over $\mathfrak{n}$. Let $H_\gamma = H_\beta$. All of the desired conditions are clear except (3). Assume that $c_iH_\gamma d_i$ for all $i < \omega$, that φ is any formula, and that $\{\delta < \mathfrak{n} : \mathfrak{A} \models \varphi[\langle c_i\delta : i < \omega\rangle]\} \in D_\gamma$.

By (4) for β it is clear that $\{\delta < \mathfrak{n} : \mathfrak{A} \models \varphi[\langle c_i\delta : i < \omega\rangle]\} \in D_\beta$. Hence by (3) for β, $\{\delta < \mathfrak{n} : \mathfrak{B} \models \varphi[\langle d_i\delta : i < \omega\rangle]\} \in D_\beta \subseteq D_\gamma$, as desired.

This completes the inductive definition. Let $\mathfrak{p} = 2^{\mathfrak{n}}$ and let $E = D_{\mathfrak{p}}$. By (1) and (7), E is an ultrafilter over $\mathfrak{n}$. Let $K = \{([a]_E, [b]_E) : aH_{\mathfrak{p}}b\}$. By (5) and (6), K has domain ${}^{\mathfrak{n}}A/E$ and range ${}^{\mathfrak{n}}B/E$. If $aH_{\mathfrak{p}}b$ and $a'H_{\mathfrak{p}}b'$, then $aH_\gamma b$ and $a'H_\gamma b'$ for some $\gamma < \mathfrak{p}$, so

$$\begin{aligned}
[a]_E = [a']_E \quad & \text{iff } \{\delta < \mathfrak{n} : a_\delta = a'_\delta\} \in E \\
& \text{iff } \{\delta < \mathfrak{n} : \mathfrak{A} \models (v_0 = v_1)[a_\delta, a'_\delta]\} \in E \\
& \text{iff } \{\delta < \mathfrak{n} : \mathfrak{A} \models (v_0 = v_1)[a_\delta, a'_\delta]\} \in D_\gamma \quad \text{by (4)} \\
& \text{iff } \{\delta < \mathfrak{n} : \mathfrak{B} \models (v_0 = v_1)[b_\delta, b'_\delta]\} \in D_\gamma \quad \text{by (3)} \\
& \text{iff } [b]_E = [b']_E \text{ similarly.}
\end{aligned}$$

Thus K is a one-one function. In an exactly analogous way it is shown that K preserves relations and operations. For example, if $\mathbf{R}$ is an m-ary relation symbol, $a_0, \ldots, a_{m-1} \in {}^{\mathfrak{n}}A$, and $a_iH_{\mathfrak{p}}b_i$ for all $i < m$, then $a_iH_\gamma b_i$ for some $\gamma < \mathfrak{p}$, and hence, with φ the formula $\mathbf{R}v_0 \cdots v_{n-1}$, $\mathfrak{C} = {}^{\mathfrak{n}}\mathfrak{A}/E$, $\mathfrak{D} = {}^{\mathfrak{n}}\mathfrak{B}/E$,

$$\begin{aligned}
\langle [a_0], \ldots, [a_{m-1}]\rangle \in \mathbf{R}^{\mathfrak{C}} \quad & \text{iff } \mathfrak{C} \models \varphi[[a_0], \ldots, [a_{m-1}]] \\
& \text{iff } \{\delta < \mathfrak{n} : \mathfrak{A} \models \varphi[a_0\delta, \ldots, a_{m-1}\delta]\} \in E \\
& \text{iff } \{\delta < \mathfrak{n} : \mathfrak{A} \models \varphi[a_0\delta, \ldots, a_{m-1}\delta]\} \in D_\gamma \quad \text{by (4)} \\
& \text{iff } \{\delta < \mathfrak{n} : \mathfrak{B} \models \varphi[b_0\delta, \ldots, b_{m-1}\delta]\} \in D_\gamma \quad \text{by (3)} \\
& \text{iff } \langle [b_0], \ldots, [b_{m-1}]\rangle \in \mathbf{R}^{\mathfrak{D}} \text{ similarly.}
\end{aligned}$$

□

As mentioned at the beginning of this section, a characterization as in 26.42 immediately gives a characterization of the notion of elementary class:

Theorem 26.43 (Keisler, Shelah). *The following conditions are equivalent:*

(*i*) $\mathbf{K}$ *is an elementary class;*
(*ii*) $\mathbf{UpK} = \mathbf{K}$ *and if* ${}^I\mathfrak{A}/F \in \mathbf{K}$, *then* $\mathfrak{A} \in \mathbf{K}$.

Proof. The implication (*i*) ⇒ (*ii*) is obvious. Now assume (*ii*). By 26.2 it suffices to show that $\mathbf{K}$ is elementarily closed. Assume, then, that $\mathfrak{A} \equiv_{\mathrm{ee}} \mathfrak{B} \in \mathbf{K}$. By Theorem 26.42 there exist I and F with ${}^I\mathfrak{A}/\mathfrak{F} \cong {}^I\mathfrak{B}/\mathfrak{F}$. Since $\mathbf{UpK} = \mathbf{K}$ and $\mathfrak{B} \in \mathbf{K}$, it follows that ${}^I\mathfrak{A}/F \in \mathbf{K}$. Hence by (*ii*), $\mathfrak{A} \in \mathbf{K}$, as desired. □

Thus we may say that a class $\mathbf{K}$ is an elementary class iff it is closed under ultraproducts and ultra*roots*.

EXERCISES

26.44. If $\mathbf{K}$ is compact and $\mathbf{L} = \{\mathfrak{A} : \text{for some } \mathfrak{B}, \mathfrak{A} \equiv_{\mathrm{ee}} \mathfrak{B} \in \mathbf{K}\}$, then $\mathbf{L}$ is elementary.

26.45. For any class $\mathbf{K}$, let $\mathbf{K}^{\preccurlyeq} = \{\mathfrak{A} : \text{for some } \mathfrak{B}, \mathfrak{A} \preccurlyeq \mathfrak{B} \in \mathbf{K}\}$. The following conditions are equivalent:
(*i*) $\mathbf{K}$ is an elementary class;
(*ii*) $\mathbf{UpK}^{\preccurlyeq} \subseteq \mathbf{K}$.

26.46. $\mathfrak{A} \equiv_{ee} \mathfrak{B}$ iff $\mathfrak{A}$ can be elementarily embedded in some ultrapower of $\mathfrak{B}$. Prove directly, without using 26.42.

26.47. The converse of 26.5 fails in general. *Hint:* consider $\mathfrak{A} = \langle A, R_i \rangle_{i \in \omega}$ and $\mathfrak{B} = \langle B, S_i \rangle_{i \in \omega}$, with $A = B = \omega$, $R_0 = S_0 = \omega$, $R_{i+1} = R_i \sim \{i + 1\}$, $S_{i+1} = S_i \sim \{i\}$.

26.48. Use 26.14 to show that the following two structures are elementarily equivalent: $\mathfrak{A} = (\omega, <)$, $\mathfrak{B} = (\omega + \omega^* + \omega, <)$ (see 20.2 for notation).

26.49. Show that the conditions (1)–(5) of 26.19(v) are independent of each other.

26.50. $\Theta\rho\mathbf{K} \cup \Theta\rho\mathbf{L}$ is consistent iff $\mathbf{UpK} \cap \mathbf{UpL} \neq 0$.

26.51. $\mathbf{K}$ is closed under elementary equivalence iff both $\mathbf{K}$ and $\mathbf{S}_{\mathscr{L}} \sim \mathbf{K}$ are closed under ultrapowers and isomorphisms.

26.52. $\mathbf{K}$ is a finitely axiomatizable elementary class iff both $\mathbf{K}$ and $\mathbf{S}_{\mathscr{L}} \sim \mathbf{K}$ are closed under ultraproducts.

26.53. If L is a general logic in which every $L, \mathscr{L}$-elementary class is closed under ultraproducts, then $L \subseteq L_{\text{fo}}$.

26.54. If $\mathbf{K} = \mathbf{UpK}$, $\mathbf{L} = \mathbf{UpL}$, and $\mathbf{K} \cap \mathbf{L} = 0$, then there is a finitely axiomatizable elementary class $\mathbf{M}$ such that $\mathbf{K} \subseteq \mathbf{M}$ and $\mathbf{L} \cap \mathbf{M} = 0$.

26.55. Derive Craig's interpolation theorem from the result of Exercise 26.54.

Types 27

So far in this part we have spoken about satisfaction of *sentences* and methods for finding models for sets of *sentences*. We now want to consider similar general questions concerning sets of *formulas*. Thus instead of considering global questions, about classes of structures, we shall be considering local questions, about sets of elements, or sets of finite sequences of elements. Generally speaking, we shall be concerned with conditions on a set Δ of formulas for there to exist a structure $\mathfrak{A}$ and an $x \in {}^{\omega}A$ such that $x \in \varphi^{\mathfrak{A}}$ for all $\varphi \in \Delta$. The basic definitions with which we shall be working in this chapter are given in

Definition 27.1.

(*i*) For any $n \in \omega$, $\mathrm{Fmla}^n_{\mathscr{L}}$ is the set of all formulas φ of $\mathscr{L}$ such that $\mathrm{Fv}\,\varphi \subseteq \{v_0, \ldots, v_{n-1}\}$.

(*ii*) If $\Delta \subseteq \mathrm{Fmla}^n_{\mathscr{L}}$, $\mathfrak{A}$ is an $\mathscr{L}$-structure, and $a \in {}^nA$, then we say that a *realizes* Δ *in* $\mathfrak{A}$ provided that $\mathfrak{A} \models \varphi[a]$ for all $\varphi \in \Delta$.

(*iii*) If $\Delta \subseteq \mathrm{Fmla}^n_{\mathscr{L}}$ and $\mathfrak{A}$ is an $\mathscr{L}$-structure, then we say that $\mathfrak{A}$ *realizes* Δ provided that some $a \in {}^nA$ realizes Δ in $\mathfrak{A}$. If no $a \in {}^nA$ realizes Δ, then we say that $\mathfrak{A}$ *omits* Δ.

(*iv*) If $\mathfrak{A}$ is an $\mathscr{L}$-structure and $a \in {}^nA$, then the *n-type* of a in $\mathfrak{A}$ is the set $\{\varphi \in \mathrm{Fmla}^n_{\mathscr{L}} : \mathfrak{A} \models \varphi[a]\}$. A set $\Delta \subseteq \mathrm{Fmla}^n_{\mathscr{L}}$ is an *n-type* if it is the n-type of a in $\mathfrak{A}$ for some $a \in {}^nA$ and some $\mathscr{L}$-structure $\mathfrak{A}$.

(*v*) Let $(\Gamma, \mathscr{L})$ be a theory, and let $\Delta \subseteq \mathrm{Fmla}^n_{\mathscr{L}}$. We say that Δ is *consistent over* Γ provided that for every finite $\Delta' \subseteq \Delta$, $\Gamma \cup \{\exists v_0 \cdots \exists v_{n-1} \bigwedge \Delta'\}$ is consistent. We say that Δ is an *n-type over* Γ provided that it is the n-type of a in $\mathfrak{A}$ for some a and some model $\mathfrak{A}$ of Γ.

Note that if $\Delta \subseteq \mathrm{Fmla}^n_{\mathscr{L}}$ is realized by a in $\mathfrak{A}$, then Δ is contained in an

n-type. It is clear that a set $\Delta \subseteq \mathrm{Fmla}^n_{\mathscr{L}}$ is consistent over a theory Γ iff every finite subset of Δ is realized in a model of Γ. The possibility of omitting sets of formulas can be illustrated by the structure $\mathfrak{A} = (\omega, 0, \mathscr{s}, <)$. Consider the set Δ of all formulas $\mathbf{\Delta} n < v_0$ for $n \in \omega$ (recall the definition of $\mathbf{\Delta}$ from 14.1). Obviously Δ is omitted in $\mathfrak{A}$. Some natural questions which will occupy us for much of this and the next chapter are: does a theory Γ always have a model which omits many sets; or omits few sets; does Γ have a model with many sequences which admit a given set?

The following equivalent formulations of n-types are frequently useful.

Proposition 27.2. *Let* $(\Gamma, \mathscr{L})$ *be a theory and let* $\Delta \subseteq \mathrm{Fmla}_{\mathscr{L}}$. *The following conditions are equivalent:*

(*i*) Δ *is an n-type over* Γ;

(*ii*) Δ *is maximal consistent over* Γ;

(*iii*) *there is a model* $\mathfrak{A}$ *of* Γ *with* $|A| \leq |\mathrm{Fmla}_{\mathscr{L}}|$ *and an* $a \in {}^nA$ *such that* Δ *is the n-type of a in* $\mathfrak{A}$.

PROOF. Obviously (*i*) $\Rightarrow$ (*ii*) and (*iii*) $\Rightarrow$ (*i*). Now assume (*ii*). Let us expand the language $\mathscr{L}$ to $\mathscr{L}'$ by adjoining new individual constants $\mathbf{c}_0, \ldots, \mathbf{c}_{n-1}$. Let

$$\Delta' = \{\varphi(\mathbf{c}_0, \ldots, \mathbf{c}_{n-1}) : \varphi \in \Delta\}.$$

Our assumption that Δ is consistent over Γ clearly implies that every finite subset of $\Delta' \cup \Gamma$ is consistent. But $\Delta' \cup \Gamma$ is a set of *sentences*, so we can apply the compactness theorem to obtain a model $(\mathfrak{A}, a_0, \ldots, a_{n-1})$ of $\Delta' \cup \Gamma$ with $|A| \leq |\mathrm{Fmla}_{\mathscr{L}}|$. Clearly Δ is contained in the n-type of a over $\mathfrak{A}$. Since Δ is maximal consistent over Γ, it must actually equal this n-type. □

By applying Zorn's lemma we obtain a simple criterion for realizability:

Corollary 27.3. *Let* $(\Gamma, \mathscr{L})$ *be a theory and* $\Delta \subseteq \mathrm{Fmla}^n_{\mathscr{L}}$. *Then the following conditions are equivalent:*

(*i*) Δ *is realizable in a model of* Γ;

(*ii*) Δ *is consistent over* Γ.

It is more difficult to give a simple criterion for omitting types. The following general theorem turns out to be very useful.

Theorem 27.4 [Omitting types theorem (Henkin, Orey)]. *Let* Γ *be a consistent theory in a countable language* $\mathscr{L}$. *Suppose that* N *is a non-empty subset of* ω *and that for each* $n \in N$ *a set* $\Delta_n \subseteq \mathrm{Fmla}^n_{\mathscr{L}}$ *is given. Then the following conditions are equivalent:*

(*i*) *there is a countable model* $\mathfrak{A}$ *of* Γ *which omits each* Δ_n;

(*ii*) *there is a consistent theory* $\Gamma' \supseteq \Gamma$ *which satisfies the following condition: for any* $n \in N$ *and any* $\psi \in \mathrm{Fmla}^n_{\mathscr{L}}$, *if* $\Gamma' \cup \{\exists v_0 \cdots \exists v_{n-1}\psi\}$ *has a model, then there is a* $\varphi \in \Delta_n$ *such that* $\Gamma' \cup \{\exists v_0 \cdots \exists v_{n-1}(\psi \wedge \neg\varphi)\}$ *has a model.*

PROOF

(*i*) ⇒ (*ii*). Let $\mathfrak{A}$ be a model of Γ which omits each Δ_n. Set $\Gamma' = \Theta_{\rho}\mathfrak{A}$. To verify (*ii*), suppose $n \in N$, $\psi \in \text{Fmla}^n_{\mathscr{L}}$, and $\Gamma' \cup \{\exists v_0 \cdots \exists v_{n-1}\psi\}$ has a model. Since Γ' is complete, it follows that $\exists v_0 \cdots \exists v_{n-1}\psi$ holds in $\mathfrak{A}$. Hence there is an $a \in {}^nA$ such that $\mathfrak{A} \models \psi[a]$. Since $\mathfrak{A}$ omits Δ_n, we also have $\mathfrak{A} \models \neg\varphi[a]$ for some $\varphi \in \Delta_n$. Hence $\mathfrak{A} \models \exists v_0 \cdots \exists v_{n-1}(\psi \wedge \neg\varphi)$, as desired.

(*ii*) ⇒ (*i*). We shall apply the model existence theorem 18.9, in the form of Corollary 18.11. Let $\mathscr{L}'$ be an expansion of $\mathscr{L}$ rich by C. Let S consist of all finite $\Theta \subseteq \text{Sent}_{\mathscr{L}'}$ such that $\Gamma' \cup \Theta$ has a model. Now S satisfies (C0)–(C9). To check this, only (C6) and (C9) require any thought. For (C6), assume that $\exists\alpha\varphi \in \Theta \in S$. Choose $\mathbf{c} \in C$ so that it does not occur in any sentence of Θ. Obviously then any model of Θ can be modified to become a model of $\Theta \cup \{\text{Subf}^{\alpha}_{\mathbf{c}}\varphi\}$. Hence clearly $\Theta \cup \{\text{Subf}^{\alpha}_{\mathbf{c}}\varphi\} \in S$. Condition (C9) is similarly checked. So, S satisfies (C0)–(C9).

It also has the following property:

(1) if $\Theta \in S$, $n \in N$, and $\mathbf{c} \in {}^nC$, then there is a $\varphi \in \Delta_n$ such that $\Theta \cup \{\neg\varphi(\mathbf{c}_0, \ldots, \mathbf{c}_{n-1})\} \in S$.

For, let $\mathbf{d}_0, \ldots, \mathbf{d}_{m-1}$ be all of the members of $C \sim \{\mathbf{c}_0, \ldots, \mathbf{c}_{n-1}\}$ which occur in $\bigwedge \Theta$. Let ψ be obtained from $\bigwedge \Theta$ by first changing all bound variables of $\bigwedge \Theta$ to variables v_i with $i \geq m + n$, and then replacing $\mathbf{c}_0, \ldots, \mathbf{c}_{n-1}, \mathbf{d}_0, \ldots, \mathbf{d}_{m-1}$ respectively by $v_0, \ldots, v_{n-1}, v_n, \ldots, v_{n+m-1}$. Since $\Gamma' \cup \Theta$ has a model, clearly $\Gamma' \cup \{\exists v_0 \cdots \exists v_{n+m-1}\psi\}$ has a model. Therefore, by (*ii*), there is a $\varphi \subset \Delta_n$ such that the set $\Gamma' \cup \{\exists v_0 \cdots \exists v_{n-1}(\exists v_n \cdots \exists v_{n+m-1}\psi \wedge \neg\varphi)\}$ has a model, call it $\mathfrak{A}$. Choose $a \in {}^{n+m}A$ such that $\mathfrak{A} \models \neg\varphi[a_0, \ldots, a_{n-1}]$ and $\mathfrak{A} \models \psi[a_0, \ldots, a_{n+m-1}]$. Now we expand $\mathfrak{A}$ to an $\mathscr{L}'$-structure $\mathfrak{A}'$ by assigning $a_0, \ldots, a_{n-1}$ to $\mathbf{c}_0, \ldots, \mathbf{c}_{n-1}$, $a_n, \ldots, a_{n+m-1}$ to $\mathbf{d}_0, \ldots, \mathbf{d}_{m-1}$, and extending on in any fashion. Clearly $\mathfrak{A}'$ is a model of $\Gamma' \cup \Theta \cup \{\neg\varphi(\mathbf{c}_0, \ldots, \mathbf{c}_{n-1})\}$, so $\Theta \cup \{\neg\varphi(\mathbf{c}_0, \ldots, \mathbf{c}_{n-1})\} \in S$, as desired in (1).

Now the set $\{(n, \mathbf{c}) : n \in N \text{ and } \mathbf{c} \in {}^nC\}$ has power $\aleph_0$, so we can enumerate it in a simple infinite sequence $\langle(n_i, \mathbf{c}_i) : i < \omega\rangle$. For each $i < \omega$ we define a function $f_i\colon S \to S$: for each $\Theta \in S$,

$$f_i\Theta = \Theta \cup \{\neg\varphi(\mathbf{c}_{i0}, \ldots, \mathbf{c}_{i,ni-1})\},$$

where φ is minimum (in some fixed well-ordering of $\text{Fmla}_{\mathscr{L}}$) such that $\Theta \cup \{\neg\varphi(\mathbf{c}_{i0}, \ldots, \mathbf{c}_{i,ni-1})\} \in S$ and $\varphi \in \Delta_n$; this is possible by (1). Clearly each function f_i is admissible over S. Clearly $\Gamma \in S$, so by 18.11 we obtain a countable model $(\mathfrak{A}, a_{\mathbf{c}})_{\mathbf{c}\in C}$ of Γ such that $A = \{a_{\mathbf{c}} : \mathbf{c} \in C\}$ and for each $i < \omega$, there is a $\Phi \in S$ with $\Gamma \subseteq f_i\Phi \subseteq \{\varphi : (\mathfrak{A}, a_{\mathbf{c}})_{\mathbf{c}\in C} \models \varphi\}$. To see that $\mathfrak{A}$ omits each Δ_n with $n \in N$, take $n \in N$ and $x \in {}^nA$. Say $x = a \circ \mathbf{d}$ with $\mathbf{d} \in {}^nC$. Choose $i \in \omega$ with $(n, \mathbf{d}) = (n_i, \mathbf{c}_i)$, and choose $\Phi \in S$ with $\Gamma \subseteq f_i\Phi \subseteq \{\varphi : (\mathfrak{A}, a_{\mathbf{c}})_{\mathbf{c}\in C} \models \varphi\}$. By the definition of f_i it follows that there is a $\varphi \in \Delta_n$ such that $(\mathfrak{A}, a_{\mathbf{c}})_{\mathbf{c}\in C} \models \neg\varphi(\mathbf{d}_0, \ldots, \mathbf{d}_{n-1})$, which shows that $\mathfrak{A}$ omits Δ_n. □

We shall derive some consequences of the omitting types theorem which are most easily motivated using terminology of Boolean algebras, although no results of Chapter 9 will be used.

Definition 27.5. Let Γ be a theory and $m \in \omega$.

(*i*) We define $\approx_m^\Gamma = \{(\varphi, \psi) : \varphi, \psi \in \mathrm{Fmla}_{\mathscr{L}}^m \text{ and } \Gamma \models \varphi \leftrightarrow \psi\}$.

(*ii*) A formula $\varphi \in \mathrm{Fmla}_{\mathscr{L}}^m$ is *m-atomic over* Γ provided that $\Gamma \cup \{\exists v_0 \cdots \exists v_{m-1}\varphi\}$ has a model and for any $\psi \in \mathrm{Fmla}_{\mathscr{L}}^m$, either $\Gamma \models \varphi \rightarrow \psi$ or $\Gamma \models \varphi \rightarrow \neg\psi$.

(*iii*) A formula $\varphi \in \mathrm{Fmla}_{\mathscr{L}}^m$ is *m-atomless over* Γ provided that $\Gamma \cup \{\exists v_0 \cdots \exists v_{m-1}\varphi\}$ has a model but there is no formula $\psi \in \mathrm{Fmla}_{\mathscr{L}}^m$ m-atomic over Γ such that $\Gamma \models \psi \rightarrow \varphi$.

Using the procedure of 9.56, $\mathrm{Fmla}_{\mathscr{L}}^m/\approx_m^\Gamma$ can be made into a Boolean algebra $\mathfrak{A}_m^\Gamma$. A formula $\varphi \in \mathrm{Fmla}_{\mathscr{L}}^m$ is m-atomic over Γ iff $[\varphi]$ is an atom of $\mathfrak{A}_m^\Gamma$, and it is m-atomless over Γ iff $[\varphi]$ is an atomless element of $\mathfrak{A}_m^\Gamma$. Let $m \in \omega$ and let $\mathscr{L}'$ be an expansion of $\mathscr{L}$ by adjoining m new individual constants $\mathbf{c}_0, \ldots, \mathbf{c}_{m-1}$. Then a formula $\varphi \in \mathrm{Fmla}_{\mathscr{L}}^m$ is m-atomic over Γ iff $\Gamma \cup \{\varphi(\mathbf{c}_0, \ldots, \mathbf{c}_{m-1})\}$ is a consistent complete set, while φ is m-atomless over Γ iff $\Gamma \cup \{\varphi(\mathbf{c}_0, \ldots, \mathbf{c}_{m-1})\}$ is consistent but is not contained in any finitely axiomatizable complete consistent extension of Γ.

To illustrate these notions by concrete examples, first take $\Gamma = \Theta\rho(\omega, 0, \mathfrak{s}, <)$. Since this is a complete theory, it is clear that $v_0 = \mathbf{0}$ is 1-atomic over Γ. On the other hand, consider the theory P of Peano arithmetic. For each $m \in \omega$, there are *no* formulas m-atomic over P, for if ψ were m-atomic over P, clearly $\mathrm{P} \cup \{\exists v_0 \cdots \exists v_{m-1}\psi\}$ would be complete and axiomatizable, contradicting 16.2. Thus every $\psi \in \mathrm{Fmla}_{\mathscr{L}(\mathrm{nos})}^m$ is m-atomless over P.

The following theorem is a weaker kind of omitting types theorem.

Theorem 27.6 (Ehrenfeucht). *Let Γ be a consistent theory in a countable language $\mathscr{L}$, and suppose $0 \neq N \subseteq \omega$. For each $n \in N$ let Δ_n be an n-type over Γ, and assume that no set Δ_n contains a formula n-atomic over Γ. Then Γ has a countable model that omits each type Δ_n.*

Proof. We shall apply 27.4. To this end, assume that $n \in N$, $\psi \in \mathrm{Fmla}_{\mathscr{L}}^n$, and $\Gamma \cup \{\exists v_0 \cdots \exists v_{n-1}\psi\}$ has a model. If $\psi \notin \Delta_n$, then $\neg\psi \in \Delta_n$ since Δ_n is an n-type. So in this case $\Gamma \cup \{\exists v_0 \cdots \exists v_{n-1}(\psi \wedge \neg\neg\psi)\}$ has a model, as desired in 27.4(*ii*). Assume, on the other hand, that $\psi \in \Delta_n$. Now by hypothesis ψ is not n-atomic over Γ, so there is a $\varphi \in \mathrm{Fmla}_{\mathscr{L}}^n$ such that not $(\Gamma \models \psi \rightarrow \varphi)$ and not $(\Gamma \models \psi \rightarrow \neg\varphi)$. Since Δ_n is an n-type, either $\varphi \in \Delta_n$ or $\neg\varphi \in \Delta_n$, say $\varphi \in \Delta_n$. From not $(\Gamma \models \psi \rightarrow \varphi)$ it follows that there is a model $\mathfrak{A}$ of Γ such that $\mathfrak{A} \models \exists v_0 \cdots \exists v_{n-1}(\psi \wedge \neg\varphi)$, as desired in 27.4(*ii*). □

We can apply Ehrenfeucht's theorem to the theory P to obtain further incompleteness properties of it. For example, if Δ is any 1-type there is a countable model of P which omits Δ. Thus the infinitely long formula

$\exists v_0 \bigwedge \Delta$ is not a consequence of P. Another example: let $\mathfrak{A}$ be any denumerable model of P, say $A = \{a_n : n \in \omega\}$. For each $n \in \omega \sim 1$ let $\Delta_n = \{\varphi \in \mathrm{Fmla}^n_{\mathscr{L}} : \mathfrak{A} \models \varphi[a_n, \ldots, a_n, \ldots]\}$. Clearly Δ_n is an n-type over P. Let $\mathfrak{B}$ be a countable model of P omitting each type Δ_n. Then each element of B has type different from each element of A, i.e., first-order distinguishable from each element of A.

Using the omitting types theorem 27.4 we can prove the following useful theorem of Vaught.

Theorem 27.7 (Vaught). *Let Γ be a consistent theory in a countable language. Then Γ has a countable model $\mathfrak{A}$ such that for every $n \in \omega$, every n-tuple of elements of A satisfies either an n-atomic formula over Γ or an n-atomless formula over Γ.*

PROOF. For each $n \in \omega$ let

$$\Delta_n = \{\varphi \in \mathrm{Fmla}^n_{\mathscr{L}} : \neg\varphi \text{ is } n\text{-atomic or } n\text{-atomless over } \Gamma\}.$$

To check the hypothesis of 27.4, assume that $n \in \omega$, $\psi \in \mathrm{Fmla}^n_{\mathscr{L}}$, and that $\Gamma \cup \{\exists v_0 \cdots \exists v_{n-1}\psi\}$ has a model. If ψ is n-atomless over Γ, then $\neg\psi \in \Delta_n$ and the desired conclusion is obvious. Otherwise there is a formula φ n-atomic over Γ such that $\Gamma \models \varphi \rightarrow \psi$. Thus $\neg\varphi \in \Delta_n$, and any model of $\Gamma \cup \{\exists v_0 \cdots \exists v_{n-1}\varphi\}$ is a model of $\Gamma \cup \{\exists v_0 \cdots \exists v_{n-1}(\psi \wedge \neg\neg\varphi)\}$, as desired.

Thus 27.4 applies, and this obviously gives the desired model. □

We now apply these results to characterize the following important notion (cf. the notion of a *prime* model in 21.28):

Definition 27.8. A structure $\mathfrak{A}$ is an *elementary prime* model of Γ provided that $\mathfrak{A}$ is a model of Γ which can be elementarily embedded in each model of Γ.

Note that if Γ has an elementarily prime model, then Γ is automatically complete. If Γ is model complete, then prime and elementarily prime have the same meaning.

Theorem 27.9 (Vaught). *Let $\mathscr{L}$ be countable and Γ a complete theory in $\mathscr{L}$. Then for any $\mathscr{L}$-structure $\mathfrak{A}$ the following conditions are equivalent:*

(i) $\mathfrak{A}$ is an elementarily prime model of Γ;

(ii) $\mathfrak{A}$ is a countable model of Γ, and for every $n \in \omega$ and $a \in {}^nA$, a satisfies some n-atomic formula over Γ in $\mathfrak{A}$.

PROOF

(i) ⇒ *(ii)*. Let $\mathfrak{A}$ be an elementarily prime model of Γ. By the downward Lowenheim–Skolem theorem, Γ has a countable model $\mathfrak{B}$. Since $\mathfrak{A}$ can be elementarily embedded in $\mathfrak{B}$, $\mathfrak{A}$ is also countable. Now let $n \in \omega$ and $a \in {}^nA$. Set $\Delta_n = \{\varphi \in \mathrm{Fmla}^n_{\mathscr{L}} : \mathfrak{A} \models \varphi[a]\}$; thus Δ_n is an n-type over Γ. Now

(1) Δ_n contains an n-atomic formula over Γ.

For, if not, by 27.6 with $N = \{n\}$ there is a countable model $\mathfrak{B}$ of Γ which omits Δ_n. Let $f\colon \mathfrak{A} \rightarrowtail \mathfrak{B}$ be an elementary embedding. For any $\varphi \in \Delta_n$ we have $\mathfrak{A} \models \varphi[a]$ and hence $\mathfrak{B} \models \varphi[f \circ a]$. Thus $\mathfrak{B}$ admits Δ_n, contradiction. Hence (1) holds. Therefore a satisfies an n-atomic formula over Γ in $\mathfrak{A}$.

$(ii) \Rightarrow (i)$. Assume (ii). Let $A = \{a_i : i < \omega\}$, and let $\mathfrak{B}$ be any model of Γ. For each $n \in \omega$ let φ_n be an n-atomic formula over Γ such that $\mathfrak{A} \models \varphi_n[a_0, \ldots, a_{n-1}]$. We now define a sequence $\langle b_i : i < \omega \rangle$ of elements of B by recursion. Since $\mathfrak{A} \models \varphi_0$, $\mathfrak{A}$ is a model of Γ, and Γ is complete, we have $\Gamma \models \varphi_0$ and hence $\mathfrak{B} \models \varphi_0$. Suppose that $n \in \omega$, $b_0, \ldots, b_{n-1}$ have been defined, and $\mathfrak{B} \models \varphi_n[b_0, \ldots, b_{n-1}]$. Now $\mathfrak{A} \models \varphi_n[a_0, \ldots, a_{n-1}]$ and $\mathfrak{A} \models \varphi_{n+1}[a_0, \ldots, a_n]$, so $\mathfrak{A} \models (\varphi_n \wedge \exists v_n \varphi_{n+1})[a_0, \ldots, a_{n-1}]$. Since $\mathfrak{A}$ is a model of Γ, therefore it is not the case that $\Gamma \models \varphi_n \rightarrow \neg\exists v_n \varphi_{n+1}$. Since φ_n is n-atomic over Γ, it follows that $\Gamma \models \varphi_n \rightarrow \exists v_n \varphi_{n+1}$. Hence, since $\mathfrak{B}$ is a model of Γ and $\mathfrak{B} \models \varphi_n[b_0, \ldots, b_{n-1}]$, choose $b_n \in B$ such that $\mathfrak{B} \models \varphi_{n+1}[b_0, \ldots, b_n]$. This completes the definition of the sequence $\langle b_i \rangle_{i \in \omega}$.

Now let $f = \{(a_i, b_i) : i \in \omega\}$. We claim that f is the desired elementary embedding of $\mathfrak{A}$ into $\mathfrak{B}$. First we show that f is a function. For, suppose $a_i = a_j$, say $i < j$. Let ψ be the formula $v_i = v_j$. Thus $\mathfrak{A} \models \psi[a_0, \ldots, a_j]$, and $\psi \in \mathrm{Fmla}_{\mathscr{L}}^{j+1}$. Since also $\mathfrak{A} \models \varphi_{j+1}[a_0, \ldots, a_j]$, and φ_{j+1} is n-atomic over Γ, it follows easily that $\Gamma \models \varphi_{j+1} \rightarrow \psi$. Since $\mathfrak{B} \models \varphi_{j+1}[b_0, \ldots, b_j]$, hence $\mathfrak{B} \models \psi[b_0, \ldots, b_j]$, i.e., $b_i = b_j$. If we suppose $a_i \neq a_j$, take ψ to be $\neg(v_i = v_j)$ and apply the same argument, we see that $b_i \neq b_j$. Thus f is a one-one function mapping A into B.

To show that f is an elementary embedding, let ψ be any formula, let $x \in {}^\omega A$, and suppose that $\mathfrak{A} \models \psi[x]$. Say $x_i = a_{ji}$ for all $i < \omega$. Now for any formula χ, say with $\mathrm{Fv}\,\chi \subseteq \{v_0, \ldots, v_{n-1}\}$, and $\mathscr{L}$-structure $\mathfrak{C}$, and any $u, w \in {}^\omega C$ with $u_i = w_{ji}$ for all $i < \omega$ we have

$$(2) \qquad \mathfrak{C} \models \chi[u_0, \ldots, u_{n-1}] \qquad \text{iff } \mathfrak{C} \models \chi(v_{j0}, \ldots, v_{j(n-1)})[w_0, \ldots, w_m]$$

where $m = \max\{j_0, \ldots, j_{n-1}\}$. (2) is easily established by induction on χ. Now we return to our formula ψ. By (2), $\mathfrak{A} \models \psi(v_{j0}, \ldots, v_{j(n-1)})[a_0, \ldots, a_m]$. But also $\mathfrak{A} \models \varphi_{m+1}[a_0, \ldots, a_m]$ so, as above, $\Gamma \models \varphi_{m+1} \rightarrow \psi(v_{j0}, \ldots, v_{j(n-1)})$. Therefore, $\mathfrak{B} \models \psi(v_{j0}, \ldots, v_{j(n-1)})[b_0, \ldots, b_m]$. Hence, again by (2), $\mathfrak{B} \models \psi[fx_0, \ldots, fx_{n-1}]$, since $fx_i = fa_{ji} = b_{ji}$ for all $i < n$. □

Theorem 27.10 (Vaught). *Let Γ be a complete theory in a countable language $\mathscr{L}$. Then the following conditions are equivalent:*

(i) Γ has an elementarily prime model:

(ii) for every $n \in \omega$ there are no n-atomless formulas over Γ.

Proof

$(i) \Rightarrow (ii)$. Let Γ have an elementarily prime model $\mathfrak{A}$. Let $n \in \omega$, and let $\varphi \in \mathrm{Fmla}_{\mathscr{L}}^n$, with $\Gamma \cup \{\exists v_0 \cdots \exists v_{n-1}\varphi\}$ consistent. Thus $\mathfrak{A}$ is a model of $\exists v_0 \cdots \exists v_{n-1}\varphi$, so choose $a \in {}^nA$ such that $\mathfrak{A} \models \varphi[a_0, \ldots, a_{n-1}]$. By 27.9, there is a n-atomic formula θ over Γ such that $\mathfrak{A} \models \theta[a_0, \ldots, a_{n-1}]$. Thus

$\mathfrak{A} \models (\theta \wedge \varphi)[a_0, \ldots, a_{n-1}]$. Since θ is n-atomic over Γ, it follows easily that $\Gamma \models \theta \rightarrow \varphi$. Thus φ is not n-atomless, as desired.

$(ii) \Rightarrow (i)$. Assume (ii). Obviously 27.7 and 27.9 yield an elementarily prime model of Γ. □

Theorem 27.11 (Vaught). *Any two elementarily prime models of a theory* Γ *in a countable language* $\mathscr{L}$ *are isomorphic.*

Proof. We shall apply 26.20; so we need to construct an ω-sequence for the elementarily prime models $\mathfrak{A}$ and $\mathfrak{B}$ of Γ. For each $m \in \omega$, let I_m consist of all pairs $(x, y) \in {}^mA \times {}^mB$ such that there is an m-atomic formula φ over Γ such that both $\mathfrak{A} \models \varphi[x]$ and $\mathfrak{B} \models \varphi[y]$. Note that the 0-atomic formulas over Γ are just the φ with $\Gamma \models \varphi$, since Γ is complete. Thus $0I_00$. Now suppose that xI_my and $a \in A$. Say φ is m-atomic over Γ and $\mathfrak{A} \models \varphi[x]$, $\mathfrak{B} \models \varphi[y]$. By 27.9 choose an $(m + 1)$-atomic formula ψ over Γ such that $\mathfrak{A} \models \chi[x_0, \ldots, x_{m-1}, a]$. Hence, easily, $\Gamma \models \varphi \rightarrow \exists v_m\psi$, so $\mathfrak{B} \models \exists v_m\psi[y]$. Hence choose $b \in B$ so that $\mathfrak{B} \models \psi[y_0, \ldots, y_{m-1}, b]$. Thus we have $\langle x_0, \ldots, x_{m-1}, a\rangle I_{m+1}\langle y_0, \ldots, y_{m-1}, b\rangle$, as desired. The proof is similar with $\mathfrak{A}$ and $\mathfrak{B}$ interchanged. Finally, suppose that $m \in \omega$, xI_my, φ is an atomic formula with $\mathrm{Fv}\,\varphi \subseteq \{v_0, \ldots, v_{m-1}\}$, and (say) $\mathfrak{A} \models \varphi[x]$. Say ψ is m-atomic and both $\mathfrak{A} \models \psi[x]$ and $\mathfrak{B} \models \psi[y]$. Clearly $\Gamma \models \psi \rightarrow \varphi$, so $\mathfrak{B} \models \varphi[y]$, as desired. □

One of the most striking theorems concerning types is the following characterization of $\aleph_0$-categoricity. Recall the definition of $\approx_n^\Gamma$ from 27.5(i).

Theorem 27.12 (Engeler, Ryll–Nardzewski, Svenonius). *Let* Γ *be a complete theory in a countable language* $\mathscr{L}$. *Assume that* Γ *has only infinite models. Then the following conditions are equivalent:*

(i) Γ *is* $\aleph_0$*-categorical:*

(ii) *for all* $n \in \omega$, $\mathrm{Fmla}_{\mathscr{L}}^n/\approx_n^\Gamma$ *is finite.*

Proof

$(i) \Rightarrow (ii)$. Assume that (ii) is false; fix $n \in \omega$ such that $\mathrm{Fmla}_{\mathscr{L}}^n/\approx_n^\Gamma$ is infinite. Let

$$\Delta = \{\varphi : \varphi \in \mathrm{Fmla}_{\mathscr{L}}^n \text{ and } \neg\varphi \text{ is } n\text{-atomic over } \Gamma\}.$$

We claim that Δ is consistent over Γ. If not, choose a finite $\Delta' \subseteq \Delta$ such that

(1) there is no model $\mathfrak{A}$ of Γ having an $x \in {}^nA$ with $\mathfrak{A} \models \varphi[x_0, \ldots, x_{n-1}]$ for all $\varphi \in \Delta'$.

Let $\Delta'' = \Delta' \cup \{\exists v_0(v_0 = v_0)\}$. Now we claim:

(2) for any $\psi \in \mathrm{Fmla}_{\mathscr{L}}^n$, there is a subset Δ''' of Δ'' such that $\Gamma \models \psi \leftrightarrow \bigvee_{\varphi\in\Delta'''} \neg\varphi$.

[Of course, (2) is a contradiction]. To prove (2), let $\psi \in \mathrm{Fmla}_{\mathscr{L}}^n$. Let $\Delta''' = \{\varphi \in \Delta'' : \Gamma \models \neg\varphi \rightarrow \psi\}$. Clearly $\Gamma \models \bigvee_{\varphi\in\Delta'''} \neg\varphi \rightarrow \psi$. Suppose $\Gamma \not\models \psi \rightarrow \bigvee_{\varphi\in\Delta'''} \neg\varphi$.

Let $\mathfrak{A}$ be a model of Γ such that there is an $x \in {}^nA$ with $\mathfrak{A} \models \psi[x_0, \ldots, x_{n-1}]$ and $\mathfrak{A} \models \varphi[x_0, \ldots, x_{n-1}]$ for all $\varphi \in \Delta'''$. By (1), choose $\varphi \in \Delta'$ such that $\mathfrak{A} \models \neg\varphi[x_0, \ldots, x_{n-1}]$. Thus $\varphi \notin \Delta'''$. Since $\mathfrak{A} \models (\psi \wedge \neg\varphi)[x_0, \ldots, x_{n-1}]$ and $\neg\varphi$ is n-atomic over Γ, it follows easily that $\Gamma \models \neg\varphi \rightarrow \psi$. Thus $\varphi \in \Delta'''$. This is impossible. Thus (2) holds. Hence Δ is consistent over Γ after all. Also note:

(3) if φ is n-atomic over Γ, then $\neg\varphi$ is not n-atomic over Γ.

For otherwise both φ and $\neg\varphi$ would be in Δ, contradicting the consistency of Δ.

Now extend Δ to an n-type Δ^* over Γ. By 27.2, there is a denumerable model of Γ which realizes Δ^*. Note by construction and (3) that Δ^* does not contain an n-atomic formula over Γ. By 27.6 there is a denumerable model of Γ which omits Δ^*. Thus Γ is not $\aleph_0$-categorical.

(*ii*) $\Rightarrow$ (*i*). By 27.11 it suffices to show under the assumption (*ii*) that any denumerable model $\mathfrak{A}$ of Γ is elementarily prime. We shall do this by applying 27.9. (From the point of view of the theory of Boolean algebras the desired conclusion is obvious, since $\mathrm{Fmla}^n_{\mathscr{L}}/\approx^\Gamma_n$ is finite. The remainder of this proof is just to work out in first-order logic some elementary Boolean algebra.) To this end, let $n \in \omega$ and $a \in {}^nA$. Let Δ have exactly one member in common with each equivalence class under $\approx^\Gamma_n$. By (*ii*), Δ is finite. Now $\exists v_0(v_0 = v_0) \in \mathrm{Fmla}^n_{\mathscr{L}}$, so $\Gamma \models \exists v_0(v_0 = v_0) \leftrightarrow \varphi$ for some $\varphi \in \Delta$. Hence there is a $\varphi \in \Delta$ with $\mathfrak{A} \models \varphi[a]$, and we can form $\psi = \bigwedge \{\varphi \in \Delta : \mathfrak{A} \models \varphi[a]\}$. Thus $\mathfrak{A} \models \psi[a]$. Furthermore, ψ is n-atomic over Γ (as desired). For, let φ be any member of $\mathrm{Fmla}^n_{\mathscr{L}}$. Say $\Gamma \models \varphi \leftrightarrow \varphi'$ with $\varphi' \in \Delta$, and $\Gamma \models \neg\varphi \leftrightarrow \varphi''$ with $\varphi'' \in \Delta$. If $\mathfrak{A} \models \varphi[a]$, then $\mathfrak{A} \models \varphi'[a]$ and clearly $\Gamma \models \psi \rightarrow \varphi$. If $\mathfrak{A} \models \neg\varphi[a]$ it is similarly clear that $\Gamma \models \psi \rightarrow \neg\varphi$. Thus ψ is n-atomic over Γ. □

The last topic of this section is *indiscernibles*. We want to find models in which many n-tuples realize exactly the same n-type and hence cannot be distinguished from each other by first-order means. For the main theorem we need a purely set-theoretical result known as *Ramsey's theorem*.

Definition 27.13. $\mathbf{S}_m X = \{Y : Y \subseteq X, |Y| = m\}$.

Theorem 27.14 (Ramsey). *If X is an infinite set, $n \in \omega \sim 1$, and $\mathbf{S}_n X = A_0 \cup A_1$, then there is an infinite $Z \subseteq X$ and an $i < 2$ such that $\mathbf{S}_n Z \subseteq A_i$.*

Proof. We proceed by induction on n. The case $n = 1$ is trivial. Assume the result for n, and assume that X is an infinite set and $\mathbf{S}_{n+1} X = A_0 \cup A_1$. We now define three sequences $\langle x_i : i < \omega\rangle$, $\langle Y_i : i < \omega\rangle$, $\langle m_i : i < \omega\rangle$ by recursion. Let f be a choice function for non-empty subsets of X. Set $x_0 = fX$, $Y_0 = X$, $m_0 = 0$. Now suppose that x_i, Y_i, m_i have been defined so that Y_i is an infinite subset of X and $x_i \in Y_i$. Let $A'_j = \{Z : Z \in \mathbf{S}_n(Y_i \sim \{x_i\})$, and $Z \cup \{x_i\} \in A_j\}$ for each $j \in 2$. Thus $\mathbf{S}_n(Y_i \sim \{x_i\}) = A'_0 \cup A'_1$, so there is an infinite subset Y_{i+1} of $Y_i \sim \{x_i\}$ and an $m_{i+1} \in 2$ such that $\mathbf{S}_n Y_{i+1} \subseteq A'_{m(i+1)}$.

Let $x_{i+1} = fY_{i+1}$. This completes the recursive definition. Now $\{i : m_i = 0\}$ or $\{i : m_i = 1\}$ is infinite; say, by symmetry, that $\{i : m_i = 0\}$ is infinite. Let $Z = \{x_i : m_i = 0\}$. Now note:

(1) $Y_{i+1} \subseteq Y_i$;
(2) $x_i \in Y_i \sim Y_{i+1}$;
(3) x is one-one;
(4) $\forall i \in \omega[m_i = 0 \Rightarrow \forall W \in \mathbf{S}_n Y_{i+1}(W \cup \{x_i\} \in A_0)]$.

By (3), Z is infinite. We claim that $\mathbf{S}_{n+1}Z \subseteq A_0$. For, let $W \in \mathbf{S}_{n+1}Z$. Choose i minimum such that $x_i \in W$. Then $m_i = 0$ and by (1) and (2) $W \sim \{x_i\} \subseteq Y_{i+1}$. Hence by (4), $W \in A_0$, as desired. □

Definition 27.15

(*i*) A *simple ordering structure* is a pair $\mathfrak{A} = (A, \leq)$ such that $\leq$ is a simple ordering with field A.

(*ii*) *Let* $\mathfrak{A} = (A, \leq)$ be a simple ordering structure, $\mathfrak{B}$ an $\mathscr{L}$-structure ($\mathscr{L}$ is an arbitrary first-order language). We say that $\mathfrak{A}$ is *homogeneous for* $\mathfrak{B}$, or is a set of *indiscernibles* for $\mathfrak{B}$ provided that $A \subseteq B$ and the following condition holds:

(*) $\forall n \in \omega \forall x \in {}^nA \forall y \in {}^nA(\forall i, j < n(x_i < x_j \text{ iff } y_i < y_j)$
$\Rightarrow x$ and y realize the same n-type in $\mathfrak{B}$).

Perhaps the archtype of indiscernibles is furnished by the transcendentals in $\mathbb{C}$. If $x_0, \ldots, x_{n-1}$ are distinct transcendentals and $y_0, \ldots, y_{n-1}$ are distinct transcendentals, then there is an automorphism f of $\mathbb{C}$ such that $fx_i = y_i$ for each $i < n$. Hence, clearly, x and y realize the same n-type in $\mathbb{C}$. Thus *any* ordering of the transcendentals gives a set of indiscernibles for $\mathbb{C}$ in the sense of 27.15. Perhaps the notion in 27.15 should be termed order-indiscernibles, since indiscernibles in the more natural sense do exist. The following slight reformulation of the notion will be useful in what follows.

Proposition 27.16. *Let* $\mathfrak{A} = \langle A, \leq \rangle$ *be a simple ordering structure,* $A \subseteq B$, $\mathfrak{B}$ *an* $\mathscr{L}$*-structure. The following conditions are equivalent:*

(*i*) $\mathfrak{A}$ *is homogeneous for* $\mathfrak{B}$;
(*ii*) $\forall n \in \omega \forall x \in {}^nA \forall y \in {}^nA[\forall i, j < n(i < j \Rightarrow x_i < x_j$ *and* $y_i < y_j) \Rightarrow x$ *and* y *realize the same n-type in* $\mathfrak{B}$].

PROOF. (*i*) $\Rightarrow$ (*ii*): trivial. (*ii*) $\Rightarrow$ (*i*). Assume (*ii*), and assume that $n \in \omega$, $\varphi \in \mathrm{Fmla}^n_{\mathscr{L}}$, $x, y \in {}^nA$, and

(1) $$\forall i, j < n(x_i < x_j \text{ iff } y_i < y_j).$$

Say $|\{x_i : i < n\}| = m$. There is an order-preserving function x' mapping m one-one onto $\{x_i : i < n\}$. Choose $s : n \twoheadrightarrow m$ such that $x_i = x'_{si}$ for all $i < n$. For any $j < m$ choose $i < n$ with $si = j$ and set $y'_j = y_i$. This does not depend on the particular choice of such an i, by (1). Thus $y_i = y'_{si}$ for all $i < n$.

If $j < k < m$, then $x'_j < x'_k$ by the choice of x'; and if $si = j$, $sl = k$, then $x_i < x_l$, hence $y_i < y_l$ by (1), hence $y'_j < y'_k$. Hence

(2) $$\forall i, j < m(i < j \Rightarrow x'_i < x'_j \text{ and } y'_i < y'_j).$$

Hence, using (*ii*),

$$\begin{aligned}\mathfrak{B} \models \varphi[x_0, \ldots, x_{m-1}] \quad &\text{iff } \mathfrak{B} \models \varphi(v_{s0}, \ldots, v_{s(n-1)})[x'_0, \ldots, x'_{m-1}]\\ &\text{iff } \mathfrak{B} \models \varphi(v_{s0}, \ldots, v_{s(n-1)})[y'_0, \ldots, y'_{m-1}]\\ &\text{iff } \mathfrak{B} \models \varphi[y_0, \ldots, y_{n-1}].\end{aligned}$$ □

The basic existence theorem for indiscernibles is as follows. This theorem has found wide use in logic.

Theorem 27.17 (Ehrenfeucht, Mostowski). *Let Γ be a theory with an infinite model, and let $\mathfrak{A} = \langle A, \leq \rangle$ be a simple ordering structure. Then Γ has a model $\mathfrak{B}$ such that:*

(*i*) *$\mathfrak{A}$ is homogeneous for $\mathfrak{B}$;*

(*ii*) *any automorphism of $\mathfrak{A}$ extends to an automorphism of $\mathfrak{B}$;*

(*iii*) $|B| \leq |\mathrm{Fmla}_{\mathscr{L}}| + |A|$.

Proof. Let $\mathfrak{C}$ be an infinite model of Γ, and let $\leq$ be a simple ordering of C. Let $\mathscr{L}'$ be a Skolem expansion of $\mathscr{L}$ (see 11.33), and let Γ' be Γ together with the Skolem set for $\mathscr{L}$, $\mathscr{L}'$. By Lemma 11.36 let $\mathfrak{C}'$ be an expansion of $\mathfrak{C}$ to a model of Γ'.

Next, expand $\mathscr{L}'$ to $\mathscr{L}''$ by adjoining new individual constants $\mathbf{d}_a$ for $a \in A$. Let Δ be the set of all sentences of the forms

$$\neg(\mathbf{d}_a = \mathbf{d}_b) \qquad \text{for } a \neq b,$$
$$\varphi(\mathbf{d}_{a0}, \ldots, \mathbf{d}_{an}) \leftrightarrow \varphi(\mathbf{d}_{b0}, \ldots, \mathbf{d}_{bn}),$$

where $\varphi \in \mathrm{Fmla}^{n+1}_{\mathscr{L}'}$ for some $n \in \omega$, $a, b \in {}^{n+1}A$, and for all $i, j \leq n$, if $i < j$ then $a_i < a_j$ and $b_i < b_j$. We claim that $\Gamma' \cup \Delta$ has a model. To prove this we apply the compactness theorem. Let Θ be a finite subset of Δ; say $\Theta = \{\psi_0, \ldots, \psi_m\} \cup \Theta'$, where for each $i \leq m$ the sentence ψ_i is, as above,

(1) $$\varphi_i(\mathbf{d}_{ai0}, \ldots, \mathbf{d}_{ain(i)}) \leftrightarrow \varphi_i(\mathbf{d}_{bi0}, \ldots, \mathbf{d}_{bin(i)}),$$

while Θ' is a finite set of sentences of the form $\neg(\mathbf{d}_a = \mathbf{d}_b)$ where $a \neq b$. We now define subsets $D_0, \ldots, D_{m+1}$ of C by induction. Let $D_0 = C$. Having defined D_i as an infinite subset of C, where $i \leq m$, we define D_{i+1} as follows. Let

$$\begin{aligned}E_0 &= \{F \in \mathbf{S}_{ni+1}D_i : \text{if } x \text{ is the unique order-preserving map from}\\ &\qquad n_i + 1 \text{ onto } F, \text{ then } \mathfrak{C}' \models \varphi_i[x_0, \ldots, x_{ni}]\},\\ E_1 &= \mathbf{S}_{ni+1}D_i \sim E_0.\end{aligned}$$

By Ramsey's theorem, there is an infinite $D_{i+1} \subseteq D_i$ and a $j \in 2$ such that $\mathbf{S}_{ni+1}D_{i+1} \subseteq E_j$. This completes the definition of the sequence $D_0, \ldots, D_{m+1}$.

Now let $A' = \{x \in A : x = a_{ij} \text{ or } x = b_{ij} \text{ for some } i \leq m \text{ and } j \leq n_i\} \cup \{x \in A : \mathbf{d}_x \text{ occurs in some sentence of } \Theta'\}$. Thus A' is finite. Let f be an order preserving map from A' into D_{m+1}. Set $\mathfrak{C}'' = (\mathfrak{C}', l_x)_{x \in A}$, where $l: A \to C$ is arbitrary subject only to the restriction that $f \subseteq l$. We claim that $\mathfrak{C}''$ is the desired model of $\Gamma' \cup \Theta$. It is obviously a model of $\Gamma' \cup \Theta'$. Now suppose that $i \leq m$; we must check that the sentence (1) holds in $\mathfrak{C}''$. Now $a_{i0} < \cdots < a_{in(i)}$, so $fa_{i0} < \cdots < fa_{in(i)}$, and similarly $fb_{i0} < \cdots < fb_{in(i)}$. With E_0 and E_1 as above, let us suppose that $\mathbf{S}_{ni+1}D_{i+1} \subseteq E_0$, the case $\mathbf{S}_{ni+1}D_{i+1} \subseteq E_1$ being treated similarly. Since $\text{Rng} f \subseteq D_{m+1} \subseteq D_{i+1}$, we have both $\{fa_{i0}, \ldots, fa_{in(i)}\} \in E_0$ and $\{fb_{i0}, \ldots, fb_{in(i)}\} \in E_0$. Thus $\mathfrak{C}' \models \varphi_i[fa_{i0}, \ldots, fa_{in(i)}]$ and $\mathfrak{C}' \models \varphi_i[fb_{i0}, \ldots, fb_{in(i)}]$. Since $f \subseteq l$, it follows that (1) holds in $\mathfrak{C}''$. Thus, indeed, $\mathfrak{C}''$ is a model of $\Gamma' \cup \Theta$.

Hence, by the compactness theorem, $\Gamma' \cup \Delta$ has a model, which we may assume has the form $\mathfrak{D}$ with $A \subseteq D$, and with $\mathbf{d}_a$ interpreted by a for each $a \in A$. Let $\mathfrak{B}'$ be the substructure of $\mathfrak{D}$ generated by A, and set $\mathfrak{B} = \mathfrak{B}' \upharpoonright \mathscr{L}$. We now proceed to check the conditions (*i*)–(*iii*) of the theorem. Clearly $|B| \subseteq |\text{Fmla}_{\mathscr{L}}| + |A|$, i.e., (*iii*) holds. To check (*i*), first note that $\mathfrak{B}' \subseteq \mathfrak{D}$ and $\mathfrak{D}$ is a model of the Skolem set for $\mathscr{L}$, $\mathscr{L}'$, so by Proposition 19.20, $\mathfrak{B} \preccurlyeq \mathfrak{D} \upharpoonright \mathscr{L}$. Hence if $n \in \omega$, $x, y \in {}^nA$, and $x_i < x_j$, $y_i < y_j$ whenever $i < j < n$, then for any $\varphi \in \text{Fmla}^n_{\mathscr{L}}$ we have

$$\begin{aligned} \mathfrak{B} \models \varphi[x] \quad & \text{iff } \mathfrak{D} \models \varphi[x] \text{ iff } \mathfrak{D} \models \varphi[\mathbf{d}_{x0}, \ldots, \mathbf{d}_{x(n-1)}] \\ & \text{iff } \mathfrak{D} \models \varphi[\mathbf{d}_{y0}, \ldots, \mathbf{d}_{y(n-1)}] \\ & \text{iff } \mathfrak{B} \models \varphi[y]. \end{aligned}$$

Thus by 27.16 $\mathfrak{A}$ is homogeneous for $\mathfrak{B}$, i.e., (*i*) holds.

Finally, let f be an automorphism of $\mathfrak{A}$. Clearly

(2) $$B = \{\sigma^{\mathfrak{D}}x : \sigma \text{ is a term of } \mathscr{L}' \text{ and } x \in {}^{\omega}A\}.$$

Now let $f^+ = \{(\sigma^{\mathfrak{D}}x, \sigma^{\mathfrak{D}}(f \circ x)) : \sigma \text{ is a term of } \mathscr{L}' \text{ and } x \in {}^{\omega}A\}$. Then by (2), $\text{Dmn} f^+ = \text{Rng} f^+ = B$. Obviously $f \subseteq f^+$. Now f^+ is a function. For, suppose that $\sigma^{\mathfrak{D}}x = \tau^{\mathfrak{D}}y$, where σ and τ are terms of $\mathscr{L}'$ and $x, y \in {}^{\omega}A$. Say the variables of σ are among $v_0, \ldots, v_m$ and those of τ are among $v_0, \ldots, v_n$. Let τ' be obtained from τ by replacing $v_0, \ldots, v_n$ by $v_{m+1}, \ldots, v_{m+n+1}$ respectively. Let $z_i = x_i$ for $i \leq m$ and $z_i = y_{i-m-1}$ for $i > m$. Clearly then $\sigma^{\mathfrak{D}}x = \sigma^{\mathfrak{D}}z$ and $\tau^{\mathfrak{D}}y = \tau'^{\mathfrak{D}}z$. Thus $\mathfrak{D} \models (\sigma \equiv \tau')[z_0, \ldots, z_{m+n+1}]$. Note from 27.16 that $\mathfrak{A}$ is homogeneous for $\mathfrak{D} \upharpoonright \mathscr{L}'$. Hence $\mathfrak{D} \models (\sigma \equiv \tau')[fz_0, \ldots, fz_{m+n+1}]$, so $\sigma^{\mathfrak{D}}(f \circ x) = \sigma^{\mathfrak{D}}(f \circ z) = \tau'^{\mathfrak{D}}(f \circ z) = \tau^{\mathfrak{D}}(f \circ y)$, showing that f^+ is a function. In an entirely analogous way one shows that f^+ is one-one and that it preserves relations and operations. □

We shall present one corollary of this important theorem. It depends on the following result from the theory of ordered sets.

Theorem 27.18. *For any infinite cardinal* $\mathfrak{m}$ *there is a simple ordering structure* $\mathfrak{A} = \langle A, \leq \rangle$ *with* $|A| = \mathfrak{m}$ *such that* $\mathfrak{A}$ has $2^{\mathfrak{m}}$ *automorphisms.*

Proof. Let $A = \mathfrak{m} \times T$, where T is the set of all rational numbers r such that $0 \leq r < 1$. We define a linear order on A as follows:

$$(\alpha, r) < (\beta, s) \text{ iff } \alpha < \beta, \text{ or } \alpha = \beta \text{ and } r < s.$$

In the notation of the theory of ordered sets, this linear order has type $(1 + \eta)\cdot\mathfrak{m}$; one replaces each ordinal $\alpha < \mathfrak{m}$ by a copy of the rationals in $[0, 1)$. Let $F = \{f \in {}^{\mathfrak{m}}2 : f\alpha = 0$ for every limit ordinal $\alpha < \mathfrak{m}\}$. Clearly $|F| = 2^{\mathfrak{m}}$. With every $f \in F$ we shall associate an automorphism g_f of $\mathfrak{A} = \langle A, \leq\rangle$, as follows. Take any $(\alpha, r) \in A$. Write $\alpha = \beta + m$, where β is a limit ordinal and $m \in \omega$. We define $g_f(\alpha, r)$ as follows:

Case 1. $f\alpha = 0, f(\alpha + 1) = 0$. Let

$$g_f(\alpha, r) = (\beta + 2m, 2r) \quad \text{if } 0 \leq r < 1/2,$$
$$g_f(\alpha, r) = (\beta + 2m + 1, 2(r - 1/2)) \quad \text{if } 1/2 \leq r < 1.$$

Case 2. $f\alpha = 0, f(\alpha + 1) = 1$. Let

$$g_f(\alpha, r) = (\beta + 2m, 3r) \quad \text{if } 0 \leq r < 1/3,$$
$$g_f(\alpha, r) = (\beta + 2m + 1, 3(r - 1/3)) \quad \text{if } 1/3 \leq r < 2/3,$$
$$g_f(\alpha, r) = (\beta + 2m + 2, 3(r - 2/3)) \quad \text{if } 2/3 \leq r < 1.$$

Case 3. $f\alpha = 1, f(\alpha + 1) = 0$. Let

$$g_f(\alpha, r) = (\beta + 2m + 1, r).$$

Case 4. $f\alpha = 1, f(\alpha + 1) = 1$. Let

$$g_f(\alpha, r) = (\beta + 2m + 1, 2r) \quad \text{if } 0 \leq r < 1/2,$$
$$g_f(\alpha, r) = (\beta + 2m + 2, 2(r - 1/2)) \quad \text{if } 1/2 \leq r < 1.$$

Obviously g_f maps A into A. Also, g_f is onto, for let $(\gamma, s) \in A$. Write $\gamma = \beta + n$, β a limit ordinal, $n \in \omega$.

Case 1. n even, say $n = 2m$.

Subcase 1. $f(\beta + m) = 0, f(\beta + m + 1) = 0$. Then $g_f(\beta + m, s/2) = (\gamma, s)$.

Subcase 2. $f(\beta + m) = 0. f(\beta + m + 1) = 1$. Then $g_f(\beta + m, s/3) = (\gamma, s)$.

Subcase 3. $f(\beta + m) = 1$. Then by our assumption on f, $m \neq 0$.

Subsubcase 1. $f(\beta + m - 1) = 0$. Then $g_f(\beta + m - 1, (s + 2)/3) = (\beta + 2(m - 1) + 2, 3((s + 2)/3 - 2/3)) = (\gamma, s)$.

Subsubcase 2. $f(\beta + m - 1) = 1$. Then $g_f(\beta + m - 1, (s + 1)/2) = (\beta + 2(m - 1) + 2, 2((s + 1)/2 - 1/2)) = (\gamma, s)$.

Case 2. n odd, say $n = 2m + 1$.

Subcase 1. $f(\beta + m) = 0, f(\beta + m + 1) = 0$. Then $g_f(\beta + m, (s + 1)/2) = (\gamma, s)$.

Subcase 2. $f(\beta + m) = 0, f(\beta + m + 1) = 1$. Then $g_f(\beta + m, (s + 1)/3) = (\gamma, s)$.

Subcase 3. $f(\beta + m) = 1, f(\beta + m + 1) = 0$. Then $g_f(\beta + m, s) = (\gamma, s)$.

Subcase 4. $f(\beta + m) = 1, f(\beta + m + 1) = 1$. Then $g_f(\beta + m, s/2) = (\gamma, s)$.

Thus g_f maps A onto A. To show that g_f is an automorphism of $\mathfrak{A}$ it remains only to show that it preserves $<$. So, assume that $(\alpha, r) < (\beta, s)$.

Case 1. $\alpha < \beta$. Write $\alpha = \gamma + m$, $\beta = \delta + n$, where γ and δ are limit ordinals and $m, n \in \omega$. If $\gamma < \delta$, clearly $g_f(\alpha, r) < g_f(\beta, s)$. Hence assume that $\gamma = \delta$. Clearly it suffices now to take the case $n = m + 1$, i.e., $\beta = \alpha + 1$.

Subcase 1. $f\alpha = 0, f\beta = 0$. Then 1st coord $g_f(\alpha, r) \leq \gamma + 2m + 1$, 1st coord $g_f(\beta, s) \geq \gamma + 2(m + 1) = \gamma + 2m + 2$. So $g_f(\alpha, r) < g_f(\beta, s)$.

Subcase 2. $f\alpha = 0, f\beta = 1$. Then 1st coord $g_f(\alpha, r) \leq \gamma + 2m + 2$, 1st coord $g_f(\beta, s) \geq \gamma + 2(m + 1) + 1 = \gamma + 2m + 3$. So $g_f(\alpha, r) < g_f(\beta, s)$.

Subcase 3. $f\alpha = 1, f\beta = 0$. Then 1st coord $g_f(\alpha, r) = \gamma + 2m + 1$,
1st coord $g_f(\beta, s) \geq \gamma + 2(m + 1) = \gamma + 2m + 2$. So $g_f(\alpha, r) < g_f(\beta, s)$.
Subcase 4. $f\alpha = 1, f\beta = 1$. Then 1st coord $g_f(\alpha, r) \leq \gamma + 2m + 2$,
1st coord $g_f(\beta, s) \geq \gamma + 2(m + 1) + 1 = \gamma + 2m + 3$. So $g_f(\alpha, r) < g_f(\beta, s)$.

Case 2. $\alpha = \beta$, $r < s$. The desired conclusion is clear.

Thus g_f is an automorphism of $\mathfrak{A}$. Now suppose $f, f' \in F$ and $f \neq f'$. Say $\alpha < \mathfrak{m}$ and $f\alpha \neq f'\alpha$. Say $f\alpha = 0, f'\alpha = 1$. Let $\alpha = \beta + m$, β a limit ordinal, $m \in \omega$. Then 1st coord $g_f(\alpha, 0) = \beta + 2m$. 1st coord $g_{f'}(\alpha, 0) = \beta + 2m + 1$, so $g_f \neq g_{f'}$. □

This theorem together with the Ehrenfeucht–Mostowki theorem gives the following interesting property of first-order theories:

Corollary 27.19 *If* Γ *is a theory with an infinite model and* $\mathfrak{m} \geq |\text{Fmla}_{\mathscr{L}}|$, *then* Γ *has a model of power* $\mathfrak{m}$ *with* $2^{\mathfrak{m}}$ *automorphisms.*

EXERCISES

27.20. Let $\mathscr{L}$ be a language whose only nonlogical constants are individual constants $\mathbf{c}_i$ for $i < \omega$. Let Γ be the theory in with axioms $\neg(\mathbf{c}_i \equiv \mathbf{c}_j)$ for all $i \neq j$. Show that Γ is complete.

27.21. Continuing 27.20, find the elementarily prime model of Γ.

27.22. Give an example of a complete theory in a countable language, having only infinite models, which has a prime model but no elementarily prime model.

27.23. Give an example of a complete theory in a countable language with only infinite models, $\aleph_0$-categorical, and with exactly one 1-atomic formula (up to equivalence of formulas).

27.24. Suppose $\mathscr{L}$ is a countable language, $\mathfrak{A}$ is an $\mathscr{L}$-structure, and $\varphi \in \text{Fmla}^2_{\mathscr{L}}$ is such that $\varphi^{\mathfrak{A}}$ well-orders A. Show that $\Theta\rho\mathfrak{A}$ has an elementarily prime model.

27.25. Under the hypotheses of 27.12, the conditions (*i*) and (*ii*) are also equivalent to:
(*) for every model $\mathfrak{A}$ of Γ, every $n \in \omega$, and every $a \in {}^nA$, a satisfies some n-atomic formula over Γ in $\mathfrak{A}$ (cf. 27.9).

27.26. If $\mathfrak{A}$ is any Boolean algebra, then the atoms of $\mathfrak{A}$ form a set of indiscernibles for $\mathfrak{A}$, under any ordering of them.

27.27. Does $\Theta\rho(\omega, \mathfrak{s}, 0)$ have an elementarily prime model?

27.28. Suppose a group $\mathfrak{G} = (G, \cdot, 1)$ has the property that for every $\varphi \in \text{Fmla}^1$, if $\mathfrak{G} \models \exists v_0\varphi(v_0)$, then there is an $n \in \omega \sim 1$ such that $\mathfrak{G} \models \exists v_0(v_0^{\mathbf{n}} \equiv \mathbf{1} \wedge \varphi(v_0))$. Show that $\mathfrak{G}$ is elementarily equivalent to a group in which every element has finite order.

28 Saturated Structures

Saturated structures are structures in which all possible types are realized (roughly speaking; see below for an exact formulation). In this chapter we prove their existence and uniqueness and show how they can be used in logical investigations.

Definition 28.1. An $\mathscr{L}$-structure $\mathfrak{A}$ is $\mathfrak{m}$-*saturated* iff for every $X \subseteq A$ with $|X| < \mathfrak{m}$, *if* $\mathscr{L}'$ is an expansion of $\mathscr{L}$ obtained by adding a new individual constant $\mathbf{c}_x$ for each $x \in X$, and if $\Delta \subseteq \mathrm{Fmla}^1_{\mathscr{L}'}$ is such that for each finite subset Δ' of Δ there is an $a \in A$ with $(\mathfrak{A}, x)_{x\in X} \models \bigwedge \Delta'[a]$, *then* there is an $a \in A$ such that $(\mathfrak{A}, x)_{x\in X} \models \varphi[a]$ for all $\varphi \in \Delta$.

An $\mathscr{L}$-structure $\mathfrak{A}$ is *saturated* iff it is $|A|$-saturated.

Given that are are trying to make exact the notion of a structure in which the maximum amount of formulas can be simultaneously satisfied, Definition 28.1 exhibits two peculiarities. First, constants are taken from A, rather than dealing with just formulas of the original language $\mathscr{L}$. Second, only formulas with just one free variable are considered rather than arbitrary formulas. The first restriction is essential; without mentioning constants from A, an essentially weaker notion is obtained (see Exercise 28.34). It is from the point of view of applications that the stronger notion given in 28.1 is to be preferred. The second restriction in 28.1 is inessential, as the following proposition shows.

Proposition 28.2. *Let $\mathfrak{A}$ be an $\mathscr{L}$-structure and $\mathfrak{m}$ an infinite cardinal. Then the following conditions are equivalent*:

(*i*) *$\mathfrak{A}$ is $\mathfrak{m}$-saturated*;
(*ii*) *for every $X \subseteq A$ with $|X| < \mathfrak{m}$, if $\mathscr{L}'$ is an X-expansion of $\mathscr{L}$, and if Δ is a set of formulas of $\mathscr{L}'$ such that for each finite subset Δ' of Δ*

there is an $a \in {}^{\omega}A$ *with* $(\mathfrak{A}, x)_{x \in X} \models \bigwedge \Delta'[a]$, then *there is an* $a \in {}^{\omega}A$ *such that* $(\mathfrak{A}, x)_{x \in X} \models \varphi[a]$ *for all* $\varphi \in \Delta$.

Proof. Obviously $(ii) \Rightarrow (i)$. Now assume (i) and the hypothesis of (ii). We may assume that Δ is closed under conjunction. For each $\varphi \in \Delta$ choose m_φ such that $\mathrm{Fv}\,\varphi \subseteq \{v_0, \ldots, v_{m\varphi}\}$. We define $a \in {}^{\omega}A$ by recursion. Suppose that $a \restriction i$ has been defined in such a way that each formula

$$\exists v_i \cdots \exists v_{m\varphi}\varphi(\mathbf{c}_{a0}, \ldots, \mathbf{c}_{a(i-1)}), \qquad \varphi \in \Delta,$$

holds in $(\mathfrak{A}, x, a_j)_{x \in X, j < i}$. This is clearly true for $i = 0$. Now let Θ be the set of all formulas $\exists v_{i+1} \cdots \exists v_{m\varphi}\varphi(\mathbf{c}_{a0}, \ldots, \mathbf{c}_{a(i-1)})$ with $\varphi \in \Delta$. Thus

$$|X \cup \{a_j : j < i\}| < \mathfrak{m},$$

and Θ is a set of formulas ψ with $\mathrm{Fv}\,\psi \subseteq \{v_i\}$. If Θ' is a finite subset of Θ, we may write $\Theta' = \{\exists v_{i+1} \cdots \exists v_{m\varphi}\varphi(\mathbf{c}_{a0}, \ldots, \mathbf{c}_{a(i-1)}) : \varphi \in \Delta'\}$, where Δ' is a finite subset of Δ. Now $\bigwedge \Delta' \in \Delta$, so $\exists v_i \cdots \exists v_{mx} \bigwedge \Delta'(\mathbf{c}_{a0}, \ldots, \mathbf{c}_{a(i-1)})$ holds in $(\mathfrak{A}, x, a_j)_{x \in X, j < i}$. Thus there is a $b \in A$ such that b assigned to v_i satisfies each $\psi \in \Theta'$ in $(\mathfrak{A}, x, a_j)_{x \in X, j < i}$. It follows from (i) that there is an $a_i \in A$ such that each formula

$$\exists v_{i+1} \cdots \exists v_{m\varphi}\varphi(\mathbf{c}_{a0}, \ldots, \mathbf{c}_{ai}), \qquad \varphi \in \Delta,$$

holds in $(\mathfrak{A}, x, a_j)_{x \in X, j \le i}$. This finishes the construction of a. Clearly the conclusion of (ii) then holds. □

We are really interested in the notion "the $\mathscr{L}$-structure $\mathfrak{A}$ is $\mathfrak{m}$-saturated" only when $|\mathrm{Fmla}_{\mathscr{L}}| \le \mathfrak{m} \le |A|$. The next few remarks indicate why these restrictions are reasonable. Since, clearly, $\mathfrak{n} < \mathfrak{m}$ and $\mathfrak{A}$ $\mathfrak{m}$-saturated imply $\mathfrak{A}$ $\mathfrak{n}$-saturated, the assumption that $|\mathrm{Fmla}_{\mathscr{L}}| \le \mathfrak{m}$ does no harm when we prove the existence of $\mathfrak{m}$-saturated structures. Since the assumption $|\mathrm{Fmla}_{\mathscr{L}}| \le \mathfrak{m}$ is needed for our constructions, we are justified in making it. The assumption $\mathfrak{m} \le |A|$ is resonable in view of the following two results.

Proposition 28.3. *If* $\mathfrak{A}$ *is* $\mathfrak{m}$*-saturated, then either* $\mathfrak{A}$ *is finite or* $|A| \ge \mathfrak{m}$.

Proof. Suppose $\mathfrak{A}$ is infinite and $|A| < \mathfrak{m}$. Let $\mathscr{L}'$ be an A-expansion of $\mathscr{L}$, and let Δ be $\{\neg(v_0 = \mathbf{c}_a) : a \in A\}$. Clearly every finite subset of Δ can be realized in $(\mathfrak{A}, a)_{a \in A}$, but Δ itself cannot be, so $\mathfrak{A}$ is not $\mathfrak{m}$-saturated. □

Proposition 28.4. $\mathfrak{A}$ *is finite iff* $\mathfrak{A}$ *is* $\mathfrak{m}$*-saturated for all* $\mathfrak{m}$.

Proof. The second condition implies the first, by 28.3. Now suppose that $\mathfrak{A}$ is finite, $X \subseteq A$, $|X| < \mathfrak{m}$, $\mathscr{L}'$ is an X-expansion of $\mathscr{L}$, and $\Delta \subseteq \mathrm{Fmla}^1_{\mathscr{L}'}$. Suppose that Δ cannot be realized in $(\mathfrak{A}, x)_{x \in X}$. Then for each $a \in A$ there is a $\varphi_a \in \Delta$ such that $(\mathfrak{A}, x)_{x \in X} \models \neg\varphi_a[a]$. Thus $\{\varphi_a : a \in A\}$ is a finite subset of Δ which cannot be realized in $(\mathfrak{A}, x)_{x \in X}$. Hence $\mathfrak{A}$ is $\mathfrak{m}$-saturated. □

Before coming to the main existence and uniqueness results we want to give a few more trivial properties of $\mathfrak{m}$-saturated structures, some of which are useful in applications.

Proposition 28.5

(i) *$\mathfrak{A}$ is $\mathfrak{m}$-saturated iff $\mathfrak{A}$ is $\mathfrak{n}$-saturated for every $\mathfrak{n} \leq \mathfrak{m}$;*

(ii) *if $\mathfrak{m}$ is a limit cardinal, then $\mathfrak{A}$ is $\mathfrak{m}$-saturated iff $\mathfrak{A}$ is $\mathfrak{n}$-saturated for every $\mathfrak{n} < \mathfrak{m}$;*

(iii) *$\mathfrak{A}$ is $\mathfrak{m}$-saturated iff for every $X \subseteq A$ with $|X| < \mathfrak{m}$, the structure $(\mathfrak{A}, x)_{x \in X}$ is 0-saturated;*

(iv) *if $\mathfrak{A}$ is $\mathfrak{m}$-saturated, $X \subseteq A$, then $|X| < \mathfrak{m}$, then $(\mathfrak{A}, x)_{x \in X}$ is $\mathfrak{m}$-saturated.*

We now consider the existence of $\mathfrak{m}$-saturated structures.

Theorem 28.6. *Let $\mathscr{L}$, $\mathscr{L}'$, $\mathscr{L}''$ be first-order languages, $\mathfrak{m}$ and $\mathfrak{n}$ cardinals, and Γ and Γ' theories in $\mathscr{L}$ and $\mathscr{L}'$ respectively, subject to the following conditions:*

(i) $\aleph_0 \leq \mathfrak{m}, \mathfrak{n}$;

(ii) *$\mathscr{L}'$ is an expansion of $\mathscr{L}$, and $|\mathrm{Fmla}_{\mathscr{L}'}| = \mathfrak{n}$;*

(iii) *$\mathscr{L}''$ is an expansion of $\mathscr{L}'$ rich by C;*

(iv) *if $\mathfrak{p}$ is any nonzero cardinal $< \mathfrak{m}$, then $\mathfrak{n}^{\mathfrak{p}} \leq \mathfrak{n}$;*

(v) *$\Gamma \subseteq \Gamma'$, and Γ' is consistent;*

(vi) *for each $B \subseteq C$ with $|B| < \mathfrak{m}$, let $\mathscr{L}_B$ be the reduct of $\mathscr{L}''$ to the symbols of $\mathscr{L}$ and B; then there are at most $\mathfrak{n}$ 1-types over Γ in the language $\mathscr{L}_B$.*

Under all of these assumptions it follows that there is a model $\mathfrak{A}$ of Γ' of power $\leq \mathfrak{n}$ such that $\mathfrak{A} \upharpoonright \mathscr{L}$ is $\mathfrak{m}$-saturated.

Proof. Let a well-ordering of C be given. We shall apply the model existence Theorem 18.9 in its full form. Let S be the set of all $\Theta \subseteq \mathrm{Sent}_{\mathscr{L}''}$ such that $\Gamma \cup \Theta$ is consistent and $\{\mathbf{c} \in C : \mathbf{c}$ occurs in some $\varphi \in \Theta\}$ has power $< \mathfrak{n}$. It is a straightforward matter to check that S is a consistency family; cf. the proofs of 22.1 and 27.4. The family S also has this additional property:

(1) if $\Theta \in S$, $B \subseteq C$, $|B| < \mathfrak{m}$, and Δ is a 1-type over Γ in $\mathscr{L}_B$ which is consistent over Θ, then $\Theta \cup \{\varphi(\mathbf{c}) : \varphi \in \Delta\} \in S$, where $\mathbf{c}$ is the first member of C not in B or in any sentence of Θ.

For, if Δ' is any finite subset of Δ, then by the hypothesis of (1), $\Theta \cup \{\exists v_0 \bigwedge \Delta'\}$ has a model, $\mathfrak{A}$. We can modify $\mathfrak{A}$ by assigning to $\mathbf{c}$ an element $a \in A$ such that $\mathfrak{A} \models \bigwedge\Delta'[a]$, and we then obtain a model of $\Theta \cup \{\varphi(\mathbf{c}) : \varphi \in \Delta'\}$. Thus $\Theta \cup \{\varphi(\mathbf{c}) : \varphi \in \Delta\}$ is consistent, so it follows that it is in S, and (1) holds.

Let $T = \{(B, \Delta) : B \subseteq C,\ |B| < \mathfrak{m}$, and Δ is a 1-type over Γ in $\mathscr{L}_B\}$. The assumptions of the theorem imply that $|T| \leq \mathfrak{n}$, and obviously $T \neq 0$. Hence we may write $T = \{(B_\beta, \Delta_\alpha) : \alpha < \mathfrak{n}\}$. For each $\alpha < \mathfrak{n}$ we define a function f_α with domain S:

$f_\alpha\Theta = \Theta \cup \{\varphi(\mathbf{c}) : \varphi \in \Delta_\alpha\}$ if this set is in S, where $\mathbf{c}$ is the first member of C not in B_α or in any sentence of Θ;

$f_\alpha\Theta = \Theta$ otherwise.

Obviously f_α is admissible over S.

By the model existence theorem, since $\Gamma' \in S$, let $(\mathfrak{A}, a_\mathbf{c})_{\mathbf{c}\in C}$ be a model of Γ' such that $A = \{a_\mathbf{c} : \mathbf{c} \in C\}$, hence $|A| \leq \mathfrak{n}$, and:

(2) if $\alpha < \mathfrak{n}$, then there is a $\Theta \in S$ such that $\Gamma' \subseteq f_\alpha\Theta \subseteq \{\varphi : (\mathfrak{A}, a_\mathbf{c})_{\mathbf{c}\in C} \models \varphi\}$.

To show that $\mathfrak{A} \upharpoonright \mathscr{L}$ is $\mathfrak{m}$-saturated, let $B \subseteq C$ with $|B| < \mathfrak{m}$, and let $\Delta \subseteq \mathrm{Fmla}^1_{\mathscr{L}B}$ be such that $(\mathfrak{A} \upharpoonright \mathscr{L}, a_\mathbf{c})_{\mathbf{c}\in B} \models \exists v_0 \bigwedge \Delta'$ for each finite subset Δ' of Δ. Thus Δ is consistent over $\Theta\rho(\mathfrak{A} \upharpoonright \mathscr{L}, a_\mathbf{c})_{\mathbf{c}\in B}$, so we can extend Δ to a 1-type Δ^* over $\Theta\rho(\mathfrak{A} \upharpoonright \mathscr{L}, a_\mathbf{c})_{\mathbf{c}\in B}$. Thus Δ^* is a 1-type over Γ in $\mathscr{L}_B$, so $(B, \Delta^*) \in T$. Write $(B, \Delta^*) = (B_\alpha, \Delta_\alpha)$, and choose Θ by (2) so that $\Theta \in S$ and $\Gamma \subseteq f_\alpha\Theta \subseteq \Theta\rho(\mathfrak{A}, a_\mathbf{c})_{\mathbf{c}\in C}$. Since Δ^* is consistent over Θ because $\Theta \subseteq f_\alpha\Theta \subseteq \Theta\rho(\mathfrak{A}, a_\mathbf{c})_{\mathbf{c}\in C}$, it follows from (1) and the definition of f_α that $(\mathfrak{A} \upharpoonright \mathscr{L}, a_\mathbf{c})_{\mathbf{c}\in C} \models \varphi(\mathbf{d})$ for each $\varphi \in \Delta^*$, where $\mathbf{d}$ is a suitable element of C. Thus $a_\mathbf{d}$ realizes Δ in $(\mathfrak{A} \upharpoonright \mathscr{L}, a_\mathbf{c})_{\mathbf{c}\in B}$, as desired. □

Theorem 28.6 is our main existence theorem for saturated structures. We shall apply it in two ways: to $\aleph_0$-saturation on the one hand, and $\mathfrak{m}$-saturation for $\mathfrak{m} > \mathrm{Fmla}_{\mathscr{L}}$ on the other hand. In the case of countable languages, we can actually characterize when a theory has an $\aleph_0$-saturated countable model:

Theorem 28.7 (Vaught). *Let Γ be a theory in a countable language $\mathscr{L}$. Then the following conditions are equivalent:*

(*i*) *Γ has a countable $\aleph_0$-saturated model;*

(*ii*) *for each $m \in \omega$ there are only countably many m-types over Γ.*

PROOF

(*i*) ⇒ (*ii*). Let $\mathfrak{A}$ be a countable $\aleph_0$-saturated model of Γ, and let $m \in \omega$. For each m-type Δ over Γ let $f\Delta = \{a \in {}^mA : a \text{ realizes } \Delta \text{ in } \mathfrak{A}\}$. Since $\mathfrak{A}$ is $\aleph_0$-saturated, by Proposition 28.2 we see that $f\Delta \neq 0$ for each m-type Δ over Γ. Clearly $f\Delta \cap f\Delta' = 0$ for distinct m-types Δ, Δ' over Γ. Hence (*ii*) holds since $\mathfrak{A}$ is countable.

(*ii*) ⇒ (*i*). We apply 28.6 with $\mathfrak{m} = \mathfrak{n} = \aleph_0$, $\mathscr{L} = \mathscr{L}'$, and $\Gamma = \Gamma'$. Obviously $\mathfrak{n}^\mathfrak{p} = \mathfrak{n}$ for $0 < \mathfrak{p} < \mathfrak{m}$. Let $\mathscr{L}''$ be as in 28.6, and let $B \subseteq C$ with $|B| < \aleph_0$, say $B = \{\mathbf{b}_0, \ldots, \mathbf{b}_{m-1}\}$ with $m < \omega$. For each $\varphi \in \mathrm{Fmla}^1_{\mathscr{L}B}$, let φ' be obtained from φ by first suitably replacing bound variables and then replacing $\mathbf{b}_0, \ldots, \mathbf{b}_{m-1}$ by $v_1, \ldots, v_m$ respectively. For each 1-type Δ over Γ

in $\mathscr{L}_B$, let $f\Delta = \{\varphi' : \varphi \in \Delta\}$. We claim that f establishes a one-one correspondence between 1 types over Γ in $\mathscr{L}_B$ and $(m + 1)$-types over Γ in $\mathscr{L}$. It is a routine matter to check this claim. Hence the assumption (*ii*) yields the fact that there are only countably many 1-types over Γ in $\mathscr{L}_B$. Thus the hypotheses of 28.6 are fulfilled, and (*i*) follows. □

From 28.7 and 27.12 we obtain many examples of theories with countable $\aleph_0$-saturated models:

Corollary 28.8. *If Γ is a consistent $\aleph_0$-categorical theory with only infinite models, in a countable language, then Γ has a countable $\aleph_0$-saturated model.*

Of course, actually all the countable models of Γ in this case are $\aleph_0$-saturated. Recall from Chapter 21 our many examples of $\aleph_0$-categorical theories. Actually 27.12 is not needed to establish 28.8. A more direct argument can be given which generalizes to prove

Corollary 28.9. *If Γ is a consistent theory in a countable language and Γ has up to isomorphism only countably many countable models, then Γ has a countable $\aleph_0$-saturated model.*

Proof. Let $m \in \omega$. Each m-type over Γ is realized in some countable model of Γ, by 27.2. Hence there are only countably many m-types over Γ (cf. the proof (*i*) ⇒ (*ii*) in 28.7). So by 28.7, Γ has a countable $\aleph_0$-saturated model. □

Corollary 28.9 makes evident the existence of $\aleph_0$-saturated models in several more theories. For example, consider the theory Γ of algebraically closed fields of characteristic 0. The countable models of Γ are, up to isomorphism, the algebraically closed fields $\mathfrak{A}_\alpha$ for $\alpha \leq \omega$, where $\mathfrak{A}_\alpha$ has transcendence degree α over $\mathbb{Q}$. Hence by Corollary 28.9, one of these fields is $\aleph_0$-saturated. In fact, only $\mathfrak{A}_\omega$ is $\aleph_0$-saturated. For, let $m < \omega$, and let X be a transcendence base for $\mathfrak{A}_m$. Thus $|X| = m$. For each $n \in \omega$ one has a formula φ_n of $\mathscr{L}'$ (with $\mathscr{L}'$ as in 28.1) such that $\mathfrak{A}_m \models \varphi_n[a]$ iff a is not algebraic of degree n over $\mathbb{Q}(X)$. For each n there is such an a, by elementary field theory, but $\{\varphi_n : n \in \omega\}$ is *not* realized in $(\mathfrak{A}_m, x)_{x \in X}$. Thus $\mathfrak{A}_m$ is not $\aleph_0$-saturated. So $\mathfrak{A}_\omega$ is $\aleph_0$-saturated by 28.9.

The following interesting result connects $\aleph_0$-saturation with the existence of elementarily prime models.

Theorem 28.10. *Let Γ be a complete theory in a countable language. If Γ has a countable $\aleph_0$-saturated model, then Γ has an elementarily prime model.*

Proof. Suppose that Γ has no elementarily prime model. Then by 27.10 there is an $n \in \omega$ for which there is an n-atomless formula φ over Γ. We shall show that there are $2^{\aleph_0}$ n-types over Γ, so that by 28.7 Γ has no countable $\aleph_0$-saturated model.

For each $f \in \bigcup_{m\in\omega} {}^m 2$ we define a formula ψ_f. Let $\psi_0 = \varphi$. Suppose that ψ_f has been defined so that $\psi_f \in \mathrm{Fmla}^n_{\mathscr{L}}$ and ψ_f is n-atomless over Γ. Then there is a $\chi \in \mathrm{Fmla}^n_{\mathscr{L}}$ such that $\Gamma \nvDash \psi_f \to \chi$ and $\Gamma \nvDash \psi_f \to \neg\chi$. Let $\psi_{f0} = \psi_f \wedge \chi$ and $\psi_{f1} = \psi_f \wedge \neg\chi$. Clearly ψ_{f0} and ψ_{f1} are again n-atomless over Γ. This completes the definition. For each $f \in {}^\omega 2$ let $\Delta_f = \{\psi_g : g \subseteq f\}$. Clearly Δ_f is consistent over Γ and so it can be extended to an n-type Θ_f. For $f \neq h$ obviously $\Theta_f \neq \Theta_h$, as desired. □

In the case of $\aleph_0$-categorical theories with only infinite models, the denumerable model is both $\aleph_0$-saturated and elementarily prime. If Γ is the theory of algebraically closed fields of characteristic 0, its elementarily prime model, the field of algebraic numbers, is different from its $\aleph_0$-saturated model.

Now we turn to the existence problem for $\mathfrak{m}$-saturated structures with $\mathfrak{m} > |\mathrm{Fmla}_{\mathscr{L}}|$. The main theorem is as follows:

Theorem 28.11. *Let $\mathscr{L}$, $\mathscr{L}'$ be the first-order languages, $\mathfrak{m}$ and $\mathfrak{n}$ cardinals, and Γ' a theory in $\mathscr{L}'$, subject to the following conditions:*

(*i*) $|\mathrm{Fmla}_{\mathscr{L}}| < \mathfrak{m}$;
(*ii*) *$\mathscr{L}'$ is an expansion of $\mathscr{L}$, and* $|\mathrm{Fmla}_{\mathscr{L}'}| \leq \mathfrak{n}$;
(*iii*) *if $\mathfrak{p}$ is any nonzero cardinal $< \mathfrak{m}$, then $\mathfrak{n}^{\mathfrak{p}} \leq \mathfrak{p}$*;
(*iv*) *Γ' is consistent.*

Under all of these assumptions it follows that there is a model $\mathfrak{A}$ of Γ' of power $\leq \mathfrak{n}$ such that $\mathfrak{A} \upharpoonright \mathscr{L}$ is $\mathfrak{m}$-saturated.

Proof. Let $\Gamma = \Gamma' \cap \mathrm{Sent}_{\mathscr{L}}$. By extending $\mathscr{L}'$ and Γ' we may assume that $|\mathrm{Fmla}_{\mathscr{L}'}| = \mathfrak{n}$. Let $\mathscr{L}''$ be as in 28.6. Clearly 28.6 will give the desired result, and only 28.6(*vi*) remains to be verified. Suppose that $B \leq C$ with $|B| < \mathfrak{m}$ and with $\mathscr{L}_B$ as in 28.6. Now by (*i*), $|\mathrm{Fmla}_{\mathscr{L}B}| < \mathfrak{m}$. The number of 1-types over Γ in $\mathscr{L}_B$ is clearly $\leq 2^{|\mathrm{Fmla}_{\mathscr{L}B}|} \leq \mathfrak{n}$ by (*iii*), so 28.6(*vi*) holds. □

The most useful corollaries of this theorem are as follows.

Corollary 28.12. *Suppose $|\mathrm{Fmla}_{\mathscr{L}}| \leq \mathfrak{m}$, and $\mathfrak{A}$ is an $\mathscr{L}$-structure with $\aleph_0 \leq |A| \leq 2^{\mathfrak{m}}$. Then there is an $\mathfrak{m}^+$-saturated elementary extension $\mathfrak{B}$ of $\mathfrak{A}$ of power $2^{\mathfrak{m}}$.*

Proof. Let $\mathscr{L}^*$ be an A-expansion of $\mathscr{L}$, with new individual constants $\mathbf{c}_a$ for $a \in A$, and let $\mathscr{L}'$ be an expansion of $\mathscr{L}^*$ by adjoining new individual constants $\mathbf{d}_\alpha$, $\alpha < 2^{\mathfrak{m}}$. Let $\Gamma' = \{\varphi \in \mathrm{Sent}_{\mathscr{L}'} : (\mathfrak{A}, a)_{a\in A} \models \varphi\} \cup \{\neg \mathbf{d}_\alpha = \mathbf{d}_\beta : \alpha < \beta < 2^{\mathfrak{m}}\}$. The conditions of 28.11 are clearly met with $\mathfrak{m}$, $\mathfrak{n}$ replaced by $\mathfrak{m}^+$, $2^{\mathfrak{m}}$ respectively. □

Corollary 28.13. *Suppose that $\mathfrak{m}$ is strongly inaccessible, $|\mathrm{Fmla}_{\mathscr{L}}| < \mathfrak{m}$, and $\mathfrak{A}$ is an $\mathscr{L}$-structure with $\aleph_0 \leq |A| \leq \mathfrak{m}$. Then $\mathfrak{A}$ has an $\mathfrak{m}$-saturated elementary extension $\mathfrak{B}$ of power $\mathfrak{m}$.*

The proof of 28.13 is similar to that of 28.12. Using GCH we obtain

Corollary 28.14 (GCH). *Every theory in $\mathscr{L}$ with infinite models has a saturated model of each regular cardinality* $>|\mathrm{Fmla}_{\mathscr{L}}|$.

We now want to give different proofs for these corollaries which are perhaps more natural than the above. The proofs are based on the following lemma, which is of course, weaker than 28.12:

(*) *Suppose* $|\mathrm{Fmla}_{\mathscr{L}}| \leq \mathfrak{m}$, *and* $\mathfrak{A}$ *is an* $\mathscr{L}$*-structure with* $\aleph_0 \leq |A| \leq 2^{\mathfrak{m}}$. *Then there is an elementary extension* $\mathfrak{B}$ *of* $\mathfrak{A}$ *such that* $|B| = 2^{\mathfrak{m}}$, *and if* X *is a subset of* A *of power* $\leq \mathfrak{m}$, $\mathscr{L}_X$ *is an* X*-expansion of* $\mathscr{L}$, *and* Δ *is a* 1*-type over* $\Theta\rho(\mathfrak{A}, x)_{x\in X}$ *in* $\mathscr{L}_X$, *then* Δ *is realized in* $(\mathfrak{B}, x)_{x\in X}$.

We prove (*) by applying the compactness theorem; it can also be proved using ultraproducts (Exercise 28.39). We consider each language $\mathscr{L}_X$ as a reduct of a certain A-expansion $\mathscr{L}_A$ of $\mathscr{L}$. For each pair (X, Δ) such that $X \subseteq A$, $|X| \leq \mathfrak{m}$, and Δ is a 1-type over $\Theta\rho(\mathfrak{A}, x)_{x\in X}$ in $\mathscr{L}_X$, we introduce a new constant $\mathbf{c}_{X\Delta}$, thus expanding $\mathscr{L}_A$ to a new language $\mathscr{L}'$. Let Γ be the set

$$\Theta\rho(\mathfrak{A}, a)_{a\in A} \cup \{\varphi(\mathbf{c}_{X\Delta}) : X \subseteq A, |X| < \mathfrak{m}, \Delta \text{ a } \\ \text{1-type over } \Theta\rho(\mathfrak{A}, x)_{x\in X} \text{ in } \mathscr{L}_X, \varphi \in \Delta\}.$$

Clearly every finite subset of Γ has a model, so Γ has a model $\mathfrak{B}$ of power $\leq |\mathrm{Fmla}_{\mathscr{L}'}|$. Now the number of pairs (X, Δ) of the above sort is clearly at most $2^{\mathfrak{m}}$, so $|B| \leq 2^{\mathfrak{m}}$. Also, $\mathfrak{A}$ is elementarily embeddable in $\mathfrak{B}$, so B is infinite. By the upward Löwenheim–Skolem theorem we can assume that $|B| = 2^{\mathfrak{m}}$. Clearly $\mathfrak{B} \upharpoonright \mathscr{L}$ is as desired in (*) (up to isomorphism).

On the basis of (*), Corollary 28.12 is established as follows: Assume the hypothesis of 28.12. We define a sequence $\langle \mathfrak{B}_\alpha : \alpha \leq \mathfrak{m}^+ \rangle$ of $\mathscr{L}$-structures. Let $\mathfrak{B}_0 = \mathfrak{A}$. Suppose that $\alpha < \mathfrak{m}^+$ and $\mathfrak{B}_\alpha$ has been defined so that $\aleph_0 \leq |B_\alpha| \leq 2^{\mathfrak{m}}$. By (*), choose $\mathfrak{B}_{\alpha+1}$ to be an elementary extension of $\mathfrak{B}_\alpha$ of power $2^{\mathfrak{m}}$ such that if X is any subset of B_α of power $\leq \mathfrak{m}$, $\mathscr{L}_X$ is an X-expansion of $\mathscr{L}$, and Δ is a 1-type over $\Theta\rho(\mathfrak{B}_\alpha, x)_{x\in X}$, then Δ is realized in $(\mathfrak{B}_{\alpha+1}, x)_{x\in X}$. For λ a limit ordinal $\leq \mathfrak{m}^+$, let $\mathfrak{B}_\lambda = \bigcup_{\alpha<\lambda} \mathfrak{B}_\alpha$. Then $\mathfrak{B}_{\mathfrak{m}+}$ satisfies the conclusion of 28.12. In fact, it is obviously an elementary extension of $\mathfrak{A}$ of power $2^{\mathfrak{m}}$. Now suppose that $X \subseteq B_{\mathfrak{m}+}$, $|X| \leq \mathfrak{m}$, $\mathscr{L}_X$ is an X-expansion of $\mathscr{L}$, and Δ is a subset of $\mathrm{Fmla}^1_{\mathscr{L}X}$ such that $(\mathfrak{B}_{\mathfrak{m}+}, x)_{x\in X} \models \exists v_0 \bigwedge \Delta'$ for every finite subset Δ' of Δ. Extend Δ to a 1-type Δ^* over $\Theta\rho(\mathfrak{B}_{\mathfrak{m}+}, x)_{x\in X}$. There is an $\alpha < \mathfrak{m}^+$ such that $X \subseteq B_\alpha$, since $|X| \leq \mathfrak{m}$ and $\mathfrak{m}^+$ is regular. Clearly Δ^* is also a 1-type over $\Theta\rho(\mathfrak{B}_\alpha, x)_{x\in X}$. From our construction it follows that Δ^* is realized in $(\mathfrak{B}_{\alpha+1}, x)_{x\in X}$. Since $\mathfrak{B}_\alpha \preccurlyeq \mathfrak{B}_{\mathfrak{m}+}$, Δ^* is also realized in $(\mathfrak{B}_{\mathfrak{m}+}, x)_{x\in X}$. Thus $\mathfrak{B}_{\mathfrak{m}+}$ is $\mathfrak{m}$-saturated.

Corollary 28.13 can also be easily established in case $|A| < \mathfrak{m}$, on the basis of (*) or 28.12: one simply extends the above construction all the way to $\mathfrak{m}$. We leave the details for an exercise.

Another way of establishing these corollaries, and one which is very interesting from the point of view of general algebra, is via the notions of universal and homogeneous structures.

Definition 28.15. Let **K** be a class of $\mathscr{L}$-structures and let $\mathfrak{m}$ be a cardinal. An $\mathscr{L}$-structure $\mathfrak{A}$ is $\mathfrak{m}$-*universal over* **K** provided that every member of **K** of power $<\mathfrak{m}$ can be embedded in $\mathfrak{A}$. This is the general algebraic notion, which we modify as follows for the present purposes. An $\mathscr{L}$-structure $\mathfrak{A}$ is $\mathfrak{m}$-*universal* provided that every model $\mathfrak{B} \equiv_{ee} \mathfrak{A}$ of power $<\mathfrak{m}$ can be elementarily embedded in $\mathfrak{A}$. $\mathfrak{A}$ is *universal* iff it is $|A|$-universal.

An $\mathscr{L}$-structure $\mathfrak{A}$ is *algebraically* $\mathfrak{m}$-*homogeneous over* **K** provided that for every substructure $\mathfrak{B}$ of $\mathfrak{A}$ and for every isomorphism f of $\mathfrak{B}$ into $\mathfrak{A}$, if $|B| < \mathfrak{m}$ and $\mathfrak{B} \in \mathbf{K}$, then f can be extended to an automorphism of $\mathfrak{A}$. Again, we modify this for logical purposes. An $\mathscr{L}$-structure $\mathfrak{A}$ is $\mathfrak{m}$-*homogeneous* iff for all $\alpha < \mathfrak{m}$ and all $x \in {}^{\alpha}A$ and $y \in {}^{\alpha}A$, if $(\mathfrak{A}, x_\xi)_{\xi<\alpha} \equiv_{ee} (\mathfrak{A}, y_\xi)_{\xi<\alpha}$, then for any $c \in A$ there is a $d \in A$ such that $(\mathfrak{A}, x_\xi, c)_{\xi<\alpha} \equiv_{ee} (\mathfrak{A}, y_\xi, d)_{\xi<\alpha}$. $\mathfrak{A}$ is *homogeneous* iff it is $|A|$-homogeneous.

We will not be working with the purely algebraic notions defined in 28.15. We mention them only in order to be able to informally describe the third method for constructing saturated structures. In fact, for all classes **K** and cardinals $\mathfrak{m}$ having some natural properties one can construct in a straightforward algebraic fashion a structure $\mathfrak{m}$-universal and $\mathfrak{m}$-homogeneous over **K** and of power $\mathfrak{m}$. Applying this general existence theorem to certain very special classes **K** one obtains the existence of universal-homogeneous structures (in the above, logical, sense). We shall now establish that these structures are just the saturated structures. We break the proof into several steps, some of which are independently interesting. First of all it is convenient to slightly reformulate the definition of $\mathfrak{m}$-saturation.

Proposition 28.16. *Let $\mathfrak{A}$ be an $\mathscr{L}$-structure. The following conditions are equivalent:*

(*i*) *$\mathfrak{A}$ is $\mathfrak{m}$-saturated;*

(*ii*) *for all $\alpha < \mathfrak{m}$, if $\mathscr{L}'$ is an expansion of $\mathscr{L}$ obtained by adding new individual constants $\mathbf{c}_\xi$ for each $\xi < \alpha$, if $a \in {}^{\alpha}A$, and if $\Delta \subseteq \mathrm{Fmla}^1_{\mathscr{L}'}$ is such that each finite subset of it can be realized in $(\mathfrak{A}, a_\xi)_{\xi<\alpha}$, then Δ can be realized in $(\mathfrak{A}, a_\xi)_{\xi<\alpha}$.*

Proof. Assume (*i*) and the hypothesis of (*ii*). Let $X = \{a_\xi : \xi < \alpha\}$, and let $\mathscr{L}''$ be an X-expansion of $\mathscr{L}$, obtained by adding new individual constants $\mathbf{d}_x$ for each $x \in X$. For each formula φ of $\mathscr{L}'$ let φ'' be obtained from φ by replacing each individual constant $\mathbf{c}_\xi$ by $\mathbf{d}_{a\xi}$. Then for any $y \in {}^{\omega}A$ and any formula φ of $\mathscr{L}'$, we have $(\mathfrak{A}, a_\xi)_{\xi<\alpha} \models \varphi[y]$ iff $(\mathfrak{A}, x)_{x\in X} \models \varphi''[y]$. It follows that if we let $\Delta'' = \{\varphi'' : \varphi \in \Delta\}$, then each finite subset of Δ'' can be realized in

$(\mathfrak{A}, x)_{x \in X}$. Hence by (*i*), Δ'' can be realized in $(\mathfrak{A}, x)_{x \in X}$, so Δ can be realized in $(\mathfrak{A}, a_\xi)_{\xi < \alpha}$.

Now assume (*ii*), and let X, $\mathscr{L}'$, and Δ be as 28.1. Then there is an $\alpha < \mathfrak{m}$ and a one-one $a: \alpha \twoheadrightarrow X$. Let $\mathscr{L}''$ be an expansion of $\mathscr{L}$ be adjoining new individual constants $\mathbf{c}_\xi$ for $\xi < \alpha$. Say $\mathscr{L}'$ is obtained from $\mathscr{L}$ by adjoining new individual constants $\mathbf{d}_x$ for $x \in X$. For each formula φ of $\mathscr{L}'$ let φ'' be obtained from φ by replacing each individual constant $\mathbf{d}_x$ by $\mathbf{c}a^{-1}x$. Then for any $y \in {}^\omega A$, $(\mathfrak{A}, x)_{x \in X} \models \varphi[y]$ iff $(\mathfrak{A}, a_\xi)_{\xi < \alpha} \models \varphi''[y]$. We proceed further as in the proof of (*i*) $\Rightarrow$ (*ii*). □

Lemma 28.17. *If $\mathfrak{A}$ is $\mathfrak{m}$-saturated, then $\mathfrak{A}$ is $\mathfrak{m}$-homogeneous.*

PROOF. Assume that $\mathfrak{A}$ is an $\mathfrak{m}$-saturated $\mathscr{L}$-structure, $\alpha < \mathfrak{m}$, $x \in {}^\alpha A$, $y \in {}^\alpha A$, $(\mathfrak{A}, x_\xi)_{\xi < \alpha} \equiv_{\mathrm{ee}} (\mathfrak{A}, y_\xi)_{\xi < \alpha}$, and $c \in A$. Let $\mathscr{L}'$ be the expansion of $\mathscr{L}$ to a language for $(\mathfrak{A}, x_\xi)_{\xi < \alpha}$. Set $\Delta = \{\varphi \in \mathrm{Fmla}^1_{\mathscr{L}'} : (\mathfrak{A}, x_\xi)_{\xi < \alpha} \models \varphi[c]\}$. Since $(\mathfrak{A}, x_\xi)_{\xi < \alpha} \equiv_{\mathrm{ee}} (\mathfrak{A}, y_\xi)_{\xi < \alpha}$, each finite subset of Δ can be realized in $(\mathfrak{A}, y_\xi)_{\xi < \alpha}$. Hence by 28.16, Δ can be realized in $(\mathfrak{A}, y_\xi)_{\xi < \alpha}$. Say $(\mathfrak{A}, y_\xi)_{\xi < \alpha} \models \varphi[d]$ for each $\varphi \in \Delta$. Clearly then $(\mathfrak{A}, x_\xi, c)_{\xi < \alpha} \equiv_{\mathrm{ee}} (\mathfrak{A}, y_\xi, d)_{\xi < \alpha}$. □

Lemma 28.18. *If $\mathfrak{A}$ is $\mathfrak{m}$-saturated, $\mathfrak{A} \equiv_{\mathrm{ee}} \mathfrak{B}$, and $b \in {}^{\mathfrak{m}}B$, then there is an $a \in {}^{\mathfrak{m}}A$ such that $(\mathfrak{A}, a_\xi)_{\xi < \mathfrak{m}} \equiv_{\mathrm{ee}} (\mathfrak{B}, b_\xi)_{\xi < \mathfrak{m}}$.*

PROOF. Assume the hypothesis. We define the sequence $a \in {}^{\mathfrak{m}}A$ by induction so that for each $\alpha \leq \mathfrak{m}$ the following condition holds:

$$(1) \qquad (\mathfrak{A}, a_\xi)_{\xi < \alpha} \equiv_{\mathrm{ee}} (\mathfrak{B}, b_\xi)_{\xi < \alpha}.$$

Thus (1) is given for $\alpha = 0$. Suppose a_ξ has been defined for each $\xi < \alpha$ so that (1) holds, where $\alpha < \mathfrak{m}$. Let $\Delta = \{\varphi : \mathrm{Fv}\varphi \subseteq \{v_0\}, \varphi$ is in the language of $(\mathfrak{B}, b_\xi)_{\xi < \alpha}$, and $(\mathfrak{B}, b_\xi)_{\xi < \alpha} \models \varphi[b_\alpha]\}$. From (1) it is clear that each finite subset of Δ can be realized in $(\mathfrak{A}, a_\xi)_{\xi < \alpha}$. Hence, by 28.16, choose a_α so that a_α realizes Δ in $(\mathfrak{A}, a_\xi)_{\xi < \alpha}$. Thus (1) holds with α replaced by $\alpha + 1$, the definition of a is complete, and (1) holds for $\alpha = \mathfrak{m}$. □

Lemma 28.19. *If $\mathfrak{A}$ is $\mathfrak{m}$-saturated, then $\mathfrak{A}$ is $\mathfrak{m}^+$-universal.*

PROOF. Let $\mathfrak{A}$ be $\mathfrak{m}$-saturated, and assume that $\mathfrak{A} \equiv_{\mathrm{ee}} \mathfrak{B}$ and $|B| \leq \mathfrak{m}$. Let $b: \mathfrak{m} \twoheadrightarrow B$ be an onto map. Then by 28.18 there is an $a \in {}^{\mathfrak{m}}A$ such that $(\mathfrak{A}, a_\xi)_{\xi < \mathfrak{m}} \equiv_{\mathrm{ee}} (\mathfrak{B}, b_\xi)_{\xi < \mathfrak{m}}$.

Let $f = \{(b_\xi, a_\xi) : \xi < \mathfrak{m}\}$. Clearly f is a one-one function mapping B into A. Now let φ be any formula of $\mathscr{L}$, say $\varphi \in \mathrm{Fmla}^n_{\mathscr{L}}$, and let $x \in {}^nB$. Say $x_i = b_{\xi i}$ for each $i < n$. Then

$$\begin{aligned}
\mathfrak{B} \models \varphi[x] \quad & \text{iff } \mathfrak{B} \models \varphi[b_{\xi 0}, \ldots, b_{\xi(n-1)}] \\
& \text{iff } (\mathfrak{B}, b_\xi)_{\xi < \mathfrak{m}} \models \varphi(\mathbf{c}_{\xi 0}, \ldots, \mathbf{c}_{\xi(n-1)}) \\
& \text{iff } (\mathfrak{A}, a_\xi)_{\xi < \mathfrak{m}} \models \varphi(\mathbf{c}_{\xi 0}, \ldots, \mathbf{c}_{\xi(n-1)}) \\
& \text{iff } \mathfrak{A} \models \varphi[f \circ x].
\end{aligned}$$
□

Lemma 28.20. *Let $\mathscr{L}$ be a first-order language and assume that* $|\mathrm{Fmla}_{\mathscr{L}}| < \mathfrak{m}$. *If* $\mathfrak{A}$ *is* $\mathfrak{m}$*-universal and* $\mathfrak{m}$*-homogeneous, then* $\mathfrak{A}$ *is* $\mathfrak{m}$*-saturated.*

PROOF. Suppose that $\mathfrak{A}$ is $\mathfrak{m}$-universal and $\mathfrak{m}$-homogeneous, and assume that the hypothesis of 28.16(*ii*) holds. Expand $\mathscr{L}'$ to $\mathscr{L}''$ by adjoining a new individual constant **d**. Now consider the set

$$\Gamma = \Theta\rho(\mathfrak{A}, a_\xi)_{\xi<\alpha} \cup \{\varphi(\mathbf{d}) : \varphi \in \Delta\}.$$

Our hypothesis of 28.16(*ii*) implies that every finite subset of Γ has a model which is an expansion of $(\mathfrak{A}, a_\xi)_{\xi<\alpha}$. Hence Γ has a model $(\mathfrak{B}, b_\xi, e)_{\xi<\alpha}$ of power $<\mathfrak{m}$. Thus $\mathfrak{A} \equiv_{\mathrm{ee}} \mathfrak{B}$, so by the $\mathfrak{m}$-universality of $\mathfrak{A}$ we may assume that $\mathfrak{B} \preccurlyeq \mathfrak{A}$. Hence $(\mathfrak{A}, a_\xi)_{\xi<\alpha} \equiv_{\mathrm{ee}} (\mathfrak{B}, b_\xi)_{\xi<\alpha} \equiv_{\mathrm{ee}} (\mathfrak{A}, b_\xi)_{\xi<\alpha}$. Furthermore, e realizes Δ in $(\mathfrak{B}, b_\xi)_{\xi<\alpha}$ and hence in $(\mathfrak{A}, b_\xi)_{\xi<\alpha}$. From the $\mathfrak{m}$-homogeneity of $\mathfrak{A}$ it follows that there is an $x \in A$ with $(\mathfrak{A}, a_\xi, x)_{\xi<\alpha} \equiv_{\mathrm{ee}} (\mathfrak{A}, b_\xi, e)_{\xi<\alpha}$. Hence x realizes Δ in $(\mathfrak{A}, a_\xi)_{\xi<\alpha}$. □

Combining the essential content of the preceding lemmas, we obtain

Theorem 28.21. *Let $\mathscr{L}$ be a first-order language,* $\mathfrak{A}$ *an* $\mathscr{L}$*-structure,* $\mathfrak{m}$ *a cardinal, and assume that* $|\mathrm{Fmla}_{\mathscr{L}}| < \mathfrak{m}$. *Then the following conditions are equivalent:*

(*i*) $\mathfrak{A}$ *is* $\mathfrak{m}$*-saturated;*
(*ii*) $\mathfrak{A}$ *is* $\mathfrak{m}^+$*-universal and* $\mathfrak{m}$*-homogeneous;*
(*iii*) $\mathfrak{A}$ *is* $\mathfrak{m}$*-universal and* $\mathfrak{m}$*-homogeneous.*

Now we turn to the question of uniqueness of saturated structures.

Theorem 28.22. *If* $\mathfrak{A}$ *and* $\mathfrak{B}$ *are elementarily equivalent saturated structures of the same power, then* $\mathfrak{A} \cong \mathfrak{B}$.

PROOF. Let $|A| = |B| = \mathfrak{m}$, and let $a: \mathfrak{m} \rightarrowtail\!\!\!\!\twoheadrightarrow A$ and $b: \mathfrak{m} \rightarrowtail\!\!\!\!\twoheadrightarrow B$. We use the by now very familiar back-and-forth method to show that $\mathfrak{A} \cong \mathfrak{B}$; otherwise the proof is similar to that for 28.18. We define sequences $x \in {}^{\mathfrak{m}}A$, $y \in {}^{\mathfrak{m}}B$ by induction so that for each $\alpha \leq \mathfrak{m}$ the following condition holds:

(1) $$(\mathfrak{A}, x_\xi)_{\xi<\alpha} \equiv_{\mathrm{ee}} (\mathfrak{B}, y_\xi)_{\xi<\alpha}.$$

Thus (1) is given for $\alpha = 0$. Suppose x_ξ and y_ξ have been defined for each $\xi < \alpha$ so that (1) holds, where $\alpha < \mathfrak{m}$. If α is even, we proceed as follows. Choose ξ minimum such that $a_\xi \notin x^*\alpha$, and set $x_\alpha = a_\xi$. Let $\Delta = \{\varphi : \mathrm{Fv}\,\varphi \subseteq \{v_0\}$, φ is in the language of $(\mathfrak{A}, x_\xi)_{\xi<\alpha}$, and $(\mathfrak{A}, x_\xi)_{\xi<\alpha} \models \varphi[x_\alpha]\}$. From (1) it is clear that each finite subset of Δ can be realized in $(\mathfrak{B}, y_\xi)_{\xi<\alpha}$ so, since $\mathfrak{B}$ is $\mathfrak{m}$-saturated, by 28.16 we see that there is a $y_\alpha \in B$ which realizes Δ in $(\mathfrak{B}, y_\xi)_{\xi<\alpha}$. If α is odd, we interchange the roles of a and b, x and y, $\mathfrak{A}$ and $\mathfrak{B}$ above. Clearly (1) now holds with α replaced by $\alpha + 1$, the definition of x and y is complete, and (1) holds for $\alpha = \mathfrak{m}$.

Let $f = \{(x_\xi, y_\xi) : \xi < \alpha\}$. We claim that f is the desired isomorphism. To prove this, first note:

(2) if $\alpha < \mathfrak{m}$, and $\alpha = \beta + n$ where β is a limit ordinal and $n \in \omega$, then $a_\alpha \in x^*(\beta + 2n + 1)$ and $b_\alpha \in y^*(\beta + 2n + 2)$.

This condition is easily proved by induction on α. Hence f has domain A and range B. It is then clear from (1) for $\alpha = \mathfrak{m}$ that f is the desired isomorphism. □

We now want to give a couple of applications of saturated models. Saturated models have been extensively used in logic, and these two applications just give a very small sample of their effectiveness for logical problems. The first, rather simple, application is a corollary of the following property of universal structures.

Theorem 28.23. *If $\mathfrak{A}$ is an $\mathfrak{m}^+$-universal $\mathscr{L}$-structure of power $\mathfrak{m} \geq |\text{Fmla}_{\mathscr{L}}|$, then $\mathfrak{A}$ is isomorphic to a proper elementary substructure of itself.*

Proof. Assume the hypothesis. Let $\mathfrak{B}$ be a proper elementary extension of $\mathfrak{A}$. Choose $b \in B \sim A$. Let $(\mathfrak{C}, a, b)_{a \in A}$ be an elementary substructure of $(\mathfrak{B}, a, b)_{a \in A}$ of power $\mathfrak{m}$. Thus $A \subseteq C$, and in fact $\mathfrak{A} \preccurlyeq \mathfrak{C}$. For, if φ is any formula and $x \in {}^\omega A$, then $\mathfrak{A} \models \varphi[x]$ iff $\mathfrak{B} \models \varphi[x]$ iff $\mathfrak{C} \models \varphi[x]$. Also, note that $b \notin A$, so $\mathfrak{A}$ is a proper elementary substructure of $\mathfrak{C}$. Since $\mathfrak{A}$ is $\mathfrak{m}^+$-universal and $\mathfrak{A} \equiv_{ee} \mathfrak{C}$, there is an elementary embedding f of $\mathfrak{C}$ into $\mathfrak{A}$. Clearly $f \restriction A$ is an isomorphism from $\mathfrak{A}$ onto a proper elementary substructure of $\mathfrak{A}$. □

Corollary 28.24 (GCH). *If $\mathfrak{A}$ is any infinite $\mathscr{L}$-structure and $|\text{Fmla}_{\mathscr{L}}| \leq \mathfrak{m}$, then there is a $\mathfrak{B} \equiv_{ee} \mathfrak{A}$ of power $\mathfrak{m}$ such that $\mathfrak{B}$ is isomorphic to a proper elementary substructure of itself.*

Proof. By Corollary 28.12, let $\mathfrak{C}$ be a saturated elementary extension of $\mathfrak{A}$ of any power $\mathfrak{n} \geq \mathfrak{m}$. By Theorem 28.21, $\mathfrak{C}$ is $\mathfrak{n}^+$-universal, so by Theorem 28.23, there is an isomorphism f of $\mathfrak{C}$ onto a proper elementary substructure $\mathfrak{D}$ of $\mathfrak{C}$. Expand $\mathscr{L}$ to $\mathscr{L}'$ by adjoining a unary relation symbol $\mathbf{P}$ and a unary operation symbol $\mathbf{O}$. Then for any $\varphi \in \text{Fmla}^n_{\mathscr{L}}$, the following sentence holds in $(\mathfrak{C}, D, f)$:

$$(1) \qquad \forall v_0 \cdots \forall v_{n-1}\Big(\bigwedge_{i<n} \mathbf{P}v_i \rightarrow [\varphi \leftrightarrow \varphi^{\mathbf{P}}(\mathbf{O}v_0, \ldots, \mathbf{O}v_{n-1})]\Big),$$

where $\varphi^{\mathbf{P}}$ is obtained from φ by relativizing quantifiers to $\mathbf{P}$. Let $(\mathfrak{B}, E, g)$ be an elementary substructure of $(\mathfrak{C}, D, f)$ of power $\mathfrak{m}$. Since all of the sentences (1) hold in $(\mathfrak{C}, D, f)$, they also hold in $(\mathfrak{B}, E, g)$, and it follows easily that g is an isomorphism of $\mathfrak{B}$ onto a proper elementary substructure of itself. □

We will see at the end of this chapter that GCH can be eliminated in 28.24. Our second application depends on the following two important properties of saturated structures. The first proposition is obvious.

Proposition 28.25. *If $\mathfrak{A}$ is an $\mathfrak{m}$-saturated $\mathscr{L}$-structure and $\mathscr{L}'$ is a reduct of $\mathscr{L}$, then $\mathfrak{A} \upharpoonright \mathscr{L}'$ is $\mathfrak{m}$-saturated.*

Proposition 28.26. *Let $\mathscr{L}$ be a language having a unary relation symbol* **P**, *and suppose that $\mathfrak{A}$ is an $\mathfrak{m}$-saturated $\mathscr{L}$-structure satisfying the following conditions:*

(*i*) $|\mathbf{P}^{\mathfrak{A}}| \geq \aleph_0$;
(*ii*) *for each operation symbol* **O** *of $\mathscr{L}$, the set $\mathbf{P}^{\mathfrak{A}}$ is closed under $\mathbf{O}^{\mathfrak{A}}$.*

Let $\mathfrak{B}$ be the $\mathscr{L}$-structure such that $B = \mathbf{P}^{\mathfrak{A}}$, $\mathbf{R}^{\mathfrak{B}} = \mathbf{R}^{\mathfrak{A}} \cap {}^m\mathbf{P}^{\mathfrak{A}}$ for each m-ary relation symbol **R** *of $\mathscr{L}$, and $\mathbf{O}^{\mathfrak{B}} = \mathbf{O}^{\mathfrak{A}} \upharpoonright {}^m\mathbf{P}^{\mathfrak{A}}$ for each m-ary operation symbol* **O** *of $\mathscr{L}$.*

Then $\mathfrak{B}$ *is $\mathfrak{m}$-saturated, and $|B| \geq \mathfrak{m}$.*

Proof. Suppose $X \subseteq \mathbf{P}^{\mathfrak{A}}$ with $|X| < \mathfrak{m}$, and let $\mathscr{L}'$ be an X-expansion of $\mathscr{L}$. Assume that $\Delta \subseteq \mathrm{Fmla}^1_{\mathscr{L}'}$ is such that each finite subset of Δ can be realized in $(\mathfrak{B}, x)_{x\in X}$. For each formula φ of $\mathscr{L}'$ let φ^* be obtained from φ by relativizing quantifiers to **P**. Then for any $b \in \mathbf{P}^{\mathfrak{A}}$,

(1) $$(\mathfrak{A}, x)_{x\in X} \models \varphi^*[b] \quad \text{iff } (\mathfrak{B}, x)_{x\in X} \models \varphi[b].$$

Now let $\Delta^* = \{\varphi^* : \varphi \in \Delta\} \cup \{\mathbf{P}v_0\}$. Then each finite subset of Δ^* can be realized in $(\mathfrak{A}, x)_{x\in X}$, by (1), so Δ^* is realized by some element b. Clearly $b \in B$ and, by (1) b realizes Δ in $(\mathfrak{B}, x)_{x\in X}$. Thus $\mathfrak{B}$ is $\mathfrak{m}$-saturated. Obviously $|B| \geq \mathfrak{m}$. □

The construction of 28.26 is of general interest; see, e.g., 25.6–25.8.

Theorem 28.27 (Specker) (GCH). *Let $\mathscr{L}$ be a first-order language having unary relation symbols* **P** and **Q**. *Suppose that $\mathscr{L}'$ is another first-order language, and that f and g are isomorphisms from $\mathscr{L}'$ into $\mathscr{L}$ (see 26.19). Let $\mathfrak{A}$ be an $\mathscr{L}$-structure, and assume the following:*

(*i*) $|\mathbf{P}^{\mathfrak{A}}|, |\mathbf{Q}^{\mathfrak{A}}| \geq \aleph_0$;
(*ii*) *for each operation symbol* **O** *of $\mathscr{L}'$, $\mathbf{P}^{\mathfrak{A}}$ is closed under $(f\mathbf{O})^{\mathfrak{A}}$, and $\mathbf{Q}^{\mathfrak{A}}$ is closed under $(g\mathbf{O})^{\mathfrak{A}}$.*

Now let $\mathfrak{B}$ and $\mathfrak{C}$ be the $\mathscr{L}'$-structures such that $B = \mathbf{P}^{\mathfrak{A}}$, $C = \mathbf{Q}^{\mathfrak{A}}$, while if **R** *is an m-ary relation symbol of $\mathscr{L}'$ then $\mathbf{R}^{\mathfrak{B}} = (f\mathbf{R})^{\mathfrak{A}} \cap {}^m\mathbf{P}^{\mathfrak{A}}$ and $\mathbf{R}^{\mathfrak{C}} = (g\mathbf{R})^{\mathfrak{A}} \cap {}^m\mathbf{Q}^{\mathfrak{A}}$, and if* **O** *is an m-ary operation symbol of $\mathscr{L}'$ then $\mathbf{O}^{\mathfrak{B}} = (f\mathbf{O})^{\mathfrak{A}} \upharpoonright {}^m\mathbf{P}^{\mathfrak{A}}$ and $\mathbf{O}^{\mathfrak{C}} = (g\mathbf{O})^{\mathfrak{A}} \upharpoonright {}^m\mathbf{Q}^{\mathfrak{A}}$. Assume that $\mathfrak{B} \equiv_{ee} \mathfrak{C}$.*

Under all of these assumptions there is a structure $\mathfrak{A}' \equiv_{ee} \mathfrak{A}$, such that if $\mathfrak{B}'$ and $\mathfrak{C}'$ are formed from $\mathfrak{A}'$ similarly to the formation of $\mathfrak{B}$ and $\mathfrak{C}$ from $\mathfrak{A}$, then $\mathfrak{B}' \cong \mathfrak{C}'$.

Proof. By Corollary 28.12 choose $\mathfrak{A}'$ to be a saturated structure elementarily equivalent to $\mathfrak{A}$. By 28.25 and 28.26, $\mathfrak{B}'$ and $\mathfrak{C}'$ are saturated, and $|B'| = |C'| = |A|$. Hence by 28.22 (since clearly $\mathfrak{B}' \equiv_{ee} \mathfrak{C}'$ because $\mathfrak{B} \equiv_{ee} \mathfrak{C}$), $\mathfrak{B}' \cong \mathfrak{C}'$. □

Again, we shall see shortly how to eliminate GCH. These two applications of saturated structures are typical in the need to assume GCH. For many applications, one can show by delicate considerations concerning models of set theory that the GCH is unnecessary. Even in such cases the problem remains to devise a proof within ordinary set theory without GCH for the result in question. A small modification of the notion of saturated structure turns out to be sufficient for most problems:

Definition 28.28. An $\mathscr{L}$-structure $\mathfrak{A}$ is *special* if and only if either $|A|$ is a successor cardinal, and $\mathfrak{A}$ is saturated, or else $|A|$ is a limit cardinal and there is a sequence $\langle \mathfrak{B}_{\mathfrak{m}} : \mathfrak{m} < |A| \rangle$ such that:

(*i*) each $\mathfrak{B}_{\mathfrak{m}}$ is $\mathfrak{m}^+$-saturated;
(*ii*) $\mathfrak{B}_{\mathfrak{m}} \preccurlyeq \mathfrak{B}_{\mathfrak{n}}$ whenever $\mathfrak{m} < \mathfrak{n} < |A|$;
(*iii*) $\mathfrak{A} = \bigcup_{\mathfrak{m} < |A|} \mathfrak{B}_{\mathfrak{m}}$.

We shall not go into the theory of special models very much; we restrict ourselves to an existence theorem and a uniqueness theorem.

Theorem 28.29. *Suppose $\mathfrak{A}$ is an $\mathscr{L}$-structure and $\mathfrak{m}$ is a cardinal. Assume that $2^{\mathfrak{n}} \leq \mathfrak{m}$ whenever $\mathfrak{n} < \mathfrak{m}$, that $|\mathrm{Fmla}_{\mathscr{L}}| < \mathfrak{m}$, and that $\aleph_0 \leq |A| \leq \mathfrak{m}$. Then $\mathfrak{A}$ has a special elementary extension of power $\mathfrak{m}$.*

Proof. If $\mathfrak{m} = \mathfrak{n}^+$ for some $\mathfrak{n}$, then the assumptions imply that $2^{\mathfrak{n}} = \mathfrak{n}^+$, so the desired result is clear from 28.12. So assume that $\mathfrak{m}$ is a limit cardinal. We define the desired sequence $\langle \mathfrak{B}_{\mathfrak{n}} : \mathfrak{n} < \mathfrak{m} \rangle$, leading up to the desired structure, by induction on $\mathfrak{n}$. Set $\mathfrak{p} = |\mathrm{Fmla}_{\mathscr{L}}| + |A|$. Thus $2^{\mathfrak{q}} \leq \mathfrak{m}$. By 28.12, let $\mathfrak{B}_0$ be a $\mathfrak{p}^+$-saturated elementary extension of $\mathfrak{A}$ of power $2^{\mathfrak{q}}$, and for $0 < \mathfrak{q} < \mathfrak{p}$, let $\mathfrak{B}_{\mathfrak{q}} = \mathfrak{B}_0$. Now suppose that $\mathfrak{p} \leq \mathfrak{n} < \mathfrak{m}$ and $\mathfrak{B}_{\mathfrak{q}}$ has been defined for all $\mathfrak{q} < \mathfrak{n}$ so that $\mathfrak{B}_{\mathfrak{q}} \preccurlyeq \mathfrak{B}_{\mathfrak{r}}$ for $\mathfrak{q} < \mathfrak{r} < \mathfrak{n}$, and $|B_{\mathfrak{q}}| \leq 2^{\mathfrak{p}+\mathfrak{q}}$. Then $|\bigcup_{\mathfrak{q}<\mathfrak{n}} \mathfrak{B}_{\mathfrak{q}}| \leq 2^{\mathfrak{n}}$, so by 28.12 let $\mathfrak{B}_{\mathfrak{n}}$ be an $\mathfrak{n}^+$-saturated elementary extension of $\bigcup_{\mathfrak{q}<\mathfrak{n}} \mathfrak{B}_{\mathfrak{q}}$ of power $2^{\mathfrak{n}}$. This completes the definition of $\langle \mathfrak{B}_{\mathfrak{n}} : \mathfrak{n} < \mathfrak{m} \rangle$. Clearly $\bigcup_{\mathfrak{n}<\mathfrak{m}} \mathfrak{B}_{\mathfrak{n}}$ is as desired in the theorem. □

Note that, by this theorem, without GCH, any infinite $\mathscr{L}$-structure $\mathfrak{A}$ has a special elementary extension. In fact, pick any cardinal $\mathfrak{m} > |A| + |\mathrm{Fmla}_{\mathscr{L}}|$ such that $\mathfrak{n} < \mathfrak{m}$ implies $2^{\mathfrak{n}} \leq \mathfrak{m}$. For example, $\mathfrak{m}$ can be the supremum of the sequence

$$|A| + |\mathrm{Fmla}_{\mathscr{L}}|, \exp(|A| + |\mathrm{Fmla}_{\mathscr{L}}|), \exp\exp(|A| + |\mathrm{Fmla}_{\mathscr{L}}|), \ldots.$$

Then apply 28.28. On the other hand, without GCH one cannot always find a saturated elementary extension of a structure; see exercise 28.40.

Theorem 28.30. *If $\mathfrak{A}$ and $\mathfrak{C}$ are elementary equivalent special $\mathscr{L}$-structures of the same non-denumerable cardinality, then $\mathfrak{A} \cong \mathfrak{C}$.*

PROOF. We may assume that $\mathfrak{m} = |A|$ is a limit cardinal. Let $\langle \mathfrak{B}_{\mathfrak{n}} : \mathfrak{n} < \mathfrak{m} \rangle$ be a sequence of the sort described in 28.28 for $\mathfrak{A}$, and let $\langle \mathfrak{D}_{\mathfrak{n}} : \mathfrak{n} < \mathfrak{m} \rangle$ be a similar sequence for $\mathfrak{C}$. Let $\mathfrak{n} : \alpha \rightarrowtail\!\!\twoheadrightarrow \{\mathfrak{p} : \aleph_0 \leq \mathfrak{p} < \mathfrak{m}\}$, where $\mathfrak{n}$ is strictly increasing and hence α is a limit ordinal. Choose $a : \mathfrak{m} \rightarrowtail\!\!\twoheadrightarrow A$ and $c : \mathfrak{m} \rightarrowtail\!\!\twoheadrightarrow C$. We now modify the sequence a; more precisely, we define a new sequence $a' : \mathfrak{m} \to A$ by recursion. Suppose that $\xi < \mathfrak{m}$ and $a' \restriction \xi$ has been defined. Choose $\gamma < \alpha$ minimum such that $\xi < \mathfrak{n}_\gamma$. Then choose η minimum such that $a_\eta \in B_{\mathfrak{n}\gamma} \sim \{a'_\gamma : \gamma < \xi\}$, which is possible since $|B_{\mathfrak{n}\gamma}| \geq \mathfrak{n}_\gamma^+$ by 28.28 and 28.3, and set $a'_\xi = a_\eta$. Thus clearly

(1) $$a' \restriction \mathfrak{n}_\gamma \in {}^{\mathfrak{n}\gamma}B_{\mathfrak{n}\gamma} \qquad \text{for each } \gamma < \alpha.$$
(2) $$a' : \mathfrak{m} \rightarrowtail\!\!\twoheadrightarrow A.$$

For, obviously a' is one-one. Suppose $\operatorname{Rng} a' \neq A$, and choose η minimum such that $a_\eta \notin \operatorname{Rng} a'$. Choose $\gamma < \alpha$ minimum such that $a_\eta \in B_{\mathfrak{n}\gamma}$. For each $\tau < \eta$ there is a $\theta_\tau < \mathfrak{m}$ such that $a_\tau = a'_{\theta\tau}$. Let $\xi = (\bigcup_{\tau < \eta} \theta_\tau + 1) \cup \mathfrak{n}_\gamma$. Let γ' be minimum such that $\xi < \mathfrak{n}_{\gamma'}$. Thus $\gamma < \gamma'$, and η is minimum such that $a_\eta \in B_{\mathfrak{n}\gamma'} \sim \{a'_\mu : \mu < \xi\}$, since $a_\tau = a'_{\theta\tau}$ and $\theta_\tau < \xi$ for each $\tau < \eta$. Hence $a'_\xi = a_\eta$, contradiction. So (2) holds. Similarly we can define $c' : \mathfrak{m} \rightarrowtail\!\!\twoheadrightarrow C$ so that

(4) $$c' \restriction \mathfrak{n}_\gamma \in {}^{\mathfrak{n}\gamma}D_{\mathfrak{n}\gamma} \qquad \text{for each } \gamma < \alpha;$$
(5) $$c' : \mathfrak{m} \rightarrowtail\!\!\twoheadrightarrow C.$$

Now we define two sequences $\langle x_\gamma : \gamma < \alpha \rangle$ and $\langle y_\gamma : \gamma < \alpha \rangle$ by recursion; we want $x_\gamma : \mathfrak{n}_\gamma \to B_{\mathfrak{n}\gamma}$ and $y_\gamma : \mathfrak{n}_\gamma \to D_{\mathfrak{n}\gamma}$ for each $\gamma < \alpha$. Suppose we have already defined x_γ and y_γ for all $\gamma < \beta$, where $\beta < \alpha$, in such a way that

(6) $$(\mathfrak{A}, x_{\gamma\nu})_{\gamma<\beta,\nu<\mathfrak{n}\gamma} \equiv_{\mathrm{ee}} (\mathfrak{C}, y_{\gamma\nu})_{\gamma<\beta,\nu<\mathfrak{n}\gamma};$$
(7) $$x_\gamma : \mathfrak{n}_\gamma \to B_{\mathfrak{n}\gamma} \text{ and } y_\gamma : \mathfrak{n}_\gamma : \mathfrak{n}_\gamma \to D_{\mathfrak{n}\gamma} \qquad \text{for each } \gamma < \beta.$$

We now consider two cases

Case 1. $\beta = \delta + 2n$ for some δ, a limit ordinal or 0, and some $n \in \omega$. Let $x\beta$ be any function mapping $\mathfrak{n}_\beta$ onto $\{a'\gamma : \gamma < \mathfrak{n}_\beta\}$. Thus by (1), $x\beta : \mathfrak{n}_\beta \to B_{\mathfrak{n}\beta}$. Now

$$|\{y_{\gamma\nu} : \gamma < \beta, \nu < \mathfrak{n}_\gamma\}| \leq \sum_{\gamma<\beta} \mathfrak{n}_\gamma \leq \mathfrak{n}_\beta;$$
$$y_{\gamma\nu} \in D_{\mathfrak{n}\gamma} \subseteq D_{\mathfrak{n}\beta} \text{ for } \gamma < \beta \text{ and } \nu < \mathfrak{n}_\gamma;$$
$$(\mathfrak{C}, y_{\gamma\nu})_{\gamma<\beta,\nu<\mathfrak{n}\gamma} \equiv_{\mathrm{ee}} (\mathfrak{D}_{\mathfrak{n}\beta}, y_{\gamma\nu})_{\gamma<\beta,\nu<\mathfrak{n}\gamma}.$$

Since $\mathfrak{D}_{\mathfrak{n}\beta}$ is $\mathfrak{n}_\beta^+$-saturated, clearly $(\mathfrak{D}_{\mathfrak{n}\beta}, y_{\gamma\nu})_{\gamma<\beta,\nu<\mathfrak{n}\gamma}$ is also. Hence from (6) and all of the above we easily infer from 28.18 that there is a $y\beta : \mathfrak{n}\beta \to D_{\mathfrak{n}\beta}$ such that $(\mathfrak{A}, x_{\gamma\nu})_{\gamma\leq\beta,\nu<\mathfrak{n}\gamma} \equiv_{\mathrm{ee}} (\mathfrak{D}_{\mathfrak{n}\beta}, y_{\gamma\nu})_{\gamma\leq\beta,\nu<\mathfrak{n}\gamma}$. Thus (6) and (7) now hold with β replaced by $\beta + 1$.

Case 2. β is odd. We reverse the roles of x and y, etc.

The rest of the proof now follows a familiar pattern; cf. the proof of 28.22. □

Now we return to the consideration of our two applications of saturated structures. The first one can now clearly by proved without GCH using special structures as soon as we prove

Lemma 28.31. *If $\mathfrak{A}$ is an uncountable special structure, then $\mathfrak{A}$ is $|A|^+$ universal.*

The proof of this lemma is, however, just a "one-sided" version of the proof of 28.30; see the relationship between the proofs of 28.18 and 28.22.

Turning to the second application of saturated structures, we first note that the analog of 28.25 is clear:

Proposition 28.32. *If $\mathfrak{A}$ is a special $\mathscr{L}$-structure and $\mathscr{L}'$ is a reduct of $\mathscr{L}$, then $\mathfrak{A} \upharpoonright \mathscr{L}'$ is special.*

The analog of 28.26 also holds. Since it is not quite so clear, we sketch the proof of it.

Proposition 28.33. *Let $\mathscr{L}$ be a language having a unary relation symbol* **P**, *and suppose that $\mathfrak{A}$ is a special $\mathscr{L}$-structure satisfying the following conditions:*

(*i*) $|\mathbf{P}^{\mathfrak{A}}| \geq \aleph_0$;
(*ii*) *for each operation symbol* **O** *of $\mathscr{L}$, the set $\mathbf{P}^{\mathfrak{A}}$ is closed under $\mathbf{O}^{\mathfrak{A}}$.*

Let $\mathfrak{B}$ be the $\mathscr{L}$-structure such that $B = \mathbf{P}^{\mathfrak{A}}$, $\mathbf{R}^{\mathfrak{B}} = \mathbf{R}^{\mathfrak{A}} \cap {}^m\mathbf{P}^{\mathfrak{A}}$ for each m-ary relation symbol **R** *of $\mathscr{L}$, and $\mathbf{O}^{\mathfrak{B}} = \mathbf{O}^{\mathfrak{A}} \upharpoonright {}^m\mathbf{P}^{\mathfrak{A}}$ for each m-ary operation symbol* **O** *of $\mathscr{L}$.*

Then $\mathfrak{B}$ is special, and $|B| = |A|$.

Proof. We take only the case in which $\mathfrak{m} = |A|$ is a limit cardinal. Let $\langle \mathfrak{C}_\mathfrak{n} : \mathfrak{n} < \mathfrak{m} \rangle$ be a sequence as indicated in 28.28 for $\mathfrak{A}$. For each $\mathfrak{n} < \mathfrak{m}$ let $\mathfrak{D}_\mathfrak{n}$ be obtained from $\mathfrak{C}_\mathfrak{n}$ like $\mathfrak{B}$ was obtained from $\mathfrak{A}$. Then by 28.26, each structure $\mathfrak{D}_\mathfrak{n}$ is $\mathfrak{n}^+$-saturated and of power $\geq \mathfrak{n}^+$. Clearly $\mathfrak{D}_\mathfrak{n} \preccurlyeq \mathfrak{D}_\mathfrak{p}$ whenever $\mathfrak{n} < \mathfrak{p} < \mathfrak{m}$, and $\mathfrak{B} = \bigcup_{\mathfrak{n}<\mathfrak{m}} \mathfrak{D}_\mathfrak{n}$. □

From 28.32 and 28.33 one can prove 28.27 without using GCH, by replacing saturated structures by special structures in the proof of 28.27.

EXERCISES

28.34. Let $\mathfrak{m} \geq \aleph_0$. Give an example of a 0-saturated structure of power $\mathfrak{m}$ which is not $\mathfrak{m}$-saturated.

28.35. Every complete theory with infinite models in a countable language has 0-saturated models of each power $\geq \exp \aleph_0$.

28.36. Let Γ be a theory in a language $\mathscr{L}$, and let $\mathscr{L}'$ be an expansion of $\mathscr{L}$ by adjoining m new individual constants, where $m \in \omega$. Show that there is a one-one correspondence between 1-types over Γ in $\mathscr{L}'$ and $(m + 1)$-types over Γ in $\mathscr{L}$. (See the proof of 28.7.)

28.37. Assume the hypotheses of 27.12. Then the following condition is equivalent to (*i*) and (*ii*) of 27.12:

(*) Γ has a model which is both $\aleph_0$-saturated and elementarily prime.

28.38. Let Γ be a complete theory in a countable language. Then Γ cannot have exactly two nonisomorphic countable models.

28.39. Prove the lemma (*) following 28.14 using ultraproducts instead of the compactness theorem.

28.40. Find a denumerable structure $\mathfrak{A}$ such that $\mathfrak{A}$ has a saturated elementary extension of power $\aleph_1$ iff $\exp \aleph_0 = \aleph_1$.

28.41. If $\mathfrak{A}_\alpha$ is $\mathfrak{m}$-saturated for each $\alpha < \mathfrak{n}$, $\mathfrak{A}_\alpha \preccurlyeq \mathfrak{A}_\beta$ whenever $\alpha < \beta < \mathfrak{n}$, and $\mathfrak{n}$ is regular, then $\bigcup_{\alpha < \mathfrak{n}} \mathfrak{A}_\alpha$ is $\mathfrak{m}$-saturated.

28.42. Find the denumerable $\aleph_0$-saturated model of the theory of 27.20.

PART V

Unusual Logics

In this part we shall treat several kinds of logic which differ from first-order logic in standard formalization, which has been the almost exclusive topic in Parts II, III, and IV. Our aim is not to give as comprehensive a treatment as for first-order logic, but just to introduce various new forms of logic, prove one or two important theorems about the new logic, and state without proof some further results. To avoid consideration of trivialities, we shall not formulate the foundations of the new logics as carefully as we have for first-order logic. The one or two results we select usually show strong connections, or strong differences, of the new logic when compared with first-order logic.

Inessential Variations 29

We begin by considering four logics which are usually considered to be inessential variations of first-order logic with equality: first-order logic without equality, description operators, Hilbert's ε-operator, and many-sorted logic.

Logic without Equality

Here we modify the notion of a first-order language only by eliminating the equality symbol. Throughout this section, unless otherwise mentioned, the languages referred to will be equality-free. Most of the facts concerning logic with or without equality are similar, and we shall not state all the important results even without proof. The notion of an $\mathscr{L}$-structure remains the same, and the relation $\mathfrak{A} \models \varphi[x]$ and related concepts are defined as before. For logical axioms we can take the following formulas, where φ and ψ are arbitrary formulas and α is any variable:

φ, for φ a tautology;
$\forall\alpha(\varphi \to \psi) \to (\varphi \to \forall\alpha\psi)$ if α does not occur free in φ;
$\forall\alpha\varphi \to \mathrm{Subf}_\tau^\alpha\varphi$, if no free occurrence of α in φ is within the scope of a quantifier on a variable appearing in τ.

This leads to a notion $\Gamma \vdash_{\mathrm{ne}} \varphi$ as in Chapter 10. Various elementary facts about $\vdash_{\mathrm{ne}}$ are proved just as in Chapter 10. Of course we still have the notion $\Gamma \models \varphi$ as in Chapter 11. We can prove the model existence theorem for logic without equality, and it even becomes simpler. We shall carry this through in detail for relational languages.

Definition 29.1. Let $\mathscr{L}'$ be a rich expansion of $\mathscr{L}$ by C, and let S be a family of sets of sentences of $\mathscr{L}'$. Then S is a *consistency family* (*for* $\mathscr{L}, \mathscr{L}'$)

iff for each $\Gamma \in S$ all of the following hold, for all sentences φ, ψ of $\mathscr{L}'$:

(C0) if $\Delta \subseteq \Gamma$, then $\Delta \in S$;
(C1) $\varphi \notin \Gamma$ or $\neg\varphi \notin \Gamma$;
(C2) if $\neg\varphi \in \Gamma$, then $\Gamma \cup \{\varphi^{\rightarrow}\} \in S$;
(C3) if $\varphi \wedge \psi \in \Gamma$, then $\Gamma \cup \{\varphi\} \in S$ and $\Gamma \cup \{\psi\} \in S$;
(C4) if $\varphi \vee \psi \in \Gamma$, then $\Gamma \cup \{\varphi\} \in S$ or $\Gamma \cup \{\psi\} \in S$;
(C5) if $\forall\alpha\varphi \in \Gamma$, then for all $\mathbf{c} \in C$, $\Gamma \cup \{\mathrm{Subf}_{\mathbf{c}}^{\alpha}\varphi\} \in S$;
(C6) if $\exists\alpha\varphi \in \Gamma$, then for some $\mathbf{c} \in C$, $\Gamma \cup \{\mathrm{Subf}_{\mathbf{c}}^{\alpha}\varphi\} \in S$;
(C7) if α is a limit ordinal $< |\mathrm{Fmla}_{\mathscr{L}}|$, $\Theta_\beta \in S$ for each $\beta < \alpha$, $\Theta_\beta \subseteq \Theta_\gamma$ whenever $\beta < \gamma < \alpha$, and $|\{\mathbf{c} \in C : \mathbf{c}$ occurs in some $\varphi \in \Theta_\beta$ for some $\beta < \alpha\}| < |C|$, then $\bigcup_{\beta<\alpha} \Theta_\beta \in S$;
(C8) $|\{\mathbf{c} \in C : \mathbf{c}$ occurs in some $\varphi \in \Gamma\}| < |C|$.

Theorem 29.2. *Let $\mathscr{L}$ be a relational language, and let $\mathscr{L}'$ be a rich expansion of $\mathscr{L}$ by C. Let S be a consistency family, and let $\Gamma \in S$, Then Γ has a model $\mathfrak{A}$ with $A = C$, and hence $|A| = |\mathrm{Fmla}_{\mathscr{L}}|$.*

PROOF. Let $\mathfrak{m} = |\mathrm{Fmla}_{\mathscr{L}}|$, $\varphi: \mathfrak{m} \rightarrowtail\!\!\!\twoheadrightarrow \mathrm{Sent}_{\mathscr{L}'}$, and let a well-ordering of C be given. We now define a sequence $\langle \Theta_\alpha : \alpha \leq \mathfrak{m}\rangle$ of members of S. Let $\Theta_0 = \Gamma$. Suppose that $\Theta_\alpha \in S$ has been defined, where $\alpha < \mathfrak{m}$. Let

$$\Theta'_\alpha = \Theta_\alpha \quad \text{if } \Theta_\alpha \cup \{\varphi_\alpha\} \notin S;$$
$$\Theta'_\alpha = \Theta_\alpha \cup \{\varphi_\alpha\} \quad \text{otherwise;}$$
$$\Theta''_\alpha = \Theta'_\alpha \cup \{\varphi_s\} \quad \text{if } \Theta_\alpha \cup \{\varphi_\alpha\} \in S,\ \varphi_\alpha = \varphi_s \vee \varphi_t, \text{ and } \Theta'_\alpha \cup \{\varphi_s\} \in S,$$
$$\Theta''_\alpha = \Theta'_\alpha \cup \{\varphi_t\} \quad \text{if } \Theta_\alpha \cup \{\varphi_\alpha\} \in S,\ \varphi_\alpha = \varphi_s \vee \varphi_t, \text{ and } \Theta'_\alpha \cup \{\varphi_s\} \notin S,$$
$$\Theta''_\alpha = \Theta'_\alpha \quad \text{otherwise;}$$
$$\Theta_{\alpha+1} = \Theta''_\alpha \cup \{\mathrm{Subf}_{\mathbf{c}}^{\beta}\psi\} \quad \text{if } \Theta_\alpha \cup \{\varphi_\alpha\} \in S,\ \varphi_\alpha = \exists\beta\psi, \text{ and } \mathbf{c} \text{ is the least member of } C \text{ such that } \Theta'_\alpha \cup \{\mathrm{Subf}_{\mathbf{c}}^{\beta}\psi\} \in S,$$
$$\Theta_{\alpha+1} = \Theta'''_\alpha \quad \text{otherwise.}$$

Clearly $\Theta_{\alpha+1} \in S$. For λ a limit ordinal $\leq\mathfrak{m}$ let $\Theta_\lambda = \bigcup_{\alpha<\lambda} \Theta_\alpha$. Clearly $\Theta_\lambda \in S$ if $\lambda < \mathfrak{m}$. This completes the definition of $\langle \Theta_\alpha : \alpha \leq \mathfrak{m}\rangle$. All of the following hold, for all sentences φ, ψ of $\mathscr{L}'$.

(1) $\varphi \notin \Theta_\mathfrak{m}$ or $\neg\varphi \notin \Theta_\mathfrak{m}$;
(2) if $\neg\varphi \in \Theta_\mathfrak{m}$, then $\varphi^{\rightarrow}\ \Theta_\mathfrak{m}$;
(3) if $\varphi \wedge \psi \in \Theta_\mathfrak{m}$, then $\varphi, \psi \in \Theta_\mathfrak{m}$;
(4) if $\varphi \vee \psi \in \Theta_\mathfrak{m}$, then $\varphi \in \Theta_\mathfrak{m}$ or $\psi \in \Theta_\mathfrak{m}$;
(5) if $\forall\alpha\varphi \in \Theta_\mathfrak{m}$, then for all $\mathbf{c} \in C$, $\mathrm{Sub\ f}_{\mathbf{c}}^{\alpha}\varphi \in \Theta_\mathfrak{m}$;
(6) if $\exists\alpha\varphi \in \Theta_\mathfrak{m}$, then for some $\mathbf{c} \in C$, $\mathrm{Sub\ f}_{\mathbf{c}}^{\alpha}\varphi \in \Theta_\mathfrak{m}$;

Let $A = C$, and for each m-ary relation symbol $\mathbf{R}$ of $\mathscr{L}$ let $\mathbf{R}^{\mathfrak{A}} = \{\mathbf{c} \in {}^m C : \mathbf{R}\mathbf{c}_0 \cdots \mathbf{c}_{m-1} \in \Theta_\mathfrak{m}\}$. For $\mathbf{c} \in C$ let $\mathbf{c}^{\mathfrak{A}} = \mathbf{c}$. This defines an $\mathscr{L}'$-structure $\mathfrak{A}$.

(7) for any sentence φ of $\mathscr{L}'$, if $\varphi \in \Theta_\mathfrak{m}$ then $\mathfrak{A} \models \varphi$, while if $\neg\varphi \in \mathbf{Q}_\mathfrak{m}$ then $\mathfrak{A} \models \neg\varphi$.

The proof of (7), by induction on φ, is straightforward using (1)–(6); see the proof of 18.9. Hence the theorem follows. □

From 29.2 we immediately obtain the completeness theorem for logic without equality:

Theorem 29.3. *Let $\mathscr{L}$ be a relational language, Γ a set of sentences of $\mathscr{L}$ consistent in the sense $\vdash_{\mathrm{ne}}$. Then Γ has a model of power $|\mathrm{Fmla}_{\mathscr{L}}|$.*

Corollary 29.4. $\Gamma \vdash_{\mathrm{ne}} \varphi$ *iff* $\Gamma \models \varphi$ *iff* $\Gamma \vdash \varphi$.

Corollary 29.5. $\vdash_{\mathrm{ne}} \varphi$ *iff* $\models \varphi$ *iff* $\vdash \varphi$.

Let us go into a little more detail about the relationship between logic with and without equality.

Definition 29.6. If $\mathscr{L}$ is a first-order language with equality, we let $\mathscr{L}^{\mathrm{e}}$ be the first-order language without equality in which equality is treated as a new binary relation symbol.

Thus $\mathscr{L}^{\mathrm{e}}$ is obtained from $\mathscr{L}$ by moving the equality symbol from the list of logical constants to the list of nonlogical constants. If $\mathfrak{A}$ is an $\mathscr{L}$-structure, then $(\mathfrak{A}, =)$ is a natural kind of $\mathscr{L}^{\mathrm{e}}$-structure. But of course in $\mathscr{L}^{\mathrm{e}}$-structures equality can be interepreted as any binary relation whatsoever. The relationship between $\mathscr{L}$ and $\mathscr{L}^{\mathrm{e}}$ can be expressed more fully using the following concepts.

Definition 29.7. Let $\mathscr{L}$ be a relational first-order language without equality, and let $\mathfrak{A}$ and $\mathfrak{B}$ be $\mathscr{L}$-structures. A *two-way homomorphism from $\mathfrak{A}$ onto $\mathfrak{B}$* is a function f mapping A onto B such that for every m-ary relation symbol $\mathbf{R}$ and every $a \in {}^{m}A$, $a \in \mathbf{R}^{\mathfrak{A}}$ iff $f \circ a \in \mathbf{R}^{\mathfrak{B}}$.

If $\mathscr{L}$ is a relational first-order language with equality and $\Gamma \subseteq \mathrm{Fmla}_{\mathscr{L}}$, then $\mathrm{E}\Gamma$ is the following set of sentences of $\mathscr{L}$:

$\forall v_0(v_0 = v_0)$
$\forall v_0 \forall v_1(v_0 = v_1 \rightarrow v_1 = v_0)$
$\forall v_0 \forall v_1 \forall v_2(v_0 = v_1 \wedge v_1 = v_2 \rightarrow v_0 = v_2)$
$\forall v_0 \cdots \forall v_{2m-1}(\mathbf{R} v_0 \cdots v_{m-1} \wedge \bigwedge_{i<m} v_i = v_{m+i}$
$\rightarrow \mathbf{R} v_m \cdots v_{2m-1})$ for any m-ary relation symbol $\mathbf{R}$ which occurs in some formula of Γ.

The following proposition is easily established by induction on φ:

Proposition 29.8. *Let $\mathscr{L}$, $\mathfrak{A}$, $\mathfrak{B}$ and f be as in the first part of 29.7. Let φ be a formula and $x \in {}^{\omega}A$. Then $\mathfrak{A} \models \varphi[x]$ iff $\mathfrak{B} \models \varphi[f \circ x]$.*

From the very definition of satisfaction we obtain the following proposition:

Proposition 29.9. *Let $\mathscr{L}$ be a relational first-order language with equality, φ an $\mathscr{L}$-sentence, and $\mathfrak{A}$ an $\mathscr{L}$-structure. Then $\mathfrak{A} \models_{\mathscr{L}} \varphi$ iff $(\mathfrak{A}, =) \models_{\mathscr{L}^e} \varphi$.*

Proposition 29.10. *Let $\mathscr{L}$ be a relational first-order language with equality, $\varphi \in \mathrm{Sent}_{\mathscr{L}}$, and Γ a set of $\mathscr{L}$-sentences such that equality does not appear in any sentence of Γ.*

Then $\Gamma \models_{\mathscr{L}} \varphi$ iff $\Gamma \models_{\mathscr{L}^e} \bigwedge E\{\varphi\} \rightarrow \varphi$.

Proof. First assume that $\Gamma \models_{\mathscr{L}} \varphi$. Let $(\mathfrak{A}, E)$ be any model in the $\mathscr{L}^e$-sense of $\Gamma \cup \mathrm{E}\{\varphi\}$, where $\mathfrak{A}$ is the underlying $\mathscr{L}$-structure. Then E is an equivalence relation on A. We form an $\mathscr{L}$-structure $\mathfrak{A}/E$ with universe A/E by setting $\mathbf{R}^{\mathfrak{A}/E} = \{([a_0], \ldots, [a_{m-1}]) : (a_0, \ldots, a_{m-1}) \in \mathbf{R}^{\mathfrak{A}}\}$ for each m-ary relation symbol $\mathbf{R}$ of $\mathscr{L}$. For each $a \in A$ let $fa = [a]$. It is easily verified that f is a two-way homomorphism of $\mathfrak{A}$ onto $\mathfrak{A}/E$. From 29.8 we see, then, that $(\mathfrak{A}/E, =)$ is a model of Γ in the $\mathscr{L}^e$-sense, so obviously $\mathfrak{A}/E$ is a model of Γ in the $\mathscr{L}$-sense. Hence by assumption $\mathfrak{A}/E \models_{\mathscr{L}} \varphi$. Thus 29.9 yields $(\mathfrak{A}/E, =) \models_{\mathscr{L}^e} \varphi$, and so by 29.8 we get $(\mathfrak{A}, E) \models_{\mathscr{L}^e} \varphi$, as desired.

Conversely, suppose $\Gamma \models_{\mathscr{L}^e} \bigwedge \mathrm{E}\{\varphi\} \rightarrow \varphi$, and let $\mathfrak{A}$ be any model of Γ. Then $(\mathfrak{A}, =)$ is a model of $\Gamma \cup \mathrm{E}\{\varphi\}$ in the $\mathscr{L}^e$-sense, so it is a model of φ. By 29.9 this means that $\mathfrak{A} \models_{\mathscr{L}} \varphi$. □

If we recall 16.51, we obtain the following lemma.

Lemma 29.11. *Let $\mathscr{L}$ be a first-order language without equality, and with exactly two non-logical constants, both binary relation symbols. Then $\mathscr{g}^{+*}\{\varphi : \varphi \in \mathrm{Sent}_{\mathscr{L}}, \models_{\mathscr{L}} \varphi\}$ is not recursive.*

We can apply the method of proof of 16.51 to improve 29.11:

Lemma 29.12. *The theory of one binary relation, in logic without equality, is undecidable.*

Proof. Let $\mathscr{L}$ be as in 29.11, with $\mathbf{R}$ and $\mathbf{S}$ the nonlogical constants, and let $\mathscr{L}'$ be a language without equality with single nonlogical constant $\mathbf{T}$, a binary relation symbol. With each formula φ of $\mathscr{L}$ we associate a formula φ^0 of $\mathscr{L}'$; we define φ^0 by recursion on φ. For $\varphi = \mathbf{R}v_iv_j$, let v_k and v_l be the first two new variables, and set

$$\varphi^0 = \exists v_k[\mathbf{T}v_kv_k \wedge \mathbf{T}v_kv_i \wedge \exists v_l(\mathbf{T}v_kv_l \wedge \neg\mathbf{T}v_lv_k \wedge \mathbf{T}v_lv_j)].$$

For $\varphi = \mathbf{S}v_iv_j$, let v_k and v_l be the first two new variables, and set

$$\varphi^0 = \exists v_k[\exists v_l(\mathbf{T}v_kv_l \wedge \mathbf{T}v_lv_k \wedge \neg\mathbf{T}v_lv_l) \wedge \mathbf{T}v_kv_i \wedge \exists v_l(\mathbf{T}v_kv_l \wedge \mathbf{T}v_lv_j \wedge \neg\mathbf{T}v_lv_k)].$$

Further, we let

$$(\varphi \wedge \psi)^0 = \varphi^0 \wedge \psi^0, (\varphi \vee \psi)^0 = \varphi^0 \vee \psi^0, (\neg\varphi)^0 = \neg\varphi^0,$$
$$(\forall\alpha\varphi)^0 = \forall\alpha(\neg\exists\beta(\mathbf{T}\alpha\beta) \rightarrow \varphi^0),$$

where β is the first variable not occurring in $\forall\alpha\varphi$. Now:

(1) if $\mathfrak{A}$ is any $\mathscr{L}$-structure, then there is an $\mathscr{L}'$-structure $\mathfrak{B}$ such that $\mathfrak{A} \models \varphi[x]$ iff $\mathfrak{B} \models \varphi^0[x]$ for every formula φ and every $x \in {}^{\omega}A$.

For, given $\mathfrak{A}$ we introduce new elements S_{ab0}, S_{ab1} for each $(a, b) \in \mathbf{R}^{\mathfrak{A}}$ and new elements t_{ab0}, t_{ab1}, t_{ab2} for each $(a, b) \in \mathbf{S}^{\mathfrak{A}}$. These elements are to lie outside A and should be distinct from each other for different choices of (a, b). Let B be A together with all of the new elements. For each $(a, b) \in \mathbf{R}^{\mathfrak{A}}$ the following set is to be a part of $\mathbf{T}^{\mathfrak{B}}$:

$$\{(S_{ab0}, S_{ab0}), (S_{ab0}, a), (S_{ab0}, S_{ab1}), (S_{ab1}, b)\}.$$

For each $(a, b) \in \mathbf{S}^{\mathfrak{A}}$ the following is a part of $\mathbf{T}^{\mathfrak{B}}$:

$$\{(t_{ab0}, t_{ab1}), (t_{ab1}, t_{ab0}), (t_{ab0}, a), (t_{ab0}, t_{ab2}), (t_{ab2}, b)\}.$$

This completes the definition of $\mathfrak{B}$. It is routine to check (1). Next,

(2) if $\mathfrak{B}$ is an $\mathscr{L}'$-structure for which $\{b \in B : \forall c(b, c) \notin \mathbf{T}^{\mathfrak{B}}\} \neq 0$, then there is an $\mathscr{L}$-structure $\mathfrak{A}$ such that $\mathfrak{A}$ and $\mathfrak{B}$ are related as in (1).

For, let $A = \{b \in B : \forall c(b, c) \notin \mathbf{T}^{\mathfrak{A}}\}$, and set

$$\mathbf{R}^{\mathfrak{A}} = \{(a, b) : \mathfrak{B} \models (\mathbf{R}v_0v_1)^0[a, b]\},$$
$$\mathbf{S}^{\mathfrak{A}} = \{(a, b) : \mathfrak{B} \models (\mathbf{S}v_0v_1)^0[a, b]\}.$$

Then (2) is easily checked. From (1) and (2) we easily get

(3) for any sentence φ of $\mathscr{L}$, $\models \varphi$ iff $\models \exists v_0\forall v_1 \neg \mathbf{T}v_0v_1 \rightarrow \varphi^0$.

Our lemma now follows. □

Theorem 29.13. *Let $\mathscr{L}$ be a relational first-order language without equality and with at least one relation symbol of rank ≥ 2. Then $\{\varphi : \models \varphi\}$ is undecidable.*

Proof. Let $\mathbf{R}$ be a relation symbol of $\mathscr{L}$ of rank ≥ 2. Let $\mathscr{L}'$ be a first-order language with sole non-logical constant the binary relation symbol $\mathbf{S}$. With each formula φ of $\mathscr{L}'$ we associate the formula φ^0 of $\mathscr{L}$ obtained by replacing each atomic part $\mathbf{S}v_iv_j$ of φ by $\mathbf{R}v_iv_jv_j \cdots v_j$. Clearly $\models \varphi$ iff $\models \varphi^0$, so 29.12 yields our theorem. □

Now we turn to model-theoretic matters. The compactness theorem is, of course, an immediate consequence of 29.3:

Theorem 29.14. *If Γ is a set of sentences, every finite subset of which has a model, then Γ has a model of power $|\mathrm{Fmla}_{\mathscr{L}}|$ (in a relational language).*

The notions of elementary substructure and elementary extension carry-over in an obvious way to logic without equality, and the proof of 19.17 yields:

Theorem 29.15. Let $\mathscr{L}$ *be a relational language without equality,* $\mathfrak{B}$ *an* $\mathscr{L}$-*structure,* $\mathfrak{m}$ *a cardinal such that* $|\mathrm{Fmla}_{\mathscr{L}}| \leq \mathfrak{m} \leq |B|$, *and* C *a subset of* B *such that* $|C| \leq \mathfrak{m}$. Then *there is an elementary substructure* $\mathfrak{A}$ *of* $\mathfrak{B}$ *such that* $C \subseteq A$ *and* $|A| = \mathfrak{m}$.

Concerning the upward Löwenheim–Skolem theorem, we can prove here an even stronger theorem than 19.24. This is perhaps surprising, since our second proof of 19.24, essentially involved equality. Finite structures can be elementarily extended to infinite structures!

Theorem 29.16. *Let* $\mathfrak{A}$ *be an* $\mathscr{L}$-*structure,* $\mathscr{L}$ *a relational language without equality, and let* $A \subseteq B$. *Then there is an* $\mathscr{L}$-*structure* $\mathfrak{B}$ *with universe* B *such that* $\mathfrak{A} \preccurlyeq \mathfrak{B}$.

PROOF. Fix $a \in A$ and define $f: B \to A$ by setting $fx = x$ for $x \in A$ and $fx = a$ for $x \in B \sim A$. For each m-ary relational symbol $\mathbf{R}$, let $\mathbf{R}^{\mathfrak{B}} = \{x \in {}^mB : f \circ x \in \mathbf{R}^{\mathfrak{A}}\}$. This defines our $\mathscr{L}$-structure $\mathfrak{B}$. One easily checks

(1) for any $x \in {}^\omega B$ and any formula φ of $\mathscr{L}$ we have $\mathfrak{B} \models \varphi[x]$ iff $\mathfrak{A} \models \varphi[f \circ x]$.

Thus $\mathfrak{A} \preccurlyeq \mathfrak{B}$. □

Our final theorem for logic without equality is a version of Craig's interpolation theorem; cf. Exercise 25.35, where an even stronger result is stated.

Theorem 29.17. *Let* $\mathscr{L}$ *be a relational language without equality, and let* φ *and* ψ *be sentences of* $\mathscr{L}$ *such that* $\models \varphi \to \psi$ *while neither* $\models \neg \varphi$ *nor* $\models \psi$. *Then there is a sentence* χ *such that* $\models \varphi \to \chi$, $\models \chi \to \psi$, *and every nonlogical constant occurring in* χ *occurs in both* φ *and* ψ.

PROOF. First note:

(1) there is a nonlogical constant common to φ and ψ.

For, suppose not. Let $\mathfrak{A}$ be a model of φ and $\mathfrak{B}$ a model of $\neg\psi$. By taking elementary extensions, by 29.16 we may assume that $A = B$. Let $C = A$, and for $\mathbf{R}$ a relation symbol occurring in φ let $\mathbf{R}^{\mathfrak{C}} = \mathbf{R}^{\mathfrak{A}}$, and let $\mathbf{R}^{\mathfrak{C}} = \mathbf{R}^{\mathfrak{B}}$ otherwise. Clearly $\mathfrak{C}$ is a model of $\varphi \wedge \neg\psi$, contradiction.

Making use of (1), one can now apply the model existence theorem 29.2 just like 18.12 was applied in the proof of 22.1. Instead of $\exists v_0(v_0 = v_0)$, used in the proof of 22.1, one can use any sentence $\theta \vee \neg\theta$ where θ involves only nonlogical constants common to φ and ψ. □

The Description Operator

We now imagine a first-order language *with* equality $\mathscr{L}$ endowed with a new symbol $\boldsymbol{\tau}$ and a distinguished individual constant $\mathbf{O}$. Intuitively, we want to consider expressions like $\boldsymbol{\tau} v_i \varphi$, where φ is any formula, and interpret it as "the unique v_i such that φ, or $\mathbf{O}$ if there is no v_i, or several v_i, such that φ." We want to define for such a language the precise meaning of terms and formulas and their precise interpretation in $\mathscr{L}$-structures. Then we prove the basic fact about descriptions, which for clear reasons is analogous to a similar fact about definitions: description operators can be eliminated. A triple $(\mathscr{L}, \boldsymbol{\tau}, \mathbf{O})$ as above is called a *descriptive triple*.

Definition 29.18. Let $(\mathscr{L}, \boldsymbol{\tau}, \mathbf{O})$ be a descriptive triple. We define simultaneously the notions of terms and formulas for $(\mathscr{L}, \boldsymbol{\tau}, \mathbf{O})$:

(*i*) any variable is a term;
(*ii*) if $\mathbf{O}$ is an operation symbol of rank m and $\sigma_0, \ldots, \sigma_{m-1}$ are terms, then $\mathbf{O}\sigma_0 \cdots \sigma_{m-1}$ is a term;
(*iii*) if σ and ρ are terms, then $\sigma \equiv \rho$ is a formula;
(*iv*) if $\mathbf{R}$ is a relation symbol of rank m and $\sigma_0, \ldots, \sigma_{m-1}$ are terms, then $\mathbf{R}\sigma_0 \cdots \sigma_{m-1}$ is a formula;
(*v*) if φ and ψ are formulas, then so are $\neg\varphi$, $\varphi \vee \psi$, and $\varphi \wedge \psi$;
(*vi*) if α is a variable and φ is a formula, then $\forall\alpha\varphi$ is a formula;
(*vii*) if α is a variable and φ is a formula, then $\boldsymbol{\tau}\alpha\varphi$ is a term.

Some examples of terms are v_i, $\boldsymbol{\tau} v_i(v_i \equiv \mathbf{O})$, $v_i \mathbin{\mathbf{+}} \boldsymbol{\tau} v_j(v_j \equiv \boldsymbol{\tau} v_k(v_k \mathbin{\mathbf{<}} v_i))$, in a language with appropriate operation and relation symbols $\mathbf{O}$, $\mathbf{+}$, $\mathbf{<}$. The notions of free and bound occurrences of variables in formulas or terms (!) is defined as usual. In this regard, v_i is to be regarded as bound wherever it occurs in $\boldsymbol{\tau} v_i \varphi$, i.e., $\boldsymbol{\tau}$ is a variable-binding operator.

Definition 29.19. Let $(\mathscr{L}, \boldsymbol{\tau}, \mathbf{O})$ be a descriptive triple, let $\mathfrak{A}$ be an $\mathscr{L}$-structure, and let $x \in {}^{\omega}A$. We define simultaneously the notions $\sigma^{\mathfrak{A}}x$ (for σ a term) and $\mathfrak{A} \models \varphi[x]$ (for φ a formula):

$v_i^{\mathfrak{A}}x = x_i$;
$(\mathbf{O}\sigma_0 \cdots \sigma_{m-1})^{\mathfrak{A}}x = \mathbf{O}^{\mathfrak{A}}(\sigma_0^{\mathfrak{A}}x, \ldots, \sigma_{m-1}^{\mathfrak{A}}x)$;
$\mathfrak{A} \models (\sigma \equiv \rho)[x]$ iff $\sigma^{\mathfrak{A}}x = \rho^{\mathfrak{A}}x$;
$\mathfrak{A} \models \mathbf{R}\sigma_0 \cdots \sigma_{m-1}[x]$ iff $(\sigma_0^{\mathfrak{A}}x, \ldots, \sigma_{m-1}^{\mathfrak{A}}x) \in \mathbf{R}^{\mathfrak{A}}$;
$\mathfrak{A} \models \neg\varphi[x]$ iff not $(\mathfrak{A} \models \varphi[x])$;
$\mathfrak{A} \models (\varphi \vee \psi)[x]$ iff $\mathfrak{A} \models \varphi[x]$ or $\mathfrak{A} \models \psi[x]$;
$\mathfrak{A} \models (\varphi \wedge \psi)[x]$ iff $\mathfrak{A} \models \varphi[x]$ and $\mathfrak{A} \models \psi[x]$;
$\mathfrak{A} \models \forall v_i\varphi[x]$ iff for all $a \in A$, $\mathfrak{A} \models \varphi[x_a^i]$;
$(\boldsymbol{\tau} v_i\varphi)^{\mathfrak{A}}x$ is the unique $a \in A$ such that $\mathfrak{A} \models \varphi[x_a^i]$, if such an a exists and is unique, and it is $\mathbf{O}^{\mathfrak{A}}$ otherwise.

Before proceeding to our main theorem, we may mention that the main place in formalized mathematics where descriptions are useful is in axiomatic set theory. In fact, the class $\{x : \varphi(x)\}$ is most naturally interpreted as

$$\boldsymbol{\tau} y \forall x[x \in y \leftrightarrow \varphi(x)]$$

in formalized set theory. In precise metamathematical investigations of axiomatic set theory the following theorem is then quite useful.

Theorem 29.20. *Let $(\mathscr{L}, \boldsymbol{\tau}, \mathbf{O})$ be a descriptive triple, Γ a set of sentences in $(\mathscr{L}, \boldsymbol{\tau}, \mathbf{O})$, σ a term of $(\mathscr{L}, \boldsymbol{\tau}, \mathbf{O})$ and φ a formula of $(\mathscr{L}, \boldsymbol{\tau}, \mathbf{O})$. Then*

(*i*) *if v_k is a variable not occurring in σ, then there is a formula ψ not involving τ and with $\mathrm{Fv}\, \psi \subseteq \mathrm{Fv}\, \sigma \cup \{v_k\}$, such that $\Gamma \models \sigma \equiv v_k \leftrightarrow \psi$;*

(*ii*) *there is a formula φ^* not involving τ and with $\mathrm{Fv}\, \varphi^* \subseteq \mathrm{Fv}\, \varphi$ such that $\Gamma \models \varphi \leftrightarrow \varphi^*$.*

Proof. We proceed by simultaneous induction on σ and φ. If σ is v_i, then for ψ in (*i*) we may obviously take the formula $v_i \equiv v_k$. If σ is $\mathbf{O}\rho_0 \cdots \rho_{m-1}$ and v_k does not occur in σ, first choose distinct new variables $v_{l0}, \ldots, v_{l(m-1)}$ not occurring in σ and different from v_k. By the induction assumption there are formulas ψ_i for $i < m$ not involving τ such that

$$\Gamma \models \rho_i \equiv v_{li} \leftrightarrow \psi_i \qquad \text{for each } i < m.$$

Clearly then

$$\Gamma \models \sigma \equiv v_k \leftrightarrow \exists v_{l0} \cdots \exists v_{l(m-1)} \Big(\bigwedge_{i<m} \psi_i \wedge \mathbf{O} v_{l0} \cdots v_{l(m-1)} \equiv v_k \Big).$$

Next, suppose φ is the formula $\sigma \equiv \rho$. Let v_k be a variable not occurring in φ, and by the induction hypothesis let ψ and χ be formulas not involving τ such that $\Gamma \models \sigma \equiv v_k \leftrightarrow \psi$ and $\Gamma \models \rho \equiv v_k \leftrightarrow \chi$. Clearly then

$$\Gamma \models \sigma \equiv \rho \leftrightarrow \exists v_k(\psi \wedge \chi).$$

In all of the above cases the free variable restriction in the theorem can clearly be met. Atomic formulas $\mathbf{R}\sigma_0 \cdots \sigma_{m-1}$ are treated similarly. The induction steps involving $\neg$, $\vee$, $\wedge$, $\forall$ are trivial. Finally, suppose σ in the term $\boldsymbol{\tau} v_i \varphi$, and v_k does not occur in σ. By the induction assumption let φ^* be a formula not involving $\boldsymbol{\tau}$ with $\mathrm{Fv}\, \varphi^* \subseteq \mathrm{Fv}\, \varphi$ and $\Gamma \models \varphi \leftrightarrow \varphi^*$. Let v_l be a new variable. Clearly then

$$\begin{aligned}\Gamma \models \sigma \equiv v_k \leftrightarrow \exists v_l[\forall v_i(\varphi^* \leftrightarrow v_i \equiv v_l) \wedge v_l \equiv v_k] \\ \vee [\neg \exists v_l \forall v_i(\varphi^* \leftrightarrow v_i \equiv v_l) \wedge v_k \equiv \mathbf{O}]. \quad \square\end{aligned}$$

The correspondence from φ to φ^* in 29.20 is clearly effective. A particular case of 29.20 is that $\models \varphi$ iff $\models \varphi^*$, for any sentence φ of $(\mathscr{L}, \boldsymbol{\tau}, \mathbf{O})$. Thus we obtain a weak completeness theorem by 11.21 and 6.10:

Theorem 29.21. *For a descriptive triple* $(\mathscr{L}, \tau, \mathbf{O})$ *the set* $\mathscr{g}^{+*}\{\varphi : \varphi$ *a sentence of* $(\mathscr{L}, \boldsymbol{\tau}, \mathbf{O})$ *and* $\models \varphi\}$ *is recursively enumerable.*

This gives rise to a possibility of giving logical axioms for $\boldsymbol{\tau}$, and this has been done, leading to a notion $\Gamma \vdash_D \varphi$. The completeness theorem can thus be proved for this notion.

The Hilbert ϵ-operator

A *choice triple* is a triple $(\mathscr{L}, \boldsymbol{\varepsilon}, \mathbf{O})$ of the same kind as a descriptive triple, and all the syntactical definitions above are the same. But now we want to interpret $\boldsymbol{\varepsilon} v_i \varphi$ differently, namely ambiguously as any v_i such that φ holds. It is not completely clear how to make this rigorous, but the following way has seemed appropriate.

Definition 29.22. Let $(\mathscr{L}, \boldsymbol{\varepsilon}, \mathbf{O})$ be a choice triple. A *choice structure* for $(\mathscr{L}, \boldsymbol{\varepsilon}, \mathbf{O})$ is a pair $(\mathfrak{A}, f)$ such that $\mathfrak{A}$ is an $\mathscr{L}$-structure and f is a choice function for nonempty subsets of A. The definition of $\sigma^{\mathfrak{A}f}x$ and $(\mathfrak{A}, f) \models \varphi[x]$ for $x \in {}^{\omega}A$ goes just as for the description operator in 29.19 except for the term $\boldsymbol{\varepsilon} v_i \varphi$, where we set

$$(\boldsymbol{\varepsilon} v_i \varphi)^{\mathfrak{A}f} x = f\{a \in A : (\mathfrak{A}, f) \models \varphi[x_a^i]\} \qquad \text{if this set is nonempty;}$$
$$(\boldsymbol{\varepsilon} v_i \varphi)^{\mathfrak{A}f} x = \mathbf{O}^{\mathfrak{A}} \qquad \text{otherwise.}$$

The analog of 29.20 does *not* hold for the $\boldsymbol{\varepsilon}$-operator. A counterexample can be obtained as follows. Let $\mathscr{L}$ be a language with just one nonlogical constant, an individual constant $\mathbf{O}$, and consider a choice triple $(\mathscr{L}, \boldsymbol{\varepsilon}, \mathbf{O})$. Suppose that ψ is a formula not involving ε such that $\models \boldsymbol{\varepsilon} v_0(v_0 = v_0) = v_0 \leftrightarrow \psi$; such a formula would have to exist if the analog of 29.20 holds. Let $\mathfrak{A}$ be any $\mathscr{L}$-structure with $|A| > 1$, and fix $a \in A$. If $\mathfrak{A} \models \psi[a]$, let f be a choice function such that $fA \neq a$. Then $(\boldsymbol{\varepsilon} v_0(v_0 = v_0))^{\mathfrak{A}f} = fA \neq a$, so $(\mathfrak{A}, f) \not\models (\boldsymbol{\varepsilon} v_0(v_0 = v_0) = v_0 \leftrightarrow \psi)[a]$. If $\mathfrak{A} \not\models \psi[a]$, we let f be a choice function such that $fA = a$ and get a similar contradiction.

As is to be expected, however, there *is* a close connection between the $\boldsymbol{\varepsilon}$-operator and Skolem expansions. We describe this in the next definition and results.

Definition 29.23. *Let* $(\mathscr{L}, \boldsymbol{\varepsilon}, \mathbf{O})$ be a choice triple, and let $\mathscr{L}'$ be a Skolem expansion of $\mathscr{L}$, with notation as in 11.33. With terms σ and formulas φ of $(\mathscr{L}, \boldsymbol{\varepsilon}, \mathbf{O})$ we shall associate terms σ^* and formulas φ^* of $\mathscr{L}'$, by recursion:

$$v_i^* = v_i;\ (\mathbf{O}\sigma_0 \cdots \sigma_{m-1})^* = \mathbf{O}\sigma_0^* \cdots \sigma_{m-1}^*;$$
$$(\sigma = \rho)^* = \sigma^* = \rho^*;\ (\mathbf{R}\sigma_0 \cdots \sigma_{m-1})^* = \mathbf{R}\sigma_0^* \cdots \sigma_{m-1}^*;$$
$$(\neg\varphi)^* = \neg\varphi^*,\ (\varphi \vee \psi)^* = \varphi^* \vee \psi^*,\ (\varphi \wedge \psi)^* = \varphi^* \wedge \psi^*,$$
$$(\forall v_i \varphi)^* = \forall v_i \varphi^*;$$

the only nontrivial part of the definition is in dealing with $\varepsilon v_i\varphi$. Let $\mathrm{Fv}\,\exists v_i\varphi = \{v_{j0}, \ldots, v_{j(m-1)}\}$ with $j_0 < \cdots < j_{m-1}$. Choose k minimal such that φ^* is a formula of $\mathscr{L}_k$. Then we set

$$(\varepsilon v_i\varphi)^* = S^k_{\exists vi\varphi^*} v_{j0} \cdots v_{j(m-1)}.$$

Next, let $\mathfrak{A}$ be an $\mathscr{L}$-structure, and let f be a choice function for nonempty subsets of A. We shall define an $\mathscr{L}'$-structure $\mathfrak{B}_{\mathfrak{A}f}$ which is to be an expansion of $\mathfrak{A}$. Thus we must interpret in $\mathfrak{B}_{\mathfrak{A}f}$ all of the new operation symbols $S^k_{\exists\alpha\psi}$. For each term $\varepsilon v_i\varphi$ of $(\mathscr{L}, \varepsilon, \mathbf{O})$ with notation as above, and for each $a_0, \ldots, a_{m-1} \in A$, let $x \in {}^\omega A$ be any sequence with $x_{jt} = a_t$ for each $t < m$ and set

$$S^{k\mathfrak{B}\mathfrak{A}f}_{\exists vi\varphi^*}(a_0, \ldots, a_{m-1}) = f\{a : (\mathfrak{A}, f) \models \varphi[x^i_a]\} \qquad \text{if this is nonempty,}$$
$$S^{k\mathfrak{B}\mathfrak{A}f}_{\exists vi\varphi^*}(a_0, \ldots, a_{m-1}) = \mathbf{O}^{\mathfrak{A}} \qquad \text{otherwise.}$$

Since our mapping * is clearly one-one, this is possible. For $S^k_{\exists\alpha\psi}$ not of the above form, let $S^{k\mathfrak{B}\mathfrak{A}f}_{\exists\alpha\psi}$ be the constant function with value $\mathbf{O}^{\mathfrak{A}}$.

Proposition 29.24. *Let the notation be as above. Suppose that σ is a term of $(\mathscr{L}, \varepsilon, \mathbf{O})$, φ is a formula of $(\mathscr{L}, \varepsilon, \mathbf{O})$, and $x \in {}^\omega A$. Then $\sigma^{\mathfrak{A}f}x = \sigma^{*\mathfrak{B}\mathfrak{A}f}x$ and $(\mathfrak{A}, f) \models \varphi[x]$ iff $\mathfrak{B}_{\mathfrak{A}f} \models \varphi^*[x]$.*

PROOF. The simultaneous inductive proof on σ and φ is straightforward. □

From this proposition we again obtain a weak completeness theorem:

Theorem 29.25. *The set $g^{+*}\{\varphi : \varphi$ is a sentence $(\mathscr{L}, \varepsilon, \mathbf{O})$ and $\models \varphi\}$ is recursively enumerable.*

PROOF. We claim that $\models \varphi$ iff $\Gamma \models \varphi^*$, using the notation of 29.23, where Γ is the following set of sentences of $\mathscr{L}'$:

the Skolem set of $\mathscr{L}'$ over $\mathscr{L}$;
$\chi_{ij\varphi\psi}$ for $i, j < \omega$ and φ, ψ formulas of $\mathscr{L}'$,

where $\chi_{ij\varphi\psi}$ is formed as follows. Let $\mathrm{Fv}\,\exists v_i\varphi = \{v_{k0}, \ldots, v_{k(m-1)}\}$ with $k_0 < \cdots < k_{m-1}$, and $\mathrm{Fv}\,\exists v_j\psi\{v_{l0}, \ldots, v_{l(n-1)}\}$ with $l_0 < \cdots < l_{n-1}$. Choose s minimum such that φ is a formula of $\mathscr{L}_s$, and t minimum such that ψ is a formula of $\mathscr{L}_t$. Let $v_{u0}, \ldots, v_{u(m+n)}$ be the first $m + n + 1$ distinct variables not occurring in $\exists v_i\varphi$ or $\exists v_j\psi$, and set

$$\varphi' = \mathrm{Subf}^{vk0}_{vu0} \cdots \mathrm{Subf}^{vk(m-1)}_{vu(m-1)}\varphi,$$
$$\psi' = \mathrm{Subf}^{vl0}_{vum} \cdots \mathrm{Subf}^{vl(n-1)}_{vu(m+n-1)}\psi.$$

Then let $\chi_{ij\varphi\psi}$ be the sentence

$$\forall v_{u0} \cdots \forall v_{u(m+n-1)}[\forall v_{u(m+n)}(\mathrm{Subf}^{vi}_{vu(m+n)}\varphi' \leftrightarrow \mathrm{Subf}^{vj}_{vu(m+n)}\psi') \rightarrow S^s_{\exists vi\varphi}v_{u0} \cdots v_{u(m-1)} = S^t_{\exists vj\psi}v_{um} \cdots v_{u(m+n-1)}]$$

Our claim can now be routinely checked. Since Γ is clearly effective, the theorem follows. □

As in the case of description operators, Theorem 29.25 implies the possibility of developing a proof theory based upon the ε-operator. This has been done, and the completeness theorem has been proved for the resulting notion $\vdash_\varepsilon$. Two of the major results here are: (1) ("second ε-theorem") if $\Gamma \cup \{\varphi\}$ is an ε-free set of sentences and $\Gamma \vdash_\varepsilon \varphi$, then $\Gamma \models \varphi$; (2) ("first ε-theorem") a formulation of Herbrand's theorem in the ε-language. Note that for any formula φ we have $\models \exists v_0\varphi \leftrightarrow \varphi(\varepsilon v_0\varphi)$. This gives rise to the possibility of founding logic using the ε-symbol and no quantifiers, taking the above as a definition of $\exists$; this has been carefully worked out. Another interesting use of the ε-calculus is in axiomatic set theory. Although the second ε-theorem above implies that nothing is gained by introducing the ε-symbol after setting out the usual axioms for ZF (set theory without the axiom of choice), the situation is different if ε-formulas are allowed in the schema of set formation of ZF. Then the axiom of choice in its usual formulation becomes provable. This is worked out carefully in Bourbaki's treatment of set theory.

Many-Sorted Logic

This variant of first-order logic is considerably more important than the ones above. The basic idea is to allow several universes in the structure instead of only one. As far as syntax is concerned, this is expressed in the following definition:

Definition 29.26. A many sorted language $\mathscr{L}$ is determined by specifying the following. There is a nonempty set $\mathscr{S}$ of sorts. For each $s \in \mathscr{S}$ we have individual variables of sort s:

$$v_0^s, v_1^s, \ldots.$$

The logical symbols are the usual ones: $\neg, \vee, \wedge, \forall, =$. The nonlogical constants are some relation and operation symbols. Each relation symbol and each operation symbol has a rank which is a finite non-empty sequence of members of $\mathscr{S}$.

Given such a language, we define terms and formulas as follows:

(*i*) v_i^s is a term of sort s;
(*ii*) if $\mathbf{O}$ is an operation symbol of rank $(s_0, \ldots, s_m)$ and $\sigma_0, \ldots, \sigma_{m-1}$ are terms of sorts $s_0, \ldots, s_{m-1}$ respectively, then $\mathbf{O}\sigma_0 \cdots \sigma_{m-1}$ is a term of sort s_m;
(*iii*) if σ and τ are terms of the same sort, then $\sigma = \tau$ is a formula;
(*iv*) if $\mathbf{R}$ is a relation symbol of rank $(s_0, \ldots, s_m)$ and $\sigma_0, \ldots, \sigma_m$ are terms of sorts $s_0, \ldots, s_m$ respectively, then $\mathbf{R}\sigma_0 \cdots \sigma_m$ is a formula;
(*v*) if φ and ψ are formulas and α is a variable, all of the following are formulas: $\neg\varphi$, $\varphi \vee \psi$, $\varphi \wedge \psi$, $\forall\alpha\varphi$;
(*vi*) terms and formulas can only be formed in these ways.

The notion of an $\mathscr{L}$-structure for such a language is clear:

Definition 29.27. Let $\mathscr{L}$ be a many sorted language as above. An $\mathscr{L}$-structure is a triple $\mathfrak{A} = (A, f, R)$ such that:

(*i*) A is a function which assigns to each $s \in \mathscr{S}$ a nonempty set A_s;
(*ii*) f is a function whose domain is the set of operation symbols of $\mathscr{L}$; if $\mathbf{O}$ is an operation symbol of rank $(s_0, \ldots, s_m)$, then $f_{\mathbf{O}} : A_{s0} \times \cdots \times A_{s(m-1)} \to A_{sm}$;
(*iii*) $\mathbf{R}$ is a function whose domain is the set of relation symbols of $\mathscr{L}$; if $\mathbf{R}$ is a relation symbol of rank $(s_0, \ldots, s_m)$, then $R_{\mathbf{R}} \subseteq A_{s0} \times \cdots \times A_{sm}$.

Let $\mathscr{L}$ and $\mathfrak{A}$ be as in 29.27. Given $x \in \mathsf{P}_{s\in\mathscr{S}}\, {}^{\omega}A_s$, it is clear how to define the notions $\sigma^{\mathfrak{A}}x$ and $\mathfrak{A} \models \varphi[x]$. This, then, defines the fundamental notions for many-sorted logic.

There is a natural way of relating many sorted logic to ordinary logic. Let $\mathscr{L}$ be a many sorted language, as above. With $\mathscr{L}$ we associate an ordinary first-order language $\mathscr{L}^*$ as follows. The language $\mathscr{L}^*$ is to have the relation and operation symbols of $\mathscr{L}$, and additional unary relation symbols $\mathbf{P}_s$ for $s \in \mathscr{S}$. If $\mathbf{O}$ is an operation symbol of $\mathscr{L}$ of rank $(s_0, \ldots, s_m)$, then as a symbol of $\mathscr{L}^*$ the operation symbol $\mathbf{O}$ will have rank m. For a relation symbol $\mathbf{R}$ of rank $(s_0, \ldots, s_m)$, the symbol $\mathbf{R}$ will have rank $m + 1$. We shall treat the variables v_i^s as the variables of $\mathscr{L}^*$ also. Note that this is, strictly speaking, impossible when $|\mathscr{S}| > \aleph_0$, but for almost all logical purposes it makes no difference. Now with each formula φ of $\mathscr{L}$ we associate the formula φ^* of $\mathscr{L}^*$ obtained by replacing "$\forall v_i^s$" by "$\forall v_i^s(\mathbf{P}_s v_i^s \to$" throughout φ. Let Γ be the set of all of the following sentences of $\mathscr{L}^*$:

$$\exists v_0^s \mathbf{P}_s v_0^s \qquad \text{for each } s \in \mathscr{S};$$

$$\forall v_0^{s0} \forall v_1^{s1} \cdots \forall v_{m-1}^{s(m-1)} \left[\bigwedge_{i<m} \mathbf{P}_{si} v_i^{si} \to \mathbf{P}_{sm} \mathbf{O} v_0^{s0} \cdots v_{m-1}^{s(m-1)} \right]$$

for each operation symbol $\mathbf{O}$ of $\mathscr{L}$ of rank $(s_0, \ldots, s_m)$.

Next, let $\mathfrak{A}$ be any $\mathscr{L}$-structure. Fix $a \in A$. We convert it into an $\mathscr{L}^*$-structure $\mathfrak{A}_a^*$ as follows. Set

$$A^* = \bigcup_{s\in\mathscr{S}} A_s;$$

$$\mathbf{R}^{\mathfrak{A}*} = \mathbf{R}^{\mathfrak{A}};$$

$$\mathbf{O}^{\mathfrak{A}*}(a_0, \ldots, a_{m-1}) = \mathbf{O}^{\mathfrak{A}}(a_0, \ldots, a_{m-1}) \qquad \text{if } a_i \in A_{si} \text{ for each } i > m,$$

$$\mathbf{O}^{\mathfrak{A}*}(a_0, \ldots, a_{m-1}) = a_0 \qquad \text{otherwise},$$

where $\mathbf{O}$ has rank $(s_0, \ldots, s_{m-1})$. The following proposition is then easy to establish.

Proposition 29.28. *Let $\mathscr{L}$ be a many sorted language, $\mathscr{L}^*$ the associated first-order language, $\mathfrak{A}$ an $\mathscr{L}$-structure. Then $\mathfrak{A}^*$ is a model of Γ above. Furthermore, let $x \in \mathsf{P}_{s \in \mathscr{S}} {}^{\omega}A_s$. Then $\sigma^{\mathfrak{A}}x = \sigma^{\mathfrak{A}^*}x$ and $\mathfrak{A} \models \varphi[x]$ iff $\mathfrak{A}^* \models \varphi^*[x]$.*

Corollary 29.29. *For any many-sorted sentence φ, $\models \varphi$ iff $\Gamma \models \varphi^*$.*

Corollary 29.30. *$g^{+*}\{\varphi : \varphi$ is a many-sorted sentence and $\models \varphi\}$ is r.e.*

Again, explicit axiom systems for many-sorted logic have been developed. The model theory for many-sorted logic is rather well developed. Some of the details are outlined in the exercises.

BIBLIOGRAPHY

These references form a more extensive introduction to logic without equality, description operators, ε-operators, and many-sorted logic respectively.

1. Church, A. *Introduction to Mathematical Logic*. Princeton: Princeton Univ. Press (1956).
2. Kalish, D. and Montague, R. Remarks on descriptions and natural deduction. *Arch. Math. Logik u. Grundl.*, *3* (1957), 50–73.
3. Leisenring, A. C. *Mathematical Logic and Hilbert's ε-Symbol*. New York: Gordon and Breach (1969).
4. Mal′cev, A. I. Model correspondences. In *The metamathematics of Algebraic Systems: Collected Papers 1936–1967*. North-Holland (1971), 66–94.

EXERCISES

29.31. Let $\mathscr{L}$ be a relational language without equality. Let $\langle \mathfrak{A}_i : i \in I \rangle$ be a system of $\mathscr{L}$-structures. Let F be an ultrafilter on I. We define a structure $\mathfrak{B} = \mathsf{P}^F_{i \in I} \mathfrak{A}_i$, which serves the role of ultraproducts in logic without equality. Its universe is $B = \mathsf{P}_{i \in I} A_i$. Given an m-ary relation symbol $\mathbf{R}$, we set

$$\mathbf{R}^{\mathfrak{B}} = \{x \in {}^mB : \{i : (x_{0i}, \ldots, x_{m-1,i}) \in \mathbf{R}^{\mathfrak{A}}\} \in F$$

Prove the following version of the fundamental theorem on ultraproducts: If $x \in {}^{\omega}B$ and φ is a formula of $\mathscr{L}$, then the following two conditions are equivalent:

(*i*) $\mathsf{P}^F_{i \in I} \mathfrak{A}_i \models \varphi[x]$;
(*ii*) $\{i \in I : \mathfrak{A}_i \models \varphi[\mathrm{pr}_i \cdot x]\} \in F$.

29.32. Find a set Γ of sentences in a suitable language without equality such that Γ has no finite model.

29.33. Let $\mathscr{L}$ be a relational language without equality, $\mathfrak{A}$ an $\mathscr{L}$-structure. A relation $E \subseteq A \times A$ is a *congruence* relation in $\mathfrak{A}$ if it is an equivalence relation on A, and for any m-ary relation symbol $\mathbf{R}$ of $\mathscr{L}$, if $a, b \in {}^mA$,

$a_i E b_i$ for all $i < m$, and $a \in \mathbf{R}^{\mathfrak{A}}$ then $b \in \mathbf{R}^{\mathfrak{A}}$. Given such an E, we let $\mathfrak{A}/E$ be the $\mathscr{L}$-structure with universe A/E and with

$$\mathbf{R}^{\mathfrak{A}/E} = \{x \in {}^m(A/E) : \text{there exists } a \in \mathbf{R}^{\mathfrak{A}} \text{ with } a_i \in x_i \text{ for all } i < m\}.$$

Show that the mapping f such that $fa = [a]_E$ for all $a \in A$ is a two-way homomorphism from $\mathfrak{A}$ onto $\mathfrak{A}/E$.

29.34. With $\mathscr{L}$ and $\mathfrak{A}$ as in 29.33, there is a maximal congruence on $\mathfrak{A}$.

29.35. For any relational language $\mathscr{L}$ without equality there is a denumerable $\mathscr{L}$-structure $\mathfrak{A}$ and a congruence E on $\mathfrak{A}$ such that $|A/E| = 1$.

29.36. Let $\mathscr{L}$ be a relational language without equality. An $\mathscr{L}$-structure $\mathfrak{A}$ is *primitive* if there is no congruence on $\mathfrak{A}$ except the identity. If $|A| > 1$ and $\mathfrak{A}$ is elementarily equivalent (without equality) to a one-element $\mathscr{L}$-structure, then $\mathfrak{A}$ is not primitive.

29.37. Let $\mathscr{L}$ be a relational language without equality and with only finitely many non logical constants. Suppose that $\mathfrak{A}$ is a finite $\mathscr{L}$-structure, $\mathfrak{A}$ is elementarily equivalent (without equality) to $\mathfrak{B}$, and $|A| < |B|$. Then $\mathfrak{B}$ is not primitive.

29.38. Formulate Definition 29.18 more precisely. *Hint*: let Trmfmla be the set of all pairs $(\tau, 0)$, $(\varphi, 1)$ with τ a term and φ a formula, and define Trmfmla in a standard set-theoretic way.

29.39. Let $(\mathscr{L}, \boldsymbol{\tau}, \mathbf{0})$ be a descriptive triple. Take as logical axioms the schemes 10.23(1)–(5) as well as the following two schemes:

(6) $\forall v_i(v_i \equiv v_j \leftrightarrow \varphi) \rightarrow v_j \equiv \boldsymbol{\tau} v_i \varphi$, if j is minimum such that v_j does not occur in φ;

(7) $\neg \exists v_j \forall v_i(v_i \equiv v_j \leftrightarrow \varphi) \rightarrow \boldsymbol{\tau} v_i \varphi \equiv \mathbf{0}$, with j as in (6).

This gives rise to a notion $\Gamma \vdash_{\mathrm{d}} \varphi$ and the attendant notions, such as d-consistency. Prove the usual versions of the completeness theorem, analogous to 11.19–11.20. *Hint*: reprove 29.20, using $\vdash_{\mathrm{d}}$ instead of $\models$.

29.40. Show that with descriptions the usual validity 10.61, that $\models \forall \alpha \varphi \rightarrow \mathrm{Subf}^{\alpha}_{\sigma} \varphi$ under certain conditions, no longer holds.

29.41. The description operator can be made precise in several non-equivalent ways. We indicate briefly another possibility different from the one expounded in this section. It applies to an arbitrary first-order language, without an individual constant $\mathbf{O}$ being distinguished. But satisfaction is modified as follows. If $\mathfrak{A}$ is an $\mathscr{L}$-structure and $a \in A$, we define $(\mathfrak{A}, a) \models \varphi[x]$ for $x \in {}^m A$ just as in 29.19, except that $(\boldsymbol{\tau} v_i \varphi)^{\mathfrak{A}} x$ is a if there is no unique $b \in A$ such that $(\mathfrak{A}, a) \models \varphi[x^i_b]$. Show that 29.20 then fails, in general.

29.42. The compactness theorem holds for many-sorted logic. *Hint*: Use 29.28.

29.43. Define the ultraproduct of many-sorted structures and prove the fundamental theorem on ultraproducts.

29.44. Formulate and prove a general downward Löwenheim–Skolem theorem for many-sorted logic.

29.45. Formulate and prove a general upward Löwenheim–Skolem theorem for many-sorted logic.

29.46. Let Γ be the theory of vector spaces of infinite dimension over algebraically closed fields, formulated in a two-sorted language. Show how this can be done in a precise way, and show that Γ is complete.

30 Finitary Extensions

In this chapter we want to consider various essential extensions of first-order languages in which the formulas are still of finite length. In particular, we shall consider higher-order logic, ω-logic, and cardinality quantifiers.

Weak Second-Order Logic

Let a first-order language $\mathscr{L}$ be augmented by new quantifiable variables $F_0, F_1, \ldots$ which in an $\mathscr{L}$-structure are interpreted to range over finite subsets of the universe: this gives us a formulation of *weak second-order logic*. Among the atomic formulas are new ones of the form $F_i\sigma$, where σ is a term.

In a weak second-order logic the compactness theorem fails to hold. Perhaps the simplest example is given by the following set of sentences:

$$\exists F_0 \forall v_0 F_0 v_0;$$

there are at least n things (a sentence for each $n \in \omega$).

Various other of our theorems in first-order logic fail in weak second-order logic. Two important theorems are the upward Löwenheim–Skolem theorem and the recursive enumerability of the set of validities. To prove that these do not hold here, consider the following set Q^w of weak second-order sentences, formulated in $\mathscr{L}_{nos}$:

$$Q \qquad \text{(recall 14.17)}$$

$$\forall v_0 \exists F_0 [F_0 v_0 \wedge \forall v_1 (F_0 \mathbf{s} v_1 \rightarrow F_0 v_1)]$$

It is clear that $\mathfrak{A}$ is a model of Q^w if and only if $\mathfrak{A}$ is isomorphic to $(\omega, +, \cdot, 0, \mathscr{s})$. In fact, let $\mathfrak{A} = (A, +, \cdot, 0, f)$ be a model of Q^w. Define $g: \omega \rightarrow A$ recursively by setting $g0 = 0$ and $g(m + 1) = fgm$ for all $m \in \omega$. Thus $gm = (\mathbf{\Delta} m)^{\mathfrak{A}}$

for each $m \in \omega$ (proof by induction on m). The function g preserves $+$ and $\cdot$ because R is a subtheory of Q. Suppose that g does not map onto A, and choose $a \in A \sim g^*\omega$. By axiom 14.17(i)(c) of Q we see that it is possible to define a sequence $\langle x_i : i \in \omega \rangle$ such that $x_0 = a$ and $\mathfrak{s}x_{i+1} = x_i$ for every $i \in \omega$. This clearly contradicts our one weak second-order axiom for Q^w. Hence g is onto, so it is an isomorphism from $(\omega, +, \cdot, \mathfrak{s}, 0)$ onto $\mathfrak{A}$. Thus the upward Löwenheim–Skolem theorem fails. Furthermore, we can repeat the proofs of 15.20 and 15.21 to obtain

Theorem 30.1. *The set $\{\mathscr{g}^+\psi : \psi$ is a sentence of $\mathscr{L}_{\text{nos}}$ in weak second-order logic and $Q^w \models \psi\}$ is not elementarily definable in $(\omega, +, \cdot, \mathfrak{s}, 0)$, in weak second-order logic.*

Now $Q^w \models \psi$ iff $\models \bigwedge Q^w \rightarrow \psi$. Hence $\{\mathscr{g}^+\psi : \models\psi\}$ is not weak second order elementarily definable in $(\omega, +, \cdot, 0, \mathfrak{s})$ either. Thus by 14.16, $\{\mathscr{g}^+\psi : \models \psi\}$ is not recursively enumerable, so there is no hope of finding a reasonable proof theory for this logic.

In view of all of these negative results, it is interesting that the downward Löwenheim–Skolem theorem, 19.17, goes over verbatim. The proof of this fact depends upon the following lemma analogous to 19.16 (where we use a rather obvious notation for satisfaction):

Lemma 30.2. *Let $\mathfrak{A}$ and $\mathfrak{B}$ be $\mathscr{L}$-structures, with $\mathfrak{A} \subseteq \mathfrak{B}$. Then the following two conditions are equivalent:*

(i) *$\mathfrak{A} \preccurlyeq \mathfrak{B}$ in the weak second-order sense;*
(ii) *for every weak second-order formula φ, every $k \in \omega$, every $x \in {}^{\omega}A$, and every $y \in {}^{\omega}\{F : F \subseteq A, |F| < \aleph_0\}$, if $\mathfrak{B} \models \exists v_k\varphi[x; y]$ then there is an $a \in A$ such that $\mathfrak{B} \models \varphi[x^k_a; y]$, and if $\mathfrak{B} \models \exists F_k\varphi[x; y]$ then there is a finite $G \subseteq A$ such that $\mathfrak{B} \models \varphi[x; y^k_G]$.*

The proof is just like for 19.16.

Theorem 30.3 (Tarski). Let *$\mathfrak{B}$ be an $\mathscr{L}$-structure, let $\mathfrak{m}$ be a cardinal such that $|\text{Fmla}_{\mathscr{L}}| \leq \mathfrak{m} \leq |B|$, and let C be a subset of B of power $\leq \mathfrak{m}$.* Then *there is an elementary substructure $\mathfrak{A}$ of $\mathfrak{B}$ (in the weak second-order sense) such that $C \subseteq A$ and $|A| = \mathfrak{m}$.*

PROOF. The proof is very similar to that for 19.17, but for completeness we shall sketch it. Let $\mathscr{C}$ be a choice function for nonempty subsets of $B \cup SB$. We now define a sequence $\langle D_m : m \in \omega \rangle$ by recursion. Let D_0 be any subset of B such that $C \subseteq D_0$ and $|D_0| = \mathfrak{m}$. Fix $d \in D_0$. Now suppose D_m has been defined. Let J_m be the set of all sextuples (G, φ, k, l, x, y) such that G is a finite subset of D_m, φ is a weak second-order formula of $\mathscr{L}$, $k, l \in \omega$, $x \in {}^{\omega} G$, and $x_j = d$ whenever v_j does not occur in φ, and $y \in {}^{\omega}$ SG and $y_j = 0$ whenever F_j does not occur in φ. Clearly equality (5) in the proof of 19.17 still holds. Now let

$$D_{m+1} = D_m \cup \{\mathscr{C}\{a : \mathfrak{B} \models \varphi[x^k_a; y]\} : (G, \varphi, k, l, x, y) \in J_m \text{ and } \mathfrak{B} \models \exists v_k\varphi[x; y]\}$$
$$\cup \{\mathscr{C}\{H : \mathfrak{B} \models \varphi[x; y^l_H]\} : (G, \varphi, k, l, x, y) \in J_m \text{ and } \mathfrak{B} \models \exists F_l\varphi[x; y]\}.$$

The rest of the proof proceeds as before. □

We may mention without proof the following decidability results of Läuchli [4] and Rabin [6]: the weak second-order theory of linear ordering is decidable; the weak second-order theory of one function is decidable.

Monadic Second-Order Logic

In monadic second-order logic we strengthen a first-order language by adjoining variables $P_0, P_1, \ldots$ which are to range over *all* subsets of a given universe. The negative facts mentioned above for weak second-order logic still are true for the monadic case. We need new proofs now, however. Let Q^m consist of the theory Q together with the sentence

$$\forall P_0[P_0\mathbf{O} \wedge \forall v_0(P_0v_0 \rightarrow P_0\mathbf{s}v_0) \rightarrow \forall v_0 P_0 v_0].$$

Then up to isomorphism $(\omega, +, \cdot, 0, \mathfrak{s})$ is the only model of Q^m. Hence the upward Löwenheim–Skolem theorem fails to hold. Thus just as for 30.1 we obtain the important

Theorem 30.4. *The set $\{\mathscr{g}^+\psi : \psi$ is a sentence of $\mathscr{L}_{nos}$ in monadic second-order logic and $Q^m \models \psi\}$ is not elementarily definable in $(\omega, +, \cdot, \mathfrak{s}, 0)$, in monadic second-order logic.*

Thus again $\{\mathscr{g}^+\psi : \models \psi\}$ is not r.e., and there is no reasonable proof theory for this logic. Theorem 30.4 stands in contrast with the following result:

Theorem 30.5. *The set $\{\mathscr{g}^+\psi : \psi$ is a sentence of $\mathscr{L}_{nos}$ in ordinary first-order logic which holds in $(\omega, +, \cdot, 0, \mathfrak{s})$ is elementarily definable in $(\omega, +, \cdot, 0, \mathfrak{s})$ in monadic second-order logic.*

Proof. We wish to formalize in monadic second-order logic the following definition of truth of a sentence in $(\omega, +, \cdot, 0, \mathfrak{s})$. First define a function f mapping the variable-free terms of $\mathscr{L}_{nos}$ into ω as follows:

$$f0 = 0; \qquad f\mathbf{s}\sigma = f\sigma + 1;$$
$$f(\sigma + \tau) = f\sigma + f\tau; \qquad f(\sigma \cdot \tau) = f\sigma \cdot f\tau.$$

Then we can say that a sentence φ holds in $(\omega, +, \cdot, 0, \mathfrak{s})$ if and only if it is a member of the unique set A of sentences such that a sentence ψ is in A iff one of the following conditions holds:

(1) ψ has the form $\sigma \equiv \tau$, and $f\sigma = f\tau$;
(2) ψ has the form $\neg\chi$, and $\chi \notin A$;
(3) ψ has the form $\chi \wedge \theta$, and $\chi \in A$ and $\theta \in A$;

(4) ψ has the form $\chi \vee \theta$, and $\chi \in A$ or $\theta \in A$;
(5) ψ has the form $\forall v_i \chi$, and $\mathrm{Subf}^{vi}_{\Delta m}\chi \in A$ for every $m \in \omega$.

In fact, $\{\varphi \in \mathrm{Sent}_{\mathscr{L}_{\mathrm{nos}}} : (\omega, +, \cdot, \mathfrak{s}, 0) \models \varphi\} = B$ clearly satisfies (1)–(5), and if A satisfies (1)–(5), it is easily seen by induction on the sentence φ that $\varphi \in A$ iff $\varphi \in B$. To formalize this definition, consider the following relations:

$g^{+*}\mathrm{Sent}_{\mathscr{L}_{\mathrm{nos}}}$;
$\{m : m = g^+(\sigma = \tau)$ for some variable-free terms σ, τ such that $f\sigma = f\tau\}$;
$\{(m, n) : m = g^+\varphi$ and $n = g^+(\neg\varphi)$ for some sentence $\varphi\}$;
$\{(m, n, p) : m = g^+\varphi,\ n = g^+\psi$, and $p = g^+(\varphi \wedge \psi)$ for some sentences $\varphi, \psi\}$;
$\{(m, n, p) : m = g^+\varphi,\ n = g^+\psi$, and $p = g^+(\varphi \wedge \psi)$ for some sentences $\varphi, \psi\}$;
$\{(m, n, p) : m = g^+\varphi$ and $p = g^+\forall v_n\varphi$ for some sentence $\forall v_n\varphi\}$;
$\{(m, n, p, q) : m = g^+\varphi$ and $q = g^+\ \mathrm{Sub}f^{vn}_{\Delta p}\varphi$ for some formula φ with $\mathrm{Fv}\varphi \subseteq \{v_n\}\}$.

Clearly all of these relations are recursive. Hence by 14.15 they are elementarily definable in $(\omega, +, \cdot, \mathfrak{s}, 0)$, say by formulas $\varphi_0, \varphi_1, \varphi_2, \varphi_3, \varphi_4, \varphi_5, \varphi_6$, respectively. Hence $\{g^+\psi : \psi$ is a sentence of $\mathscr{L}_{\mathrm{nos}}$ in ordinary first-order logic which holds in $(\omega, +, \cdot, \mathfrak{s}, 0)\}$ is elementarily defined in $(\omega, +, \cdot, \mathfrak{s}, 0)$ by the following formula of monadic second-order logic:

$$\begin{aligned}&\forall P_0[\forall v_0(\varphi_0 \rightarrow \{P_0v_0 \leftrightarrow \varphi_1 \vee \exists v_1[\varphi_2(v_1, v_0)] \wedge P_0v_1\\&\vee \exists v_1\exists v_2[\varphi_3(v_1, v_2, v_0) \wedge P_0v_1 \wedge P_0v_2] \vee \exists v_1\exists v_2[\varphi_4(v_1, v_2, v_0)\\&\wedge (P_0v_1 \vee P_0v_2)] \vee \exists v_1\exists v_2[\varphi_5(v_1, v_2, v_0)\\&\wedge \forall v_3\forall v_4(\varphi_6(v_1, v_2, v_3, v_4) \rightarrow P_0v_4)]\}) \rightarrow P_0v_0].\end{aligned}$$

□

The compactness theorem fails in monadic second-order logic, since we can add to $\mathscr{L}_{\mathrm{nos}}$ a new individual constant $\mathbf{c}$ and consider $\mathrm{Q^m}$ together with all sentences $\neg(\mathbf{c} = \Delta m)$. The downward Löwenheim–Skolem theorem fails, since, for example, we can fully express, by finitely many sentences, the usual definition of the field of reals as a Dedekind complete ordered field; in fact, the ordered field axioms are finitely many first order sentences, and completeness is expressed by the following sentence:

$$\begin{aligned}&\forall P_0(\exists v_0P_0v_0 \wedge \exists v_1\forall v_0(P_0v_0 \rightarrow v_0 < v_1) \rightarrow \exists v_1\{\forall v_0(P_0v_0 \rightarrow v_0 < v_1)\\&\wedge \forall v_2[\forall v_0(P_0v_0 \rightarrow v_0 < v_2) \rightarrow v_1 < v_2 \vee v_1 = v_2]\}).\end{aligned}$$

One of the strongest decidability results in all of logic is Rabin's theorem that the monadic second-order theory of two successor functions is decidable (see the table at the beginning of Chapter 13).

Second-Order Logic

Now we expand the apparatus of monadic second-order logic by adding to first-order logic for each $i \in \omega \sim 1$ variables $R^i_0, R^i_1, \ldots$ ranging over i-ary relations of the universe, and variables $O^i_0, O^i_1, \ldots$ ranging over i-ary operations on the universe; the formation rules and satisfaction and truth rules for this language are obvious. For simplicity we shall delete from this natural definition the variables ranging over operations, retaining only those which range over relations (plus all of the usual first-order set-up). As for first-order logic, this is not an essential restriction, since m-ary operations can be considered as special $(m + 1)$-ary relations. Naturally enough, the negative results concerning monadic second-order logic extend automatically to this case. There are a couple of results about this logic that are worth mentioning in this short survey, however.

Theorem 30.6. *For any formula φ of second-order logic there is a formula ψ in the following very particular prenex normal form such that $\models \varphi \leftrightarrow \psi$:*

$$\psi = \mathbf{Q}_0\alpha_0 \cdots \mathbf{Q}_{m-1}\alpha_{m-1}\forall\beta_0 \cdots \forall\beta_{n-1}\exists\gamma_0 \cdots \exists\gamma_{p-1}\chi,$$

where the α's are all relation variables, the $\mathbf{Q}$'s are quantifiers $\forall$ or $\exists$, the β's and γ's are all individual variables, and χ is quantifier-free.

Proof. The usual transformations yield a prenex normal form for φ. To get the quantifier prefix in the above specific form, we need some facts which will allow us to interchange types of quantifiers. The first fact is obvious:

(1) $$\models \exists v_i \exists R_k^j \psi \leftrightarrow \exists R_k^j \exists v_i \psi.$$
(2) $$\models \forall v_i \exists R_k^j \psi \leftrightarrow \exists R_k^{j+1} \forall v_i \psi',$$

where R_k^{j+1} is new and ψ' is obtained from ψ by replacing each atomic part $R_k^j \sigma_0 \cdots \sigma_{j-1}$ of ψ by $R_k^{j+1}\sigma_0 \cdots \sigma_{j-1} v_i$. We prove (2) somewhat informally, in particular ignoring assignment of values to the variables other than those mentioned in (2). Take a structure $\mathfrak{A}$ and first suppose $R^{(j+1)\mathfrak{A}} \subseteq {}^{j+1}A$ so that for all $a \in A$, $\mathfrak{A} \models \psi'$ under the obvious assignment. Given any $a \in A$, let $R_k^{j\mathfrak{A}} = \{x \in {}^jA : x\langle a \rangle \in R_k^{(j+1)\mathfrak{A}}\}$. Clearly, then $\mathfrak{A} \models \psi$ under this assignment. Conversely, suppose $\mathfrak{A} \models \forall v_i \exists R_k^j \psi$. For each $a \in A$ choose $R_{ka}^j \subseteq {}^jA$ so that ψ will hold under this assignment. Let $R_k^{j+1} = \{x\langle a \rangle : a \in A \text{ and } x \in R_{ka}^j\}$. Clearly then $\mathfrak{A} \models \forall v_i \psi'$ under this assignment.

Taking negations in (2) we easily obtain

(3) $$\models \exists v_i \forall R_k^j \psi \leftrightarrow \forall R_k^{j+1} \exists v_i \psi'.$$

with assumptions similar to the above. Another obvious fact is

(4) $$\models \forall v_i \forall R_k^j \psi \leftrightarrow \forall R_k^j \forall v_i \psi.$$
(5) $$\models \exists v_i \forall v_j \psi \leftrightarrow \exists R_k^1[\exists v_i R_k^1 v_i \wedge \forall v_i \forall v_j (R_k^1 v_i \rightarrow \psi)],$$

where R_k^1 is new. Again we give an informal proof. Assume that $\mathfrak{A} \models \exists v_i \forall v_j \psi$, and choose $a \in A$ so that for all $b \in A$ we have $\mathfrak{A} \models \psi$ under the obvious assignment. Let $R_k^{1\mathfrak{A}} = \{a\}$. Obviously then the right hand side of (5) holds in $\mathfrak{A}$. Now suppose the right-hand side of (5) holds in $\mathfrak{A}$, and choose $R_k^{1\mathfrak{A}} \subseteq A$ accordingly. Fix $a \in R_k^{1\mathfrak{A}}$. Then for any $b \in A$, $\mathfrak{A} \models \psi$ under the natural assignment, as desired. By standard arguments for shifting quantifiers around we obtain the following from (5):

if ψ is in prenex normal form, $\psi = \mathbf{Q}_0\alpha_0 \cdots \mathbf{Q}_{m-1}\alpha_{m-1}\chi$ with χ quantifier free, each $\mathbf{Q}_i$ either $\forall$ or $\exists$, and the α_i's individual or relation variables, then

(6) $$\models \exists v_i \forall v_j \mathbf{Q}_0\alpha_0 \cdots \mathbf{Q}_{m-1}\alpha_{m-1}\chi \leftrightarrow \exists R_k^1 \forall v_i \forall v_j \exists v_l \mathbf{Q}_0\alpha_0 \cdots \mathbf{Q}_{m-1}\alpha_{m-1}(R_k^1 v_l \wedge (R_k^1 v_i \rightarrow \chi)),$$

where R_k^1 and v_l are new.

Now using (1)–(6) we can easily carry out the proof of our theorem. In fact, we can use (1), (3), and (6) to shift all existential quantifiers $\exists v_i$ to the far right of the quantifier prefix of φ. Then we can use (2) and (4) to shift the second-order quantifiers to the far left of the prefix. □

Another important fact about second-order logic is that there is a reasonable proof theory with respect to a weakened form of validity. We shall go into this in some detail for the more complicated theory of types below. In conclusion, note that for second order logic it is natural to expand our notion of structure, by also admitting various second-order relations. For example we might admit a binary relation constant between elements and subsets of the universe. Again, more details will be found below in the discussion of type theory.

Theory of Types

If we imagine iterating the process which led from first-order logic to second-order logic we arrive at third-, fourth-, etc., order logic. The union of them all is the theory of types, a version of which we now proceed to describe fairly precisely.

First we need the notion of a *type* (*symbol*). The type symbols are built up from symbols (**o**) as follows:

(1) **o** is a type symbol;
(2) if σ is a finite nonzero sequence of type symbols, then (σ) is a type symbol;
(3) type symbols can only be formed in these ways.

Now a *type language* consists of the following parts:

logical constants: $\neg$, $\vee$, $\wedge$, $\forall$, $=$; for each type $\tau \neq \mathbf{o}$, relation variables $P_0^\tau, P_1^\tau, \ldots$ and for each type τ not of the form (σ), operation variables $v_0^\tau, v_1^\tau, \ldots$;

nonlogical constants, namely relation constants of types $\neq \mathbf{o}$ and operation constants of various types.

For such a language, we define the notion of a term:

(4) operation variables of type **o** and operation constants of type **o** are terms of type **o**;
(5) a relation variable or constant of type $\sigma \neq \mathbf{o}$ is a term of type σ;
(6) If **O** is an operation variable or constant of type $(\sigma_0, \ldots, \sigma_m)$ with $m > 0$ and $\tau_0, \ldots, \tau_{m-1}$ are terms of types $\sigma_0, \ldots, \sigma_{m-1}$ respectively, then $\mathbf{O}\tau_0, \ldots, \tau_{m-1}$ is a term of type σ_m;
(7) terms can only be formed in these ways.

Then we define *formulas*:

(8) If σ and τ are terms of the same type, then $\sigma = \tau$ is a formula;
(9) if **R** is a relation variable or constant of type $(\sigma_0, \ldots, \sigma_m)$, and $\tau_0, \ldots, \tau_m$ are terms of types $\sigma_0, \ldots, \sigma_m$ respectively, then $\mathbf{R}\tau_0 \cdots \tau_m$ is a formula;
(10) if φ and ψ are formulas, so are $\neg\varphi$, $\varphi \vee \psi$ and $\varphi \wedge \psi$;
(11) if φ is a formula and α is a variable, then $\forall\alpha\varphi$ is a formula;
(12) formulas can only be formed in these ways.

Now we turn to the semantic notions. After the precise definitions we shall give a few concrete examples of formulas and satisfaction of them in structures.

Let A be a nonempty set. By induction on σ we now define the set A_σ of *relations of type* σ *on* A;

(13) $A_{\mathbf{o}} = A$;
(14) for a type $\sigma = (\tau_0, \ldots, \tau_m)$, set $A_\sigma = \{R : R \subseteq A_{\tau 0} \times \cdots \times A_{\tau m}\}$.

Next, we define the set A^σ of *operations of type* σ *on* A:

(15) for a type $\sigma = (\tau_0, \ldots, \tau_m)$, $m > 0$, set $A^\sigma = \{f : f$ is a function and $f : A_{\tau 0} \times \cdots \times A_{\tau(m-1)} \to A_{\tau m}\}$.

Now let a type language $\mathscr{L}$ be given, as above. An $\mathscr{L}$*-structure* is a pair $\mathfrak{A} = (A, D)$ such that $A \neq 0$ and D is a function which assigns to each relation and operation constant of $\mathscr{L}$ a relation or operation respectively of the same type on A. Instead of $D\mathbf{R}$ or $D\mathbf{O}$ we might write $\mathbf{R}^{\mathfrak{A}}$ or $\mathbf{O}^{\mathfrak{A}}$. An *assignment for* $\mathfrak{A}$ is a function x assigning to each relation and operation variable of $\mathscr{L}$ a relation or operation on $\mathfrak{A}$ of the appropriate type. It is then obvious how to define $\tau^{\mathfrak{A}}x$ and $\mathfrak{A} \models \varphi[x]$, and we omit these details.

A formula in which the constants have only the types **o** or $(\mathbf{o} \cdots \mathbf{o})$ and the variables are only the operation variables of type **o** is essentially just a first-order formula. Similarly, the second-order formulas can be easily identified. It follows that the negative results for second-order logic carry over to this case. Type theory has enormous expressive power; indeed, it really encompasses a substantial portion of set theory. To illustrate the basic concepts, let us consider a type language whose sole non-logical constant is a relation constant $<$ of type $(\mathbf{oo})$. The following sentence φ holds in a structure $(A, <)$ iff $(A, <) \simeq (\omega, <)$:

($<$ is a linear order) $\wedge$
($<$ has a first element) $\wedge$
(in A every element has an immediate successor) $\wedge$
$\forall P_0^{(\mathbf{o})}$ (if the least element under $<$ is in $P_0^{(\mathbf{o})}$ and $P_0^{(\mathbf{o})}$ is closed under taking immediate successors, then $\forall v_0^{\mathbf{o}} P_0^{(\mathbf{o})} v_0$).

[The actual sentence φ is easily constructed from this informal description.]

The following sentence holds in $(A, <)$ iff $<$ well-orders A in a type $\geq \omega_1$:

($<$ is a well ordering of A) $\wedge$ $\exists v_0^{\mathbf{o}}[v_0^{\mathbf{o}}$ has no immediate predecessor, but every element $< v_0^{\mathbf{o}}$ which is not the first element of A has an immediate predecessor, and $\neg\exists v_0^{(\mathbf{oo})}$ ($v_0^{(\mathbf{oo})}$ is a one-one function onto A when it is restricted to $\{v_i^{\mathbf{o}} : v_i^{\mathbf{o}} < v_0^{\mathbf{o}}\}$].

Now we want to give a weakened kind of validity, for which it is possible to develop a proof theory and prove a completeness theorem. Let $\mathscr{L}$ be a type language, as above, and $\mathfrak{A}$ any $\mathscr{L}$-structure. A *frame* for $\mathfrak{A}$ consists of two mappings F, G whose domains are the set of all appropriate type symbols and which satisfy the following conditions:

(16) $F^{\mathbf{o}} = A = G^{\mathbf{o}}$;
(17) for a type $\sigma = (\tau_0, \ldots, \tau_m)$, with $m > 0$ we have $0 \neq F^\sigma \subseteq \{f : f$ is a function and $f: G^{\tau 0} \times \cdots \times G^{\tau(m-1)} \to G^{\tau m}\}$;
(18) for a type $\sigma = (\tau_0, \ldots, \tau_m)$, we have $0 \neq G^\sigma \subseteq \{R : R \subseteq G^{\tau 0} \times \cdots \times G^{\tau m}\}$;
(19) if $\mathbf{O}$ is an operation constant of $\mathscr{L}$ of type σ, then $\mathbf{O}^{\mathfrak{A}} \in F^\sigma$;
(20) if $\mathbf{R}$ is a relation constant of $\mathscr{L}$ of type $\sigma \neq 0$, then $\mathbf{R}^{\mathfrak{A}} \in G^\sigma$.

Note that frames exist; for example let $F^\sigma = A^\sigma$ for all types σ and $G^\sigma = A_\sigma$ for all types $\sigma \neq \mathbf{o}$. It is not immediately clear that frames other than this natural one exist, but that is true, and it follows from the theorem below.

Let (F, G) be a frame for A. An *assignment* for (A, F, G) is a function x such that for each type σ, $xP_i^\sigma \in G^\sigma$ for $\sigma \neq \mathbf{o}$ and $xv_i^\sigma \in F^\sigma$. Given such an assignment, we define $\tau^{\mathfrak{A}FG}x$ and $(\mathfrak{A}, F, G) \models \varphi[x]$ as follows.

$$v_i^{\mathbf{o}\mathfrak{A}FG}x = xv_i^{\mathbf{o}};$$
$$P_i^{\sigma\mathfrak{A}FG}x = xP_i^\sigma;$$

for $\mathbf{R}$ a relation constant of type $\sigma \neq 0$, $\mathbf{R}^{\mathfrak{A}FG}x = \mathbf{R}^{\mathfrak{A}}$; if $\mathbf{O}$ is an operation constant or variable of type $(\sigma_0, \ldots, \sigma_m)$ and $\tau_0, \ldots, \tau_{m-1}$ are terms of types $\sigma_0, \ldots, \sigma_{m-1}$ respectively, then we let $L = \mathbf{O}^{\mathfrak{A}}$ (for $\mathbf{O}$ a constant), or $L = x\mathbf{O}$ (for $\mathbf{O}$ a variable), and set

$$(\mathbf{O}\tau_0 \cdots \tau_{m-1})^{\mathfrak{A}FG}x = L(\tau_0^{\mathfrak{A}FG}x, \ldots, \tau_{m-1}^{\mathfrak{A}FG}x).$$

If σ and τ are terms of the same type, then we define

$$(\mathfrak{A}, F, G) \models (\sigma = \tau)[x] \qquad \text{iff } \sigma^{\mathfrak{A}FG}x = \tau^{\mathfrak{A}FG}x.$$

Let $\mathbf{R}$ be a relation variable or constant of type $\sigma \neq 0$. If $\mathbf{R}$ is a variable, let $S = x\mathbf{R}$, and if $\mathbf{R}$ is a constant, let $S = \mathbf{R}^{\mathfrak{A}}$. If $\sigma = (\tau_0 \cdots \tau_m)$ and $\rho_0, \ldots, \rho_m$ are terms of types $\tau_0, \ldots, \tau_m$ respectively, we let

$$(\mathfrak{A}, F, G) \models \mathbf{R}\rho_0 \cdots \rho_m[x] \qquad \text{iff } (\rho_0^{\mathfrak{A}FG}x, \ldots, \rho_m^{\mathfrak{A}FG}x) \in S.$$

Further,

$(\mathfrak{A}, F, G) \models \neg\varphi[x]$ iff not$[(\mathfrak{A}, F, G) \models \varphi[x]]$;
$(\mathfrak{A}, F, G) \models (\varphi \vee \psi)[x]$ iff $(\mathfrak{A}, F, G) \models \varphi[x]$ or $(\mathfrak{A}, F, G) \models \psi[x]$;
$(\mathfrak{A}, F, G) \models (\varphi \wedge \psi)[x]$ iff $(\mathfrak{A}, F, G) \models \varphi[x]$ and $(\mathfrak{A}, F, G) \models \varphi[x]$;
$(\mathfrak{A}, F, G) \models \forall P_i^\tau\varphi[x]$ iff for every $R \in G^\tau$ we have $(\mathfrak{A}, F, G) \models \varphi[x_R^Q]$, where $Q = P_i^\tau$;
$(\mathfrak{A}, F, G) \models \forall v_i^\tau\varphi[x]$ iff for every $O \in F^\tau$ we have $(\mathfrak{A}, F, G) \models \varphi[x_O^Q]$, where $Q = v_i^\tau$.

Thus we have a new notion of validity, which we shall denote by $\models_{\mathrm{f}}$. We shall now show that this notion reduces, in a certain sense, to first-order validity, and that (hence) there is a reasonable proof theory for $\models_{\mathrm{f}}$. The basic idea is as for many sorted logic in Chapter 29. For simplicity we shall assume that there are no operation constants, and no operation variables except those of type $\mathbf{o}$. As for first-order logic, this is not an essential restriction.

Let $\mathscr{L}$ be a type language as above, with these restrictions. Say the relation constants are $\langle R_i : i \in I\rangle$. Let $\mathscr{L}'$ be the associated first-order language, in which:

(21) the variables $P_0^\tau, P_1^\tau, \ldots, \tau \neq \mathbf{o}, v_0^{\mathbf{o}}, v_1^{\mathbf{o}}, \ldots$ are all treated as individual variables;
(22) A relation constant of type $(\sigma_0 \cdots \sigma_m)$ is treated as an $(m + 1)$-ary relation symbol.

Now we expand $\mathscr{L}'$ to $\mathscr{L}''$ by adjoining new unary relation symbols $\mathbf{T}^\tau$ for each type τ, and also for each type $\tau = (\sigma_0 \cdots \sigma_m)$ an $(m + 2)$-ary relation symbol $\mathbf{S}^\tau$. Now with each formula φ of $\mathscr{L}$ we associate a formula φ^* of $\mathscr{L}''$:

$(v_i^{\mathbf{o}} \equiv v_j^{\mathbf{o}})^* = v_i^{\mathbf{o}} \equiv v_j^{\mathbf{o}}$;
$(P_i^\tau \equiv P_j^\tau)^* = P_i^\tau \equiv P_j^\tau$;
$(P_i^\tau \equiv \mathbf{R}_j)^* = \forall P_0^{\sigma 0} \cdots \forall P_m^{\sigma m}[\mathbf{S}^\tau P_0^{\sigma 0} \cdots P_m^{\sigma m} P_i^\tau \leftrightarrow \mathbf{R}_j P_0^{\sigma 0}, \ldots, P_m^{\sigma m}]$,
where $\tau = (\sigma_0 \cdots \sigma_m)$ is the type of $\mathbf{R}_j$;
$(\mathbf{R}_j \equiv P_i^\tau)^* = (P_i^\tau \equiv \mathbf{R}_j)^*$, τ the type of $\mathbf{R}_j$;
$(\mathbf{R}_i \equiv \mathbf{R}_j)^* = \forall P_0^{\sigma 0} \cdots \forall P_m^{\sigma m}(\mathbf{R}_i P_0^{\sigma 0} \cdots P_m^{\sigma m} \leftrightarrow \mathbf{R}_j P_0^{\sigma 0} \cdots P_m^{\sigma m})$,
where $\mathbf{R}_i$ and $\mathbf{R}_j$ are of type $(\sigma_0 \cdots \sigma_m)$;
$(P_i^\tau M_0 \cdots M_m)^* = \exists P_k^{\sigma 0} \cdots \exists P_{k+m}^{\sigma m}[\bigwedge_{i \leq m} (P_{k+i}^{\sigma i} \equiv M_i)^*$
$\wedge\, \mathbf{S}^\tau P_k^{\sigma 0} \cdots P_{k+m}^{\sigma m} P_i^\tau]$, where $\tau = (\sigma_0 \cdots \sigma_m)$ and k is minimal
such that $P_k^{\sigma 0}, \ldots, P_{k+m}^{\sigma m} \notin \{M_0, \ldots, M_m\}$;
$(\mathbf{R}_i M_0 \cdots M_m)^* \equiv \exists P_k^{\sigma 0} \cdots \exists P_{k+m}^{\sigma m}[\bigwedge_{i \leq m} (P_{k+i}^{\sigma i} \equiv M_i)^*$
$\wedge\, \mathbf{R}_i P_k^{\sigma 0} \cdots P_{k+m}^{\sigma m}]$, where $\mathbf{R}_i$ has type $\tau = (\sigma_0 \cdots \sigma_m)$ and k is
as above;
$(\neg\varphi)^* = \neg\varphi^*$; $(\varphi \vee \psi)^* = \varphi^* \vee \psi^*$; $(\varphi \wedge \psi)^* = \varphi^* \wedge \psi^*$;
$(\forall v_i^{\mathbf{o}}\varphi)^* = \forall v_i^{\mathbf{o}}(\mathbf{T}^{\mathbf{o}} v_i^{\mathbf{o}} \to \varphi^*)$; $(\forall P_i^\tau\varphi)^* = \forall P_i^\tau(\mathbf{T}^\tau P_i^\tau \to \varphi^*)$.

With each $\mathscr{L}$-structure $\mathfrak{A}$ and each frame G for A (since there are no operation symbols except v_i^0, F is omitted) we associate an $\mathscr{L}''$-structure $\mathfrak{A}_G^*$. Let

$$A_G^* = \bigcup_{\tau \text{ a type}} G^\tau;$$

$$\mathbf{R}_i^{\mathfrak{A}*G} = \mathbf{R}_i^{\mathfrak{A}};$$

$$\mathbf{T}^{\tau\mathfrak{A}*G} = G^\tau;$$

$$\mathbf{S}^{\tau\mathfrak{A}*G} = \{(U_0, \ldots, U_m, V) : U_0, \ldots, U_m, V \in A_G^*, (U_0, \ldots, U_m) \in V \in G^\tau, \text{ where } \tau = (\sigma_0, \ldots, \sigma_m).$$

Now the following lemma is easily checked, by induction on φ:

Lemma 30.7. *Let G be a frame for A, $\mathfrak{A}$ an $\mathscr{L}$-structure, x an assignment over A, i.e., $xP_i^\sigma \in G^\sigma$ for each type $\sigma \neq 0$, and $xv_i^0 \in A$ for each $i \in \omega$. Then $(\mathfrak{A}, G) \models \varphi[x]$ iff $\mathfrak{A}_G^* \models \varphi^*[x]$.*

Now let Γ be the following set of sentences of $\mathscr{L}''$:

$\exists v_0^0 \mathbf{T}^\tau v_0^0$ for each type τ;
$\forall P_0^{\sigma 0} \cdots \forall P_m^{\sigma m} \forall P_0^\tau (\mathbf{S}^\tau P_0^{\sigma 0} \cdots P_m^{\sigma m} P_0^\tau \rightarrow \bigwedge_{i \leq m} \mathbf{T}^{\sigma i} P_i^{\sigma i} \wedge \mathbf{T}^\tau P_0^\tau)$
for a type $\tau = (\sigma_0 \cdots \sigma_m)$;
$\forall P_0^\tau \forall P_1^\tau [\forall P_0^{\sigma 0} \cdots \forall P_m^{\sigma m} (\mathbf{S}^\tau P_0^{\sigma 0} \cdots P_m^{\sigma m} P_0^\tau$
$\leftrightarrow \mathbf{S}^\tau P_0^{\sigma 0} \cdots P_m^{\sigma m} P_1^\tau) \rightarrow P_0^\tau = P_1^\tau]$, where $\tau = (\sigma_0 \cdots \sigma_m)$;
$\exists P_0^\tau \forall P_0^{\sigma 0} \cdots \forall P_m^{\sigma m} (\mathbf{R}_i P_0^{\sigma 0} \cdots P_m^{\sigma m} \leftrightarrow \mathbf{S}^\tau P_0^{\sigma 0} \cdots P_m^{\sigma m} P_0^\tau)$,
where $\mathbf{R}_i$ has rank $\tau = (\sigma_0 \cdots \sigma_m)$.

Clearly then:

Lemma 30.8. *Under the hypotheses of* 30.7, $\mathfrak{A}_G^*$ *is a model of* Γ.

Now we can prove the main theorem:

Theorem 30.9. *For any sentence φ of $\mathscr{L}$, $\models_f \varphi$ iff $\Gamma \models_{\mathscr{L}''} \varphi^*$.*

Proof. The direction $\Leftarrow$ is given by the preceding two lemmas. For $\Rightarrow$, assume that $\models_f \varphi$, and let $\mathfrak{B}$ be any $\mathscr{L}''$-structure which is a model of Γ. Set $A = \mathbf{T}^{o\mathfrak{B}}$. We now define by induction functions f^τ with domain $\mathbf{T}^{\tau\mathfrak{B}}$ for each type τ. We let f^o be the identity on A. For $\tau = (\sigma_0 \cdots \sigma_m)$ and $y \in \mathbf{T}^{\tau\mathfrak{B}}$ we set

$$f^\tau y = \{(f^{\sigma 0}x_0, \ldots, f^{\sigma m}x_m) : (x_0, \ldots, x_m, y) \in \mathbf{S}^{\tau\mathfrak{B}}\}.$$

Now let G^τ be the range of f^τ for each type τ. For each relation constant $\mathbf{R}_i$ of type $\tau = (\sigma_0 \cdots \sigma_m)$ let

$$\mathbf{R}_i^{\mathfrak{A}} = \{(f^{\sigma 0}x_0, \ldots, f^{\sigma m}x_m) : (x_0, \ldots, x_m) \in \mathbf{R}_i^{\mathfrak{B}}\}.$$

Then it is easily verified that G is a frame for $\mathfrak{A}$. The following is easily verified:

(23) Suppose $xP_i^\sigma \in \mathbf{T}^{\sigma\mathfrak{B}}$ and $xv_i^0 \in A$ for all σ and i; and let $yv_i^0 = xv_i^0$, $yP_i^\sigma = f^\sigma xP_i^\sigma$ for all i and σ; then for any formula of $\mathscr{L}$, $\mathfrak{A}_G^* \models \psi^*[x]$ iff $\mathfrak{B} \models \psi^*[y]$.

Now $\mathfrak{A} \models_f \varphi$ by assumption, so by Lemma 30.7, $\mathfrak{A}_G^* \models \varphi^*$ and hence by (23), $\mathfrak{B} \models \varphi^*$, as desired. □

From 30.9 we see that $\{\mathscr{g}^+\varphi : \varphi$ is a sentence of $\mathscr{L}$ and $\models_f\varphi\}$ is r.e. if the language is effective. Thus there is a reasonable proof theory for this form of validity; it has actually been worked out in detail. See Henkin [2] for the details, where a more comprehensive type theory is used.

ω–logic

Let $\mathscr{L}$ be any first-order language. An *ω-expansion* of $\mathscr{L}$ is an expansion $\mathscr{L}'$ of $\mathscr{L}$ in which a new unary relation symbol $\mathbf{N}$ is introduced as well as new individual constants $\mathbf{c}_0, \mathbf{c}_1, \ldots$. The term *$\omega$-logic* is applied in a loose sense to considerations about the pair $(\mathscr{L}, \mathscr{L}')$. An $\mathscr{L}'$-structure $(\mathfrak{A}, \mathbf{N}^{\mathfrak{A}}, a_i)_{i<\omega}$ is an *ω-structure* if $\mathbf{N}^{\mathfrak{A}} = \{a_i : i < \omega\}$. Note that if the a_i's are pairwise distinct, then the ω-structure is isomorphic to one of the form $(\mathfrak{B}, \omega, i)_{i<\omega}$. Thus in $\mathscr{L}'$ one has available, so to speak, the set of natural numbers, constants for each of them, and variables to range over ω (by restricting quantifiers to $\mathbf{N}$). Now with the above notation, let Γ be the set of all sentences $\mathbf{Nc}_i$ for $i < \omega$. Let Δ be the set of all formulas $\mathbf{N}v_0$ and $\neg(v_0 \equiv \mathbf{c}_i)$ for $i \in \omega$. Then clearly the ω-structures are exactly those models of Γ which omit Δ. Now let Θ be a theory in $\mathscr{L}'$. We say that Θ is *ω-complete* provided that for any formula $\varphi \in \mathrm{Fmla}^1_{\mathscr{L}'}$, if $\varphi(\mathbf{c}_i) \in \Theta$ for each $i \in \omega$, then $\forall v_0(\mathbf{N}v_0 \rightarrow \varphi) \in \Theta$. Then the following basic theorem follows easily from the omitting types theorem 27.4:

Theorem 30.10 (ω-completeness theorem). *With the above notation, let Θ be a theory in $\mathscr{L}'$ which includes Γ, has a model, and is ω-complete. Then Θ has an ω-model, i.e., it has a model which is an ω-structure.*

Proof. From the above remarks we see that it suffices to find a model of Θ which omits Δ. We apply 27.4 with $N = \{1\}$. Suppose that $\psi \in \mathrm{Fmla}^1_{\mathscr{L}'}$ and $\Theta \cup \{\exists v_0\psi\}$ has a model. To obtain a contradiction, suppose that $\Theta \cup \{\exists v_0(\psi \wedge \neg\varphi)\}$ is inconsistent for every $\varphi \in \Delta$. Thus $\Theta \models \neg\varphi \rightarrow \neg\psi$ for each $\varphi \in \Delta$, and hence $\Theta \models \neg\psi(\mathbf{c}_i)$ for every $i < \omega$. Hence by ω-completeness, $\Theta \models \forall v_0(\mathbf{N}v_0 \rightarrow \neg\psi)$. But also, since $\mathbf{N}v_0 \in \Delta$, $\Theta \models \forall v_0(\neg\mathbf{N}v_0 \rightarrow \neg\psi)$, so $\Theta \models \forall v_0 \neg \psi$, which contradicts the supposition that $\Theta \cup \{\exists v_0\psi\}$ has a model. □

Many other logics can be reduced to ω-logic, in a certain sense. We shall illustrate this with weak second-order logic. Let $\mathscr{L}$ be a first-order language, and let **K** be a class of $\mathscr{L}$-structures. We say that **K** is *ω-projective* provided that there is an expansion $\mathscr{L}'$ of $\mathscr{L}$, an ω-expansion $\mathscr{L}''$ of $\mathscr{L}'$, and a theory Θ in $\mathscr{L}''$, such that **K** is the class of all $\mathscr{L}$-reducts of ω-models of Θ.

Theorem 30.11. *Let $\mathscr{L}$ be any first-order language, and let Θ be a set of weak second-order sentences over $\mathscr{L}$ with only infinite models. Then* Mod Θ *is ω-projective.*

PROOF. Let $\mathscr{L}'$ be an expansion of $\mathscr{L}$ by adding a new binary relation symbol **R** and a new ternary relation symbol S, and let $\mathscr{L}''$ be an ω-expansion of $\mathscr{L}'$. Now we shall associate with each second-order formula φ of $\mathscr{L}$ a formula φ^* of $\mathscr{L}''$; we shall assume that $\mathscr{L}''$ is supplied with certain new individual variables F_{i0}, F_{i1} for $i \in \omega$:

$$(\sigma = \tau)^* = \sigma = \tau;$$
$$(\mathbf{P}\sigma_0 \cdots \sigma_{m-1})^* = \mathbf{P}\sigma_0 \cdots \sigma_{m-1};$$
$$(F_i\sigma)^* = \mathbf{S}F_{i0}F_{i1}\sigma;$$
$$(\neg\varphi)^* = \neg\varphi^*; \qquad (\varphi \vee \psi)^* = \varphi^* \vee \psi^*; \qquad (\varphi \wedge \psi)^* = \varphi^* \wedge \psi^*;$$
$$(\forall v_i\varphi)^* = \forall v_i\varphi^*;$$
$$(\forall F_i\varphi)^* = \forall F_{i0}\forall F_{i1}(\mathbf{R}F_{i0}F_{i1} \rightarrow \varphi^*).$$

Now let Ω consist of the following sentences of $\mathscr{L}''$;

$$\varphi^*, \text{ for } \varphi \in \Theta;$$
$$\mathbf{N}\mathbf{c}_i, \text{ for each } i < \omega;$$
$$\neg(\mathbf{c}_i = \mathbf{c}_j) \text{ for } i < j < \omega;$$
$$\forall v_0 \cdots \forall v_{n-1}\exists v_n\forall v_{n+1}[\mathbf{S}(\mathbf{c}_n, v_n, v_{n+1}) \leftrightarrow \bigvee_{i<n} (v_i = v_{n+1}), \text{ for each } n \in \omega;$$
$$\forall v_0\forall v_1(\mathbf{R}v_0v_1 \rightarrow \mathbf{N}v_0);$$
$$\forall v_0\forall v_1\forall v_2(\mathbf{S}v_0v_1v_2 \rightarrow \mathbf{R}v_0v_1);$$
$$\forall v_0[\neg v_0 = \mathbf{c}_0 \rightarrow \exists v_1\exists v_2(\mathbf{S}v_0v_1v_2)];$$
$$\forall v_0\exists v_1\mathbf{R}v_0v_1;$$
$$\forall v_0\forall v_1\forall v_2\{\forall v_3[\mathbf{S}(v_0, v_1, v_3) \leftrightarrow \mathbf{S}(v_0, v_2, v_3)] \rightarrow v_1 = v_2\}.$$

We claim that Mod Θ is exactly the class of all $\mathscr{L}$-reducts of ω-models of Ω. To see this, we have to establish a relationship between satisfaction of φ and φ^*. Let $\mathfrak{A}$ be any infinite $\mathscr{L}$-structure. Let $f_{\mathfrak{A}}$ be a one-to-one function mapping A onto the set of all finite subsets of A. Let $a_{\mathfrak{A}}: \omega \rightarrowtail\!\!\!\!\!\twoheadrightarrow A$. Set

$$\mathbf{R}^{\mathfrak{A}*} = \{(a_{\mathfrak{A}}n, b) : n \in \omega, b \in A, |f_{\mathfrak{A}}b| \leq n\};$$
$$\mathbf{S}^{\mathfrak{A}*} = \{(a_{\mathfrak{A}}n, b, c) : n \in \omega, b \in A, |f_{\mathfrak{A}}b| \leq n, c \in f_{\mathfrak{A}}b\}.$$

Then with $a = a_{\mathfrak{A}}$, as is easily seen, $\mathfrak{A}^* = (\mathfrak{A}, \mathbf{R}^{\mathfrak{A}*}, \mathbf{S}^{\mathfrak{A}*}, \text{Rng}\, a, a_i)_{i<\omega}$ is a model of Ω except possibly for the sentences φ^*. Furthermore, if $x \in {}^{\omega}A$, $y \in {}^{\omega}\{X : X \subseteq A,\ X \text{ finite}\}$, and if for each $i < \omega$ we choose $n_i \in \omega$ and $b_i \in A$ so that $|fb_i| \leq n_i$ and $fb_i = y_i$, and if φ is any second-order formula of $\mathscr{L}$, then, with $zF_{i0} = a_{ni}$ and $zF_{i1} = b_i$ for all $i < \omega$, and with a natural satisfaction notation,

(1) $$\mathfrak{A} \models \varphi[x; y] \quad \text{iff } \mathfrak{A}^* \models \varphi^*[x; z].$$

This is easily shown by induction on φ. Now it follows that if $\mathfrak{A}$ is a model of Θ, then $\mathfrak{A}^*$ is a model of Ω. Conversely, it is easily seen that any ω-model $\mathfrak{B}$ of Ω is isomorphic to $\mathfrak{A}^*$ for some $\mathscr{L}$-structure $\mathfrak{A}$, so the $\mathscr{L}$-reduct of $\mathfrak{B}$ is a model of Θ. □

Cardinality Quantifiers

Given a first-order language $\mathscr{L}$, we can convert $\mathscr{L}$ into a **Q**-*language* by adjoining a new symbol **Q** which is used syntactically just like the universal quantifier $\forall$. Thus to the rules of formulas is added the stipulation that $\mathbf{Q}v_i\varphi$ is a formula whenever φ is. Of course, **Q** can be given many interpretations; some reasonable limitations on the interpretations of **Q** have been set down in Mostowski [5]. We consider here certain cardinality interpretations. Given an infinite cardinal $\mathfrak{m}$, the $\mathfrak{m}$-*interpretation* of $\mathbf{Q}v_i$ is "there are at least $\mathfrak{m}$ v_i such that." Thus to the other satisfaction rules we add:

$$\mathfrak{A} \models_{\mathfrak{m}} \mathbf{Q}v_i\varphi[x] \qquad \text{iff } |\{a \in A : \mathfrak{A} \models \varphi[x^i_a]\}| \geq \mathfrak{m}.$$

First we consider the case $\mathfrak{m} = \aleph_0$. Then $\mathbf{Q}v_i$ is interpreted as "there are infinitely many v_i such that" and hence "$\neg\mathbf{Q}v_i$" means "there are only finitely many v_i such that." Hence in this interpretation the **Q**-language resembles weak second-order logic. Various negative facts are seen just as in that case. The compactness theorem fails, as is seen from the set

$$\neg\mathbf{Q}v_0(v_0 = v_0);$$
there are at least n things (a sentence for each $n \in \omega$).

The system $(\omega, +, \cdot, \mathfrak{s}, 0)$ is characterized up to isomorphism by the finite set

the theory Q;
$$\forall v_0 \neg \mathbf{Q}v_1\exists v_2(v_1 + v_2 = v_0).$$

Hence in the familiar way we see that $\{\mathscr{g}^+\varphi : \varphi$ is a **Q**-sentence in $\mathscr{L}_{\text{nos}}$ and $\models_\omega\varphi\}$ is not r.e., so that there is no reasonable proof theory for the ω-valid **Q**-sentences.

In spite of these discouraging facts, the situation improves considerably for $\mathfrak{m} > \aleph_0$. First consider the compactness theorem. It still fails, for any $\mathfrak{m} \geq \aleph_0$. This can be seen from the following set of sentences (where we deal with a language with individual constants $\mathbf{c}_\alpha$ for each $\alpha < \mathfrak{m}$):

$$\neg(\mathbf{c}_\alpha = \mathbf{c}_\beta) \qquad \text{for } \alpha < \beta < \mathfrak{m};$$
$$\neg\mathbf{Q}v_0(v_0 = v_0).$$

However, a weaker kind of compactness sometimes occurs, depending on $\mathfrak{m}$. To encompass some other kinds of weak compactness introduced in the next section, we give a general definition:

Definition 30.12. Let $\mathfrak{m}$ and $\mathfrak{n}$ be infinite cardinals with $\mathfrak{m} < \mathfrak{n}$. We say that a logic $\mathscr{L}$ is $(\mathfrak{m}, \mathfrak{n})$-*compact* provided that if Γ is a set of sentences with $|\Gamma| < \mathfrak{n}$, and if every subset of Γ of power $<\mathfrak{m}$ has a model, then Γ has a model. And Γ is $(\mathfrak{m}, \infty)$-*compact* if it is $(\mathfrak{m}, \mathfrak{n})$-compact for all $\mathfrak{n} > \mathfrak{m}$.

Thus ordinary first-order logic is (ω, ∞)-compact. To formulate our main result on compactness for **Q**-quantifiers we need a certain cardinal number function. For any $\mathfrak{m} \geq \aleph_0$ let

$\pi\mathfrak{m}$ = the least $\mathfrak{n}$ such that there is a family $\langle \mathfrak{p}_\alpha : \alpha < \mathfrak{n}\rangle$ of cardinals $<\mathfrak{m}$ such that $\prod_{\alpha<n} \mathfrak{p}_\alpha \geq \mathfrak{m}$.

Thus $\aleph_0 \leq \pi\mathfrak{m} \leq \mathfrak{m}$ for any cardinal $\mathfrak{m}$. We have $\pi\aleph_0 = \aleph_0$; $\pi\aleph_1 = \aleph_0$; $\pi\mathfrak{m}^+ \leq \mathfrak{m}$; $\pi\mathfrak{m} = \mathfrak{m}$ if $\mathfrak{m}$ is strongly inaccessible; $\pi\mathfrak{m} \leq \mathrm{cf}\,\mathfrak{m}$ if $\mathfrak{m}$ is singular; and $\pi(\exp \aleph_0)^+ \geq \aleph_1$.

Theorem 30.13 (Fuhrken). *Let $\langle \mathfrak{A}_i : i \in I\rangle$ be a system of $\mathscr{L}$-structures, with $|I| < \pi\mathfrak{m}$, where $\mathfrak{m} \geq \aleph_0$. Let φ be a* **Q**-*formula and $x \in {}^\omega\mathsf{P}_{i\in I} A_i$. Then $\mathsf{P}_{i\in I} \mathfrak{A}_i/\bar{F} \models_\mathfrak{m} \varphi[[\;] \circ x]$ iff $\{i \in I : \mathfrak{A}_i \models_\mathfrak{m} \varphi[\mathrm{pr}_i \circ x]\} \in F$.*

Proof. The proof is by induction on φ, and goes just as in Chapter 18 except for the step from φ to $\mathbf{Q}v_j\varphi$. Let

$$J = \{i \in I : \mathfrak{A}_i \models_\mathfrak{m} \mathbf{Q}v_j\varphi[\mathrm{pr}_i \circ x]\},$$
$$T_i = \{a \in A_i : \mathfrak{A}_i \models_\mathfrak{m} \varphi[(\mathrm{pr}_i \circ x)^j_a]\} \qquad \text{for each } i \in I.$$

First suppose $J \in F$. Then $|T_i| \geq \mathfrak{m}$ for each $i \in J$, so we can choose $y_i : \mathfrak{m} \rightarrowtail T_i$. For each $\alpha < \mathfrak{m}$ let $f_\alpha \in \mathsf{P}_{i\in I} A_i$ be such that $f_{\alpha i} = y_{i\alpha}$ whenever $i \in J$. Then $\langle [f_\alpha] : \alpha < \mathfrak{m}\rangle$ is one-one. For suppose that $\alpha < \beta < \mathfrak{m}$. Then $J \subseteq \{i \in I : f_{\alpha i} = y_{i\alpha}$ and $f_{\beta i} = y_{i\beta}\} \subseteq \{i : f_{\alpha i} \neq f_{\beta i}\}$, so $[f_\alpha] \neq [f_\beta]$. Furthermore, $J \subseteq \{i \in I : \mathfrak{A}_i \models_\mathfrak{m} \varphi[\mathrm{pr}_i \circ x^j_{f\alpha}]\}$, so by the induction hypothesis $\mathsf{P}_{i\in I} \mathfrak{A}_i/\bar{F} \models_\mathfrak{m} \varphi[([\;] \circ x)^j_{[f\alpha]}]$. Thus we have shown that $\mathsf{P}_{i\in I} \mathfrak{A}_i/\bar{F} \models_\mathfrak{m} \mathbf{Q}v_j\varphi[[\;] \circ x]$.

Second, suppose that $J \notin F$. Thus $I \sim J \in F$. Let $K = \{c \in \mathsf{P}_{i\in I} A_i/\bar{F} : \mathsf{P}_{i\in I} \mathfrak{A}_i/\bar{F} \models_\mathfrak{m} \varphi[([\;] \circ x)^j_c]\}$, and let L consist of one representative from each member of K. Now for each $f \in L$ we define $Gf \in \mathsf{P}_{i\in I\sim J} (T_i \cup \{0\})$ as follows:

$$(Gf)_i = f_i \qquad \text{if } f_i \in T_i,$$
$$(Gf)_i = 0 \qquad \text{otherwise.}$$

Now we claim that G is one-one. For, if f and g are distinct members of L, then $\mathsf{P}_{i\in I} \mathfrak{A}_i/\bar{F} \models_\mathfrak{m} \varphi[[\;] \circ x^j_f]$ and so by the induction hypothesis $\{i \in I : \mathfrak{A}_i \models_\mathfrak{m} \varphi[\mathrm{pr}_i \circ x^j_f]\} \in F$, and hence, arguing similarly for g, the set

$$(I \sim J) \cap \{i \in I : f_i \neq g_i\} \cap \{i \in I : \mathfrak{A}_i \models_\mathfrak{m} \varphi[\mathrm{pr}_i \circ x^j_f]\}$$
$$\cap \{i \in I : \mathfrak{A}_i \models_\mathfrak{m} \varphi[\mathrm{pr}_i \circ x^j_g]\}$$

is in F. But for any member of this set we have $f_i, g_i \in T_i$, and $i \in I \sim J$, so $(Gf)_i \neq (Gg)_i$. Thus, indeed, G is one-one. Hence

$$|L| \leq \prod_{i\in I\sim J} (|T_i| + 1) < \mathfrak{m},$$

since $|T_i| < \mathfrak{m}$ for $i \in I \sim J$ and $|I| < \pi\mathfrak{m}$. Therefore not($\mathrm{P}_{i \in I}\, \mathfrak{A}_i/\bar{F} \models_{\mathfrak{m}} \mathbf{Q}v_j\varphi[x]$). □

Recalling from p. 323 our proof of the compactness theorem using ultraproducts, we obtain

Corollary 30.14. *In the* $\mathfrak{m}$*-interpretation, any* **Q***-language is* $(\aleph_0, \pi\mathfrak{m})$*-compact.*

Thus in the $(\exp \aleph_0)^+$-interpretation, any **Q**-language is $(\aleph_0, \aleph_1)$-compact: any countable set of **Q**-sentences such that every finite subset has a model, also has a model. Since $\pi\aleph_1 = \aleph_0$, this corollary does not give any information about the $\aleph_1$-interpretation. Fuhrken, however, has shown that in the $\aleph_1$-interpretation, any **Q**-language is $(\aleph_0, \aleph_1)$-compact, which is clearly a "best possible" result. Vaught has shown that the valid **Q**-sentences in the $\aleph_1$-interpretation are r.e., and so there is a reasonable proof theory for these sentences. Keisler has given an elegant axiom system for them. These results make the $\aleph_1$-interpretation of **Q**-languages one of the most interesting extensions of first-order logic. For results on **Q**-languages, see Bell, Slomson [1].

BIBLIOGRAPHY

1. Bell, J. L. and Slomson, A. B. *Models and Ultraproducts.* Amsterdam: North-Holland (1969).
2. Henkin, L. Completeness in the theory of types. *J. Symb. Logic, 15* (1950), 81–91.
3. Keisler, H. J. Some model-theoretic results for ω-logic. *Israel J. Math., 4* (1966), 249–261.
4. Läuchli, H. A decision procedure for the weak second-order theory of linear order. In *Contrib. math. logic.* Amsterdam: North-Holland (1968), 189–197.
5. Mostowski, A. On a generalization of quantifiers. *Fund. Math.,* **44** (1957), 12–36.
6. Rabin, M. Decidability of second-order theories and automata on infinite trees. *Trans. Amer. Math. Soc., 141* (1969), 1–35.

EXERCISES

30.15. Let $\mathscr{L}$ be a countable first-order language and let Γ be an $\aleph_0$-categorical weak second-order theory in $\mathscr{L}$ with only infinite models. Show that Γ is complete in the weak second-order sense.

30.16. Let $\mathscr{L}$ be a first-order language whose only nonlogical constant is a binary relation symbol $<$. Show that there is a weak second order sentence φ whose models are exactly all $\mathscr{L}$-structures isomorphic to $(\omega + \omega, <)$.

30.17. Modify the equivalence relations $\equiv_m$ of 26.4 as follows. First, $\mathfrak{A} \equiv_0^w \mathfrak{B}$ iff $\mathfrak{A} \equiv_0 \mathfrak{B}$ as in 26.4. Second, $\mathfrak{A} \equiv_{m+1}^w \mathfrak{B}$ iff for every $n \in \omega$ and every $a \in {}^nA$ there is a $b \in {}^nB$ such that $(\mathfrak{A}, a_i)_{i<n} \equiv_m (\mathfrak{B}, b_i)_{i<n}$, and similarly with the

roles of $\mathfrak{A}$ and $\mathfrak{B}$ interchanged. Show that if $\mathfrak{A} \equiv_m^w \mathfrak{B}$ for every $m \in \omega$ then $\mathfrak{A}$ is weak second order elementarily equivalent to $\mathfrak{B}$.

30.18. Using 30.17, show that $(\omega^\omega \cdot \omega \cdot 2, <)$ and $(\omega^\omega \cdot (\omega + \omega^*), <)$ are weak second-order elementarily equivalent. Hence well-order is not definable in weak second-order logic.

30.19. Show that in 30.6 we may instead take the existential quantifiers on individual variables to occur before the universal quantifiers on individual variables.

30.20. Let Γ be a collection of *sentences* with the special form of 30.6, where no existential quantifiers on individual variables occur. Show that Mod Γ is a universal class in the usual, first-order, sense.

30.21. Let Γ be a set of sentences in a **Q**-language, and let **K** be the class of all models of Γ in the $\aleph_0$-interpretation. Show that **K** is ω-projective.

30.22. There is a set Γ of sentences in ω-logic such that every finite subset of Γ has an ω-model while Γ does not have an ω-model.

30.23. Take a **Q**-language $\mathscr{L}$ with the $\mathfrak{m}$-interpretation. Then every set Γ of sentences of $\mathscr{L}$ which has a model of power $\geq \mathfrak{m} + |\mathrm{Fmla}_{\mathscr{L}}|$ has a model of power $\mathfrak{m} + |\mathrm{Fmla}_{\mathscr{L}}|$.

31 Infinitary Extensions

In this final chapter we want to consider the presently most popular extensions of first-order logic, namely languages in which the expressions are of infinite length.

The Languages $L_{\omega_1\omega}$

Consider an ordinary first order language $\mathscr{L}$, and enlarge it by allowing countable conjunctions and disjunctions $\bigwedge_{i\in\omega}\varphi_i$ and $\bigvee_{i\in\omega}\varphi_i$. The satisfaction rules are clear for such a language. Many natural things can be expressed in an $L_{\omega_1\omega}$-language which require circumlocutions in the other languages we have discussed. For example, the natural numbers are characterized by adding to the theory Q the sentence

$$\forall v_0(v_0 = \mathbf{0} \vee v_0 = \mathbf{1} \vee v_0 = \mathbf{2} \vee \cdots).$$

The notion of an Archimedean-ordered field is expressed by the following sentence added to the axioms for ordered fields:

$$\forall v_0(v_0 < \mathbf{1} \vee v_0 < \mathbf{1} + \mathbf{1} \vee v_0 < \mathbf{1} + \mathbf{1} + \mathbf{1} \vee \cdots).$$

As is to be expected, the compactness theorem fails in $L_{\omega_1\omega}$, as is seen from the following set of sentences:

$$\exists v_0[\neg(v_0 = \mathbf{c}_0) \wedge \cdots \wedge \neg(v_0 = \mathbf{c}_n)] \qquad (\text{for all } n \in \omega);$$
$$\forall v_0(v_0 = \mathbf{c}_0 \vee v_0 = \mathbf{c}_1 \vee \cdots).$$

It turns out, however, that much of the model-theoretic machinery of first-order logic is still available in the $L_{\omega_1\omega}$ case, and new methods can also be used. We give two illustrations: Scott's isomorphism theorem and the model existence theorem.

Theorem 31.1 (D. Scott). *Let $\mathscr{L}$ be a countable language, and $\mathfrak{A}$ a countable $\mathscr{L}$-structure. Then there is an $L_{\omega_1\omega}$-sentence ψ in $\mathscr{L}$ whose countable models are exactly the $\mathscr{L}$-structures isomorphic to $\mathfrak{A}$.*

PROOF. For each finite sequence a of length n of elements of A and each $\beta < \omega_1$ we define a formula $\varphi_a^\beta \in \mathrm{Fmla}_{\mathscr{L}}^n$. Let

$$\varphi_a^0 = \bigwedge \{\psi \in \mathrm{Fmla}_{\mathscr{L}} : \mathfrak{A} \models \psi[a] \text{ and } \psi \text{ is atomic or the negation of an atomic formula}\}.$$

For λ a limit ordinal $<\omega_1$ let

$$\varphi_a^\lambda = \bigwedge_{\alpha<\lambda} \varphi_a^\alpha.$$

Finally, for $\alpha < \omega_1$ let

$$\varphi_a^{\alpha+1} = \varphi_a^\alpha \wedge \bigwedge_{b\in A} \exists v_n \varphi_{a\langle b\rangle}^\alpha \wedge \forall v_n \bigvee_{b\in A} \varphi_{a\langle b\rangle}^\alpha.$$

Thus we have, by induction on α,

(1) $$\mathfrak{A} \models \varphi_a^\alpha[a];$$

(2) $$\models \varphi_a^\alpha \rightarrow \varphi_a^\beta \qquad \text{if } \beta < \alpha < \omega_1.$$

Now by (2), $\langle\{x : \mathfrak{A} \models \varphi_a^\alpha[x]\} : \alpha < \omega_1\rangle$ is a sequence of subsets of ${}^nA$ non-increasing under $\subseteq$. Since $|{}^nA| = \aleph_0$, it follows that there is an $\alpha < \omega_1$ such that for all $\beta \geq \alpha$, $\mathfrak{A} \models \varphi_a^\alpha \leftrightarrow \varphi_a^\beta$. Thus to each finite sequence a of elements of A there is such an ordinal α_a. If we let β be the sup of all ordinals α_a, we have $\beta < \omega_1$ and

(3) for all finite sequences a of elements of A and all $\alpha \geq \beta$, $\mathfrak{A} \models \varphi_a^\alpha \leftrightarrow \varphi_a^\beta$.

Now let the desired sentence ψ be

$$\varphi_0^\beta \wedge \bigwedge \{\forall v_0 \cdots \forall v_{n-1}(\varphi_a^\beta \rightarrow \varphi_a^{\beta+1}) : a \text{ is a finite sequence of elements of } A, \text{ say of length } n\}.$$

From (1) and (3) it is clear that $\mathfrak{A}$ is a model of ψ. Now let $\mathfrak{B}$ be any countable model of ψ. We shall show that $\mathfrak{A} \cong \mathfrak{B}$ by constructing an ω-sequence for $\mathfrak{A}$ and $\mathfrak{B}$, and applying 26.20. For each $m \in \omega$, let

$$I_m = \{(a, b) \in {}^mA \times {}^mB : \mathfrak{B} \models \varphi_a^\beta[b]\}.$$

Since φ_0^β is a part of ψ, it is immediate that $0I_00$. Thus conditions (1) and (2) of 26.6(*ii*) hold. Next, suppose that aI_mb and $x \in A$. Thus $\mathfrak{B} \models \varphi_a^\beta[b]$. Since $\mathfrak{B}$ is a model of ψ, it follows that $\mathfrak{B} \models \varphi_a^{\beta+1}[b]$, so, by the definition of $\varphi_a^{\beta+1}$ we have $\mathfrak{B} \models \exists v_n \varphi_{a\langle x\rangle}^\beta[b]$. Hence there is a $y \in B$ such that $\mathfrak{B} \models \varphi_{a\langle x\rangle}^\beta[b\langle y\rangle]$. Thus $a\langle x\rangle I_m b\langle y\rangle$. Now suppose that $y \in B$ and aI_mb. Thus $\mathfrak{B} \models \varphi_a^\beta[b]$, so $\mathfrak{B} \models \varphi_a^{\beta+1}[b]$ and hence $\mathfrak{B} \models \varphi_{a\langle x\rangle}^{\beta+1}[b\langle y\rangle]$ for some $x \in A$, so again $a\langle x\rangle I_m b\langle y\rangle$. Thus 26.6(*ii*)(3) holds. Since $\models \varphi_a^\beta \rightarrow \varphi_a^0$, condition 26.6(*ii*)(4) is clear. □

The model existence theorem carries over almost verbatim to $L_{\omega_1\omega}$. We sketch here the necessary changes. First of all, the notation $\varphi^{\rightarrow}$ must be extended to infinite conjunctions and disjunctions:

$$\left(\bigwedge_{i<\omega} \varphi_i\right)^{\rightarrow} = \bigvee_{i<\omega} \varphi_i^{\rightarrow};$$

$$\left(\bigvee_{i<\omega} \varphi_i\right)^{\rightarrow} = \bigwedge_{i<\omega} \varphi_i^{\rightarrow}.$$

The other parts of the definition of $\varphi^{\rightarrow}$ remain as before. Let $\mathscr{L}$ be an $L_{\omega_1\omega}$-language, and let $\mathscr{L}'$ be an expansion of $\mathscr{L}$ by adjoining a countable set C of new individual constants. A *consistency family for* $(\mathscr{L}, \mathscr{L}')$ is a family S of *countable* sets of sentences of $\mathscr{L}'$ such that for each $\Gamma \in S$, conditions (C0)–(C9) of Chapter 18 hold, as well as:

(C3′) If $\bigwedge_{i<\omega} \varphi_i \in \Gamma$, then $\Gamma \cup \{\varphi_i\} \in S$ for each $i < \omega$.
(C4′) If $\bigvee_{i<\omega} \varphi_i \in \Gamma$, then $\Gamma \cup \{\varphi_i\} \in S$ for some $i < \omega$.

The following lemma is proved just like 18.7:

Lemma 31.2. *If S satisfies* (C1)–(C9), (C3′), *and* (C4′), *then* $S' = \{\Gamma : \Gamma \subseteq \Delta$ *for some* $\Delta \in S\}$ *satisfies* (C0)–(C9), (C3′), *and* (C4′).

For our present purposes the following weak kind of model existence theorem will prove useful.

Theorem 31.3. *Let $\mathscr{L}$ have only countably many nonlogical constants. If S is a consistency family for $(\mathscr{L}, \mathscr{L}')$ and $\Gamma \in S$, then Γ has a model.*

Proof. Let Δ be the closure of Γ under the following operations: passage to subformulas; passage from φ to $\mathrm{Subf}^{\alpha}_{\mathbf{c}}\varphi$, where $\mathbf{c} \in C$; passage from $\neg\varphi$ to $\varphi^{\rightarrow}$; formation of $\mathbf{c} = \mathbf{d}$ for any $\mathbf{c}, \mathbf{d} \in C$. Clearly Δ is countable. Let φ map ω onto all of the sentences in Δ, let τ enumerate all primitive terms of $\mathscr{L}'$, and let a well-ordering of C be given. Now we define a sequence $\langle \Theta_i : i \leq \omega\rangle$. Let $\Theta_0 = \Gamma$. Θ_{i+1} is formed from Θ_i as in the proof of 18.9, so that:

(1) if $\Theta_i \cup \{\varphi_i\} \in S$, then $\varphi_i \in \Theta_{i+1}$;
(2) if $\Theta_i \cup \{\varphi_i\} \in S$ and $\varphi_i = \varphi_s \vee \varphi_t$, then $\varphi_s \in \Theta_{i+1}$ or $\varphi_t \in \Theta_{i+1}$;
(3) if $\Theta_i \cup \{\varphi_i\} \in S$ and $\varphi_i = \exists\alpha\psi$, then $\mathrm{Subf}^{\alpha}_{\mathbf{c}}\psi \in \Theta_{i+1}$ for some $\mathbf{c} \in C$;
(4) for some $\mathbf{c} \in C$, $\mathbf{c} = \tau_i \in \Theta_{i+1}$;
(5) if $\Theta_i \cup \{\varphi_i\} \in S$ and $\varphi_i = \bigvee_{j<\omega} \psi_j$, then $\psi_j \in \Theta_{i+1}$ for some $j < \omega$.

At this point the remainder of the proof of 18.9 continues with practically no changes. □

As an application of 31.3 we give Craig's interpolation theorem for $L_{\omega_1\omega}$:

Theorem 31.4. *Let φ and ψ be sentences in an $L_{\omega_1\omega}$-language such that $\models \varphi \rightarrow \psi$. Then there is a sentence χ such that $\models \varphi \rightarrow \chi$, $\models \chi \rightarrow \psi$, and any nonlogical constant occurring in χ occurs in both φ and ψ.*

The proof is almost identical with that of 22.1, and we omit it.

Of the infinitary languages, $L_{\omega_1\omega}$ is perhaps the most interesting. The above results give some indications that many interesting methods and results are still available in $L_{\omega_1\omega}$, and it is this fact that has led to its extensive study among the logics much more expressive than ordinary first-order logic.

Languages $L_{\mathfrak{m}\mathfrak{n}}$

Let $\mathfrak{m}$ and $\mathfrak{n}$ be infinite cardinal numbers. An $L_{\mathfrak{m}\mathfrak{n}}$-language is a first-order language augmented by allowing conjunction and disjunction of fewer than $\mathfrak{m}$ formulas, and universal quantification on fewer than $\mathfrak{n}$ variables. Thus the new formation and satisfaction rules are codified as follows:

If $\langle \varphi_\alpha : \alpha < \mathfrak{p} \rangle$ is a system of formulas with $\mathfrak{p} < \mathfrak{m}$, then $\bigwedge_{\alpha<\mathfrak{p}} \varphi_\alpha$ and $\bigvee_{\alpha<\mathfrak{p}} \varphi_\alpha$ are formulas. We allow, instead of just $\aleph_0$ variables, $\mathfrak{n}$ variables $\langle v_\alpha : \alpha < \mathfrak{n} \rangle$ and add the formation rule:

If φ is a formula and $\alpha \in {}^{\mathfrak{p}}\mathfrak{n}$ is one–one, where $\mathfrak{p} < \mathfrak{n}$, then $\forall v_{\alpha 0} \cdots v_{\alpha\xi} \cdots \varphi$ is a formula.

We say that $\mathfrak{A} \models \bigwedge_{\alpha<\mathfrak{p}} \varphi_\alpha[x]$ iff for all $\alpha < \mathfrak{p}$, $\mathfrak{A} \models \varphi_\alpha[x]$. And we write $\mathfrak{A} \models \forall v_{\alpha 0} \cdots v_{\alpha\xi} \cdots \varphi[x]$ iff $\mathfrak{A} \models \varphi[y]$ for every $y \in {}^{\mathfrak{n}}A$ such that $y_\xi = x_\xi$ for all $\xi \notin \operatorname{Rng} \alpha$. Clearly these definitions correspond to the intuitive notions. Note that one could conceive of more general quantification than the above. For example,

$$\forall v_0 \exists v_1 \forall v_2 \exists v_3 \cdots \varphi$$

clearly has an intuitive sense. For development of such more general quantification see, e.g., [2].

The languages $L_{\mathfrak{m}\mathfrak{n}}$ have been extensively developed. We shall restrict ourselves here to just one aspect of this development, the question of extending the compactness theorem. If $\mathfrak{m} > \aleph_0$, the compactness theorem always fails, as indicated above for $L_{\omega_1\omega}$, by an example that obviously extends to any language $L_{\mathfrak{m}\mathfrak{n}}$ with $\mathfrak{m} > \aleph_0$. The languages $L_{\omega\mathfrak{n}}$ are clearly just slight variants of ordinary first-order logic. Despite the failure of the compactness theorem in general, there are two modifications of it whose study has been very fruitful in the discussion of numerous foundational questions:

Definition 31.5 (Recall Definition 30.12). A cardinal $\mathfrak{m} \geq \aleph_0$ is *strongly compact* provided that any $L_{\mathfrak{m}\mathfrak{m}}$-language is $(\mathfrak{m}, \infty)$-compact. The cardinal is *weakly compact* if any $L_{\mathfrak{m}\mathfrak{m}}$-language is $(\mathfrak{m}, \mathfrak{m}^+)$-compact.

Obviously $\aleph_0$ is both strongly and weakly compact, and every strongly compact cardinal is weakly compact. Our object now is to show that uncountable strongly and weakly compact cardinals are very big, and to give some purely mathematical equivalents of these notions.

Lemma 31.6. *Any strongly compact cardinal is a strong limit cardinal.*

PROOF. Suppose that $\mathfrak{m}$ is not a strong limit cardinal. We shall show that it is not strongly compact. Clearly there is a cardinal $\mathfrak{n} < \mathfrak{m}$ such that $\mathfrak{m} \leq 2^{\mathfrak{n}}$. Let $\mathscr{L}$ be a first-order language with unary relation symbols $\mathbf{P}_{\xi\varepsilon}$ for each $\xi < \mathfrak{n}$ and $\varepsilon < 2$. Let Γ consist of the following sentences:

(1) $\bigwedge_{\xi<\mathfrak{n}} \bigvee_{\varepsilon<2} \forall v_0 \mathbf{P}_{\xi\varepsilon} v_0$;
(2) $\forall v_0 \forall v_1 (v_0 = v_1)$;
(3) $\neg \bigwedge_{\xi<\mathfrak{n}} \forall v_0 \mathbf{P}_{\xi, f\xi} v_0$ (for each $f \in {}^{\mathfrak{n}}2$).

Suppose Γ has a model, $\mathfrak{A} = \langle \{a\}, \mathbf{P}^{\mathfrak{A}}_{\xi\varepsilon} \rangle_{\xi<\mathfrak{n}, \varepsilon<2}$. By (1), there is a function $f \in {}^{\mathfrak{n}}2$ such that $\mathbf{P}^{\mathfrak{A}}_{\xi, f\xi} = \{a\}$ for every $\xi < \mathfrak{n}$. This contradicts (3).

Now let Δ be any subset of Γ of power $< \mathfrak{m}$. Since $\mathfrak{m} \leq 2^{\mathfrak{n}}$, there is then a function $f \in {}^{\mathfrak{n}}2$ such that the corresponding sentence of type (3) is not in Δ. Let $A = \{0\}$, and for $\xi < \mathfrak{n}$, $\varepsilon < 2$, let

$$\mathbf{P}^{\mathfrak{A}}_{\xi\varepsilon} = \{0\} \qquad \text{if } f\xi = \varepsilon,$$
$$\mathbf{P}^{\mathfrak{A}}_{\xi\varepsilon} = 0 \qquad \text{if } f\xi \neq \varepsilon.$$

Clearly then $\mathfrak{A} = \langle \{0\}, \mathbf{P}^{\mathfrak{A}}_{\xi\varepsilon} \rangle_{\xi<\mathfrak{n}, \varepsilon<2}$ is a model of Δ. □

In the next few results we shall be working with expansions of a first-order language which has a binary relation symbol $<$. In this language, in any $L_{\mathfrak{m}\mathfrak{n}}$ with $\mathfrak{m}, \mathfrak{n} \geq \aleph_1$, the notion of well-order can be defined:

$$< \text{ is a simple order,}$$
$$\neg \exists v_0 \cdots v_i \cdots_{i<\omega} \bigwedge_{i<\omega} (v_{i+1} < v_i).$$

We also need certain formulas $\varphi_\alpha \in \mathrm{Fmla}^1_{\mathscr{L}}$, formulated in any language $L_{\mathfrak{m}\mathfrak{m}}$ such that $\alpha < \mathfrak{m}$:

$$\exists v_1 \cdots v_{1+\xi} \cdots_{\xi<\alpha} \Bigg\{ \bigwedge_{\xi<\eta<\alpha} (v_{1+\xi} < v_{1+\eta}) \wedge$$
$$\forall v_{1+\alpha} \Bigg[v_{1+\alpha} < v_0 \leftrightarrow \bigvee_{\xi<\alpha} (v_{1+\alpha} = v_{1+\xi}) \Bigg] \Bigg\}.$$

If β is a nonzero ordinal and $\gamma < \beta$, then $(\beta, <) \models \varphi_\alpha[\gamma]$ iff $\alpha = \gamma$, as is easily seen.

Lemma 31.7. *Any weakly compact cardinal is regular.*

PROOF. Assume that $\mathfrak{m}$ is singular, say $\mathfrak{m} = \bigcup_{\alpha<\mathfrak{n}} \mathfrak{p}_\alpha$, where $\mathfrak{n} < \mathfrak{m}$ and each $\mathfrak{p}_\alpha < \mathfrak{m}$. Let $\mathscr{L}$ be a first-order language with two nonlogical constants, a binary relation symbol $<$ and a unary relation symbol $\mathbf{P}$. Consider the following set Γ of sentences:

(1) $<$ is a well-ordering,
(2) $\bigvee_{\alpha<\mathfrak{n}} \exists v_0 \cdots v_\xi \cdots_{\xi<\mathfrak{p}\alpha} \forall v_{\mathfrak{p}\alpha}[\mathbf{P}v_{\mathfrak{p}\alpha} \rightarrow \bigvee_{\xi<\mathfrak{p}\alpha}(v_{\mathfrak{p}\alpha} = v_\xi)]$,
(3) $\exists v_0(\varphi_\alpha \wedge \mathbf{P}v_0)$ for each $\alpha < \mathfrak{m}$.

Thus there are $\mathfrak{m}$ sentences altogether. Suppose Γ has a model; we may assume that it has the form $(\beta, <, S)$ for some nonzero ordinal β, where $<$ is the natural ordering on β. The sentences of type (3) together with the remarks preceding this lemma show that $\mathfrak{m} \subseteq S$. But the sentence (2) says that $|S| \leq \mathfrak{p}_\alpha$ for some $\alpha < \mathfrak{n}$, contradiction. So, Γ has no model. Take any subset Δ of Γ of power $<\mathfrak{m}$. Let $S = \{\alpha < \mathfrak{m} : \exists v_0(\varphi_\alpha \wedge \mathbf{P}v_0) \in \Delta\}$. Thus $|S| < \mathfrak{m}$, so $|S| \leq \mathfrak{p}_\alpha$ for some $\alpha < \mathfrak{n}$. Hence $(\mathfrak{m}, <, S)$ is a model of Δ. □

The two preceding lemmas give us the following important result concerning strongly compact cardinals:

Theorem 31.8. *Every strongly compact cardinal is strongly inaccessible.*

Recall that the existence of strongly inaccessible uncountable cardinals is not provable in ordinary set theory. Thus it is not provable that any of the languages $L_{\mathfrak{m}\mathfrak{m}}$ for $\mathfrak{m} > \aleph_0$ are $(\mathfrak{m}, \infty)$-compact. Actually, strongly compact cardinals are extremely inaccessible. If there are arbitrarily large inaccessibles, and all inaccessibles are enumerated in an increasing sequence $\langle \theta_\alpha : \alpha \in \text{Ord} \rangle$, then the strongly compact cardinals are certain fixed points of θ, i.e., they satisfy $\mathfrak{m} = \theta_\mathfrak{m}$. Even stronger properties of the strongly compact cardinals are known. We shall not prove any very general facts here, but to give an idea of the flavor of the arguments we will at least show that if inaccessibles exist, not all of them are strongly compact:

Theorem 31.9 (Hanf). *If* $\mathfrak{m}$ *is an uncountable weakly compact cardinal, then* $\mathfrak{m}$ *is not the first uncountable strongly inaccessible cardinal.*

Proof. Let $\mathfrak{m}$ be the first uncountable strongly inaccessible cardinal; we show that it is not weakly compact. Take a first-order language $\mathscr{L}$ with nonlogical constants $<$, $\mathbf{F}$ (ternary), $\mathbf{G}$ (ternary), $\mathbf{P}$ (unary), and $\mathbf{Q}$ (unary). Let Γ be the following set of sentences of $L_{\mathfrak{m}\mathfrak{m}}$:

(1) $<$ is a well ordering;
(2) $\exists v_0 \varphi_\alpha$ (for each $\alpha < \mathfrak{m}$);
(3) $\forall v_0 \forall v_1 \forall v_2 \forall v_3(\mathbf{F}v_0v_1v_2 \wedge \mathbf{F}v_0v_1v_3 \rightarrow v_2 = v_3)$;
(4) $\forall v_0 \exists v_1[v_1 < v_0 \wedge \forall v_2 \forall v_3(\mathbf{F}v_0v_2v_3 \rightarrow v_2 < v_1 \wedge v_3 < v_0)]$;
(5) $\forall v_0 \forall v_1[\mathbf{P}v_0 \wedge v_1 < v_0 \rightarrow \exists v_2 \exists v_3(\mathbf{F}v_0v_2v_3 \wedge v_1 < v_3)]$;
(6) $\forall v_0 \forall v_1 \forall v_2[v_1 < v_0 \wedge v_2 < v_0 \wedge \forall v_3(\mathbf{G}v_0v_1v_3 \leftrightarrow \mathbf{G}v_0v_2v_3) \rightarrow v_1 = v_2]$;
(7) $\forall v_0\{\mathbf{Q}v_0 \rightarrow \exists v_1[v_1 < v_0 \wedge \forall v_2 \forall v_3(\mathbf{G}v_0v_2v_3 \rightarrow v_3 < v_1)]\}$;
(8) $\exists v_0 \forall v_1 \neg(v_0 < v_1)$;
(9) $\forall v_0(\mathbf{P}v_0 \vee \mathbf{Q}v_0 \vee \exists v_1\{v_1 < v_0 \wedge \forall v_2[v_1 < v_2 \rightarrow \neg(v_2 < v_0)]\} \vee \varphi_\omega)$.

Thus $|\Gamma| = \mathfrak{m}$. Suppose Γ has a model, $\mathfrak{A}$. We may assume that $\mathfrak{A}$ has the form $(\alpha + 1, <, \mathbf{F}^{\mathfrak{A}}, \mathbf{G}^{\mathfrak{A}}, \mathbf{P}^{\mathfrak{A}}, \mathbf{Q}^{\mathfrak{A}})$, by virtue of (1) and (8). Because of the sentences (2), it follows that $\mathfrak{m} \subseteq \alpha + 1$ hence $\mathfrak{m} \leq \alpha + 1$ and actually $\mathfrak{m} \leq \alpha$ since $\mathfrak{m}$ is an infinite cardinal. Thus $\mathfrak{m} \in \alpha + 1$. We shall obtain a contradiction by showing that each uncountable infinite cardinal $\mathfrak{n} \in \alpha + 1$ is accessible, i.e., is either singular or else satisfies $\mathfrak{n} \leq 2^{\mathfrak{p}}$ for some $\mathfrak{p} < \mathfrak{n}$. By (9) we have $\mathfrak{n} \in \mathbf{P}^{\mathfrak{A}}$ or $\mathfrak{n} \in \mathbf{Q}^{\mathfrak{A}}$.

Case 1. $\mathfrak{n} \in \mathbf{P}^{\mathfrak{A}}$. Let $f = \{(\beta, \gamma) : (\mathfrak{n}, \beta, \gamma) \in \mathbf{F}^{\mathfrak{A}}\}$. Thus by (3), f is a function. By (4) we can choose $\delta < \mathfrak{n}$ such that f maps a subset of δ into $\mathfrak{n}$. By (5), $\bigcup \operatorname{Rng} f = \mathfrak{n}$. Thus $\mathfrak{n}$ is singular.

Case 2. $\mathfrak{n} \in \mathbf{Q}^{\mathfrak{A}}$. By (7), choose $\beta < \mathfrak{n}$ such that $\delta < \beta$ whenever $(\mathfrak{n}, \gamma, \delta) \in \mathbf{G}^{\mathfrak{A}}$. Now for each $\gamma < \mathfrak{n}$ let $f\gamma = \{\delta : (\mathfrak{n}, \gamma, \delta) \in \mathbf{G}^{\mathfrak{A}}\}$. Thus $f\gamma \subseteq \beta$. And by (6), f is one–one. Hence $f\colon \mathfrak{n} \rightarrowtail \mathbf{S}\beta$, as desired.

Since $\mathfrak{m}$ is strongly inaccessible, we have reached a contradiction. Thus Γ has no model.

Now let Δ be a subset of Γ of power $<\mathfrak{m}$. Let β be the supremum of all ordinals $\alpha < \mathfrak{m}$ such that $\exists v_0\varphi_\alpha \in \Delta$. We now build up a model $\mathfrak{A}$ of Δ with universe $A = \beta + 1$ and with the natural ordering of $\beta + 1$ as the interpretation of $\mathbf{<}$. Thus (1), (2), and (8) automatically hold for formulas in Δ. Let $\mathbf{P}^{\mathfrak{A}}$ consist of all singular cardinals $\leq \beta$. For each $\mathfrak{n} \in \mathbf{P}^{\mathfrak{A}}$ choose $\mathfrak{p}_\mathfrak{n} < \mathfrak{n}$ and a function $f_\mathfrak{n}\colon \mathfrak{p}_\mathfrak{n} \to \mathfrak{n}$ such that $\bigcup \operatorname{Rng} f_\mathfrak{n} = \mathfrak{n}$. Set

$$\mathbf{F}^{\mathfrak{A}} = \{(\mathfrak{n}, \gamma, \delta) : \mathfrak{n} \in \mathbf{P}^{\mathfrak{A}}, (\gamma, \delta) \in f_\mathfrak{n}\}.$$

Clearly, then, (3)–(5) will hold. Next, let $\mathbf{Q}^{\mathfrak{A}} = \{\mathfrak{n} : \mathfrak{n}$ is a cardinal $\leq \beta$, and $\mathfrak{n} \leq 2^{\mathfrak{p}}$ for some $\mathfrak{p} < \mathfrak{n}\}$. For each $\mathfrak{n} \in \mathbf{Q}^{\mathfrak{A}}$ choose $\mathfrak{p}_\mathfrak{n} < \mathfrak{n}$ and $g_\mathfrak{n}\colon \mathfrak{n} \rightarrowtail \mathbf{S}\mathfrak{p}_\mathfrak{n}$. Set

$$\mathbf{G}^{\mathfrak{A}} = \{(\mathfrak{n}, \delta, \gamma) : \mathfrak{n} \in \mathbf{Q}^{\mathfrak{A}}, \delta < \mathfrak{n}, \text{ and } \gamma \in g_\mathfrak{n}\delta\}.$$

Then (6) and (7) are clear. Finally, (9) holds since $\beta + 1 < \mathfrak{m}$. □

The above results give a fairly good picture of the rarity of strongly compact cardinals. Now we want to add a few results which indicate that weakly compact cardinals are rare also. (But perhaps they are not as rare as strongly compact cardinals.) By the proof of 31.6 we have

Theorem 31.10. *If* $\mathfrak{m} \geq \aleph_0$, *then* $2^{\mathfrak{m}}$ *is not weakly compact.*

Theorem 31.11. *Every weakly compact cardinal is weakly inaccessible.*

PROOF. By 31.7 it suffices to show that any successor cardinal $\mathfrak{m}^+ > \aleph_0$ is not weakly compact. Let Γ be the following set of sentences:

(1) $\mathbf{<}$ is a well-ordering;
(2) $\exists v_0\varphi_\alpha$ for each $\alpha < \mathfrak{m}^+$;
(3) $\exists v_0 \cdots v_\xi \cdots_{\xi<\mathfrak{m}} \forall v_\mathfrak{m} \bigvee_{\xi<\mathfrak{m}} (v_\mathfrak{m} \mathbf{=} v_\xi)$.

It is easily checked that Γ violates $(\mathfrak{m}^+, \mathfrak{m}^{++})$-compactness. □

Theorem 31.12 (Hanf). *If* $\mathfrak{m}$ *is an uncountable weakly compact cardinal, then there is a cardinal* $\mathfrak{n} < \mathfrak{m}$ *which is uncountable, weakly inaccessible, and not a power of* 2.

Proof. Assume the hypothesis holds but the conclusion fails. By 31.10 and 31.11, $\mathfrak{m}$ is weakly inaccessible but not a power of 2. We take a language $\mathscr{L}$ with nonlogical constants $<$, **F** (ternary), **G** (ternary), **H** (4-ary), **P**, **Q**, **R** (unary). Let Γ be the following set of sentences of $L_{\mathfrak{m}\mathfrak{m}}$:

(1) $<$ is a well-ordering;
(2) $\exists v_0\varphi_\alpha$ (for each $\alpha < \mathfrak{m}$);
(3) $\forall v_0\forall v_1\forall v_2\forall v_3(\mathbf{F}v_0v_1v_2 \wedge \mathbf{F}v_0v_1v_3 \rightarrow v_2 = v_3)$;
(4) $\forall v_0\exists v_1[v_1 < v_0 \wedge \forall v_2\forall v_3(\mathbf{F}v_0v_2v_3 \rightarrow v_2 < v_1 \wedge v_3 < v_0)]$;
(5) $\forall v_0\forall v_1[\mathbf{P}v_0 \wedge v_1 < v_0 \rightarrow \exists v_2\exists v_3(\mathbf{F}v_0v_2v_3 \wedge v_1 < v_3)]$;
(6) $\forall v_0\forall v_1\forall v_2\forall v_3[(\mathbf{G}v_0v_1v_2 \wedge \mathbf{G}v_0v_1v_3 \rightarrow v_2 = v_3) \wedge (\mathbf{G}v_0v_2v_1 \wedge \mathbf{G}v_0v_3v_1 \rightarrow v_2 = v_3)]$;
(7) $\forall v_0[\mathbf{Q}v_0 \rightarrow \exists v_1(v_1 < v_0 \wedge \forall v_2\{v_2 < v_0 \rightarrow \forall v_3[v_3 < v_2 \rightarrow \exists v_4(v_4 < v_1 \wedge \mathbf{G}v_2v_3v_4)]\})]$;
(8) $\forall v_0\forall v_1\forall v_2\forall v_3\{\forall v_4[\mathbf{H}(v_0, v_1, v_4, v_3) \leftrightarrow \mathbf{H}(v_0, v_2, v_4, v_3)] \rightarrow v_1 = v_2\}$;
(9) $\forall v_0\{\mathbf{T}v_0 \rightarrow \exists v_3[v_3 < v_0 \wedge \forall v_1\forall v_2(v_1 < v_0 \wedge \mathbf{H}(v_0, v_1, v_2, v_3) \rightarrow v_2 < v_3)]\}$;
(10) $\forall v_0\forall v_3[\mathbf{T}v_0 \wedge v_3 < v_0 \rightarrow \forall v_4\cdots v_\xi\cdots_{\xi<\mathfrak{n}}(\bigwedge_{4\le\xi\le\mathfrak{n}}(v_\xi < v_3) \rightarrow \exists v_1\{v_1 < v_0 \wedge \forall v_2[\mathbf{H}(v_0, v_1, v_2, v_3) \leftrightarrow \bigvee_{4\le\xi\le\mathfrak{n}} v_2 = v_\xi]\})]$, a sentence for each $\mathfrak{n} < \mathfrak{m}$;
(11) $\exists v_0\forall v_1 \neg(v_0 < v_1)$;
(12) $\forall v_0(\mathbf{P}v_0 \vee \mathbf{Q}v_0 \vee \mathbf{T}v_0 \vee \exists v_1\{v_1 < v_0 \wedge \forall v_2[v_1 < v_2 \rightarrow \neg(v_2 < v_0)]\} \vee \varphi_\omega)$.

Clearly Γ has $\mathfrak{m}$ elements. Suppose Γ has a model, $\mathfrak{A}$.

We may assume that $\mathfrak{A}$ has the form $(\alpha + 1, <, \mathbf{F}^{\mathfrak{A}}, \mathbf{G}^{\mathfrak{A}}, \mathbf{H}^{\mathfrak{A}}, \mathbf{P}^{\mathfrak{A}}, \mathbf{Q}^{\mathfrak{A}}, \mathbf{R}^{\mathfrak{A}})$, by (1) and (11). We have $\mathfrak{m} \subseteq \alpha + 1$ by (2). Thus $\mathfrak{m} \in \alpha + 1$. By (12) we have three cases.

Case 1. $\mathfrak{m} \in \mathbf{P}^{\mathfrak{A}}$. As in the proof of 31.9, we see that $\mathfrak{m}$ is singular, so it is not weakly compact (see 31.7), contradiction.

Case 2. $\mathfrak{m} \in \mathbf{Q}^{\mathfrak{A}}$. By (7), choose $\beta < \mathfrak{m}$ as indicated there. For each $\gamma < \mathfrak{m}$ let $f_\gamma = \{(\delta, \varepsilon) : (\gamma, \delta, \varepsilon) \in \mathbf{G}^{\mathfrak{A}}\}$. Thus f_γ is a one–one function by (6). By (7), $f_\gamma \restriction \gamma : \gamma \rightarrowtail \beta$. Thus $\mathfrak{m}$ is the least cardinal $>\beta$, so $\mathfrak{m}$ is a successor cardinal and hence by 31.11, $\mathfrak{m}$ is not weakly compact, contradiction.

Case 3. $\mathfrak{m} \in \mathbf{T}^{\mathfrak{A}}$. Choose $\beta < \mathfrak{m}$ in accordance with (9). For each $\gamma < \mathfrak{m}$ let $f_\gamma = \{\delta : (\mathfrak{m}, \gamma, \delta, \beta) \in \mathbf{H}^{\mathfrak{A}}\}$. Thus by (9) $f: \mathfrak{m} \rightarrow \mathbf{S}\beta$. By (8), f is one–one. From this we know that β is infinite. Let Δ be any subset of β and set $|\Delta| = \mathfrak{n}$. From (10) we easily infer that $\Delta \in \operatorname{Rng} f$. Thus $\mathfrak{m}$ is a power of 2, so by 31.10, $\mathfrak{m}$ is not weakly compact, contradiction.

Thus Γ does not have a model. Now let Δ be a subset of power $<\mathfrak{m}$. Let β, $\mathbf{P}^{\mathfrak{A}}$, $\mathbf{F}^{\mathfrak{A}}$ be as in the proof of 31.9. $\mathbf{Q}^{\mathfrak{A}}$ and $\mathbf{G}^{\mathfrak{A}}$ are easily defined to satisfy (6), (7). Let $\mathbf{T}^{\mathfrak{A}}$ be the set of all uncountable powers of 2 less than $\beta + 1$. For each $\mathfrak{n}$ with $\aleph_0 < 2^{\mathfrak{n}} < \beta + 1$ let $h_{\mathfrak{n}}: 2^{\mathfrak{n}} \twoheadrightarrow \mathbf{S}\mathfrak{n}$. Set $\mathbf{H}^{\mathfrak{A}} = \{(2^{\mathfrak{n}}, \alpha, \gamma, \mathfrak{n}) : 2^{\mathfrak{n}} \in \mathbf{T}^{\mathfrak{A}}, \gamma < 2^{\mathfrak{n}}, \alpha \in h_{\mathfrak{n}}\gamma\}$. This yields, as is easily seen, a model $\mathfrak{A}$ of Δ. □

To hint at the importance of strongly and weakly compact cardinals we shall now give some results which show their relationships with properties of cardinals defined in purely mathematical ways. The following definition will play a role here; it gives one of the most important notions of large cardinals.

Definition 31.13. An infinite cardinal $\mathfrak{m}$ is *measurable* if there is a mapping $\mu: \mathbf{S}\mathfrak{m} \to \{0, 1\}$ satisfying the following conditions:

(*i*) $\mu\mathfrak{m} = 1$;
(*ii*) $\mu 0 = \mu\{\alpha\} = 0$ for each $\alpha \in \mathfrak{m}$;
(*iii*) if X is a collection of pairwise disjoint nonempty subsets of $\mathfrak{m}$ with $|X| < \mathfrak{m}$, then $\mu \bigcup_{x \in X} x = \Sigma\{\mu x : x \in X\}$ (thus $\mu \bigcup_{x \in X} x = 1$ iff for some unique $x \in X$, $\mu x = 1$).

We will give its connections with compact cardinals presently. For now we give this important result of Ulam:

Theorem 31.14 (Ulam). *Every measurable cardinal is strongly inaccessible.*

Proof. Let $\mathfrak{m}$ be measurable, and suppose $\mathfrak{n} < \mathfrak{m}$ and $\mathfrak{p}_\alpha < \mathfrak{m}$ for each $\alpha < \mathfrak{n}$; we shall show that $\Pi_{\alpha<\mathfrak{n}} \mathfrak{p}_\alpha < \mathfrak{m}$ (this is one of the equivalent conditions for $\mathfrak{m}$ to be strongly inaccessible). Let μ be a measure on $\mathbf{S}(\mathfrak{m})$ (as in 31.13). Suppose $\mathfrak{m} \leq \Pi_{\alpha<\mathfrak{n}} \mathfrak{p}_\alpha$, and let $f: \mathfrak{m} \rightarrowtail \mathsf{P}_{\alpha<\mathfrak{n}} \mathfrak{p}_\alpha$. For each $\alpha < \mathfrak{n}$ and $\beta < \mathfrak{p}_\alpha$ let

$$A_{\alpha\beta} = \{\gamma \in \mathfrak{m} : f_\gamma\alpha = \beta\}.$$

Clearly $A_{\alpha\beta} \cap A_{\alpha\delta} = 0$ if $\beta \neq \delta$, and $\bigcup_{\beta<\mathfrak{p}_\alpha} A_{\alpha\beta} = \mathfrak{m}$. It follows that for each $\alpha < \mathfrak{n}$ there is a unique β, call it g_α, such that $\mu A_{\alpha,g\alpha} = 1$. Hence $\mu(\mathfrak{m} \sim A_{\alpha,g\alpha}) = 0$ and so

$$\mu \bigcup_{\alpha<\mathfrak{n}} (\mathfrak{m} \sim A_{\alpha,g\alpha}) = 0.$$

But $\bigcup_{\alpha<\mathfrak{n}} (\mathfrak{m} \sim A_{\alpha,g\alpha}) = \mathfrak{m} \sim \bigcap_{\alpha<\mathfrak{n}} A_{\alpha,g\alpha}$. Thus $\mu \bigcap_{\alpha<\mathfrak{n}} A_{\alpha,g\alpha} = 1$. But if $\gamma \in \bigcap_{\alpha<\mathfrak{n}} A_{\alpha,g\alpha}$, then for all $\alpha < \mathfrak{n}$, $f_\gamma\alpha = g_\alpha$. Thus $f_\gamma = g$. Hence $|\bigcap_{\alpha<\mathfrak{n}} A_{\alpha,g\alpha}| \leq 1$, contradiction. □

The following equivalences for strongly compact cardinals are due to Tarski, Monk, Scott, and Mycielski.

Theorem 31.15. *For any infinite cardinal* $\mathfrak{m}$, *the following conditions are equivalent:*

(*i*) $\mathfrak{m}$ *is strongly compact;*

(*ii*) *if* $\mathfrak{A}$ *is an* $\mathfrak{m}$*-complete,* $\mathfrak{m}$*-distributive Boolean algebra, then* $\mathfrak{A}$ *has an* $\mathfrak{m}$*-complete ultrafilter (recall that* $\mathfrak{A}$ *is* $\mathfrak{m}$*-complete provided that* ΣX *exists whenever* $X \subseteq A$ *and* $|X| < \mathfrak{m}$; $\mathfrak{A}$ *is* $\mathfrak{m}$*-distributive provided that the conditions* $\mathfrak{n}, \mathfrak{p} < \mathfrak{m}$ *and* $a : \mathfrak{n} \times \mathfrak{p} \to A$ *imply that* $\Sigma\{\Pi_{\alpha<\mathfrak{n}} a_{\alpha, f\alpha} : f \in {}^{\mathfrak{n}}\mathfrak{p}\}$ *exists and equals* $\Pi_{\alpha<\mathfrak{n}} \Sigma_{\beta<\mathfrak{p}} a_{\alpha\beta}$; *a filter* F *is* $\mathfrak{m}$*-complete provided that* $\Pi X \in F$ *whenever* $X \subseteq F$ *and* $|X| < \mathfrak{m}$*);*

(*iii*) *in every* $\mathfrak{m}$*-complete field of sets, any proper* $\mathfrak{m}$*-complete filter can be extended to an* $\mathfrak{m}$*-complete ultrafilter (a field* F *of sets is* $\mathfrak{m}$*-complete if* $\bigcup X \in F$ *whenever* $X \subseteq F$ *and* $|X| < \mathfrak{m}$*);*

(*iv*) *in the field of all subsets of a set, any proper* $\mathfrak{m}$*-complete filter can be extended to an* $\mathfrak{m}$*-complete ultrafilter;*

(*v*) *any* $\mathfrak{m}$*-product of* $(\mathfrak{m}, \infty)$*-compact Hausdorff spaces is* $(\mathfrak{m}, \infty)$*-compact (a space* X *is* $(\mathfrak{m}, \infty)$*-compact provided that every cover of* X *by open sets has a subcover with* $<\mathfrak{m}$ *elements; the* $\mathfrak{m}$*-product topology is the closure of the usual product topology under taking intersections of* $<\mathfrak{m}$ *things);*

(*vi*) *any product of* $(\mathfrak{m}, \infty)$*-compact Hausdorff spaces is* $(\mathfrak{m}, \infty)$*-compact.*

Proof. We shall prove (*ii*) ⇒ (*iii*) ⇒ (*iv*) ⇒ (*v*) ⇒ (*vi*) ⇒ (*ii*) and (*iv*) ⇒ (*i*) ⇒ (*iv*).

(*ii*) ⇒ (*iii*). Assume (*ii*), and let $\mathfrak{A}$ be an $\mathfrak{m}$-complete field of sets and F a proper $\mathfrak{m}$-complete filter over $\mathfrak{A}$. Then $\mathfrak{A}/F$ is an $\mathfrak{m}$-complete, $\mathfrak{m}$-distributive Boolean algebra. In fact, if $a \in {}^{\mathfrak{n}}A$ where $\mathfrak{n} < \mathfrak{m}$, it is easily seen that $\Sigma_{\alpha<\mathfrak{n}} [a_\alpha] = [\bigcup_{\alpha<\mathfrak{n}} a_\alpha]$. And if $\mathfrak{n}, \mathfrak{p} < \mathfrak{m}$ and $a: \mathfrak{n} \times \mathfrak{p} \to A$, then

$$\bigcap_{\alpha<\mathfrak{n}} \bigcup_{\beta<\mathfrak{p}} a_{\alpha\beta} = \bigcup\Big\{\bigcap_{\alpha<\mathfrak{n}} a_{\alpha, f\alpha} : f \in {}^{\mathfrak{n}}\mathfrak{p}\Big\},$$

and it follows easily that $\mathfrak{A}/F$ is $\mathfrak{m}$-distributive. By (*ii*), let G be an $\mathfrak{m}$-complete ultrafilter over $\mathfrak{A}/F$. Then $\{a : [a] \in G\}$ is easily seen to be an $\mathfrak{m}$-complete ultrafilter over $\mathfrak{A}$ which extends F.

(*iii*) ⇒ (*iv*): obvious.

(*iv*) ⇒ (*v*). Here we generalize one of the usual proofs of the Tychonoff product theorem. Assume (*iv*), and suppose that $\langle X_i : i \in I \rangle$ is a family of $(\mathfrak{m}, \infty)$-compact spaces. Recall that a subbase for the ordinary topology on $Y = \mathsf{P}_{i \in I} X_i$ consists of all sets of the form $\{f : f_i \in U\}$ where $i \in I$ and U is open in X_i. To proceed further we need to show

(1) $\qquad \mathfrak{m}$ is regular.

For let $J = \{G \subseteq {}^{\mathfrak{m}}2 : |G| < \mathfrak{m}\}$. Let F be the $\mathfrak{m}$-complete ideal in $\mathbf{S}({}^{\mathfrak{m}}2)$ generated by J. Since $\mathrm{cf}(2^{\mathfrak{m}}) > \mathfrak{m}$, it is clear that F is an $\mathfrak{m}$-complete proper ideal in $\mathbf{S}({}^{\mathfrak{m}}2)$. By (*iv*), let K be an $\mathfrak{m}$-complete prime ideal in $\mathbf{S}({}^{\mathfrak{m}}2)$ extending

F. Suppose $\mathfrak{m}$ is singular, say $\mathfrak{m} = \bigcup_{\alpha<\mathfrak{n}} \mathfrak{p}_\alpha$ where $\mathfrak{n} < \mathfrak{m}$ and $\mathfrak{p}_\alpha < \mathfrak{m}$ for each $\alpha < \mathfrak{n}$. Then K is $\mathfrak{m}^+$-complete. For, if $a : \mathfrak{m} \to K$, then

$$\bigcup_{\alpha<\mathfrak{m}} a_\alpha = \bigcup_{\alpha<\mathfrak{n}} \bigcup_{\beta<\mathfrak{p}_\alpha} a_\beta \in K.$$

Now for each $\alpha < \mathfrak{m}$ we have ${}^{\mathfrak{m}}2 = \{f : f\alpha = 0\} \cup \{f : f\alpha = 1\}$, so we can choose $g_\alpha \in 2$ such that $\{f : f\alpha = g\alpha\} \notin K$. By the $\mathfrak{m}^+$-completeness of K, $\{g\} = \bigcap_{\alpha<\mathfrak{m}} \{f : f\alpha = g\alpha\} \notin K$, contradiction.

Now we come back to the main part of the argument. From (1) it follows that a base for the $\mathfrak{m}$-topology on Y is the family of all sets of the form

$$\{f : f \restriction J \in \mathsf{P}_{i\in J}\, U_i\}, \tag{2}$$

where J is a subset of I of power $<\mathfrak{m}$ and U_i is open in X_i for each $i \in J$. To check (v), it suffices to take a family $\mathscr{B}$ of basic open sets of Y, assume that no subset of $\mathscr{B}$ of power $<\mathfrak{m}$ covers Y, and show that $\mathscr{B}$ does not cover Y. Let K be the $\mathfrak{m}$-complete ideal in $\mathbf{S}Y$ generated by $\mathscr{B}$. By (1), K is a proper ideal. Hence by (iv), there is an $\mathfrak{m}$-complete prime ideal L in $\mathbf{S}Y$ which extends K. For each $i \in I$ let

$$\mathscr{C}_i = \{U : U \text{ is an open set in } X_i \text{ and } \{f \in Y : f_i \in U\} \in L\}.$$

Since L is $\mathfrak{m}$-complete and $Y \notin L$, no subfamily of $\mathscr{C}_i$ of power $<\mathfrak{m}$ covers X_i. Hence by the $(\mathfrak{m}, \infty)$-compactness of X_i, $\mathscr{C}_i$ does not cover X_i. Thus for each $i \in I$ we can choose $f_i \in X_i \sim \bigcup \mathscr{C}_i$. We claim that $f \notin \bigcup \mathscr{B}$. For, suppose $f \in \bigcup \mathscr{B}$, say $f \in V \in \mathscr{B}$, where V has the form (2). Now

$$V = \bigcap_{i\in J} \{g \in Y : g_i \in U_i\},$$

and $|J| < \mathfrak{m}$. Since $V \in L$ (because $\mathscr{B} \subseteq K \subseteq L$), it follows that there is an $i \in J$ with $\{g \in Y : g_i \in U_i\} \in L$. Hence $U_i \in \mathscr{C}_i$ and $f_i \in \bigcup \mathscr{C}_i$, contradiction.

(v) $\Rightarrow$ (vi): obvious.

(vi) $\Rightarrow$ (ii). Assume (vi), and let $\mathfrak{A}$ be an $\mathfrak{m}$-complete, $\mathfrak{m}$-distributive Boolean algebra. Let B be the set of all subsets of A of power $<\mathfrak{m}$. Now we need to show

$$\mathfrak{m} \text{ is measurable.} \tag{3}$$

To prove (3), let $C = \bigcup_{\alpha<\mathfrak{m}} {}^{\mathfrak{m}}\alpha$. Thus $|C| = 2^{\mathfrak{m}}$. Let $f : 2^{\mathfrak{m}} \twoheadrightarrow C$. For each $\xi < 2^{\mathfrak{m}}$ we have $|\mathrm{Rng} f_\xi| < \mathfrak{m}$, and hence as a discrete space $\mathrm{Rng} f_\xi$ is $(\mathfrak{m}, \infty)$-compact. Hence by (vi), the space $X = \mathsf{P}_{\xi<\exp\mathfrak{m}} \mathrm{Rng} f_\xi$ is $(\mathfrak{m}, \infty)$-compact. For each $\alpha < \mathfrak{m}$, let t_α be the member of X such that $t_\alpha\xi = f_\xi\alpha$ for all $\xi < 2^{\mathfrak{m}}$. Set $D = \{t_\alpha : \alpha < \mathfrak{m}\}$. For each $\alpha < \mathfrak{m}$ let ξ_α be the unique ordinal $<2^{\mathfrak{m}}$ such that $f_{\xi\alpha} : \mathfrak{m} \to 2, f_{\xi\alpha}\alpha = 1$, while $f_{\xi\alpha}\beta = 0$ for all $\beta \in \mathfrak{m} \sim \{\alpha\}$. Let $U_\alpha = \{g \in X : g_{\xi(\alpha)} = 1\}$. Thus U_α is open in X, and $D \cap U_\alpha = \{t_\alpha\}$. It follows that D is not closed, since otherwise $\{X \sim D\} \cup \{U_\alpha : \alpha < \mathfrak{m}\}$ would be an open cover of X with no subcover of power $<\mathfrak{m}$. So, let u be in the closure of D but not in D. Now we are ready to define our "measure" μ on $\mathbf{S}(\mathfrak{m})$. For any $\Gamma \subseteq \mathfrak{m}$ we set

$$\begin{aligned} \mu\Gamma = 1 &\quad \text{if there is a } \xi < 2^{\mathfrak{m}} \text{ such} \\ &\quad\quad \text{that } \Gamma = \{\alpha < \mathfrak{m} : t_\alpha\xi = u\xi\}, \\ \mu\Gamma = 0 &\quad \text{otherwise.} \end{aligned}$$

To check 31.13(*i*), choose $\xi < 2^{\mathfrak{m}}$ such that $f_\xi\colon \mathfrak{m} \to \{0\}$. Clearly then $\mathfrak{m} = \{\alpha < \mathfrak{m} : t_\alpha\xi = u\xi\}$, since $u\xi = 0$ and $t_\alpha\xi = 0$ for any $\alpha < \mathfrak{m}$. For 31.13(*ii*), for any $\xi < 2^{\mathfrak{m}}$ let $U = \{g \in X : g\xi = u\xi\}$. Thus U is a neighborhood of u, so there is an $\alpha < \mathfrak{m}$ with $t_\alpha \in U$. Hence $t_\alpha\xi = u\xi$. This shows that $\mu 0 = 0$. Next, for any $\alpha < \mathfrak{m}$, suppose $\mu\{\alpha\} = 1$, and let $\xi < 2^{\mathfrak{m}}$ be such that $\{\alpha\} = \{\beta < \mathfrak{m} : t_\beta\xi = u\xi\}$. Let U be a neighborhood of u such that $t_\alpha \notin U$. Let $V = \{g \in X : g\xi = u\xi\}$. Thus V is also a neighborhood of u. Choose $t_\beta \in U \cap V$. Then $\beta \neq \alpha$ and $t_\beta\xi = u\xi$ contradiction. Hence $\mu\{\alpha\} = 0$ after all. Finally, suppose that Y is a collection of pairwise disjoint nonempty subsets of $\mathfrak{m}$ with $|Y| < \mathfrak{m}$. First suppose $\mu \bigcup_{y\in Y} y = 1$. Choose $\xi < 2^{\mathfrak{m}}$ accordingly so that $\bigcup_{y\in Y} y = \{\alpha < \mathfrak{m} : t_\alpha\xi = u\xi\}$. Since $|Y| < \mathfrak{m}$, we can choose $\zeta < 2^{\mathfrak{m}}$ such that for each $y \in Y$ there is a $\beta \in \operatorname{Rng} f_\zeta$ with $f_\zeta^{-1}\{\beta\} = y$. Let $U = \{g \in X: g\xi = u\xi \text{ and } g\zeta = u\zeta\}$. Then U is a neighborhood of u, so we can choose $t_\alpha \in U$. Since $t_\alpha\xi = u\xi$, by choice of ξ there is a $y \in Y$ with $\alpha \in Y$. Thus by choice of ζ, $f_\zeta\alpha = t_\alpha\zeta = u\zeta$ implies that $y = f_\zeta^{-1}\{u\zeta\} = \{\alpha : t_\alpha\zeta = u\zeta\}$, and $\mu y = 1$. Thus

(4) if Y is a collection of pairwise disjoint nonempty subsets of $\mathfrak{m}$ with $|Y| < \mathfrak{m}$ and $\mu \bigcup_{y\in Y} y = 1$, then $\mu y = 1$ for some $y \in Y$.

(5) If y and z are pairwise disjoint subsets of X with $\mu y = 1$, then $\mu z = 0$.

Indeed, suppose that $\mu z = 1$ also. Thus there are $\xi, \zeta < 2^{\mathfrak{m}}$ with $y = \{\alpha < \mathfrak{m} : t_\alpha\xi = u\xi\}$ and $z = \{\alpha < \mathfrak{m} : t_\alpha\zeta = u\zeta\}$. Choose t_α in the neighborhood $\{g \in X : g\xi = u\xi, g\zeta = u\zeta\}$ of u. Then $\alpha \in y \cap z$, contradiction. Thus (5) holds. In particular, the y in (4) is unique. If $\mu \bigcup_{y\in Y} y = 0$, then by (4) $\mu(\mathfrak{m} \sim \bigcup_{y\in Y} y) = 1$, and hence by (5) $\mu y = 0$ for each $y \in Y$. We have now verified 31.13(*iii*), and with it (3) has now been checked. Actually we need not (3), but the following consequence of it (by 31.14):

(6) $\mathfrak{m}$ is strongly inaccessible.

For each $C \in B$ let X_C be the set $\mathsf{P}_{a\in C}\{-a, a\}$, with the discrete topology. By (6), $|X_C| < \mathfrak{m}$, so X_C is $(\mathfrak{m}, \infty)$-compact. Hence by our assumption (*vi*), so is $Y = \mathsf{P}_{C\in B} X_C$. For each $C \in B$, set

$$D_C = \{f \in \mathsf{P}_{a\in C}\{-a, a\} : \Pi_{a\in C} fa \neq 0\}.$$

Now by $\mathfrak{m}$-distributivity we have

$$1 = \Pi_{a\in C}(a + -a) = \Sigma\{\Pi_{a\in C} fa : f \in \mathsf{P}_{a\in C}\{-a, a\}\},$$

from which it follows that $D_C \neq 0$. Now for $C, E \in B$ we set

$$K_{CE} = \{f \in Y : f_C \upharpoonright C \cap E = f_E \upharpoonright C \cap E \text{ and } f_C \in D_C \text{ and } f_E \in D_E\}.$$

Each such set is closed; this follows easily from the fact that

$$\begin{aligned}\{f \in Y : f_C \upharpoonright C \cap E = f_E \upharpoonright C \cap E\} \\ = \bigcap_{x \in C \cap E} [(\{f \in Y : f_C x = x\} \cap \{f \in Y : f_E x = x\}) \\ \cup (\{f \in Y : f_C x = -x\} \cap \{f \in Y : f_E x = -x\})].\end{aligned}$$

Now if $F \subseteq B$ and $|F| < \mathfrak{m}$, then $\bigcap_{C,E \in F} K_{CE} \neq 0$. In fact $\bigcup F \in B$ since $\mathfrak{m}$ is regular. Choose $f \in D_{\cup F}$, and let $g \in Y$ be any element such that $g_C = f \upharpoonright C$ for each $C \in F$. Clearly then $g \in \bigcap_{C,E \in F} K_{CE}$. From the $(\mathfrak{m}, \infty)$-compactness of Y it now follows that there is an $h \in \bigcap_{C,E \in B} K_{CE}$. Set $F = \bigcup_{C \in B} \text{Rng } h_C$. Then F is the desired $\mathfrak{m}$-complete ultrafilter on $\mathfrak{A}$. For, clearly $F \neq 0$. Suppose $0 \in F$, say $0 \in \text{Rng } h_C$. This contradicts $h_C \in D_C$. Suppose $x \in F$ and $x \leq y$. Say $x \in \text{Rng } h_C$. Let $E = C \cup \{y\}$. Since $h \in K_{CE}$, it follows that $x \in \text{Rng } h_E$, and $h_E \in D_E$ implies that $h_E y = y$, so $y \in F$. Next, suppose $x \in A$ is arbitrary. Then $h_{\{x\}} x = x$ or $h_{\{x\}} x = -x$, so $x \in F$ or $-x \in F$. Finally, suppose $Z \subseteq F$ with $|Z| < \mathfrak{m}$; we must show that $\Pi Z \in F$. For each $z \in Z$ choose $C_z \in B$ with $z \in \text{Rng } h_{C_z}$. Let $E = \bigcup_{z \in Z} C_z \cup \{\Pi Z\}$. Thus $E \in B$ since $\mathfrak{m}$ is regular. Since $h \in K_{Cz,E}$ for each $z \in Z$, we have $z \in \text{Rng } h_E$. Now $\Pi Z \in \text{Rng } h_E$ or $-\Pi z \in \text{Rng } h_E$. If the latter is the case, then

$$\Pi_{x \in E}\, h_E x \leq \Pi_{z \in Z}\, z \cdot -\Pi Z = 0,$$

contradicting $h_E \in D_E$. Thus $\Pi Z \in \text{Rng } h_E \subseteq F$, as desired.

(iv) ⇒ *(i)*. This is a straight-forward generalization of the ultraproduct proof of compactness. Indeed, it is routine to check that the proof of 18.29 extends to $L_{\mathfrak{m}\mathfrak{m}}$-languages if F is assumed to be $\mathfrak{m}$-complete. In the proof of the compactness theorem on p. 323, we take $I = \{\Delta : \Delta \subseteq \Gamma, |\Delta| < \mathfrak{m}\}$, and for each $\Delta \in I$ set $G_\Delta = \{\Theta : \Delta \subseteq \Theta \in I\}$. The rest of the proof is similar, replacing "finite" by "cardinality $<\mathfrak{m}$". But here it is necessary to show that $\mathfrak{m}$ is regular; this was done above in the proof that *(iv)* ⇒ *(v)*.

(i) ⇒ *(iv)*. Let A be the field of all subsets of X, and let F be an $\mathfrak{m}$-complete filter over A. Consider a first-order language with a unary relation symbol $\mathbf{P}_a$ for each $a \in A$ and an individual constant $\mathbf{c}$. Let Γ be the following set of sentences of $L_{\mathfrak{m}\mathfrak{m}}$:

all sentences holding in $(X, a)_{a \in A}$,
$\mathbf{P}_a\mathbf{c}$ for each $a \in F$.

If Δ is a subset of Γ of power $<\mathfrak{m}$, then there is an $x \in X$ such that $x \in \bigcap \{a \in F : \mathbf{P}_a\mathbf{c} \text{ is in } \Delta\}$. Hence $(X, a, x)_{a \in A}$ is a model of Δ. It follows from $(\mathfrak{m}, \infty)$-compactness that Γ has a model $\mathfrak{B} = (B, b_a, y)_{a \in A}$. Let $G = \{a \in A : y \in b_a\}$. Our claim is that G is the desired $\mathfrak{m}$-complete ultrafilter extending F. If $a \in F$, then $\mathbf{P}_a\mathbf{c} \in \Gamma$ and so $y \in b_a$ since $\mathfrak{B}$ is a model of Γ. Thus $F \subseteq G$.

Suppose $a \in G$ and $a \leq d \in A$. Then $\forall v_0(\mathbf{P}_a v_0 \rightarrow \mathbf{P}_d v_0)$ holds in $(X, u)_{u \in A}$ and hence in $\mathfrak{B}$. Since $y \in b_a$, it follows that $y \in b_d$ and so $d \in G$. Next, $0 \notin G$, since $\forall v_0 \neg \mathbf{P}_0 v_0$ holds in $(X, u)_{u \in A}$ and hence in $\mathfrak{B}$, so that $y \notin b_0$ and hence $0 \notin G$. Now let $a \in A$ be given. Then $\forall v_0(\mathbf{P}_a v_0 \vee \mathbf{P}_{-a} v_0)$ holds in $(X, u)_{u \in A}$ and hence in $\mathfrak{B}$, so $y \in b_a$ or $y \in b_{-a}$, and hence $a \in G$ or $-a \in G$. Finally, suppose $Y \subseteq G$ and $|Y| < \mathfrak{m}$. Let $d = \bigcap Y$. Then $\forall v_0(\bigwedge_{a \in Y} \mathbf{P}_a v_0 \rightarrow \mathbf{P}_d v_0)$ holds in $(X, u)_{u \in A}$, and hence in $\mathfrak{B}$. Since $y \in b_a$ for each $a \in Y$, it follows that $y \in b_d$ and $d \in G$, as desired. □

Recalling the proof of (*vi*) ⇒ (*ii*), above, we have

Theorem 31.16. *Every strongly compact cardinal is measurable.*

Now we turn to equivalences for weakly compact cardinals, which we shall only carry through for strongly inaccessible cardinals. One of them concerns a generalization of Ramsey's theorem. Recall that $\mathbf{S}_2 S = \{X \subseteq S : |X| = 2\}$. We shall write $\mathfrak{m} \rightarrow (\mathfrak{n})^2$ provided that whenever $\mathbf{S}_2\mathfrak{m} = A \cup B$ it follows that there is a $\Gamma \subseteq \mathfrak{m}$ with $|\Gamma| = \mathfrak{n}$ such that $\mathbf{S}_2\Gamma \subseteq A$ or $\mathbf{S}_2\Gamma \subseteq B$. Thus Ramsey's theorem says that $\mathfrak{m} \rightarrow (\aleph_0)^2$ for any $\mathfrak{m} \geq \aleph_0$. Another equivalence involves the notion of a tree. A *tree* is a partially ordered set $(P, \leq)$ in which $\{x : x < y\}$ is well-ordered for each $y \in P$; its order type is the *level* of y. A cardinal $\mathfrak{m}$ has the *tree property* provided that if $(T, \leq)$ is a tree of power $\mathfrak{m}$ and every level has $<\mathfrak{m}$ elements, then there is a simply ordered subset of T of power $\mathfrak{m}$.

The following characterizations of weakly compact cardinals are due to Keisler, Tarski, Erdös, Parovičenko, Monk, Scott, and Hanf.

Theorem 31.17. *Assume that* $\mathfrak{m}$ *is strongly inaccessible. Then the following conditions are equivalent;*

(*i*) $\mathfrak{m}$ *is weakly compact;*
(*ii*) $\mathfrak{m}$ *has the tree property;*
(*iii*) *if* $\mathfrak{A}$ *is an* $\mathfrak{m}$*-complete,* $\mathfrak{m}$*-distributive Boolean algebra of power* $\leq \mathfrak{m}$ *but* >1, *then* $\mathfrak{A}$ *has an* $\mathfrak{m}$*-complete ultrafilter;*
(*iv*) *if* $\mathfrak{A}$ *is an* $\mathfrak{m}$*-complete field of subsets of* $\mathfrak{m}$ *of power* $\mathfrak{m}$, *then any* $\mathfrak{m}$*-complete proper filter on* $\mathfrak{A}$ *can be extended to an* $\mathfrak{m}$*-complete ultrafilter;*
(*v*) $\mathfrak{m} \rightarrow (\mathfrak{m})^2$;
(*vi*) *if* $(T, \leq)$ *is a linear ordering with* $|T| = \mathfrak{m}$, *then* T *has a subset of power* $\mathfrak{m}$ *which is either well-ordered under* $\leq$ *or under* $\geq$.

PROOF. We shall show (*i*) ⇒ (*ii*) ⇒ (*iii*) ⇒ (*iv*) ⇒ (*v*) ⇒ (*vi*) ⇒ (*ii*) ⇒ (*i*).

(*i*) ⇒ (*ii*). Assume (*i*), and let $(T, \leq)$ be a tree such that every level L_α for $\alpha < \mathfrak{m}$ has $<\mathfrak{m}$ elements, while $|T| = \mathfrak{m}$. Clearly then $L_\alpha \neq 0$ for all $\alpha < \mathfrak{m}$. Let $\mathscr{L}$ be a language with a binary relation symbol $\mathbf{<}$, unary relation symbols $\mathbf{P}_\alpha$ for each $\alpha < \mathfrak{m}$, individual constants $\mathbf{c}_t$ for each $t \in T$, and one more individual constant $\mathbf{d}$. Consider the following set of sentences:

$\exists v_0(\mathbf{P}_\alpha v_0 \wedge v_0 < \mathbf{d})$ each $\alpha < \mathfrak{m}$,
all $L_{\mathfrak{mm}}$-sentences holding in $\mathfrak{B} = (T, <, L_\alpha, t)_{\alpha<\mathfrak{m},t\in T}$.

Clearly every subset of Γ of power $<\mathfrak{m}$ has a model. Hence Γ has a model, $\mathfrak{A}$. For each $\alpha < \mathfrak{m}$ choose $a_\alpha \in \mathbf{P}_\alpha^{\mathfrak{A}}$ with $a_\alpha <^{\mathfrak{A}} \mathbf{d}^{\mathfrak{A}}$. Now the following sentence holds in $\mathfrak{B}$, hence in $\mathfrak{A}$:

$$\forall v_0\left[\mathbf{P}_\alpha v_0 \leftrightarrow \bigvee_{t\in L_\alpha} v_0 = \mathbf{c}_t\right].$$

Hence for each $\alpha < \mathfrak{m}$ choose $t_\alpha \in T$ so that $a_\alpha = \mathbf{c}_{t\alpha}^{\mathfrak{A}}$. Now the sentence

$$\forall v_0\forall v_1\forall v_2(v_0 < v_2 \wedge v_1 < v_2 \rightarrow v_0 < v_1 \vee v_0 = v_1 \vee v_1 < v_0)$$

holds in $\mathfrak{B}$, hence in $\mathfrak{A}$. Thus $\mathbf{c}_{t\alpha}^{\mathfrak{A}}$ and $\mathbf{c}_{t\beta}^{\mathfrak{A}}$ are comparable under $<^{\mathfrak{A}}$ for any $\alpha, \beta < \mathfrak{m}$. Hence t_α and t_β are comparable (otherwise the sentence saying they aren't would hold in $\mathfrak{B}$, hence in $\mathfrak{A}$). Hence (*ii*) holds.

(*ii*) $\Rightarrow$ (*iii*). Assume (*ii*), and let $\mathfrak{A}$ be as in the hypothesis of (*iii*). Say $\langle a_\alpha : \alpha < \mathfrak{m}\rangle$ has range A. Let $T = \{g : \exists\alpha < \mathfrak{m}(g \in \mathsf{P}_{\beta<\alpha}\{a_\alpha, -a_\alpha\}$ and $\Pi_{\beta<\alpha} g_\beta \neq 0)\}$. Thus $(T, \subseteq)$ is a tree; the level of $g \in T$ is its domain. Since $\mathfrak{m}$ is strongly inaccessible, the set of elements of level $\alpha < \mathfrak{m}$ has power $\leq 2^{|\alpha|} < \mathfrak{m}$. Furthermore, for each $\alpha < \mathfrak{m}$ we have

$$1 = \Pi_{\beta<\alpha}(a_\beta + -a_\beta) = \Sigma\{\Pi_{\beta<\alpha} g_\beta : g \in \mathsf{P}_{\beta<\alpha}\{a_\beta, -a_\beta\}\},$$

and so there is an element of T of level α. Thus $|T| = \mathfrak{m}$. So now we can apply (*ii*) to obtain $g \in \mathsf{P}_{\beta<\mathfrak{m}}\{a_\beta, -a_\beta\}$ such that $\Pi_{\beta<\alpha} g_\beta \neq 0$ for all $\alpha < \mathfrak{m}$. Let $F = \operatorname{Rng} g$. Obviously $F \neq 0$, $1 \in F$, and $x \in F$ or $-x \in F$ for all $x \in A$. Suppose $x \in F$ and $x \leq y$. Then $-y \notin F$, otherwise with α such that $x, -y \in g^*\alpha$ we would have $\Pi_{\beta<\alpha} g_\beta = 0$. So $y \in F$. Finally, suppose that $Y \subseteq F$ and $|Y| < \mathfrak{m}$. Then there is an $\alpha < \mathfrak{m}$ such that $Y \subseteq g^*\alpha$ and $\Pi Y \in g^*\alpha$ or $-\Pi Y \in g^*\alpha$. In the second case, $\Pi_{\beta<\alpha} g_\beta \leq \Pi Y \cdot -\Pi Y = 0$, which is impossible. Thus $\Pi Y \in F$ and F is the desired $\mathfrak{m}$-complete ultrafilter.

(*iii*) $\Rightarrow$ (*iv*). Assume (*iii*) and the hypothesis of (*iv*). Let F be an $\mathfrak{m}$-complete proper filter on $\mathfrak{A}$. Then $\mathfrak{A}/F$ is an $\mathfrak{m}$-complete, $\mathfrak{m}$-distributive Boolean algebra of power $\leq\mathfrak{m}$. By (*iii*), let G be an $\mathfrak{m}$-complete ultrafilter over $\mathfrak{A}/F$. Then $\{a : [a] \in G\}$ is an $\mathfrak{m}$-complete ultrafilter on $\mathfrak{A}$ which extends F.

(*iv*) $\Rightarrow$ (*v*). Assume (*iv*), and let $\mathbf{S}_2\mathfrak{m} = A \cup B$. For each $\alpha \in \mathfrak{m}$ let $M_\alpha = \{\beta : \{\alpha, \beta\} \in A\}$. Let $\mathfrak{A}$ be the $\mathfrak{m}$-complete field of subsets of $\mathfrak{m}$ generated by all the sets M_α as well as the singletons $\{\beta\}$ for $\beta < \mathfrak{m}$. Let F be any $\mathfrak{m}$-complete ultrafilter over $\mathfrak{A}$ containing all the complements of singletons. Thus each member of F has power $\mathfrak{m}$. Define $\{x_\alpha : \alpha < \mathfrak{m}\}$ by induction by letting x_α for each $\alpha < \mathfrak{m}$ be the least element of

$$\bigcap\{M_{x\beta} : \beta < \alpha, M_{x\beta} \in F\}$$
$$\cap \bigcap\{\mathfrak{m} \sim M_{x\beta} : \beta < \alpha, M_{x\beta} \notin F\} \cap (\mathfrak{m} \sim \{x_\beta : \beta < \alpha\}).$$

Thus x is one–one. Set $S = \{x_\alpha : M_{x\alpha} \in F\}$. If $\alpha < \beta$ and $x_\alpha, x_\beta \in S$, then $x_\beta \in Mx_\alpha$ and so $\{\alpha, \beta\} \in A$. Thus $\mathbf{S}_2 S \subseteq A$. Similarly, $[\text{Rng}\, x \sim S]^2 \subseteq B$. Since $|S| = \mathfrak{m}$ or $|\text{Rng}\, x \sim S| = \mathfrak{m}$, (*v*) holds.

(*v*) $\Rightarrow$ (*vi*). Assume (*v*), and let $(T, \leq)$ be a linear ordering with $|T| = \mathfrak{m}$. Let $\preccurlyeq$ be a well-ordering of T. Set $A = \{\{s, t\} : s \prec t \text{ and } s < t\}$, and let $B = [T]^2 \sim A$. An application of (*v*) yields the desired result.

(*vi*) $\Rightarrow$ (*ii*). Assume (*vi*), and let $(T, \leq)$ be a tree of power $\mathfrak{m}$ such that every level has $<\mathfrak{m}$ elements. For each element x of T and each $\alpha < \mathfrak{m}$, let $x^\alpha = x$ if $\alpha \geq$ level of x, $x^\alpha =$ the unique y of level α with $y < x$ if $\alpha <$ level of x. Clearly then

(1) if x and y are incomparable, then there is a unique α such that $x^\alpha \neq y^\alpha$ and the predecessors of x^α and y^α are the same.

Let $<'$ be any well-order of T. We now define another order $<''$ on T. For any $x, y \in T$, $x <'' y$ iff $x < y$, or else x and y are incomparable and $x^\alpha <' y^\alpha$, where α is as in (1). We leave it to the reader to check that $<''$ is a linear order of T such that

(2) if $u < x$, $u < z$, and $x <'' y <'' z$, then $u < y$.

Now by (*vi*), say that L is a subset of T of power $\mathfrak{m}$ well-ordered by $<''$. We may assume that L has order type $\mathfrak{m}$. Let

$$B = \{t \in T : \exists l \in L\, \forall a \in L(l <'' a \Rightarrow t \leq a)\}.$$

We claim that B is the desired linearly ordered subset of T of power $\mathfrak{m}$. Suppose $t_0, t_1 \in B$, and choose l_0, l_1 correspondingly. Let l_i be the maximum of l_0, l_1 under $<''$. Then by the definition of B, $t_0 \leq l_i$ and $t_1 \leq l_i$. Hence t_0 and t_1 are comparable. Now let $\alpha < \mathfrak{m}$; we show that B has an element of level α. Let E be the collection of all sets of the form $V_x = \{l \in L : x \leq l\}$ where x is an element of T of level α. Then $\{l \in L : \text{level of } l \geq \alpha\} = \bigcup \{V_x : x \text{ of level } \alpha\}$, and since $|L| = \mathfrak{m}$ and $\{x : x \text{ is of level } <\alpha\}$ has power $<\mathfrak{m}$, this union has power $\mathfrak{m}$. But there are fewer than $\mathfrak{m}$ elements of level α, so $|V_x| = \mathfrak{m}$ for some x of level α. Thus V_x is cofinal in L (in the $<''$ sense). We claim $x \in B$. Choose any $l \in V_x$ with $x \neq l$ and suppose $l <'' a$ with $a \in L$. Then there is an $l' \in V_x$ with $a <'' l'$. Thus $x < l$ and $x < l'$, so by (2), $x < a$, as desired.

(*ii*) $\Rightarrow$ (*i*). Assume (*ii*). We shall prove (*i*) by modifying the proof of the completeness theorem, 11.19. Let Γ be a set of sentences of power $\mathfrak{m}$ such that every subset Δ of Γ of power $<\mathfrak{m}$ has a model $\mathfrak{A}_\Delta$. First we modify the proof of 11.18. Let Θ be the collection of all subformulas of sentences in Γ. Expand our language by adjoining individual constants $\mathbf{c}_\alpha$ for each $\alpha < \mathfrak{m}$. Clearly $|\Theta| \leq \mathfrak{m}$. Let $C = \{\mathbf{c}_\alpha : \alpha < \mathfrak{m}\}$. Let $\langle \varphi_\alpha : \alpha < \mathfrak{m}\rangle$ be a list of all sentences of the form $\varphi_\alpha = \exists v_{i\alpha 0} \cdots v_{i\alpha\xi} \cdots_{\xi < \beta\alpha} \Psi_\alpha$ which are obtainable from members of Θ by replacing variables by members of C. Then we define a sequence $\langle \mathbf{d}_\alpha : \alpha < \mathfrak{m}\rangle$; each $\mathbf{d}_\alpha$ is to be a sequence of members of C, of length $<\mathfrak{m}$. Suppose $\mathbf{d}_\beta$ defined for all $\beta < \alpha$, where $\alpha < \mathfrak{m}$. Then

$$\bigcup_{\beta < \alpha} \text{Rng}\, \mathbf{d}_\beta \cup \{\mathbf{c} \in C : \mathbf{c} \text{ occurs in } \varphi_\beta \text{ for some } \beta < \alpha\}$$

has power $<\mathfrak{m}$. We let $\mathbf{d}_\alpha$ be a β_α-termed one–one sequence of members of C not in this set. This defines $\langle \mathbf{d}_\alpha : \alpha < \mathfrak{m} \rangle$. Next, for each $\alpha < \mathfrak{m}$ let

$$\Omega_\alpha = \Gamma \cup \{\exists v_{i\gamma 0} \cdots v_{i\gamma\xi} \cdots_{\xi < \beta\gamma} \Psi_\gamma \rightarrow \Psi'_\gamma : \gamma < \alpha\},$$

where Ψ'_γ is obtained from Ψ_γ by replacing all free occurrences of $v_{i\gamma\xi}$ ($\xi < \beta_\gamma$) by $\mathbf{d}_{\gamma\xi}$. Then it is easy to expand each structure $\mathfrak{A}_\Delta$ to be a model of Δ together with the set in braces. Hence $\Omega_\mathfrak{m}$ still has the property that each subset Δ of it of power $<\mathfrak{m}$ has a model $\mathfrak{A}'_\Delta$.

Now we shall use (*ii*) to extend $\Omega_\mathfrak{m}$. Let $\langle \Psi_\alpha : \alpha < \mathfrak{m} \rangle$ be an enumeration of all sentences obtained from members of Θ by replacing variables by members of C. Let T consist of all functions f such that $f \in \mathsf{P}_{\beta < \alpha}\{\Psi_\beta, \neg\Psi_\beta\}$ for some $\alpha < \mathfrak{m}$, Rng$f$ has a model, and $\forall \beta < \alpha[(\Psi_\beta \in \Omega_\mathfrak{m} \Rightarrow f_\beta = \Psi_\beta) \wedge (\neg\Psi_\beta \in \Omega_\mathfrak{m} \Rightarrow f_\beta = \neg\Psi_\beta)]$. Under $\subseteq$, T is a tree. There is an element of T of any level $\alpha < \mathfrak{m}$. For, let $\Delta = \{\Psi_\beta : \beta < \alpha, \Psi_\beta \in \Omega_\mathfrak{m}\} \cup \{\neg\Psi_\beta : \beta < \alpha, \neg\Psi_\beta \in \Omega_\mathfrak{m}\}$. Then Δ has a model, $\mathfrak{A}'$, and if we let $f_\beta = \Psi_\beta$ or $\neg\Psi_\beta$ according as $\mathfrak{A}' \models \Psi_\beta$ or $\mathfrak{A}' \models \neg\Psi_\beta$, we see that $f \in T$. The other hypotheses of (*ii*) clearly hold, so T has a branch F of power $\mathfrak{m}$. Let $\Xi = \{f_\alpha : f \in F, \alpha < \mathfrak{m}\}$. Then $\Xi \supseteq \Theta_m$ and $\Psi_\beta \in \Xi$ or $\neg\Psi_\beta \in \Xi$ for each $\beta < \mathfrak{m}$. Note that every subset of Ξ of power $<\mathfrak{m}$ has a model. We are now in a position to repeat the proof of 11.12, except that instead of (4) one checks

(3) for any sentence Ψ_β, $\Psi_\beta \in \Gamma$ iff $\mathfrak{A} \models \Psi_\beta$.

We leave the details to the reader. □

BIBLIOGRAPHY

1. Dickmann, M. A. *Large Infinitary Languages*. Amsterdam: North-Holland (1976).
2. Henkin, L. Some remarks on infinitely long formulas. In *Infinitistic Methods*. Warsaw (1961), 167–183.
3. Keisler, H. J. *Model Theory for Infinitary Logic*. Amsterdam: North-Holland (1971).

EXERCISES

31.18. Give examples of structures $\mathfrak{A}$, $\mathfrak{B}$ in a countable language $\mathscr{L}$ such that $\mathfrak{A} \equiv_{ee} \mathfrak{B}$ in the $L_{\omega_1\omega}$-sense, $\mathfrak{A}$ is countable, but $\mathfrak{B}$ is uncountable.

31.19. If Γ is a countable set of sentences of $L_{\omega_1\omega}$ and Γ has a model, then Γ has a countable model.

31.20. Show that for any set a there is a formula φ with one free variable v_0, in some language $L_{\mathfrak{m}\mathfrak{m}}$ with only ϵ as a nonlogical constant, such that for any transitive set A, $(A, \epsilon) \models \varphi[b]$ iff $b = a$.

31.21. Every measurable cardinal is weakly inaccessible.

31.22. Let $\mathfrak{A}$ be an $\mathscr{L}$-structure. Suppose $\mathfrak{A} \equiv_{ee} \mathfrak{B}$ in the $L_{\mathfrak{m}\mathfrak{m}}$ sense, where $|A| < \mathfrak{m}$ and $\mathscr{L}$ has fewer than $\mathfrak{m}$ nonlogical constants. Show that $\mathfrak{A} \cong \mathfrak{B}$.

31.23. Give an example in which 31.17(*vi*) fails if $\mathfrak{m} = \aleph_1$.

Index of symbols

Set theory

Recursive function theory

Roman alphabet

Logic

Roman alphabet

Other

Index of Symbols

Index of names and definitions

A

B

C

D

E

P

Q

R

S

V

W